PATTY'S INDUSTRIAL HYGIENE

Fifth Edition

Volume 1

I INTRODUCTION TO INDUSTRIAL HYGIENE

II RECOGNITION AND EVALUATION OF CHEMICAL AGENTS

PATTY'S INDUSTRIAL HYGIENE

Fifth Edition

Volume 1

I INTRODUCTION TO INDUSTRIAL HYGIENE

II RECOGNITION AND EVALUATION OF CHEMICAL AGENTS

ROBERT L. HARRIS

Editor

CONTRIBUTORS

H. E. Ayer
W. E. Babcock
S. Cha
G. D. Clayton
D. E. Cohen
L. J. Cralley
L. V. Cralley
M. R. Cullen
B. D. Dinman
J. D. Dinman
W. Eschenbacher
C. Gomberg
R. L. Harris
G. J. Kullman
R. G. Lieckfield, Jr.
M. Lippman
J. R. Lynch
A. Merlo
S. M. Rappaport
P. C. Reist
C. H. Rice
J. Singh
R. D. Soule
L. A. Todd
P. E. Tolbert
A. Turk

A Wiley-Interscience Publication

JOHN WILEY & SONS, INC.

New York / Chichester / Weinheim / Brisbane / Singapore / Toronto

This book is printed on acid-free paper. ♾

Published simultaneously in Canada.

For ordering and customer service, call 1-800-CALL-WILEY.

Library of Congress Cataloging in Publication Data:

Patty's industrial hygiene / [edited by] Robert L. Harris. — 5th ed.
[rev.]
v. ⟨ ⟩ cm.
Fourth ed. published as: Patty's industrial hygiene and toxicology.
Includes bibliographical references and index.
ISBN 0-471-29756-9 (Vol. 1) (cloth : alk. paper); 0-471-29784-4 (set)
1. Industrial hygiene. I. Harris, Robert L., 1924– .
II. Patty, F. A. (Frank Arthur), 1897– Industrial hygiene and toxicology.
RC967.P37 2000
613.6′2—dc21 99-32462

Printed in the United States of America.

10 9 8 7 6 5 4 3 2 1

Contributors

Howard E. Ayer, CIH, Cincinnati, Ohio

William E. Babcock, OSHA Salt Lake Technical Center, Salt Lake City, Utah

Samuel Cha, MS, TRC Environmental Group, Windsor, Connecticut

George D. Clayton, CIH, Retired, Formerly Chairman of the Board, Clayton Environmental Consultants Inc. San Luis Rey, California.

David E. Cohen, MD, MPH, Director of Occupational and Environmental Dermatology NYU Medical Center, Department of Dermatology New York, New York

Lester V. Cralley, Ph.D Fallbrook, California

Lewis J. Cralley, Ph.D Cincinnati, Ohio

Mark R. Cullen, MD, Yale Occupational and Environmental Medical Program New Haven, Connecticut

Bertram D. Dinman, MD, Sc.D, Graduate School of Public Health, University of Pittsburgh Pittsburgh, Pennsylvania

Jonathan D. Dinman, Ph.D Department of Molecular Genetics and Microbiology, Robert Wood Johnson School of Medicine, University of Medicine and Dentistry of New Jersey Piscataway, New Jersey

William Eschenbacher, MD, NIOSH Morgantown, West Virginia

Christina Gomberg, RN, MS NIOSH Morgantown, West Virginia

Robert L. Harris, Ph.D, CIH Raleigh, North Carolina

Greg J. Kullman, Ph.D, CIH NIOSH Morgantown, West Virginia

Robert G. Lieckfield Jr., CIH, Clayton Environmental Consultants Novi, Michigan

Morton Lippmann, Ph.D, CIH NY University Medical Center Tarrytown, New York

Jeremiah R. Lynch, CIH
Rumson, New Jersey

Angela Merlo, MD, Ph.D
Lawrenceville, New Jersey

Stephen M. Rappaport, Ph.D, CIH
Department of Environmental Sciences and Engineering, University of North Carolina
Chapel Hill, North Carolina

Parker C. Reist, Ph.D, PE, Department of Environmental Sciences and Engineering, University of North Carolina, Chapel Hill, North Carolina

Carol Rice, Ph.D, CIH, University of Cincinnati Medical Center
Cincinnati, OH

Jaswant Singh, Ph.D, CIH, Managing Director, Liberty Risk Services
Malaysia

Robert D. Soule, CIH, CSP
Indiana, Pennsylvania

Lori A. Todd, Ph.D, CIH
Department of Environmental Sciences and Engineering, University of North Carolina, Chapel Hill, North Carolina

Paige E. Tolbert, Ph.D, Environmental and Occupational Health Department, Rollins School of Public Health of Emory University, Atlanta, Georgia

Amos Turk, Ph.D
Danbury, Connecticut

Preface

Industrial hygiene is an applied science and a profession. Like other applied sciences such as medicine and engineering, it is founded on basic sciences such as biology, chemistry, mathematics, and physics. In a sense it is a hybrid profession because within its ranks are members of other professions—chemists, engineers, biologists, physicists, physicians, nurses, and lawyers. In their professional practice all are dedicated in one way or another to the purposes of industrial hygiene, to the anticipation, recognition, evaluation, and control of work-related health hazards. All are represented among the authors of chapters in these volumes.

Although the term "industrial hygiene" used to describe our profession is probably of twentieth century origin, we must go further back in history for the origin of its words. The word "industry," which has a dictionary meaning, "systematic labor for some useful purpose or the creation of something of value," has its English origin in the fifteenth century. For "hygiene" we must look even earlier. Hygieia, a daughter of Aesklepios who is god of medicine in Greek mythology, was responsible for the preservation of health and prevention of disease. Thus, Hygieia, when she was dealing with people who were engaged in systematic labor for some useful purpose, was practicing our profession, industrial hygiene.

Industrial Hygiene and Toxicology was originated by Frank A. Patty with publication of the first single volume in 1948. In 1958 an updated and expanded Second Edition was published with his guidance. A second volume, Toxicology, was published in 1963. Frank Patty was a pioneer in industrial hygiene; he was a teacher, practitioner, and manager. He served in 1946 as eighth President of the American Industrial Hygiene Association. To cap his professional career he served as Director of the Division of Industrial Hygiene for the General Motors Corporation.

At the request of Frank Patty, George and Florence Clayton took over editorship of the ever-expanding *Industrial Hygiene and Toxicology* series for the Third Edition of Volume I, *General Principles*; published in 1978, and Volume II, *Toxicology*, published in

1981–1982. The First Edition of Volume III, *Theory and Rationale of Industrial Hygiene Practice,* edited by Lewis and Lester Cralley, was published in 1979 with its Second Edition published in 1984. The ten-book, two-volume Fourth Edition of *Patty's Industrial Hygiene and Toxicology,* edited by George and Florence Clayton, was published in 1991–1994, and the Third Edition of Volume III, *Theory and Rationale of Industrial Hygiene Practice,* edited by Robert Harris, Lewis Cralley, and Lester Cralley, was published in 1994. With the agreement and support of George and Florence Clayton, and Lewis and Lester Cralley, it is a signal honor for me to follow them and Frank A. Patty as editor of the Industrial Hygiene volumes of *Patty's Industrial Hygiene and Toxicology.*

Industrial hygiene has been dealt with very broadly in past editions of *Patty's Industrial Hygiene and Toxicology.* Chapters have been offered on sampling and analysis, exposure measurement and interpretation, absorption and elimination of toxic materials, occupational dermatoses, instrument calibration, odors, industrial noise, ionizing and nonionizing radiation, heat stress, pressure, lighting, control of exposures, safety and health law, health surveillance, occupational health nursing, ergonomics and safety, agricultural hygiene, hazardous wastes, occupational epidemiology, and other vital areas of practice. These traditional areas continue to be covered in this new edition. Consistent with the past history of *Patty's,* new areas of industrial hygiene concerns and practices have been addressed as well: aerosol science, computed tomography, multiple chemical sensitivity, potential endocrine disruptors, biological monitoring of exposures, health and safety management systems, industrial hygiene education, and other areas not covered in earlier editions.

Although industrial hygiene has been practiced in one guise or another for centuries, the most systematic approaches and the most esoteric accomplishments have been made in the past fifty or sixty years—generally in the years since Frank Patty published his first book. This accelerated progress is due primarily to increased public awareness of occupational health and safety issues and need for environmental control as is evidenced by Occupational Safety and Health, Clean Air, and Clean Water legislation at both federal and state levels.

Industrial hygienists know that variability is the key to measurement and interpretation of workers' exposures. If exposures did not vary, exposure assessment could be limited to a single measurement, the results of which could be acted upon, then the matter filed away as something of no further concern. We know, however, that exposures change. But not only do exposures change—change is characteristic of the science and practice of our profession as well. We must be alert to recognize new hazards, we must continue to evaluate new and changing stresses, we must evaluate performance of exposure controls and from time to time upgrade them. These volumes represent the theory and practice of industrial hygiene as they are understood by their chapter authors at the time of writing. But, as observed by the Greek philosopher Heracleitus about 2500 years ago, "There is nothing permanent except change." Improvements and changes in theory and practice of industrial hygiene take place continuously and are generally reported in the professional literature. Industrial hygienists, the practitioners, the teachers, and the managers, must stay abreast of the professional literature. Furthermore, when an industrial hygienist develops new knowledge, he/she has what almost amounts to an ethical obligation to share it in our journals.

One cannot ponder the rapid changes and advancements made in recent decades in science and technology, and in our own profession as well, without wondering at what the next two or three decades will bring. Developments in computer technology and information processing and exchange have greatly influenced manufacturing (robotics, computer controlled machining) and the general conduct of commerce and business in the past one or two decades. This change will only accelerate with computer speeds and capacities doubling every 18 months or so, and processing units approaching microsize. The possibility for continuously monitoring and computer storage of exposures of individual workers may become reality within the next decade. The human genome project holds promise for prevention and cure of many diseases, including some associated with conditions of work. World population continues to increase geometrically and is expected to be about eight billion in the year 2020; with improvements in preventive health care this will be an increasingly older population. Genetic engineering and highly effective pesticides are already improving yields of agricultural commodities; if all goes well in this area, and if we can avoid set-backs as might be associated with potential endocrine disruptors, feeding the expanding human population may not be a limiting factor. Globalization of manufacturing and commerce has already begun to reduce manufacturing employment in the United States and in Europe, and to expand opportunities for expanding populations in some developing nations. The United States and other developed nations are on their way to becoming world centers of information and innovation.

How will all of this affect the future practice of industrial hygiene? In the Preface to the Fourth Edition of *Patty's,* George and Florence Clayton suggested that the future of industrial hygiene is limited only by the narrowness of vision of its practitioners. More recently, Lawrence Birkner, past president of the American Academy of Industrial Hygiene, and his co-worker and spouse, Ruth McIntyre Birkner, in writing about the future of the occupational and environmental hygiene profession, say much the same thing. (See "The Future of the Occupational and Environmental Hygiene Profession" in *A.I.H.A.* Journal, pp. 370–374, 1997) Larry and Ruth report that we must be aware of the changes likely to take place in the next couple of decades, and must develop strategies now to assure the profession's full participation in protecting the health and safety of workers, and the environment, of tomorrow.

ROBERT L. HARRIS

Raleigh, North Carolina

Contents

USEFUL EQUIVALENTS AND CONVERSION FACTORS

1 kilometer = 0.6214 mile
1 meter = 3.281 feet
1 centimeter = 0.3937 inch
1 micrometer = 1/25,4000 inch = 40 microinches
  = 10,000 Angstrom units
1 foot = 30.48 centimeters
1 inch = 25.40 millimeters
1 square kilometer = 0.3861 square mile (U.S.)
1 square foot = 0.0929 square meter
1 square inch = 6.452 square centimeters
1 square mile (U.S.) = 2,589,998 square meters
  = 640 acres
1 acre = 43,560 square feet = 4047 square meters
1 cubic meter = 35.315 cubic feet
1 cubic centimeter = 0.0610 cubic inch
1 cubic foot = 28.32 liters = 0.0283 cubic meter
  = 7.481 gallons (U.S.)
1 cubic inch = 16.39 cubic centimeters
1 U.S. gallon = 3,7853 liters = 231 cubic inches
  = 0.13368 cubic foot
1 liter = 0.9081 quart (dry), 1.057 quarts
  (U.S., liquid)
1 cubic foot of water = 62.43 pounds (4°C)
1 U.S. gallon of water = 8.345 pounds (4°C)
1 kilogram = 2.205 pounds
1 gram = 15.43 grains
1 pound = 453.59 grams
1 ounce (avoir.) = 28.35 grams
1 gram mole of a perfect gas ≎ 24.45 liters
  (at 25°C and 760 mm Hg barometric pressure)
1 atmosphere = 14.7 pounds per square inch
1 foot of water pressure = 0.4335 pound per
  square inch
1 inch of mercury pressure = 0.4912 pound per
  square inch
1 dyne per square centimeter = 0.0021 pound per
  square foot
1 gram-calorie = 0.00397 Btu
1 Btu = 778 foot-pounds
1 Btu per minute = 12.96 foot-pounds per second
1 hp = 0.707 Btu per second = 550 foot-pounds
  per second
1 centimeter per second = 1.97 feet per minute
  = 0.0224 mile per hour
1 footcandle = 1 lumen incident per square foot
  = 10.764 lumens incident per square meter
1 grain per cubic foot = 2.29 grams per
  cubic meter
1 milligram per cubic meter = 0.000437 grain per
  cubic foot

To convert degrees Celsius to degrees Fahrenheit: °C (9/5) + 32 = °F
To convert degrees Fahrenheit to degrees Celsius: (5/9) (°F − 32) = °C
For solutes in water: 1 mg/liter ≎ 1 ppm (by weight)
Atmospheric contamination: 1 mg/liter ≎ 1 oz/1000 cu ft (approx)
For gases or vapors in air at 25°C and 760 mm Hg pressure:
  To convert mg/liter to ppm (by volume): mg/liter (24,450/mol. wt.) = ppm
  To convert ppm to mg/liter: ppm (mol. wt./24,450) = mg/liter

CONVERSION TABLE FOR GASES AND VAPORS[a]

(Milligrams per liter to parts per million, and vice versa; 25°C and 760 mm Hg barometric pressure)

Molecular Weight	1 mg/liter ppm	1 ppm mg/liter	Molecular Weight	1 mg/liter ppm	1 ppm mg/liter	Molecular Weight	1 mg/liter ppm	1 ppm mg/liter
1	24,450	0.0000409	39	627	0.001595	77	318	0.00315
2	12,230	0.0000818	40	611	0.001636	78	313	0.00319
3	8,150	0.0001227	41	596	0.001677	79	309	0.00323
4	6,113	0.0001636	42	582	0.001718	80	306	0.00327
5	4,890	0.0002045	43	569	0.001759	81	302	0.00331
6	4,075	0.0002454	44	556	0.001800	82	298	0.00335
7	3,493	0.0002863	45	543	0.001840	83	295	0.00339
8	3,056	0.000327	46	532	0.001881	84	291	0.00344
9	2,717	0.000368	47	520	0.001922	85	288	0.00348
10	2,445	0.000409	48	509	0.001963	86	284	0.00352
11	2,223	0.000450	49	499	0.002004	87	281	0.00356
12	2,038	0.000491	50	489	0.002045	88	278	0.00360
13	1,881	0.000532	51	479	0.002086	89	275	0.00364
14	1,746	0.000573	52	470	0.002127	90	272	0.00368
15	1,630	0.000614	53	461	0.002168	91	269	0.00372
16	1,528	0.000654	54	453	0.002209	92	266	0.00376
17	1,438	0.000695	55	445	0.002250	93	263	0.00380
18	1,358	0.000736	56	437	0.002290	94	260	0.00384
19	1,287	0.000777	57	429	0.002331	95	257	0.00389
20	1,223	0.000818	58	422	0.002372	96	255	0.00393
21	1,164	0.000859	59	414	0.002413	97	252	0.00397
22	1,111	0.000900	60	408	0.002554	98	249.5	0.00401
23	1,063	0.000941	61	401	0.002495	99	247.0	0.00405
24	1,019	0.000982	62	394	0.00254	100	244.5	0.00409
25	978	0.001022	63	388	0.00258	101	242.1	0.00413
26	940	0.001063	64	382	0.00262	102	239.7	0.00417
27	906	0.001104	65	376	0.00266	103	237.4	0.00421
28	873	0.001145	66	370	0.00270	104	235.1	0.00425
29	843	0.001186	67	365	0.00274	105	232.9	0.00429
30	815	0.001227	68	360	0.00278	106	230.7	0.00434
31	789	0.001268	69	354	0.00282	107	228.5	0.00438
32	764	0.001309	70	349	0.00286	108	226.4	0.00442
33	741	0.001350	71	344	0.00290	109	224.3	0.00446
34	719	0.001391	72	340	0.00294	110	222.3	0.00450
35	699	0.001432	73	335	0.00299	111	220.3	0.00454
36	679	0.001472	74	330	0.00303	112	218.3	0.00458
37	661	0.001513	75	326	0.00307	113	216.4	0.00462
38	643	0.001554	76	322	0.00311	114	214.5	0.00466

CONVERSION TABLE FOR GASES AND VAPORS (*Continued*)

(Milligrams per liter to parts per million, and vice versa; 25°C and 760 mm Hg barometric pressure)

Molecular Weight	1 mg/liter ppm	1 ppm mg/liter	Molecular Weight	1 mg/liter ppm	1 ppm mg/liter	Molecular Weight	1 mg/liter ppm	1 ppm mg/liter
115	212.6	0.00470	153	159.8	0.00626	191	128.0	0.00781
116	210.8	0.00474	154	158.8	0.00630	192	127.3	0.00785
117	209.0	0.00479	155	157.7	0.00634	193	126.7	0.00789
118	207.2	0.00483	156	156.7	0.00638	194	126.0	0.00793
119	205.5	0.00487	157	155.7	0.00642	195	125.4	0.00798
120	203.8	0.00491	158	154.7	0.00646	196	124.7	0.00802
121	202.1	0.00495	159	153.7	0.00650	197	124.1	0.00806
122	200.4	0.00499	160	152.8	0.00654	198	123.5	0.00810
123	198.8	0.00503	161	151.9	0.00658	199	122.9	0.00814
124	197.2	0.00507	162	150.9	0.00663	200	122.3	0.00818
125	195.6	0.00511	163	150.0	0.00667	201	121.6	0.00822
126	194.0	0.00515	164	149.1	0.00671	202	121.0	0.00826
127	192.5	0.00519	165	148.2	0.00675	203	120.4	0.00830
128	191.0	0.00524	166	147.3	0.00679	204	119.9	0.00834
129	189.5	0.00528	167	146.4	0.00683	205	119.3	0.00838
130	188.1	0.00532	168	145.5	0.00687	206	118.7	0.00843
131	186.6	0.00536	169	144.7	0.00691	207	118.1	0.00847
132	185.2	0.00540	170	143.8	0.00695	208	117.5	0.00851
133	183.8	0.00544	171	143.0	0.00699	209	117.0	0.00855
134	182.5	0.00548	172	142.2	0.00703	210	116.4	0.00859
135	181.1	0.00552	173	141.3	0.00708	211	115.9	0.00863
136	179.8	0.00556	174	140.5	0.00712	212	115.3	0.00867
137	178.5	0.00560	175	139.7	0.00716	213	114.8	0.00871
138	177.2	0.00564	176	138.9	0.00720	214	114.3	0.00875
139	175.9	0.00569	177	138.1	0.00724	215	113.7	0.00879
140	174.6	0.00573	178	137.4	0.00728	216	113.2	0.00883
141	173.4	0.00577	179	136.6	0.00732	217	112.7	0.00888
142	172.2	0.00581	180	135.8	0.00736	218	112.2	0.00892
143	171.0	0.00585	181	135.1	0.00740	219	111.6	0.00896
144	169.8	0.00589	182	134.3	0.00744	220	111.1	0.00900
145	168.6	0.00593	183	133.6	0.00748	221	110.6	0.00904
146	167.5	0.00597	184	132.9	0.00753	222	110.1	0.00908
147	166.3	0.00601	185	132.2	0.00757	223	109.6	0.00912
148	165.2	0.00605	186	131.5	0.00761	224	109.2	0.00916
149	164.1	0.00609	187	130.7	0.00765	225	108.7	0.00920
150	163.0	0.00613	188	130.1	0.00769	226	108.2	0.00924
151	161.9	0.00618	189	129.4	0.00773	227	107.7	0.00928
152	160.9	0.00622	190	128.7	0.00777	228	107.2	0.00933

CONVERSION TABLE FOR GASES AND VAPORS (*Continued*)

(Milligrams per liter to parts per million, and vice versa; 25°C and 760 mm Hg barometric pressure)

Molecular Weight	1 mg/liter ppm	1 ppm mg/liter	Molecular Weight	1 mg/liter ppm	1 ppm mg/liter	Molecular Weight	1 mg/liter ppm	1 ppm mg/liter
229	106.8	0.00937	253	96.6	0.01035	277	88.3	0.01133
230	106.3	0.00941	254	96.3	0.01039	278	87.9	0.01137
231	105.8	0.00945	255	95.9	0.01043	279	87.6	0.01141
232	105.4	0.00949	256	95.5	0.01047	280	87.3	0.01145
233	104.9	0.00953	257	95.1	0.01051	281	87.0	0.01149
234	104.5	0.00957	258	94.8	0.01055	282	86.7	0.01153
235	104.0	0.00961	259	94.4	0.01059	283	86.4	0.01157
236	103.6	0.00965	260	94.0	0.01063	284	86.1	0.01162
237	103.2	0.00969	261	93.7	0.01067	285	85.8	0.01166
238	102.7	0.00973	262	93.3	0.01072	286	85.5	0.01170
239	102.3	0.00978	263	93.0	0.01076	287	85.2	0.01174
240	101.9	0.00982	264	92.6	0.01080	288	84.9	0.01178
241	101.5	0.00986	265	92.3	0.01084	289	84.6	0.01182
242	101.0	0.00990	266	91.9	0.01088	290	84.3	0.01186
243	100.6	0.00994	267	91.6	0.01092	291	84.0	0.01190
244	100.2	0.00998	268	91.2	0.01096	292	83.7	0.01194
245	99.8	0.01002	269	90.9	0.01100	293	83.4	0.01198
246	99.4	0.01006	270	90.6	0.01104	294	83.2	0.01202
247	99.0	0.01010	271	90.2	0.01108	295	82.9	0.01207
248	98.6	0.01014	272	89.9	0.01112	296	82.6	0.01211
249	98.2	0.01018	273	89.6	0.01117	297	82.3	0.01215
250	97.8	0.01022	274	89.2	0.01121	298	82.0	0.01219
251	97.4	0.01027	275	88.9	0.01125	299	81.8	0.01223
252	97.0	0.01031	276	88.6	0.01129	300	81.5	0.01227

[a] A. C. Fieldner, S. H. Katz, and S. P. Kinney, "Gas Masks for Gases Met in Fighting Fires," *U.S. Bureau of Mines, Technical Paper No. 248,* 1921.

PATTY'S INDUSTRIAL HYGIENE

Fifth Edition

Volume 1

I INTRODUCTION TO INDUSTRIAL HYGIENE

II RECOGNITION AND EVALUATION OF CHEMICAL AGENTS

CHAPTER ONE

Industrial Hygiene: Retrospect and Prospect

George D. Clayton, CIH*

1 INTRODUCTION

History is the systematic recording of past events. A decade has passed since publication of the 4th Edition of the Patty series on industrial hygiene and toxicology. This decade has had an explosion of technical advances in our profession, and this mountain of new information provides a solid platform for future development.

Prior to the 1950s, industrial hygienists were frequently asked "What is an industrial hygienist?" As we reach the end of this century, recognition of the profession has improved. In both government and industry identity of the profession is well established. In fact, it is one of the highest paid professions in the health related field (1).

The current definition of industrial hygiene states that it is "the anticipation, recognition, evaluation, and control of workplace environmental factors that may affect the health, comfort, or productivity of the worker." Earlier definition defined it as both a science and art, and also included discomfort and inefficiency among workers or citizens of the community. The early goal of the industrial hygiene profession, to educate the public, unions and workers, concerning the importance of protecting the health of mankind has been achieved. Although progress was at a creeping pace prior to the Occupational Safety and Health Act of 1970, it has accelerated as a result of that Act and the communication explosion.

In the past, the introduction of modern production of chemicals was blamed for the resulting occupational diseases. Some people believed that prior to 1950 there was very

**Retired.*

Patty's Industrial Hygiene, Fifth Edition, Volume 1. Edited by Robert L. Harris.
ISBN 0-471-29756-9

little activity in the field of industrial hygiene. Among other fallacious beliefs was that unhealthful conditions are inherent in certain trades. Time has proven that with the proper design of facilities and equipment, along with correct ventilation, design, and good practice, no industry need be considered "unhealthy."

2 RETROSPECT

2.1 Early History

It is true that prior to 1900 there was very little concern expressed for the health of workers. At the beginning of civilization, people struggled for existence, and survival itself was an occupational disease. As stratification of social classes evolved, common labor was performed by slaves. This practice continued until the nineteenth century. The victories of war provided a steady supply of slaves. Manual labor by others than slaves was scorned. At one period in their culture Egyptians were prohibited by law from performing manual labor. With such an attitude toward the working man, it is not surprising that no efforts were made to control the work environment or to provide a healthful, comfortable workplace.

Early in the fourth century B.C., lead toxicity in the mining industry was recognized and recorded by Hippocrates, although no effort was made to provide protection for the workers. Some 500 years later, Pliny the Elder, a Roman scholar, referred to the dangers imminent in dealing with zinc and sulfur. He described a bladder-derived protective mask to be used by laborers subjected to large amounts of dust or lead fumes. However, the Romans were more concerned with engineering and military achievements than with any type of occupational medicine.

The writings of the Greek physician Galen, who resided in Rome in the second century, present many theories on anatomy and pathology. Galen was authoritative and assertive in his writings, but even though he recognized the dangers of acid mists to copper miners, his writings gave no incentive to the solution of the problem.

During the Middle Ages feudalism made its appearance, and little improvement was made in work standards. However, one advancement of this period was the provision of assistance to ill members and their families by the feudal guilds. During the twelfth and thirteenth centuries, observation and experimentation flourished in the great universities; however, the study of occupational disease was virtually ignored. Thus little was achieved in the field of industrial hygiene until 1473, with the publication of a pamphlet on occupational disease by Ulrich Ellenbog, which included notable hygiene instructions. This was followed in 1556 by the writing of a German scholar, Georgius Agricola, who effectively described hazards associated with the mining industry. His *De Re Metallica* was translated into English in 1912 by Herbert and Lou Henry Hoover. Agricola's 12-section treatment included suggestions for mine ventilation and protective masks for miners, a discussion of mining accidents, descriptions of what is referred to today as "trench foot" (effects on the extremities caused by lengthy exposure to the cold water of damp mines), and silicosis (disease of the lungs caused by inhalation of silica or quartz dust).

In the sixteenth century industrial hygiene still abounded with the mystical. Many believed that demons lived in the mines and could be controlled by fasting and prayer. The published observations of Paracelsus, the alchemist son of a Swiss doctor, are based on the 10 years he worked in a smelting plant and as a laborer in the mines of Tyrol. The book makes many erroneous conclusions: for example, he attributes miners' "lung sickness' to a vapor of mercury, sulfur, and salt. However, his warnings about the toxicity of certain metals and outline of mercury poisoning were quite advanced. He pointed out the fallacies of many medical theories then current, taught the use of specific remedies instead of indiscriminate bleeding and purging, and introduced new medicines.

It is generally agreed that the first comprehensive treatise on occupational disease, *De Morbis Artificum Diatriba* by Bernardo Ramazzini, an Italian physician, was published in 1700. The book described silicosis in pathological terms, as observed by autopsies on miners' bodies. He presented cautions that he felt would alleviate many industrial hazards. Unfortunately, his warnings for vigilance with these hazards were ignored for centuries. Nevertheless, this book had a prodigious effect on the future of public hygiene. Ramazzini, believing the work environment affected health, asked of his patients, "Of what trade [or occupation] are you?" Most physicians to this day include this same question in recording case histories of their patients.

In the eighteenth century more industrial hygiene problems were being recognized and uncovered. Sir George Baker correctly attributed "Devonshire colic" to lead in the cider industry, and was instrumental in its removal from use. Percival Pott recognized soot as one of the causes of scrotal cancer and was a major force in the passage of the Chimney-Sweepers Act of 1788. Charles Thackrah, both a political and a medical influence, wrote a 200-page treatise dealing with occupational medicine. A scientist with vision, Thackrah stated "Each master . . . has in great measure the health and happiness of his workpeople in his power. . . . Let benevolence be directed to the prevention, rather than to the relief of the evils." Others who were beacons of this age were Thomas Beddoes and Sir Humphry Davy, who collaborated in describing occupations that were prone to cause "phthisis" (tuberculosis). Sir Davy also aided in the development of the miner's safety lamp.

It has been reported that health problems suffered by the artists Rubens, Renoir, Dufy, and Klee may have been caused by toxic heavy metals in bright paints they used. These four artists might have been heavily exposed to such poisonous metals as lead and mercury when they used yellow, red, white, green, blue, and violet paints. Rubens (1577–1640), Renoir (1841–1919), and Dufy (1877–1953) had rheumatoid arthritis, and Klee (1879–1940) suffered from sclerosis. It is believed that they were more heavily exposed to pigments containing salts of heavy metals because they used significantly brighter and clearer colors than their contemporaries who did not suffer from rheumatic disease.

Another early example of toxicity of lead has been reported in the *National Geographic* (September, 1990). The episode occurred in 1845, when 129 members of Sir John Franklin's fateful voyage, which was searching for the Northwest Passage, perished. It had been assumed they died because of a combination of starvation, scurvy, and bad luck. It is now suspected they died of lead poisoning. A study of the bones of Franklin's sailors on Canada's King William Island, and of the bodies of three others from Beechey Island, shows that their remains contained higher than normal levels of lead, and that the lead in the skeletons came from a single source, which matched lead in solder used to seal food tins found in a Beechey Island cache. An anthropologist who participated in the study stated that if the lead did not kill the men, it probably affected their judgment, leading to poor decisions that contributed to the death of the entire crew. Interestingly, it was the invention of tinned foods in 1810 that made possible such long voyages. The degree of the toxicity of lead was not recognized until the 1880s; and as a result, new methods of sealing food tins were devised.

Although developments of the eighteenth century surpassed any of the previous centuries, safeguards for workers' safety appeared to be in a suspended state until the English factory acts of 1833 were passed, which indicated an interest of the government in the health of the working man. These acts are considered the first effective legislative acts in the field of industry; they required that some concern be shown to the working population. This concern, however, was in practice directed more toward providing compensation for accidents than controlling the *causes* of these accidents. Various European nations followed the lead of England and developed workmen's compensation acts. These laws were instrumental in stimulating the adoption of increased factory safety precautions and the inauguration of medical service in industrial plants. *Community responsibility* was developing, which was illustrated by the interest of newspapers and magazines, in efforts to control the environment. One of the most popular publications of the nineteenth century, the *London Illustrated News*, affixed the blame in a mine explosion to negligence in proper gas-testing methods. The same article made a point of the fact that no safety lamps had been provided the workers. In 1878 the last of the English "factory acts" centralized the inspection of factories by creating a post for this purpose in London.

2.2 Development of Industrial Hygiene in the United States

There were probably fewer than 50 industrial hygienists in the United States during the early 1930s, devoting their efforts to protecting the health of workers. The general public had little interest in industrial hygiene. Some managers in industry believed it was not only a source of trouble, but an economic waste. A few physicians viewed it as an intolerable invasion of the doctor's domain, and insisted that only medical doctors were qualified to express opinions regarding the effects of any material or any stress on the human body. Unions were more interested in getting "hazardous pay" than in controlling the environment. Workers were kept in ignorance of the actual hazards.

In the early twentieth century, a champion of social responsibility for workers' health and welfare in the United States was Alice Hamilton, a physician. She not only presented substantiated evidence of a relationship between illness and exposure to toxins, but proposed concrete solutions to these problems. This was the start of an "occupational medicine renaissance." The public was becoming increasingly aware of problems that could be encountered in certain industrial environments, and of the need for legislation. In 1908 the federal government passed a compensation act for certain civil service employees, and in 1911 the first state compensation laws were passed. By 1948 all the states had passed such legislation. These workmen's compensation laws significantly influenced the development of industrial hygiene in the United States, for management began to recognize and appreciate that controlling the environment was less costly than paying huge sums in compensation.

In June of 1939 a number of professionals involved with providing health care for workers through the control of their environment formed the American Industrial Hygiene

Association (AIHA) to provide a means of evaluating mutual problems. The members of this new organization consisted of all persons interested in industrial hygiene, whether from industry, academia, or government. These founders came from a variety of backgrounds, such as chemistry, engineering, and medicine. Today the typical industrial hygienist, although specializing in one of these sciences, will have a "working knowledge" of the others.

Through the association's efforts, management began to realize that a *healthy* worker is a *productive* worker. Unions began to understand that the health of their members should be of uppermost concern to them, supplanting hazardous pay. Government passed a series of workmen's compensation laws, culminating in 1970 with the passage of the Williams–Steiger Occupational Safety and Health Act. Today laws require industry to provide a healthful environment; unions assist in supplying information to their members regarding the hazards of the environment; and employees must be informed of the hazards of the products and materials with which they work. As these developments occurred instrumentation was developed, from crude beginnings to the present sophisticated, automated instrumentation.

The astute founders of the AIHA realized the importance of disseminating technical information, and only a few months after they organized (January, 1940) began publishing articles on instrumentation, toxicity, and so on, in *Industrial Medicine* (see Section 2.2.2), as a section of the medical journal. This endeavor was expanded to a variety of publications authored by committees of the membership, which were influential in developing the profession.

Currently thousands of professionals are dedicated to protecting people's health. Today AIHA is the leading technical association in this field. Its membership currently numbers 12,000. At the present time, membership depends on certain qualifications. For current eligibility the association should be contacted.

In addition to full membership, there are satellite groups, referred to as "local sections," which had their origin almost simultaneously as the national group. Individuals whose responsibilities were diversified but included industrial hygiene problems, who felt the need for a local means of communicating with other industrial hygienists, found that local sections satisfied their needs. The first such section was the Michigan Industrial Hygiene Society which was organized prior to 1939. In 1990, the number of such sections was 75, and included 8 foreign groups. There are presently 76 local section and 5 foreign sections, with some of the earlier foreign groups becoming independent industrial hygiene organizations. Table 1.1 lists the independent foreign organizations.

In the late 1980s student membership was introduced at the universities, and has grown from about 60 members the first year, to approximately 500 at the universities, where their program includes industrial hygiene/occupational subjects.

In 1993, the classification of "fellow membership" was created. Fellows of the American Industrial Hygiene Association are individuals who have been full members in good standing for a minimum of 15 years. These individuals have been nominated by a local section or standing committee for their recognized achievement in the field of industrial hygiene, either through research, leadership, publications, education or service to AIHA. Individuals are approved as fellows by the AIHA Board of Directors after recommendation

Table 1.1. Former Foreign Local Sections

Name	Approximate Dates as an AIHA Local Section	Converted to
Southern Ontario	1964–1968	Occupational Hygiene Association of Ontario (OHAO)
Ottawa	1964–1978	Combined with OHAO
Quebec	1965–1976	Quebec Industrial Hygiene Association
Spanish	1983–1990	Spanish Industrial Hygiene Association
Swiss	1983–1987	Swiss Occupational Hygienists Society

by the Membership Committee. The fellow classification is limited to no more than 5 percent of the membership.

A year before the formation of the AIHA, a group of industrial hygienists working under the aegis of the U.S. Public Health Service in Washington, D.C., organized the Conference of Governmental Industrial Hygienists, under the guidance of Jack Bloomfield, the Director of State Programs for the Public Health Service. Membership in this group was limited to professional personnel in governmental agencies or educational institutions, who were engaged in occupational safety and health programs, and who desired a medium for the free exchange of ideas and experiences, and for promotion of standards and techniques in industrial health. This was not an official government agency.

Several attempts have been made during the current decade to merge the AIHA and ACGIH (the Conference), but have not succeeded. One of the outstanding contributions of that organization to the field of industrial hygiene has been the formation of a committee that annually evaluates and establishes threshold limit values and biological exposure indexes for various chemicals used in the workplace. These standards are utilized throughout the world. The ACGIH has a worldwide reputation; its membership is now 4,612.

2.2.1 Significant and Historical Events

Most events, whether or not significant, are not isolated incidents by themselves. They may inspire, incite, or cause future occurrences. The following historical events illustrate this axiom.

In 1910 the first national conference on industrial diseases was called in Chicago by the American Association for Labor Legislation. A commission consisting of representatives of medicine, engineering, and chemistry was assigned the task of investigating the magnitude of the problem, and of proposing a method of attack in waging war against industrial disease. About this time several other groups began the study of occupational diseases. The U.S. Bureau of Mines was created in 1919. The U.S. Bureau of Labor, established in 1885, became the federal Department of Labor in 1913, with the charge of collecting "information upon the subject of labor, its relation to capital, the hours of labor, and the earnings of laboring men and women, and also upon the means of promoting their material, social, intellectual, and moral prosperity." The Department of Labor had the responsibility of collecting and disseminating information in the field of industrial hygiene, and following the establishment of the Occupational Safety and Health Administration in 1970, which is under the jurisdiction of the Labor Department, its role was greatly expanded.

The American Museum of Safety was created in New York in 1911 and later became known as the Safety Institute of America. The National Safety Council was organized in 1913. The American Public Health Association organized a section on industrial hygiene in 1914, The U.S. Public Health Service organized a Division of Industrial Hygiene and Sanitation in 1915. The American Association of Industrial Physicians and Surgeons was organized in 1916. Accelerated production of munitions and other war materials for World War I resulted in increased mortality and ill health, making many persons conscious of the necessity for technical guidance in the recognition and control of occupational diseases.

Upon its formation in 1939, the AIHA participated in a joint conference sponsored by the Industrial Medical Association, headquartered in Chicago. Other participants were the Industrial Nurses and the Dental Association, as well as ACGIH. In 1960, AIHA decided to develop its own program and invited ACGIH to join them in sponsoring their first conference, which was held in Detroit, Michigan in 1961. They accepted the invitation, and from this first joint conference which had approximately 600 attendees, a very successful joint conference has emerged which currently has an attendance of over 12,000 registrants.

2.2.2 Significant Publications

The first publication of importance in the field of industrial hygiene was the *Journal of Industrial Hygiene* established in 1919, and for some years it was the leading publication in this field. In 1949 it was acquired by the American Medical Association, underwent editorial policy changes, and was then published as the *AMA Archives of Industrial Health.*

Several months after it was organized, the AIHA, recognizing the importance of disseminating knowledge concerning workers' health and safety, made arrangements with the Industrial Medicine Association to have a section of their publication devoted to industrial hygiene. This section made its debut in the January 1940 issue. Interest in this section grew rapidly, and in 1945 it was published as a supplement to the magazine. In June 1946, AIHA began publishing the magazine as the *AIHA Quarterly*, during the months of March, June, September, and December. Further growth of interest and availability of papers inspired another change to a bimonthly publication schedule, and the name became the *AIHA Journal* in 1958. Approximately 12 years elapsed before it became a monthly publication, and for many years it has been, and still is today, the leading source of information in this field.

The current decade has seen many changes in the profession, and in an effort to recognize these changes the *AIHA Journal* added a tagline—"A Publication for the Science of Occupational and Environmental Health." Society has moved from the industrial revolution era to the technology phase (2). Global economy has become important, and industrial hygienists are very suitably qualified to meet these challenges, having the skills and knowledge required in the occupational, environmental, and safety areas. Industrial hygiene professionals will continue to analyze the issues and environment of the workplace, to work with multidisciplinary teams and management style, and to approach problem solving in a flexible mode, capable of responding to changes and challenges in the workplace as well as the community. The *AIHA Journal*, along with other publications, is working to provide industrial hygienists with the tools they need in the changing environment.

In 1998 the *AIHA Journal* was made available totally online at the AIHA Home Page web site on the Internet. The current and past articles and/or abstracts are being placed in an archival format with extensive search engine capability for members and others. Effective in 1999, members and subscribers are able to receive the Journal online, in print, or in both formats.

In 1986 the American Conference of Governmental Industrial Hygienists (ACGIH) began to publish a journal called *Applied Industrial Hygiene*. The name was later changed to *Applied Occupational and Environmental Hygiene*.

To respond to members' needs, both AIHA and ACGIH have developed publication distribution programs, which offer a variety of publications and information in a wide variety of formats: print, CD-ROM, software programs, and other multimedia options. Both organizations have established home page sites on the World Wide Web. The AIHA Home Page won awards and recognition as one of the top 50 sites in occupational health and safety, and as the best association web page. For both associations, the Internet World Wide Web sites provide an alternative means of delivering information, and enhancing communications both with and among the members of the profession, also with those outside the profession (http://www.aiha.org or http://www.acgih.org).

Both associations also have developed magazine/newsletter publication for their memberships. ACGIH publishes *Today!*, its membership newsletter. AIHA has *The Synergist*, a monthly magazine for members and subscribers.

Within their publication programs, both provide wide selections of products to meet the diverse needs of the profession. Although ACGIH's catalog concentrates primarily on the works of other publishers, they also publish some key publications themselves, such as their *Threshold Limit Values (TLV)* book, and *Biological Exposure Indices (BEI)*. Other valuable publications are *Industrial Ventilation: A Manual of Recommended Practice* and *Air Sampling Instruments*.

The publication program of AIHA takes a different approach. The AIHA Press actively pursues the development of new publications, some coming from the 37 plus technical committees; others from individual authors. In 1999 AIHA Press published in excess of 100 publications. Resale publications and AIHA Press publications are carefully selected, based on member needs identified in market research.

In 1973 the National Institute for Occupational Safety and Health (NIOSH) published the *Industrial Environment: Its Evaluation and Control*, referred to as the *White Book* by most industrial hygienists. In 1997 AIHA published a new rendition of this book and entitled it *The Occupational Environment: Its Evaluation and Control*. It is still referred to as the *White Book*, and is taking its place as a leading reference and education tool for the profession. AIHA Press publications, *The Occupational Environment: Its Evaluation and Control* and another *The Quiet Sickness: A Photographic Chronicle of Hazardous Work in America*, received several publishing awards in 1998 and 1999. Other publications of note from the AIHA Press include: the *Emergency Response Planning Guidelines* series, the *Workplace Environmental Exposure Level Guide* series, *A Strategy for Assessing and*

Managing Occupational Exposures, Second Edition, and the *Noise and Hearing Conservation Manual*, Fourth Edition. Both AIHA and ACGIH have had selected publications translated into other languages, primarily Spanish; both have international distribution arrangements with other countries. The opening up of communications links via phone, fax, and the Internet have changed and enhanced methods of sharing information and creating knowledge banks for the profession. Utilization of these viable resources by industrial hygienists and others in the occupational health field will enhance the quality of their performance.

3 INDUSTRIAL HYGIENE WORLDWIDE

As membership in the AIHA grew, it was recognized as the leading technical source concerning the practice of industrial hygiene. Simultaneously, communication between countries accelerated, and among the first to join AIHA were industrial hygienists from Canada. From this nucleus, foreign membership increased, with members from England and other European countries. In 1977, there were 127 members from Canada, 20 from England, 10 from Germany, 7 from Italy and 7 from Australia, and 4 from France. In that year 34 foreign countries were represented. In 1999, 45 foreign countries were represented. The Association has become international in scope, and currently has more than 800 members outside the United States.

During the period 1980 to 1990 as the number of professional industrial hygienists in some countries grew, local sections were organized as sections of the AIHA. In five of those sections, the membership and country's interest in industrial hygiene grew and inspired them to become separate organizations (See Table 1.1).

3.1 International Efforts by AIHA Presidents (3)

As communications and transportation technology brought the world closer together, and U.S. headquartered-corporations became multinational in the early 1980s, AIHA presidents began an organized effort to establish formal relationships between AIHA and industrial hygiene organizations abroad. The first effort was made in 1981 by AIHA President Newell E. Bolton who represented the AIHA at the British Occupational Hygiene Society conference. The AIHA presidents who followed continued their contacts with various foreign industrial hygiene organizations. In 1986 President Alice Farrar served as co-chair of the International Congress on Industrial Hygiene held in Rome, Italy. This international meeting, gathering speakers and attendees from more than 15 countries, was one of the first of its kind. During the congress, presidents of several industrial hygiene organizations in attendance, including AIHA, ACGIH, BOHS, AIDII, the French, the Belgians, and the Spanish, gathered to discuss interest in forming an international occupational hygiene association. This culminated in a decision to hold a workshop during the 1987 AIHC in Montreal to develop an international occupational hygiene association. At this workshop, co-sponsored by AIHA and ACGIH, representatives from 10 industrial hygiene organization from eight countries, signed an agreement to form the International Occupational Hygiene Association (IOHA). Among the objectives of the newly organized IOHA in 1987 were to promote and develop occupational hygiene throughout the world, and to promote the exchange of occupational hygiene information among organizations and individuals.

4 PROFESSIONAL CERTIFICATION

During the mid-1950s the AIHA Board of Directors explored the feasibility of certification and registration as a means of improving and maintaining professionalism in the field of industrial hygiene. Two separate committees were established with these charges. Although it was not successful in demonstrating the imperativeness of registration at that time, at the 1989 May meeting of the AIHA Law Committee round table the Committee on Registration cited a critical need to determine what action is necessary to prevent industrial hygienists from being displaced as the professionals within their area of expertise. The problem had been intensified because of the increasing number of new and proposed laws and regulations that limit who can practice in areas traditionally believed to be within the scope of industrial hygiene. The committee had distributed a questionnaire in 1988 to determine how the membership felt about this subject. Over 2000 replies were received. In summary, they indicated that if the association did not act in a timely way the profession was in danger of being legislated out of existence, and further, that the survey conducted was about five years late. Regrettably the efforts of the 1957 ad hoc committee did not come to fruition, and in 1999 industrial hygiene registration still is not a reality.

The efforts of the ad hoc committee on certification were more successful. The committee recommended to the AIHA board that the association endorse and initiate establishment of a voluntary certification program for qualified industrial hygienists. The Board approved this recommendation and appointed an ad hoc committee on certification standards, which in 1958 recommended to AIHA that the ACGIH should be invited to join with AIHA in initiating a certification program, and should be invited to delegate six members to join with the ad hoc committee in planning. In March 1959 the joint committee recommended to the two associations that voluntary certification should be conducted by an *independent* incorporated board, and that the two associations should sponsor this board and advance funds for initial expenses. Both AIHA and ACGIH accepted these recommendations. Each organization thereupon delegated six of its members to join in organizing the board. A charter as a nonprofit corporation under the laws of the Commonwealth of Pennsylvania was approved in September 1960, and the first annual meeting of the American Board of Industrial Hygiene was held in Pittsburgh on October 28, 1960, and the process of professional certification for industrial hygienists was begun.

Eligibility for certification as an industrial hygienist includes five years of practice in the field of industrial hygiene. To become certified one must take a comprehensive two-day exam which is similar in form and intensity to the CPA exam. The American Board of Industrial Hygiene (ABIH) conducts the examination and certification process. If the industrial hygienist meets all the requirements, he or she becomes a Certified Industrial Hygienist (CIH) and must maintain certification by participating in continuing education. More than 50% of the Association's members are CIHs.

In 1966 the diplomates activated the American Academy of Industrial Hygiene as a voluntary professional society. The Academy has conducted a meeting each autumn in various locations in the United States and Canada since 1977. At present (1999) there are 5748 diplomates, 513 retired diplomates, and 550 in training.

All diplomates of the American Board of Industrial Hygiene become members of the American Academy of Industrial Hygiene. They are members in good standing as long as they pay the annual dues assessed by the Board within the time specified by the Board.

5 ACADEMIC PROGRAMS

In 1918 the Harvard Medical School established a Department of Applied Physiology, which in 1922 became a part of the present Harvard School of Public Health as the Department of Physiology and the Department of Industrial Hygiene. This was the first time that instruction and research in industrial hygiene leading to advanced degrees had been offered anywhere in the world. This school cooperates with the graduate school of engineering: any of the courses offered in the School of Public Health may be elected by students working for a degree of master or doctor of science in engineering. The School of Public Health, which is open to graduates of schools of medicine and graduates in arts and sciences with training in basic medical sciences or specialized training and experience in an important phase of public health work, offers the degree of master of public health and, to especially qualified persons, the degree of doctor of public health. The Harvard School of Public Health was the first place in the world where a qualified person could obtain scheduled, broad instruction in industrial hygiene, regardless of whether his undergraduate training had been in medicine or in the sciences. Many of the pioneers of the AIHA were Harvard graduates with backgrounds of medicine, engineering and other sciences who were active in developing and promoting industrial hygiene.

With the passage of the Occupational Safety and Health Act of 1970, the demand for professionals required to accomplish the mandate of the act soared; this included disciplines of occupational medicine, occupational health nursing, occupational safety, and industrial hygiene.

Many more schools offer undergraduate programs in industrial hygiene today than before, however few schools offer an undergraduate "major" in industrial hygiene. Consequently industrial hygienists usually prepare for their careers by pursuing an undergraduate degree in one of the sciences, such as engineering, chemistry, biology, etc., then continue on to attain a master's and/or doctoral degree in industrial hygiene, occupational health, occupational health and safety, environmental health, or other variations.

Information on academic sources in this field was first made available by the National Institute for Occupational Safety and Health (NIOSH). The NIOSH report was first published in 1979 and revised in 1987. The first report listed 201 programs, the later report included only 106 programs.

In 1985 the National Safety Council published a Directory of College and University Safety Courses. It listed 405 schools, offering safety education. A total of 145 degrees were listed, including 38 entitled "environmental health" or "environmental safety and Health."

In 1988 the American Society of Safety Engineers published a survey that listed 129 offered degrees, including 30 in the "environmental" category.

In 1989 AIHA published a report *Education Opportunities in Industrial Hygiene*, which lists 82 institutions offering degrees or certificates in industrial hygiene. Currently, the AIHA identifies 107 institutions offering industrial hygiene-related programs. The list is available in the AIHA Careerworks booklet revised in 1998. More updated lists can be obtained by contacting the AIHA, either by phone at (703) 849-8888, or fax: (703) 207-3561. It can also be obtained by going to the AIHA web site at http://www.aiha.org.

Other key web sites offering the most up-to-date information about training and education in industrial hygiene are National Institutes on Occupational Safety and Health (NIOSH) at: *http://www.cdc.gov/niosh/training.html*. This site provides full information about Training Project Grants and Education & Research Center (ERC) training grants, and about these university-based ERCs. As of April 21, 1998 there were 15 ERCs across the country.

In 1997 the AIHA conducted a member survey of its 12,000 plus professionals, and according to the results, industrial hygienists are a highly educated group, with 98% being college graduates, 60% having master's degrees, and 11% having doctoral degrees. According to the membership data reports, these numbers continue in 1999.

As a group, industrial hygienists generally take an active role in expanding their knowledge by participating in continuing education programs offered by various groups, such as the AIHA, government programs, and educational institutions.

6 PROSPECT

In the decade now nearing termination, the industrial hygiene profession has achieved many of its goals previously set forth, for protecting the health and safety of workers and their environment. The profession has succeeded in "designing out" the health problems before they are incorporated into the workplace. As an example, the "sick building syndrome" that has frequently been a problem because of the lack of input from industrial hygienists in the design of buildings and factories, is now more rare, as architects and builders consult industrial hygienists in their planning stages. The objective of eliminating physical and mental hazards in the workplace and its environment has been successfully promoted. Standards for noise have been formulated; however emission rates for individual machines are still in the developmental stage. Gaseous and particulate emission rates should be essentially zero. Standards for lighting and color are in the making to adhere to the fundamentals of ergonomics.

The problem of lack of "popular" recognition of the industrial hygiene role in ecology still exists, although awareness on the part of the general public and governmental bodies has greatly improved, especially since the AIHA headquarters moved to the Washington, DC, area, the seat of government operations. A more aggressive effort will bring the profession to the attention of policy makers and various publications.

There has been an increasing awareness by the general public of the adverse effects of chemical hazards on their health and wellbeing, and this has created a proliferation of new organizations interested in the environment. They include professional technical organizations, civic groups, and governmental groups at all levels. Their activity has created expanding interest in the profession that we recognize as industrial hygiene.

BIBLIOGRAPHY

1. *U.S. News and World Report*, 98 (Nov. 11, 1991).
2. Personal communication, courtesy of A. Dees, Director Communications, AIHA.
3. Personal communications, courtesy of Alice C. Farrar, past president of AIHA, and currently Senior Vice President of Clayton Environmental Consultants.

CHAPTER TWO

Rationale for Industrial Hygiene Practice

Robert L. Harris, Ph.D., Lewis J. Cralley, Ph.D.
and Lester V. Cralley, Ph.D.

1 BACKGROUND

The professional field of industrial hygiene has its roots in the profession of public health. *Webster's* dictionary defines *public health* as "the art and science dealing with the protection and improvement of community health by organized community effort and including preventive medicine and sanitary and social science." The term *public health* was first used early in the seventeenth century. The term *industrial hygiene* to identify a profession or field of work is probably of twentieth century origin. One must go further back in history, however, to find the origin of its words. The term "industry," with a dictionary meaning, "systematic labor for some useful purpose or the creation of something of value," has its English origin in the fifteenth century. The word "hygiene" goes back to much earlier times. Hygieia, a daughter of Aesklepios who is god of medicine in Greek mythology, was responsible for the preservation of health and prevention of disease. Thus, the roots of the term *industrial hygiene* mean *preservation of health and prevention of disease among people engaged in systematic labor for some useful purpose or the creation of something of value*. In the public health context, then, the purpose of industrial hygiene is to protect the health of communities of workers. The modern definition of industrial hygiene includes protection of the health of persons living around a place of work from hazards that may arise from that place of work.

That working in mines and other workplaces caused diseases and death among workers has been known for more than two thousand years. The association between work and

Patty's Industrial Hygiene, Fifth Edition, Volume 1. Edited by Robert L. Harris.
ISBN 0-471-29756-9

disease was recognized, but knowledge on the toxicity of materials, the hazards of physical and biologic agents, and of ergonomic stressors encountered in industry, did not exist in those very early times. With few exceptions, then, the earliest attention given to worker health was in applying the knowledge at hand, which was primarily the recognition and treatment of illnesses associated with a job.

Thus, although diseases caused by conditions of work have been recognized and treated for many hundreds of years, it was not until around the turn of the twentieth century that major effort begin to be directed toward what we now think of as fundamental industrial hygiene—the anticipation, recognition, evaluation, and control of workplace environmental health stresses in the prevention of occupational diseases.

The aim of this chapter is not to document or present chronologically the major past contributors to worker health and their relevant works, or the events and episodes that gave urgency to the development of industrial hygiene as a science and a profession. Rather, the purpose of the chapter is to place in perspective the many factors involved in relating environmental stresses to health and the rationale upon which the practice of industrial hygiene is based, including the anticipation, recognition, evaluation, and control of workplace stresses, the biological responses to these stresses, the body defense mechanisms involved, and their interrelationships.

2 INSEPARABILITY OF ENVIRONMENT AND HEALTH

Knowledge is constantly being developed regarding the ecological balance that exists between the earth's natural environmental forces and the existing biological species, and how the effects of changes in either may affect the other. In the earth's early biologic history this balance was maintained by the natural interrelationships of stresses and accommodations between the environment and the existing biological species at each particular site. This system related as well to the ecological balance within species, both plant and animal.

Studies of past catastrophic events, such as the ice ages, have shown the effects that changes in this balance can have on the existing species. The forces that brought on the demise of the dinosaurs that lived during the Mesozoic Era are uncertain. Most probably major geological events were involved. Studies have also shown that in the earth's past history a great many other animal species, as well, have originated and disappeared.

The human species, however, has been an exception to the ecological balance that existed between the natural environment and the evolving biological species in the earth's earlier history. The human ability to think, create, and change the natural environment has brought on changes above and beyond those of the existing natural forces and environment. These changes have had, and continue to have, an ever increasing impact upon the previous overall ecologic balance.

The capacity of humans to alter the environment to serve their purposes is beyond the bounds of anticipation. In early human history these efforts predictably addressed themselves to better means of survival, that is to food, shelter, and protection. As these efforts succeeded, humankind was able to devote some of its energy to gaining knowledge concerning factors affecting human health and well-being. Thus evolved the medical sciences, including public health. In some instances, these efforts resulted in intervention in the ecological balance in the control of disease. In other situations the environment may have been altered to make desirable resources available, e.g., in the damming of streams for flood control and developing hydroelectric power. This type of alteration of the localized natural environment and the associated ecological systems may have an impact by developing additional stresses in readjusting the existing ecological balance.

Of more recent impact on health have been the stresses of living brought on by activities associated with personal gratifications such as life-styles as well as those associated with an ever more complex and advancing technology in almost all areas of human endeavor.

The quality of the indoor environment is receiving increasing attention in relation to good health. This applies to the study and control of factors giving rise to psychological stresses associated with living or working in enclosed spaces, as well as air pollution arising from life-styles, building designs, and materials and activities.

The advantages associated with changes in the environment for human benefit and improving the essential quality of living should be assessed for their cost-effectiveness as well as their potential to produce deterioration of the environment and concomitant new stresses. That humans, for optimum health, must exist in harmony with their surrounding 24-hour daily environment and its stresses is self-evident.

For better understanding of the significance of on-the-job environmental health hazards, an overview of the 24-hour daily stress patterns of workers is helpful. This permits a perspective in which the overall component stresses are related to the whole of workers' health.

The human habitat, the earth and its flora and fauna, is in reality a chemical one, that is, an entity that can be described in terms of tens of millions of related elements and compounds. It is the habitat in which the many species have evolved and in which a sort of symbiosis exists that supports the survival of individual species. The intricacy of this relationship is illustrated in the recognition that copper, chromium, fluorine, iodine, molybdenum, manganese, nickel, selenium, silicon, vanadium, and zinc, in trace amounts, are essential to human health and well-being. All, however, are toxic when absorbed in excess, and all are listed in standards relating to permissible exposure limits in working environments. Some compounds of several of these elements are classified as carcinogens. It is most revealing that some trace elements essential for survival are, under some circumstances, capable of causing our destruction. Thus, the matter of need or hazard is a question of "How much?"

The environment is both friendly and hostile. The friendly milieu provides the components necessary for survival; oxygen, food, and water. On the other hand, the hostile environment constitutes a combination of stresses in which survival is constantly challenged.

Although numerous factors are obviously involved in the optimal health of an individual, stresses arising out of the overall environment, that is, the workplace, life-style, and off-the-job activities, are substantial. The stresses encountered over full 24-hour daily periods have an overall impact on an individual's health. Any activity over the same period of time that can be stress relieving will have a beneficial effect in helping the body to adjust to the remaining insults of the day.

The inseparability of the environment and its relation to health is presented graphically in Figure 2.1.

An environmental health stress may be thought of as any agent in the environment capable of significantly diminishing a sense of well-being, causing severe discomfort, interfering with proper body organ functions, or causing illness or disease. These stresses may be chemical, physical, biologic, psychologic, or ergonomic in nature. They may arise from natural or created sources.

2.1 Macrocosmic Sources

Macrocosmic sources of environmental stress agents, such as the hemispheric or global quality or state of air, soil, or water, are those emanating from the sun or from extensive geographical pertubations and are capable of affecting large geographical areas. Examples of natural stress agents are ultraviolet, thermal, and other radiations from the sun, major volcanic eruptions which release huge quantities of gases and particulate material into the upper atmosphere, the changing of the upper air jetstream, and other factors which influence climate, and the movement of the earth's tectonic plates resulting in earthquakes and tidal waves.

Examples of human-created macrocosmic stresses include interference, through release to the atmosphere of some synthetic organic compounds, with the upper atmosphere ozone layer that shields the earth from excessive ultraviolet radiation; the burning of fossil fuels that increases the carbon dioxide level in the atmosphere and alters the earth's heat balance and surface temperature; and the destruction of forests and pollution of oceans which inhibit the biosphere's oxygen producing capability.

These macrocosmic stresses may act directly upon individuals through such conditions as excessive exposure to ultraviolet and thermal radiation, or indirectly by influencing the earth's climate—sunshine, rain, and temperature—thus, affecting vegetation and habitability.

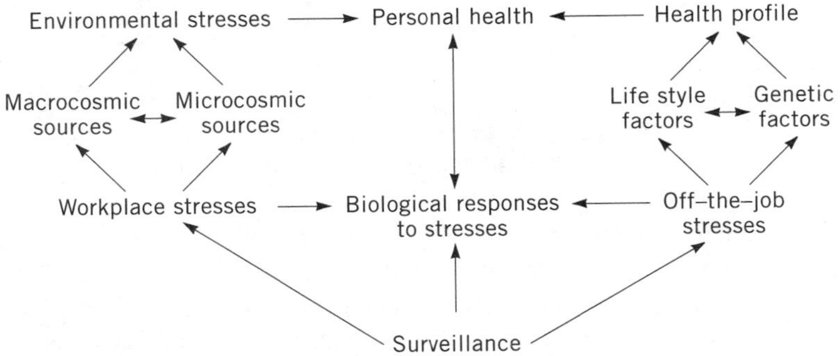

Figure 2.1. Inseparability of environment and health.

2.2 Microcosmic Sources

Microcosmic sources of stress agents are those emanating from localized areas and generally affecting a single region. These are most commonly at the regional or community level and may also include the home and work environments.

Examples of natural sources of microcosmic stress agents are pollen, which gives rise to sensitization, allergy, and hay fever; water pollution from ground sources having high mineral or salt content; and the release of methane, radon, sulfur gases and other air contaminants from underground and surface areas. It is noteworthy that human evolution has taken place in the presence of natural macrocosmic and microcosmic sources.

Human created microcosmic sources of health stress agents at the community level include noise from everyday activities such as lawn mowing, motorcycle and truck traffic, and loud music; air pollution from motor vehicle exhaust, release of industrial emissions into the air, emissions from refuse and garbage landfills, toxic waste disposal sites, spraying of crops, and life-style; surface water pollution through the release of contaminants from home, community, agricultural, and industrial activities into streams; and seepage into groundwater of contaminants from land fills, agricultural activities, and from industrial and other waste disposal sites.

Stress agents from these sources may cause direct responses, such as the effects of noise on hearing and toxic exposures on health, or indirect responses such as acid smog and rain affecting vegetation and soil quality.

Regarding microcosmic sources, it is noteworthy that segments of industry are taking seriously their obligations and opportunities for protection of the health and well-being of both their employees and the communities in which their plants operate. In late 1988 the Chemical Manufacturers Association adopted an initiative identified as Responsible Care: A Public Commitment. Each CMA member company, as a condition of membership, will manage its business according to the following listed principles (1):

> To recognize and respond to community concerns about chemicals and our operations.
>
> To develop and produce chemicals that can be manufactured, transported, used and disposed of safely.
>
> To make health, safety and environmental considerations a priority in our planning for all existing and new products and processes.
>
> To report promptly to officials, employees, customers and the public, information on chemical related health or environmental hazards and to recommend protective measures.
>
> To counsel customers on the safe use, transportation, and disposal of chemical products.
>
> To operate our plants and facilities in a manner that protects the environment and the health and safety of our employees and the public.
>
> To extend knowledge by conducting or supporting research on the health, safety and environmental effects of our products, processes and waste materials.
>
> To work with others to resolve problems created by past handling and disposal of hazardous substances.

To participate with government and others in creating responsible laws, regulations and standards to safeguard the community, workplace and environment.

To promote the principles and practices of Responsible Care by sharing experiences and offering assistance to others who produce, handle, use, transport, or dispose of chemicals.

These principles pledge good practices in controlling microcosmic sources, and through those principles applying to employees, endorse good industrial hygiene practices. The Responsible Care initiative bodes well for community and worker health, as well as for the long-term financial health of the industry.

2.3 Life-Style Stresses

The life-style of individuals, including habits, nutrition, off-the-job activities, recreation, exercise, and rest may have beneficial effects as well as stresses that exert a profound influence on health. Extensive knowledge has been developed, and continues to expand, on the influence of habits such as smoking, alcohol consumption, and use of drugs on health. Stresses from such activities may have additive, accumulative, and synergistic actions, or may exert superimposed responses on the effects of other exposures arising in places of work. These off-the-job agents can be causes of respiratory, cardiovascular, renal, or other diseases, and may create grave health problems in individuals over and above any effects of exposures encountered in workplaces.

Knowledge is available on the deleterious effects on the health of their offspring caused by smoking, alcohol consumption, and drug use by women during pregnancy; effects include malformation and improper functioning of body organs and systems, low birth weight, etc.

The benefits to health of good, adequately balanced nutrition, i.e., vitamins, minerals, and other essential food intake, are gaining increased attention in relation to general fitness, weight control, prevention of disease, supporting the natural body defense mechanisms, recovery from exposure to environmental stresses, and aging. Conversely, malnutrition and obesity are associated with many diseases or dysfunctions, and may have synergistic effects with exposures to other stresses.

Recreational activities are important aspects of good health practices, releasing tension brought on through both off- and on-the-job stresses. Conversely, many recreational activities may be harmful, such as listening to excessively loud music which may lead to hearing decrement, frequent engagements in events and schedules that interfere with the body's internal rhythmic functions, failing to observe needed precautions while using toxic agents in hobby activities, and pursuing activities to the point of exhaustion.

Both exercise and rest are important for maintaining good health. Exercise helps in maintaining proper muscular tone as well as weight control. Exercise, however, should be designed for specific purposes, maintained on a regular basis, and structured to accommodate the body physique and health profile of the performing individual. Otherwise more harm than benefit may result. Rest provides time for the body to recuperate from physical and psychological stresses.

The hours between work shifts and during weekends provide time for the body to excrete substantial portions of some chemical agents absorbed during a work shift or a work week. The significance of these nonexposure recovery periods, of course, depends upon the biologic half-time of the chemical agent of interest. Work patterns that disturb this recovery period, as in moonlighting, may have an especially deleterious effect if similar stresses are encountered on the second job. Likewise, smoking, alcohol consumption, and use of drugs may impair the body's proper recuperation from previous stresses.

2.4 Off-the-Job Stresses

Workers may encounter a host of stresses outside places of work. These may be chemical, physical, biologic, or psychologic in nature and usually are encountered during the 5- to 6-hour period between the end of the work period and the beginning of sleep time, and on weekends. Many off-the-job activities, if performed in excess and without regard to necessary precautions, are capable of producing stress and injury. Hobby and recreational activities may account for a substantial portion of this time. Hobby participants may encounter environmental exposures from activities such as soldering, welding, cleaning, gluing, woodworking, grinding, sanding, and painting, which are experienced by workers on the job, but which on the job are well controlled. The home hobbyist, however, often does not have available the appropriate protective devices such as local exhaust ventilation, protective clothing, goggles, and respirators. The same hobbyist may neglect other good safety practices; he or she may ignore precautionary labels, take shortcuts to save time, fail to use the basic principles of keeping toxic materials from the skin, neglect thorough washing after skin contact, and may smoke or eat while working with such materials.

The off-the-job gamut of health stresses is wide and formidable. To deal with these stresses satisfactorily requires that a degree of accommodation be reached based on judgment, feasibility, personal options, objectives, and other factors. It is evident that although the components of the total sum of environmental health stresses must be considered on an individual basis, no single component can stand alone and apart from the others.

2.5 Workplace Stresses

Places of work may be the most important sources of health stresses if workplace operations have not been studied thoroughly and the associated health hazards have not been eliminated or controlled. This was evidenced during the time of early industrial development when little information was available on methods for identifying, measuring, evaluating, and controlling work related stresses. During this period workplace exposures were often severe, leading to high prevalence of diseases and excess deaths.

Beginning about the turn of the twentieth century, and especially since the 1950s, management, labor unions, government, academic, and other groups have taken substantial interest in worker health and in the control of exposures to stresses in places of work. This, along with the setting of standards of exposure limits, has created a broad support for expanding knowledge and practice to prevent illnesses and diseases associated with on-the-job exposures. These volumes focus on the theory and practice for anticipating, recognizing, evaluating, and controlling on-the-job health hazards, and on those professions which are allied with industrial hygiene in protection of the health of persons exposed to workplace and community stresses.

2.6 Biological Response to Environmental Health Stresses

The human body consists of a number of discrete organs and systems derived from the embryonic state, encased in a dermal sheath, and developed to perform specific functions necessary for the overall functioning of the body as an integral unit. These organs are interdependent so that a malfunction in one may affect the functioning of many others. As an example, in the alveoli, a malfunction which hinders the passage of oxygen into the blood transport system may have a direct effect on other organs through their diminished oxygen supply. Similarly, any hormonal imbalance or enzyme aberration may affect the functioning of many other body organs. Once absorption has occurred, toxic agents may selectively target one or more of the organs. Each organ has its own means for accommodating to, or resisting, stress and adjusting to injury, and its own propensity for repair.

Although the organ structures and functions of the several experimental animal species have many similarities to those of the human body, with some more similar than others to those of humans, care has to be taken in extrapolating research data on any one experimental animal species to the human. Similarly, research data obtained in isolated systems such as cell cultures have to be cautiously interpreted when applied to even the same cells in the whole integrated human body organ system.

2.7 Body Protective Mechanisms Against Environmental Stresses

The human body, in coexisting with the hostile stresses of the external environment, has developed a formidable system of protection against many of these stresses. This is accomplished in a remarkable manner by the ectodermal and endodermal barriers resisting absorption of noxious agents through inhalation, skin contact, and ingestion, and supported by the backup mesodermal and biotransformation mechanisms once absorption has occurred. These external and internal protective mechanisms, however, are not absolute and can be overwhelmed by a stress agent to the extent that they are ineffective, with resultant disease and death. Also, these protective mechanisms may become impaired in various degrees from insults associated with life-styles and other daily activities. In studying the effects of specific stresses on health it is important to be aware of the body's protective mechanisms. Suitable control of a specific stress should supplement the body's protective response to that stress.

2.8 Coaption of Health and Environmental Stresses

To survive, the human body must live in balance with the surrounding environment and its concomitant stresses. Since these stresses, singly or combined, are not constant in value even for short periods of time or over limited geographical areas, the body must have built-in mechanisms for adjusting to differing levels of stresses through adaptation, acclimatization, and other accommodating mechanisms. There is a limit, however, to which the body can protect itself against these stresses without a breakdown occurring in its protective systems.

An extremely thought-provoking concept on associations between environmental stresses and health decrements has been presented by Professor Theodore Hatch (2, 3). His concept examines associations between stresses and the human body's adjustments, compensations, and finally breakdown and failure, in response to them. The concept is particularly useful in understanding the effects of multiple risk factors when they include those of both occupational and nonoccupational origin.

If humans are to have freedom of geographic movement and of living in highly hostile environments which produce environmental insults greater than can be handled by the body's protective and adaptive mechanisms, some means of protection other than accommodation by the body must be provided. Extreme examples of the need for such protection are living and working in confined spaces where the immediate human environment is under absolute control against outside catastrophic stresses as in the cases of space travel and submarine activities.

More typical of this coaptive relationship is exposure to ultraviolet radiation from the sun. It is obvious that avoidance of all ultraviolet radiation from this source is impractical. In addition to the ozone layer of the upper atmosphere shielding the earth from major levels of ultraviolet radiation emitted by the sun, and the body's own protective mechanisms such as skin pigmentation, further accommodation is reached through the use of special clothing, skin barriers, eye protection, and a managed limitation to exposure.

At high altitudes where the partial pressure of oxygen is diminished from that to which the human body may be accustomed at lower elevations, the body acclimatizes in time by increasing the number of red blood cells and hemoglobin that carry oxygen to the tissues. When the availability of atmospheric oxygen decreases below the limit of acclimatization, further accommodation may be provided externally through the use of supplemental oxygen supply.

The human body has a number of regulating mechanisms to keep its temperature within normal limits when it is exposed to excessively high or low environmental temperatures. This permits living in a limited but wide range of environmental temperatures. Further accommodation to extremes in environmental temperatures may be provided through special clothing, protective equipment, and living and working in climate controlled structures.

Where excessive exposures to environmental stresses exist in a workplace, emphasis is placed on their elimination or on lowering them to acceptable levels. Stress levels should be lowered through recognized engineering or administrative control procedures to the point that the body defense mechanisms can adequately prevent injury to health. Some situations may not be amenable to engineering or administrative control and may properly require the use of personal protective equipment or other control strategies.

There is a limit to what can be done to alter existing environmental stresses from natural macrocosmic sources. Thus, these become ubiquitous background stresses upon which other exposures from microcosmic sources such as community and industrial pollution, off-the-job activities, life-styles, and on-the-job activities are added. Rationally, then, it is primarily the created stresses from predominantly microcosmic and a few macrocosmic sources that are amenable to control. These must be kept within acceptable limits to permit humans to avoid health damage from the stresses of an increasingly complex technologic age.

3 INDUSTRIAL TECHNOLOGICAL ADVANCES

In the early history of mankind, the many activities associated with living were at the tribal level where emphasis was placed upon survival, i.e., on procuring adequate food, protection, and shelter. The tribes, many of whom were nomadic, were undoubtedly aware that they lived in accommodation with their environments. This would have been evidenced through the appropriate use of clothing, safe use of fire, and observing climatic patterns.

It was inevitable that the nomadic way of life would give way in most instances to a more settled life-style in which cooperative efforts for food, shelter, and protection were more dependable than those based on individual or small group effort and ingenuity. During this transition period accommodations to the natural elements were made easier through more permanent shelter and and a more organized pattern of living.

The next advancement in accommodation came through the realization that increased production could be attained through specialization of work pursuits wherein a designated work group devoted its principal effort to the making of a single commodity such as clothing, pottery, or tools, or the growing of foods. Each group shared its commodity in exchange for the products of other groups. This type of trade evolved to cottage type industries that related primarily to the community level of commerce. Even at that level of production many of the health stresses associated with different pursuits were intensified over those of nomadic living in which every person was a sort of jack-of-all-trades. This was especially true where the operations tended to be restricted to confined and crowded spaces.

As means for communications and transportation improved, trade increased between adjoining communities, and the search continued for ways to produce commodities in increased volume with less manpower. Similar types of production operations tended to expand and to be concentrated within the same housing structure. This led to increases in the health stresses for the whole workforce.

This trend toward industrialization continued and intensified with the advent of the steam engine. Developments such as the steam engine gave rise to the industrial revolution which brought larger factories and new production techniques, along with increased health risks. Where in earlier times of cottage industries there were relatively few health risks in any one workplace, the new technology and industrialization led to more complex patterns of exposures.

Since the advent of the industrial revolution technological advances and their application to production have expanded at an ever increasing pace. In the latter half of the twentieth century the application in industry of knowledge gained through space technology and other such research has rapidly expanded into the current high-technology electronic age.

The advent of this high technology and its application to production is having its effect upon both the nature of employment and the concomitant health stresses. While in the past workers needed only specific instructions to perform most job operations—and this will continue for some time—the move into higher technology has created a demand for highly trained employees, rather than persons merely instructed in job requirements, for many job positions. This trend can be expected to increase dramatically. Computers, word processors, video display terminals, lasers, microwave, and other electronic equipment are becoming commonplace in industry. Robots, which require sophisticated management and control, are taking over many repetitive operations such as painting and metal working.

The nature and extent of associated health stresses are becoming more complex with the advent of high-technology industry. At the same time, the health effects of these stresses are becoming more detectable with more sophisticated measurement and diagnostic tools.

The urgent need for knowledge concerning the effects of exposures to health stresses associated with an ever expanding industrial technology, along with methods for their recognition, evaluation, and control, gave rise in the twentieth century to the science of industrial hygiene. This science must keep attuned to the ever changing applications of technology in industry.

4 EMERGENCE OF INDUSTRIAL HYGIENE AS A SCIENCE AND PROFESSION

Science may be defined as an organized body of knowledge and facts established through research, observations, and hypotheses. As such, science may be basic, as exemplified by the fundamental physical sciences, or it may be applied in the sense that the principles of other sciences are brought to bear in developing facts and knowledge in a specific area of application.

During the early history of industrial development the lack of knowledge on the effects of health stresses associated with industrial operations and how these could be controlled, with the concomitant massive exposures to toxic materials in places of work, led to many serious episodes of illness, disease, and death among workers. An example is the high incidence of silicosis and silicotuberculosis that existed a century ago among workers in hard rock mines, the granite industry, and in tunneling operations, wherever the dust had a high free silica content.

During this early period the major effort on behalf of workers' health was to apply the knowledge at hand, which related primarily to the recognition and treatment of occupational illnesses.

It was not until around the beginning of the twentieth century that specific attention began to be devoted to the preventive aspects of occupational illnesses. Scientists and practitioners, including engineers, chemists, and physicists, began to apply their knowledge and skills toward the development of methods and procedures for identifying, measuring, and controlling exposures to harmful airborne dusts and other chemical agents in workplaces. At that time there were no recognized procedures for carrying out these activities.

Choosing means for assessing levels of chemical agents in air early in the twentieth century is a case in point. Many air sampling procedures were tried by various investigators; the impingement method was judged the most adaptable one at that time for collecting many of the airborne contaminants such as particulates, mists, some fumes, and gases. The light-field microscopic dust counting technique was developed for enumerating levels of mineral dust in the air; and conventional wet chemical analytical methods available at the time were adapted for measuring quantities of chemical agents in these samples.

Even during this early period of development of airborne sample collection and analytical procedures, scientists realized that the exposure patterns that existed were more complex and complicated than the instruments for sample collection and analysis could define. These scientists also knew of many of the deficiencies associated with the measurement procedures being developed. They were aware that the data being collected represented only segments of the overall exposure patterns. They believed, however, that these segments could be used as indices which would represent overall exposure patterns so long as the production techniques and other operational factors remained the same. It must be remembered that at that time information was not available on respiratory deposition and dust size. It was imperative to investigators, and rightly so, that some method, with whatever deficiencies that may have been incumbent, be developed for indexing airborne levels of contamination in workplaces, both for estimating levels of exposure and for use as benchmarks in determining degrees of air quality improvement after controls had been established.

Since the earliest instruments for collecting airborne contaminant samples were generally nonportable during operation, the collected samples represented general room levels of contamination and the results depended on where in the workplace the samples were taken.

The procedures described above for airborne sampling and analysis in workrooms, as primitive as they may seem today, served well during that period of time. They accounted for the drastic reduction in massive exposures that existed in many work sites and were the methods and procedures upon which future refinements would be made.

These early scientists showed that it was feasible to lower the massive workplace airborne contamination levels that often existed at that time, and by relating the data to the health profiles of workers they observed that the lowering of exposure level also lowered the incidence of the associated diseases. Thus, began the first field studies that were to have a profound influence on the development of the earliest permissible exposure limits and in developing the rationale upon which the practice of industrial hygiene is predicated. The insights developed in these field studies were to be further substantiated through laboratory and clinical research.

The development of the hand-operated midget impinger pump during the 1930s was a decided improvement over the standard impinger pump since it was portable and permitted movement about a workplace while airborne samples were being taken. Samples of particulate or gaseous agents could be taken with glass impingers or fritted-glass bubblers near workers as they moved about their tasks. These new worker exposure data demonstrated that workers often had higher levels of exposure than those indicated by the general room airborne levels.

Other instruments, such as the electrostatic precipitator and evacuated containers, came into use during this period. In the late 1940s, the paper and membrane filter methods for collecting some airborne particulate samples came into use. The filter sampling procedure was found to be superior in many respects to the impingement method. The method did not fracture particles or disperse agglomerates, which often accompanied impingement collection, and could be performed in ways that permitted direct microscopic observation of particles and gravimetric measurement of samples.

Insights also began to emerge on the importance of particle size in relation to deposition and retention of particulate material in the respiratory tract. Electron microscopes and membrane filters made it possible to study particles of submicron sizes.

A surge of improved and sophisticated techniques for quantifying workers' exposures to health stress agents took place in the 1960s. This applies both to sample collection and analytical techniques in which much lower levels of exposure to specific agents could be quantified. There also began a dramatic increase in toxicologic and epidemiological studies by government, industry, universities, and foundations, directed to obtaining data upon which to base exposure standards as well as good industrial hygiene practices.

Another major advancement in developing better methods for studying occupational diseases occurred at midcentury. Toxicologic and other studies had revealed that lowering the level of exposure and extending the exposure duration changed the dose–response pattern of many toxic agents. As an example, a high level of exposure to airborne lead produces an acute response over a relatively short period of time. In contrast, lowering the exposure level of this agent and extending the exposure time shows a different dose–response pattern, a chronic form of lead poisoning. Thus, in studying the effects of exposure to health stress agents it is important to obtain relevant dose–response data over an extended period of exposure time and exposure intensity.

One method of obtaining relevant health profile data on workers is through study of their medical records. Another method is through the study of causes of death among worker populations using death certificates located through Social Security, management, retirement system, and labor union records. Such studies have revealed that a lifetime of work exposure to an agent, or an extended observation period of twenty or more years from time of initial exposure, may be necessary to fully define the wide range of dose–response relationships. This may be especially true for carcinogenic and other long latency types of exposure response.

The establishment of professional associations to support the interests and growth of the profession has played an important role in developing industrial hygiene as a science. In the United States in the 1930s the American Public Health Association had a section on industrial hygiene that supported the early growth of the profession. The American Conference of Governmental Industrial Hygienists was organized in 1938. The American Industrial Hygiene Association was organized in 1939. The American Board of Industrial Hygiene was created and held its first meeting in 1960. This Board certifies qualified industrial hygienists in the comprehensive practice of industrial hygiene, and in the past has certified in six additional industrial hygiene specialties as well. Industrial hygienists certified by the Board have the status of Diplomates and as such are eligible for membership in the American Academy of Industrial Hygiene.

A number of years ago the American Academy of Industrial Hygiene developed and adopted a 15-point Code of Ethics for the professional practice of industrial hygiene. At the initiative of the Academy, representatives of Academy, the American Board of Industrial Hygiene, the American Industrial Hygiene Association, and the American Conference of Governmental Industrial Hygienists drafted, with legal guidance, a revised Code which was adopted by the Boards of the four organizations in late 1994 and early 1995. The

Code addresses industrial hygienists' responsibilities to the profession, to workers, to employers and clients, and to the public.

The American Industrial Hygiene Association administers a laboratory accreditation program with the objective of assisting those laboratories engaged in analyses of industrial hygiene samples in achieving and maintaining performance levels within acceptable ranges.

The American Industrial Hygiene Foundation was established in 1979 under the auspices of the American Industrial Hygiene Association. The functions of the Foundation are carried out by an independent Board of Trustees. The Foundation provides fellowships to a limited number of industrial hygiene graduate students, encourages qualified science students to enter the industrial hygiene profession, and stimulates major universities to establish and maintain industrial hygiene graduate programs.

The Social Security Act of 1935 and the Walsh-Healy Act of 1936, had an immense impact in giving increased stability, incentive, and expanded concepts in the practice of industrial hygiene. These Acts stimulated industry to incorporate industrial hygiene programs as an integral part of management. They also stimulated broad base programs in industry, foundations, educational institutions, insurance carriers, labor unions, and government which address the causes, recognition, and control of occupational diseases. These Acts established the philosophy that the worker had a right to earn a living without endangerment to health, and were the forerunners for the passage of the Occupational Safety and Health Act of 1970.

Passage of the Occupational Safety and Health Act of 1970, which has the purpose of assuring " . . . so far as possible every man and woman in the nation safe and healthful working conditions . . ." had a very broad bearing on the further development and practice of the industrial hygiene profession. These enabling Acts, and the regulations deriving from them, have been substantial factors in the broad recognition of industrial hygiene as a science and a profession. It has been necessary to expand the profession in all of its concepts and technical aspects to meet its expanded responsibilities.

Other industrialized countries have had similar experiences in the professional recognition and growth of the science of industrial hygiene.

4.1 Definition of Industrial Hygiene

The American Industrial Hygiene Association defines industrial hygiene as " . . . the science and practice devoted to the anticipation, recognition, evaluation, and control of those environmental factors or stresses, arising in or from the workplace, that may cause sickness, impaired health and well-being, or significant discomfort among workers and may also impact the general community." Because the science and practice of industrial hygiene continues to evolve the Association, from time to time, reviews and revises its definition.

By any definition, however, industrial hygiene is an applied science encompassing the application of knowledge from a multidisciplinary profession including the sciences and professions of chemistry, engineering, biology, mathematics, medicine, physics, toxicology, and other specialties. Industrial hygiene meets the criteria for the definition as a science since it brings together in context and practice an organized body of knowledge necessary for the anticipation, recognition, evaluation, and control of health stresses in the work environment.

In the early 1900s, the major thrust in the control of workplace health stresses was directed toward those areas in industry having massive exposures to highly toxic materials. The professional talents of engineers, chemists, physicians, physicists, and statisticians were those largely used in these programs. As industrial technology advanced, the complexity of workers' exposures also increased, along with an increase in the professional knowledge and skills needed to study the new health effects and to develop the methods for anticipation, recognition, evaluation, and control of the new environmental stresses. The need for new knowledge and skills continues now with the advent of high technology in the electronic and allied industries. Factors such as improper lighting and contrast, glare, posture, fatigue, need for intense concentration, tension, and many other stresses arise in the operation of computers, word processors, video display terminals, and laser, microwave, and ionizing radiation equipment, which are becoming commonplace in industry. Thus, concerns for the health of employees above and beyond that of toxicity response arise. The study and control of these nonchemical stress agents point to the need for the occupational health nurse, psychologist, human factors engineer, ergonomist, and others to join the professional team in studying the effects and control of the ever widening list of health stress agents in places of work.

In the early practice of occupational health nursing, emphasis was placed on such activities as the emergency treatment of traumatic injuries stipulated in written orders of a physician, and in maintaining records and information relating to physical examinations and the like. With the current advanced training of occupational health nurses, this limited role has been found to be wasteful of professional talent and resources. Occupational health nurses are often the first interface between workers and pending health problems, and are in a position to gain information on situations and health stresses both on and off the job that may, unless addressed, lead to more serious responses. Occupational health nurses have increasingly become members of the multidisciplinary team needed in the recognition of job associated health stresses. Similarly, psychologists, in the study of effects of strain, tension, and similar stresses, and human factors engineers and ergonomists in designing machines, tools and equipment compatible with physical and morphologic limitations of workers, are examples of other professionals joining the multidisciplinary team studying the effects and control of the increasingly complex health stresses associated with advanced industrial technology.

The complexity of the multifaceted professional effort needed for carrying out the responsibilities of professional practice in the protection of worker health is further illustrated in the 28 technical committees of the American Industrial Hygiene Association and the six different specialty areas of certification which have been used in the past by the American Board of Industrial Hygiene.

4.2 Rationale of Industrial Hygiene Practice

The practice of industrial hygiene is based upon the following observations, experiences, and rationales:

1. Environmental health stresses in the workplace can be quantitatively measured and expressed in terms that relate to the degree of stress.

2. Stresses in the workplace, in main, show a dose–response relationship. The dose can be expressed as a value integrating the concentration or intensity, and the time duration, of the exposure to the stress agent. In general, as the dose increases the severity of the response also increases. As the dose decreases the biological response decreases and may at some time and dose value exhibit a different kind of response, chronic versus acute, depending on the time duration of the stress.

3. The human body has intricate mechanisms of protection, both in preventing the invasion of hostile stresses into the body and in dealing with stress agents once invasion has occurred. For most stress agents there is some point above zero level of exposure which the body can tolerate over a working lifetime without injury to health.

4. Levels of exposure of workers to specific stress agents should always be kept within prescribed safe limits. Regardless of their type, all exposures in the workplace should be kept at such levels lower than prescribed levels as are reasonably attainable through good industrial hygiene and work practices.

5. Some stress agents may cause serious biological responses among a few workers at such low levels that exposures should be controlled to levels as low as reasonable achievable regardless of any higher regulatory limit. Examples of such an agent are one having genotoxic properties, or one which may cause hypersensitivity reaction brought on by very low exposure to some agent for which some earlier sensitizing dose has occurred.

6. The elimination of health hazards through process design and/or the use of nonhazardous substitute materials should be the first objective in maintaining a safe workplace. When this is not feasible, recognized engineering or administrative controls should be used to keep exposures within acceptable limits. In some cases, when feasible engineering and administrative controls are insufficient, supplemental programs such as the use of personal protective equipment or other control practices have application.

7. Surveillance of both the work environment and workers should be maintained to assure a healthful workplace.

4.3 Elements of an Industrial Hygiene Program

The purpose of an industrial hygiene program is to assure a healthful workplace for employees. It should include all the functions needed in the anticipation, recognition, evaluation, and control of occupational health hazards associated with production, office, and other work. This requires a comprehensive program designed around the nature of the operations, documented to preserve a sound retrospective record, and executed in a professional manner.

The basic components of a comprehensive program include the following:

1. Coordinated technical activities capable of anticipating health stresses which may result from changes in materials or processes, and of detecting occupational health stresses in any part or process of the establishment.

2. Capability to conduct or obtain measurement and evaluation activities for the assessment of occupational health stresses anywhere in the establishment.

3. Capability to determine the need for and to obtain and maintain effective engineering and administrative controls necessary for safe and healthful workplaces throughout the establishment.

4. Participation in the periodic review of worker exposure and health records to detect the emergence of insufficiently controlled health stresses in the workplace.

5. Participation in research, including toxicological and epidemiological studies designed to generate data useful in establishing safe levels of exposure.

6. Maintaining a data storage system that permits appropriate retrieval of information necessary for the study of long term effects of occupational exposures.

7. Assuring the relevancy of the data being collected.

An integrated program is capable of responding to the need for the establishment of appropriate exposure controls, both for current needs and for those which may result from technological advances and associated process changes.

The almost universal availability of high capacity and powerful desk-top computers which has taken place over the past one or two decades has greatly facilitated the conduct of industrial hygiene programs. A great amount of industrial hygiene related software for record keeping, technical reference (e.g., regulations, safety data sheets, etc.), sampling data analysis, exhaust ventilation design, and other such industrial hygiene functions has become available from commercial sources, or through professional journals, professional associations, and individual industrial hygienists. The American Industrial Hygiene Association has a Computer Applications Committee whose mission is to provide a forum for advancing the use of computer applications by occupational and environmental health professionals. Among other activities this Committee reviews and reports on available software.

The industrial hygienist at the corporate or equivalent level should report to top management. His/her responsibility involves appropriate input whenever product, technological, operational, or process changes, or other corporate considerations, may have an influence on the nature and extent of associated health hazards. When new plants or processes are in prospect, the corporate industrial hygienist should assure that adequate controls are incorporated at the design stage.

5 HEALTH HAZARD RECOGNITION

An important aspect of a responsive industrial hygiene program is that it is capable of recognizing potential health hazards or, when new materials and operations are encountered, to exercise expert judgment in maintaining an adequate surveillance program until any associated health hazards have been defined. This should not be a problem in cases involving operations, procedures, or materials for which adequate knowledge is available. In such cases it is primarily a matter of application of available knowledge and techniques. In operations and procedures involving a new substance or material for which relevant information is limited or unavailable, it may be necessary to extrapolate information from other kindred sources, to use professional judgment in setting up a control program with a reasonable factor of safety, and to incorporate an ongoing surveillance program to further define health hazards that may emerge. In some instances it may be necessary to undertake toxicological research prior to the production stage to define parameters needed in setting up the control and surveillance program.

One of the basic concepts of industrial hygiene is that the environmental health stresses of the workplace can be quantitatively measured and recorded in terms that relate to the degree of stress.

The recognition of potential health hazards is dependent on such relevant basic information as:

1. Detailed knowledge of the industrial process and any resultant emissions that may be harmful.
2. The toxicological, chemical, and physical properties of these emissions.
3. An awareness of the sites in the process that may involve exposure of workers.
4. Job work patterns with energy requirements (i.e., metabolic levels) of workers.
5. Other coexisting stresses that may be important.

This information may be expressed in a number of ways depending upon its ultimate use. A very effective form is a material-process flow chart that lists each step in the process along with the appropriate information just noted. This permits the pinpointing of areas of special concern. The effort in whatever form it may take, however, remains only a tool for the use of the industrial hygienist in the actual identification of the stresses in the workplace. In the quantification itself, many approaches may be taken depending on the information sought, its intended use, the required sensitivity of measurement, the level of effort and instruments available, and the practicality of the procedures.

Aside from the production workplace with the attendant toxicological, physical, and other related health stresses, a new area of concern is rapidly gaining special attention where employees may be subjected to a high degree of stress from tension, physical and mental strain, fatigue, excessive concentration, and distraction such as may exist for operators of computers, word processors, video display terminals and other operator intensive equipment. Off-the-job stresses, life-style factors, and the immediate room environment may become increasingly important for such workers. The recognition of associated health stresses and their evaluation require a special battery of psychological and physiological body reaction and response tests to define and measure factors of fatigue, tension, eyestrain, deficits in the ability to concentrate, and the like.

6 EXPOSURE MEASUREMENTS

Both direct and indirect methods may be used to measure exposures to stress agents. Table 2.1 illustrates direct and indirect measurements of chemical agents.

Table 2.1. Methods for Measuring Workers' Exposures to Absorbed Chemical Stress Agents

Direct	Indirect
Body Dosage	*Environment*
Tissues	Ambient air
Fluids	Interface of body and stress
Blood	
Serum	*Physiological Response*
Excreta	Sensory
	Pulse rate and recovery pattern
	Body temperature and recovery pattern
	Voice masking, etc.
Urine	
Feces	
Sweat	
Saliva[a]	
Hair[a]	
Nails[a]	
Mother's milk[a]	
Alveolar air	

[a]Not usually considered to be excreta.

6.1 Direct Measurements

To measure directly the quantity of a chemical agent actually absorbed by the body, fluids, tissues, expired air, excreta, etc., must be analyzed for the agent *per se* or for a biotransformation product. The quantification of a body burden of the agent requires information regarding the biological half-time of the agent or its metabolite as well. Such procedures may be quite involved, since the evaluation of the data at times depends on previous information gathered through epidemiological studies and animal research. Studies on animals, moreover, may have used indirect methods for measuring exposure to the stress agent, necessitating appropriate extrapolation in the use of such values. The current adopted list of Biological Exposure indices published by the American Conference of Governmental Industrial Hygienists lists 62 determinants (the agent or a metabolite) for 39 compounds, and gives notice of intent to adopt 6 more determinants for 5 compounds. Twenty-one additional compounds are under study by the Committee for establishment of biological exposure indices.

One decided advantage of biological monitoring coupled with information on the biological half-time of an agent or its metabolite, is that exposure can be integrated on a time-weighted basis. Such integration is difficult to estimate through ambient air sampling when the exposure is highly intermittent or involves peak exposures of varying duration. Conversely, biological monitoring may fail to reflect adequately the influence of peak concentrations *per se* that may have special meaning for acute effects. Urine analysis may also provide valuable data on body burden in addition to current exposures when the samples are collected at specific time intervals after exposure. In general, quantitative body-burden interpretation of analytic values for a biological specimen requires knowledge of biologic half-time for the agent of interest and an appropriate exposure-sampling time sequence and schedule.

Sampling of alveolar air may be an appropriate procedure for monitoring exposures to organic vapors and gases. An acceleration of research in this area can be anticipated because of the ease with which samples can be collected.

The use of biologic specimens, particularly for purposes of research, ordinarily requires the informed consent of each study subject who provides a sample. This is clearly necessary for an invasive procedure such as blood sampling, and may apply even to the collection of excreta and exhaled air. Invasion of privacy may be at issue, e.g., detection of alcohol consumption in the analysis of exhaled breath or other excreta.

6.2 Indirect Measurement

The most widely used technique for the evaluation of occupational health hazards is indirect in that the measurement is made at the interface of the body and the stress agent, e.g., the breathing zone or skin surface. In this approach the stress level actually measured may differ appreciably from the actual body dose. For example, all the particulates of an inhaled dust are not deposited in the lower respiratory tract. Some are exhaled and others are entrapped in the mucous lining of the upper respiratory tract and eventually are expectorated or swallowed. The same is true of gases and vapors of low water solubility. Thus, the target site for inhaled chemicals is scattered along the entire respiratory tract, depending on their chemical and physical properties. Another example is skin absorption of a toxic material. Many factors, such as the source and concentration of the contaminant, i.e., airborne or direct contact, and its characteristics, body skin location, and skin physiology, relate to the amount of the contaminant that reacts with or is absorbed through the skin. For chemical agents, the sampling and analytical procedures must relate appropriately to the chemical and physical properties such as particle size, solubility, and limit of sensitivity of analytical procedures, for the agent being assessed. Other factors of importance are average values, peak exposures, and the job energy requirements which are directly related to respiratory volume and pulmonary deposition characteristics.

Exposures to physical agents such as noise and ionizing radiation are almost always measured by indirect methods such as dosimetry or assessment of work area intensity levels.

The indirect method of health hazard assessment is, nevertheless, a valid one when the techniques used are the same or equivalent to those relied on in the studies that established the standards.

7 ENVIRONMENTAL EXPOSURE QUANTIFICATION

Procedures for measuring airborne exposure levels of a stress agent depend to a great degree upon the reasons for making the measurements. Some of these are: (*1*) Obtaining worker exposure levels over a long period of time on which to base permissible exposure limits; (*2*) compliance with standards; and (*3*) performance of process equipment and controls. It is essential that the data be valid regardless of the purpose for which they were collected and that they be capable of duplication. This is a key factor in establishing exposure limits to be used in standards and in fact-finding related to compliance. Since judgment and action will in some way be passed on the data, validity is paramount if the data are to be used as a *bona fide* basis for action.

7.1 Long-Term Exposure Studies

In epidemiologic studies in which the relationship between a stress agent and the body response is sought, ideally the stress factor would be characterized in great detail. This could require massive volumes of data suitable for statistical analysis and a comprehensive data collecting procedure so that a complete exposure picture may be accurately constructed. The sampling procedures and strategy should be fully documented, including number and length of time of samples, their locations, their types, i.e., personal or area samples, and should be adequate to cover the full work-shift activities of the workers. Any departure from normal activities should be noted. These are important since the data may be used at a later date for a purpose not anticipated at the time of sample collection.

This ideal situation is seldom the case in epidemiologic studies. In research on long-term health effects the typical epidemiologic study involves use of surrogates for exposures, or efforts to retrospectively reconstruct exposures. The collection of valid retrospective data may be extremely difficult. If available at all, actual sampling data may be scanty, the sample collection and analytical procedures used in the past may not have been well documented as to precise methodology, and may have been less sensitive and efficient, or may have measured different parameters, than do current procedures; sampling locations and types may not be well defined; and the job activities of the workers may have changed considerably even though the job designations may be the same. Other factors which need to be considered in securing retrospective data relate to contrasting past and current plant operations, including changes in technology and raw materials, effectiveness of exposure control procedures and their surveillance, and housekeeping and maintenance practices. In many instances an attempt to accommodate these differences has been made through broad assumptions and extrapolations with an unknown degree of validity and without expressing the limitations of such derived data. On the other hand, some extraordinarily good exposure reconstruction work has been accomplished in the past decade (4–6).

The effect of national emergencies may significantly change the nature and extent of workers' exposures to stress agents. The experience during World War II is an example. The number of hours worked per week were increased substantially in many industries. Control equipment was allocated to specified industries and denied to others. Local exhaust ventilation systems at times became ineffective or completely inoperable due to lack of maintenance and replacement parts. Less attention was given to plant maintenance, housekeeping, and monitoring procedures than had been given in earlier years. Substitute or lower quality raw materials had to be used in many instances.

Although the major impact of World War II upon levels of exposure to harmful agents occurred from around 1940 to the early 1950s, the effects of many of these exposures may not have shown up among members of that work force and its retirees until decades later.

Thus, expressing exposure levels in the past for more than a few years may be only extrapolated guesses unless factors such as the above can be clearly examined and the data validity established.

7.2 Compliance With Standards

In contrast to the collection of data for epidemiological studies, data collected for the purpose of compliance with standards, as they are now interpreted, may require relatively few samples if the values are clearly above or below the designated value for the agent of interest. If the values are borderline, evaluation may call for a more comprehensive sampling exercise and may be a matter for legal interpretation. When exposures are borderline between compliance and noncompliance, the most effective response, whenever practicable, is to further reduce exposures to well below the standard so one is assured of compliance. The nature and type of samples taken should meet the criteria upon which the standards were based. Scientifically, though, the data should be adequate to establish a clear pattern with no one single value being given undue weight and should meet data analysis requirements.

7.3 Spot Sampling

The exposure of a worker may arise from a number of sources, including the ambient levels of the agent in the general room air which in turn may be influenced by ambient levels of the agent in the community air, leaks from improperly maintained operating equipment such as from pump seals and flanges, the inadequate performance of engineering control equipment, and the care which workers take in performing job operations. Spot sampling can easily detect the effects of any one of these factors on the overall worker exposure level and point the direction for further exposure control action.

8 DATA EVALUATION

The evaluation of the intensity of a physical agent or of airborne levels of a chemical agent to determine compliance with a standard or to determine specific sources of the stress agent are generally uncomplicated and straightforward. The evaluation of environmental exposure data that serve as a basis for determining whether a health hazard exists is more complicated and requires a denominator that characterizes a satisfactory workplace. Similarly, the use of environmental exposure data for establishing safe levels of exposure or a permissible exposure level, as in epidemiologic studies, requires their correlation with other parameters such as the health profile of the workforce.

As pointed out earlier, the early field studies of the 1920s and 1930s showed that when the very high exposures of workers were lowered, there was an corresponding lowering of the related disease incidence in workers. These and other studies gave support to the dose–response rationale upon which the practice of industrial hygiene is primarily based, i.e., there is a dose–response relationship between the extent of exposure and severity of biological response to most stress agents and in which the response is negligible at some point above zero level.

There is great difficulty, however, in determining lower levels of exposure to a specific agent and its effect on the health of workers over a working lifetime. Often this is done through extrapolation of other data or by trying to estimate past exposures. In the lower range of the dose–response region, the incidence of disease from exposure to an agent may be so low that it approaches the level for that disease in the community outside the industry under study. This results in part from exposure of the general population to stress agents such as smoking, alcohol consumption, drug use, hobby activities, community and in-house pollution, and the like. These incidental stresses may be similar in magnitude to those on the job, or may be additive to, accumulative, or synergistic with stress from on-the-job exposures. For various reasons, often including limited study population size, even well controlled studies may not be sensitive enough to give data which can be reliably extrapolated to the lowest dose-response region for lifetime exposures.

The problems of estimating dose–response of human populations at low levels of exposure to hazards has given rise in recent years to a new scientific field of endeavor called risk analysis. There is not yet unanimity of opinion among scientist in the field on the most appropriate models and estimating procedures for all types of risk situations, including those involving lifetime exposures to low levels of health stressors.

It is known that the body has protective mechanisms to guard against the effects of low levels of exposure to many environmental agents. For a great number of agents encountered in the industrial environment, data on dose-response relationships support the industrial hygiene rationale that the level of exposure does not have to be zero over a lifetime of work to avoid injury to workers' health. Some agents, however, such as those having genotoxic properties (i.e., being able to directly damage genetic material in cells), and perhaps some associated with hypersensitivity, may not have a threshold of biologic response, and the lowest achievable level of exposure for workers should be required.

Evaluation of data from exposure stresses relating to tension, fatigue, annoyances, irritation, decrements in ability to concentrate and discern, and the like are often subjective and may also involve the personal background, traits, habits, etc., of those being stressed. Such stresses may require that attention be given to personal behavior for proper definition and control. Evaluation of such situations generally must be done by specialists other than industrial hygienists.

9 ENVIRONMENTAL CONTROL

The cornerstones of an effective industrial hygiene program can be described as:

1. Proper identification of on-the-job health hazards.
2. The measurement of levels of exposure to such hazards.
3. Evaluation of all exposure data in context with work schedules and job demands.
4. Environmental control.

In essence, the success of the entire program depends upon the success of the control effort. The technical aspects of the program must encompass sound practices and must be related both to workers and to the medical preventive program.

The heart of a control program must rest with process and/or engineering controls properly designed and properly operated to protect workers' health. The most effective and economic control is that which has been incorporated at the stage of process design and production planning, and which has been made part of the process. With new processes this can be accomplished by bringing industrial hygiene input into the bench level, pilot plant, and final stages of process development. It is neither good industrial hygiene practice nor sound economics to neglect exposure control in process design with the intention of adding supplemental control hardware piecemeal as indicated by future production, or to comply with regulations.

Although engineering and administrative means should be used wherever feasible to achieve control of exposures, the need may exist for the judicious use of personal protective equipment under unique circumstances, for example during breakdowns, spills, accidental releases, and some repair jobs. Personal protective equipment should be used only sparingly and under appropriate circumstances, and never as a substitute for more reliable and effective engineering or administrative controls.

Increasingly, engineering controls are being supported with automatic alarm systems to give an alert when controls are malfunctioning and excessive air contamination or physical agent intensity is occurring.

The control of stresses associated with high technology in the operation of equipment such as computers, word processors, video display terminals, microwave equipment, and the like requires a different engineering approach from that used in the control of toxic stresses. Providing optimal lighting, adjusting equipment to the operator's stature, and maintaining an overall general room compatibility with tasks are required when such stresses are encountered. Additional considerations including special rest periods and designated exercises may be appropriate.

10 EDUCATIONAL INVOLVEMENT

In the late 1930s very few universities in the United States offered programs leading to degrees in industrial hygiene at either the undergraduate or graduate level. In 1992 more than thirty colleges and universities offer programs leading to undergraduate or graduate degrees in industrial hygiene. This reflects the enormous growth in the profession that has taken place over the past 50 years. The passage of the Occupational Safety and Health Act of 1970 had a major impact on this growth.

The American Industrial Hygiene Association has increased in membership from 160 in 1940 to almost 12,000 in 1998. The number of Diplomates in the American Academy of Industrial Hygiene was more than 5700 in 1998; in the spring of 1998 there were also more than 500 Industrial Hygienists in Training.

Professional organizations such as the American Industrial Hygiene Association, the American Conference of Governmental Industrial Hygienists, and the American Academy of Industrial Hygiene offer excellent opportunities for the interchange of professional knowledge and the continuing education of industrial hygienists. These professional organizations invite participation through technical publications, lectures, committee activities, seminars, and refresher courses. As an example, the American Industrial Hygiene Conference and Exhibition of 1999 listed more than 480 technical papers covering a wide range of subjects. The same Conference offered 112 professional development courses for the purpose of increasing knowledge and skills in the practice of industrial hygiene. This participation by experts in the many facets of the profession enhances the overall performance of practitioners and permits industrial hygienists to keep abreast of new technology in the anticipation, recognition, measurement, and control of workplace stresses. These associations and the Academy support the profession in its fullest concept. In return, practicing industrial hygienists are obliged to keep involved in the educational and knowledge sharing process by making professional information available to others who have interest in, and responsibility for, the health and well-being of workers.

Industrial hygienists should take active roles in educating management concerning environmental stresses in places of work and the means for their control. Alert management can bring pending processes and production changes to the attention of industrial hygienists for study and follow-up, thus avoiding inadvertent occurrences of health problems.

The educational involvement of workers is extremely important. Workers have the right to know the status of their working environments, the stresses that may be deleterious to their health if excessive exposures occur, and of the control systems that have been instituted for their protection. Knowledgeable workers are in a position to enhance their own protection through the proper use of control equipment, the proper response to administrative controls, and, where it is needed, to the proper use of personal protective equipment. When a control system malfunctions a worker is often the first to observe it, and can call it to the attention of management. In cases of spills and leaks, or equipment breakdown, a worker who is knowledgeable of the hazardous nature of the materials involved can better follow prescribed emergency procedures. Industrial hygienists are in an excellent position to participate in worker protection education programs.

11 SUMMARY

Gigantic strides have been made during the past four or five decades in characterizing and controlling environmental health hazards in places of work. In many industrial plants where comprehensive industrial hygiene programs are in effect and exposures to work hazards are well controlled, the off-the-job stresses such as smoking, alcohol consumption, drug use, and hobby activities may have greater effect on workers' health than do their on-the-job stresses.

Industrial technology changes rapidly. As technology changes and new technology is applied new and different on-the-job health stresses emerge. Industrial hygienists must stay abreast of these changes and with the procedures for their recognition, evaluation, and control. It is vital that the techniques used in measuring occupational health stresses cover all the relevant components of each stress and that these are incorporated into any resulting control program. The practice of industrial hygiene rests on having valid data, on proper judgment in evaluating these data, and on effective follow-through.

The science and profession of industrial hygiene has a vital role in industry. A well implemented comprehensive industrial hygiene program leads to a healthful workplace.

The chapters which follow in these volumes are devoted to the theoretical basis and the practice of the science and profession of industrial hygiene.

BIBLIOGRAPHY

1. *CMA News* **16**(8) (Nov. 1988)
2. T. Hatch, Changing Objectives in Occupational Health, *AIHAJ* **23**, 1–7, 1962
3. T. F. Hatch, "The Role of Permissible Limits for Hazardous Airborne Substances in the Working Environment in the Prevention of Occupational Disease," *Bull. World Health Organization, Geneva* **47**, 151 (1972).
4. S. K. Hammond, C. J. Hines, M. F. Hallock, S. R. Woskie, S. Abdollahzareh, C. R. Iden, R. Ramsey, E. Anson, and M. B. Schenker, "The Tiered Exposure-Assessment Strategy in the Semiconductor Health Study," *Amer. Jour. Ind. Med.* **28**, 661 (1995).
5. S. K. Hammond, C. J. Hines, M. F. Hallock, S. R. Woskie, E. Kenyon, and M. B. Schrenker, "Glycol Ether Exposures in the Semiconductor Industry," *Occup. Hyg.* **2**, 355 (1996).
6. P. A. Stewart, D. Zaebst, J. N. Zey, R. Herrick, M. Dosemeci, R. Hornung, T. Bloom, L. Pottern, B. A. Miller, and A. Blair, "Exposure Assessment for a Study of Workers Exposed to Acrylonitrile" *Scand. J. Work Environ. Health* **24**(supl. 2), 42 (1998).

CHAPTER THREE

The Mode of Absorption, Distribution, and Elimination of Toxic Materials

Bertram D. Dinman, MD, Sc.D. and Jonathan D. Dinman, Ph.D.

1 INTRODUCTION

The toxicity of materials and the hygienic standards governing their use are of obvious importance in the practice of health protection; however, they should not be confused with the *hazard* of the usage of such materials. As this chapter will attempt to demonstrate, there are multiple physical and biological factors which will under various conditions of exposure determine the risk associated with workplace chemicals.

Multiple physical factors will determine whether chemical agents ever reach target tissues to produce toxic sequella. Gases may replace all the air in the atmosphere, but the amount of vapor in the atmosphere at any time is limited by the vapor pressure of the liquid or solid from which it arises. By contrast, the amount of suspended particulate matter in the atmosphere (e.g., dust, mist, or fume) is not limited by vapor pressure. It is well established that particulates may adsorb gases and vapors, thereby altering, and in some instances increasing, their physiological action. This follows because gas or vapors sorbed on particulates must follow the same physical principles governing the deposition of such particles in the pulmonary tree. Because particulates are more likely than gases to impinge on various portions of respiratory tract surfaces, the physiological consequences of particulate inhalation may be more significant than would be the case for gas or vapor molecules that less readily deposit and are exhaled. In addition to these purely physical considerations outside the body, other biophysical factors in the body, for example, pH, pK_a, lipid/water

Patty's Industrial Hygiene, Fifth Edition, Volume 1. Edited by Robert L. Harris.
ISBN 0-471-29756-9

partition coefficient, and dissociation characteristics, discussed in this chapter also determine whether the xenobiotic and tissue target ever have the opportunity to interact.

Similarly, a multiplicity of biological variables determine whether a material will ever reach a target site. In the course of a material's absorption into the body, transport to, and entering a specific target organ or being eliminated, numerous conditions must be met before the cellular target site may be reached. The blood supply, conditions of vascular and cell permeability, integrity of the mucociliary escalator system, actions of neurohumoral mediators, and so on all determine whether cellular target molecules ever are at risk. Finally, even if the extensive complex of preconditions are met so as to permit the occurrence of xenobiotic-target cell interactions, biosystems possess intrinsic capacities for the obviation of temporary and/or persistent dysfunction, viz., cellular repair or removal of aberrant cellular transformations. For example, repair of the consequences of ionizing radiation upon DNA base sequences once believed improbable have been extensively studied. At present, a number of specific, individual mechanisms for functional restitution of such alterations have been described. In addition, where damaged cells have been functionally altered such as to cause subsequent dysfunctional behavior, events leading to the programmed death, viz., apoptosis, of such cells effectively terminate their ability to reproduce such defective progeny. Further discussion of these reparative and damage control mechanisms are briefly noted in Section 3.1.4 and more extensively in 6.2.

To plan the prevention of injury from toxic materials in industry, it is essential to have a clear understanding of the dynamics of how materials enter the body, how they are handled therein, and how they are eliminated. To comprehend these processes more clearly, one need understand the mechanics of respiration and circulation and their role in the kinetics of absorption, transport, body cellular uptake, and body elimination. Many of these processes require comprehension of the gas laws and the ability to apply them to the solution of gases in liquids and specifically in the body fluids.

On the basis of such understanding it will become apparent why the occurrence of exposure to a toxic material is not per se tantamount to the risk of an untoward effect. Indeed, after review of this chapter, it would seem that, given all the conditions required before toxicity occurs, such an outcome is less probable than no event at all.

2 CLASSIFICATION OF CONTAMINANTS

The earth is surrounded by a gaseous atmosphere of rather fixed composition: 78.09% nitrogen, 20.95% oxygen, 0.93% argon, 0.03% carbon dioxide, insignificant amounts of neon, helium, and krypton, and traces of hydrogen, xenon, radioactive emanations, oxides of nitrogen, and ozone, with which may be mixed up to 5.0% water vapor. Any of these gases in greater proportions than usual, or any other substance present in the atmosphere, may be regarded as a contaminant or as atmospheric pollution. The possibilities of contamination are legion, but they may be classified according to their physical state, their chemical composition, or their physiological action.

2.1 Physical Classifications

2.1.1 Gases and Vapors

Although strictly speaking a gas is defined as a substance above its critical temperature and a vapor as the gaseous phase of a substance below its critical temperature, the term "gas" is usually applied to any material that is in the gaseous state at 25°C and 760 mmHg pressure; "vapor" designates the gaseous phase of a substance that is ordinarily liquid or solid at 25°C and 760 mmHg pressure. The usage distinction between gas and vapor is not sharp, however. For example, hydrogen cyanide, which boils at 26°C, is always referred to as a gas, but hydrogen chloride, which boils at −83.7°C, is sometimes referred to as an acid vapor.

2.1.2 Particulate Matter

There are at least seven forms of particulate matter:

1. *Aerosol.* A dispersion of solid or liquid particles of microscopic size in a gaseous medium, for instance, smoke, fog, and mist.
2. *Dust.* A term loosely applied to solid particles predominantly larger than colloidal and capable of temporary suspension in air or other gases. Derivation from larger masses through the application of physical force is usually implied.
3. *Fog.* A term loosely applied to visible aerosols in which the dispersed phase is liquid. Formation by condensation is implied.
4. *Fume.* Solid particles generated at condensation from the gaseous state, generally after volatilization from melted substances and often accompanied by a chemical reaction, such as oxidation. Popular usage sometimes loosely includes any type of contaminant.
5. *Mist.* A term loosely applied to dispersion of liquid particles, many of which are large enough to be individually visible without visual aid.
6. *Smog.* A term derived from "smoke" and "fog" and applied to extensive atmospheric contamination by aerosols arising from a combination of natural and man-made sources.
7. *Smoke.* Small gas-borne particles resulting from incomplete combustion and consisting predominantly of carbon and other combustible materials.

2.2 Chemical Classifications

Chemical classifications are variously based on the chemical composition of the air contaminants and may vary widely depending on the aspect of the composition to be emphasized.

2.3 Pathological Classifications

The physiological classification of air contaminants is not entirely satisfactory because with many gases and vapors, the type of physiological action depends on concentration. For instance, a vapor at one concentration may exert its principal action as an anesthetic, whereas a lower concentration of the same vapor may, with no anesthetic effect, injure the nervous system, the hematopoietic system, or some visceral organ. Although it is frequently impossible to place a material in a single class correctly, a pathophysiological classification may be suggested.

2.3.1 Irritants

Irritant materials are corrosive or vesicant in their action. They inflame moist or mucous surfaces. They have essentially the same effect on animals as on humans, and the concentration factor is of far greater significance than the time (duration of exposure) factor. Some representative irritants are as follows:

1. Irritants affecting chiefly the upper respiratory tract: aldehydes (acetaldehyde, acrolein, formaldehyde), alkaline dusts and mists, ammonia, chromic acid, ethylene oxide, hydrogen chloride, hydrogen fluoride, sulfur dioxide, sulfur trioxide.
2. Irritants affecting both the upper respiratory tract and the lung tissues: bromine, chlorine, chlorine oxides, cyanogen bromide, cyanogen chloride, dimethyl sulfate, diethyl sulfate, iodine, ozone, sulfur chlorides, phosphorus trichloride, phosphorus pentachloride, and toluene diisocyanate.
3. Irritants affecting primarily the terminal respiratory passages and air sacs: arsenic trichloride, nitrogen dioxide and nitrogen tetroxide, and phosgene. (To the extent that their action frequently terminates in asphyxial death, lung irritants are related to the chemical asphyxiants.)

2.3.2 Asphyxiants (Anoxia-Producing Agents)

Strictly speaking, "asphyxia" should be restricted to descriptions of agents that produce oxygen lack and concomitant increases of carbon dioxide tension in the blood and tissues. Concern with effects of chemicals on oxygen availability to the body requires that agents producing this effect are referred to as producing anoxia (i.e., lack of oxygen). Many of the chemical agents noted below produced effects at multiple loci (e.g., carbon monoxide affects hemoglobin as well as various tissue respiratory catalysts such as cytochrome P-450). However, the classification to follow considers the major sites of action of such chemical agents.

This group can be subdivided into three classes, depending on how they cause anoxia within the body.

2.3.2.1 Anoxic Anoxia. Lack of oxygen availability to the lungs and blood stems from simple displacement or dilution of atmospheric oxygen; this may occur even in the case of physiologically inert gases. This, in turn, results in reduction of the partial pressure of oxygen required to maintain oxygen saturation of the blood sufficient for normal cellular respiration. Agents producing anoxic anoxia include ethane, helium, hydrogen, nitrogen, and nitrous oxide.

"Anoxemic anoxia" has been frequently subsumed under the term "anoxic anoxia," being defined as a lack or diminution of oxygen delivery to the blood. It is seen with defective pulmonary uptake or obstruction or oxygen transfer across the alveolar membrane.

2.3.2.2 Anemic Anoxia. Anemic anoxia implies a total or partial lack of availability of hemoglobin for oxygen carriage by the red blood cell. Whereas simple hemorrhage reduces the total loading of blood oxygen in proportion to red blood cell loss, arsine has an effect similar to hemorrhage by producing red blood cell breakdown, making it unavailable for oxygen carriage. In addition, numerous chemical agents can produce a similar effect by impairing or blocking the oxygen uptake and carrying capacity of hemoglobin in the lungs. Examples of this group are carbon monoxide, which combines with hemoglobin to form carboxyhemoglobin, and aniline, dimethylaniline, and toluidine, which form methemoglobin.

2.3.2.3 Histotoxic Anoxia. Histotoxic anoxia results from the action of agents that impair or block the action of cellular catalysts necessary for tissue oxidative metabolism. Because the ability of hemoglobin to take up oxygen is not necessarily altered, and because the tissues cannot avail themselves of this oxygen, the capillary and venous oxygen saturation of the blood usually is higher than normal. Examples in this group are cyanogen, hydrogen cyanide, and nitrites. Hydrogen sulfide blocks cellular oxidation at the respiratory center controlling respiration and directly stops pulmonary air moving action. The ultimate effect of this agent is to cause anoxic anoxia secondarily.

2.3.3 Anesthetics and Narcotics

The anesthetics and narcotics group exerts its principal action as simple anesthesia without serious systemic effects, and the members have a depressant action on the central nervous system governed by their partial pressure on the blood supply to the brain. The following examples are arranged in the order of their decreasing anesthetic action compared with other actions: (a) acetylene hydrocarbons (acetylene, allylene, crotonylene), (b) olefin hydrocarbons (ethylene to heptylene), (c) ethyl ether and isopropyl ether, (d) paraffin hydrocarbons (propane to decane), (e) aliphatic ketones (acetone to octanone), (f) aliphatic alcohols (ethyl, propyl, butyl, and amyl), and (g) esters. Although the last are not particularly anesthetic, they are placed here for want of a better classification.

2.3.4 Systemic Poisons

The following substances are classified as systemic poisons:

1. Materials that cause injury to one or more of the visceral organs: the majority of the halogenated hydrocarbons.
2. Materials damaging the hematopoietic system: benzene, phenols, inorganic lead, warfarin.
3. Nerve poisons: carbon disulfide, methyl alcohol, methyl *n*-butyl ketone.
4. Toxic metals: lead, mercury, cadmium, antimony, manganese, beryllium.
5. Toxic nonmetal inorganic materials: compounds of arsenic, phosphorus, selenium, and sulfur; fluorides.

6. Toxic organic/biologically active materials: pesticides, phorbol esters, ribosome inactivating proteins, antibiotics.

2.3.5 *Immune System Sensitizers*

Strictly speaking, materials capable of inducing the immune response are referred to as sensitizers. Sensitizing materials, i.e. antigens, are typically large molecular weight proteins and carbohydrates of bacterial, plant, or animal origin. Although most industrial compounds are in themselves of insufficient molecular weight to act as antigens, relatively small reactive molecules have been shown to be capable of inducing classical immune responses. Small molecules demonstrating this capacity are referred to as haptens. In these cases the relatively low molecular weight haptens are capable of covalently conjugating with a host protein or other macromolecules. The ability to act as a hapten is largely determined by the chemical reactivity of the compound (electrophilicity) and its molecular conformation; slight changes in shape can markedly alter antigenicity. The net result of this combination of a low molecular weight compound with a macromolecule is formation of a complete antigen capable of initiating the first step in the typical sensitization sequence. The chemical structure of the foreign compound may play a role in determining its bioreactivity toward appropriate macromolecules. The highly reactive chemical cross-linker used in plastics manufacture, toluene diisocyanate, is an example of a small molecule that can function as an immune system sensitizer.

The next step in the process of sensitization is "antigen uptake and presentation." Cells of the immune system, typically macrophages, dendritic cells and B-cells ingest the antigen. Antigen is processed into small peptide fragments in the lysosomal vacuoles, and the peptide fragments are exported to the cell surface in combination with HLA class II molecules where they are displayed to antigen-specific T-cells. The interaction between T-cell and antigen presenting cell serves to induce an antigen-specific immune response. Typically, two classes of immune response are observed with sensitizers.

The humoral response involves the production of antigen-specific antibodies by B-cells. Antibodies are large, complex protein molecules of the immunoglobulin (Ig) series which all consist of heavy and light polypeptide chains. There are five basic classes of antibodies: IgM, IgD, IgG, IgA, and IgE. Structurally, antibodies are typically divided into two distinct regions: the common and variable regions. The common regions are specific to a class of antibody, and are capable of high affinity interactions with receptors on specific cells of the immune system. The variable regions are capable of high affinity interactions with the eliciting antigen. There are thousands of gene fragments which encode the variable region, and variety is further enhanced by their hypermutability. Thus, the amino acid content of the variable region varies with the specificity of the antibody, that is, its ability to recognize and combine with antigen. Some antibodies are found in plasma, namely, IgG and IgM, whereas others, notably IgE, frequently are associated with cells fixed in specific tissues. Owing to the size and complexity of the antigen more than one antibody may result from this interaction with a single antigen. On subsequent exposures to the specific antigen such antibodies combine or complex with this antigen. Such complexes are responsible for many of the clinical immunologic response syndromes.

Antigen–antibody complexes can promote the fixation of complement, a complex cascade-type system of serum proteins. Induction of the complement cascade promotes a series of inflammatory responses including increase phagocytosis by leukocytes, cause smooth muscle contractions (see below), and damage cell surfaces, cell lysis, and death. Additionally, antigen-mediated cross linking of cell surface-associated IgE can promote the release of histamine from mast cells and platelets. Such complexing of antigen and antibody results in cell damage and release of heparin, histamine, or serotonin. Another destructive interaction results when the antibody complexes with antigens that have become bound to the surface of specific cells, for example, red blood cells. This complexing leads to the agglutination of such elements; if complement is released here the cells will lyse. In a third type of reaction circulating immune complexes may initiate blood cell damage; complement mediates this response. Kidney damage is the most common sequelae of immune complex disease.

The second arm of the immune response is the cellular response. Interactions between antigen-specific T-cells and antigen presenting cells can result in the release of mediators of inflammation, e.g. interleukins, interferons, porins, tumor necrosis factor, leukotrienes and prostaglandins. The large scale release of these molecules can result in acute inflammatory episodes, which in turn may cause severe internal tissue damage. This can lead to chronic inflammation, hypersensitivity, and possible autoimmune sequalae.

The clinical result of the immune or allergic response may be localized or widespread, immediate or variably delayed reactions. The severity of response may be so localized as to affect only a few square millimeters of skin, or may extend to involve the whole body as in severe anaphylactic type response. The dose–response relationship governs induction of the magnitude of elicited response. However, there are individual variations in responsiveness and complex modulating mechanisms regulating production of immunologic cells and antibodies.

The most severe immunologic reaction, anaphylaxis, has infrequently been associated with industrial exposures. By contrast, less life threatening but serious reactions are seen in industry. These most commonly affect the skin and less frequently the respiratory tract. Anaphylaxis may reflect topical direct contact with sensitizer or, less commonly, result from absorption via other portals of entry. The effect may be local or widespread, independent of how much dermal contact occurs.

3 PORTALS OF ENTRY AND ABSORPTION OF XENOBIOTICS

If a foreign material is to produce a biological effect, it must breach some structural and/or biochemical barrier(s) interposed between the external environment and a susceptible locus within the body. In the course of transit of these barriers exogenous materials may be (*1*) chemically altered (e.g., hydrolyzed by gastric acids), (*2*) biotransformed (i.e., altered by enzymatic action into more or less toxic moieties), (*3*) stored in depots where they usually produce little effect, or (*4*) ultimately excreted, primarily by the kidney and biliary tree. But regardless of such alterations, before any toxic event can occur the material-in whatever form or transformate, must usually pass through multiple barriers. The nature of these barriers and how they may be breached are the focus of this section.

3.1 Principles of Cell Membrane Transit

3.1.1 Cell Membrane Structure

On a gross level cells may be arranged in a variety of fashions; they may consist of layers of multiple cell units, for example, the skin, or a single layer of cells, for example, alveolar walls. However, all individual cell membranes are strikingly similar at the molecular level.

The cell membrane at the molecular level of organization consists of phospho(glyco) lipid (Figure 3.1) as well as protein and cholesterol molecules organized in a bilayer arrangement (Fig. 3.2) major components of the plasma membrane consist of phosphoglycerides and fatty acids (Fig. 3.1). The phosphoglycerides consist of glycerol esterified to phosphoric acid further esterified to an amino alcohol, namely, choline or ethanolamine. The amino alcohol, glycerol, and phosphoric acid components are hydrophilic; this assures membrane stability within the hydrophilic environment with which they are in contact, that is, the intracellular cytosol or the extracellular water. The cholesterol molecules present in the membrane are also believed to impart a degree of stiffness to this fluidlike lipid bilayer. This "head" oriented arrangement (Fig. 3.1) contrasts with the twin "tails" of the two 16- to 18-carbon fatty acids. Their lipophilic nature is consistent with their orientation within the mass of the membrane. The membrane is thus made up in greater part by a lipid bimolecular layer with "heads" oriented toward both the inside and outside of the cell (Fig. 3.2 and Fig. 3.3). The fluidity of the cell membrane is in large part a reflection of the proportion and structure of unsaturated fatty acids. As unsaturated fatty acid content increases the membranes become more fluid, resulting in more rapid active transport. The nuclear membrane contains a higher proportion of saturated fatty acids than the cell surface membrane, and is thus a more rigid structure. It is the predominance of the lipid cell wall components that contributes to the relative ease with which lipid xenobiotics penetrate into the cell.

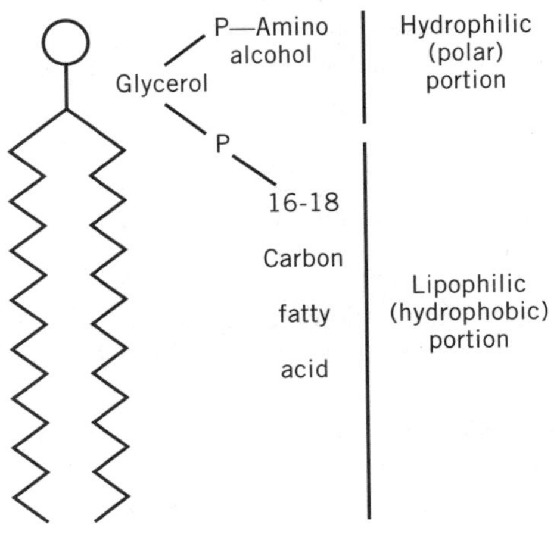

Figure 3.1. Structure of phospholipid molecular component of the bimolecular cell membrane model.

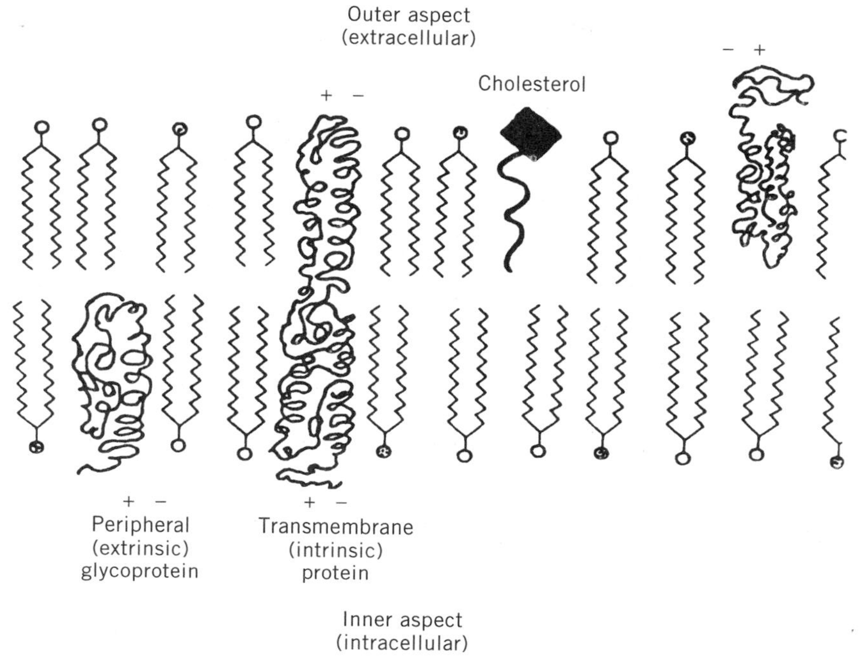

Figure 3.2. Schematic diagram of lipid bimolecular cell membrane model.

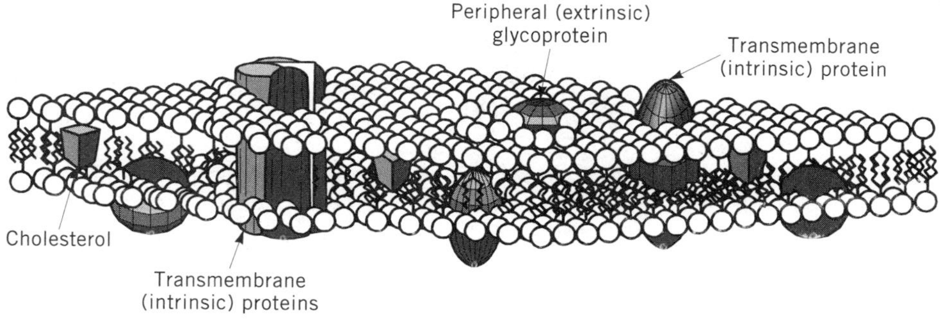

Figure 3.3. Three-dimensional proposed model of a typical cell membrane. White or stippled spheres indicate two different types of polar head groups of the membrane phospholipid with associated fatty acid chains. Both intrinsic and extrinsic membrane proteins are indicated by large hatched and mosaic topped bodies.

In addition, crystalline water containing various ions (e.g., Ca^{2+}, Mg^{2+}) is located within this two-molecule-thick layer; these also influence the permeability and the gel state of the membrane.

Immersed in this bimolecular structure approximately 6 to 10 nm thick are holo-, glyco- and lipoproteins (Fig. 3.3). The proteins may penetrate the entire bimolecular structure, that is, the intrinsic proteins, or may occur in either half of the bilayer (Fig. 3.2 and Fig. 3.3). These extrinsically oriented proteins are in contact with both aqueous and lipid media. Accordingly, they tend to demonstrate amphipathic properties. The intrinsic proteins, found largely in a lipid region, consist largely of hydrophobic amino acids; that is, they contain hydrophilic amino acids where they are in contact with water and hydrophobic amino acids within the bimolecular phospholipid components of the membrane. These latter proteins constitute the molecular bases for the presence of pores and receptors in the cell membrane. The pores are how ions and large molecules can nonspecifically enter and exit the cell by simple diffusion (see below), while trafficking of specific molecules into and out of the cell is a receptor-mediated process. Because it is the nature of proteins to (*1*) immobilize adjacent lipid molecules and (*2*) perform translational or rotational motion in the lipid bilayer, they play important roles in the maintenance of rigidity or fluidity of membrane structures. Within constraints of ambient temperature and "tail" fatty acid saturation, the number of extrinsic and intrinsic protein molecules markedly affect the physical structural properties of cell walls.

The physical chemical activities of these structures are highly dynamic. The relative mix of the chemical components varies both within an individual cell wall and from cell to cell. The membrane is constantly renewing these components; thus the lipid molecules have a 3- to 5-day half-life, the large proteins a ½- to 5-day half-life, and small proteins a 7- to 13-day half-life. Some membrane protein molecules drift laterally in the plane of the membrane whereas others are anchored to the cytoskeleton. It should thus be apparent that these cell-limiting membranes are dynamic and can be degraded and rebuilt rapidly in keeping with needs of the organism.

3.1.2 Passive Transport

Passive transport depends upon three primary determinants, namely, (*1*) concentration gradients across membranes, (*2*) liposolubility of a compound, and (*3*) ionization of the compound.

3.1.2.1 Determinants of Passive Transport. The first of these largely reflects the difference in concentration of the compound in question on both sides of the membrane. Diffusion rates also will reflect the thickness and area of the membrane as well as the diffusion constant for the material in keeping with Fick's law. This phenomenon is essentially a first-order rate process.

The interplay of these variables represents a dynamic process, because their rates of occurrence change as concentrations vary on either side of the membrane. Such variations may reflect blood flow rates, concentration changes exterior to the cell, or rates of metabolic transformation occurring within the cell. The rate of simple diffusion also will reflect the rate of dissolution of a compound in the largely lipid domains of the cell membrane.

Accordingly, the rate of transfer by this process will in turn depend upon the lipid-water partition coefficient. Thus small hydrophobic molecules, such as carbon tetrachloride, rapidly pass through cell walls of the gastric mucosa and pulmonary alveoli, respectively.

Although lipid solubility is an important determinant of the degree of passive diffusion, the extent of ionization also plays a large role in this type of transmembrane movement. The pH partition theory in effect states that only lipid-soluble, nonionized chemical species will passively diffuse over plasma membranes.

Depending upon the pH of its environment a compound may exist in a ionized or nonionized state. In addition, each compound may be characterized by its own intrinsic dissociation constant (i.e., pK_a), the pH at which an acid or basic compound is 50% associated. Knowledge of the pH of the environment and the pK_a of a compound permits use of the Henderson-Hasselbach equation for calculation of the percentage in the nonionized form and hence availability for passive diffusion across the largely lipoprotein cell membrane.

By contrast polar materials, for example, sugars, transit passively at a slow rate, if at all, at these sites. However, it should not be implied that polar compounds do not transit across cell barriers by simple diffusion. Such a process does occur via aqueous channels, that is, the transmembrane protein pores of the cell wall. However, the relative paucity of such channels in what is mainly a lipid structure makes such occurrence less frequent, that is, transiting at a lower rate relative to lipophilic compounds.

Nevertheless, movement of hydrophilic and large lipophilic molecules as well as ions readily occurs through these pores. The membrane proteins can gain both positive and negative charges; thus small-diameter pores will permit passage only to ions whose charge is opposite to that of the membrane protein. This selectivity decreases as pore diameter increases so that both cations and anions can pass; in addition, penetration is increased with ion-pair formation. The potential gradient, that is, the electrochemical potential between outer and inner membrane surfaces, also induces diffusion of ions through pores.

The number and size of pores vary in any one membrane, because these attributes can be altered by numerous extrinsic factors, for example, hormones. Thus, for example, if the number and diameter increase, hydrodynamic flow (governed by Poiseuille's law) increases as the relative proportion of diffusion decreases. As the pore diameter approaches 20 nm, hydrodynamic flow should account for 90 to 95% of total transmembrane movement. In view of this high degree of biophysical plasticity, it is apparent that the cell membrane can exhibit considerable dynamics and a remarkable potential for responsiveness to changing cell and body needs.

A dynamic reality applies *in vivo*; ionization continuously occurs as within the cell nonionized moieties are altered or removed through metabolic events. On the other hand, changes in concentration of a nonionized compound outside the cell determine how much of it is available for passive transport, as a reflection of associated changes in transmembrane concentration gradient.

3.1.2.2 Filtration. Filtration, or the "solvent drag" effect reflects the fact that water flowing through a porous membrane may carry solutes. Obviously, this form of transport involves uniquely hydrophilic compounds. If such solutes are of sufficiently small molecular weight they will pass through pores in the cell membrane as determined by osmotic or hydrostatic forces. There is a wide range in pore sizes in various cell membranes of the body, that is, from 4 nm in most cells to as large as 40 to 70 nm in blood capillaries. Thus filtration only allows passage of small molecules (100 to 200 Da) from the usual capillary; however, molecules as large as 60,000 Da can be passed through the glomerular capillaries in the kidney (1).

3.1.3 Carrier-Mediated Modes of Transport

3.1.3.1 Active Transport. Active transport occurs across concentration or electrochemical gradients. Polar compounds, electrolytes (e.g., Na^+, K^+), nonelectrolytes (e.g., sugars), and zwitterions (e.g., amino acids), which do not diffuse passively, require this mode of transport. This process of transport requires metabolic work, and is typically receptor mediated. Receptors can be highly specific (e.g., for glucose), or relatively nonspecific (e.g., multiple drug resistance pumps), and movement of molecules can be either into, or out of the cell. Active transport postulates that macromolecular carriers form a complex at one side of the membrane with the substances to be carried. This complex moves to the other side of the cell membrane at which place the substance carried is released and the carrier returns to the surface of origin, repeating once more the cycle of transport. The specificity of each carrier varies with regard to the nature of the basic structural characteristics which must exist before complex formation and subsequent transport occur. Thus among compounds with similar critical structural characteristics competitive inhibition of transport may occur for one of the several compounds present at the membrane's exterior. In addition, such systems are subject to saturation at high substrate concentrations; at this point zero-order transfer rates obtain regardless of concentration gradients. Finally, because this process requires expenditure of energy, metabolic inhibitors have the potential for blocking this form of transport. These active transport systems are frequently specific for certain forms of exogenous compounds, for example, amino acids and sugars. Active transport is also relevant to toxicologic processes because it represents an important method for transport of xenobiotics from the system. This process is also particularly important in the removal of organic acids and bases from, for example, the central nervous system, liver, and kidney.

3.1.3.2 Facilitated or Exchange Diffusion. In contrast to the type of active transport just described, this type of transport is not against an electrochemical or concentration gradient and does not require expenditure of metabolic energy. However, this type of carrier-mediated transport occurring down a concentration gradient plays a significant role in the transport of important nutrients such as glucose. It occurs at the gastrointestinal epithelial surface, the red cell membrane, and the capillaries of the blood–brain barrier.

A protein molecule acting as the carrier is formed from two subunits whose new configuration facilitates binding and transport across the plasma membrane. This carrier-substance complexation process resembles the interaction between enzyme and substrate. As with active transport, similar compounds may compete for a carrier site; irreversible binding by and of such molecules can block this form of transport.

3.1.4 Special Transport Processes, Endocytosis

The processes referred to as pinocytosis and phagocytosis represent primitive forms of transport across cell membranes. In these processes the cell wall invaginates or flows around a foreign material in order to engulf it and bring it into the cell interior. When the engulfed foreign material is a liquid, the process is referred to as pinocytosis, and when particulates are involved, phagocytosis. The former process occurs in the transport of a variety of compounds, for example, proteins and glycoproteins, hormones, and lipids. It occurs in a variety of cells, for example, leukocytes, hepatocytes, and gastrointestinal cells. Phagocytosis is an important process for the removal of foreign particulates by cells involved in the immune system, e.g., alveolar macrophages in the lung and hepatic Kuipfer cells.

The encapsulation of a xenobiotic within such an endo- or pinocytotic vesicle places it within an environment that is topologically equivalent to extracellular space. Such vesicles are typically trafficked to lysosomes, where low pH typically renders most molecules biologically inert. An alternative trafficking route is back out of the cell via active transport through the multiple drug receptor apparatus.

In the rare event that a xenobiotic manages to escape encapsulation and enters the cytoplasm, the odds that it will wander free to do unlimited damage are extremely low. It is now believed that rather than being an amorphous bag of enzymes, the cytoplasm is a highly ordered, dynamic apparatus, the integrity of which is constantly monitored. Modification of an intracellular cytoplasmic components are readily detected. Such modified intracellular molecules are rapidly sequestered and removed from the cytoplasm for disposal. For example, inappropriately modified proteins are rapidly detected and tagged for disposal by the cellular ubiquitin (2) conjugation apparatus. These ubiquinated molecules are subsequently directed to the proteosome for degradation and resultant elimination. Thus, if a xenobiotic were to enter the cytoplasm, the odds are very high that interaction with proteinaceous cytoplasmic components would rapidly result in its destruction and removal.

3.1.5 A General Overview of Membranes and Toxic Interactions

Xenobiotics first contact living organisms at their interface with the environment, namely, the cell membrane. Given the delicately balanced, dynamic state of this structure it would be expected that its interaction with toxic compounds would almost inevitably produce deleterious consequences. But this inherent dynamic also suggests, within dose limits at the cell membrane, that a finite degree of interaction and alteration can be tolerated with subsequent restitution to a previous functional and structural state. Within such limits, Table 3.1 summarizes a few known relationships between toxic agents and the specific membrane effects they may precipitate.

3.2 Penetration and Absorption Via the Respiratory Tract

Although most xenobiotics reach the interior milieu through the gastrointestinal tract, in occupational settings the pulmonary portal of entry is the more usual pathway for uptake of environmental agents. Because of the (*1*) very large surface area of the lungs (namely, varying between 30 and 100 m^2, with a mean of 70 m^2 in adult males), (*2*) extremely large blood flow to the organ carried by a capillary bed approximately 2000 km in length, and (*3*) the extremely thin physical chemical barrier between air and blood (approximately

Table 3.1. Membrane Function/Structure Alterations Associated with Toxic Agents

Membrane Alteration	Causal Agent
Structural Change	
Extraction of lipid elements	Alcohols
Altered protein configuration	Anesthetics
Alteration of ligands maintaining integrity, e.g., SH	Hg, X-rays, ozone
Transport Interference	
Diffusion impairment, secondary to altered steric relationships	Detergents
Change of terminal charge on membrane protein ligands	Mercury, lipid solvents
Competition for carrier substrate	5-Fluorouracil

0.2–0.5 μm), the potential for the ready access of ambient air and its contents to the internal environment is self-evident. To an important extent, how body penetration by toxicants occurs in the lungs largely depends on the physical state of such xenobiotics, namely, whether particulate, aerosol, gas, or vapor. In addition, because the air conducting tracheobronchial tree carries warmed air which is saturated with water vapor, hygroscopic particulate growth will occur increasing both their size and deposition potential.

3.2.1 Lung Structure and Function

Both elements are important determinants of xenobiotic penetration and absorption in the respiratory system. As a general principle, the basic purpose of the respiratory system is to bring oxygen laden air into intimate contact with capillary blood. All other structures in the lung are ancillary to this end. However, in the course of evolution the lungs have developed, extending from the nares to the ultimate gas-transfer site, viz., the alveoli, multiple lines of defense against the intrusion of chemical, physical and biologic xenobiotics.

Over the course of evolution air-breathing species have developed functional and structural elements in response to the reality that ambient air inevitably contains gases and particulates with pathogenic potentials. Clearly, their transport into the intimate regions of the environment–blood interface presents a threat to viability. Accordingly, the various defensive structural and functional modifications developed in this system.

The first of these physical defenses are encountered at the entrance of the nasal passages, viz., the nasal hairs which mechanically filter larger particles. Yet another physically based defensive structure is seen in the evolution of the nasal turbinates. These bony, rounded out-foldings covered by the nasal mucus membranes are found directly after air enters the nares; these provide a large, convoluted surface area for both warming, humidifying and abruptly changing the direction of air flow. As a consequence of these directional flow changes, larger particles with relatively larger kinetic energy change direction less and readily impact. This structures and the nasal hair succeed in preventing respiratory tract ingress of particles larger than 10–20 μm beyond the nose and throat. These physically based mechanisms, viz., mechanical filtration continues to act for progressively smaller and smaller particles as they change direction with each branching of the airways. Another mechanical defense is presented by the action of smooth muscle which encircle the tracheobronchial tree. These muscles contract when receptors in this tree are stimulated by contact with chemical or mechanical irritants in the inspired air. The result may be bronchoconstriction which decreases net air flow and penetration of irritant gases and vapors into the lungs. Another consequence of receptor stimulation is coughing and/or sneezing which can expell such xenobiotics.

A second biological line of defense is presented by the liquids which coat the intrapulmonary air conducting passages. This layer provides a physical barrier to body contact by particulates, gases, etc., while acting as a chemical and biological buffer containing detoxifying and antibacterial systems. For contained in this liquid layer are elements of immune system activity, represented by antibodies produced via the humoral immune response and lymphocyte mediated antigen elaboration. By virtue of its physical form, this fluid layer provides a convenient milieu for the movement and ultimate elimination of foreign particulates by macrophages of the respiratory tract. Finally, this aqueous layer is of special importance as a mechanical medium for transport of particulates from the supra-alveolar segments of the lungs via the mucociliary escalator (see Sec. 3.2.5.2)

Examination of the cellular components of the lung (Fig. 3.4) reveals major differences among cells over the course of the airways that are in contact with the ambient environment. The site of the primary oxygen-blood interface is the pulmonary alveoli found at the terminus of the air-conducting pathways. The alveoli consist of very thin-walled sacs about which is intimately juxtaposed a rich capillary network. The alveoli are made up of mainly flattened epithelial cells (type I) and fewer granular cuboid cells (type II) with a very few brush cells (type III). Thus the capillary network, the more common flattened cells (type I)—approximately 0.2 mm thick—and a basement membrane about 300 to 350 Å thick constitute this basic unit of respiration, the alveolus. Upon the air surface of these alveoli is found a fluid-containing surfactant that is produced by these cells. It is at this very thin alveolar sac wall site that gases are exchanged between ambient air and the blood; here the internal environment makes its closest approximation to the ambient world. Blind terminal vessels containing lymph surround the terminal bronchioles. This lymphatic system drains the secretions of the alveoli, preventing fluid accumulation and thus facilitating gas exchange. This especially protein-rich lymph drains into the lymph nodes at the lung hila or root. Thus the lymphatic system not only serves as a recovery system for protein, but also provides a channel for fluid-media based movement (active and passive) of phagocytes carrying exogenous and endogenous materials out of the lung.

3.2.2 Penetration and Absorption of Gases and Vapors into the Lungs

Gases and vapors come into equilibrium with blood passing through the alveolar capillaries practically instantaneously. This reflects Dalton's law of partial pressures, Henry's law (solubility of gases over a liquid), and the gases' lipid-aqueous partition coefficient.

Because air contained in the tracheobronchial tree is essentially saturated with warm water vapor, and as the alveolar surface is covered with an aqueous film, the water solubility of a gas or vapor (as well as of a particulate or aerosol) assists in determining the

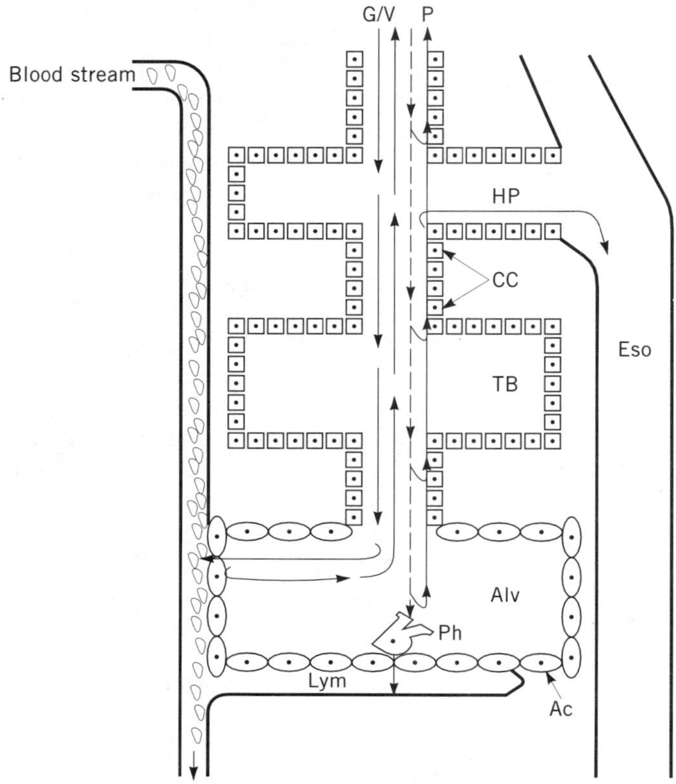

Figure 3.4. Schematic diagram of gas, vapor, and particulate movement in the respiratory system. Gases and vapors generally are exhaled as readily as they diffuse into the bloodstream at the alveolar compartment. To an insignificant extent they may also diffuse into the lymphatics. Particulates (depending upon particle aerodynamics or mass median diameter) deposit at various locations on surfaces of the respiratory tract. Those sufficiently small (<5–10 μm) to be deposited upon the alveolar membrane are generally phagocytosed. Phagocytes may transport particulates into the lung interstitium via lymphatics to lymph nodes or may carry them back up to the mucociliary escalator (see text). Physical penetration by particulates occurs less frequently. HP = hypopharynx; Eso = esophagus; TB = tracheobronchial compartment; Alv = alveolar compartment; Lym = lymphatics; AC = alveolar cells (flattened type I); cc = ciliated columnar epithelium; Ph = phagocyte; G/V = gas/vapor; P = particulate.

sites of deposition therein as well as the rate of alveolar penetration and diffusion. Uptake will also vary directly as a function of temperature, ventilation, and blood flow rates.

After passage through the alveolar membrane and capillary wall xenobiotics reach and enter the blood. This process can occur as the result of simple diffusion–dissolution in the blood or by chemical binding or absorption with blood solid or liquid components. Water constitutes 75% of the blood volume; however, although water-soluble compounds, for example, methanol, are readily taken up by the blood, other nonaqueous soluble materials can be equally rapidly taken up. Recalling the highly lipid nature of the cell membrane, simple passive diffusion of lipophilic substances should be expected. Indeed, because the blood also contains lipoidal elements, fat-soluble gases or vapors may be rapidly absorbed by the blood. In addition, because nonpolar compounds are rapidly removed from the blood to fat storage depots, the blood returning to the lung will contain low lipid concentrations and/or vapor pressures so that a gradient from alveolar air to blood is maintained.

In addition, the presence of salts, proteins, lipoproteins, and formed membrane elements may variably take up and bind a wide variety of chemical classes independent of lipid–water partitioning.

It should be clear that the uptake of gases and vapors represents the interaction of many variables, physical and chemical. Ultimately, pulmonary gas absorption will largely depend upon the concentration or partial pressure gradient between alveolar gas and blood. Because this gradient depends as much upon the partial pressure of a compound in the venous blood returned to the lung, it should be evident that penetration and absorption rates in the lung will indirectly reflect blood saturations at tissue sites resulting from various extrapulmonary biological, physical, and chemical processes.

3.2.3 Absorption of Aerosols by the Lung

Consistent with general principles of membrane passage discussed previously, the absorption of xenobiotics through the lung can be predicted in part. Brown and Schanker (1) indicated that for nonlipid soluble compounds, in the course of a nonsaturable passive diffusion process, the rate of absorption varies inversely with molecular size. Such hydrophilic chemicals pass over as expected through aqueous paths or pores between alveolar cells. Lipophilic aerosols cross the pulmonary epithelium even more rapidly by a similar passive, nonsaturable transfer phenomenon. The higher the lipid–water partition coefficient, the more rapid the passage. In addition, there are energy-dependent, high-affinity active transport systems in type I and II alveolar cells. The active transport mechanisms appear to be cytochrome P-450 dependent. Paraquat appears to be absorbed through alveolar epithelia by this method. By contrast, alveolar phagocytes do not appear to be as active in the uptake of aerosols by such processes.

3.2.4 Particulate Penetration and Absorption in the Lung

3.2.4.1 Principles of Particulate Movement and Deposition in the Lung. The general aerodynamic properties of particulates and aerosols determine if and where they may be deposited in the lungs (3, 4). Thus, depending upon diameter, mass, and shape, particulates may either preferentially deposit in various regions in the airways or not deposit at all. Accordingly, particles may be *sedimented* (depending largely upon gravity), *impacted* (mainly reflecting inertial forces acting with flow-direction change), or subject to *diffusion* (largely caused by Brownian movement). In addition, when particulates are formed by spray or mechanical comminution an electrostatic charge may be imparted; accordingly these particles are deposited on surfaces by electrostatic forces. As regards irregularly shaped particles (especially fibers), these tend to tumble and as their relatively larger lengths approximate the diameter of an airway deposition results from interception by airway walls.

Sedimentation is determined by density and diameter (density × diameter^2) and the density of the medium; thus in accordance with Stokes' law a particle will accelerate until it reaches a velocity at which the force of gravity is balanced by the viscous resistive force exerted by air and the shape of the particle. Particles demonstrating deposition as a result of this essentially gravitational force usually have aerodynamic diameters of 2 μm or more. They usually are deposited in the nasopharynx and the upper reaches of the tracheobronchial tree. Because settling of fibrous particles occurs mainly as a function of the square of their diameter, their gravitational deposition occurs mainly in large airways. Fibers capable of reaching the smaller airways have aerodynamic diameters of less than 3 μm. Thus length and shape rather than falling speed determine their deposition.

Inertial impaction occurs as a fluid stream changes direction and entrained particulates are likewise forced to change their flow direction. In effect such particulates continue their direction of flow because of inertial energy imparted by movement. Accordingly, depending upon the size of the particle, its mass, diameter, and the change of direction of the airway, as well as the velocity of flow, particles either (*1*) change their direction or (*2*) do not change and consequently impact upon the wall. This mode of deposition is particularly relevant to large particulates, that is, larger than 5 μm; such particles generally are impacted in the nasopharynx and tracheobronchial region of the lungs. By contrast, particles less than this size tend to reach the smallest airways and alveoli.

The diffusional behavior of very small particles is independent of density; rather it reflects their collision with vibrating gas molecules, that is, Brownian movement. Thus in regions of laminar air flow (alveoli, terminal bronchioles), particles less than 0.1 to 0.5 μm will tend to contact surfaces by simple diffusion. Another set of determinants governs the deposition of fibers (particulates with length/diameter ratio greater than 3:1), viz., their length and shape are more important than falling speed. As the length of a particle becomes greater than its diameter (i.e., more fibrous in configuration), impaction and sedimentation are less relevant to the probability of tracheobronchial contact. Rather, such impaction occurs at bifurcations of terminal and respiratory ducts. The curled fibers of chrysotile tend to exhibit this behavior, whereas the straight fibers of amphibole asbestos follow the aerodynamic flow lines parallel to the airway axis and tend to avoid impact at bifurcations. Thus these latter fibers penetrate more deeply into the lung than those of chrysotile asbestos.

Other physical factors (e.g., electrical charge, radius, shape, and retention time may also govern pulmonary deposition, particularly because by the agglomeration process these can affect aerodynamic size. Thus aerosols containing solutes may act as condensation centers which continuously grow in size owing to water absorption in humid air of the respiratory tract. Such particle growth requires that sizing and deposition characterizations integrate over the particle size-versus-humidity function.

Finally, physiological events must be considered in assessing deposition potentials. With increases in respiratory rate, the volume, air stream velocity, and inertial forces are increased. Airflow rate changes can affect deposition by alteration of the various combination of laminar and turbulent flows; these will have important effects upon deposition probabilities. Conversely, increased respiratory rates will decrease retention time of inhaled air streams, thus potentially decreasing diffusional deposition in the finer radicles of the respiratory tract.

The net effect of these many physical and physiological considerations as they determine loci of respiratory tract deposition of particles is shown in Figure 3.5. The curves are specific for a 1450-mL tidal volume (moderate to heavy activity) at the rate of 15 respirations per minute. As the respiratory tidal volume increases, the predicted gravimetric deposition shifts upward; conversely, at lower tidal volumes, predicted deposition decreases. Caution must be exercised in using these predictions, for the following factors militate against too-literal application of the data: variations in the mode of breathing (e.g., mouth, nose), the need for idealized models of lung anatomy and airflow patterns in making these estimates, and the occurrence of nonuniform ventilatory distribution. Nevertheless, the use of these relationships to provide approximations of pulmonary deposition patterns is warranted insofar as their approximate nature is appreciated.

3.2.4.2 Physical Penetration of the Alveolar Wall and Some Biologic Consequences. The smaller of the two main cell types making up the alveolar epithelium normally extend multiple cytoplasmic extensions into the alveolar space. Electron microscopic studies have revealed withdrawal of such extensions with alveolar wall irritation by particulate matter. Penetration of the alveolar wall into its interstitium and lymphatics may be actively facilitated by these villous withdrawals.

Fibers reaching the bronciolar-alveolar duct and further may interact with sialic acid moieties and integrins, binding with cell surfaces (7). In addition to the internalization by alveolar cells, some of the fibers are translocated across the epithelium by actin-containing microfiliments (8).

3.2.5 Clearance from the Respiratory Tract (See Fig. 3.4)

3.2.5.1 From the Nasopharynx. Particles having an aerodynamic diameter of 10 µm or more are usually deposited in the nasopharynx; the probability of deposition here increases directly with particle diameter. Nasal hairs and changes in flow direction over the nasal turbinates enhance entrapment and impaction. Particles deposited in the anterior nasopharynx are swept forward to the nonciliated nares and are ejected by nose blowing and sneezing. Ciliated epithelium present in the posterior nasopharynx direct the mucus flow downward to the pharynx so that entrapped particles are swallowed, making them eventually available for possible gastrointestinal tract ab sorption. Thus even though a particle may be too large to reach airways below the level of the nasopharynx, it is carried in this mucus to the lower throat (i.e., the hypopharynx) and swallowed.

3.2.5.2 From the Tracheobronchial Tract. Ciliated columnar epithelial cells lining the air-conducting tubes are found from the level of the seventeenth branching upward to the nose. Each of these cells, bear about 200 fine, 5–7 µm whiplike extensions of the cell membrane viz., the cilia, which simultaneously beat at a rate of approximately 1000 cycles per minute upward during their rapid phase and then restitute relatively slowly to their previous state. These projections are immersed in and covered by thin, sticky mucus formed by mucus and serous secretory cells of the airways. A lesser amount of mucus is

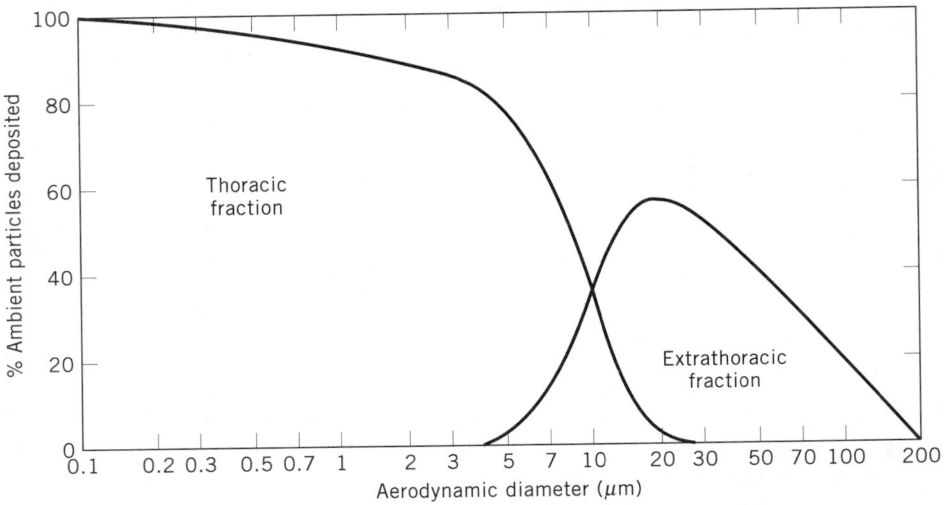

(**a**)

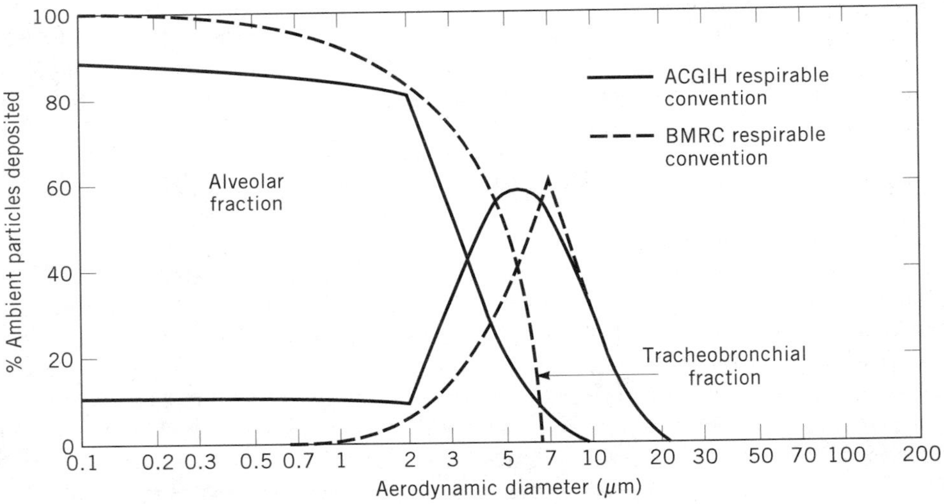

(**b**)

Figure 3.5. Deposition of inspirable aerosols in the extrathoracic (head) and thoracic segments of the respiratory system according to the ISO Model (5). (**a**) Division of inspirable fraction into extrathoracic (head) and thoracic subfractions, based primarily on oral inhalation (6). (**b**) Further division of thoracic subfraction into tracheobronchial and alveolar inspirable particulate fractions. Both the American Conference of Governmental Industrial Hygienists (ACGIH) and British Medical Research Council (BMRC) definitions of respirable fractions are provided in the lower figure.

contributed by the goblet cells which are mainly found in the large proximal airways. Clara cells mainly found in the terminal bronchioles also contribute fluids. By such repeated cycles of coordinated, upward beating of the cilia this overlying mucus film is moved headward up the tracheobronchial tree, eventually reaching the hypopharynx. Particles landing upon this sticky mucus film covering the interior of the airways are gradually carried in this mucus much as they would be transported by an escalator.

Extensive studies mucociliary escalator transport rates suggest that there are considerable individual variations in local transport rates. Such individual variation reflects topographic or regional variations in pulmonary secretory cell distributions and activity levels, individual cellular rheologic properties, autonomic nervous system activity and early pathologic change.

Accordingly, estimates of transport rates suggest they are almost individually unique. Within such limitations, study of mucociliary escalator transport rates indicate that this mucus blanket moves particles headward from as rapidly as 5 to 10 mm/min. in the trachea, to as slowly as 10 μm/min. in the smallest ciliated airways. It has been found that clearance from the ciliated nasopharynx is accomplished within 2 to 4 hours; 5 to 8 hours to clear the tracheobronchial tree and a range of 2 to 24 hours to effectuate overall clearance of the traceobronchiolar system. With large particles the rate of transport is more rapid. Depending upon individual sensitivity, cigarette smoke may cause a temporary increase in mucociliary clearance; more impairment of clearance rates with chronic exposures indubitably exacerbates the subsequent pulmonary damage that ensues in chronic smokers. On the other hand, acute irritant gas and vapor exposures decrease the ciliary beat rate, which in turn lessens mucociliary escalator flow.

3.2.5.3 Clearance by Phagocytosis. Pulmonary phagocytes are similar to mononuclear phagocytes found elsewhere in the body, for example, the liver and peritoneal cavity. There are at least three major types of pulmonary phagocytes: (*1*) alveolar phagocytes found in the seromucous, surfactant-rich fluid on the surface of the alveolar cells, (*2*) airway phagocytes found either on the surface or beneath the mucus layer of the conducting airways, and (*3*) macrophages found in various interstitial tissue compartments, for example, alveolar wall, peribronchial, and perivascular spaces and lymph vessels and nodes. Although they demonstrate variability in their specific structure and function, they all appear to play an important role in defending the airway and alveolar surfaces from toxic particles, gases, and pathogens, keeping these surfaces clean and sterile. Additionally, the placement of these cells in the organism ensure that they perform their essential roles in immune surveillance.

In the event of particle or pathogen deposition, pools of macrophages are tapped by (*1*) an active cell removal system in the interstitium, (*2*) increased cell precursor production, (*3*) release from lung reservoirs, or (*4*) increased influx of blood monocytes. Such macrophages effectively intercept foreign particles either: (*1*) by chance, (*2*) because of chemotactic substances associated with the foreign particle, (*3*) particle coatings containing antibodies, viz., IgG, or, (*4*) coatings containing antibody–antigen complexes, or (*5*) complement derived from alveolar fluids. Amplification of macrophage elaboration can occur as a result of of chemotactic substances released by other responding macrophages or lymphocytes.

Within each phagocyte are found subcellular vesicles (diameter 0.5 μm or less) rich in lytic enzymes, for example, ribonucleases, glucuronidases, phosphatases, and phospholipases. After a phagocyte engulfs a particulate and incorporates it into the cell, lysosomes attach themselves to the phagosomal membrane. These membranes become continuous and the lytic enzymes kill the pathogens.

For particulates it has been found in vivo that among dusts not known to be reactive 50 to 75% of particles were taken up in 2 hr, more than 95% by 10 hr, and essentially all by 24 hr. By contrast, when alpha quartz was tested *in vivo*, clearance was cut by 50% in the early phases of exposure (9).

However, these usually protective tissue elements may also play a deleterious role in the development of lung disease, for example, emphysema and fibrosis. In addition, certain agents may be inappropriately handled by phagocytes [e.g., silica, lead (10)] or may lead to local segregation and concentration of toxic particles (11).

The deposition of fibrous particles such as asbestos can lead to inflammatory alterations initially as an acute alveolitis. Although this response is primarily mediated via macrophage response, lymphocytes, neutrophils and eosinophils also participate. This array of inflammatory cells further respond by secreting a cascade of multiple cytokines and growth factors associated with interstitial fibrogenic processes. It is in response to such factors (e.g., interleukins, tumor necrosis agents, prostiglandins) that interstitial fibrosis and extracellular matrix deposition may occur.

In addition, particularly as regards asbestos fibers, integral to this fibrosis inducing cascade of inflammatory mediators, there is induction of reactive oxygen species. Simultaneously, interactions between cell surfaces and asbestos fibers induce further activation of phagocytic cytokines and proteases. Such agents, particularly those generating reactive oxygen species, are presently believed to play an important role in induction of unscheduled DNA synthesis, DNA strand breaks and other chromosomal aberrations, leading to neoplasia.

Despite such untoward consequences ensuing from contact with peculiarly toxic agent, in most cases when phagocytosis is effective, it provides a useful removal system for transport of toxic compounds or particles from lungs. The phagocytic route of egress is mainly via the mucociliary escalator and toward the pharynx or by direct carriage in the pulmonary lymphatic vessels and system. Ultimately these cells disintegrate with the particulates (and cell debris) reaching the pharynx and eventual gastrointestinal ingestion. A smaller proportion of the phagocytes move through alveolar walls to the lymphatic drainage system.

3.2.5.4 Transport of Particulates in the Lymphatic System. Because lymphatic capillary vessel walls are extremely thin with loose, intercellular junctions and interrupted basement membranes they are highly permeable to plasma proteins, particles and cells. The flow of lymph from alveoli and lung interstitium is unidirectional due to funnel shaped one-way valves; this flow transports xenobiotic or products of cell breakdown, etc. for activation of more centrally located humoral and cellular mechanisms of immune defense. Small particles that succeed by direct penetration, or phagocytosis, in reaching the lymphatics slowly make their way to the pulmonary lymph nodes where they can accumulate. After protracted periods such particulates may ultimately find their way to the reticuloendothelial system elements found in the various organs, namely, liver, bone marrow, and spleen.

3.3 Gastrointestinal Absorption

The gastrointestinal tract is usually considered a less important portal of entry than the lungs for the toxic agents encountered in the occupational setting. Although in general this is true, it is too frequently forgotten that ambient particles or aerosols too large to reach the alveoli eventually are available for absorption via the mucociliary escalator. Thus it should be remembered that those particulates considered irrespirable may nonetheless be presented for absorption in the gastrointestinal tract. Actually these particles are of a size range (i.e., 10 to 100 mm) that is quite readily absorbed from the gut. This consideration is especially important in industrial hygiene because sampling strategies designed only to measure respirable particle sizes are appropriate only for those materials that produce only pulmonary pathology. By contrast, measurement of only respirable airborne chemicals that are (*1*) capable of producing systemic toxicity and, (*2*) that can be absorbed both via alveoli and the gastrointestinal tract will underestimate potential occupational dose, for example, for particulate lead and fluorides.

In addition, for work environments where personal hygiene is poor, hand to mouth ingestion and absorption via the gastrointestinal tract is important. For the nonoccupational physician, and especially the pediatric clinician, this portal of entry is undoubtedly highly significant.

3.3.1 Structure and Absorptive Function of the Gastrointestinal Tract

Basically the gastrointestinal tract (GIT) can be considered as a tube going through the body; essentially, until xenobiotics transit this tube's wall, they remain outside the body's economy or its internal environment. Accordingly, poisons orally ingested and swallowed may produce no systemic effect until they traverse this tube, that is, are absorbed across the GIT wall.

This process of transepithelial absorption may occur from mouth (e.g., nitroglycerin) to anus (e.g., chlorpromazines); however, most absorption takes place in the small intestine. Until recently, the stomach was considered to be a significant site of absorption only for water and ethanol. In the past this was believed to reflect a short passage time in the stomach as well as its low pH. However, in keeping with the Henderson-Hasselbach equation, strong acids with a low pK_a would be expected to exist in a nondissociated, nonpolar form in the stomach and thus will readily be absorbed. Nevertheless, in a larger perspective, gastric absorption is relatively less important in GIT absorptive processes.

Digestion is accomplished by enzymes excreted in a mucous medium by the GIT and its associated structures (e.g., liver bile acids, pancreatic enzymes); these catalysts accomplish the breakdown of dietary protein, carbohydrate, and lipid. Mechanical mixing is achieved by autonomic system controlled pulsatile contraction of the digestive tube effected by successive contractions of muscle layers beneath the epithelial lining of the GIT lumen. Interposed between these two elements (i.e., the epithelia and muscle layers) is an extremely rich capillary network. Thus absorption of GIT contents proceeds through the epithelial cells lining the inner surface of this tube, across a cell basement membrane and subepithelial layer into the capillary network. The surface of the GIT is characterized by numerous upfoldings providing a markedly increased surface area for absorption. In contrast to the intestine, the stomach's increased surface area provides for more cells of a secretory type that produce enzymes, mucus, and hydrochloric acid.

Microscopic examination of the intestinal wall reveals that its surface consists of numerous, fingerlike projections covered largely by a single layer of columnar cells. On the surface of each cell in turn are microvillous projections that actively "pump in" materials for presentation to a bed of capillaries associated with each cell. These capillaries all drain into the portal veins leading to the liver where multiple biotransformation systems are available for further chemical alteration of absorbed compounds. By virtue of these systems of villi and microvilli, which increase surface area by about 600 times, that is, to about 2000 ft^2, the absorptive capacity of the small intestine is markedly enhanced. Almost one-half of the total GIT mucosal area is found in the upper quarter of the small intestine. The lower ileum also contains lymphoid tissues called 'Peyers patches', which serve an immune-system function. Peyers patches react with antigens (see above) which have entered via the surface mucosal barriers.

Absorption of ingested material proceeds by passive transfer in the proximal small intestine; in the distal portion both passive and active transport occur. Most essential nutrient moieties (e.g., amino acids, sugars) depend upon active transit through the gut wall. However, active transport and absorption of xenobiotics appear to be relatively uncommon in the GIT. But for toxic materials that are structurally or electrochemically similar to such nutrients, active transport can facilitate absorption of what normally would not be absorbed, for example, 5-fluorouracil by the pyrimidine transport system, or cobalt, manganese, and thallium by the iron transport system.

The microvilli are covered by a coat of mucopolysaccharide; this probably acts as a barrier against (*1*) bacteria, (*2*) macromolecules, and (*3*) ionized compounds. At the intestinal surface is found a thin acid layer at pH 5. Although this layer is found in what is generally a basic environment of pH 7.3, such an acidic layer affects the extent of ionization of compounds in the GIT. This layer is immersed in a relatively thicker second layer of water that will restrict passage of nonpolar compounds that might otherwise penetrate into the lipophilic components of the walls of the microvilli.

As previously noted the pH of the gut contents plays a critical but not limiting role in absorption. The Henderson-Hasselbach equation predicts that nonionizing weak acids would be minimally absorbed in the small intestine. However, the small amount that is absorbed into the pH 7.4 environment of the plasma is transported away, maintaining the concentration gradient across the membranes. In addition, because of the huge surface area of the small intestines, the capacity for net absorption is great although local absorption rates are low.

Physical factors further determine absorption in the GIT. Although nonpolar, nonionizable compounds (e.g., lipophilic organic solvents) most readily penetrate cell membranes, such dissolve poorly in the largely aqueous GIT fluids with consequent minimal absorption. Similarly, solid particulates of a low water solubility in such an essentially aqueous medium dissolve poorly and thus do not effectively contact the absorptive membranes. If such particulates are large, less is absorbed by diffusion, because the rate of dissolution is directly proportional to the surface area; it is upon this basis that ingested low polarity, non-ionizable metallics demonstrate little if any toxicity potential, e.g., mercury metal.

The presence of multiple metal salts may alter the absorption of other individual metallic compounds. For example, cadmium decreases zinc and copper absorption (12). The presence of phosphate and/or calcium may inhibit absorption of aluminum by formation of less soluble metallic-salt complexes. Conversely, the formation of citrate–aluminum complexes appears to enhance GIT absorption. Although elemental metals or their nonpolar forms are not readily absorbed from the GIT, EDTA chelation in the gut increases their absorbability and bio-availability.

Multiple physiological factors directly affect GIT absorption. Enzymes, microflora, and biliary emulsifiers all may alter or degrade toxic compounds, minimizing or enhancing their toxicity. For example, although oral ingestion of snake venoms is innocuous, the pH of the stomach promotes formation of carcinogenic amines when native secondary amines present in food interact with nitrites used as food additives. Intestinal flora readily convert aromatic nitro compounds to carcinogenic aromatic amines, and other flora can degrade DDT to DDE.

Other physiological factors may alter GIT absorptive capacity. In general decreased peristalsis will increase absorption; however, for a compound that may have an inherently low absorption rate, decreased GIT motility can markedly increase its residency time and thus its net absorption. Variations in local blood flow may alter absorptive rates, especially for lipids. Because these readily penetrate cell membranes, decreased blood flow will decrease local lipid uptake. Age alone may markedly affect GIT absorption; thus newborn rats may absorb cadmium 24 times as rapidly as adults (13). Finally, it should be recalled that various endocrines, e.g., parathormone, directly enhances GIT absorption of aluminum; vitamins, viz., vitamin D, facilitate the absorption of lead, strontium, magnesium, barium, beryllium, zinc and cadmium.

Although the foregoing has considered xenobiotics occurring either as solutes or otherwise in liquid form, it has also been found that particulates may move across the epithelial cell walls of the GIT. Thus colloids and small particles can move through aqueous cell membrane "pores." In addition, such particles as large as 23 μm have been observed to be taken up by pinocytosis into cellular vesicles that are subsequently discharged in the connective tissues beneath these cells. Ultimately such particles are either ingested by phagocytic cells or directly taken up in lymphatic fluids.

As should now be readily apparent, the wide variety of food components, their hydrolysis and other breakdown products, and the potentials for binding to such variegated diet-derived components provides a highly variable and rapidly changing GIT mileau. Accordingly, prediction of absorption rates under these complex conditions all too frequently approaches conjecture.

Regardless of the effects of this multiplicity of variables, once a xenobiotic traverses the epithelial wall of the GIT it is subject to one of two fates. For most materials absorbed, carriage in the portal circulation to the liver is the rule; here multiple biotransformations take place. Such metabolic transformation may produce a less toxic product so that little of the parent toxic compound reaches the systemic organs. If hepatic metabolism over the specifically required pathway were saturated or detoxifying rate limits were approached, extrahepatic target tissues may not be protected. Conversely, compounds of high cytotoxicity may selectively damage the liver by exposing it to high portal vein concentrations whereas other organs are spared such relatively higher, toxic concentrations.

For some small percentage of GIT-absorbed materials lymphatic drainage bypassing the liver is the rule. In such manner ingestion particularly of lipophilic material might avoid a first pass through the liver's detoxifying systems, thus allowing toxic consequences to occur elsewhere in the body.

3.4 Absorption and Penetration Through the Skin

The skin, a very important portal of entry in the occupational setting, is specifically considered in Chapter 6.

4 TRANSPORT AND DISTRIBUTION OF XENOBIOTICS

4.1 Introduction

After foreign materials have penetrated the epithelial barriers, the next medium encountered is the blood or the lymphatic or interstitial fluid. The liquid phase of the blood, in which are suspended various formed elements (i.e., red blood cells, platelets, white blood cells, and other reticuloendothelial derived cells in various forms) is referred to as the plasma. Humans possess from 45 to 68 mL blood/kg body weight, or from 5 to 5.5 L in a 70-kg man. Materials carried in this medium may leave blood vessels via physical "pores" in the capillary wall, by diffusion or by special transport mechanisms previously discussed. In effect, the major determinant of capillary bed egress is molecular size. With capillary wall pore sizes as large as 30 Å, compounds of up to 6000 Da, regardless of electrical charge, may directly exit from capillaries into the extracellular compartment. Larger molecules may directly exit via pinocytosis but at a very much lower rate.

After leaving the vascular system, compounds enter the extracellular compartment whose constituents differ slightly from the plasma because of variations among plasma components' ability to leave the blood system. This extracellular compartment serves as the final liquid medium before a xenobiotic encounters a tissue cell membrane. The processes governing transit of that membrane have been described in Section 3.1.

4.2 Blood Elements and Their Transport of Xenobiotics

4.2.1 Transport by Formed Elements

Because only red blood cells play a role in transport, this discussion does not consider leukocytes or platelets in this context. Mature erythrocytes as they are found in the blood possess no nucleus. The carriage of the essential element oxygen occurs because of the presence of a divalent iron contained in the hemoglobin molecule; this particular molecule's configuration results in an extremely strong affinity for oxygen. If this ferrous iron is oxidized to the trivalent state, methemoglobin is formed and oxygen transport is severely restricted.

Each 100 mL of arterial blood contains about 15 g of hemoglobin, which combines with 19 or 20 mL of oxygen. In blood that is in equilibrium with air at normal atmospheric pressure, about 1% of the total oxygen is in solution in the plasma. Normal venous blood carries 55 to 60 vol % carbon dioxide; it drops off about 10% of this in the lungs where the blood is once more oxygenated. The amount of carbon dioxide in simple solution is about 3 vol % in the venous blood and 2.5 vol % in the arterial. Blood flows through the capillaries of the lungs in approximately 1 sec (14), and through active tissue in a similar time. At such a flow rate it is evident that mere solution of oxygen and carbon dioxide cannot account for the exchange of these gases, even though the equilibration in the lungs is highly efficient. The speed of exchange is due to the reversible combination of these gases with hemoglobin in the oxyhemoglobin and carbamino reactions, respectively, as well as reflecting the presence of catalysts in the red cells that play an important part in the oxygen and carbon dioxide exchange. Carbonic anhydrase greatly increases in either direction the reversible chemical reaction $H_2CO_3 \rightleftarrows CO_2 + H_2O$, and oxidation and reduction in the tissues are greatly accelerated by oxygen-activating catalysts or by dehydrogenases.

It must be recognized that the transport of oxygen or other gases is strictly a passive, partial pressure gradient-dependent phenomenon. Transport by diffusion, either from alveolar air sacs to blood or from tissue cells to blood, involves passage through cell membranes. Because oxygen and gases other than carbon dioxide never move counter to partial pressure gradients, it may be said that such exchanges are governed simply by physical laws of diffusion and solubility.

Although the oxygen binding affinity is about 1000 times greater than its dissociation constant, carbon monoxide's binding affinity is 242,000 times greater than its dissociation constant. Accordingly, this 242-fold greater affinity favors not only a greater binding capacity for CO, but also produces a considerable reluctance for CO dissociation in the capillary bed.

The aggregate surface area of red blood cells (namely, 3000 to 4000 m^2) favors transport by adsorption (van der Waals forces) and/or stromal binding of physiologically necessary as well as xenobiotic materials. Most such molecules are in equilibrium with their fraction bound to plasma. Because the stromal density of the erythrocyte is low and readily permeable, xenobiotics can have ready access at multiple sites to the constituents of the cell, namely, the protein fibrils (albuminoids) surrounded by lipids (cholesterol and esters, neutral fats, cephalin, lecithin, and cerebrosides) as well as its other contents, for example, hemoglobin. Thus, whereas O_2, HCN, CO, and Se are bound to heme, As and Sb bind with the sulfhydryl ligands of the globin protein. Other toxic or essential compounds are bound to the red cell although their locus is not well defined. This group includes organic mercury salts, Cr, Pb, Zn, Th, Cs, and steroids.

Although the erythrocyte has no nucleus it must be considered a metabolically active entity. Accordingly, this mature, aneucleated cell contains the enzymes lipase, catalase, and anhydrase, as well as the others required for porphyrin synthesis. It is this latter activity and changes in D-aminolevulinic acid dehydrase activity that has proved to be such a highly sensitive indicator of lead exposure.

4.2.2 *Transport of Xenobiotics in Plasma*

4.2.2.1 General. Numerous electrolytes found in plasma as ions are in equilibrium with nondissociated molecules. The dissociated ions readily move from blood to interstitial fluid and into cells by diffusion. For example, Be and the alkaline earths Ca and Sr exist in this form; others, for example, Yt and Pu, exist as microcolloids in association with plasma proteins. Gases and vapors are essentially physically dissolved in the plasma.

It should be noted that very few substances are transported in the blood in association with only one blood component or element. For example, the uranyl ion is approximately 40% bound to plasma protein and the other 60 percent occurs as a diffusible bicarbonate. In such fashion, most elements exist in a state of dynamic equilibrium between various formed and/or liquid plasma components.

4.2.2.2 Transport by Plasma Proteins. Notwithstanding the foregoing, transport by these proteins represents a most important vehicle for carriage of exogenous materials through the body. This reflects in part the finding that most plasma proteins possess surface areas of 600,000 to 800,000 m^2 for physical interactions, for example, electrostatic attraction, van der Waal forces, and ion radius, all of which affect binding behavior. Thus divalent cations, for example, Ca^{2+} and Ba^{2+}, form unstable complexes with proteins, whereas heavy element polyvalent cations form more tightly bound and less dissociable complexes.

Protein binding sites may be saturated; competition for binding sites may result in the displacement or inavailability to another compound. Such an inavailability of binding sites result in marked increases in plasma concentrations of the nonbound compound, with consequent cellular availability and effect potentials.

Albumin, with a molecular weight of 6800, is the most abundant and important plasma protein involved in transport. This molecule possesses 109 cationic and 120 anionic ligands for binding, distributed over a relatively large surface areas. The most common of the ligands are carboxylic binding groups associated with the asparaginic and glutaminic amino acids. There appear to be six binding regions on albumin (15) where protein-ligand interactions occur. This binding results mainly from van der Waals and hydrophobic forces, the latter reinforced by hydrogen bonds. Many physiologically important compounds are carried in the albumin fraction, for example, bilirubin, porphyrins, ascorbic acid, fatty acids, and cholinesterase. Many cations (e.g., Cu and Cd) and anions (e.g., Br, I, and CN) are partially bound to albumin, the former via imidazole or carboxyl ligands.

Generally, metal ions are bound by sulfhydryl, amino, carboxyl, and imidazole groups of albumin amino acids; compounds with an *o*-carboxylic or *o*-hydroxylic group on a benzene ring (e.g., *o*-cresols, nitro- and halo-substituted aromatic hydrocarbons, and phenols) demonstrate considerable binding affinities for albumin.

By contrast with albumin, the globulin fractions are believed to participate in relatively fewer transport-related interactions with other compounds. The transport of mono- and bivalent copper is accomplished by ceruloplasmin found in the α-2 globulin fraction whereas divalent iron is carried by transferrin found in the β-1 globulin fraction. Steroid hormones are transported by interactions with α-1 globulins , as are vitamin B_{12} and thyroxine. The γ-globulin fraction is essentially largely associated with the immune mechanisms.

The binding of multiple toxicants with plasma protein has been extensively reported; however, specific binding site locations are less clear. Phosgene and nitrous oxide react with amino groups of proteins, whereas methyl bromide reacts with sulfhydryl binding groups on amino acid proteins. Isocyanate, aromatic amines, and carbon disulfide react with proteins, the latter binding to peptides forming dithiocarbanates and also cyclic mercaptothioazolinone.

Because lipoproteins are found in the globulin fraction, many lipophilic compounds may be transported by these proteins, for example, benzopyrene.

Variations in sites and strengths of binding in the blood affect the bioavailability and hence metabolism and elimination from the body of xenobiotics, for increase in binding diminishes diffusibility. For example, lead bound to red blood cells has a biological half-life of 30 hr, in contrast to lead binding to plasma proteins, with a half-life at that site of only 30 to 40 min. Thus while the protein-bound fraction has a greater bioavailability, lead bound to erythrocytes will tend to spare target organs from exposure to peaking dosages. If a substance binds with higher affinities to some sites more strongly than to others, its bioavailability to other target loci is likewise diminished. Protein binding by a toxic substance may, by such successful competition for such sites, cause the release of other protein-bound physiological substances. This in turn can produce other untoward effects, for example, as in the release of histamine or serotonin. Finally, binding by toxic substances may pathologically alter protein structure, as is believed to be the case for ozone, isocyanates, aromatic amines, and fresh zinc fume.

4.2.2.3 Transport by Blood Organic Acids. The organic acids found in plasma are potent complexing agents and hence important transport media for xenobiotics. The complexing of xenobiotics by blood organic acids depends upon the physicochemical properties of both reactants (e.g., valence, ionic radius) as well as the metabolism of these acids.

Although citric acid is the most effective of this class of compounds, lactic acid, although less active, is present at four-to six-fold greater concentrations in blood. Glutaminic acid and the amino acid α-glutamic acid are also effective complexing agents. The anions of organic oxy- and amino acids readily complex with alkaline earths and some heavy elements present in blood as cations.

The consequences of such bindings are variable, depending upon diffusibility, stability, and metabolism of these complexes. Thus although citric acid may complex cations of Bi, Po, and Y, such complexes may ultimately enhance the bioavailability of such toxicants as the organic acid is metabolically broken down within the cell. In the case of bone-seeking elements, for example, Pu, which complex with citric acid, the breakdown of such complexes in the region of bone minerals effectively makes such osteotrophic elements readily available for bone incorporation. Conversely, complexing of Yt by glutamic acid to form a diffusible soluble form allows excretion of this element in the urine.

4.3 Transport by the Lymphatic Vascular System

In contrast to the high-pressure blood circulatory system, the lymphatics are a low-pressure aspirating system for fluid lost from capillaries and tissues. Because such vessels do not possess basement membranes, they are extremely permeable, permitting ingress of relatively large molecules. The role of alveolar macrophages and dendritic cells, and their access to the lymphatics are discussed in Section 3.2.5.3. As the lymphatics ultimately drain these antigen presenting cells into the circulatory system, such particulates are thus made available for metabolism, particularly by the reticuloendothelial system.

5 XENOBIOTIC DISTRIBUTION TO AND DEPOSITION IN ORGANS AND TISSUES

5.1 Introduction

In terms of its constituents, the body is largely water, namely, 70% by weight. For this reason, in considering where xenobiotics are distributed within the body, it is convenient first to divide the body into three fluid compartments, that is, the vascular or plasma, the interstitial or extracellular, and the intracellular compartments.

It must be emphasized, however, that this construct simply represents a convenient model for analyzing the distribution and movement of xenobiotics throughout the body. As discussed in Section 4, these transfers operate under constraints imposed by the physicochemical nature of the compound and the bodily environment in which it is found.

The vascular system that brings each organ its supply of blood provides a unifying channel from the external environment, across the interstitial compartment, and finally to each cell membrane. The priority the body places upon the needs of each organ is suggested by comparisons between the mass of each organ and their relative blood flow rates (Table 3.2). The concept that considers the body as a liquid continuum "separated" into permeable compartments allows for kinetic analyses of body xenobiotic distributions.

Further breakdown of the intracellular compartment into organs or regions of interest, each supplied and drained by blood flow, permits more specific multicompartmental modeling and toxicokinetic analysis of xenobiotic distribution and movement.

5.2 General Factors Affecting Distribution

As noted previously, physicochemical factors as well as the state in which a material exists in blood or plasma (e.g., ion, molecule, binding affinity for blood elements, molecular weight) affect transfer potentials and thus distribution. Variations in blood supply and flow rates may affect how a toxic compound is distributed. Such variations may result from the relative blood flow to specific organs (see Table 3.2) neurohumoral-induced flow altera-

Table 3.2. Relative Body Mass and Blood Flow for a 70-kg Man

Tissue or Organ	% Total Body Mass[a,b]	% Total Cardiac Output[c]
Brain	2	15
Liver	2.6	30
Kidneys	0.4	20
Heart	0.4	20
Skin, muscle	43.7	10
Bone, connective tissue	17	
Fat	18	3

[a]Tissues here constitute 70% of total body weight.
[b]Data from Reference 16.
[c]Data from Best and Taylor (17).

tions, and so on. In addition, concentration gradients between compartments, the presence of other substances in the cell, affecting cellular permeability or intracellular distribution all may have an impact upon how a xenobiotic distributes itself within the body.

When a xenobiotic is ultimately conveyed to a tissue cell membrane, numerous factors determine its transmembrane movement. Thus nonionized compounds that are lipid soluble readily diffuse through cell membranes. In the event the cell or tissue has a high fat content, such nonpolar compounds readily dissolve in such tissue, often remaining there for long time periods. Such is the case for a large group of compounds of environmental interest, for example, PBB and DDT, in which case they are sequestered in fat depots for months and years. Other materials accumulate in association with specific macromolecules, for example, binding of CO with hemoglobin. In still other cases physical dimensions of xenobiotics may be similar to those of physiological components found in specific tissues. Such similarities may result in exchange with these specific cellular elements, for example, alkaline earths and fluoride incorporated by ion exchange with calcium of bone.

In the process of biotransformation multiple events occurring within the cell will affect body distribution. Reactive intermediates of biotransformation may combine with cell macromolecules, impeding their excretion from that site. Any reaction within the cell resulting in altered permeability of cellular and intracellular membranes, for example, endoplasmic reticulum, or which competes for binding sites within the cell, or alters pH, pK_a; or blood flow can affect permeability, transfer rates, and ultimately distribution. Depending upon whether such actions take place at a target cell or at a storage site, a toxic outcome or attenuation of damage potentials, respectively, may result.

5.3 Specific Structures Limiting Distribution

5.3.1 The Blood–Brain Barrier

This structural concept is of major importance among the several structural barriers to transfer of substances from blood to tissue. The blood–brain barrier, found at the capillary–glial cell junction, is relatively less permeable than all other such membranes. (However, because of the usual high lipid content of membranes the transit of fat-soluble compounds are relatively less impeded than ionized, dissociated compounds.) This lesser degree of permeability results in part from the capillary–endothelial cell anatomical junctions being in unusually close apposition, which leaves fewer intercellular pores. In addition, brain capillaries are closely surrounded by the processes of glial brain cells; accordingly, at least two cells must be traversed if a blood-borne chemical is to flow from blood to brain cells. Another impediment to movement reflects the relatively lower protein content of the brain interstitial fluid. Although this barrier is highly effective in controlling extravascular transfer, it is not uniformly distributed throughout this organ. Finally, this barrier is not as effective in infants as in adults.

5.3.2 The Placental Barrier

The so-called placental barrier has been considered to represent an effective impediment to transfer of toxins from the maternal bloodstream to that of the fetus. However, it has become clear that many biopathogens (e.g., viruses, spirochetes), large molecules (e.g., antibodies), and even whole cells (e.g., erythrocytes) can cross this barrier.

Although toxic chemicals can readily cross this barrier, because biotransformation systems are present in this organ, some chemical substances can be detoxified *in situ* here. Further, if some toxicants have a high affinity for specific maternal tissues (e.g., PCB in fat depots), little may be available for transfer to the fetus. Furthermore, the behavior of various organs of the fetus differs from that of the adult. Thus the immature blood–brain barrier more readily permits passage and deposition of lead or methylmercury into the fetal brain. By contrast, xenobiotics concentrated in the human liver may not be accumulated in that fetal organ.

Accordingly, although the placental barrier may alter the potential for transfer from maternal to fetal blood, this movement is carried out in a highly variable and compound specific manner. Any toxicant in question cannot be presumed to place the fetus at more or less hazard; each compound presented to the placenta can be considered only on the basis of its individual chemical, biological, or physical properties.

5.4 Redistribution of Xenobiotics

Xenobiotic concentrations in most organs will change with the passage of time. This may occur in order to maintain intercompartmental equilibrium as tissue/blood concentration gradients change. Blood flow rates to various organs (see Table 3.2) initially will determine tissue extracellular concentrations; the permeability of the tissue and relative availability of cell binding sites will determine subsequent deposition. The latter two factors will change with tissue uptake and subsequent biotransformation. Redistribution of lipophilic substances occurs regardless of biotransformation, with such xenobiotics transferring readily to and from adipose tissue and brain.

With the passage of time and uptake by specific tissues, blood carriage becomes relatively less important. At this point tissue-specific binding affinities are more important in determining tissue concentration of xenobiotics. Thus although the liver takes up 50 percent of a dose of lead within 2 hr of administration, 30 days later 90% of the remaining lead is found in bone, having substituted for calcium in the hydroxyapatite crystal lattice of that organ.

5.5 Specific Body Compartments of Deposition and Accumulation

5.5.1 *Liver and Kidneys*

Although these two organs have a high binding capacity for chemicals and metals, the mechanisms responsible for this behavior have not been clearly elucidated. On the basis of more recent studies the role of active transport and site of specific cell component binding has become clearer. Thus cadmium, mercury, and bismuth have been found to bind preferentially with the cytoplasmic soluble fraction of hepatocytes; the latter two elements are also bound to nuclear soluble protein and to a lesser extent with mitochondrial fractions of liver cells. Upon further investigation of hepatocellular protein binding, it has been found that a number of heavy metals are bound to a specific protein, metallothionine.

Biosynthesis of this protein is induced by increased metal loading of the liver and kidney. This low molecular weight protein (i.e., 10,000 to 12,000 Da) consists of a dimer of equal size; being rich in cysteine it possesses many sulfhydryl groups that readily bind certain metals. It was found that a metallothionine molecule can bind to metal ions, 3 SH groups being required to bind each ion of such metals as Zn, Cd, Hg, Bi, Co, and Cu. This binding of free metal ions represents a form of detoxification of xenobiotics. However, any benefit may be transient, because mobilization of metal ions can occur owing to their displacement as a consequence of unsuccessful competition for such binding sites.

Another hepatocellular cytoplasmic protein, ligandin (18), one of the forms of glutathione reductase, readily binds a number of organic ions. This protein has been shown to bind azo dye carcinogens, steroids, and other xenobiotic organic acids. Such binding suggests a role for this protein in facilitating transfer from plasma to liver, thus effectively removing such compounds from further circulation.

5.5.2 Deposition and Accumulation in Lipid-Rich Organs and Tissues

In general terms, nonpolar, nonelectrolyte substances that minimally dissociate and that demonstrate a high Overton-Meierhof partition coefficient for lipids will readily permeate lipid components of the cell membrane. Also, as previously noted, changes in pH and pK_a will also alter dissociation. In addition, chemical structural characteristics can change liposolubility; for example, substitution of an alkyl group greatly increase the liposolubility of hexobarbital.

With these caveats in mind one would thus expect that such substances, which would dissolve in neutral body fats, would also readily accumulate in large quantities in the body, because such fats constitute from 20 to as high as 50% of total body weight. It should therefore be predicted that compounds of industrial importance such as industrial solvents (e.g., alcohols, ketones, glycoesters) and chlorinated hydrocarbons (e.g., chlorinated aliphatics, pesticides, PCBs) all would tend to accumulate in the large lipid compartment.

Of particular importance in this regard is the central nervous system, which is rich in various lipids. As previously noted, the blood–brain barrier does not provide a completely effective blockade to nonpolar, nondissociated compounds. Thus compounds such as organic solvents (e.g., halogenated aliphatics, alcohols) and organometallics (e.g., alkyl mercury and leads) readily pass through this barrier.

Once this constraint to their movement is overcome, the liposolubility of this class of compound can further affect brain structure and nervous system functional integrity. Neurofibers acting as electrical conductors require an intact, electrically functional "insulation." The lipid myelin sheath, or a layer of Schwann cells rich in lipid that serve such function, can be disrupted by uptake of liposoluble xenobiotics. For yet other reasons, compounds such as CS_2, Pb, Hg, and Mn appear specially to have an effect on the brain's cellular elements. Yet other phenomena may precipitate neurotoxicity. For example, tertiary configuration and consequent successful competitive inhibition of acetylcholine esterase activity at the neural synapses appear to be responsible for the toxicity of the organophosphate pesticides.

In addition to the relatively high lipid concentrations of the nervous system, body fat depots that act as body stores of readily mobilized energy-rich triglycerides make up from

10% to as much as 50% of total body mass. Thus a large compartment for uptake and storage of xenobiotics is represented by such fat depots. Accordingly, accumulation at such sites can temporarily remove liposoluble xenobiotics from circulation. It may thus minimize the available substrate for biotransformations occurring elsewhere and the toxicity that might follow. Further, because of the relatively smaller blood supply (cf. Table 3.2) in such storage sites, compounds once deposited here tend to turn over less rapidly from these depots. For these reasons, biopsy of fat stores can provide an indication of previous exposures long after exposures have ceased. However, although such deposition sites are considered capacious and turnover is less rapid, these loci do not necessarily imply sequestration or protection of the body from the effects of their xenobiotic contents. For example, fat—like all other body components—undergoes continuous turnover of its chemical elements. In the event of sudden needs for energy-rich substrates (e.g., as with starvation), fat stores, and their associated xenobiotics, can be rapidly mobilized to the general circulation. This process under neurohumoral control is mediated via multiple enzyme systems found in adipose tissue, for example, lipase, diastase, phosphatase, and lecithinase, as well as catalysts involved in intermediate carbohydrate metabolism, for example, hexokinases and dehydrogenases. In addition, this tissue, which is now understood to be metabolically active, is known to carry out biotransformations not directly related to these energy mobilizing activities. Experimental evidence indicates that oxidation of benzene to phenol or hydrolysis of esters may take place in fat storage sites.

Finally, deposition in other tissues that contain lipids in somewhat higher concentrations than in general may produce direct effects at such places, for example, the hematopoietic disruptions of the bone marrow and in the thymus caused by benzene.

5.5.3 Deposition in Bone

Because bone constitutes 10 to 15% of total body mass, these structures represent a significant potential site of deposition for osteotrophic compounds or elements. Minerals are the major components of bone, representing 72% by weight of this tissue. The primary bone mineral is constituted by hydroxyapatite, $Ca_{10}(PO_4)_6(OH)_2$, which occurs as a crystal 25 to 30 Å thick and approximately 400 Å long and wide. The surface area of this crystalline structure averages 100 m^2/g of bone, totaling 7×10^6 m^2 for a standard 70-kg person. This mineral portion of bone contains primarily (in descending order of concentration) calcium, phosphate, hydroxyl, carbonate, and citrate plus lesser amounts of sodium, magnesium, and fluoride. Structurally, bone consists of a variably dense matrix, made up of (*1*) active osteocytes continually breaking down and reforming bone and (*2*) a stable, relatively inert, dense mineral condensation. Depending upon systemic (e.g., calcium regulatory, growth, and other hormones) and local factors (e.g., mechanical stress or loadings) there is variation in the percentage of bone metabolically active. The extremes of activity are presented by age, that is, almost 100% of bone is active in the young, whereas only 30% is active in adults.

Xenobiotics that demonstrate osteotrophic proclivities, though occurring as ions, molecules, or colloids, usually are transported as complexes, for example, with organic acids. Ultimately, such complexes dissociate, so that newly freed ions can exchange with calcium or phosphate ions in bone mineral. Colloids react with bone by chemisorption or adsorption. All these reactions take place only in metabolically active areas of bone.

Bone mineral structurally and functionally can be considered to consist of three layers: (*1*) a deep layer consisting of the hexagonal hydroxyapatite crystal, (*2*) an intermediate hydration layer or "shell," and (*3*) an outer shell consisting of nonspecific cations and anions in solution. All these three layers are in dynamic equilibrium with each other, the outermost ultimately in equilibrium with body interstitial fluid. Ions must diffuse through each layer or shell; diffusion through the hydration shell is restricted to specific ions. Ultimately ions reach the crystal, which can act as an anion or cation exchanger. In such fashion calcium cations or phosphate anions are exchanged for ions of similar size. Calcium ions can be exchanged for other calcium ions (i.e., by iso-ionic exchange) or with other ions of a similar ionic radius (i.e., by hetero-ionic exchange). In such a fashion phosphate ion can be exchanged for phosphate, or citrate and carbonate can be exchanged. The hydroxyl anion of the hydroxyapatite crystal can be exchanged for fluoride ion. Ultimately, as ions are incorporated at the surface of the hydroxyapatite crystal, they are repeated, overlaid by another layer of crystal. Thus with time and repeated crystallization such incorporated ions are deposited "deeper" into the crystalline bone structure and thus less readily accessible for subsequent exchange with body fluids.

The rate of diffusion and ultimate crystal deposition is complex, determined by multiple factors. The relative ionic concentration (Guldberg-Waage law) in interstitial fluid at the two outer shells and the crystal surface dynamically determine the net direction and velocity of flow. The chemical characteristics and valency of ions, pH, fluid volume, temperature, and so forth also determine velocity of flow and direction. Although all these principles are applied to ionic transfer, the rate of transfer seems to diminish as an ion moves from the outer shell toward the inner crystalline layer. Within the limits of the number of exchange sites at the outermost shell, exchange takes place more rapidly with interstitial fluid than with the immediately deeper hydration shell. Similarly, exchange between the hydration shell and the outermost layer takes place more rapidly than does exchange between hydration shell and the deeper hydroxyapatite crystal surface. About 30% of bone ion is loosely bound; exchange takes place with a biological half-life of 15 days. Because the other 70% is firmly bound by crystallization, mobilization of this fraction has a biological half-life of about 2.5 years.

In view of these dynamics, the behavior of small, large, or repeated doses of osteotrophic elements is predictable. For example, a single large does of fluoride mainly appears in the urine and is not deposited in bone as the finite number of exchange sites on the outer shell are saturated. By contrast, repeated smaller doses of fluoride are more readily accumulated deeper within this three-shell system. With cessation of fluoride exposure this ion diffuses slowly out of the bone. This diffusion out of bone proceeds over scores of months and is manifested by a slight but persistent increase in urinary fluoride concentrations. Similar behavior is generally applicable to a large range of osteotrophic elements. Indeed, calcium, citrate, fluoride, radium, and strontium being deposited in bone by ion exchange should follow similar dynamic transfer mechanisms. However, there are wide variations in the kinetics of such movements; for example, the biological half-life of barium in bone is 35 days, whereas that of radium is approximately 60 years.

Colloidal absorption represents another important mechanism by which osteotrophic elements become incorporated into bone structure. After being carried by organic acids in plasma, these complexes dissociate so that the resulting free ions form colloidal particles that ultimately reach the bone surface. Because of the enormous surface area per gram of bone (i.e., 100 m^2/mg), adsorption occurs there due to van der Waals forces and chemisorption reflecting electrostatic or covalent binding. The irreversible absorption as a mono- or multilayer film on the crystal surface is covered here by successive layers of bone crystal. By cyclic repetition these colloids are progressively buried deeper in the bone. Such colloid deposition appears to be the mechanism whereby transuranics and lanthanides accumulate in bone.

It should be apparent that bone is not a metabolically inert tissue. The dynamics of these exchange reactions makes it self-evident that ionic flow may move either toward the crystalline depths of the bone mineral or to the outer shell for ready exchange and possible excretion. Accordingly, bone deposition cannot be regarded either as permanent or irrevocable, although the dynamics of such movements is usually relatively slow and highly variable.

6 METABOLIC TRANSFORMATION OF XENOBIOTICS

There is a broad diversity of reactions whereby xenobiotics are transformed in order to facilitate their elimination from the body. Substances that exist in the plasma as polar, dissociated, and hydrophilic forms are readily eliminated by the kidneys. By contrast, most lipophilic, nonpolar compounds are not easily excreted via the renal route. Although some of this class of compounds can be eliminated either into the bile or by volatilization from the lungs, most such nonpolar materials must undergo transformation to water-soluble metabolites, which are then readily excreted into the urine.

In the larger perspective, a two-stage process affects the alteration of foreign materials required for their excretion. The first phase, that is, a "nonsynthetic" stage, largely involves lipophilic, nonpolar substances and consists largely of enzymatically mediated oxidation, reduction, or hydrolysis. In most cases this step alters molecular structure so that a "conjugation" or linkage site (i.e., a carboxyl, amino, sulfhydryl, or halogen group) is available for subsequent reaction. In most instances this process succeeds in making possible subsequent steps leading to detoxification and elimination. Occasionally this first phase leads to the opposite situation, "lethal synthesis," that is, the formation of a more toxic product.

The second basic step in the detoxification process consists of a synthesis phase. At this juncture an endogenous substance is linked at the conjugation center of the previously altered xenobiotic. This step almost always results in detoxification, largely because the newly conjugated molecule is more water soluble and thus more easily excreted. (In an evolutionary context, these processes, which all lead to water solubilization, reflect the aquatic origin of most terrestrial animal life.)

However, conjugation may also result in the formation of a larger molecule that displaces or blocks access of other toxic molecules at receptor sites. Alternatively, a newly conjugated- or altered-molecular structure may not sterically "fit" at a target cell's receptor site. In both cases the end result is prevention of toxic sequelae, that is, detoxification.

Multiple thermodynamically driven chemical processes, participant tissues, and biological mediators are involved in these complex activities referred to as detoxification and biotransformation. These are more extensively discussed in the toxicology volumes of the publication.

7 ELIMINATION AND EXCRETION OF XENOBIOTICS

7.1 Introduction

The rate and effectiveness of the elimination of xenobiotics or toxics from the body is a major determinant of the severity and duration of their possible actions. Thus if they are rapidly eliminated any effect tends to be minimized; if residence in the body is protracted they may act for longer periods of time and possibly with greater severity.

Xenobiotics are eliminated from the body via several routes. The onset of elimination reflects in part the route of absorption, so that inhaled vapors and gases may be immediately desorbed from plasma and exhaled almost simultaneously with exposure. By contrast, material absorbed from other portals of entry will first be transported, distributed, and transferred to organs, where they can be transformed and ultimately excreted. Additionally, at the cellular level, xenobiotics can be readily eliminated via active transport via the multiple drug resistance transporter system before they have an opportunity to interact with components inside of the cell.

Because excretion is the primary function of the kidneys, the majority of absorbed chemicals are eliminated via that organ. Because compounds must be converted to water-soluble materials for kidney excretion, this process is usually less rapid than is pulmonary elimination. The bile and gastrointestinal tract are also important modes of elimination for a number of xenobiotics of industrial significance, for example, lead.

Although several other organs are capable of excretion, namely, the skin via sweat and the salivary, lachrymal, and mammary glands, these are of lesser significance. In addition, via deposition and shedding of body parts (e.g., hair, fingernails), several xenobiotics of industrial significance can exit the body, for example, mercury and arsenic.

The factors that determine a material's "mobility" in the body have been discussed previously. Such factors also play a role in the elimination process by influencing transferability and hence accessibility to the organs of excretion.

7.2 The Kinetics of Elimination and Excretion

The study of the kinetics of elimination of xenobiotics requires quantitative data describing their concentrations in blood and/or urine. Such information collected over a time course permits study of the rate and character of xenobiotic elimination, in addition to providing valuable insights into their body uptake, distribution, and mobilization.

After a single dose of a compound is rapidly administered intravenously, a logarithmic plot of its concentration in plasma over time will generally demonstrate one of at least two patterns, namely, (*1*) a linear decrease or (*2*) a nonlinear decrease. Derivation of a linear excretion curve (Fig. 3.6) suggests rapid and relatively uniform distribution of the compound throughout the body, that is, as if the body was acting as one homogeneous vessel or compartment. The shape of this curve also indicates that a compound is eliminated using all transformation and excretion routes by first-order kinetics.

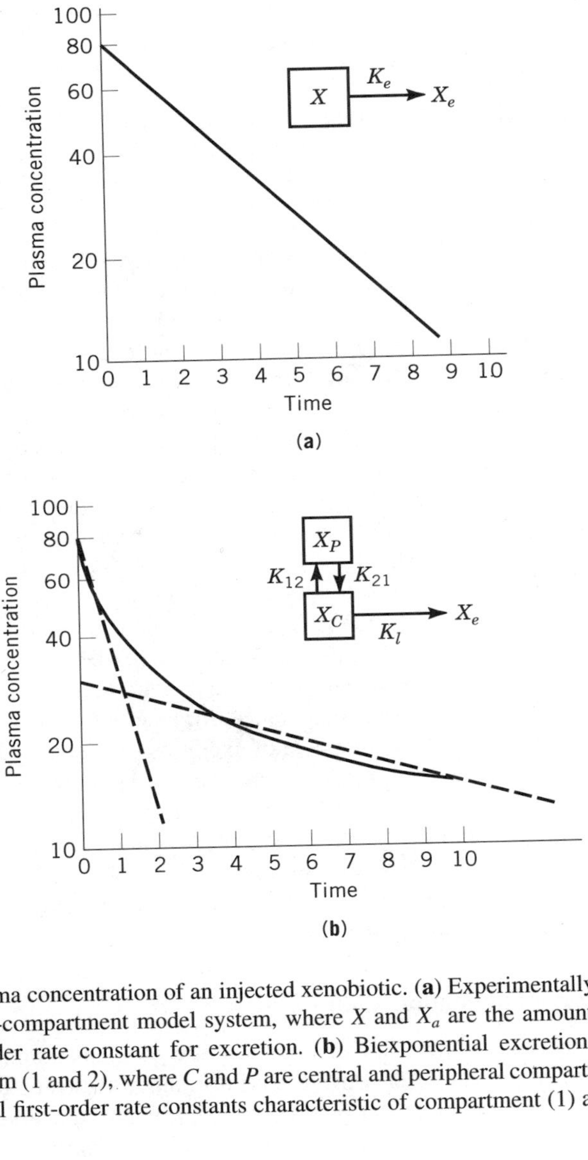

Figure 3.6. Plasma concentration of an injected xenobiotic. (**a**) Experimentally derived first-order excretion in a one-compartment model system, where X and X_a are the amount of the compound, and k_c = first-order rate constant for excretion. (**b**) Biexponential excretion curves for a two-compartment system (1 and 2), where C and P are central and peripheral compartments, respectively, and k = individual first-order rate constants characteristic of compartment (1) and (2).

The time necessary for the compound to be eliminated from the body is known as the biological half-life, or $t_{1/2}$. It can be determined visually by inspection of the derived curve in Figure 3.6 or by calculation using the compound's elimination rate over time. Among chemicals showing first-order elimination kinetics the half-life is independent of dose up to individual limits. In a single open-compartment model tissue concentrations decrease at similar half-lives as in the plasma. Because the ratio of tissue to plasma concentration is constant within limits, a determination of such a ratio can permit subsequent tissue concentration to be calculated from an assay of plasma levels.

The slope of the derived curve (Fig. 3.6) represents a first-order constant of elimination; it can be derived from the relationship:

$$t_{1/2} = \frac{\ln 2}{k_{el}} \quad k_{el} = \frac{0.693}{t_{1/2}}$$

so that the elimination rate constant demonstrates units of reciprocal time. Thus the half-life of the compound in Figure 3.6 is expressed by $k_{el} = 0.693/t_{1/2}$ because 20% of the element is excreted each hour. This open model and its first-order kinetics assume (*1*) the compound is excreted by essentially one route of elimination, (*2*) the substance is not biotransformed, and (*3*) it is distributed to tissues that are all similar in their exchange rates.

These relationships and the elimination rate hold constant, regardless of dosages, except when the dose exceeds certain limits. But as these limits are exceeded, the initial rates of elimination do not hold constant, indicating that biotransformation, protein binding, active transport, and so on all operate within finite limits. This indicates that some essential pathway(s) has become saturated, in keeping with Michaelis-Menten kinetics. Thus below the Michaelis-Menten constant (*Km*) for the process first-order kinetics apply, but as the *Km* for the process is exceeded, nonlinear kinetics apply. Such an event indicates that a change from first-order to saturation kinetics is occurring; that is, the body's processing mechanisms are operating in a different fashion than seen with low doses. Such points of departure may also indicate the attainment of a level of dosage compatible with toxicity.

Elimination curves that are not linear on a logarithmic plot but rather are exponential (Fig. 3.6) suggest the need for a multicompartment analysis. By the analysis and determination of residuals the observed curve can be distinguished as consisting of two different slopes, that is, one having a rapid and the other a slow rate of excretion, each curve having its own elimination rate constant. Such suggests that the body consists of a two-compartment system. One compartment manifests rapid equilibration, biotransformation, and excretion components, whereas a second compartment performs these activities less rapidly.

Regarding the first of these compartments, this could represent the liver and kidneys, which have a high perfusion rate and thus rapidly encounter the effect of xenobiotic dosing. In toxicokinetic modeling these may be referred to as a "central compartment." Conversely, given the relatively low blood perfusion rates and long-term deposition characteristic of fat or bone, the slow excretion component of the biphasic excretion curve may reasonably represent the essentially similar behavior of these organs. These organs with slower uptake and release in contrast to central compartment organs, for example, kidney, are referred to in multicompartment models as "peripheral compartments."

The comparison of such a biexponential system with the linear pharmacokinetics of a single compartment system reveals characteristic differences. Noting that the former has a relatively slow component, one would expect it to demonstrate (*1*) increased half-life with increasing dose, (*2*) that the composition of excretion products may quantitatively and qualitatively change, (*3*) that saturation of processes in the slow compartment may show a marked change in dose–response relationships as dose increases, and (*4*) that when operating at or near rate limits, competitive inhibition of metabolic or transport processes by other chemicals can more likely occur. Thus the elements of Michaelis-Menten saturation kinetics can reveal valuable information. Chronic dosing with compounds that produce small cumulative insults will eventually reveal indications of excretory organ damage by increases in biological half-life. Likewise with such small dosings, if half-life is less than the dosing interval, such schedule can be predicted as not leading to bioaccumulation.

In the case of chemicals whose metabolic transformation mechanisms are readily saturated, zero-order kinetics are followed. In these cases an arithmetic plot of plasma concentration versus time will produce a straight line. In effect, this indicates that a rate-limiting step is occurring at some point in the metabolic process so that a constant amount is being transformed per unit time, independent of the amount found in the organ or the body. For such compounds a true half-life does not exist, because complete excretion rather than logarithmic kinetics apply. A prime example of such a compound is ethyl alcohol.

7.3 Specific Routes of Excretion

7.3.1 The Kidneys

From the viewpoint of the body's interface with xenobiotics the excretory role of the kidney represents the most important of several functions. While this emphasis on excretion is focus of interest, it must be remembered that this organ participates in a number of other important activities aimed at maintenance of homeostasis. Thus it is involved in regulation of: (*1*) electrolyte balance, (2) body fluid osmolarity and volumes, (*3*) acid–base balance, and (*4*) production and secretion of hormones, e.g., renin, calcitrol and erythropoietin.

Nonetheless, the kidneys have as their prime function the removal of nonessential metabolites from the body; by these same mechanisms xenobiotics are also removed. By receiving 20 to 25% of the total cardiac output, large quantities of blood are brought to the kidneys for elimination of materials contained therein. This is accomplished by three basic processes: (*1*) glomerular filtration, (2) passive tubular diffusion, and (*3*) active tubular secretion.

While nominally 180 L of the noncellular elements of blood are delivered each day to the kidneys for filtration, essentially 179 L of this filtrate is re-absorbed by the tubular elements. Given such large blood volumes brought to the kidneys, the efficiency of the organ is such that only slightly less than 1% of the fluid passing through the glomerular filtration segment reaches the urinary bladder.

Passage across the single-cell glomerular capsule (see Fig. 3.7) by molecules of less than 60,000 Da, or with a molecular radius of less than 20 Å, is readily accomplished. Proteins of approximately 20 to 40 Å may be less readily passed through these pores or slits with changes in their size and/or electrical charge. This transfer is enhanced by the presence of pores of up to 700 Å diameter in the glomerular capillary walls. In this filtration process, as the unbound forms of plasma elements (e.g., metal ions) decrease in concentration, the equilibrium between free and bound forms is altered. This leads to dissociation of the protein-bound forms and their subsequent availability for passive filtration.

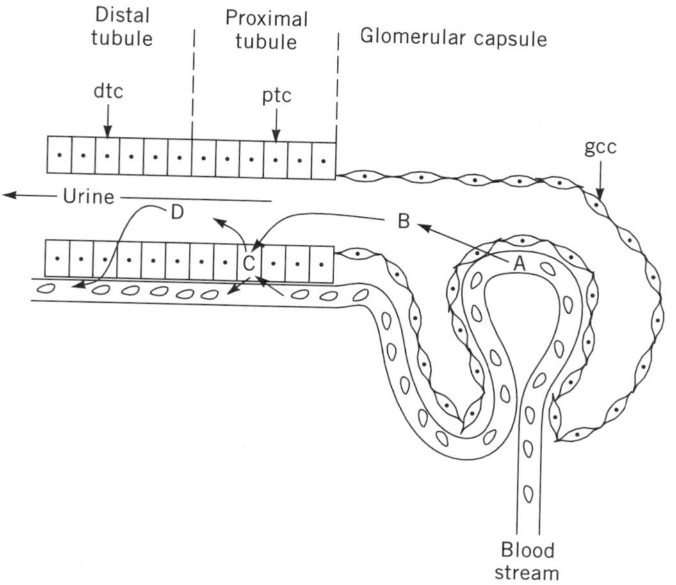

Figure 3.7. Schematic diagram, structural and functional elements of the renal glomerulus and tubules. (A) Formed elements and plasma enter glomerular capillary. (B) Ionic or free small molecules (<60,000 Da) are filtered through capillary and glomerular walls into the uriniferous lumen of the glomerular capsule. As ionic or free form concentrations decrease in plasma, the equilibrium with the bound components is altered, causing further dissociation of these components and thus an increase in their availability for filtration. (C) Energy-dependent active transport at the proximal tubules allows either (1) excretion from the blood to the glomerular filtrate, or (2) extraction from the glomerular filtrate for recovery to the blood. These shifts may occur in the presence or absence of protein binding. (D) Passive diffusion allows transfer of nonionized, nonpolar substances, consistent with basic membrane diffusion theory. gcc = glomerular capsule membrane; ptc = proximal tubule cells; dtc = distal tubular cells.

Subsequently, this filtrate (i.e., the postglomerular capsule) passes down-stream to the area of the proximal tubules and are subjected to two transfer forces, that is, passive diffusion and active tubular transport (Fig. 3.5). Both these processes may act to move compounds in either direction, that is, back across the tubular cell from the glomerular filtrate toward the tubular blood supply (i.e., reabsorption), and vice versa. Passive filtration of xenobiotics through the tubules to the urine is of relatively minor significance. Although many of such compounds are bound to plasma proteins and hence not available for glomerular filtration, nevertheless reabsorption and excretion follow basic principles of cell membrane transfer. Thus in the acidic environment of the glomerular filtrate, weak organic acids move toward a nonionized state, which leads to consequent facilitated passage across the tubular cell wall, that is, passive reabsorption. Such a principle also applies to the fate of hydrophilic, dissociated ions, which will readily be filtered through the glomerulus. That is, materials in this state are not passively transferred across cell membranes and thus are not recovered by passive transport across renal tubules, subject to variations in pK_a and pH.

Active transport is accomplished largely in the proximal tubules. As with any active process it is an energy-dependent process. As has been previously seen, similar configurations or binding constants may compete at binding sites involved in the tubular transport system involved in the secretory process. In the region of the proximal tubules there is passive reabsorption of approximately two-thirds of the water, and active reabsorption, particularly of electrolytes passively filtered through the glomerular capsule, as well for example, amino acids, ions, small peptides and hormones. In contrast to filtration at the glomerulus, protein-bound compounds are accessible for partial degradation by enzymes on the surfaces of tubular cells and subsequent pinocytosis by these cells. Further intracellular enzymatic degradation in such tubular cells produce amino acids which are then are available to general body stores of these vital protein constituents. As regards xenobiotics their active secretion by tubular cells into urine represents a major route for their elimination from the internal environment.

7.3.2 Biliary Excretion

Study of the gross and microscopic anatomy of the liver reveals its central role in xenobiotic metabolism. Because it receives a large portion of its blood supply from the GIT, it has the first opportunity to sequester, bind, or excrete xenobiotics brought into the internal milieu from the oral portal of entry. Microscopic examination of the liver demonstrates that its individual cells on one side are in direct contact with mixed portal and systemic blood and on its opposite side with the biliary vascular system. Thus materials are presented from either source to hepatocytes for biotransformation and excretion to either the blood or biliary stream. Transfer from either stream to within the hepatocyte occurs by active transport; a number of specific transport mechanisms have been identified. A xenobiotic must have either a strongly polar group or a large molecular weight (>300 Da) for biliary excretion to occur. Organic xenobiotics excreted in the bile are conjugated, for example, with glucuronic acids, glycine or with glutathione. Such conjugation takes place in the smooth endoplastic reticulum; such conjugative processes facilitate renal excretion of xenobiotics or their products by increasing their water solubility. Multiple factors, e.g., cholecystokinin., appear to modulate hepatocytic excretory capacity. Control of such hepatocytic activities is governed by autoregulatory, feedback-based control systems which are activated by detection of increases in intracellular binding by proteins, increased conjugating activities, and stimulation of microsomal biotransformation. Additionally, specific transport systems for organic acids, bases, and neutral organics have been identified.

Xenobiotics absorbed via the GIT may be eliminated after hepatic biotransformation via the blood, and ultimately the kidney, or in the bile to the GIT. However, nonpolar or lipophilic metabolites excreted in bile are most likely to be reabsorbed and carried back via the portal system for representation to the liver. By such recirculation by the enterohepatic circulatory cycle (i.e., bile to GIT with reabsorption and transport via the portal vein back to the liver), compounds that are hepatotoxic may be concentrated in the liver with resultant damage to that organ. On the other hand, many organic compounds are biotransformed to polar metabolites or conjugates before excretion in the bile. Such forms are less susceptible to reabsorption and entry into the enterohepatic circulation.

7.3.3 The Lungs as an Organ of Elimination

Although some investigators in the past have attempted to ascribe to the lung an active excretory capability for gases, vapors, and volatile metabolites, it is now universally agreed these are eliminated by simple diffusion. In addition, volatile liquids in equilibrium with their gas phase may be similarly excreted regardless of their route of uptake. Thus volatile organic solvents absorbed via the GIT may be eliminated through the lungs. Gases are eliminated at a rate that is inversely related to their solubilities; thus chloroform, which is highly soluble in blood, is more slowly excreted than ethylene. However, part of the prolonged retention of the highly soluble gases (e.g., chloroform) reflects lipid depot deposition, that is, representing the expression of a two compartment system.

Other factors determining the rate of pulmonary elimination are related to pulmonary physiological factors, for example, ventilation and blood perfusion rates in the lungs. In addition, it has become clear that the lung contains, although at generally lower concentrations, almost all of the hepatic pathways required for biotransformation of xenobiotics. However, it is generally considered that this represents simply a metabolic capacity directed toward ends unrelated to excretory functions.

Previous discussions (cf. Section 3.2.5) have considered the elimination of particulate matter by the lung.

7.3.4 The Gastrointestinal Tract as an Organ of Excretion

Because the gastrointestinal tract actively secretes about 3 L/day in humans, most xenobiotics are excreted here by passive diffusion. Thus transfer from blood via the GIT could and does occur for highly lipophilic, nonpolar substances, for example, polychlorinated biphenyls (PCBs) (19) and organochloro pesticides.

However, although other xenobiotics may be found in the feces, such does not necessarily represent or imply GIT excretion. Rather their presence can be the result of (*1*) excretion in the bile, (*2*) secretion in the saliva, stomach, pancreatic, or intestinal secretory fluids, (*3*) failure of absorption following oral intake, or (*4*) clearance from the respiratory tract (cf. Section 3.2.5) followed by swallowing.

7.3.5 Excretion via Perspiration and Saliva

Both these routes are of minor significance; they largely are associated with simple diffusion of nonionized lipophilic forms. Thus xenobiotics excreted via sweat may accumulate at sudorific and pilocarpal glands sufficient to produce dermatitis; the chloracne associated with PCB exposure may be a case in point. Evidence exists that fluoride is excreted in sweat, as are Hg, Bi, Pb, and As. The secretion of lead in saliva in chronic lead absorption may be associated with the gingival lead line. Some other diffusible substances found in saliva are bromides, iodides, alkaloids, and ethanol.

7.3.6 Excretion by Milk

Because xenobiotics may be conveyed by cows to humans, and from maternal milk to infants, such secretions are of some significance as an excretory medium. Because of its important lipid content, many foreign compounds may be concentrated in milk. Thus lipophilic organics such as DDT, PCB, and PBB concentrated in milk can be an interesting source of dietary contamination. In addition, metals whose physicochemical characteristics are similar to calcium, for example, lead, can be found in this medium.

8 MECHANISMS FOR THE CONTROL, REPAIR, AND REPLACEMENT OF DNA DAMAGE

In the preceding sections we have dealt with mechanisms whereby xenobiotics may be: (*1*) absorbed by the body; (*2*) moved by the body both inwardly and outwardly; (*3*) transported and distributed to various cellular and organ systems; (*4*) metabolically transformed into less noxious materials; (*5*) stored in various other organs, and (*6*) ultimately excreted from the internal mileau. In the course of these discussions, it has become apparent that these various cellular and organ systems have evolved systems which permit ingress of environmentally derived needs, viz., nutrients, oxygen, etc. However, integral to these mechanisms whereby the body deals with those ambient agents necessary to its vital activities it has simultaneously provided a multiplicity of means for the mitigation of those xenobiotics similarly extant in its environment. It is hoped that these realizations can help to underscore an understanding of the premise that *simply the presence of xenobiotics in the human environment is insufficient* to determine whether individual or group hazard can be posited. Even aside from dose considerations, it should be apparent at this point that ultimately, hazard to human health and well-being depends upon the operations of these multiple protective capabilities possessed by the intact body.

This chapter has in the main dealt with organs and whole cellular systems and by necessity some of their protective or restorative systems. Although there is an extensive panoply of bodily reparative mechanisms available which help restore body functions and even integrity, these are beyond the scope of this treatment. However, there exists at the cellular and subcellular level of organization, systems common to almost all living cells dedicated to that most critical of life activities required for survival, viz., maintenance of each cells DNA functional integrity.

In any discussions of xenobiotics and human health the importance of this activity must assume a high degree pertinence, since it is believed that environmentally attributed, and all other, cancers usually are associated with alterations of DNA structure and functional integrity. Indeed, the centrality of DNA damage to carcinogenesis and birth defects has been previously accentuated by the belief that DNA damage is irreversible, irreparable and inevitably expressed in neoplasia.

Although such previously held concerns represented a body of legitimate concern, more recent studies have revealed a plethora of physical, molecular and programmatic defenses making it unlikely that exposure to a single molecule of a xenobiotic will result in such a deleterious outcome. For example, the cellular DNA is found in the nucleus, which is surrounded by a lipid-rich membrane. Thus, the nuclear envelope presents a further physical barrier to xenobiotic entry, analogous to a second cell wall. Additionally, similar to the scenario described for cytoplasmic proteins, the nucleus is also highly ordered, and the chance that a xenobiotic will interact with a nuclear protein leading to its sequestration and elimination are also extremely high. Further, nuclear DNA is not naked; rather, it is in the form of chromatin, associated with histones and other nuclear proteins. Thus a potentially DNA damaging xenobiotic would have to overcome the strong interactions between DNA and these DNA-binding proteins in order to exact its toll on the cellular genetic material.

The notion that cellular DNA damage incurred by a xenobiotic is irreversibly harmful is incorrect. The fact that well over 80% of the human genome is composed of nonessential repetitive DNA diminishes the likelihood that any one 'hit' will be catastrophic. When the existence of intronic, noncoding sequences and the redundancy of the genetic code are factored in, the chance of a single lethal hit becomes negligible. Further, DNA is not a static molecule. Cells have evolved a large number of independent and functionally redundant DNA repair pathways. For example, different pathways have evolved to recognize double-stranded as opposed to single-stranded breaks in DNA, as well as to recognize and remove thymine dimers or deaminated cytosine. Each of these different pathways constantly monitors, identifies, removes and repairs damaged DNA. Thus, in this case a single "hit" by a xenobiotic would most usually lead to the removal of the damaged portion of the DNA.

Although the cell harbors a variety of mechanisms to repair DNA, damaged DNA can sometimes escape recognition and establish itself in the cell as a somatic mutation. If the mutation occurs in the protein coding region of a gene, it may promote the production of an altered, toxic protein. To safeguard against this however, cells have evolved machinery that monitors the quality of both messenger RNAs (mRNA) and their translated protein products. mRNAs that contain early termination codons (e.g., as a consequence of a somatic mutation) are readily recognized and degraded by the nonsense-mediated mRNA decay pathway (20). Thus these mRNAs are rapidly removed from the cell before they can be translated into potentially deleterious proteins. There is growing evidence that a translational surveillance complex monitors the quality of both the mRNA and its translated protein product in order to rapidly detect and eliminate mutant mRNAs and proteins before they have can injure the cell.

Apoptosis presents yet another cellular defense mechanism that has evolved to minimize the effects of DNA damage on an organism. Tumor suppressor genes have evolved to recognize the presence of irreparably damaged DNA, and to subsequently initiate a cellular self-destruct program. Although drastic from the point of view of the cell insofar as it results in cell death, the elimination of a potentially lethal mutation by this strategy is good for the well being of the organism as a whole.

Finally, there is growing evidence that cancers result from the accumulation of multiple mutations in cellular DNA (21). Thus, even if a xenobiotic were to overcome the multiple physical, enzymatic, and programmatic defenses presented by organisms, tissues, and cells, a single hit would still be insufficient to result in catastrophic event. Rather, it is becoming clear that mutagenesis is a stochastic process that takes multiple exposures to numerous agents over long periods of time. The body is indeed a wonderful, awe inspiring creation. It is, however, robustly designed to withstand all manner of repeated environmental insults.

ACKNOWLEDGMENTS

This work was supported in part by a grant to Jonathan D. Dinman from the National Institutes of Health (GM58859-01).

BIBLIOGRAPHY

1. R. A. Brown and L. S. Schanker, "Absorption of Aerosolized Drugs in the Rat," *Drug Metabolism and Disposition* **11**, 355–360 (1983).
2. A. Hershko and A. Ciechanover, "The Ubiquitin System," *Annual Review of Biochemistry*, **67**, 425–479 (1998)
3. D. A. Dungworth, J. L. Mauderly, and Obersdorfer, Eds., *Toxic and Carcinogenic Effects of Solid Particles in the Respiratory Tract*, ISLI Press, Washington, DC, 1994
4. M. Lippman and V. Timbrell, "Particle Loading in the Human Lung—Human Experience and Implications for Exposure Limits," *Journal of Aerosol Medicine* **3**, S155–S168 (1990).
5. ISO Technical Committee 146-Air Quality, *Particle Size Fraction Definitions for Health-Related Sampling*, International Standards Organization, ISO/TR 7708-1983 (E), 1983.
6. T. L. Chan and M. Lippman, "Experimental Measurements and Empirical Modeling of the Regional Disposition of Inhaled Particles in Humans," *J. Am. Ind. Hyg. Assoc.*, **41**, 399–409 (1980).
7. A. M. Boylan, D. A. Sanan, D. Shepard, and V. C. Broaddus, "Vitronectin Enhances Internalization of Crocidolite Asbestos by Rabbit Pleural Mesothelial Cells via the Integrin alpha V beta 5," *J. Clinical Investig.* **96**, 187–201 (1995).
8. A. R. Brody, L. H. Hill, and K. B. Adler, "Actin-containing Microfilaments of Pulmonary Epithelial Cells Provide a Mechanism for Translocating Asbestos to the Interstitium," *Chest* **83**, 11–12 (1983).
9. B. D. Beck, J. D. Brain, and D. E. Bohannon, "An *in vivo* Hamster Bioassay to Assess the Toxicity of Particulates for the Lungs," *Toxicol. Appl. Pharmacol.* **66**, 9–29 (1982).
10. C. R. DeVries, P. Ingram, S. R. Walker, R. W. Linton, W. F. Gutknecht, and J. D. Shelbourne, "Acute Toxicity of Lead Particulates for the Lungs," *Laboratory Investigation* Investigation-44 (1983).
11. J. D. Brain, "Toxicological Aspects of Alterations of Pulmonary Macrophage Function," *Ann. Rev. Pharmacol. Toxicol.* **26**, 547–655 (1986).
12. C. J. Pfeiffer, "Reactions to Environmental Agents," *Handbook of Physiology*, Section 9, American Physiological Society, Bethesda, MD, 1977, pp. 349–374.
13. L. B. Sasser and G. E. Jarboe, "Intestinal Absorption and Retention of Cadmium in Neonatal Rat," *Toxicol. Appl. Pharmacol.* **41**, 423–431 (1977).
14. C. N. Davies, ed., *Inhaled Particles and Vapors*, Vol. 11, Pergamon Press, Oxford, 1967, pp. 121–131.
15. U. Kragh-Hansen, "Molecular Aspects of Ligand Binding to Serum Albumin," *Pharmacol. Rev.* **33**, 17–53 (1981).
16. International Commission on Radiological Protection, *Report on the Task Group on Reference Man*, International Commission on Radiation Protection, No. 23, Pergamon Press, Oxford, 1975, pp. 325–327.

17. J. B. West, ed., *Best and Taylor's Physiological Basis of Medical Practice*, 11th ed., Williams and Wilkins, Baltimore, MD, 1985, p. 136.
18. A. J. Levi, Z. Gatmaitan, and I. M. Arias, "Two Hepatic Cytoplasmic Fractions, Y and Z, and Their Possible Role in the Hepatic Uptake of Bilirubin, Sulfobromophthalein and Other Ions," *J. Clinical Investig.* **48**, 2156–2167 (1969).
19. T. Rozman, L. Ballhorn, K. Rozman, C. Klaassen, and H. Greim, "Effect of Cholestyramine on the Disposition of Pentachlorophenol in Rhesus Monkeys," *J. Toxicol. Environ. Health* **10**, 277–283 (1982).
20. Y. Weng, M. J. Ruiz-Echevarria, S. Zhang, Y. Cui, K. Czaplinski, J. D. Dinman, and S. W., "Characterization of the Nonsense-mediated mRNA Decay Pathway and its Effect on Modulating Translation Termination and Programmed Frameshifting," in *mRNA Metabolism and Post-transcription Gene Regulation*, Wiley-Liss, Inc., New York, 1997, pp. 241–263.
21. E. R. Fearon and B. Vogelstein, "A Genetic Model for Colorectal Tumorigenesis," *Cell* **61**, 759–767 (1990).

CHAPTER FOUR

Pulmonary Effects of Inhaled Mineral Dusts

William L. Eschenbacher, MD, Gregory J. Kullman, Ph.D., CIH and Cristina C. Gomberg, RN, MS

1 INTRODUCTION

Despite advances in knowledge of the relationships between exposure to mineral dusts (asbestos, coal dust, silica) and the respiratory diseases that can occur (pneumoconioses, obstructive lung diseases, cancer, etc.), these diseases are still present and workers are still dying with these conditions. Important insight has been gained into the understanding of the mechanisms of the injury that occurs when the cells and the tissue of the lungs are exposed to these mineral dusts. However, the primary mode of protecting workers from the development of these mineral dust-induced lung diseases is by reducing or eliminating exposure to increased concentrations of the dusts in the first place. No amount of treatment can substitute for prevention in the development of these diseases.

This chapter provides the reader with background knowledge regarding the respiratory system (anatomy, defense mechanisms, physiology, assessment, and the response of the lungs to injury) and specifics of the lung diseases caused by exposure to some of the mineral dusts.

Treatment of these disorders will not be covered in great detail in this chapter. However, it remains paramount that engineering controls and personal respiratory protection be required so that these dust-induced lung diseases can be prevented.

Patty's Industrial Hygiene, Fifth Edition, Volume 1. Edited by Robert L. Harris.
ISBN 0-471-29756-9

2 PULMONARY ANATOMY

The primary function of the lung is to effect gas exchange with the environment: transfer oxygen from outside the body to the blood within the body and remove carbon dioxide from the blood to outside the body. The anatomy of the respiratory system is set up to allow for this exchange to occur in a most efficient fashion.

The respiratory system includes the lungs, the airways leading into the lungs, the pleural covering of the lungs, the blood vessels leading to, away from, and in the lungs, the muscles of the diaphragm and chest wall that allow air movement to occur and the chest wall itself which provides support and protection of the internal structures including the lungs, the heart, and the major blood vessels. The respiratory system can be divided into these separate components for convenience and also because disease when it occurs can preferentially affect one component more than another (see Fig. 4.1).

2.1 Tracheobronchial Tree

The tracheobronchial tree or airways actually begin at the nose and mouth and continue to the level of the terminal bronchiole, the smallest conducting breathing passage. Air containing oxygen enters the body through the nose and mouth and from there is inhaled

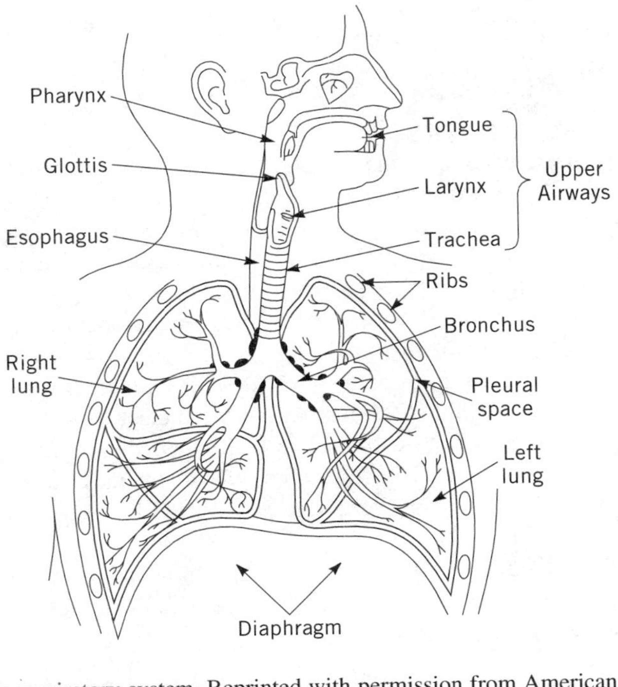

Figure 4.1. The respiratory system. Reprinted with permission from American Lung Association (1). Note figure is from 1979.

through the throat or pharynx and then through the vocal cords or larynx before entering the windpipe or trachea. From here the trachea or main airway branches into the right and left major bronchi. Air moves into these branches and then into smaller and smaller divisions of the tracheobronchial tree until the smallest conducting units or terminal bronchioles are reached. The trachea is 20–25 mm in diameter and the smaller bronchioles are less than 1 mm in diameter. However, because of the many divisions and many number of smaller bronchioles, the total cross-sectional area of the tracheobronchial tree actually increases from approximately 4 cm^2 at the trachea to 10,000 cm^2 at the level of the terminal bronchioles.

2.2 Pulmonary Parenchyma and Interstitium

The pulmonary parenchyma and interstitium refer to those parts of the lungs that are involved in functions other than air conductance. This is a network of cells and supporting matrix (including collagen) where gas exchange takes place. The next airways after the terminal bronchioles are the respiratory bronchioles which contain air sacs (alveoli) within their walls. The alveolar spaces are tiny air sacs formed by a thin cellular membrane (epithelium) abutting blood-filled capillary membranes (endothelium). There are approximately 300 million alveoli in the lungs of a human with a total surface area of a tennis court (approximately 280 sq ft). Once air reaches the bronchioles and alveoli, gas exchange occurs: oxygen diffuses across a thin cellular membrane into the blood stream and carbon dioxide is transferred from blood into the alveolar space. During exhalation, the alveolar air which now contains reduced oxygen and increased carbon dioxide moves up the tracheobronchial tree where it leaves the body at the nose and mouth.

The space between the walls of the alveoli and the walls of the capillaries is the interstitium, which is a space normally devoid of material or cells so that efficient gas exchange can occur. The interstitium can be the site for inflammatory changes. Cellular infiltrates and increased protein/collagen deposits that could occur in certain occupational diseases can make the lungs stiffer and hamper gas exchange. These diseases are varied but are collectively referred to as interstitial lung diseases.

2.3 Pulmonary Vascular Bed

Blood is carried to the lung by the pulmonary arteries from the right ventricle of the heart. The pulmonary arteries then branch and branch again, with the blood vessels paralleling the bronchi and bronchioles. Eventually the vessels become the capillaries which come into close contact with the alveolar spaces. It is thought the capillary bed is probably more a large continuous membranous surface through which blood cells and plasma flow than a conduit provided by individual vessels. After oxygen is transferred from the alveoli into the blood (actually attaching to the hemoglobin molecules within the red blood cells), the oxygenated blood leaves the capillary bed and enters the pulmonary venous system which is composed of larger and larger vessels until they become the pulmonary veins which carry the oxygenated blood to the left atrium of the heart. From here, the oxygenated blood flows to the left ventricle and then is pumped out to the rest of the body.

2.4 Pleural Surfaces

There are two adjacent layers of thin cellular coverings for the lung. In immediate contact with the lung is the mesothelium referred to as the visceral pleura. The visceral pleura is surrounded by the parietal pleura which is a thin second mesothelial covering that lines the inside of the chest wall. These two layers remain in close proximity and have smooth surfaces facing each other allowing for ease of movement of the lungs within the chest wall. The pleural surfaces can become diseased due to inflammatory changes (influx of cells and proteinaceous material) resulting in a thickened and irritated surface that can limit the expansion of the lungs. In addition, the cells present in the pleural space can mutate to dysplastic, metaplastic, or neoplastic in cases of malignancies related to the pleura (malignant mesothelioma). Also, fluid can accumulate in the pleural space between the two layers limiting lung and chest wall expansion. These pleural effusions can be benign (without malignant cells) or malignant.

2.5 Respiratory Muscles

In order for respiration (gas exchange) or ventilation (air conductance) to take place, expansion of the chest wall must occur. The respiratory muscles (primarily the diaphragm) are involved in this process. By contracting with each breath, the diaphragm descends causing the chest cavity to increase in size. This creates a negative pressure in the pleural space with a resulting expansion of the lungs. This force results in air being pulled into the lungs during inhalation as necessary for gas exchange. While inhalation is an active process requiring respiratory muscle contraction, exhalation is normally a passive process with the chest wall and the lungs relaxing back to their resting or pre-inhalation size. Diseases that affect the respiratory muscles are usually related to nerve or muscle diseases, but mineral dusts such as asbestos can result in changes to the diaphragm surface which can impact ventilation.

3 LUNG DEFENSES

3.1 Cough

The body vigilantly protects itself from foreign materials entering the tracheobronchial tree. There are many nerve fibers in the upper airways around the larynx or vocal cords and also in the trachea and major divisions of the bronchi. These nerve fibers react to the presence of any foreign object or irritant substance such as gas or particles of dust. One of the responses to stimulation of these nerve fibers in normal lungs is the cough. As part of this cough reflex, the lungs involuntarily take in a large breath, the air is then forced out with respiratory muscular effort (not passively as occurs with normal exhalation). The resulting high intrathoracic pressures generated with this forceful effort cause the major airways to narrow. The flow of air out through narrowed airways occurs at high velocities with great shear force to help expel any foreign object or substance. When cough is caused by permanent changes in the lining of the airways or in the interstitium or the lung parenchyma, then it is no longer part of a normal response but rather a symptom of a disease.

3.2 Mucociliary Escalator

The airways including the trachea, bronchi, and smaller bronchioles are covered with hairlike projections called cilia. Cells within the airways secrete mucous which layers over the cilia. The cilia beat in a rhythmic pattern moving the mucous up the airways toward the larynx or vocal cords where the mucous can then be swallowed or expectorated. A foreign substance or material that reaches the tracheobronchial tree can be trapped in this mucociliary escalator and removed from the lungs. Occasionally, however, this system can be overwhelmed by the presence of excessive amounts of foreign material such as inhaled dust. In that situation, the dust or other toxic substance can interact with airway cells and/or the alveolar spaces. Before it is adequately removed, additional damage may occur with resulting irritation and inflammation.

3.3 Macrophages

Gases, fumes and microscopic particles can eventually reach the alveolar spaces despite the cough reflex and the mucociliary escalator. The size, shape, and mass of particles determine where they are deposited within the respiratory system. Particle size can be measured in microns. A micron is 10^{-6} meters with approximately 25,400 microns in one inch. The smallest particles that can be seen by the naked eye are greater than 50 microns. To be inhaled into the lungs and into the tracheobronchial tree and the alveolar spaces, a particle is usually smaller than 5–10 microns. Particles bigger than 5 microns usually do not remain airborne long enough to be inhaled or they can be trapped on the mucosal surface of the nose. Particles of 1–5 microns are more likely to deposit in the tracheobronchial tree. Smaller particles (0.01–1 microns) are more likely to reach the smaller bronchioles and the alveolar spaces.

Once particles do reach the alveolar spaces without being coughed out or caught and removed by the mucociliary escalator, they encounter macrophages, cells that reside in the alveolar spaces. Alveolar macrophages are specialized cells that can engulf and digest particles. In time, the macrophages will then die and move up the mucociliary escalator for removal by swallowing. In this last line of lung defense, the macrophage can protect the lung (and in turn the rest of the body) from the potentially toxic effect of the particles.

However, just like the mucocilliary escalator, the macrophages can be overwhelmed by excessive quantities of particles. Particles that are not engulfed by macrophages can lead to irritation, inflammation, and disease.

4 PULMONARY PHYSIOLOGY

4.1 Ventilation

Airflow into and out of the lungs is a balance of two opposing forces intrinsic to the respiratory system: elastic recoil and flow resistance.

Elastic recoil refers to the properties of the lung that cause the lung to revert back to a resting state after the lung has been expanded. These elastic properties are produced by the networks of collagen and elastin fibers that course throughout the lungs and provide the support structure of the lungs. As lung volume increases, these fibers are stretched, and the potential elastic recoil progressively increases. The result is for the lungs to shrink back to their pre-expansion size.

In addition to the elastic forces that must be overcome with each inhalation, resistance forces related to the flow of air through the tracheobronchial tree must also be overcome. These resistance forces depend upon the flow rate of air, the characteristics of the air including density and viscosity, and the characteristics of the tracheobronchial tree.

To achieve the required oxygen uptake and carbon dioxide removal at rest, one must breathe at a rate to move 5–7 liters of air each minute into and out of the lungs. This ventilation requirement is increased in the face of lung disease because of inefficiency of diffusion of gases or in gas exchange or in inequalities of ventilation/perfusion relationships (see below). Ventilation requirements will also be increased by increasing metabolic demands such as exercise or in certain illnesses which may include fever.

4.2 Diffusion

Once air reaches the alveolar spaces, diffusion of the oxygen molecules occurs across the alveolar-capillary membranes (through the interstitium of the lung) and carbon dioxide molecules diffuse in the opposite direction to be removed from the body. This diffusion process is most efficient if the alveolar–capillary membrane is not thickened by disease and the total surface area of the alveolar–capillary bed is not reduced by disease. Interstitial disease such as the pneumoconioses caused by the inhalation of dusts can result in both the thickening of the interstitium as well as loss of the alveolar–capillary surface area by tissue destruction. Diffusion limitation can occur in these interstitial diseases. It can also occur in other lung diseases such as pulmonary emphysema where the alveolar spaces are destroyed from years of smoking. In these circumstances, the blood leaving the lungs may not be adequately oxygenated resulting in compensatory increased demands for ventilation.

4.3 Ventilation/Perfusion Relationships

In addition to the volume of air moved in and out of the lung, the distribution of air in relation to the flow of blood in the capillary bed is important. The distribution of air to the different sections of the lungs can be affected by diseases in the airways (bronchitis, emphysema, or asthma) or by interstitial disease with alterations in regional distribution of elastic recoil forces.

Once oxygen, a component of the inhaled air, reaches the alveolar surfaces and diffuses through the alveolar-capillary membrane, it will be taken up by the hemoglobin in the red blood cells that flow through the capillary bed. There must be an even matching of the air flow (5–7 liters per minute) to the alveolar spaces with the blood flow (5–6 liters per minute) through the vascular bed. In normal circumstances, 95% of the output of the heart flows through the lung from the right ventricle to the left atrium of the heart.

Again, this matching of ventilation and blood flow or perfusion can be altered by disease. When there is mismatching, there can be increased requirements for ventilation, and the energy requirements for breathing (so-called work of breathing) can be significantly increased and adequate gas exchange may not occur. In severe conditions, this can lead to respiratory failure and death.

4.4 Stress of Exercise

Any abnormality of pulmonary physiology, whether a problem with ventilation, diffusion, or ventilation/perfusion matching, can appear worsened with exercise. Normally at rest, humans consume 0.2 to 0.3 liters of oxygen per minute to satisfy the needs of the body. However, under conditions of exercise (work or play), the requirements for oxygen can be increased dramatically to as much as 5 to 6 liters of oxygen per minute. Under these circumstances, the total ventilation to support these oxygen requirements can increase to 100 to 140 liters of air per minute and the blood flow through the lungs from the heart can be increased from 5 to 6 liters a minute to 20 to 25 liters per minute. Because of the inefficiency of diseased lungs, these increased requirements of ventilation cannot be met. Therefore, any disease of the lungs even in the earliest stages can result in mild, then severe, limitation to exercise and possibly progress to limitations even at rest.

5 ASSESSMENT OF THE LUNGS

5.1 Lung Function Testing

The principal function of the lung is to exchange gas between air outside the body and blood within the body. In general, the body is quite efficient at adjusting and regulating the ventilation and the ventilation to perfusion relationships so that homeostatsis is maintained even in the presence of extensive lung disease. Additionally, the "pulmonary reserve"—the portion of lung capacity not normally used even during strenuous exercise—is quite adequate so even in the presence of mild lung disease, there may be no perception of limitation. Also, symptoms of respiratory disease (cough and shortness of breath) may not correlate with actual lung dysfunction. For these reasons, determination of lung health is best done by objective measurement of the physical properties of the lung: the size of the lung or the volume of air within the lung, expansibility or the elasticity of the lung, ventilatory ability (forces of breathing), and/or the efficiency of gas exchange. Pulmonary or lung function tests are physiologic measurements that can be used as a surrogate for direct knowledge about a physical property of the lung. A cautionary note: no single test can measure all aspects of lung function. Therefore, care should taken in generalizing from any one measurement, either normal or abnormal. The limits of interpretation are based on (1) level of knowledge of the physiologic principles on which the test is founded, (2) the biological variability inherent in the lung physiology being measured, and (3) the accuracy of the measurement itself. Improving the reliability and utility of test results can be best achieved by reducing the variability of testing. Quality control including test standardization is central to reducing variability.

5.2 Testing Standardization

Guidelines for the standardization of lung function tests have been developed by professional societies such as the American Thoracic Society (ATS) and European Pulmonary

Society (EPS). The documents that contain these guidelines are highly recommended reading for individuals involved in all aspects of lung function testing (2–5).

The goal of standardization is to decrease variability of laboratory testing. As variability with testing is reduced, sensitivity for detection of lung dysfunction and disease is increased. Or stated differently, the variability of the physiologic measures should seek to be as low as possible in order that the variation caused by the dysfunction or disease can be identified (6). It has been demonstrated that significant decrement in variability can be achieved with quality control principles applied to all aspects of spirometry (7) the most commonly used pulmonary function test.

Figure 4.2 from the 1994 ATS Spirometry Update lists spirometry standardization steps. Each of the components will be outlined to give a brief introduction to considerations to be made when critically assessing spirometry results. The most widely administered pulmonary function test—spirometry—is used as an illustrative example of how the standardization steps may be generally applied. Spirometry, covered in greater detail later in the chapter, is a "breathing" test measuring how much and how fast air can be inhaled or

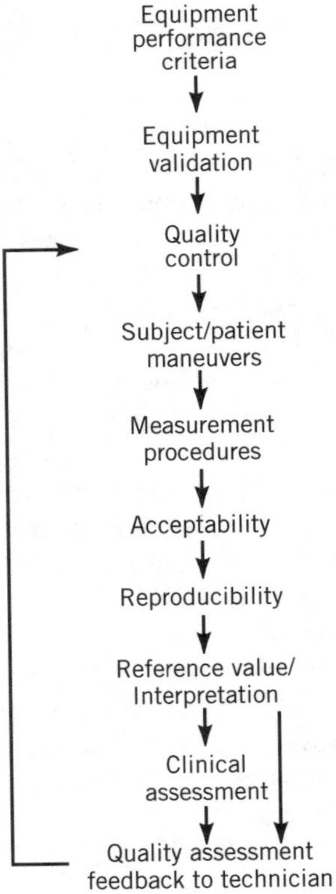

Figure 4.2. Spirometry Standardization Steps. Reprinted, with permission, from American Thoracic Society (2).

exhaled. Industrial hygienists may be asked to help design and implement screening, monitoring, or surveillance programs which include spirometry. Understanding how variability may be introduced to the measurement system is essential to obtaining acceptable results. These standardization steps with appropriate modification can be applied to any of the pulmonary function testing procedures, or any testing situation as the issue of separating the variation of interest (the signal) from the variation attributable for all other sources (the noise) is common to all testing situations.

Equipment Performance Criteria. The accuracy of spirometric equipment depends on the resolution (i.e., the minimal detectable flow) and the linearity of the entire system. Recommendations contained in ATS documents are minimal requirements. Detailed specifications for equipment to be used for spirometry are available (2).

Equipment Validation. Spirometry equipment, like all equipment, requires periodic validation to test equipment performance in meeting the pre-established equipment performance criteria.

Quality Control. Quality control is important to ensure consistency in meeting standards. Procedure manuals which include quality control plan, calibration and test performance procedures, calculations, criteria, reference values source, and emergency procedures are essential in any testing situation. Documentation of daily activities, abnormal events, continuing education and performance feedback are highly desirable. A quality control program that provides continuous feedback to the technician is increasingly seen as critical to the collection of high-quality spirometry data (2, 7).

Subject/Patient Maneuvers. Detailed description and criteria for the actual performance of maneuvers, for not only spirometry, but other pulmonary function testing as well, (i.e., methacholine challenge, diffusion capacity), is essential in obtaining consistency and standardization.

Measurement Procedures. Spirometric variables should be measured from a series of at least three acceptable forced expiratory curves (2). What should be noted is the requirement that *more than one maneuver* is needed. Additionally, the concept of acceptability and its companion concept, reproducibility, are essential and need to be defined. ATS documentation gives detailed requirements for each type of maneuver's measurement procedure.

Acceptability. Acceptability is the performance of a pulmonary function testing maneuver in accordance with criteria recommendations, usually these criteria are as set forth by the ATS (2). Acceptability criteria consists of such items as satisfactory start-of-test, minimum testing times, minimum performance variables, and end-of-test criteria. Obviously, dependent on the components and parameters to be measured, each pulmonary function test will have its specific criteria for acceptability.

Reproducibility. Reproducibility is the ability to successfully produce usually at least three acceptable maneuvers that are within a range of predefined variability, e.g., for the values of forced vital capacity (FVC) and forced expiratory volume in one second (FEV_1), the largest and the second largest, should not vary more than 0.2 L (2). Again, dependent on the parameters measured, each pulmonary function test will have its specific criteria for reproducibility.

Reference Value/Interpretation. Pulmonary function tests, like all clinical measurements, are subject to (*1*) technical variation related to instrument, procedure, observer, subject, and their interactions; (*2*) biologic variation (the focus of interest of most of the nonclinical biological sciences); (*3*) variation caused by dysfunction or disease. Interpretation of pulmonary function tests depends on establishing the variation of interest (the signal) and it relation to all other sources of variation (the noise). In assessing if particular pulmonary results are abnormal, the clinician uses reference equations to provide a context for evaluating an individual in comparison to the distribution of measurements in a reference population. There are multiple reference equations available. The choice of a particular set of reference values depends upon several factors including the population being studied. Recently published reference values for spirometry may become the standard in many different situations (8). [*Note*: There is significant statistical rationale and preference for determining abnormality using lower limits of normal (LLN) as determined from the regression model used in selected reference equations; rather than the relatively arbitrary 20% less than the predicted value (5).]

Clinical Assessment. Pulmonary function tests may be used to (*1*) describe dysfunction and assess its severity, and explain it in terms of diagnosis, prognosis, management, and assessment of trends over time, including response to treatment; (*2*) identify abnormality in individuals without known pulmonary disease; (*3*) and as part of a health assessment for a third party (e.g., insurance or government interest).

Quality Assessment and Feedback to Technician. The pulmonary function testing technician is critical to the successful performance of the pulmonary function testing maneuvers. Since many maneuvers require breathing in unusual, forceful, and sometimes stressful ways, a well-trained, knowledgeable, enthusiastic technician will affect significantly the quality of data obtained. Suboptimal technician performance is the most important factor causing inaccurate spirometry results. The maintenance of high quality data collection is critically correlated with regular written technician feedback regarding the quality of their efforts in obtaining results (7).

Lung Volume Subdivisions. The total volume capacity of the lungs is sometimes useful for understanding pulmonary pathology. See Figure 4.3 for subdivisions of lung. A reasonable estimate of total lung capacity can be obtained by combining several volume parameters. The most common parameters are

1. **Tidal Volume** (TV): during quiet, relaxed breathing, the volume of air that is exhaled with each breath.
2. **Residual Volume** (RV): the amount of air remaining in the lungs after the greatest exhalation possible.
3. **Vital Capacity** (VC): the maximum amount of air that can be exhaled after the fullest inhalation possible. The amount of air that can be exhaled with a *maximal effort* after a maximal inhalation is called the Forced Vital Capacity (FVC). The FVC is the volume that is measured in spirometry.
4. **Total Lung Capacity** (TLC): the sum of the vital capacity and the residual volume and a representation of total measurable lung volumes.

5.3 Spirometry

Spirometry is a medical test that measures the volume of air an individual inhales or exhales as a function of time. Flow, or the rate at which the volume is changing as a function of

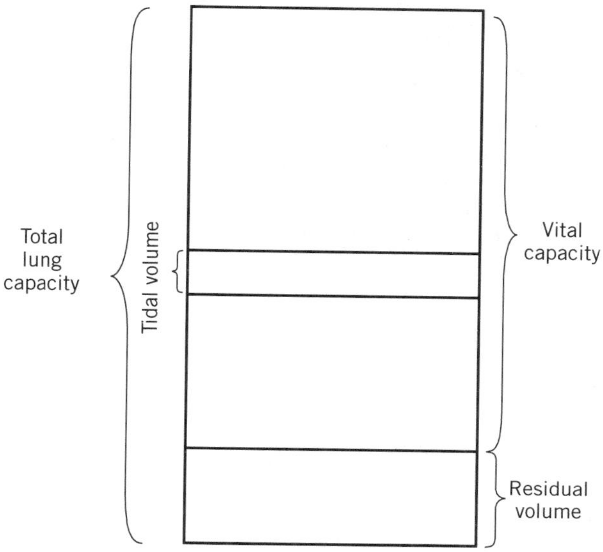

Figure 4.3. Volume subdivisions of the Lung.

time can also be measured. Spirometry is a useful screen of general health and correlates well with morbidity and life expectancy (9). Spirometry is used to gather information about respiratory function and may be used to diagnose, monitor, evaluate disability or impairment, and in public health surveillance and evaluation (10).

Spirometry refers to the measurements of exhaled air volume and flow rates from individuals who are coached by trained technicians using either volume-based or flow-based measuring equipment. The important measurements include forced vital capacity (FVC) or the greatest volume of air exhaled from a maximal inspiration to a complete exhalation; the forced expiratory volume in one second (FEV_1) or the volume of air exhaled in the first second of a FVC maneuver; and the ratio between these two values: FEV_1/FVC. These measurements can be made using either a volume-based system (such as a dry rolling-seal spirometer) or a flow-based system interfaced to a dedicated computer. All procedures should conform to standard guidelines (2). At least three maximal expiratory maneuvers or FVC maneuvers should be performed at each session. The selection and interpretation of results including the use of published reference equations should also conform to standard guidelines (5).

5.4 Bronchodilator Administration

Lung function tests such as spirometry are on occasions performed before and after the administration of an inhaled bronchodilator to measure the response to that intervention. A bronchodilator is a pharmacological agent causing the airways of the lung to dilate or become greater in diameter. Specifically, it has been proposed by the American Thoracic

Society that increases of at least 12% and 200 mL in FEV_1 or in FVC be considered to be significant (5). Bronchodilator responsiveness can be found in certain lung diseases such as asthma or chronic obstructive pulmonary disease (COPD).

5.5 Methacholine Bronchial Challenge Test

Asthma is a pulmonary condition characterized by: (*1*) airway obstruction (or narrowing) that is reversible either spontaneously or with treatment, (*2*) airway hyperresponsiveness to a variety of stimuli, and (*3*) airway inflammation (11). To assist in the diagnosis of asthma, determining the presence of reversible airway obstruction (usually to the use of a bronchodilator) or determining the presence of increased airway or bronchial responsiveness is important.

When an individual is evaluated for the possibility of asthma, baseline spirometry is first indicated. If there is presence of airways obstruction (a reduction in the ratio of FEV_1/FVC), then a bronchodilator is administered to determine the presence and extent of reversibility. If significant reversibility is present (>12% increase and a 200 ml increase in the FEV_1 or FVC) (15), then determination of bronchial responsiveness may not be necessary (4). Otherwise, the next step might be the performance of a bronchial challenge test to measure bronchial responsiveness. By far, the most commonly used nonspecific inhalation challenge agent used for such testing is methacholine.

The presence of increased bronchial responsiveness is not diagnostic of asthma but can be found in individuals with asthma. Increased bronchial responsiveness can be seen in other conditions such as smoking-induced chronic obstructive pulmonary disease (COPD), congestive heart failure (CHF), cystic fibrosis, bronchitis, and in about 5% of normal individuals (4). In addition, a few individuals with asthma may have normal bronchial responsiveness. Thus, the test for bronchial responsiveness can be useful but again is not diagnostic of any specific condition.

5.6 Lung Volume Measurement

As mentioned above (Lung volume subdivisions), the amount of air within the lungs can be represented by different volume parameters: total lung capacity, vital capacity and residual volume. Spirometry testing measures the amount of air an individual can inhale or exhale. However, even at the end of a complete exhalation, air still remains in the lung (the residual volume). Thus, spirometry is not capable of measuring all the air that comprises the total lung capacity. Knowing the total lung capacity can be useful in the evaluation of workers who may have certain lung diseases or conditions. Measurement of total lung capacity or TLC requires the direct measurement of lung volumes.

There are four methods for determining lung volumes: body plethysmography, nitrogen washout techniques, helium dilution techniques, and radiographic planimetry. Of these techniques, the one most commonly used in clinical laboratories is the body plethysmograph or the "body-box." With this method, the individual sits in an enclosed box (the size of a telephone booth) and places his or her mouth on a mouthpiece. While the individual is breathing in and out through the mouthpiece, the body-box measuring equipment senses changes in pressure surrounding the person. The resulting pressure changes are converted to volume changes which in turn provide a measure of the total volume of air within the lungs at that time. By other calculations and breathing maneuvers, the determination of TLC can be made.

As described later in this section (Classification of Lung Function Abnormalities), the determination of a restrictive ventilatory defect requires the actual measurement of TLC as by a body plethysmograph. When the TLC as measured by one of the four methods is considered to be abnormally reduced, then a restrictive ventilatory defect is present. For an obstructive ventilatory defect, the TLC can be normal, increased or reduced.

5.7 Diffusing Capacity

Diffusing capacity of the lung (DL_{CO}) is the measurement of the transfer of carbon monoxide (CO) from inspired air to pulmonary capillary blood. This lung function test is complementary to spirometry testing in that it is a crude estimate of gas exchange and the status of the pulmonary vascular bed. Procedures for performing the test conform to standard guidelines (3). A maximum of four trials can be performed to obtain at least two values that are within 10%. The mean value of at least two acceptable tests within 10% is reported. Predicted values for DL_{CO} and the ratio of DL_{CO} to the single-breath alveolar volume (V_A) or DL/V_A are based on published predicted values (2) and are the measures of interest.

5.8 Classification of Lung Function Abnormalities

Disturbances of respiratory anatomy or physiology give characteristic patterns of abnormal lung function. A lessening of ventilatory capability can either be due to a reduction in the total capacity of the lung (restrictive pattern) or obstruction in the airways of the lung (obstructive pattern).

An obstructive ventilatory defect is defined as lessening of the respiratory system's capability to cause air to flow from the lung. Usually the total amount of air the lungs are capable of holding remains relatively stable/unchanged. It is the relationship, the ratio, of the ability to force air out of the lungs (forced expiratory volume in one second, FEV_1) relative to the maximal volume that can be exhaled (forced vital capacity or vital capacity, FVC and VC, respectively) that defines an obstructive pattern. A reduced FEV_1/VC indicates airflow limitation and suggests a reduction in the cross-sectional area of the airways or airway narrowing (5). Examples of lung diseases that can have obstructive patterns include asthma, chronic obstructive pulmonary disease (COPD) (including chronic bronchitis and emphysema), cystic fibrosis, bronchitis, bronchiolitis obliterans, and bronchiectasis.

A restrictive ventilatory defect is characterized physiologically by a reduction in total lung capacity (TLC) (5). The ratio of FEV_1/VC which when reduced defines an obstructive pattern, remains normal or can even increase. Spirometry testing which measures the vital capacity might suggest a restrictive ventilatory defect but the measurement of total lung capacity (TLC) is required to determine if a true restrictive defect is present. Examples of lung diseases that can have restrictive patterns include the pneumoconioses (asbestosis, silicosis, coal workers' pneumoconiosis), interstitial fibrosis, and sarcoidosis.

5.9 Radiographic Examination

The cornerstone of the radiologic diagnosis is the plain thoracic radiograph, or the "chest x-ray" (13). Radiographic methods record anatomic structure and the changes that can occur within these structures, such as can be present with disease. The routine views of the chest are a front view and a side view. With these two visualizations, a three-dimensional inspection of the thoracic contents and chest wall can be accomplished. When the chest x-ray is used as a screening tool, the single front view may be the only one collected (14).

Other radiographic procedures may be performed to complement the simple chest x-ray, these include: digital radiography, computed tomography (CT scans), and magnetic resonance imaging (MRI).

Digital radiography techniques offer the ability to electronically transmit and compactly store images. It is possible that in the next 25 years, use of conventional radiographic film will either disappear or become dramatically reduced.

Tomography allows the selective visualization of a specified layer of tissue, excluding all others. This selective visualization is accomplished by maintaining the "focus" between object and radiographic device for one "slice" at a time. The compilation of all the slices results in a detailed three-dimensional image. Computed tomography (CT) uses the computer to process and display the resulting three-dimensional image. High resolution CT (HRCT) scans have thinner slices which can help identify certain pathological abnormalities. CT scans are particularly helpful in the evaluation of changes within the structure of lung tissue including interstitial changes or fibrosis.

MRI, or magnetic resonance imaging, is accomplished by placing certain atomic nuclei in a magnetic field. When the nuclei are excited with a particular frequency of radio waves, they emit some of the absorbed radio wave energy as radio signals. The magnetic resonance technique allows interpretation at a chemical-physiologic level while all previous imaging techniques have permitted only anatomic inspection. Since chemical–physiologic changes may occur before actual structural changes, magnetic resonance offers the potential for earlier disease detection.

5.10 ILO Classification for Presence of Pneumonconiosis

For over 50 years, the International Labour Organization (Geneva, Switzerland) has provided guidelines for the classification of chest radiographs for the presence of pneumoconioses. The most recent updated version of this classification system was released in 1999 (15). In this classification system, each chest radiograph (single front view) is scored in different categories: technical quality, presence of small opacities (representative of pneumoconioses), large opacities, pleural changes, and other abnormalities.

The most important classification scoring for the presence of pneumoconiosis is the profusion score for small opacities. The number and concentration of the small opacities is scored using a 12-point scale and the use of standard films of four categories of small opacity concentration or profusion. The four categories for the presence of profusion of small opacities are 0, 1, 2, and 3 (0 considered to essentially without significant number of small opacities to 3 which represents the greatest number or concentration or profusion of small opacities). The film to be reviewed is compared to a standard, 22-image film set.

If the reader feels that the number and concentration of the small opacities in the film being reviewed matches exactly the profusion of small opacities in the standard film, the score for the film is then recorded.

In addition to the profusion of small opacities, scoring is done for the size and shape of the small opacities; for the size and number of large opacities; for the extent, width and presence of calcium within pleural abnormalities; and for the presence of other disease conditions such as lung cancer or heart disease.

Use of the ILO classification system allows for consistency in scoring the changes on radiographs for workers that may have pneumoconioses such as asbestosis, silicosis, or coal workers' pneumoconiosis. This classification system has been used in epidemiological surveys as well as for medical-legal determination of impairment or disability.

6 INFLAMMATORY AND FIBROTIC RESPONSES OF THE LUNG

6.1 Particle Deposition

The lung diseases that occur as a result of exposure to the inhalation of mineral dusts are determined in part by the site deposition of the particles of dust. In general, airborne particles that are in the respirable range of size (<10 micron) deposit in the tracheobronchial tree through impaction, sedimentation, interception and diffusion. The physical characteristics (length, diameter, and shape of the particles) determine the distance that the particles may travel through the tracheobronchial tree. For example, the asbestos fibers of chrysotile (serpentine) and crocidolite (amphibole) are long and narrow and can impact at the bifurcation of smaller airways such as bronchioles. Once deposited in the tracheobronchial tree or within the alveolar spaces, damage then depends on the cellular response to the presence of the dust particles.

6.2 Cellular Response of Inflammation

As discussed previously in Section 3, Lung Defenses, particles that reach the terminal airways and the alveolar spaces can be engulfed by the resident alveolar macrophages. Once these cells engulf or phagocytize the particles, these cells can produce a variety of inflammatory mediators and cell-to-cell messengers depending upon the physical-chemical characteristics of the particles themselves (16, 17). For example, alveolar macrophages can release reactive oxygen species (oxygen radicals) that can be damaging to nearby cells. These macrophages can also release mediators or factors that can lead to the recruitment and proliferation of other cell types as well as the production of protein substances that are the basis of fibrosis.

For example, macrophage-released mediators such as tumor necrosis factor-alpha (TNF-α), interleukin-1 (IL-1), or interleukin-6 (IL-6) can lead to significant local responses in lung parenchyma and interstitium. TNF-α can lead to the recruitment of other inflammatory cells such as lymphocytes, eosinophils, and neutrophils. TNF-α can also cause the increased expression of adhesion molecules, which can result in the attraction and collection of other cell types in the region of lung injury. IL-1 and IL-6 are other inflammatory mediators that can alter collagen synthesis in the lung, affect the balance of protein substrates in the lung acellular matrix, and can also result in the recruitment of other cell types. All of these inflammatory mediators have been found in the lavage fluids from the lungs of workers exposed to dusts such as silica, coal dust and asbestos, as well as from the lungs of animals exposed to these mineral dusts.

Other mediators released from macrophages may be more involved with the subsequent fibrosis or scarring that occurs with long-term exposure to mineral dusts. Mediators such as transforming growth factor-beta (TGF-β), platelet-derived growth factor (PDGF), and insulinlike growth factor-1 (IGF-1) have been implicated in the recruitment and proliferation of cells involved in fibrosis such as fibroblasts. In some cases, it may be the relative proportions of these key mediators that characterizes the degree and extent of fibrotic response of the lung to the mineral dust. For example, fluids from patients' lungs with simple coal workers' pneumoconiosis (CWP) had higher concentrations of TGF-β while fluids from patients' lungs with progressive massive fibrosis (PMF) had greater concentrations of PDGF and IGF-1 (18).

In summary, particles such as asbestos fibers or coal dust or silica particles deposit in terminal airways such as respiratory bronchioles or alveolar ducts and are eaten or phagocytized by alveolar macrophages. These macrophages then release mediators that lead to inflammatory and fibrotic changes within the lung. The substances released by macrophages that may be involved include reactive oxygen species and cytokines such as interleukins (IL-1 and IL-6) as well as TNF-α, TGF-β, PDGF, and IGF-1. These mediators recruit other inflammatory cells such as lymphocytes, eosinophils, neutrophils and fibroblasts. These mediators can also damage resident cells such as the alveolar epithelial cells and endothelial cells. The activation of these recruited cells especially the fibroblasts can then lead to the production of changes within the connective tissue matrix with increased presence of collagen and resulting fibrosis.

These changes in cellular physiology, when extensive, can lead to changes in anatomy and both can be evidenced as changes in lung function.

7 MINERAL DUST-INDUCED LUNG DISEASE

7.1 Asbestos

Asbestos is a term used to describe a group of naturally occurring, fibrous magnesium silicate minerals. Asbestos is found in serpentine and amphibole mineral formations throughout the world; due to its indestructible nature and thermal insulating properties, asbestos has found many commercial applications. Asbestos has been recognized and used since ancient times in pottery, wicks for oil laps and torches, woven shrouds, and napkins or tablecloths that could be cleaned by fire (19–22). Commercial mining of asbestos began in the late 19th Century. The first asbestos textile mill in the United States began production in approximately 1896. Commercial asbestos production also began in Russia at approximately the same time. Large scale use of asbestos came with industrialization, and in particular, mass production of the steam engine which required heat resistant insulating materials for packing and seals (20–24). The production of asbestos increased rapidly during the mid 1900's as it found use in over 3000 applications.

Approximately twenty years after the start of commercial asbestos production, the first reports of pulmonary fibrosis occurred among asbestos textile workers in France (17). The considerable attention drawn to asbestos in modern times surrounds its ubiquitous use and diverse nature of attributable human health effects. Asbestos related diseases and conditions include asbestosis, small airways disease, pleural plaques, lung cancer, cancer of the larynx, cancers of the gastrointestinal tract, mesothelioma, and others (see Table 4.1) (22).

These diverse, significant health outcomes, in the context of the modern medical and legal environments, have made asbestos related respiratory disease one of the most significant occupational health issues of the 20th Century.

7.1.1 Mineralogy of Asbestos

From a mineralogical perspective, the term asbestos refers to a specific pattern of crystal formation (called asbestiform crystalline habit) producing long thin fibers with high aspect

Table 4.1. Asbestos-related Diseases and Conditions[a,b]

Pathology	Organ(s) Affected	Diseases/Condition
Nonmalignant	Lungs	**Asbestosis** (diffuse interstitial fibrosis)
		Small airway disease[c] (fibrosis limited to the peri-bronchiolar region)
		Chronic airways disease[d]
	Pleura	**Pleural plaques**
		Viscero-parietal reactions, including benign pleural effusion, diffuse pleural fibrosis and rounded otelectosis
	Skin	Asbestos corns[e]
Malignant	Lungs	**Lung cancer** (all cell types)
		Cancer of larynx
	Pleura	**Mesothelioma of pleura**
	Other mesothelium-lined cavities	**Mesothelioma of the peritoneum**, pericardium and scrotum (in decreasing frequency of occurrence)
	Gastrointestinal tract[f]	Cancer of stomach, oesaphagus, colon, rectum
	Others[f]	Ovary, gall bladder, bile ducts, pancreas, kidney

[a]Reprinted with permission from Ref. 22. Copyright© International Labour Organization, 1998. Other sources are Refs 25–30.

[b]The diseases or conditions indicated in bold type are those most frequently encountered and the ones for which a causal relationship is well established and/or generally recognized.

[c]Fibrosis in the walls of the small airways of the lung (including the membranous and respiratory bronchioles) is thought to represent the early lung parenchymal response to retained asbestos (25) which will progress to asbestosis if exposure continues and/or is heavy, but if exposure is limited or light, the lung response may be limited to these areas (26).

[d]Included are bronchitis, chronic obstructive pulmonary disease (COPD) and emphysemo. All have been shown to be associated with work in dusty environments. The evidence for causality is reviewed in ref. 27.

[e]Related to direct handling of asbestos and of historical rather than current interest.

[f]Data not consistent from all studies (27) some of the highest risks were reported in a cohort of over 17,000 American and Canadian asbestos insulation workers (29) followed from January 1, 1967 to December 31, 1986 in whom exposure had been particularly heavy.

ratios (fiber length divided by width), high tensile strength, and flexibility. Although many minerals can crystallize in an asbestiform habit, only six have been of industrial use and commonly referred to as asbestos (31). These include the serpentine mineral chrysotile and the amphibole minerals crocidolite, amosite, anthophyllite, tremolite, and actinolite (Table 4.2).

Chrysotile accounts for approximately 95% of the asbestos used in the United States (24). These asbestos minerals can also have nonasbestiform crystalline habits as noted in Table 4.2. Amphibole minerals and, to a lesser degree, serpentine minerals occur widely distributed throughout the earth's crust in many igneous or metamorphic rocks. However, only in rare instances do these mineralogical occurrences contain sufficient quantities of asbestos to be economically suited to mining.

7.1.2 Production and Use of Asbestos

The world production of asbestos has increased from the late 19th Century until about 1976 with the production of approximately 5,708,000 tons in that year. Production dropped during the 1980s and 1990s as the adverse health effects of asbestos exposure became a matter of increasing public concern (21, 22). Approximately 2.7 million tons of asbestos were produced during 1994 and world mine producers, in order of output, included Russia, Canada, Kazakhstan, China, and Brazil (23). The United States asbestos consumption was less than 27,000 metric tons in 1994. Chrysotile accounts for approximately 95% of world production and a key market for this fibrous mineral is the production of asbestos cement (21, 23). Chrysotile is produced primarily in Russia, Canada, Swaziland, and Zimbabwe with minor production plants in the United States (California), Australia, Cyprus, Italy, Brazil, and China. Crocidolite is produced in South Africa and in Australia (now discontinued) and Bolivia. Amosite was produced uniquely in South Africa. Anthophyllite was produced only in Finland.

The physical properties of asbestos that have resulted in its widespread use include (*1*) good insulator (*2*) incombustibility, (*3*) flexibility, (*4*) high tensile strength, and (*5*) resistance to corrosion. By virtue of these properties, asbestos has found many industrial ap-

Table 4.2. Classification of Asbestos Minerals

Asbestos	Nonasbestiform Analogs	Chemical Formula
	Serpentine	
Chrysotile	Antigorite, Lizardite	$Mg_3(Si_2O_5)(OH)_4$
	Amphibole	
Crocidolite	Riebeckite	$Na_2Fe_5(Si_8O_{22})(OH)_2$
Amosite	Cummingtonite-Grunerite	$(Fe, Mg)_7(Si_8O_{22})(OH)_2$
Anthophyllite	Anthophyllite	$(Mg, Fe)_7(Si_8O_{22})(OH)_2$
Tremolite	Tremolite	$Ca_2Mg_5(Si_8O_{22})(OH)_2$
Actinolite	Actinolite	$Ca_2(Mg, Fe)_5(Si_8O_{22})(OH)_2$

plications including (*1*) use in cement products including pipes and shingles, (*2*) vinyl floor tile, (*3*) paper for insulation and filtering, (*4*) friction materials for brakes and clutches, (*5*) textile products including yarn, felt, rope, wicks, and others, (*6*) paints and coatings, and (*7*) spray-on materials for fireproofing, and thermal or acoustical insulation. The shipbuilding, automobile, railroad, and construction industries have been among the larger asbestos users nationally (21, 22, 32). A 1995 list of industrial applications for asbestos shows some redistribution in uses including asbestos cement (84%), friction materials (10%), textiles (3%), seals and gaskets (2%), and other uses (1%) (22).

7.1.3 Occupational Exposures and Exposure Limits

Most asbestos exposures are of occupational origin, particularly in industrialized nations, and the list of occupations with the potential for asbestos exposures is quite large. Primary asbestos exposures occur through direct asbestos production in mining and milling operations. Most asbestos mines are open-pit operations (chrysotile) with some underground mining (amphiboles). Workers are exposed as asbestos containing ores are removed, fragmented and screened. The milling process involves occupational exposures as the asbestos containing materials are further concentrated by crushing and screening, washing/drying, cyclone separation, sorting into commercial grades, and packaging for distribution. The handling of waste ores and transportation of asbestos products at mine and milling operations is also a source of occupational exposure (21, 22, 33). Asbestos exposures can also occur to workers involved in mining stone or other mineral products through asbestos contamination of parent ores (33, 34).

Secondary asbestos exposures occur in a wide range of industrial settings through the production/application of asbestos containing products or by contact with these materials. Workers have used raw asbestos in packing and lagging (e.g., steam engines and boilers), in spraying materials onto ventilation ducts and structural surfaces in buildings, and in shipbuilding. In the manufacture of asbestos cement and asbestos tiles, exposures can occur through the opening of bags of asbestos, mixing these materials into slurries, and in the machining of the end products. In the manufacture of asbestos yarns and textile products, exposures can occur by preparing, blending, carding, spinning, weaving, and calendaring the fiber. In addition, exposure to asbestos can occur in the construction industry with the application of asbestos-containing insulation materials and the cutting, drilling, and sanding of asbestos-containing construction materials. Maintenance and construction workers, including electricians, welders, pipefitters, carpenters, and others are exposed by working directly with asbestos containing materials or by working in proximity to other construction operations involving asbestos. Asbestos exposure can also occur in office settings through uncontrolled maintenance/repair of building surfaces or structures containing asbestos materials (35). Significant levels of domestic asbestos exposure have also been reported to occur through materials adhered to work clothes when they are taken home for laundry.

Historically, occupational exposures to asbestos are reported to have been as high as 100 fibers per cubic centimeter (fibers/cm^3); although, in most occupational settings, exposures were commonly between 2 to 20 fibers/cm^3 prior to occupational exposure control. Occupational disease resulting from asbestos exposures became evident during the early 1950s. During the 1960s and 1970s, large epidemiological studies confirmed clinical observations associating asbestos exposure with asbestosis and lung cancer (21–23, 32). Occupational asbestos exposures have been substantially reduced since regulatory enforcement began during the early 1970s. In the United States, the Occupational Safety and Health Administration (OSHA), enforces a Permissible Exposure Limit (PEL) for asbestos in general industry of 0.1 fibers/cm^3 as a time-weighted average (TWA). (For regulatory purposes, asbestos is defined to be chrysotile, amosite, crocidolite, tremolite, anthophyllite, and actinolite. Fibers, measured at the microscopic level, are defined as a particulate form of asbestos 5 microns or longer with a length-to diameter ratio of at least 3 to 1). The Mine Safety and Health Administration (MSHA) exposure limit for asbestos as a TWA is 2 fibers/cm^3. The National Institute for Occupational Safety and Health (NIOSH), Recommended Exposure Limit (REL) and the American Conference of Governmental Industrial Hygienists (ACGIH), Threshold Limit Value (TLV) are also 0.1 fibers/cm^3 as a TWA for all forms of asbestos. In modern industrial settings, the occupational exposure limits/criteria and the use of engineering control methods for asbestos containing dusts have been effective in reducing asbestos exposures (36–39).

7.1.4 Asbestos-Related Respiratory Disease

The primary toxic effects of asbestos on the respiratory system involve the ability of asbestos fibers to induce a fibrotic pulmonary response and cancers of the lung and pleural spaces (48). The level of asbestos exposure, fiber type, fiber dimensions, durability, and fiber surface properties influence the toxicity, fibrogenicity, and carcinogenicity. Inhaled asbestos fibers become aligned with the airstream and can penetrate to the lower, gas exchange region of the lung depending on fiber diameter. Asbestos fibers with diameters less than 5 microns have a greater ability to penetrate to the nonciliated regions of the lung where they can exert the greatest toxic effect. Particles deposited in the major airways are cleared by the action of mucus secretions/ciliated cells and are transported out of the respiratory system. Smaller diameter fibers, deposited beyond the major airways are engulfed by alveolar macrophages and can remain in the lungs for longer periods of time. There is limited clearance of asbestos fibers deposited in this region of the lung, and when clearance occurs, it is through the lung lymphatics to the interstitium, the pleural cavities, and the lymph nodes. Often the fibers penetrating to the non-ciliated regions of the lung are partially engulfed, or coated, by several alveolar macrophages forming an asbestos body, a characteristic marker of asbestos exposure (see Fig. 4.4).

Asbestos fibers retained in the lung usually invoke an inflammatory reaction with an accumulation of white blood cells followed by a macrophagic alveolitis. The subsequent release of fibronectin, chemotactic factors, growth factors, and superoxide ion release results in the proliferation of alveolar, epithelial, endothelial, and interstitial cells (21, 22, 32) (see Section: Cellular Response of Inflammation) The asbestos-related respiratory diseases and conditions described below include pleural disease, asbestosis, lung cancer, and mesothelioma.

7.1.4.1 Asbestos-Related Pleural Disease. Pleural disease is the most common manifestation of asbestos exposure and can present as benign pleurisy with effusions (collection of fluid), pleural plaques, diffuse pleural thickening, and rounded atelectasis. Benign pleu-

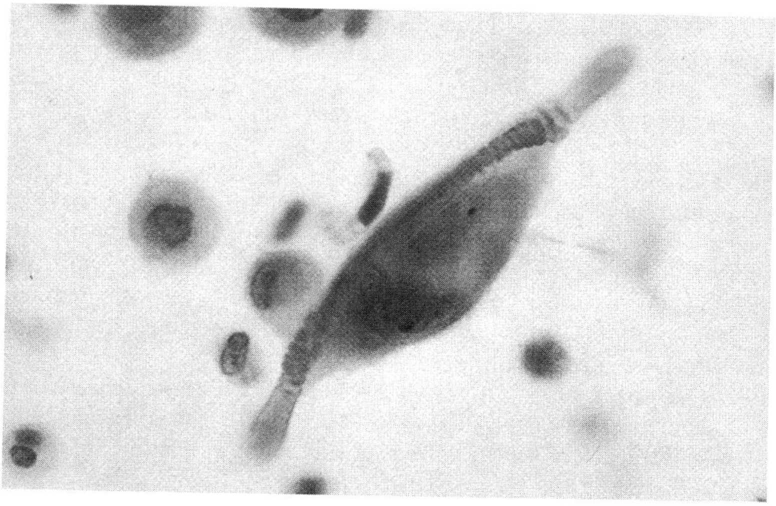

Figure 4.4. An asbestos body: an asbestos fiber coated by alveolar macrophages.

risy is defined by radiographic or thoracentesis (aspiration of fluid through the chest wall) confirmation of effusion, the absence of other causes of effusion, asbestos exposure, and the absence of tumor with clinical follow-up of at least three years (41). Benign pleurisy does not have a clear pathology although contact of asbestos fibers and the pleura is likely the initiating event. The latency period is generally less than 20 years and it is often the first manifestation of asbestos-related effects. Although it may precede other manifestations, it is not believed to be a precursor of other asbestos diseases. Benign pleurisy follows one of several courses including complete, painless regression; painful regression with minimal or no pleural scar; effusion of single or multiple instances, and, in limited instances, diffuse pleural thickening or rounded atelectasis (21–23).

Pleural plaques are raised, white, irregular lesions of hyalin fibrosis covered with mesothelium and found on the parietal (on the chest wall) pleura or on the diaphragm (22). Pleural plaques are most frequently a result of asbestos exposure and considered to be a good marker for such exposure. The occurrence and progression of pleural plaques appears in direct relation to amphibole asbestos exposures as contrasted to chrysotile asbestos (21). Pleural plaques are believed to be a local reaction to asbestos fibers reaching the pleural spaces during clearance from the lungs. Pleural plaques are largely asymptomatic in association with asbestosis; however, there is some evidence to suggest that pleural plaques may be associated with decrements in lung function as determined by spirometry (42, 43). The latency period between exposure and radiographic appearance of pleural plaques is approximately 20 years. It is possible to recognize plaques earlier than this, in much less well-defined stages, with the CT scan.

In contrast to pleural plaques, diffuse pleural thickening, also called pachypleuritis, is a fibrotic disease of the visceral (on the lung surface) pleura. Although not specific to asbestos exposure, diffuse pleural thickening is believed to result from fibers deposited in the parenchymal (interstitial) subpleural areas. Diffuse pleural thickening is attributed to the combined effects of large pleural plaques, the extension of subpleural fibrosis to the visceral pleura, and scarring from exudative (fluid forming) benign pleurisy. Due to the extent of diffuse pleural thickening resulting in restriction of the lungs to expand, dyspnea (shortness of breath) on exertion is a common symptom of this condition as is dry cough. If extensive and bilateral, lung function effects can be severe to the point of causing respiratory insufficiency and failure in some instances. Diffuse pleural thickening can be identified by chest radiograph or by CT scan, unilaterally or bilaterally. By nature of its location, diffuse pleural thickening diagnosis is facilitated by the CT scan. Since diffuse pleural thickening has etiologies other than asbestos exposure, it is not a good exposure marker as pleural plaques, and the elimination of the other disease causes should be part of the initial clinical assessment.

Rounded atelectasis, a form of asbestos-induced pleural disease, is caused by scarring of the parietal and visceral pleura, thickening of the interlobar fissure and adjacent lung tissue, and, with retraction of the scar, trapping and collapse of adjacent lung tissue. This form of pleural disease is less common than other forms. Rounded atelectasis is usually asymptomatic and detected by chest radiography as a pleural-based opacity often resembling a tumor. With the CT scan, the true nature of rounded atelectasis can often be resolved without the need for surgery. Chest pain in the area of formation is occasionally the only presenting symptom. Rounded atelectasis is usually a late condition occurring long after other pleural changes have been noted (21–23).

7.1.4.2 Asbestosis. Asbestosis is a fibrotic disease of the lungs caused by the inhalation, retention, and pulmonary reaction to asbestos fibers (44). The fibrotic changes, which give rise to asbestosis, are the result of an inflammatory process caused by asbestos fibers retained in the lung. This fibrosis is diffuse, interstitial, and tends to involve primarily the lower lobes and subpleural regions of the lung. Early lesions are characterized by discrete areas of fibrosis in the regions around the respiratory brohchioles. In advanced cases, the normal lung architecture is obliterated producing a honeycomb pattern. Fibrosis of the adjacent pleura is often seen (21). The histologic features of asbestosis do not distinguish it from interstitial fibrosis due to other fibrogenic agents; however, the presence of asbestos bodies or asbestos fibers in the lung are markers for asbestos exposure. The extent of the lung fibrosis generally relates to the measured asbestos fiber burden in the lungs.

The two most significant clinical findings of asbestosis are dyspnea on exertion and end-inspiratory crackles or rales on auscultation of the chest. Dyspnea is the earliest and most consistently reported symptom. Other symptoms can include a dry cough and chest tightness.

The chest radiograph has been one of the most important diagnostic tools for determining the presence of asbestosis. The radiographic appearance of asbestosis is characterized by irregular opacities in the lower two-thirds of the fields for both lungs. When densely profuse, these opacities can obscure the cardiac outline and the dome of the diaphragm. The reduction of lung volumes is radiographically evident (see Fig. 4.5).

The diagnosis of asbestosis is based upon several separate criteria (45). This diagnosis of asbestosis by radiography has been facilitated by the International Classification of Radiographs of the Pneumoconioses established by the International Labor Office (ILO)

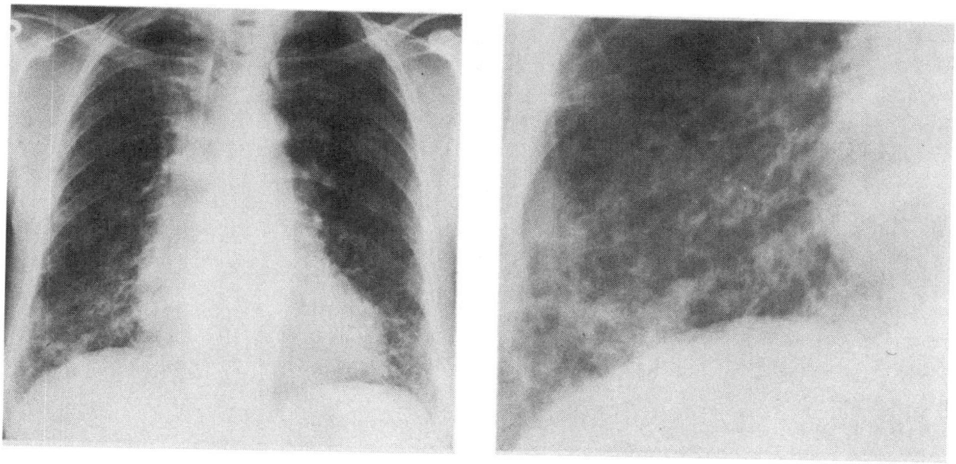

Figure 4.5. A chest radiograph of asbestosis with a close-up view of the right lower lung.

(15). The use of CT scanning and high resolution computer tomography (HRCT) have greatly increased the sensitivity for detecting asbestos-related lesions. Although, at present, no standardized reading method has been developed for HRCT. Cigarette smokers and ex-smokers have a higher prevalence of asbestos-related radiologic lesions than nonsmokers in studies of British shipyard workers. Since smokers without dust exposure have few irregular opacities on the chest radiograph, this likely reflects the impaired lung dust clearance mechanisms due to smoking (22, 23, 46). The combination of smoking and exposure has a synergistic effect.

Interstitial lung fibrosis due to asbestos is generally associated with a restrictive pulmonary function pattern. Characteristic features include reduced lung volumes, particularly vital capacity (VC); the ratio of forced expiratory volume in one second (FEV_1) to forced vital capacity (FVC) is usually preserved. Reduced lung compliance and reduced diffusing capacity are of note. In addition to spirometry, bronchoalveolar lavage (washing out) is being used more frequently as a clinical tool in the investigation of asbestos related diseases of the lung (22, 23, 32).

Medical interventions in cases of asbestosis occur on initial diagnosis, in follow-up evaluations, in the treatment of respiratory infections complicating the disease, and in the treatment of hypoxemia and heart failure associated with advanced disease stages. Other forms of support include smoking cessation programs, if needed, treatment of cough symptoms or in treatment to prevent viral or pneumococcal infections. Asbestosis has no established regimen of treatment and, due to the risks of lung cancer and mesothelioma, periodic chest radiographs are recommended. Continued work in environments with asbestos exposures should be prevented; however, disease progression is noted in some cases following exposure cessation (21–23).

7.1.4.3 Lung Cancer. During the 1930s, publication of a number of clinical case reports from the United Kingdom, Germany, and the United States first suggested a possible association between lung cancer and asbestos exposure. Today there is substantial evidence to demonstrate an association between lung cancer and asbestos exposure including exposure to all types of asbestos fibers (22, 23, 47–49). All histologic forms of lung cancer are seen in asbestos-exposed workers and the pathology of asbestos-related cancers is not distinct in type, nature, or location within the lung. There is an approximate fivefold increase in lung cancer risk for workers exposed to asbestos and lung cancer rates peak at approximately 30 years latency. Current knowledge on the pathogenesis of asbestos related cancers in incomplete. The increased risk of lung cancer in asbestos exposed smokers increases exponentially. There is some controversy that the increased risk for lung cancer exists only for those individuals where asbestosis (the presence of fibrosis related to asbestos exposure) is present (50). Others contend that sufficient, prolonged exposure to asbestos even without the development of asbestosis is associated with increased risk for lung cancer (57).

The clinical presentation of asbestos-related lung tumors is similar to that of lung tumors caused by other carcinogens, excluding the possible associated symptoms of lung fibrosis due to asbestosis. The major symptoms in lung cancer patients can include cough, dyspnea, chest pain, hemoptysis, recurrent bouts of pneumonia, localized wheezing and others. Lung cancer in asbestos exposed workers presents as a mass lesion on the chest radiograph and is indistinguishable from lung cancers associated with other carcinogens. The CT scan increases the potential for early detection of lung cancers when applied to high risk groups. The diagnosis of lung cancer is based on histopathologic grounds with cytology or tissue samples; the histologic changes are not specific to asbestos-related cancers.

The overall lung cancer survival rates remains poor with only an 8 to 10% five-year survival rate. There have been recent advances in chemotherapy, surgery, and in the understanding of the molecular pathogenesis of lung cancer. As a result of the latter, gene therapy approaches show much promise (21–23, 32).

7.1.4.4 Malignant Mesothelioma. Mesothelioma is a rare malignant tumor arising largely in the pleura and to lesser extent from the peritoneum (lining of the abdomen). In early stages of disease, mesothelioma presents as small grayish nodules on the visceral and parietal pleura that coalesce and form larger tumor masses. The tumor develops by direct extension, forming large masses that can invade adjacent structures including the chest wall, lung parenchyma, and the diaphragm (21, 23).

Malignant mesothelioma is relatively rare with an incidence of approximately two cases per million in the general population of North America. The majority of malignant mesothelioma cases have a history of exposure to asbestos through occupational or environmental sources. No other associated exposure has been found. This asbestos exposure history may be quite brief. Asbestos-exposed populations have mesothelioma incidence rates five to twentyfold higher than the general population. Wagner and colleagues (52) were the first to report a case series of mesothelioma among individuals from the crocidolite mining region and asbestos factories in South Africa. Since this report, the association between asbestos exposure and mesothelioma has been confirmed repeatedly (21–23, 52).

All types of asbestos fibers have produced mesothelioma. Amphibole asbestos fibers are considered to be the cause of most cases of mesothelioma. Mesothelioma cases have also been attributed to chrysotile exposures but much less frequently. The reasons for the increased prevalence of mesothelioma from amphibole asbestos exposures is believed associated with the geometric properties of amphibole fibers, favoring deeper penetration into the lung, and the increased resistance of amphibole fibers to degradation, favoring persistence in lung tissue. It is also accepted that mesothelioma has no association with cigarette smoking. A characteristic of mesothelioma is a long latency period of approximately 20 to 50 years (21).

Malignant mesotheliomas are diagnosed primarily in males at ages from 50 to 70 years. The most common clinical symptom is chest pain. Cough and dyspnea are also common symptoms due largely to pleural effusions. Weight loss, fever, and general malaise often present in later stages of disease. In addition to being locally invasive, approximately 50 to 80% of malignant mesotheliomas are metastatic. The radiologic findings in mesothelioma typically reveal a thick pleural peel along the lateral chest wall with extension to the apex. There is usually pleural effusion and this is a common presenting sign. Most of these changes can be detected by plain radiography; however, the use of CT and HRCT scans add sensitivity to better evaluate the extent of the tumor progression. Biopsy is usually needed to establish a firm diagnosis of malignant mesothelioma. The various forms of treatment for malignant mesothelioma can include surgery, radiotherapy, chemotherapy, and more recently immunotherapy and cytokine therapy. The clinical course of disease is usually rapid and survival is limited; the treatments generally provide only a few months of life extension. In most instances, medical intervention remains largely supportive in nature. Death by cardiorespiratory failure usually occurs within 2 years from diagnosis (21–23).

7.2 Other Silicate Minerals

Two of the most abundant elements on earth are silicon and oxygen; the combination of these two produces silicon dioxide (SiO_2). Silica or silicon dioxide (SiO_2) exists in both crystalline and amorphous forms. Amorphous (noncrystalline) silica occurs in natural glasses (e.g., volcanic tuff) as well as synthetic glasses such as mineral wool. Amorphous silica is characterized by a random, nonrepeating organization of silicon dioxide molecules. Crystalline forms of silicon dioxide are characterized by an organized, repeating pattern of molecules in a three-dimensional array commonly known as quartz (see 7.4 Silica-Crystalline). Crystalline forms of silica are produced by increased pressure and heat during formation in the earth's crust. Alpha-quartz is the most common, naturally occurring form of crystalline silica. Much of the silicon dioxide present in the earth's crust is combined with other elements to form silicate minerals. These silicate minerals contain the silicon dioxide tetrahedron in combination with various cations or anions including aluminum (Al), calcium (Ca), fluorine (Fl), hydroxide (OH), iron (Fe), magnesium, (Mg), sodium (Na) and others. Collectively, the silicates comprise abut 25% of known minerals and constitute over 90% of the earth's crust (53–55).

A wide variety of silicate minerals in addition to asbestos (chrysotile, crocidolite, amosite, anthophyllite, tremolite, and actinolite) and crystalline silica (quartz, cristobalite, and tridymite) can cause pneumoconiosis, cancers of the respiratory system, and other respiratory health problems. Other minerals, described in this section, can give rise to pulmonary responses, including radiographic changes, but are seldom associated with impairment and respiratory disability. Additionally, many mining, construction, and manufacturing materials to which workers are exposed contain a mixture of silicates, often including crystalline silica, in the form of quartz or cristobalite; this combination of mineral constituents often presents a mixture of respiratory health effects caused by specific silicate minerals. The following discussions present some of the common, commercial silicate minerals according to their mineral habitat (fibrous and nonfibrous) and respiratory health effects. Silicate minerals commonly associated with fibrous contaminants are also discussed (53–56).

7.2.1 Fibrous Silicates

7.2.1.1 Attapulgite and Sepiolite. Attapulgite $[(Mg, Al)_2Si_4O_{10}(OH) \cdot 4H_2O]$ and sepiolite are the commercial varieties of palygorskites, fibrous, crystalline magnesium–aluminum clays that can contain varying amounts of aluminum, magnesium, and iron. These silicates have a unique chain structure producing characteristic colloidal and absorptive properties, and as a result have a variety of industrial applications. Attapulgite is mined predominantly in the southeastern United States (Georgia and Florida). Spain is the predominant producer of sepiolite. Both are used in drilling muds, as paint thickeners, and as insulation substitutes for asbestos. Attapulgite and sepiolite are known as fuller's earth constituents as derived from use in finishing and thickening (or fulling) wool. Attapulgite is also used as an anti-diarrheal agent and as an absorbent in pet litter. Sepiolite is used as a material in cigarette filters (53, 54).

Studies on the carcinogenicity of attapulgite have resulted in variable results and this has been largely attributed to differences in the geological origins and mineralogical properties of attapulgite. Attapulgite from some geological deposits is reported to cause lung cancer and fibrosis on intrapleural injection and inhalation studies. Long attapulgite fibers have been reported to cause mesothelioma on intrapleural injection into rats; while short fibers (less than 4 microns in length) have not produced mesothelioma. There is one case report of attapulgite causing pulmonary fibrosis in a mining engineer exposed for over 2 years. A mortality study of approximately 2,300 workers, exposed to attapulgite mining and milling for one month or longer, showed a slight but significant increase in lung cancer; nonmalignant respiratory disease was not increased in this study (57).

7.2.1.2 Wollastonite. Wollastonite is a naturally occurring monocalcium silicate with a predominant acicular (or needlelike) crystalline habit with common fibers aspect ratios between 3 and 20. Wollastonite occurs in deposits throughout the world; although, the largest deposits are in the United States, Mexico, and Finland. Wollastonite is used in ceramics, in paint and bonding cements, as a substitute for asbestos in brake linings, wallboards, insulating materials, plastics, and fiberglass yarn (53–56).

Long term animal studies of wollastonite indicate low carcinogenic and fibrogenic potentials. These studies show that wollastonite fibers induced an alveolar macrophage response that resolves on exposure cessation. Wollastonite has been found to activate serum complement and increase pulmonary macrophage chemotaxis. Human morbidity studies of 103 wollastonite miners in New York suggest no significant relationship between exposure and pulmonary function or radiographic changes. A study of 46 Finnish wollastonite miners showed some mild profusion of irregular opacities (14 workers) and bilateral pleural thickening (13 workers); airborne fiber concentrations were high with a range from 1 to 63 fibers/cc and there was also a potential for exposures to crystalline silica. Wollastonite appears to have a relative low toxicity but limited evidence suggests some potential for pleural and interstitial changes on high levels of exposure (22, 54–59).

7.2.1.3 ***Zeolites.*** Zeolites are a group of hydrated aluminum silicates with exchangeable cations. Approximately 40 different zeolite species have been identified. The zeolites contain large pores, filled with water, which can be released on heating to temperatures of 200°C. The term zeolite is derived from the Greek word meaning boiling stone. Natural zeolites can occur in both fibrous and nonfibrous crystalline habits. The fibrous zeolite minerals, erionite and mordenite, are crystalline structures that can be cleaved into long thin, respirable fibers with high aspect ratios. In the United States, the main deposits of zeolites occur in the Western Intermountain regions, especially in the Great Basin (including Nevada and portions of Utah and neighboring states). Currently there is almost no commercial mining of zeolites and the zeolite currently used in industrial applications is produced synthetically in nonfibrous form (54–56).

Zeolites have several important commercial properties including reversible selective adsorption and ion-exchange capabilities. These properties account for much of the industrial application of the synthetic form of zeolites. The mineral is used in the wastewater treatment, petrochemical, and water filtration industries. Zeolites are also used in animal litter, as dietary supplement in animal feeds, and in certain cements (54).

The nonfibrous, synthetic forms of zeolite are not considered to be fibrogenic or carcinogenic. In contrast, the natural, fibrous forms of zeolites are fibrogenic and carcinogenic as determined by both animal and human data. The fibrous zeolite erionite has been identified as the causal agent in the increased incidence of malignant mesothelioma in the Turkish villages of Tuzkoy and Karain, with an incidence of pleural and peritoneal mesotheliomas approximately 1000 times higher than the general population. Villagers were highly exposed as they carved homes into volcanic tuffs containing fibrous erionite. The fibrous zeolite mineral erionite should be handled according to all health and safety guidelines for asbestos (54–56).

7.2.2 Minerals with the Potential for Fiber Contamination

7.2.2.1 ***Vermiculite.*** Vermiculite is the geological name for a group of nonfibrous, hydrated, ferromagnesium aluminum silicates. Vermiculite has the unique characteristic of expanding many times its original size on heating. Heat converts the water, present in the spaces between the laminar mineral plates, to steam forcing the thin, flexible plates apart into an expanded, wormlike configuration. The first significant domestic deposit of vermiculite was found in 1916 near Libby, Montana and commercial production began in 1921. In the United States, commercial quantities of vermiculite are also found South Carolina and Virginia. Outside the United States, vermiculite is also mined in Phalaborwa, South Africa, and Transvall. Vermiculite, in its expanded form, is used as insulation, as a carrier for fertilizers, herbicides, and pesticides, as a soil conditioner, as an aggregate and filler for concrete and gypsum, and in water purification. Pure vermiculite is a silicate mineral considered to be a particulate or particle not classified (PNOC) without significant pneumoconotic or carcinogenic potential. However, certain vermiculite deposits can be contaminated with amphibole asbestos fibers, thereby creating occupational exposure hazards on mining and processing. Ore from the vermiculite deposits near Libby, Montana contains actinolite and tremolite asbestos. Ore from deposits in Virginia has also been shown to contain actinolite asbestos; while, some of the vermiculite deposits in South Africa contain anthophyllite asbestos. The presence of asbestos minerals in the vermiculite presents substantially increased exposure hazards for workers. Routine monitoring of the vermiculite ores for fiber contamination is required to prevent asbestos exposures.

7.2.2.2 Talc. Talc is a layered, hydrated magnesium silicate with a chemical composition of $Mg_3Si_4O_{10}(OH)_2$. It can occur in platy, granular, and fibrous forms and occasionally as mixtures of the three forms. Commercial quantities of talc are found in several different states with the majority of commercial production from mines in New York, California, Texas, and Vermont. Outside of the United States, the major producers of talc include Australia, Austria, China, and France. Commercial talc often contains other mineral contaminants including crystalline silica as well as amphibole and serpentine forms of asbestos (54). Talc has a wide range of uses and is part of formulations for paint, paper, cosmetics, ceramics, textile materials, roofing products, rubber, fire extinguishing powders, water filtration agents, insecticides, and others. Over 500 different products are sold under the name talc and the worldwide production of talc is about 5 to 6 million tons on an annual basis (53–56).

Talc is considered to have pneumoconotic potential independent of asbestos and crystalline silica contamination. Workers exposed to the nonasbestiform varieties of talc have developed lung fibrosis. Epidemiologic studies of talc miners and millers in upstate New York revealed a progressive pneumoconiosis in older workers, disabling them with decreased vital capacity, parenchymal opacities, and pleural fibrosis. The upstate New York talc mines are contaminated with tremolite and anthophyllite (60–63). Studies of Vermont talc workers, mining ore deposits without significant asbestos and crystalline silica contamination, found evidence of pneumoconiosis on chest radiographs that correlated with exposure but not smoking (64). Some of the earlier studies of talc miners and millers suggested an association between exposure and bronchogenic carcinoma (60–63). Although, more recent studies on miners and millers working in talc deposits free of asbestos and crystalline silica have not substantiated these findings. In a study of 389 workers exposed to nonasbestiform talc, Wergeland did not observe an increase in deaths from lung cancer (65). As with vermiculite, routine monitoring of talc ore deposits is recommended to detect and prevent exposures to asbestos present as ore contaminants.

7.2.2.3 Metal Ore Deposits. Serpentine and amphibole asbestos can be present as inclusions in the ore deposits of many different metal or mineral ores including copper, gold, lead, silver, tungsten, zinc, and traprock. Careful mineralogical assessment of the ores prior to mining and milling is recommended to help prevent overexposures to asbestos minerals (54).

7.2.3 Nonfibrous Silicates

7.2.3.1 Bentonite. Bentonite is a commercial term for a group of clays formed by crystallization of vitreous volcanic ashes deposited in aqueous environments. Bentonite defines clays that contain large percentages of sodium montmorillonite or beidelite. Bentonite takes its name from the clays found at Fort Benton, Wyoming. Within the United States, bentonite is produced in approximately 12 states. Bentonite clays are taken from open pit mines, ground, dried in kilns, and bagged for shipment. Bentonite is used in drilling muds, as catalysts in petroleum refining, as a foundry sand bond, as a filler in paints, in carbonless copy paper, in filtering agents for wine, water, or other liquids, in pharmaceuticals and cosmetics, as a component in animal feeds, and others. Pneumoconiosis has been documented among bentonite miners and millers although this is believed to be due to crystalline silica present in the clays as cristobalite (54, 56).

7.2.3.2 Mica. Mica refers to a family of nonfibrous silicate minerals occurring as complex hydrated aluminum silicates having a platy structure similar to that of talc. Micas are associated with the alkaline metals of iron or magnesium. Important mica groups include the muscovites, phlogopite, lepidolites, and biotites. The only mica groups used commercially include muscovite and phlogopite. Mica was used historically in shades for oil lamps and in windows for stoves. Continued uses include liners for steam boilers, in optical instruments, as artificial snow and flocking agents for Christmas decorations, in roofing materials, in drilling muds, in asphalt, as an electrical insulator, in ceiling tiles, and in wallboard joint cements. Epidemiologic studies of mica workers demonstrate an association with pneumoconiosis. In a study of mica workers in North Carolina, 10 of 57 workers grinding mica free of crystalline silica had radiographs consistent with pneumoconiosis. Symptoms of cough and dyspnea were identified among workers consistent with radiographic evidence of pneumoconiosis. Other human epidemiologic studies support the association between mica exposures and pneumoconiosis (46, 47, 50).

7.2.3.3 Diatomaceous Earth. Diatomaceous earth is comprised of the solidified remains of skeletons of diatoms (a unicellular algae) deposited millions of years ago. The diatom has a skeleton made of silicon dioxide. In nature, diatomite is largely an amorphous, nonfibrous silicate. For industrial applications, diatomite is calcined at approximately 800 to 1000°C producing crystalline silica in the form of quartz and cristobalite. Natural diatomite usually contains approximately 1% or less of crystalline silica; following calcining, the crystalline silica content can increase to approximately 90%. Diatomaceous earth finds industrial application based on its insulating and adsorbent properties. It is used in the lining of molds in foundries, in abrasive lubricants, and as filtering agents, in pottery glazes, and other uses. High exposures to cristobalite have been described among workers in diatomaceous earth processing mills. Exposed workers may develop lung changes involving simple or complicated pneumoconiosis. The pneumoconioses observed among diatomaceous earth workers resembles that from crystalline silica. Workers are also at increased risk for lung cancer related to cumulative exposures to crystalline silica (22, 54–56).

7.2.3.4 Kaolin. Kaolin is a nonfibrous hydrated aluminum silicate referred to as China clay. Kaolin belongs to a family of silicate minerals termed the phyllosilicates based on the platy or flaky mineral habit. Kaolin [$Al_2Si_2O_5(OH)_4$], as a clay, does not have an exact chemical composition, but is comprised primarily of kaolinite with minute and variable quantities of metal oxides that can include iron and titanium. Crystalline silica is also an ancillary mineral contaminant in some kaolin deposits. The mining of sedimentary kaolin deposits is by open-pit methods using high pressure water jets to remove the clay materials. This wet mining technique does not generate high dust concentrations. By contrast, workers are exposed to higher dust levels during subsequent drying, processing, and bagging of the kaolin mine slurry. Some kaolin is calcined by heating to temperatures of 1000°C and this can increase the crystalline silica content. Kaolin has many industrial applications. It is used in paper products, ceramics, inks, paints, adhesives, insecticides, medicines, cosmetics, crayons, detergents, absorbents, cements, fertilizers, plastics, and rubber products.

Workers exposed to kaolin clays can develop pneumoconiosis. Kaolin workers show radiographic evidence of pneumoconiosis, with reductions in pulmonary function including FVC, FEV_1, and peak flow. The presence of crystalline silica in the kaolin deposits is believed largely attributable for these pneumoconioses and potential for kaolin to induce lung damage in the absence of crystalline silica contamination is not resolved.

7.2.3.5 Volcanic Ash. Volcanic ash is aerosolized as a result of volcanic eruptions. The nature of the particulates can be highly variable as determined by the geologic characteristics of the region. The eruption of the Mount St. Helens Volcano in Washington State on May 18, 1980, produced vertical clouds of ash and gases more then 20 kilometers into the air. A large percentage of this particulate was of respirable size fraction. The ash was comprised of plagioclase and aluminum silicates. The ash was found to contain crystalline silica, both quartz and cristobalite, at concentrations of 3 to 7%. Toxicological studies of the ash further suggested that it was moderately fibrogenic and should be considered a pneumoconiosis risk for those individuals more heavily exposed on a regular basis. Longitudinal studies of loggers exposed to variable concentrations of ash found significant declines in FEV_1 associated with exposures during the first year after the eruption. Symptoms of cough, phlegm, and wheeze were reported among the loggers. No radiographic changes were observed in the 4 year follow-up period (54, 56).

7.2.3.6 Feldspars. Feldspars are one of the largest groups of minerals comprising igneous rocks. The two major groups are known as orthoclase, a potassium based feldspar, and plagioclase, a sodium–calcium feldspar. Granite is an igneous rock containing large quantities of feldspars. Feldspars are used in the ceramics industry. In the absence of quartz contamination, the feldspars are believed to present a relatively low pneumoconotic potential (53, 54).

7.3 Coal Dust

7.3.1 Exposure

Workers involved in extracting and drilling operations to obtain coal are exposed to coal dust. The risk of coal dust-induced lung disease development is related to the length of time of exposure (job tenure), the concentration of dust in the environment, and the rank of the coal. Coal dust exposure in the work place in high concentrations leads to the development of disabling and deadly pneumoconiosis. Studies from the 1960s reported the presence of Coal Workers' Pneumoconioses or CWP as involving as many as 46% of coal miners. In 1969, the Coal Mine Health and Safety Act was passed which mandated a reduction in dust levels in underground mines to 2 mg/m^3. As a result of this legislation, there have been reductions in dust concentrations with corresponding reductions in coal dust-induced lung disease. However, it is still estimated that exposure to 2 mg/m^3 of coal dust over 40 years of work can lead to an approximately 9–10% risk of developing simple Coal Workers' Pneumoconiosis (CWP) and an approximately 1–2% risk of developing Progressive Massive Fibrosis (PMF) (66). In addition to length of time of exposure and dust concentrations, the rank of the coal is an independent risk factor for the development of coal dust-induced lung disease. Coal rank is related to the formation of coal and is often measured by the percentage of carbon in the coal. Anthracite coal has the highest percentage of carbon and is associated with the greatest risk for development of disease.

7.3.2 Pathophysiology

The most common coal dust-induced lung disease is Coal Workers' Pneumoconiosis or CWP (see below). The characteristic pathological lesion associated with CWP is the macule which has been defined by the College of American Pathologists as "a focal collection of coal dust-laden macrophages at the division of the respiratory bronchioles that may exist within the alveoli and extend into the peribronchiolar interstitium with associated reticulin deposits and focal emphysema" (67). Macules can range in size from 1 to 5 millimeters in diameter (see Fig. 4.6).

Progressively larger lesions, nodules, can be found in the lung of coal dust exposed workers. With time and continued exposure to coal dust, nodules can coalesce to form masslike lesions which are the hallmark of Progressive Massive Fibrosis or PMF (see

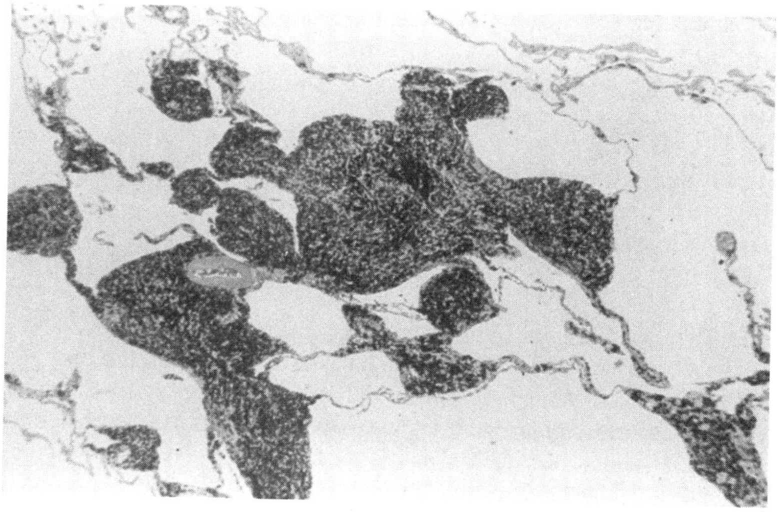

Figure 4.6. Lung histology showing a coal macule.

below). These lesions are by definition greater than 10 millimeters or 1 centimeter in diameter. These various, mostly rounded lesions usually exist within the upper zones of the lung. Nodules and the masses of PMF are seen in chest radiographs and can be characterized by classification systems such as the ILO Guidelines for the Classification of Radiographs of Pneumoconiosis (15). Macules, nodules, and masses can lead to pathological changes within the interstitium of the lung causing degenerative changes including calcification, blood vessel obliteration, and collagen deposition. The affected lung substance becomes stiff and restricted in movement. Gas exchange (the primary function of the lungs) can become altered with the development of respiratory limitation, disability and in severe cases, death.

In addition to macular changes, coal dust exposure is also associated with emphysema. Emphysema is the loss of elastic recoil of the alveoli with subsequent trapping of air and the concomitant decrease in air exchange. Furthermore, there may be airway changes characterized by increases in the number and size of mucous glands within the walls of the airways (68). These findings of emphysema and airway involvement seen in pathologic samples correlate with the physiological findings of obstructive airways disease, independent of the effects of smoking.

7.3.3 Coal Workers' Pneunoconiosis

Studies from 1960s reported the presence of Coal Workers' Pneumoconiosis or CWP in as many as 46% of coal miners (69). As stated previously, greater prevalence of CWP is related to longer times of exposure, higher rank of coal (anthracite > bituminous), and greater concentrations of dust. The diagnosis of CWP is made in the presence of an appropriate exposure history and by characterization of the radiographic presence of nodules (usually rounded and located in the upper lung zones) using a classification system like the ILO Classification. CWP is the reaction of the lung to chronic inhalation of coal dust which over time has overwhelmed the normal defense mechanisms of the lung (cough, mucociliary escalator, and alveolar macrophages) with the formation of macules, nodules and in severe cases masslike lesions (70).

The term "Black Lung" is often used in describing the lung diseases associated with coal dust exposure. Black Lung is occasionally classified as a lay definition referring to any lung disease or disorder associated with work in coal mining. The legal term as used in the Coal Mine Health and Safety Act is similar to this lay definition. CWP refers to the interstitial lung disease that occurs as a result of the accumulation of coal dust within the lung.

Using the ILO Classification System, workers with CWP can be classified as having Category 1, 2 or 3 disease. These Categories are based upon the presence of small opacities with greater number of opacities associated with a higher category of classification. The presence of large opacities represents Progressive Massive Fibrosis or PMF (see below).

Miners with CWP but without PMF may have few signs or symptoms and no physiological abnormalities such as reductions in lung function testing. However, the presence of even Category 1 CWP should be a cause for concern because with continued exposure, these individuals are at risk for progressing to higher categories of CWP and also to PMF.

7.3.4 *Progressive Massive Fibrosis*

With time and continued exposure to coal dust (especially higher concentrations [>2 mg/m^3] and higher rank coal), miners can progress from simple CWP to Progressive Massive Fibrosis (PMF). The hallmark of PMF is the presence of mass-like lesions or large opacities (>1 centimeter) seen on plain chest radiographs. These lesions (like the rounded smaller nodules) tend to be in the upper lung zones and can be unilateral or bilateral and may be rounded or irregular in shape. It may be difficult in some instances to distinguish these lesions from lung carcinoma which may have a similar radiographic appearance.

In contrast to simple CWP with minimal signs, symptoms or physiological abnormalities, workers with PMF can have cough, sputum production, shortness of breath, and chest pain. Impairment can range from minor problems to severe disability and in some cases to death.

The physiological changes associated with PMF include restrictive pulmonary physiology with reduction in lung volumes, reduction in gas transfer capability or diffusing capacity, exercise limitations, and in some cases cardiac disability of right ventricular enlargement and right ventricular dysfunction (cor pulmonale).

There is no treatment for the pneumoconiotic conditions of CWP or PMF other than avoidance of further dust exposure. In severe cases, lung transplantation has been offered for some individuals.

7.3.5 *Obstructive Lung Disease*

In addition to the interstitial pneumoconiotic diseases of CWP and PMF, coal dust exposure has been linked with the development of obstructive lung disease. Early studies showing presence of airways obstruction were criticized for not completely eliminating the confounding factors of smoking (a well-known cause of obstructive lung disease) and asthma (an obstructive lung disease that can occur in the general population). Recent reviews however, have found a significant correlation between coal dust exposure and obstructive lung defects. In one recent review (71), the statement is made that "the balance of evidence points overwhelmingly to impairment of lung function from exposure to coal mine dust, and this is consistent with the increased mortality from Chronic Obstructive Pulmonary Disease (COPD) that has been observed in miners".

7.4 Silica-Crystalline

7.4.1 *Exposure*

Silica or silicon dioxide (SiO_2) can exist in a fixed or crystalline pattern as opposed to a random molecular arrangement defined as amorphous. It is exposure to respirable concentrations of the crystalline form that is associated with the development of disease. Specifically, the crystalline forms of silica are primarily quartz, tridymite, and cristobalite. The most common forms of crystalline silica are found in common sand and sandstone. In addition, crystalline silica may exist as minute grains cemented with amorphous silica and these composites include tripoli, flint, chalceony, agate, onyx, and silica flour. Exposure to respirable crystalline silica occurs in individuals engaged in the occupations of mining, quarrying, drilling, and tunneling. Other occupations with exposure to silica include sandblasters, stonecutters, pottery workers, foundry workers, and refractory brick workers. Concern also exists for individuals exposed to silica flour (72) and for drillers in surface coal mining who are involved in removing the overburden to gain access to the coal (73).

As with exposure to other dusts, the risks involved with development of disease are the length of time of exposure and the concentration of the silica to which the worker is exposed. The current U.S. Occupational Safety and Health Administration (OSHA) exposure limit for respirable silica (Permissible Exposure Limit or PEL) is based on the measurement of respirable dust and on the percent silica content of the respirable dust (74). After chemical analysis of the respirable dust to determine the percent silica (%SiO_2), the PEL is calculated as:

$$\text{PEL} = (10\ \text{mg/m}^3)/(\%\text{SiO}_2 + 2)$$

The measured 8-hr time weighted average (TWA) concentration of respirable dust is compared to this PEL to determine if the concentration exceeds the exposure limit.

A study of gold miners from South Dakota predicted that exposure to 0.09 mg/m^3 over a 45-year period of time would result in a 47% lifetime risk of silicosis (the fibronodular interstitial disease of the lung associated with deposition of silica within the lung) (75). The Recommended Exposure Limit or REL by the National Institute for Occupational Safety and Health (NIOSH) for quartz, tridymite, and cristobalite is 0.05 mg/m^3.

Recent studies have examined alternatives to the use of silica sand for abrasive blasting. Although these studies have shown effectiveness of these alternative materials in their use as abrasives, silica-containing products are still used.

In contrast to exposure to coal dust and to asbestos fibers where the interstitial lung disease occurs only after a minimum of 15 to 20 years of exposure, some forms of silicosis (acute and accelerated silicosis [see below]) can occur in only a few years or less under circumstances of intense exposure.

7.4.2 Pathophysiology

The hallmark of silica-induced disease is the presence of the silicotic nodule (see Fig. 4.7). This nodular lesion contains a central acellular zone with extracellular silica particles surrounded by whorls of collagen and fibroblasts and an active peripheral zone composed of macrophages, fibroblasts, plasma cells, and additional extracellular silica. The formation of these nodules is thought to occur because of surface properties of the silica particles that activate lung macrophages with the subsequent release of chemotactic and inflammatory mediators in turn resulting in further cellular inflammatory responses. It is thought that freshly fractured silica is more toxic because of reactive radical groups found on the freshly fractured surface.

As with coal dust-induced disease, there is a predilection for these nodules within the upper zones of the lung. With continued exposure, there are increasing numbers of these nodules with coalescence resulting in larger masses. In addition to involvement within the lung parenchyma and interstitium, silica can also cause pathological changes within bron-

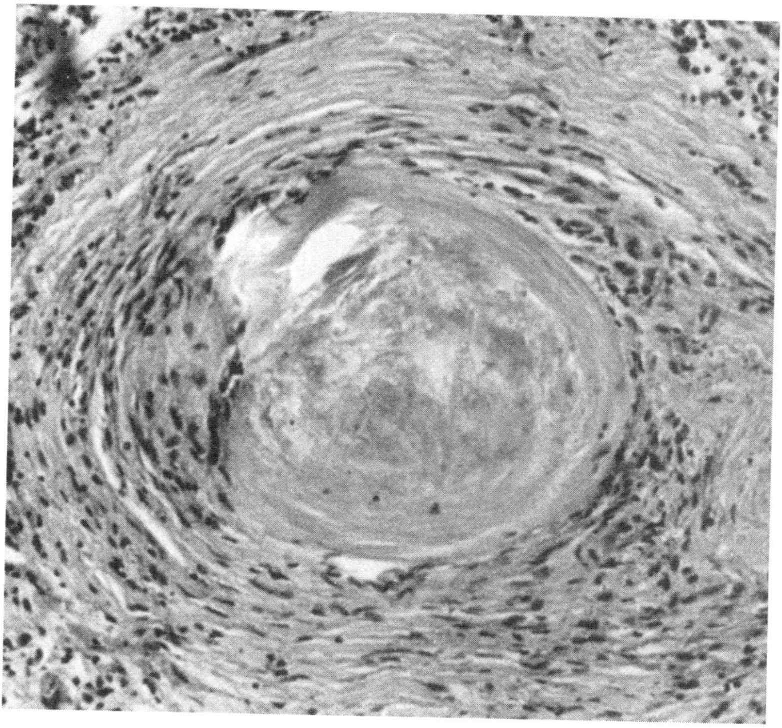

Figure 4.7. Lung histology showing a silicotic nodule.

chial associated lymphoid tissue (BALT) and within lymph nodes. Silicotic nodules can also be present in other organs within the body outside of the lung including the liver, brain, and kidney. In addition to affecting lung structure and lung physiology, silica-induced disease can alter the body's immune system increasing the likelihood of certain infectious diseases such as tuberculosis and autoimmune diseases such as progressive systemic sclerosis (76).

7.4.3 Chronic Silicosis

Chronic or Classic Silicosis occurs after 15–20 years or more of exposure to respirable crystalline silica. The presence of this form of silicosis is recognized by the appearance of the multiple small nodules on a chest radiograph. The classification of these changes seen on the chest radiograph can be made using the International Labour Organization Guidelines (15). Recent studies have examined the role of other radiographic evaluation for workers with silicosis including the use of Computed Tomography (CT) of the chest. CT examination of the chest can be more sensitive than plain chest radiography for the identification of the presence of nodules and associated emphysematous changes (77, 78). However, further studies will be required before CT examination becomes the preferred study of choice for dust-induced lung disease.

If the lesions of silica-induced lung disease remain as only individual small nodules (<5 millimeters in size), there is usually minimal if any physiological changes associated with the disease and, therefore, individuals usually have few if any symptoms. In this case, the disease is classified as Simple Silicosis.

If there is coalescence of the nodules over time with the formation of large masses (large opacities on the chest radiograph of >10 millimeters in diameter), the disease is then referred to a Progressive Massive Fibrosis or PMF (similar terminology to the changes with coal dust-induced disease) or Complicated Silicosis. In this case, individuals with PMF can have reductions in lung function with restrictive-type pathophysiology and decreases in diffusing capacity measurements. The worker may have significant respiratory and systemic symptoms of shortness of breath, productive cough, chest discomfort, with weight loss. In contrast to coal dust-induced disease (especially simple CWP), silica-induced lung disease can progress once exposure ends (79, 80).

7.4.4 *Accelerated Silicosis*

The pathologic and radiographic changes of silicosis can occur after only 5–10 years of exposure to higher concentrations of respirable crystalline silica. Progression occurs in these cases even if the worker is removed from further exposure. The association of rheumatological diseases such as Progressive Systemic Sclerosis (see below) can be found more commonly with accelerated silicosis as compared to chronic or classic silicosis. Accelerated silicosis can be fatal.

7.4.5 *Acute Silicosis*

If the concentrations of respirable crystalline silica are extremely high, disease can occur in only a few months or 1–2 years. Severe, disabling shortness of breath, weakness and weight loss can be seen with Acute Silicosis. An historical example of acute silicosis was the tragic situation of the Hawk's Nest Tunnel construction in the early 1930s near Gauley Bridge, West Virginia. Hundreds of workers died from acute and accelerated silicosis when they were exposed to extremely high concentrations of respirable silica by cutting through three miles of high-quartz sandstone without adequate respiratory protection (81).

In acute silicosis, there is filling of the alveolar spaces of the lungs with a proteinaceous material similar to that found in the disease Pulmonary Alveolar Proteinosis or PAP. This form of silicosis is associated with a relentless progressive course that is frequently fatal with death due to respiratory failure.

7.4.6 *Silico-Tuberculosis*

The association between the presence of silicosis and the development of pulmonary mycobacterial infections including pulmonary tuberculosis, has been known for many years. Recent studies have shown that those workers with chronic or classic silicosis have a 3-fold increased risk of developing both pulmonary and extra-pulmonary tuberculosis when compared to a similar control population of silica-exposed workers without silicosis (82). In addition, this report also suggested that those individuals with long exposures to silica dust were at a greater risk of developing tuberculosis even if they did not have silicosis compared to workers not exposed to silica. For those workers with silicosis, the risk of incidence of tuberculosis is in direct relationship to the radiographic presence of silicotic nodules.

7.4.7 Obstructive Lung Disease

In addition to the development of the fibro-nodular insterstitial disease that is silicosis, workers exposed to respirable crystalline silica dust can also develop chronic obstructive pulmonary disease (COPD). Increased frequency of chronic bronchitis (chronic cough and sputum production, a form of COPD) has been reported in miners of coal, gold, granite, and agate (76). It is felt that the pathophysiological changes contributing to silica-induced chronic lung disease are the occurrence of silicotic nodules in close proximity to small and medium airways with narrowing and distortion of the lumen of these airways. In addition to airway lumen compromise, there may also be emphysematous-type changes especially in association with conglomerate lesions, which can be detected with radiographic CT imaging of the chest. Thus, in many workers exposed to silica who develop silica-induced lung disease, there may be the presence of both obstructive and restrictive-type ventilatory defects as measured with pulmonary function testing.

7.4.8 Lung Cancer from Silica Dust

In October of 1996, a committee of the International Agency for Research on Cancer (IARC) reclassified silica as a Group I substance described as "carcinogenic to humans,", concluding that there is "sufficient evidence of carcinogenicity in humans" (83). Although many studies have had confounding factors such as smoking and exposure to other minerals that are carcinogenic, the balance of evidence indicates that workers with silicosis have an increased risk for lung cancer. It is less clear whether long-term exposure to silica dust without development of silicosis is also a risk factor for lung cancer.

7.4.9 Progressive Systemic Sclerosis

Silica has not only been shown to be associated with diseases of the respiratory system (silicosis, obstructive lung disease, and lung cancer), it is also been shown to be associated with Progressive Systemic Sclerosis or Scleroderma. A recent review concluded that the "systemic sclerosis" associated with silica exposure is indistinguishable from idiopathic Scleroderma (84). Although it is known that silica-exposure can lead to changes within the lymphatic and immune systems with the development of auto-antibodies, the exact mechanism for the development of this systemic disease in silica-exposed workers is unclear.

7.4.10 Comparison of Silica vs. Coal Dust-Induced Lung Disease

Similarities exist between coal dust-induced lung disease and silica-induced lung disease. Specifically, both mineral dusts can lead to the development of small rounded opacities (< 5 millimeters) located in the upper lung zones identified by plain chest radiography. In addition, both may progress to show coalescence of the smaller opacities into larger opacities classified as Progressive Massive Fibrosis or PMF. In both exposure situations, workers can also develop obstructive lung disease. In the past, it was thought that the pneumoconiosis seen in coal dust exposed workers was actually due to the silica content of the dust. However, it has been recognized in recent years that coal itself can lead to the pathological, radiographic, and physiologic changes seen in CWP and in coal dust-induced PMF.

However, despite the similarities between coal dust and silica-induced lung disease, there are significant differences. (*1*) Silica can cause changes in lymphoid tissue with involvement of mediastinal lymph nodes as well as bronchial associated lymphoid tissue (BALT); (*2*) silica can lead to systemic involvement with silicotic nodules present in different extra-pulmonary tissue; (*3*) silica has been associated with the development of systemic diseases such as Progressive systemic Sclerosis or Scleroderma; (*4*) silicosis and silica exposure can be associated with the development of mycobacterial disease such as tuberculosis; (*5*) silicosis can be associated with the development of lung carcinoma; (*6*) there are more rapidly progressive forms of silicosis (accelerated and acute silicosis) that can develop in months or a few year; (*7*) many cases of silicosis can progress after removal from further exposure. Unless the worker has PMF from coal dust exposure, it is unusual that coal dust-induced simple pneumoconiosis (CWP) will progress after removal from further exposure.

BIBLIOGRAPHY

1. *Occupational Lung Diseases: An Introduction*, American Lung Association, New York, 1979, P. 10(5).
2. American Thoracic Society, "Standardization of Spirometry—1994 Update," *Am. J. Respir. Crit. Care. Med.* **152**, 1107–1136 (1995).
3. American Thoracic Society, "Single-Breath Carbon Monoxide Diffusing Capacity (Transfer Factor)," *Am. J. Respir. Crit. Care Med.* **152**, 2185–2198 (1995).
4. American Thoracic Society, "Guidelines for Methacholine Challenge Testing," *Am. J. Respir. Crit. Care Med.* (1999), in press.
5. American Thoracic Society, "Lung Function Testing: Selection of Reference Values and Interpretative Strategies," *Am. Rev. Respir. Dis.* **144**, 1202–1218 (1991).
6. M. R. Becklake, "Concepts of Normality Applied to the Measurement of Lung Function." *Am. J. Med.* **80**, 1158–1164 (1986).
7. P. L. Enright et al., "Spirometry in the Lung Health Study. 1. Methods and Quality Control," *Am. Rev. Respir. Dis.* **143**, 1215–1223 (1991).
8. J. Hankinson et al., "Spirometric Reference Values from a Sample of the General U.S. Population," *Am. J. Respir. Crit. Care Med.* **159**, 179–187 (1999).
9. S. T. Weiss et al., "Relation of FEV_1 and Peripheral Blood Leukocyte Count to Mortality. The Normative Aging Study," *Am. J. Epidemiol.* **142**, 493–498 (1995).
10. R. O. Crapo, "Pulmonary Function Testing," *New Eng. J. Med.* **331**, 25–30 (1994).
11. National Institutes of Health, *Guidelines for the Diagnosis and Management of Asthma*. NIH Publication No. 97-4051, 1997.
12. A. Miller, "Single Breath Diffusing Capacity in a Representative Sample of the Population of Michigan, a Large Industrial State," *Am. Rev. Respir. Dis.* **127**, 270–277 (1983).

13. R. S. Fraser et al., *Synopsis of Disease of the Chest*, 2nd ed., W.B. Saunders Company, Philadelphia, 1994.
14. G. Wagner et al., "Chest Radiography in Dust-Exposed Miners: Promise and Problems, Potential and Imperfections," *Occup. Med.* **8**, 127–141 (1993).
15. International Labour Organization, *ILO 1998 International Classification of Radiographs of the Pneumoconioses*, ILO, Geneva, Switzerland, 1999.
16. W. N. Rom et al., "Cellular and Molecular Basis of the Asbestos-related Diseases," *Am. Rev. Respir. Dis.* **143**, 408–422 (1991).
17. D. Vanhee, "Cytokines and Cytokine Network in Silicosis and Coal Workers' Pneumoconiosis," *Eur. Respir. J.* **8**, 834–842 (1995).
18. D. Vanhee, "Mechanisms of Fibrosis in Coal Workers' Pneumoconiosis. Increased Production of Platelet-derived Growth Factor, Insulin-like Growth Factor Type I, and Transforming Growth Factor beta and Relationship to Disease Severity," *Am. J. Respir. Crit. Care Med.* **150**, 1049–1055 (1994).
19. NIOSH, *Workplace Exposure to Asbestos*, (1980), pp. 1–39.
20. NIOSH, *Asbestos Bibliography (Revised)*, 1997 pp. 1–212.
21. B. Begin, J. M. Samet, and R. A. Shaikh, "Asbestos," in Harber, Schenker, and Balmes, eds., *Occupational and Environmental Respiratory Disease* 1996 pp. 293–329.
22. J. M. Stellman ed., *Encyclopaedia of Occupational Health and Safety*, 4th Ed., Geneva, International Labour Office, Vol. 1. Respiratory System (D. Alois and G. R. Wagner eds., Chapt. 10), 1998 pp. 10.1–10.97.
23. W. N. Rom, "Asbestos-Related Diseases," in *Environmental and Occupational Medicine*, chapt. 24 pp. 349–375.
24. National Bureau of Standards (NBS). "Workshop on Asbestos: Definitions and Measurement Methods," U.S. Dept of Commerce/NBS Special Publication 506, 1977.
25. J. L. Wright et al., "Diseases of the Small Airways," *Am. Rev. Respir. Dis.* **146**, 240–262 (1992).
26. M. R. Becklake, "The Epidemiology of Asbestosis," in Liddell and Miller, eds., *Mineral Fibers and Health*, CRC Press, Boca Raton, FL, 1991.
27. M. R. Becklake, "Occupational Exposure and Chronic Airways Disease," in *Environmental and Occupational Medicine*, Little, Brown & Co., Boston, 1992, Chapt. 13.
28. R. Doll and J. Peto, in Antman and Aisner, eds., *Asbestos-Related Malignancy*, Grune & Stratton Orlando, FL, 1987.
29. I. J. Selikoff, "Historical Developments and Perspectives in Inorganic Fiber Toxicity in Man," *Environ Health Persp* **88**, 269–276 (1990).
30. M. R. Becklake, "Pneumoconioses." in Murray and Nadel, eds., *A Textbook of Respiratory Medicine* WB Saunders, Philadelphia, 1994, Chapt. 66.
31. Bureau of Mines, "Selected Silicate Minerals and their Asbestiform Varieties: Mineralogical Definitions and Identification-Characterization." U.S. Dept of Interior/Bureau of Mines Information Circular 8751, 1977 pp. 1–56.
32. NIOSH, *Occupational Respiratory Diseases*, 1986, pp. 287–342.
33. W. A. Burgess, *Recognition of Health Hazards in Industry*, 2nd ed., 1995, pp. 409–482,
34. G. J. Kullman. "Occupational Exposures to Fibers and Quartz at 19 Crushed Stone Mining and Milling Operations," *Am. J. Indust. Med.* 641–660, (1995).
35. EPA, *Guidance for Controlling Asbestos-Containing Materials in Buildings*, 1985, pp. 1-1 to 10-1

36. *Code of Federal Regulations*, **29** *CFR* 1910.1001, U.S. Government Printing Office, *Federal Register*, Washington DC 1995.
37. *Code of Federal Regulations*, **30** *CFR* 56.5001, U.S. Government Printing Office, *Federal Register*, Washington DC 1995.
38. NIOSH, *Recommendations for Occupational Safety and Health: Compendium of Policy Documents and Statements*, U.S. Department of Health and Human Services, Public Health Service, Centers for Disease Control, National Institute for Occupational Safety and Health, DHHS (NIOSH) Publication No. 92-100, Cincinnati, OH, 1992.
39. American Conference of Governmental Industrial Hygienists, 1998 TLVs® and BEIs®: *Threshold Limit Values for Chemical Substances and Physical Agents and Biological Exposure Indices*, Cincinnati, OH, 1998.
40. W. N. Rom, "Asbestos-related Lung Disease," *Fishman's Pulmonary Diseases and Disorders*, 3rd ed. 1998, pp. 877–891.
41. G. R. Epler, T. C. McLoud, and E. A. Gaensler, "Prevalence and Incidence of Benign Asbestos Pleural Effusion in a Working Population," *JAMA* **247**, 617–622 (1982)
42. L. Rosenstock, S. Barnhart, N. J. Heyer, D. J. Pierson, and L. D. Dudson, "The Relation Among Pulmonary Function, Chest Roentgenographic Abnormalities, and Smoking Status in an Asbestos-Exposed Cohort," *Am. Rev. Resp. Dis.* **138**(2), 272–277 (1988).
43. D. A. Schwartz, L. J. Fuortes, J. R. Galvin, L. F. Burmeister, L. E. Schmidt, B. N. Leistikow, F. P. Lamarte, and J. A. Merchant, Asbestos-induced Pleural Fibrosis and Impaired Lung Function, *Am. Rev. Respir. Dis.* **141**(2) 321–326 (1990).
44. S. A. Levy, "Asbestosis," *Occup. Med.* 179–193 (1994).
45. Consenus Report, "Asbestos, Asbestosis, and Cancer: the Helsinki Criteria for Diagnosis and Attribution," *Scand. J. Work Environ. Health* **23**, 311–316 (1997).
46. P. C. Elms, "Inorganic Dusts," *Hunter's Diseases of Occupations*, 8th ed. pp. 410–457, 1994, pp. 410–457.
47. V. R. Doll, "Mortality from Lung Cancer in Asbestos Workers," *Br. J. Ind. Med.* **12**, 81–86 (1955).
48. W. McDonald, J. C. Liddell, F. D. Gibbs, G. W. Eyssen and G. E. McDonald, "Dust Exposure and Mortality in Chrysotile Mining 1919–1975," *Br. J. Ind. Med.* **37**, 11–24 (1980).
49. I. J. Selikoff and H. Seidman, "Asbestos-associated Deaths Among Workers in the United States and Canada, 1967–1987." *Ann. N.Y. Acad. Sci.* **643**, 1–14 (1991).
50. W. Weiss, "Asbestosis: A Marker for the Increased Risk of Lung Cancer Among Workers Exposed to Asbestos," *Chest* **115**, 536–549 (1999).
51. D. E. Banks et al., "Asbestos Exposure, Asbestosis, and Lung Cancer," *Chest* **115**, 320–322 (1999).
52. J. C. Wagner, C. A. Sleggs, and P. Marchand, "Diffuse Pleural Mesothelioma and Asbestos Exposure in the Northwestern Cape Province," *Br. J. Ind. Med.* 1965; **22**, 261–266 (1965).
53. J. E. Craighead, J. Kleinerman et al., "Diseases Associated with Exposure to Silica and Nonfibrous Silicate Minerals," *Arch. Pathol. Lab. Med.* **12**, 673–720 (July 1988).
54. J. E. Lockey, "Man-Made Fibers and Nonasbestos Fibrous Silicates," in Harber, Schenker, and Balmes, eds., *Occupational and Environmental Respiratory Disease* 1996, pp. 330–343.
55. J. F. Gamble, "Silicate Pneumoconiosis," in J. A. Merchant, ed., *Occupational Respiratory Diseases*, DHHS (NIOSH) Publication No 86-102. U.S. Government Printing Office, Washington, DC, 1986.

56. W. N. Rom, "Silicates and Benign Pneumoconioses," *Environ. Occup. Med.* **16**, 345–349 (1974).
57. R. J. Waxweiler, R. D. Zumwalde, G. O. Ness et al., "A Retrospective Cohort Mortality Study of Males Mining and Milling Attapulgite Clay," *Am. J. Ind. Med.* **13**, 305–315 (1988).
58. E. E. McConnel, L. Hall and B. Adkins, "Studies on the Chronic Tonicity (Inhalation) of Wollastonite in Fischer 344 Rats," *Inhalat. Toxicol.* **3**, 323–337 (1991).
59. M. S. Huuskonen, A. Tossavainen, H. Koskinen, et al., "Wollastonite Exposure and Lung Fibrosis," *Environ. Res.* **30**, 291–304 (1983).
60. D. P. Brown and J. K. Wagoner, "Occupational Exposure to Talc Containing Asbestos: III. Retrospective Cohort Study of Mortality," U.S. Dept of HEW. (NIOSH) Pub No 80-15, 1980.
61. M. Kleinfeld, J. Messite and M. H. Zaki. "Mortality Experiences among Talc Workers. A Follow-up Study," *J. Occup. Med.* **16**, 345–349 (1974).
62. W. T. Stille, and I. R. Tabershaw, "The Mortality Experience of Upstate New York Talc Workers," *J. Occup. Med.* **24**, 480–484 (1982).
63. J. F. Gamble, W. Felner, and M. J. Dimeo. "An Epidemiologic Study of a Group of Talc Workers." *Am. Rev. Resp. Dis.* **119**, 741–753 (1979).
64. D. H. Wegman, J. M. Petes, M. G. Boundy, and T. J. Smith, "Evaluation of Respiratory Effects in Miners and Millers Exposed to Talc Free of Asbestos and Silica," *Br. J. Ind. Med.* **39**, 233–238 (1982).
65. E. Wergeland, A. Andersen and A. Baerheim, "Morbidity and Mortality in Talc-Exposed Workers," *Am. J. Ind. Med.* **17**, 505–513 (1990).
66. M. D. Attfield et al., "An Investigation into the Relationship between Coal Workers' Pneumoconiosis and Dust Exposure in U.S. Coal Miners," *Am. Ind. Hyg. Assoc. J.* **53**, 486–492 (1992).
67. J. Kleinerman et al., "Pathology Standards for Coal Workers' Pneumoconiosis. Report of the Pneumoconiosis Committee of the College of American Pathologists to the National Institute for Occupational Safety and Health," *Arch. Pathol. Lab. Med.* **103**, 375–432 (1979).
68. A. N. Douglas et al., "Bronchial Gland Dimensions in Coalminers: Influence of Smoking and Dust Exposure," *Thorax* **37**, 760–764 (1982).
69. R. E. Hyatt et al., "Respiratory Disease in Southern West Virginia Coal Miners," *Am. Rev. Respir. Dis.* **89**, 387–401 (1964).
70. J. Parker et al., "Coal Workers' Lung Disease and Silicosis," *Fishman's Pulmonary Diseases and Disorders*, 3rd ed. 1998 pp. 901–914.
71. D. Coggon et al., "Coal Mining and Chronic Obstructive Pulmonary Disease: A Review of the Evidence," *Thorax* **53**, 398–407 (1998).
72. D. E. Banks et al., "Silicosis in Silica Flour Workers," *Am. Rev. Respir. Dis.* **124**, 445–450 (1981).
73. D. E. Banks et al., "Silicosis in Surface Coalmine Drillers," *Thorax* **38**, 275–278 (1983).
74. Office of the Federal Register, **29** *CFR* (United States Code of Federal Regulations) 1910.1000. National Archives and Records Administration, U.S. Government Printing Office, Washington, DC 1998.
75. K. Steenland et al., "Silicosis among Gold Miners: Exposure–Response Analyses and Risk Assessment," *Am. J. Public Health* **85**, 1372–1377 (1995).
76. American Thoracic Society, "Adverse Effects of Crystalline Silica," *Am. J. Respir. Crit. Care Med.* **155**, 761–765 (1997).

77. C. J. Bergin et al., "CT in Silicosis: Correlation with Plain Films and Pulmonary Function Tests," *Am. J. Roentgenol.* **146**, 477–483 (1986).
78. G. Gamsu, "Computed Tomography and High-Resolution Computed Tomography of Pneumoconioses," *J. Occup. Med.* **33**, 794–796 (1991).
79. K. Kreiss et al., "Risk of Silicosis in a Colorado Mining Community," *Am. J. Ind. Med.* **30**, 529–539 (1996).
80. E. Hnizdo et al., "Risk of Silicosis in a Cohort of White South African Gold Miners," *Am. J. Ind. Med.* **24**, 447–457 (1993).
81. J. K. Corn, "Historical Aspects of Industrial Hygiene, II. Silicosis," *Am. Ind. Hyg. Assoc. J.* **41**, 125–133 (1980).
82. R. L. Cowie, "The Epidemiology of Tuberculosis in Gold Miners with Silicosis," *Am. J. Respir. Crit. Care Med.* **150**, 1460–1462 (1994).
83. International Agency for Research on Cancer, "Silica, Some Silicates, Coal Dust and *para*-Aramid Fibrils," IARC Monograph Evaluating Carcinoma Risk, Vol. 68, 1997.
84. M. H. Rustin et al., "Silica-associated Systemic Sclerosis is Clinically, Serologically and Immunologically Indistinguishable form Idiopathic Systemic Sclerosis," *Br. J. Dermatol.* **123**, 725–734 (1990).

CHAPTER FIVE

Man-Made Mineral Fibers

Jaswant Singh, Ph.D., CIH and Michael A. Coffman, CIH

1 INTRODUCTION

Concerns about the adverse health effects of exposure to asbestos have prompted widespread removal of asbestos-containing materials, resulting in the increased use of substitutes composed of both naturally occurring and synthetic materials. Man-made mineral fiber asbestos substitutes are mineral fibrous materials such as fibrous glass, rock wool, slag wool, and refractory (ceramic) fibers.

Because of the similarity of the chemical composition and morphology of these substitute fibrous materials to those of asbestos, serious questions have been raised about their health implications. In particular, there is growing concern whether such substitutes pose a carcinogenic risk similar to that of asbestos. These health concerns are even more pronounced considering that man-made mineral fibers (MMMF) have found wide applications in commerce, in addition to their use as asbestos substitutes in the building industry.

1.1 Terminology

There is a variety of commercially produced and used fibers, both naturally occurring and man-made (Fig. 5.1). Artificial fibers are fibers in which the fiber-forming material is of vegetable or animal origin. These fibers include viscose rayon, cellulose ester, and protein fibers.

Organic fibers are fibers in which the fiber-forming material is derived from monomeric organic compounds. Examples of organic fibers are polyamides, polyesters, polyacrylonitrile, and polytetrafluoroethylene (Teflon).

Carbon and graphite fibers are considered synthetic organic fibers because they are commercially produced by high-temperature processing of organic precursors such as

Patty's Industrial Hygiene, Fifth Edition, Volume 1. Edited by Robert L. Harris.
ISBN 0-471-29756-9

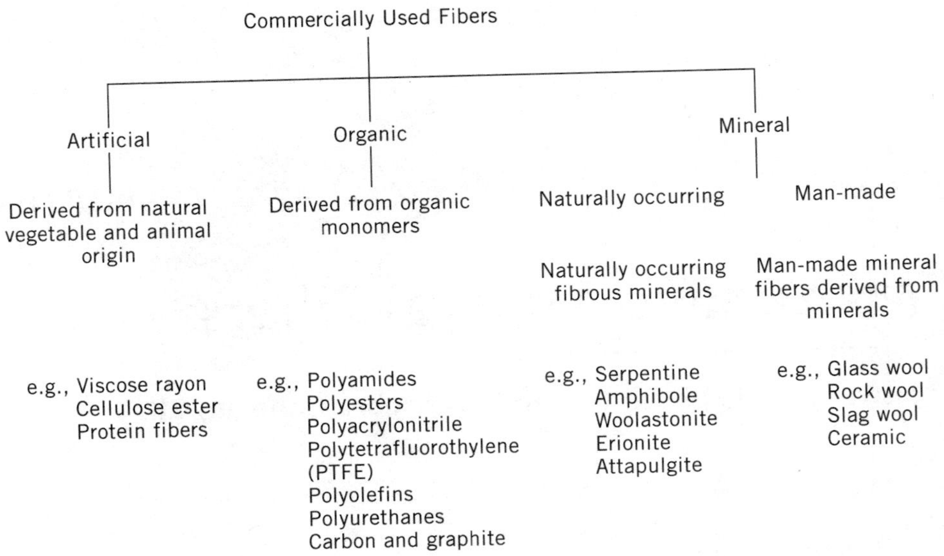

Figure 5.1. Commercially used fibers.

rayon, pitch, or polyacrylonitrile. The terms carbon and graphite are often used interchangeably but there are important differences. Graphite fibers are generally stronger and stiffer because of their polycrystalline nature. Carbon/graphite fibers are used as reinforcing materials in composites for the aerospace and automobile industries. Although manufactured to nominal diameters of 5 to 8 μm, a considerable portion of these fibers may exist as respirable fibers (less than 3 μm in diameter).

Naturally occurring mineral fibers such as serpentine (chrysotile) and amphiboles (amosite and crocidolite) have been extensively used in commerce as thermal and acoustical insulation products. Other commonly known, naturally occurring mineral fibers are woolastonite, erionite, and attapulgite. Woolastonite ($CaSiO_3$) is widely used in the ceramic industry, in paints, plastics, and abrasives, and in metallurgy. Erionite is one of the two naturally occurring zeolites known to occur in the fibrous form. The main use of attapulgite is as an absorbent for oil and grease and pet wastes. Attapulgite is also used in drilling muds, fertilizers, cosmetics and pharmaceutical products.

MMMF, the main topic of this chapter, are further divided into three categories; fibrous glass, mineral wool, and refractory/ceramic fibers.

1. *Fibrous glass* includes glass wool, continuous filament, and special-purpose fibers. These materials are typically composed of oxides of silicon, calcium, sodium, potassium, aluminum, and boron. The main raw materials are silica sand, limestone, fluorspar, boron oxide, and glass fragments (cullet). The typical composition of glass wool and other MMMF is shown in Table 5.1.

Table 5.1. Composition (% by Weight) of MMMF[a,b]

Component	Glass Wool	Rock Wool	Slag Wool	Ceramic Fiber
SiO_2	34–73	45.5–52.9	40.6–41.0	0–53.9
Al_2O_3	2.0–14.5	6.5–13.4	11.8–12.5	0–95.0
MgO	3.0–5.5	—	—	0–0.5
CaO	5.5–22.0	10.8–30.3	37.5–40.0	0–0.7
FeO	—	1.0–5.8	0.9–1.0	—
B_2O_3	3.5–8.5	—	—	0–14
Na_2O	0.5–16.0	2.3–2.5	0.2–1.45	0–0.2
K_2O	0.5–3.5	1.0–1.6	0.3–0.4	0–0.1
TiO_2	0–8.0	0.5–2.0	0.4–0.44	0–1.6
Z_1O_2	0–4.0	—	—	0–92
PbO	0–59.0	—	—	—
Fe_2O_3	—	0.5–8.2	—	0–0.97
Y_2O_3	—	—	—	0–8

[a]From Ref. 2.
[b]Some of the mineral fibers shown also contain small percentages of other components including P_2O_5, CaS, S, F, MnO.

2. *Mineral wool* includes both rock wool, derived from magma rock, and slag wool, made from molten slag produced in metallurgical processes such as the production of iron, steel, or copper. The main components of rock wool and slag wool are oxides of silicon, calcium, magnesium, aluminum, and iron (Table 5.1). Some researchers have also included glass wool under the term mineral wool.
3. *Refractory/ceramic fibers* include a wide range of amorphous or partially crystalline materials made from kaolin clay or oxides of aluminum, silicon, or other metal oxides. Some ceramic fiber products developed for special applications contain little or no aluminum oxide or silicon dioxide. For example, zirconia fibers contain mainly oxides of zirconium and yttrium. Less commonly, refractory fibers are also made from nonoxide refractory materials such as silicon carbide (SiC), silicon nitride (Si_3N_4), or boron nitride (BN).

1.2 Historical Data

Glass fibers were reportedly used as early as 2000 B.C. (1). Early Egyptians are reported to have used coarse glass fibers for decorative purposes. The commercial use of MMMF was marked by the awarding of a patent to a Russian in 1840 for the preparation of fibrous glass. During this same time, rock wool/slag wool was first produced in Wales (2). Production of fibrous glass and slag wool was accelerated during World War I when the allied blockade of Germany created an asbestos shortage. In the United States in 1938, the formation of Owens Corning Fiberglass Company resulted in increased use and production of glass fibers. In the same year, German production of mineral wool reportedly reached 15,000 tons. In 1951, a U.S. patent was issued to Carborundum Corporation for its high-melting Fiberfrax *f*. The commercial production of ceramic fibers began in the early 1970s and has seen consistent growth, reaching a level of over $100 million U.S. sales by the mid-1980s.

The chronology of the production of MMMF is presented in Table 5.2.

2 PRODUCTION AND USES OF MMMF

2.1 Mineral Wool

In the production of mineral wool, the raw material (slag and/or rock) is loaded into a cupola furnace in alternating layers with coke and small amounts of other ingredients to impart special characteristics of ductility or size to the fibers. Approximately 70% of the mineral wool sold in the United States is produced from blast furnace slag (3). Most of the remainder is produced from copper, lead, and iron smelter slag. A small amount is produced with natural rock (rock wool). Combustion of the coke in the cupola furnace generates high temperatures (about 3000°F). The slag or rock is melted, then discharged from the furnace in a stream, and fiberized. In fiberizing mineral wool, the most commonly used techniques are (*1*) the steam jet process, (*2*) the Powell process, and (*3*) the Downey process. Each of these is described below in greater detail.

Steam Jet Process. Until the early 1940s, nearly all of the mineral wool was produced using a steam jet process. In this process, the molten stream of material discharged from the cupola furnace is fiberized by passing it in front of a high-pressure steam jet (Fig. 5.2). The molten stream of slag or rock is broken up into many small droplets by the steam jets. Fiberization occurs as these droplets are swept out at high velocity in front of the jet.

Powell Process. The Powell process (Fig. 5.3) uses a group of rotors spinning at high speed to collect and distribute the molten stream of slag in a thin film on the rotor surfaces and then fiberizing it by throwing it off the rotors with centrifugal force.

Downey Process. The Downey process (Fig. 5.4) combines a spinning concave rotor with steam jets. The molten stream of slag or rock is distributed in a thin pool over the concave-surface of the rotor. The stream then flows up and over the edge of the rotor where it is swept out by the steam jets and is fiberized.

2.2 Fibrous Glass

Fibrous glass may be produced in two steps (marble melt) or in a single step (direct melt). In the marble melt process, glass is produced in a furnace that fuses the raw materials and homogenizes the melt. The glass melt exits the furnace through a forehearth to a marble making machine. The marbles are eventually remelted to make fibers. In the direct melt process, the molten glass proceeds from the forehearth directly to the fiberizer. Glass is commonly fiberized using one of four techniques: (*1*) extrusion (filimentous glass), (*2*) rotary, (*3*) flame attenuation, and (*4*) spinning (Powell) processes. Each of these processes is described in greater detail except for the spinning process, which was described previously.

Extrusion. In the production of continuous filament textile glass, the molten glass flows from the forehearth into a series of platinum tanks fitted with hundreds of small-diameter orifices. The glass flows through these orifices, and the individual filaments produced are collected together in a strand. A binder is applied and the strand is wound around a rapidly rotating drum (Fig. 5.5).

Rotary Process. Similar to the Downey process described above, glass wool is produced in the spinning process (Fig. 5.6) by flowing the molten glass stream into the bowl of a concave rotor. The molten glass is distributed by centrifugal force to the sidewall of the rotor which contains many small holes. As the glass flows through these holes, it is further fiberized by high-velocity jets around the perimeter.

Flame Attenuation Process. In the flame attenuation process, the molten glass stream is passed through high-velocity gas jets to fiberize coarse primary filaments of glass, which are then fed in front of high-velocity jet flames (Fig. 5.7).

2.3 Ceramic Fibers

Refractory or ceramic fibers are produced by (*1*) blowing and spinning, (*2*) colloidal evaporation, (*3*) continuous filamentation, and (*4*) vapor-phase deposition.

Blowing and Spinning. The blowing and spinning processes are essentially the same as those used for the production of glass and mineral wool fibers. Refractory or ceramic fibers are produced by fusing mixtures of natural minerals (such as kaolin clay), or synthetic blends of alumina, silica, or other metal oxides in an electric furnace. Fibers are formed either by (*1*) passing the molten material through high pressure steam or other hot gas jets (blowing), or (*2*) forcing the molten stream onto rapidly rotating disks that fiberize the stream by throwing it off the disks with centrifugal force.

Colloidal Evaporation. Ceramic fibers of alumina, zirconia, silica, mixtures of zirconia and silica, and thoria have been produced by evaporating a colloidal suspension of these materials.

Vapor-Phase Deposition. Special ceramic fibers composed of nonoxide materials are produced by a vapor-phase deposition technique. In this technique, a volatile compound of the desired material is reduced or decomposed on a resistively heated substrate such as tungston wire. Monocrystalline ceramic materials known as whiskers are also produced by vapor-phase deposition techniques. These materials have high strength and micron-sized diameters. A large number of fibrous materials may be produced this way by using materials such as carbides, nitrides, halides, arsenates, vanadates, and silicates, to name a few.

2.4 Secondary Processes

The secondary or finishing processes are closely linked to specific final products. The secondary processes can present significant worker exposures. These processes include:

- Dust suppression by adding agents such as mineral oils, vegetable oils, or waxes to almost all insulation wools.
- Addition of binders such as natural resins, tars, or synthetic resins. More recently, the phenol–formaldehyde or urea–formaldehyde resins have meant the introduction of curing processes and posed exposure problems to volatile components such as formaldehyde.

Table 5.2. Chronology of the Production of MMMF

	Fiber Type		
Year	Fibrous Glass	Mineral Wool	Ceramic Fibers
2000 B.C.	First reported production of glass fibers. Egyptions used coarse glass fibers for decorative purposes		
1840	Patent for method and apparatus for preparation of fibrous glass using a spinning process awarded to a Russian named Shamo	Rock/slag wool first produced in Wales	
1870		Slag wool first produced in Germany First U.S. patent issued to produce slag wool by blowing molten slag produced in a blast furnace	
1885		First successful commercial production of slag wool	
1893	Edward Drummond Libbey exhibits a dress, lamp shades, and other articles woven of glass fibers at the Columbian Exposition in Chicago		
1895		German production of slag wool reaches 50 tons/year	
1906	Pollack, Pick, Pazsicky, and Bornkessel awarded patents for glass wool manufacture. Patents for fibrous glass also issued in England and Germany during this time		

1915	Allied blockade of Germany in World War I creates asbestos shortage resulting in full scale production of fibrous glass and slag wool	
1925	Spun glass production starts in U.S. by drum-winding process	
1932	Coarse glass fibers first used in an air filter. First fibrous glass thermal insulation used in U.S. Naval vessels	
	First house installed with fibrous glass	
1938	Owens Corning Fiberglas Company formed by Owens Illinos and Owens Corning	German production of mineral wool reaches 15,000 tons/year
1941	Report on fibrous glass health hazard investigation by Walter J. Siebert with the aid of Owens Corning Fiberglass Corp. Evidence of pulmonary disease found	
	Report by Leroy Gardner concludes that exposure to glass wool dust involved "no hazard to the lungs"	
	U.S. Patent Office issues 353 patents for glass wool products	
1942	Baer Sulzberger publishes "The Effects of Fiberglass on Animal and Human Skin in Industrial Medicine." Sulzberger notes skin irritation, but no permanent damage	
1932	First use of matted fibrous glass to reinforce plastic sheet	

Table 5.2. (continued)

Year	Fiber Type		
	Fibrous Glass	Mineral Wool	Ceramic Fibers
1944		Germany has 23 cupolas producing 44,800 tons of slag and rock wool	
1951	Fibrous glass coveralls used by firefighters to withstand temperatures up to 2000°F		U.S. patent issued to Carborundum Corp. for its high melting Fiberfrax
1954			Babco and Wilson issued a patent for producing aluminoborosilicate fiber by fusing kaolin
1963	Johns–Manville assigned U.S. patent for manufacturing fibrous glass mats with a binder for insulation		
Early 1970s			Commercial production of ceramic fibers begins
Mid-1980s			Production reaches 70–90 million
1987	International Agency for Research on cancer assigns glass wool, mineral wool and ceramic fibers to category 2B. Continuous filimentous glass is assigned to category 3		

[a]From Refs. 1 and 2

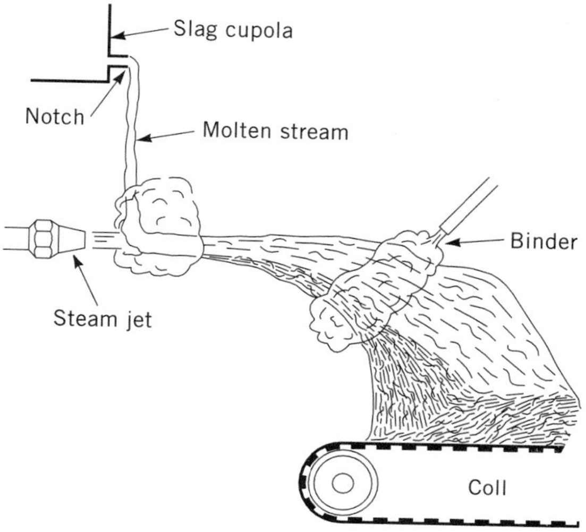

Figure 5.2. Steam-jet fiberization process.

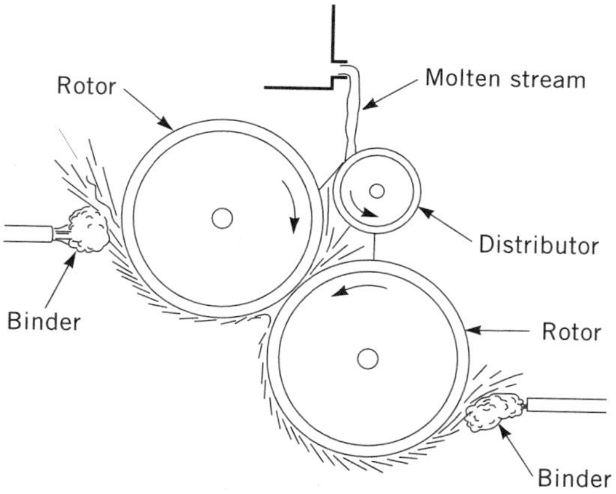

Figure 5.3. The Powell process.

• Other secondary operations include sawing, cutting, packing, and dust removal. In addition, operations such as painting or surface cutting may also impact the work environment.

2.5 Uses of MMMF

Products derived from MMMF have provided great benefits to society in terms of energy conservation, thermal comfort, factory thermal insulation, acoustic insulation, and fire

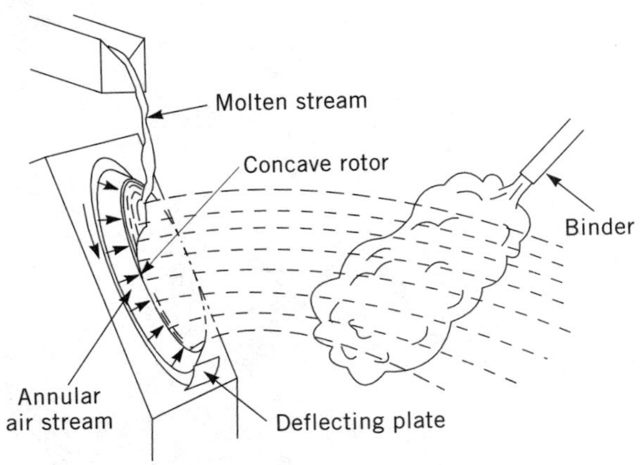

Figure 5.4. The Downey process.

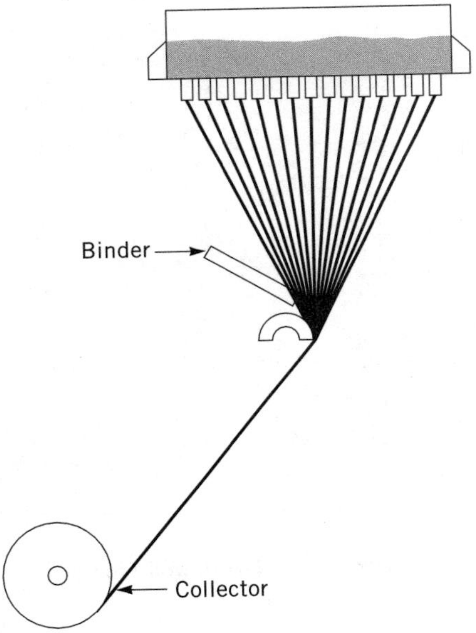

Figure 5.5. Production of continuous filamentous glass.

protection, among others. *Ceramic fibers have been extensively used for lining furnaces and kilns, because of their ability to withstand high temperatures.* Table 5.3 lists some major uses of MMMF. One researcher has identified more than 30,000 products based on MMMF (4).

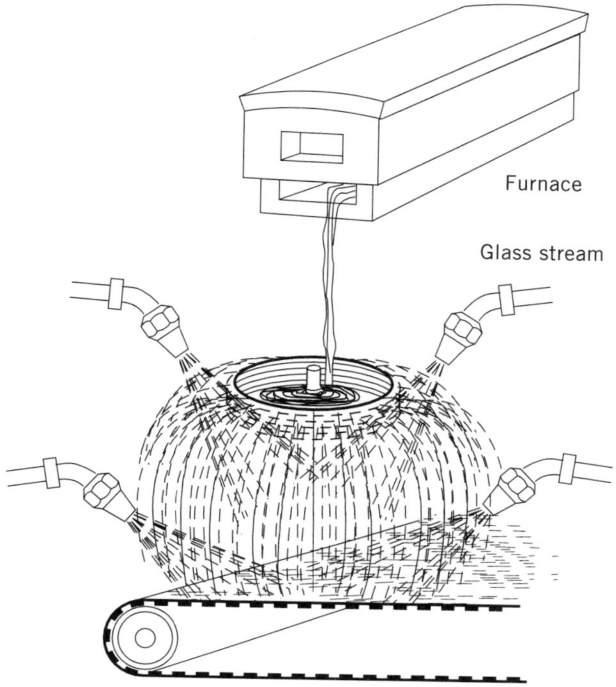

Figure 5.6. Rotary process for glass wool production.

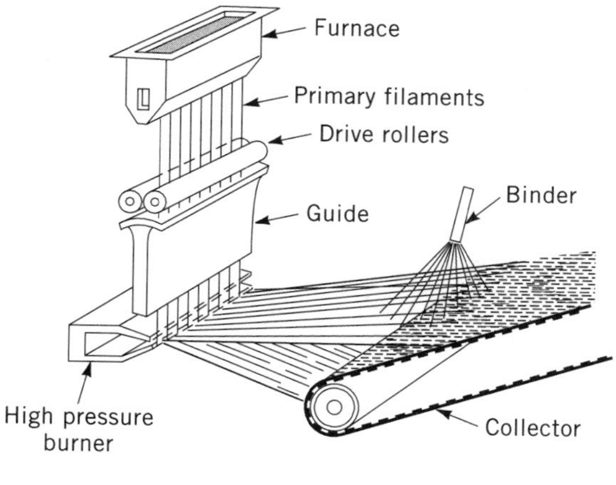

Figure 5.7. Flame attenuation process for producing glass wool.

Table 5.3. Uses of MMMF

Factory thermal insulation
Building system thermal insulation
Acoustic insulation
Fireproofing
Aerospace insulation
Reinforcing material in plastics, cement, and textiles
Automotive components
Fiber optics
Air and liquid filtration
Refractory coatings
Gaskets and seals

Because of a variety of uses to which MMMF have been put, including their use as asbestos substitutes, production of MMMF has been steadily increasing. It is estimated that over five million tons of MMMF are being produced worldwide. Rapid increase in the production of MMMF is evident in the production figures (Table 5.4) for glass fibers in the United States between 1975 and 1986 (6).

3 HEALTH EFFECTS

Since their introduction in commerce almost 100 years ago, MMMF have been known to cause irritation of the skin and, under dusty conditions, irritation of the eyes and throat. Skin irritation is generally associated with fibers that have diameters greater than 4.5 to 5.0 μm (6), sizes commonly found in insulation wools and filamentous glass. Mechanical irritation can result in irritant dermatitis which in general is not severe and does not last long.

Table 5.4. U.S. Glass Fiber Production[a]

Year	Quantity (million kg)
1975	247.88
1976	306.90
1977	357.30
1978	419.04
1979	460.36
1980	393.62
1981	472.61
1982	408.15
1983	530.27
1984	632.88

[a]From Ref. 5

3.1 Nonmalignant Respiratory Diseases

Several epidemiologic studies have addressed the issue of nonmalignant respiratory disease (NMRD) among populations exposed to MMMF. Although some studies have suggested MMMF-related effects on the respiratory system, in general, no statistically significant increase in NMRD mortality has been observed in the studied populations in comparison with local rates (6). The two largest epidemiologic investigations (7, 8) conducted in Europe and the United States showed little evidence of excess mortality from NMRD.

Similarly, in the animal studies to date, there has been little or no evidence of fibrosis of the lungs. In all cases, the tissue reaction in animals exposed to MMMF was much less than the reaction in animals exposed to equal masses of chrysotile or crocidolite asbestos. Researchers have suggested, however, that the number of asbestos fibers reaching the lung may have been greater than MMMF.

3.2 Carcinogenic Effects

3.2.1 Experimental and In Vitro Studies

Numerous studies have been conducted to evaluate the carcinogenic potency of MMMF. These studies may be categorized according to the method by which the subject animals are exposed. In many of the studies, the test animals were exposed via implantation of injection techniques. These procedures involve the direct injection or implantation of MMMF into the body of the test animal, by one of three methods:

- Intrapleural—the material is deposited adjacent to the pleural lining of the lung cavity.
- Intratracheal—the material is deposited into the windpipe (trachea).
- Intraperitoneal—the material is deposited adjacent to the lining of the abdominal cavity.

The carcinogenic potential of MMMF has also been studied using mutagenicity screening or *in vitro* studies. The induction of cancer is thought to proceed in steps. The first of these steps, initiation, involves damage to DNA resulting in heritable alterations or in rearrangements of genetic information (mutations). It is safe to assume that all chemical carcinogens are mutagens. However, not all cellular mutations result in cancer; therefore, not all mutagens are carcinogens. The use of mutagenicity screening has, however, gained widespread acceptance as a rapid and inexpensive method of identifying potential carcinogens.

Fibrous Glass. Animal studies involving the intrapleural or intraperitoneal administration of glass fibers indicate some apparent carcinogenic potential (9–10). In the in vitro studies reported in the literature, the investigators were able to detect morphological changes and chromosomal damage of mammalian cells (11–13). No mutagenicity was observed in tests using bacterial cultures (14).

Long-term inhalation studies, much better indicators of the harmful effects of occupational and environmental exposures to airborne fibers, have not shown the same positive results (15–21). In most of the inhalation studies cited, test animals were exposed to unusually high levels of the fibrous glass (as high as 400 mg/m^3). These levels are considerably higher than typical occupational exposures reported in the literature. The apparent discrepancy between the results of inhalation versus implantation studies may be due to the fact that particle deposition and retention in test animals are different than those in humans. It is, therefore, possible that the particles may never reach the target tissues in sufficient quantities in rodents via the inhalation route (22). The implantation studies are, therefore, justified to reach exposure levels (in rodents), that are possible in humans. The debate about the use of implantation studies continues within the Scientific Community, particularly because of the very different results obtained from the implantation studies versus inhalation. Review of five inhalation studies in rats by the IARC working group on MMMF showed no statistically significant increase in tumor incidence (22). The IARC working group, however, expressed concern about the adequacy of the Inhalation Studies and noted factors such as short exposure period, small number of animals used, lack of survival data and fiber dimensions.

Filamentous Glass. In animal studies involving the intraperitoneal administration of filamentous glass, no statistically significant carcinogenic response was observed (9).

Mineral Wool. In animal studies involving the intrapleural or intraperitoneal administration of mineral wool fibers, no statistically significant increases in the incidence of tumors of lung or pleura were found (16, 19, 23). Similar results were reported for inhalation studies involving mineral wool fibers (15, 16, 19).

Refractory Fibers. In animal studies involving the intrapleural or intraperitoneal administration of ceramic fibers, Smith et al. (19) and Pott et al. (9) reported increased incidence of tumors. No tumors were reported for animals exposed to ceramic fibers via intratracheal administration (19).

The results of animal studies involving the inhalation of ceramic fibers are conflicting. Davis (24) reported a statistically significant increase in lung tumors whereas in two other studies (19, 25), no increase in tumor incidence was reported.

3.2.2 Epidemiologic Studies

Although many studies have been conducted on the carcinogenic effects of MMMF on humans, much of the current state of knowledge about the carcinogenicity of MMMF is derived from three cohort studies conducted in the United States, Europe, and Canada. The three studies were based on a total of 7862 deaths among 41,185 workers in Europe and North America, and provided evidence of cancer mortality over a sufficiently long period of time. These studies indicated an excess of mortality (although not large in relative terms) caused by lung cancer among rock wool/slag wool workers but not a statistically significant increase among glass wool workers. In the United States and European studies, four cases of death from mesothelioma were reported. In view of the large cohort (involving about 800,000 person-years of observation), the incidence of mesothelioma was not considered excessive. Moreover, after further review, the researchers concluded that one of the mesothelioma cases could be attributed to that particular worker's exposure to asbestos in a naval shipyard. Table 5.5 summarizes the standardized mortality ratios (SMRs) for the three epidemiological studies (7, 8, 26). Their conclusions can be summed up as follows.

Table 5.5. Summary of Epidemiologic Studies for MMMF

Investigators	Number of Subjects	Estimated Exposure (fibers/cm^3)	Lung Cancer Mortality Rate		SMR
			Observed	Expected	
		Glass Wool			
Enterline et al. (8) (U.S.)	11,380	0.005 to 0.29	291	267	109
Simonato et al. (7) (Europe)	8,286	(Not reported)	93	91	103
Shannon et al. (26) (Canada)	2,557	<0.2	19	9.5	199
		Mineral Wool			
Enterline et al. (8) (U.S.)	1,601	0.35	60	45	134
Simonato et al. (7) (Europe)	10,115	(Not reported)	81	124	65
		Continuous Filamentous Glass			
Enterline et al. (8) (U.S.)	3,435	0.011	64	69	92
Simonato et al. (7) (Europe)	3,566	(Not reported)	15	97	15

Glass Wool. The Enterline (U.S.) study showed a slightly raised mortality rate from respiratory cancer compared to local rates. This increase, however, was not statistically significant. Moreover, mortality was neither related to the duration of exposure nor to exposure levels. In addition, it was suggested by the mineral fiber industry representatives that the small but statistically significant excess in lung cancer might be due to other factors such as smoking. A follow-up study (27) concluded that taking into account the compounding effects of cigarette smoking reduced the elevation lung cancer SMR in the population studied, to a nonsignificant level similar to that observed for local rates.

The European study also showed no overall excess mortality from lung cancer among glass wool workers as compared to local rates. There was an increase in mortality with time (since first exposure) but this was not statistically significant, and the trend was not related to the duration of exposure. The Canadian study of glass wool workers showed a statistically significant increase in lung cancer mortality but was not related to the duration of exposure or to the time since the first exposure occurred. The Canadian cohort was also much smaller compared to the U.S. and European studies and did not account for smokers in the study pcpulation.

Filamentous Glass. Both the U.S. and the European studies showed no increase in lung cancer mortality among workers exposed to filamentous glass. The observed standardized mortality ratios (SMRs) were lower than expected in both studies.

Mineral Wool. In the U.S. and European studies combined, there was a statistically significant excess of mortality from lung cancer among rock wool/slag wool workers. In the European study, a statistically significant lung cancer rate was found among workers who were first exposed to slag wool/rock wool in the early days of production when the dust levels were relatively high and before the introduction of dust suppressants and binders.

Refractory Fibers. There are no human data available on refractory fibers.

3.3 Factors Affecting Carcinogenic Potency

It is now generally agreed that it is the morphology (fiber shape and size) and *biopersistence* not the chemical composition that drives cytogenic response. Fiber dimension dictates what is inhaled, what is deposited, and which fibers reach the target tissue.

Durability of fibers and their persistence in the body also plays a significant role in determining carcinogenic response. An important question arises on how long fibers have to stay in the bronchial wall or serossa tissue to cause tumors. To persist, the fibers must be chemically durable, although the opposite may not be true; that is, durable fibers may not always be persistent. The durability of fibers depends on their chemical composition and crystal structure. Therefore, even though chemical composition by itself is not considered a direct factor in inducing carcinogenic response, indirectly it may be an important factor because it determines durability.

The morphology and durability factors have led to the suggestion (28) that the carcinogenicity of fibers depends on three Ds: *1*. Dimension; *2*. Dose; and *3*. Durability.

3.3.1 Dimension

In relation to technical products, a fiber is defined as a "long, thin filament of material, possessing the useful properties of high tensile strength and flexibility." Dimensional criteria of a fiber, as defined by the American Society for Testing and Materials (ASTM), is a length-to-width ratio of greater than 10: 1. With respect to airborne concentrations, a fiber is defined as "a particle with an aspect ratio (length to width) of at least 3: 1."

A respirable fiber according to the ILO definition (6) is "a particle with a diameter of less than 3 μ*m* and with an aspect ratio of 3: 1 or greater." In the United States, the National Institute for Occupational Safety and Health (NIOSH) recommends exposure limits for glass fibers that are less than 3.5 μ*m* in diameter and greater than 10 mm long (29). Lippman (30) concluded that mesothelioma is most closely associated with exposure to fibers longer than 5 μ*m*, and lung cancer is most closely associated with fibers longer than 10 μ*m* and with diameters greater than 0.15 μ*m*. It is generally agreed now that fibers of length greater than 5 μm and diameter greater than 1 μm are the most critical determinants of fibrogenic and carcinogenic risk (31).

For technical products, MMMF are manufactured to specific nominal diameters varying according to fiber type and use. Particle size is a major factor in imparting thermal properties to insulating wools. Table 5.6 shows the fiber size range for different fiber types.

Table 5.6 indicates that MMMF in general have much larger diameters compared to asbestos fibers. This fact is even more apparent in Table 5.7 (durability ranking of fibers). Some researchers believe that comparatively larger fiber diameters of MMMF may explain the observed differences in the carcinogenic potency between MMMF and asbestos in both human and animal inhalation studies. The larger MMMF fibers may be deposited in the respiratory tract and may not reach the lung tissue in the same quantities as finer asbestos fibers.

Table 5.6. Nominal Diameters of Some MMMF

Fiber Type	Nominal Diameter (μm)
Refractory (ceramic) fibers (2)	1.2–6.0
Fiberfrax® HSA	1.2
Fibermax® bulk	2–3.5
Fiberfrax® bulk	2–3.0
Alumina bulk	3
Zirconia bulk	3–6
Glass wool (2)	
Thermal insulation	6–15
Molded pipe insulation	7–9
Lightweight aircraft insulation	1.0–1.5
Special purpose	0.05–3.0
Continuous filaments (6)	6–15
Electrical insulation	6–9.5
Nextel® 312	8–12
Mineral wool	6–9

Table 5.7. Durability Ranking of Fibers[a]

Rank	Material	Median Diameter (μm)	Dissolution Velocity (nm/day)	Fiber "Life" (years)
1	Glass wool TEL	3.5	3.45	0.4
2	Glass wool, Superfine	0.38	1.4	1.0
3	Diabase wool	4.0	1.14	1.2
4	Glass wool 475, JM, 104	0.41	0.90	1.7
5	Glass wool E, JM, 104	0.47	0.21	6.5
6	Slag wool	4.8	0.69	2.0
7	Refractory, Fiberfrax R	1.85	0.27	5.0
8	Refractory, Fiberfrax H	1.85	0.28	4.9
9	Refractory, Silica	0.77	1.1	1.2
10	Chrysotile	(0.074)	0.005	(~100)
11	Crocidolite	(0.17)	0.011	(~110)
12	Erionite	(0.005)	0.0002	(~170)

[a]From Ref. 39.

3.3.2 *Durability and Persistence*

There appears to be a growing consensus among researchers that fibers must be durable and persist for a certain time to induce a tumor. There is, however, no precise knowledge or agreement among researchers about the length of time that fibers must persist to alter cells of the bronchial walls or the pleural or peritoneal tissues.

Durability here means the relative resistance of the fiber to dissolution through attack by biological fluids. To persist in the body, a fiber must be chemically durable. On the other hand, a chemically durable fiber may not always persist. The persistence of a fiber may be affected by dissolution, disintegration, elimination, or simple migration in the body. The importance of fiber persistence in inducing tumors has perhaps been best stated by Pott (32):

Only a few seconds of exposure to ionizing radiation can cause damage to cells which will lead to a tumor after years or even decades. With chemical carcinogens, too, a very short exposure time can cause the decisive biological alteration which will be followed by a tumor after a long latent period. A fibre has to be regarded as a physical carcinogen that works by its elongated shape. Clearly, a fiber which is both durable and persistent should have a stronger effect than a nondurable or nonpersistent one.

The concept of biopersistence as a significant factor in fiber toxicity arises from the very different lung retention of natural versus man-made mineral fibers. Studies have indicated that most biopersistant fibers are those with the highest fibrogenic or carcinogenic potential. The difference in the carcinogenicity between chrysotile fibers and the amphiboles (which show greater incidence of mesothelioma) (33), has been attributed to the greater pulmonary retention of amphibole fibres. For all fibres exceeding 10 μm in length, the longer the fibre, the more readily it dissolves.

In general, durability studies indicate that glass and mineral wool fibers are less durable than ceramic fibers and asbestos. Biopersistence has been suggested as one of the determinants of carcinogenic potential in an animal inhalation study related to refractory ceramic fibers (34). Several researchers (35–38) have investigated the relative durability of man-made mineral fibers through in vivo studies. Other researchers (39–41) conducted in vitro experiments to determine the relative resistance of MMMF to attack by biological fluids. These studies are summarized in Table 5.8.

A durability ranking (Table 5.7) by Scholze and Conradt (39) is particularly interesting. Table 5.8 shows the durability of various fiber types expressed by the number of years it will take to "dissolve" a fiber of 1 mm diameter in a simulated extracellular fluid (pH 7.6 ± 0.2) derived from Gamble's solution. These data show striking differences in durability between MMMF and natural fibers (chrysotile, crocidolite). Although MMMF were estimated to dissolve in less than 10 years, natural fibers require approximately 100 years or longer to dissolve. Also noticeable in Table 5.7 are the differences among MMMF. In general, glass wool fibers were found to be the least durable, mineral wools somewhat more durable, and some ceramic fibers the most durable.

Another noticeable feature of Table 5.7 is the difference in the median diameter of MMMF as compared to natural fibers. Because median diameters of 1 μm (used by Scholze and Conradt for comparing the durability of fiber types) are unlikely for natural fibers, the researchers used a multiple of 5 of the average diameter for the natural fibers shown in Table 5.7.

3.3.3 Dose

It is a well-established principle in toxicology that physiological response is related to the dose. Dose is perhaps the most critical factor in explaining the observed differences in the carcinogenic potency of asbestos and MMMF. Exposures of workers in MMMF industries to airborne fibers are generally lower than exposures in similar processes where asbestos

Table 5.8. Summary of Fiber Durability Studies

Investigator(s)	Type of Study	Results
Morgan and Holmes (35)	*In vivo*	Short fibers (≤10 μm) dissolved slowly and uniformly. Longer fibers (≥30 μm) dissolved much more rapidly, and less uniformly
Johnson et al. (36)	*In vivo*	Examination by electron microscopy of fibers removed from rat lung—surface etching observed
Le Bouffant et al. (37)	*In vivo*	Erosion of fiber surfaces observed using electron microscopy. Loss of sodium and calcium noted
Bellmann et al. (38)	*In vivo*	Residence times for different types of fibers: ceramic fiber > rock wool > glass wool > microfiber
		Fibers with a high calcium content dissolve most rapidly
Scholze and Conradt (39)	*In vitro*	Natural and refractory fibers more durable than glass fiber and mineral wool
Leineweber (40)	*In vitro*	High variability observed in the solubility of glass fiber. Chemical composition of the glass appears to be determinant
Klingholz and Steinkopf (41)	*In vitro*	Residence times for different types of fibers: slag wool dissolved more rapidly than glass wool, which dissolved more rapidly than rock wool

is involved. This is mostly due to large fiber diameters of MMMF compared to asbestos. Unlike asbestos fibers, MMMF do not split longitudinally into fibers of small diameter but tend to break transversely into shorter segments. It is postulated that inhaled coarser MMMF may not always reach the target issue to cause comparable (to asbestos) damage.

3.3.3.1 Industrial Workplace Exposures. In the most comprehensive epidemiologic study of MMMF workers conducted to date in the United States, Enterline et al. (8) estimated the average intensity of worker exposures to respirable (less than 3 mm) fibers at 17 glass wool and mineral plants shown in Table 5.9. The data in Table 5.9 show fibrous glass concentrations around 0.03 fibers per cubic centimeter of air (fibers/cm^3). Exposures in the mineral wool plants were higher, but less than 0.5 fibers/cm^3 in all cases.

In general, these results are consistent with those reported by Corn et al. (42) (Table 5.10) for exposure in a rock wool and a slag wool plant. Corn reported higher fiber concentrations in the rock wool plant compared with the slag wool plant. Fiber concentrations in glass wool production reported by European investigators (43, 44) (Table 5.11) also are in general agreement with exposures reported by Enterline. These exposures are much lower than the comparable historical exposures to asbestos. Such low worker exposures to glass wool and continuous filaments led Enterline to conclude that "equivalent exposures (0.03 fibers/cc or less) to asbestos may not produce detectable respiratory cancer."

3.3.3.2 End-User Exposures. Exposures of workers installing acoustical ceiling ducts, attic insulation, and aircraft insulation products have been reported by Esmen et al.(45)

Table 5.9. Worker Exposure to Fibrous Glass and Mineral Wool at Seventeen U.S. Plants[a]

Process and Plant	Average Intensity of Exposure Fibers (<3 μm/cm^3) Mean	Std. Error	Min.	Max.
All fibrous glass plants	0.039	<0.001	0.0	1.500
Plant 1	0.027	<0.001	0.0	0.032
Plant 2	0.021	<0.001	0.0	0.093
Plant 4	0.008	<0.001	0.0	0.032
Plant 5	0.003	<0.001	0.0	0.003
Plant 6	0.061	0.003	0.0	0.320
Plant 9	0.067	0.001	0.0	1.500
Plant 10	0.293	0.007	0.0	0.320
Plant 11	0.005	<0.001	0.0	0.032
Plant 14	0.023	<0.001	0.0	0.032
Plant 15	0.026	<0.001	0.0	0.032
Plant 16	0.005	<0.001	0.0	0.032
All mineral wool plants	0.353	0.006	0.0	1.413
Plant 3	0.427	0.011	0.0	1.413
Plant 7	0.195	0.011	0.0	0.344
Plant 8	0.367	0.016	0.0	1.342
Plant 12	0.215	0.009	0.0	1.074
Plant 13	0.238	0.011	0.0	0.888
Plant 17	0.391	0.013	0.0	1.355

[a]From Ref. 8.

(Table 5.12). The highest exposures were found for workers installing mineral wool insulation in attics. These exposures ranged from 0.04 to 14.8 fibers/cm^3. The exposures for fibrous glass duct installers were found to be low, ranging from 0.005 to 0.2 fibers/cm^3. Esmen found the majority of the fibers to be in the respirable range (less than 3 μm in diameter). Some end-use applications may involve significant exposure to airborne MMMF. For example, "blowing wool" involves installation of loose fiberglass or mineral wool in attics (46). This process in a confined environment may cause high concentration of airborne fibers.

3.3.3.3 Nonoccupational Exposures. Several investigators have reported ambient air concentrations in buildings in which MMMF insulation products have been applied. In one study, the concentration of glass fibers in ambient air samples taken from several sites in California ranged from nondetectable to 0.009 fibers/cm3 with an arithmetic mean of 0.0026 (47) for fibers greater than 2.5 μm diameter as determined by phase-contrast mi- croscopy.

Hohr (48) reported chrysotile, amphibole, and glass fiber concentrations in the air of several cities in the former Federal Republic of Germany (Table 5.13). In all cases, glass fiber concentrations were less than 0.002 fibers/cm^3 and, in general, lower than the concentrations of chrysotile and amphibole fibers.

Table 5.10. Average Concentrations of Total Fibers in a Rock Wool and a Slag Wool Plant

Dust Zone	Number of Samples	Average Number of Total Fibers (fibers/cm^3)
	Rock Wool	
Warehouse	3	1.4
Mixing-Fourdrinier ovens	3	0.14
Panel finishing	12	0.40
Fiber forming	10	0.20
Erection and repair	13	0.24
Tile finishings	22	0.31
All samples	63	0.34
	Slag Wool	
Maintenance	15	0.08
Block production	8	0.05
Blanket line	5	0.05
Boiler room	2	0.05
Yard	2	0.09
Ceramic block	7	0.42
Shipping	3	0.04
Main plant	11	0.01
Mold formation	19	0.03
All samples	72	0.10

[a]From Ref. 42.

Table 5.11. Respirable Glass Fiber Concentrations

	At Four European Plants[a]			At Two Swedish Plants[b]		
Exposure Process	Number of Samples	Mean	Range	Number of Samples	Mean	Range
Preproduction	23	0.01	<0.01–0.03	—	—	—
Production	153	0.04	<0.01–0.62	49	0.22	0.056–0.65
Maintenance	63	0.04	<0.01–0.17	89	0.36	0.037–5.3
General	47	0.03	<0.01–0.06	34	0.19	0.034–0.53
Secondary process 1	131	0.03	<0.01–0.21	59	0.19	0.038–0.73
Secondary process 2	70	0.43	0.02–4.02	5	0.13	0.083–0.16
Cleaning	4	0.01	0.01–0.02	76	0.21	0.026–1.0

[a]From Ref. 43.
[b]From Ref. 44.

Table 5.12. Concentrations of Respirable Fibers Installation of MMMF Insulation[a]

Product and Job Classification	No. of Samples	Fiber Concentration (fibers/cm^3) Average	Range	Average Respirable Fractions
Acoustical ceiling installer	12	0.003	0–0.006	0.55
Duct installation				
Pipe covering	31	0.06	0.007–0.38	0.82
Blanket insulation	8	0.05	0.025–0.14	0.71
Wrap around	11	0.06	0.03–0.15	0.77
Attic insulation				
Fibrous glass				
Roofer	6	0.31	0.07–0.93	0.91
Blower	16	1.8	0.67–4.8	0.44
Feeder	18	0.70	0.06–1.48	0.92
Mineral wool				
Helper	9	0.53	0.04–2.03	0.71
Blower	23	4.2	0.50–14.8	0.48
Feeder	9	1.4	0.26–4.4	0.74
Building insulation installer	31	0.13	0.013–0.41	0.91
Aircraft insulation				
Plant A				
Sewer	16	0.44	0.11–1.05	0.98
Cutter	8	0.25	0.05–0.58	0.98
Cementer	9	0.30	0.18–0.58	0.94
Isolated jobs	7	0.24	0.03–0.31	0.99
Plant B				
Sewer	8	0.18	0.05–0.26	0.96
Cutter	4	1.7	0.18–3.78	0.99
Cementer	1	0.12	—	0.93
Isolated jobs	3	0.05	0.012–0.076	0.94
Fibrous glass duct				
Duct fabricator	4	0.02	0.006–0.05	0.66
Sheet-metal worker	8	0.02	0.005–0.05	0.65
Duct installer	5	0.01	0.006–0.20	0.87

[a]From Ref. 45.

Measurements of MMMF in schools and office buildings have been reported by researchers in Denmark (49, 50) (Table 5.14). In all cases, the concentrations of respirable MMMF reported were 0.001 fibers/cm^3 or lower. These researchers also reported other respirable and nonrespirable fibers (including organics) in the buildings.

3.4 Carcinogenicity Evaluation

An important symposium on health implications of MMMF was held in Copenhagen October 28 to 29, 1986. This international symposium was organized by the Regional

Table 5.13. Fiber Concentrations in Ambient Air In the Former Federal Republic of Germany[a]

		Concentration (fibers/cm^3)			
Measuring Site	No. of Samples	Total	Chrysotile	Amphiboles	Glass
Duisburg	17	0.041	0.0022	0.0019	0.00050
Dortmund	6	0.036	0.0026	0.0019	0.00170
Dusseldorf	21	0.027	0.0014	0.0013	0.00040
Krahm (rural area)	9	0.012	0.0005	0.0007	0.00004

[a]From Ref. 48.

Table 5.14. Mean Dust and Fiber Concentrations in Schools and Office Buildings in Denmark[a]

No. of Buildings	Respirable MMMF (fibers/cm^3)	Nonrespirable MMMF (fibers/cm^3)	Other Respirable Fibers (Including Organics) (fibers/cm^3)	Other Nonrespirable Fibers (Including Organics) (fibers/cm^3)
10	0.0001	0.00002	0.18	0.013
6	0.0001	0.00004	0.15	0.011
8	0.00004	0–0.00008	0.17	0.012

[a]From Refs. 49 and 50.

Office for Europe of the World Health Organization (WHO), IARC, and the Joint European Medical Research Board in association with the Thermal Insulation Manufacturers Association of America. Many researchers discussed their findings at this conference, including the three epidemiologic studies described earlier.

The central issue that the participants in the symposium were confronted with was the interpretation of the epidemiologic studies, which showed a moderate increase in mortality rates among workers exposed to MMMF. Interpretation of these studies is made difficult not only because the excess mortality rates observed are only moderate, but because of the uncertainty of the fiber counts in the early days of the industry and the extent of other potentially carcinogenic materials present in the workplace. These extraneous exposures include arsenic in copper slag, polycyclic aromatic hydrocarbons in cupola melting operations, silica (crystobalite) in refractory fiber production and use, and asbestos in some MMMF workplaces.

In spite of these complications, participants of the Copenhagen symposium reached some definitive conclusions summarized by the symposium chairman, Sir Richard Doll, as follows (51):

- There has been a risk of lung cancer in people employed in the early days of both the rock or slag and glass wool sectors of the MMMF industry. The risk has been approximately 25% above normal for 30 years after first employment. This risk is numerically substantial, however, because lung cancer is so common.

- The risk has been greater in the rock or slag wool sector than in the glass wool sector.
- No risk has been demonstrated in the glass filament sector.
- A variety of carcinogens has contributed to the hazard.
- Uncertainty about the fiber counts in the early days of the industry and the extent of the contribution of other carcinogens make it impossible to provide a precise quantitative estimate of the likely effect of exposure to current fiber levels.
- No specific hazard other than a hazard of lung cancer has been established.

IARC also took into consideration the available research, and presented its findings in the IARC monograph on man-made mineral fibers and radon (2).

In evaluating carcinogenicity, IARC takes into account the total body of evidence and describes the chemical according to the wording of one of the categories shown in Table 5.15. Assignment of IARC category is a matter of scientific judgment reflecting the strength of evidence based on animal and human studies and other relevant data. The IARC has classified glasswool, rockwool, slagwool and refractory ceramic fibers as "possibly carcinogenic to humans" (Group 2B) (52). IARC's evaluation of the available scientific evidence and the carcinogenicity groupings for the various MMMF is summarized in Table 5.16. The National Toxicology Program (NTP) has identified respirable glasswool and ceramic fibers (respirable size) as "substances which may reasonably be anticipated to be carcinogens" (22).

4 REGULATION OF MMMF

Despite major concerns about the potential health effects, particularly the carcinogenic effect of MMMF, only a few countries have adopted regulations to limit exposure to MMMF. It appears that most countries are awaiting the outcome of *additional* epidemiologic studies to determine to what extent they should regulate MMMF. Most countries regulate MMMF either as total dust or respirable dust, and in a few cases, as fibers (1, 6). In some countries, such as Italy and Japan, the silica content of the dust determines the permissible exposure limit. Typical exposure limits for MMMF are shown in Table 5.17.

Currently in the United States, there are no regulatory limits for MMMF. NIOSH has recommended an exposure limit of 3 fibers/cm^3 for fibers with a diameter of less than 3.5 mm and a length greater than 10 mm (29). Indirectly, however, the U.S. Occupational Safety and Health Administration (OSHA) hazard communication standard (53) specifically applies to MMMF. According to OSHA's standard, all materials with an IARC carcinogenicity rating of 2B must be labelled as suspect carcinogens on Material Safety Data Sheets. The American Conference of Governmental Industrial Hygienists, (ACGIH), classifies respirable glass wool fibers as A3 animal carcinogens and has established a Threshold Limit Value (TLV) of one fiber per cc (27). The U.S. Environmental Protection Agency (EPA) has identified naturally occurring and man-made (synthetic) fibers of respirable size as priority substances for risk reduction and pollution prevention under the Toxic Sub-

Table 5.15. IARC Carcinogenicity Grouping[a]

IARC Group	Evaluation	Evidence in Humans		Evidence in Animals
I	Carcinogenic to humans	Sufficient		—
2A	Probably carcinogenic to humans	Limited		Sufficient
2B	Possibly carcinogenic to humans	Limited		Absence of sufficient evidence
			or	
		Inadequate		Sufficient
			or	
		Inadequate		Limited
3	Not classifiable as to its carcinogenicity to humans	—		—
4	Probably not carcinogenic to humans	Evidence suggests lack of carcinogenicity		Evidence suggests lack of carcinogenicity

[a]From Ref. 2.

Table 5.16. Summary of IARC Findings on the Carcinogenicity of MMMF[a]

Fiber Type	Evidence in Humans	Evidence in Animals	IARC Group	Overall IARC Evaluation
Rock wool	Limited	Limited	2B	Possibly carcinogenic to humans
Slag wool	Limited	Inadequate	2B	Possibly carcinogenic to humans
Glass wool	Inadequate	Sufficient	2B	Possibly carcinogenic to humans
Continuous filament	Inadequate	Inadequate	3	Not classifiable as to the carcinogenicity to humans
Refractory (ceramic)	No data	Sufficient	2B	Possibly carcinogenic to humans

[a] From Ref. 2.

stances Control Act (TSCA). In Canada, the man-made vitreous fibers are classified as "Controlled Product Classification. D2A, under the Canadian WHMIS (Workplace Hazardous Materials Information System) programs.

5 SURVEYS AND INSPECTIONS

5.1 Identification of MMMF Products in Buildings

The first step in a survey for MMMF is to find the presence of such materials in construction products or pipe wrap. In general, visual observation is not sufficient to determine whether the suspect material is asbestos or MMMF. Bulk sampling of the material is therefore necessary to ascertain whether the material contains asbestos or MMMF.

The most difficult identification problem is encountered in the building where it is common to find both asbestos and MMMF in insulation, pipe wrap, and other building-related products. Because the health hazards of asbestos have become better known, MMMF have been increasingly used to replace asbestos-containing materials during renovation and maintenance and repair activities. The result is that very often several types of fibrous materials are encountered in the same facility. It is therefore important to identify these materials by taking an appropriate number of bulk samples and analyzing them by polarized light microscopy, and in some cases confirmation by electron microscopy.

5.2 Air Sampling

In the past, concentrations of MMMF in air have been determined on the basis of total dust or respirable dust in the air or on the number of fibers present per unit volume of air. Samples of air are drawn through a filter at a flow rate of 2.0 L/min. The filters used are made of either mixed cellulose ester or polyvinyl chloride (PVC). For respirable mass, a cyclone sampler operated at a flow rate of 1.7 L/min. provides an adequate approximation of the respirable mass fraction. The total dust or respirable dust-determination has now been replaced by fiber counting techniques. Current practice, therefore, is to sample MMMF with an open-face filter cassette containing mixed cellulose ester filter. Air is

Table 5.17. Worldwide Exposure Limits for MMMF[a]

Country	Fiber Type	Exposure Limit: As Total Dust (mg/m³)	Exposure Limit: As Fiber Concentration (fibers/cm³)	Comments
Bulgaria	All fibers with diameters >3 μm	3	—	
	Fibers with diameters <3 μm	2		
Former Czechoslovakia	Glass	8	—	
Denmark	All	4	0.2	Proposed limits
		5	—	Nonstationary workplaces
		—	2	Stationary workplaces
Former Federal Republic of Germany	All	6	—	Fibers with diameters <1 μm are listed as suspect carcinogens
Former German Democratic Republic	All	2	—	
Italy	All	30/(% quartz + 3)	—	
Japan	All	2.9/[(0.22)(% quartz) + 1]	—	
Norway	All	5	—	
New Zealand	All	5	1	(diameter <3 μm)
Poland	All	4	2	(length <5 μm)
Sweden	All	—	1	
United Kingdom	All	5	2	
			1	Recommended for fibers with diameter <3 μm
United States				
OSHA		No limit established		
NIOSH	Fibrous glass	5	3	(diameter <3.5 μm and length >10 μm)
Former USSR	All	2 mg/m³	—	
Former Yugoslavia	All	12 mg/m³	—	

[a]From Refs. 2 and 6

drawn at a rate of 2.0 L/min. To determine fiber concentrations, care must be taken not to overload the filters. This can be accomplished by taking several samples of short duration and time-weighting the average exposures of the samples for the entire shift.

5.3 Analysis of Air Samples

5.3.1 Gravimetric Analysis.

Commonly employed in the past, the gravimetric method is easy to use, is efficient, and provides a reasonably good measure of the overall condition of the work environment.

In the past, several sampling strategies were used to characterize the work environment containing MMMF. Although most common techniques have been to measure either fiber concentrations or dust concentrations gravimetrically, some investigators have used "semi-specific analytical methods" for studies of MMMF exposure, particularly fibrous glass exposures. Johnson et al. (54) in 1987 performed chemical analysis of air samples for total silica content and estimated the amount of glass dust based on the silica content. This method has problems because silica content can differ considerably for the various types of glass. In another survey, Fowler (55) analyzed total dust samples for metals, such as cadmium (Cd), chromium (Cr), cobalt (Co), nickel (Ni), manganese (Mn), lead (Pb), and zinc (Zn), to determine the exposure of workers to mineral wool fibers.

The advisability of determining concentrations of MMMF on the basis of weight alone is questionable because it is the number and dimensions of the fibers and not the total weight that determines toxicologic response.

5.3.2 Microscopic Analysis

For fiber counting, the analytical methods employed are phase-contrast microscopy (PCM) or electron microscopy. The PCM method involves collecting samples with a membrane filter and counting the fibers by PCM at a magnification of 400 to 450 ×. The PCM method is more effective for analyzing MMMF than asbestos fibers because MMMF found in the workplace air are larger than asbestos fibers. Because most MMMF have relatively large diameters, they are easily resolved by optical microscopy. Using the PCM counting method, fiber levels as low as 0.01 fibers/cm^3 or lower can be determined. However, compared to asbestos, there has been much less standardization of MMMF techniques. In the United States, NIOSH has recommended a PCM method of fibrous glass determination. In the NIOSH-recommended method, the fibers with diameters less than 3.5 μm and length greater than 10 μm are counted. There has been, however, no evaluation of this technique through either intralaboratory or interlaboratory trials. In Europe, the MMMF counting methods have been more extensively studied and evaluated. In the mid-1970s, the World Health Organization (WHO) in Copenhagen initiated a program known as the WHO/EURO Reference Scheme (56). The program's goals were to produce reference methods for sampling and evaluating MMMF and to minimize interlaboratory variations in the results obtained with these methods. The WHO/EURO reference method (57) is also based on the membrane filter technique used to determine asbestos fiber concentrations. In the WHO/EURO reference method, the number of respirable fibers in randomly selected areas of the filter is counted using a magnification of about 500 ×. The respirable fibers counted are fibers that are longer than 5 μm, have a diameter of less than 3 μm, and have an aspect ratio of 3 to 1 or greater. In contrast to the asbestos fiber counting rules, fibers in contact with particulate or other fibers are counted provided they meet the above criteria.

Although PCM is quite suitable to identify MMMF in most cases, there is a need to distinguish MMMF positively from asbestos fibers because, increasingly, both MMMF and asbestos fibers are present in the same environment. To identify and confirm the fiber types further, both scanning electron microscopy (SEM) and transmission electron microscopy (TEM) are used.

The WHO/EURO technical committee for monitoring and evaluating MMMF also established an SEM reference method in which samples are collected on a polycarbonate filter (Nucleopore *f*) or a PVC membrane filter. After preparation, the samples are observed with an SEM at a magnification of 5000 ×. The SEM method determines fiber concentrations as well as fiber size. Most SEMs are also capable of performing energy dispersive X-ray diffraction analysis which enables the analyst to distinguish between fiber types by determining elemental composition of the fiber.

The specific identification of fiber type can be even more effectively made through TEM. Because most MMMF are amorphous fibers and are not crystalline in nature, it is possible to distinguish them from noncrystalline materials by electron diffraction of the individual fibers. The transmission electron microscope can provide very high resolution. For identification of fibers by morphology, the electron microscope is used at a magnification of 20,000 ×. A magnification of 20,000 × is also used when performing selected area electron diffraction of individual fibers.

5.4 Other Environmental Measurements

In any work environment containing MMMF, industrial hygienists should also be aware of other occupational exposures. These exposures may be to excessive concentrations of dust, heavy metals such as lead, chromium, cadmium, cobalt, and nickel, polycyclic aromatic hydrocarbons, carbon monoxide, and other contaminants. These exposures should be given appropriate consideration while evaluating the work environment. Other industrial hygiene aspects such as ventilation, housekeeping, work practices, and personal protective measures must also be evaluated.

6 CONTROL OF MMMF

The specific types of controls necessary to prevent or reduce occupational exposure to man-made mineral fibers depend heavily on the type and application of the product containing the fibers. These control measures are not unique to MMMF, but rely on methods well established in the practice of industrial hygiene. These measures include engineering controls, work practices, and the use of personal protective equipment.

6.1 Engineering Controls

Fabrication operations, including cutting, shaping, and drilling of MMMF products can generate airborne fibers by mechanical disturbance. In production operations, such exposures are best controlled through the use of local exhaust ventilation systems. In maintenance operations and construction, use of power tools, such as drills and saws may also generate airborne fibers. Such tools may be ventilated using high-efficiency particulate air (HEPA) filter-equipped vacuum cleaners. Hoods designed to fit standard power tools are commercially available.

6.2 Work Practices

The use of work practices that minimize disturbance of fibers is critical. In many instances in both industry and construction, the use of engineering controls may not be feasible. This makes it critical to minimize the amount of airborne fibers generated while handling the MMMF-containing products and during any subsequent cleanup. The following work practices are recommended:

- Use of unventilated power tools should be minimized because the high operating speed can generate greater amounts of airborne dust. Hand tools *may be* preferred *in some situations* because they generally produce less dust.
- Some products should be lightly sprayed with water, amended with a surfactant to enhance penetration prior to handling. This helps to minimize fiber release.
- Good housekeeping is an essential part of any safe construction or maintenance project. Materials that fall to the ground should be picked up as soon as possible because walking on scraps can break them into smaller pieces that can more easily become airborne or may be tracked further around the facility. Good housekeeping in the work area helps to minimize airborne dust. All external surfaces of equipment should be kept free of dust accumulation which, if dispersed, contributes to airborne concentrations. Such cleanup should be performed only using HEPA filter-equipped vacuum cleaners or wet methods, and never by dry sweeping.

6.3 Personal Protective Equipment

Persons working with loose insulation products should wear protective clothing to minimize skin contact and resulting irritation. Either disposable or launderable clothing is acceptable. Protective clothing should be laundered separately from other clothes and should never be worn or taken home by employees for laundering.

If engineering and work practice controls are not adequate to prevent exposure to airborne fibers, use of respiratory protection may be necessary. In selecting respiratory protection, guidelines established by OSHA for asbestos may be used. Safety glasses with side shields are recommended to keep fibers and dust particles out of the eyes.

6.4 Disposal

Wastes containing MMMF are not regulated as hazardous wastes under the U.S. EPA's Resource Conservation and Recovery Act (RCRA). With other regulatory agencies (state, local, etc.) inquire regarding waste classification of MMMF to make sure such wastes are disposed in accordance with all applicable regulations.

7 CONCLUSIONS

Man-made mineral fibers are extremely important commercial products. Their use as commercial insulating materials and plastic reinforcement products have provided significant benefits to society. As asbestos substitutes, and in certain "high-tech" industries, MMMF are finding wide application. The major epidemiologic studies conducted to date have shown only moderately increased mortality rates for a few specific diseases, in particular, lung cancer, for cohorts of industrial workers.

Although many questions still remain unanswered, evidence to date suggests that the carcinogenic risk from MMMF is much less when compared to asbestos. Perhaps the most definitive statement on this complex issue was made by Sir Richard Doll, Chairman of the 1986 Copenhagen symposium, in his concluding remarks (57):

If, now, I abandon the firm basis of scientific fact and express a personal judgment, I do so because I know that in the absence of such a conclusion many people may think that the symposium has been a waste of time. Let me therefore add a seventh conclusion that, taking into account also the results of animal experiments, the experience of the asbestos industry and the experience of the glass filament sector of the MMMF industry. MMMF are not more carcinogenic than asbestos fibres and exposure to current mean levels in the manufacturing industry of 0.2 Fr mL^{-1} (0.2 respirable fibers per milliliter of air) or less is unlikely to produce a measurable risk after another 20 years have passed.

Because of the commercial importance of these materials, a number of studies are underway. Until there is a better understanding of the health effects of man-made mineral fibers, it is prudent for industrial hygienists to treat these materials with the same precautions as asbestos by instituting appropriate work practices and engineering controls to minimize worker and community exposures.

BIBLIOGRAPHY

1. G. A. Peters and B. J. Peters, *Sourcebook on Asbestos Diseases: Medical, Legal and Engineering Aspects*, Vol. 2, Garlanel Law Publishing, New Toni, NY, 1986, pp. 190–2102.
2. IARC Monographs on the Evaluation of Carcinogenic Risks to Humans, Vol. 43, *Man-made Mineral Fibers (Proceedings of a WHO/IARC Conference)*, Vol. 2, World Health Organization, Copenhagen, 1984, pp. 234–252.
3. H. J. Smith, "History, Processes, and Operations in the Manufacturing and Uses of Fibrous Glass-One Company's Experience," in Occupational Exposure in Fibrous Glass, A Symposium, National Institute for Occupational Safety and Health, Cincinnati, OH, 1976, pp. 19–26.
4. J. M. Dement, *Preliminary Results of the NIOSH Industry-wide Study of the Fibrous Glass Industry*, DHEW(NIOSH) Publ. No. 1W3.35.3b; NTIS Pub. 40. PB-81-224693), National Institute for Occupational Safety and Health, Cincinnati, OH, 1973, pp. 1–5.
5. Anonymous, "Facts and Figures," *Chem. Eng. News* **64**, 32–44 (1986).

6. International Labour Organization, "Working Document on Safety in the Use of Mineral and Synthetic Fibres of a Meeting of Experts," ILO, Geneva, 1989.
7. L. Simonato, A. C. Fletcher, J. Cherrie, A. Andersen, P. Bertazzi, N. Charnay, J. Claude, J. Dodgson, J. Esteve, R. Frentzel-Beyme, M. J. Gardner, O. Jensen, J. Olsen, L. Teppo, R. Winkelmann, P. Westerholm, P. D. Winter, C. Zocchetti, and R. Saracci, "The International Agency for Research on Cancer Historical Cohort Study of MMMF Production Workers in Seven European Countries: Extension of the Follow-up," *Ann. Occup. Hyg.* **31**, 603–623 (1987).
8. P. E. Enterline, G. M. Marsh, V. Henderson, and C. Callahan, "Mortality Update of Cohort of U.S. Man-made Mineral Fibre Workers," *Ann. Occup. Hyg.* **31**, 625–656 (1987).
9. F. Pott, U. Ziem, and U. Mohr, "Lung Carcinomas and Mesotheliomas Following Intratracheal Installation of Glass and Asbestos," in W. I. Bergbau-Berufsgenossen, ed, International Pneumoconiosis Conference, Bochum, 1984, pp. 746–757.
10. M. F. Stanton, M. Layard, A. Tegeris, E. Miller, M. May, and E. Kent, "Carcinogenicity of Fibrous Glass: Pleural Response in the Rat in Relation to Fiber Dimension," *J. Nat. Cancer Inst.* **58**, 587–603 (1977).
11. A. M. Sincock, J. D. A. Delhanty, and G. Casey, "A Comparison of the Cytogenic Response to Asbestos and Glass Fibre in Chinese Hamster and Human Cell Lines," *Mutat. Res.* **101**, 257–268 (1982).
12. T. W. Hesterberg and J. C. Barrett, "Dependence of Asbestos and Mineral Dust Induced Transformation of Mammalian Cells in Culture on Fiber Dimension," *Cancer Res.* **44**, 2170–2180 (1984).
13. M. Oshimura, T. W. Hesterberg, T. Tsutsui, and J. C. Barrett, "Correlation of Asbestos-induced Cytogenetic Efforts with Cell Transformation of Syrian Hamster Embryo Cells in Culture," *Cancer Res.* **44**, 5017–5022 (1984).
14. M. Chamberlain, and E. M. Tarmy, "Asbestos and Glass Fibres in Bacterial Mutation Tests," *Mutat Res.* **43**, 159–164 (1977).
15. L. LeBouffant, J-P. Henn, J. C. Martin, C. Normand, G. Tichoux, and R. Trolarel, "Distribution of Inhaled MMMF in the Rat Lung-Long Term Effects," in *Biological Effects of Man-Made Mineral Fibers (Proceedings of a WHO/IARC Conference)*, Vol. 2, WHO, Copenhagen, 1984, pp. 143–168.
16. J. C. Wagner, G. B. Berry, R. J. Hill, D. E. Munday, and J. S. Skidmore, "Animal Experiments with MMM(V)F-Effects of Inhalation and Intrapleural Inoculation in Rats," in *Biological Effects of Man-made Mineral Fibres (Proceedings of a WHO/IARC Conference)*, Vol. 2, WHO, Copenhagen, 1984, pp. 209–233.
17. H. Muhle, F. Pott, B. Bellmann, S. Takenaka, and U. Ziem, "Inhalation and Injection Experiments in Rats to Test the Carcinogenicity of MMMF," *Ann. Occup. Hyg.* **31**, 755–764 (1987).
18. K. P. Lee, G. E. Barras, F. D. Griffith, and R. S. Warity, "Pulmonary Response to Glass Fibre by Inhalation Exposure," *Lab Invest.* **40**, 123–133 (1979).
19. D. M. Smith, L. W. Ortiz, R. F. Archuleta, and N. F. Johnson, "Long-term Health Effects in Hamsters and Rats Exposed Chronically to Man-made Vitreous Fibers," *Ann. Occup. Hyg.* **31**, 731–754 (1987).
20. B. Goldstein, I. Webster, and R. E. C. Rendall, "Changes Produced by the Inhalation of Glass Fibre in Non-human Primates," Proceedings of a WHO/IARC Conference in Association with JEMRB and TIMA, Copenhagen, April 20–22, 1982, World Health Organization, Regional Office for Europe, Copenhagen, 1984, pp. 273–286.
21. E. E. McConnell, J. C. Wagner, J. W. Skidmore, and J. A. Moore, "A Comparative Study of the Fibrogenic and Carcinogenic Effects of UICC Canadian Chrysotile B Asbestos and Glass Mi-

crofibre (JM 100)," in *Biological Effects of Man-made Mineral Fibres (Proceedings of a WHO/IARC Conference)*, Vol. 2, World Health Organization, Copenhagen, 1984, pp. 234–252.

22. National Toxicology Program, Seventh Annual Report on Carcinogens, U.S. Department of Health and Human Services, Public Health Service, 1994.
23. F. Pott, U. Ziem, F. J. Reiffer, F. Huth, H. Ernst, and U. Mohr, "Carcinogenicity Studies on Fibres, Metal Compounds and Some other Dusts in Rats," *Exp. Pathol.* **32**, 129–152 (1987).
24. J. M. G. Davis, J. Addison, R. E. Bolton, K. Donaldson, A. D. Jones, and A. Wright, "The Pathogenic Effects of Fibrous Ceramic Aluminum Silicate Glass Administered to Rats by Inhalation or Peritoneal Injection," in *Biological Effects of Man-made Mineral Fibres (Proceedings of WHO/IARC Conference)*, Vol. 2, World Health Organization, Copenhagen, 1984, pp. 303–322.
25. G. H. Pigott, and J. Ishmael, "A Strategy for the Design and Evaluation of a 'Safe' Inorganic Fibre," *Ann. Occup. Hyg.* **26**, 371–380 (1982).
26. H. S. Shannon, E. Jamieson, J. A. Julian, D. C. F. Muir, and C. Walsh, "Mortality Experience of Ontario Glass Fibre Workers-Extended Follow up," *Ann. Occup. Hyg.* **31**, 657–662 (1987).
27. Threshold Limit Values and Biological Exposure Indices for 1987–1988, American Conference of Governmental Industrial Hygienists, ISBN:1-88-2417-23-2.
28. M. Corn, personal communication, 1988.
29. NIOSH, *Criteria for a Recommended Standard Occupational Exposure to Fibrous Glass*, National Institute for Occupational Safety and Health, Publication no. 77-152, 1977.
30. M. Lippmann, "Review of Asbestos Exposure Indices," *Environ. Res.* **46**(1), (June 1988).
31. J. Bigon, R. Saracci, and J. C. Touray, *Environ. Health Perspectives*, **102** (Suppl. 5) (Oct. 1997).
32. F. Pott, "Problems in Defining Carcinogenic Fibres," *Ann. Occup. Hyg.* **31**, 799–802 (1987).
33. J. C. McDonald, "Epidemiological Significance of Mineral Fiber Persistence in Human Lung Tissue," *Environ Health Perspect.* **102** (Suppl. 5) (1994).
34. T. W. Hesterberg, R. Mast, E. E. McConnell, F. Chevalier, D. H. Bernstein, W. D. Bunn, and R. Anderson, "Chronic Inhalation Toxicity of Refractory Ceramic Fibers in Syrian Hamsters," in *Mechanism of Fibre Carcinogenesis*, Plenum Press, New York, 1991, pp. 531–538.
35. A. Morgan and A. Holmes, "Solubility of Asbestos and Man-made Mineral Fibers *in Vitro* and *in Vivo*: Its Significance in Lung Disease," *Environ. Res.* **39**, 475–484 (1986).
36. N. F. Johnson, D. M. Griffiths, and R. J. Hill, "Size Distribution Following Long-term Inhalation of MMMF," in *Biological Effects of Man-made Mineral Fibres (Proceedings of a WHO/IARC Conference)*, Vol. 2, World Health Organization, Copenhagen, 1984, pp. 102–125.
37. L. Le Bouffant, H. Daniel, J. P. Henin, J. C. Martin, C. Normand, G. Tichoux, and F. Trolard, "Experimental Study on Long-term Effects of Inhaled MMMF on the Lung of Rats," *Ann. Occup. Hyg.* **31**, 765–790 (1987).
38. B. Bellmann, H. Muhle, F. Pott, H. Konig, H. Kloppel, and K. Spurny, "Persistence of Man-made Mineral Fibres (MMMF) and Asbestos in Rat Lungs, 1987," *Ann. Occup. Hyg.* **31**, 693–709 (1987).
39. J. Scholze, and R. Conradt, "An in Vitro Study of the Chemical Durability of Siliceous Fibres," *Ann. Occup. Hyg.* **31**, 683–692 (1987).
40. J. P. Leineweber, "Solubility of Fibres *in Vitro* and *in Vivo*," in *Biological Effects of Man-made Mineral Fibres (Proceedings of a WHO/IARC Conference)*, Vol. 2, World Health Organization, Copenhagen, 1984, pp. 87–101.

41. R. Klingholz, and B. Steinkopf, The Reactions of MMMF in a Physiological Model Fluid and in Water, in *Biological Effects of Man-made Mineral Fibres (Proceedings of a WHO/IARC Conference)*, Vol. 2, World Health Organization, Copenhagen, 1986, pp. 60–86.
42. M. Corn, Y. Y. Hammad, D. Whittier, and N. Kotsko, "Employee Exposure to Airborne Fiber and Total Particle Matter in Two Mineral Wool Facilities," *Environ. Res.* **12**, 59–74 (1976).
43. J. Cherrie, J. Dodgson, S. Groat, and W. MacLaren, Environmental Surveys in the European Man-made Mineral Fiber Production Industry, *Scand. J. Work Environ. Health.* **12** (Suppl. I), 18–25 (1986).
44. Arbetarskyddsstyrelsen (National Swedish Board of Occupational Safety and Health), Measurement and Characterization of MMMF Dust (Partial Reports 3–9), Stockholm, 1981.
45. N. A. Esmen, M. J. Sheehan, M. Corn, M. Engel, and N. Kotsko, "Exposure of Employees to Man-made Vitreous Fibers: Installation of Insulation Materials," *Environ. Res.* **28**, 386–398 (1982).
46. P. S. J. Less et al. "End User Exposures to Man-Made Vitreus Fibers: I. Installation of Residual Insulation Products," *Appl. Occup. Environ. Hyg.* **8**(12), 1022–1030 (1993).
47. J. L. Balzer, "Environmental Data: Airborne Concentrations Found in Various Operations," in W. N. LeVee, and P. A. Schulte, eds., *Occupational Exposure to Fibrous Glass* (DHEW Publ. No. (NIOSH) 76-151; NTIS Publ. No. PB-258869), National Institute for Occupational Safety and Health, Cincinnati, OH, 1976, pp. 83–89.
48. D. Hohr, "Investigations by Transmission Electron Microscopy of Fibrous Particles in Ambient Air" (Ger.), *Staub. Reinhalt. Luft* **45**, 171–171 (1985).
49. T. Schneider, "Man-made Mineral Fibers and Other Fibers in the Air and in Settled Dust," *Deniron. Int.* **12**, 61–65 (1986).
50. A. Rindel, E. Bach, N. O. Breum, C. Hugod, and T. Schneider, "Correlating Health Effect with Indoor Air Quality in Kindergartens," *Int. Arch. Occup. Environ. Health* **59**, 363–373 (1987).
51. WHO/EURO Technical Committee for Monitoring and Evaluating MMMF, *The Reference Methods for Measuring Airborne Man-made Minerals Fibres (MMMF)*, WHO Regional Office for Europe, Environmental Health Report 4, 1985.
52. International Agency for Research on Cancer (IARC) *Man-made Mineral Fibers I*, IARC Monographs on the Evaluation of the Carcinogenic Risk of Chemicals to Man, Vol. 43, Lyon, France, 1988, pp. 39–171.
52. L. Chrazze, D. K. Watkins, and C. Fryar, "Adjustment for the Effect of Cigarette Smoking in a Historical Cohort Mortality Study of Workers in a Fiberglass Manufacturing Facility," *JOEM* **37** 744–748 (1995).
53. U.S. Dept. of Labor, Occupational Safety & Health Administration, Code of Federal Regulations, Title zq, 1910–1200.
54. D. L. Johnson, J. J. Healey, H. E. Ayer, and J. R. Lynch, "Exposure to Fibers in the Manufacture of Fibrous Glass," *Am. Ind. Hyg. Assoc. J.* **30**, 545–550 (1969).
55. D. P. Fowler, *Industrial Hygiene Surveys of Occupational Exposures to Mineral Wool*, National Institute for Occupational Safety & Health Publication No. 80-135, 1980.
56. WHO/EURO Technical Committee for Monitoring and Evaluating MMMF, "The WHO/EURO Man-made Mineral Fibre Reference Scheme," *Scand. J. Work Environ. Health* **11**, 123–129 (1985).
57. R. Doll, "Symposium on MMMF-Overview & Conclusions," *Ann. Occup. Hyg.* **31(4B)**, 805–819 (1987).

CHAPTER SIX

Occupational Dermatoses

David E. Cohen, MD, MPH

1 INTRODUCTION

Diseases of the skin caused by physical or chemical agents or conditions at work continue to outnumber all other work-associated illnesses. They occur in most occupations and vary from a minor dermatitides to life threatening eruptions and benign and malignant neoplasms. Descriptive titles associated with particular illnesses allow the nosology of illnesses to be easily elucidated, for example, asbestos wart, cement burn, chrome holes, fiber glass itch, hog itch, oil acne, rubber rash, and tar smarts. In view of the variety of skin lesions known to result from contactants within the workplace, the term "occupational dermatoses" is preferred because it includes any abnormality of the skin resulting directly from or aggravated by the work environment (1).

1.1 History

It is unknown in what context the first occupational skin disease occurred in. Perhaps the earliest reference occurred in the writings of Celsus about 100 A.D. (2), when he described ulcers of the skin caused by corrosive metals. During later centuries, several authors enhanced the knowledge of certain occupational skin diseases, but cutaneous ulcerations seem to have been the major skin disease of record. Ulcerations of the skin were easily recognized, especially among those handling metal salts in mining, smelting, tool and weapon making, creating objects of art, glassmaking, gold and silver coinage, casting, and similar metallics. Little was recorded about occupational skin disease until Ramazzini's historic treatise on diseases of tradesmen in 1700 (3). In this tome he described skin disorders experienced by bath attendants, bakers, guilders, midwives, millers, and miners,

Patty's Industrial Hygiene, Fifth Edition, Volume 1. Edited by Robert L. Harris.
ISBN 0-471-29756-9

among other tradespeople. Seventy-five years later Sir Percival Pott published the first account of occupational skin cancer when he described scrotal cancer among chimney-sweeps (4).

The Industrial Revolution of the eighteenth century brought change from an agricultural and guild economy to one dominated by machines and industrial expansion. As cities and industries grew, so also occurred the growth of science and the eventual discovery and use of new materials such as chromium, mercury, and petroleum, among many others. The chemical age brought enormous numbers of natural and synthetic materials into industrial and household use (5). As a result physicians began to recognize occupational dermatoses and publish their observations in England, Germany, Italy, and France. Similarly, industrialization within the United States led to the recognition of new causes of occupational skin disease. A number of updated texts and related publications are available (6–15). Today is an age of technical ingenuity brought about by atomic power, electronics, computers, precision tools, machinery, and, space (5).

1.2 Epidemiology

In the 1978 edition of *Patty's Industrial Hygiene and Toxicology* text, dermatologic diseases were shown to account for about 40% of all occupational diseases reported to the U.S. Department of Labor (15). In the 1984 Bureau of Labor Statistics, dermatologic disease had decreased to 34% of all reported occupational disease. Most of the cases were associated with manufacturing, services, wholesale-retail trade, construction, agriculture, forestry, fishing, and transportation. In this Bureau of Labor Statistics breakdown of occupational diseases by type, lung disease due to dust accounted for 1.4%, disorders due to repeated trauma 27.8%, disorders due to physical agents 7.2%, poisonings 3.6%, respiratory disease (toxic) 8.5%, and skin disease 34.1% (16). However, from 1983 to 1994, reported cases indicate an increase in the rate of occupational dermatoses from 64 to 81 per 100,000 employees. Skin disease resulting from exposures in the agriculture and manufacturing industries were responsible for the greatest number of cases with incidence rates of 86 and 41 per 10,000 workers respectively. Although the health care field has a relatively low rate of disease, the large number of workers in this industry results in almost 3,900 cases per year.

Since skin disease may not always be debilitating or life threatening, many believe that the rate at which it is reported to government agencies is underrepresented 10 to 50 fold. Hence this disparity often results in inappropriate estimates of the economic impact of occupational skin disease. It has been estimated that up to 20 to 25% of persons with occupational skin disease lose an average of 11 days of work annually. This translates to an economic loss of $222 million to $one billion annually (17).

Thermal and chemical burns, lacerations, and blunt skin trauma are extremely common occupational skin injuries. NIOSH estimates that there are approximately 1.07 to 1.65 million skin injuries per year accounting for a rate of 1.4 to 2.2 cases per 100 workers. These potentially preventable injuries are probably very common and should not be overlooked. No data are available to estimate the total economic impact of occupational skin injuries.

Marked decreases in the number of reported cases of skin diseases over the past decades have been noted but the reasons are not readily definable. It is possible that governmental regulations such as the Toxic Substances Control Act and the Occupational Safety and Health Act improved hygiene controls in industry. Better management surveillance and the "right-to-know" laws have also had a significant impact on the lowered incidence.

1.3 Structure and Function of the Skin

The skin performs a legion of homeostatic functions that are critical for survival. Protection against physical or chemical assaults can be reasonably afforded by the routine defense mechanisms that the integument possesses. However, repeated or serious acute attacks on the skin can overwhelm these defenses and result in disease. This in turn can lead to ineffective performance of other skin duties such as thermoregulation, excretory and secretory functions, immunoregulation, photoprotection, and sensory perception. Hence the body as a whole may suffer when a single organ system such as the skin is insulted.

As will be described later, the skin is not composed of a monomorphous group of cells, but rather is a complex dynamic system of differentiating cells that closely interact with almost every other organ system in the body. Injury to the skin from occupational or environmental stressors may result from toxicity to cells or through interference of normal homeostatic functions that the skin performs. Insufficient thermoregulation and electrolyte homeostasis from destruction of the integument will result in death. Chemical and thermal burns, overexposure to heat and humidity, or prolonged occlusion of the skin are capable of inducing regulatory failure.

Anatomically, the skin is composed of two main levels, epidermis and dermis. They are separated from each other by a basement membrane, which forms a wavy interface between the layers. The skin appendages, which include the hair follicle unit and sebaceous, eccrine, and apocrine sweat glands, have their respective ductal structure crossing the epidermis to the surface. The concentration of these appendageal structures varies greatly by location. For most, hair follicles abound on the scalp, sebaceous glands are concentrated on the face, and eccrine glands are on the palms, soles, and axillae. Apocrine glands, which have only an incompletely characterized function, are found in the axillae, areolae, and groin.

The outer or epidermal layer varies in thickness, being most protective on the palms and soles. Because it is contiguous with the dermis, it also acts as the outer cover of the cushion of connective and elastic tissue that guards the blood and lymph vessels, nerves, secretory glands, hair shafts, and muscles.

The epidermis has many layers, however, for the purpose of this review, it can be functionally divided into the noncornified layer and the cornified layer or stratum corneum. The principal cell of the epidermis is called the keratinocyte and is arranged as a stratified squamous epithelium. Keratinocytes begin as basal cells abutting the basement membrane and progress outward through the epidermis via a terminal differentiation pathway that lasts approximately 14 days. During this period there are substantial changes in cell surface markers and a concomitant accumulation of keratin proteins. By the end of the two-week cycle, the keratinocyte has lost its nucleus and intracellular organelles and has flattened. It is at this point that the nonviable keratinocytes enter the stratum corneum and are known as corneocytes. A tightly linked layer of fibrous proteins will largely replace the cell structure of the keratinocyte. It will take another 14 days for the original corneocyte to reach the surface of the skin, where it will be sloughed (9, 11, 18–20).

The stratum corneum is important in resisting transit of water and electrolytes. Stresses such as friction, pressure, and natural and artificial ultraviolet light can induce compensatory thickening of the stratum corneum in the form of a callous. Besides acting as a water barrier and a physical shield, the stratum corneum provides modest protection against mild acids. It is, however, quite vulnerable to the action of organic and inorganic alkaline materials. Such chemical substances attack the stratum corneum by denaturing the keratin proteins, thus altering the cohesiveness and the capacity to retain water, which is essential in the maintenance of the barrier layer. Common chemical and environmental insults to the skin eventually lead to impairment of the barrier efficiency because of water loss and dryness (21).

Also in the basal layer are melanocytes, which are neuroendocrine-derived cells that are responsible for the production of melanin pigment. Melanin is packaged into pigment granules called melanosomes and imparts the natural pigmentation to the skin. Differences in skin pigmentation occur mostly from differences in melanosome structure rather than from gross quantities of melanin. The pigment granules that arise from complex enzymatic reactions within the melanocytes are picked up by the epidermal cells and eventually are shed by way of the keratin exfoliation (18, 22). Melanin acts as the principal defense against ultraviolet light, since it acts as a broad-spectrum chromophore or light absorber. Besides the natural production of melanin, certain agents such as coal, tar, pitch, selected aromatic chlorinated hydrocarbons, petroleum products, and trauma can cause excess melanin production leading to hyperpigmentation (15, 23). In contrast, members of the quinone family and some phenolics can inhibit pigment by direct action on the melanin enzymatic system or direct toxicity to melanocytes (1, 9, 24–27). This can result in depigmentation, impacting an ivory-white appearance.

Immediately below the epidermal region lies the dermis, which is thicker than the epidermis. It is primarily composed of connective tissue such as collagen and elastin bathed in glycoproteins. These constituents are produced by resident fibroblasts that maintain the integrity of the dermis and can respond to injury and repair lost tissue. Dermal resiliency provides protection within limits against blunt trauma and its flexibility accounts for the return of stretched skin to its normal location. The dermis also houses eccrine sweat glands and ducts, hair follicles, sebaceous glands, blood and lymph vessels, and nerve endings (18).

Thermoregulation is modulated by the excretion of eccrine sweat and changes in the superficial blood flow controlled by the central nervous system. Thus, core body temperature and circulating blood are physiologically stabilized at a constant temperature despite climatic variations. Eccrine sweat is composed primarily of water and electrolytes and participates in temperature control through evaporative heat from the surface. Radiant heat loss is facilitated by the dilation of upper dermal blood vessels. The opposite occurs when a decreased core and/or surface temperature causes blood vessels of the skin to constrict and shunt warm blood away from exposed surfaces thus preserving heat (6, 9, 18, 28).

Secretory functions within the skin are relegated to sweat gland and sebaceous gland activity. Eccrine sweat is composed mostly of water and electrolytes serves primarily for thermoregulation. Inability to sweat adequately under heat stresses will result in increase core temperature and possibly heat-related illnesses. Sebaceous glands reside within the dermis as part of the hair follicle unit. Their functional product, sebum, is excreted through the hair follicle and reaches the skin surface through the follicular os. These glands are frequently targets for occupational follicular diseases caused by exposures to coal tar, heavy oils, greases, and certain aromatic chlorinated hydrocarbons (9, 15, 29–30).

Special receptors within the skin are part of a network of nerve endings and fibers that receive and conduct various stimuli, later recognized as heat, cold, pain, and other perceptions such as wet, dry, sharp, dull, smooth, and rough (18).

1.4 Percutaneous Absorption

Historically, the skin was believed to be an impervious barrier to external chemicals. More recently, however, it is understood that percutaneous absorption occurs frequently and depends on a variety of properties of the agent. It is a major avenue of entry for a number of toxic agents present in the agricultural and industrial areas (6, 9, 15, 19). The skin maintains a relatively hydrophobic character, which is imparted from the sphingolipids present in the epidermis (31). The water content, which makes up about 20% of the weight of the epidermis, is concentrated around intracellular proteins within keratinocytes. Prolonged submersion can cause a several-fold increase in skin hydration and loss of barrier function. Percutaneous absorption tends to follow the physics of typical membrane kinetics. Hence increasing the concentration gradient across the epidermis will result in increasing percutaneous absorption. However, two other significant factors must be considered in this complex biological membrane barrier. Molecular weight and hydrophobicity are important in determining or predicting percutaneous absorption. Low molecular weight molecules pass through the epidermis with greater alacrity than do larger weight compounds. Also a chemical with a high octanol:water partition coefficients (more hydrophobic) will pass more easily. Therefore low molecular weight hydrophobic compounds may pass readily through an intact epidermis and cause toxicity. Since percutaneous absorption is becoming increasingly important, pharmacokinetic models have been developed to predict percutaneous absorption of substances (32, 33).

Absorption may be accomplished via two routes: initially through the skin appendages such as hair follicles and sweat glands, then later through diffusion through the epidermis. Occlusion may serve to enhance penetration by hydrating the epidermis and by preventing evaporation or mechanical removal of a chemical from the skin surface. From a practical perspective, percutaneous absorption has been a particular problem for only certain chemicals such as pesticides, cyanides, aromatic hydrocarbons, mercury, lead, and others. Organophosphate pesticides like parathion are sufficiently absorbed through the skin to be fatal after moderate skin exposure (34). Until recently organic leads were thought to be the primary form of the heavy metal capable of percutaneous absorption. More recent work has indicated that inorganic lead in dust may be result in substantial dermal levels of lead and can cause increases in total body lead burden (35). Body site has long been described as an important variable in percutaneous absorption. The scrotum has been identified as an area where absorption is greatest, with the face as a moderately penetrable site, and the abdomen the least (36).

1.5 Metabolism

Keratinocytes in the lower epidermis are capable of performing many metabolic reactions that can biotransform toxic xenobiotics to more innocuous compounds. This detoxification process is part of the barrier function of the skin that transcends the physical obstacle of the epidermal lipids and cross-linked proteins of the stratum corneum. In fact, the skin is able to perform most of the phase 1 and phase 2 metabolic enzymes necessary for xenobiotic metabolism and biotransformation. Overall, the skin has 2% of the metabolic capacity of the liver. Induction of specific enzyme pathways may result in dramatic rises in enzyme activity. This is particularly important for biotransformation reactions when carcinogenic intermediates are formed during detoxifying metabolism of certain polycyclic hydrocarbons (37). Animal studies have demonstrated that when skin is exposed to polychlorinated biphenyls or benzo-*a*-pyrene, 20% of the total body activity of aryl hydrocarbon hydroxylase can be linked to integumentary activity. Hence, under conditions of repeated dermal exposure, the skin may be responsible for a marked amount of detoxifying and metabolism of the chemicals. Exposure to these procarcinogens on the skin, however, may result in biotransformation to active carcinogens locally, without the aide of hepatic enzymes (38, 39).

2 ETIOLOGIC FACTORS

To detect the cause of an occupational dermatosis it is fundamental to consider a spectrum of factors that can have indirect relationship to the disease (2, 7, 40). In determining the cause of an occupationally acquired disease, it is important to recognize host and environmental factors that may predispose or aggravate an already existing dermatosis. These include host factors such as genetic predisposition, preexisting dermatologic disease, age, and environmental factors such as temperature, humidity, and season (1, 2, 6, 18, 23).

2.1 Genetic Predisposition

This particular subheading represents a wide variety of potentially confounding factors with regard to occupational skin disease. They can include genetic predisposition to skin diseases as described below or extraordinary sensitivity to routinely encountered occupational and environmental stressors. Sensitivity to chemical irritants and ability to acquire allergic contact sensitivity are both likely to be genetically determined. Extreme examples are metabolic genetic diseases that predispose individuals to disease after banal exposures. An example is a person with xeroderma pigmentosa (inability to repair DNA after routine sun exposure) who works outdoors and would have an enormously high risk of developing skin cancer compared to even the fairest worker without the disease. Another genetically determined characteristic potentially influencing the development of occupational skin disease is skin type. This nomenclature refers to the sun reactivity of the skin and can be loosely correlated to the degree of skin pigmentation. It can be particularly important for workers exposed to ultraviolet light either from sunlight or from artificial sources. The susceptibility of some skin cancers may be linked to those having lower skin types. Sunburn or ultraviolet-induced skin damage will reduce the barrier function of the skin as well as increase sensitivity to irritant chemicals. Table 6.1 illustrates their characteristics (18). Skin type, however, is not a predictor of the development of irritant dermatitis from commonly encountered irritant such as sodium lauryl sulfate (41).

2.2 Age

Young workers, particularly those in the adolescent group, often incur acute contact dermatitis although age related susceptibility is not the cause. The incidence of allergic contact dermatitis is lower in childhood than in adolescence. In the elderly there is an age dependent decline in the incidence. These findings may be secondary to less exposure to allergens, lack of immune responsiveness, and limited patch test studies in these groups (42). More often, young people are placed in service jobs, for example, fast food, janitorial, or car wash, where wet work prevails and protection is difficult. At times disregard for safety and hygiene measures may be the reason. Older workers usually are more careful through experience, but aging skin is often dry and when work is largely outdoors, sunlight can cause skin cancer (2, 9).

2.3 Preexisting Skin Disease

New and old employees with preexisting skin disease are prime candidates for a supervening occupational dermatitis or an aggravation of a preexisting disease of the skin. Atopics (those with allergic tendency) are probably at a greater risk of developing irritant dermatitis from chemical and environmental stimuli. The dogmatic belief that atopy predisposes individuals to irritant and allergic contact dermatitis has not been irrefutably proven in controlled studies. Allergic contact dermatitis or irritant dermatitis from routine chemical irritants may not be a greater concern in atopic when their disease is quiescent. Certainly the presence of active or subclinical atopic dermatitis will probably lead to easier evolvement of overt dermatitis if provoked with routine chemical or physical irritants.

Other skin diseases known to be worsened by physical or chemical trauma, even though mild, are psoriasis, lichen planus, chronic recurrent eczema of the hands, and those skin conditions to which light exposure can be detrimental. The need of careful skin evaluation and job placement is clear (6, 9).

Table 6.1. Skin Characteristics

Skin Type	Characteristics
1	Always burns, never tans
	Usually burns, tans less than average (with difficulty)
3	Sometimes mild burn, tan about average
4	Rarely burns, tan more than average (easily)
5	Rarely burn, tan profusely, brown skin
6	Never burn, tan profusely, black skin

2.4 Environmental Factors: Temperature, Humidity, Season

Normal seasonal variations in temperature, humidity, wind, and incident ultraviolet light can affect the normal physiologic defense mechanisms of the body. Increased temperature coupled with high humidity can overload the normal thermoregulatory mechanisms. Anomalous thermoregulation can result in skin disease related to over-hydration from inappropriate sweating. In extreme cases temperature related diseases such as heat exhaustion and heat stroke can ensue.

Sweat can foster the trapping of dusts and suspended particles allowing percutaneous absorption in this hydrated and occluded environment. Water-soluble dusts can go into solution in sweat and cause irritation or allergy. (1, 2, 6, 9, 11, 12, 15, 40). Sweat can also lavage contaminants off the surface and keep it relatively free of irritant chemicals (43). High temperature and humidity may alone produce skin disease if pores are occluded. Miliaria (prickly heat) or occlusive eccrine sweat gland disease may produce annoying symptoms such as itching and stinging. Widespread involvement can interfere with temperature regulation secondary to altered sweat patterns.

Occupational skin diseases are generally more frequent in the warmer months for two reasons. First, hot weather discourages the use of protective clothing gear, thus allowing more unprotected skin for exposure to environmental contactants. Secondly, hot environments induce excess sweating. In addition, higher temperatures have been shown to increase the irritant potential of a particular chemical (44). Conversely, barrier function and skin hydration are reduced during the winter and susceptibility to the common irritant sodium lauryl sulfate is enhanced during the colder months (45). Also in cold weather some workers may not like to shower as frequently and hence chemical contact exposures may be lengthened (2, 7).

2.5 Personal Hygiene

Poor washing habits breed prolonged occupational contact with agents that harm the skin. Personal cleanliness is a sound preventive measure, but it depends upon the presence of readily accessible washing facilities, quality hand cleansers, and the recognition by the workers of the need to use them (1, 2, 6, 7, 9, 15, 40).

3 OCCUPATIONAL DERMATOSES

Chemical agents are unquestionably the major cutaneous hazards; however, there are multiple additional agents that are categorized as mechanical, physical, and biological causalities.

3.1 Chemicals

Chemical agents have always been and will most likely continue to be a major cause of work-incurred skin disease. Organic and inorganic chemicals are used throughout modern industrial processes and increasingly on the farm. They act as primary skin irritants, as allergic sensitizers, or as photosensitizers to induce acute and chronic contact dermatitis, which accounts for the majority of cases of occupational skin disease, probably no less than 75 to 80% (1, 2, 6, 7–15, 17, 40).

3.1.1 Primary Irritants

Most diseases of the skin caused by work result from contact with primary irritant chemicals. These materials cause dermatitis by direct action on normal skin at the site of contact if allowed to act in sufficient quantity and intensity for a sufficient time. In other words, a primary irritant can injure any normal skin and represents a nonallergic dermatitis related to characteristics of the offending agent and exposure pattern. Certain irritants, such as sulfuric, nitric, or hydrofluoric acid, can be exceedingly powerful in damaging the skin within moments. Similarly, sodium hydroxide, chloride of lime, or ethylene oxide gas can produce rapid damage. These are absolute or strong irritants and they can produce necrosis and ulceration resulting in severe scarring. More commonly encountered are the low-grade or marginal irritants that through repetitious contact produce a slowly evolving contact dermatitis. Marginal irritation is often associated with contact with soluble metalworking fluids, soap and water, and solvents such as acetone, ketone, and alcohol. Wet work in general is associated with repetitive contact with marginal irritants (2, 7).

3.1.1.1 Irritant Contact Dermatitis. Clinical manifestations produced by contact with primary irritant agents are readily recognized but not well understood (Table 6.2). Appearance of lesions may resemble other dermatoses—erythema, scaling, lichenification, vesicles, and bullae (46). Identification of the offending agent requires full understanding of the clinical picture and patient history. General behavior of many chemicals in the laboratory or in industrial processes is fairly well known and at times their chemical action can be applied theoretically to chemical action on human skin.

For example, it is known that organic and inorganic alkalies damage keratin; that organic solvents dissolve surface lipids and remove lipid components from keratin cells; that heavy

Table 6.2. Typical Primary Irritants Chemical Classes

Acids	Alkalies
Organic	Organic
Inorganic	Inorganic
Cement	Aliphatic and aromatic solvents
Soaps/detergents/surfactants	Alcohols
Metal salts	Ketones
Antimony trioxide	Petroleum
Arsenic trioxide	Coal tar
Chromium and alkaline chromates	
Cobalt sulfate	
Mercuric chloride	
Nickel sulfate	
Silver nitrate	
Zinc chloride	

metal salts, notably arsenic and chromium, precipitate protein and cause it to denature; that salicylic acid, oxalic acid, and urea, among other substances, can chemically and physically reduce keratin; that arsenic, tar, methylcholanthrene, and other known carcinogens stimulate skin to take on abnormal growth patterns. Just how these interactions take place in a molecular biological sequence remains to be explained (1, 2, 6, 7, 9, 11–15, 40).

Irritant chemicals are commonly present in agriculture, manufacturing, and service pursuits. Hundreds of these agents classed as acids, alkalies, gases, organic materials, metal salts, solvents, resins, soaps including synthetic detergents, can cause absolute or marginal irritation (15).

3.1.1.2 Chemical Burns. Under irritant dermatitis conditions, epidermal necrosis of scattered keratinocytes is typical (47). If, however, the damage to relatively large areas of epidermis is overwhelming to the defenses of the skin, necrosis of the epidermis may occur as a result of a chemical exposure. The mechanisms of such chemical burns are similar to those of irritant dermatitis, but on a more destructive scale. For many irritants, extreme exposures can result in significant chemical burns. Many of their mechanisms lie in their ability to denature epidermal proteins rapidly or cause damage by their ability to cause extreme temperature changes. A more exhaustive list can be found in de Groot's text on untoward effects of chemicals on the skin (48).

3.1.2 Allergic Contact Dermatitis

Allergic contact dermatitis represents the quintessential delayed-type hypersensitivity (type IV) allergic reaction to an external chemical. Over 3,000 allergens have thus far been identified, and allergic contact dermatitis represents approximately 20% of cases of contact dermatitis (48). Agents capable of causing contact dermatitis are generally small-molecular-weight chemicals that act as haptens. Haptens are not intrinsically allergenic but become so when they bind to an endogenous protein, usually in the epidermis or dermis, and mark the formation of a complete allergen. As with all truly allergic reactions, initial sensitization must occur before overt clinical disease can present. Sensitization occurs when an allergen is incorporated in cutaneous immune surveillance cells called Langerhans cells. They possess surface proteins (human leukocyte antigens, HLA, class II), which allow direct presentation of the digested allergen to a helper T lymphocyte (CD4 cells). This lymphocyte will become activated and migrate to regional lymph nodes, where a clone of similarly sensitized cells will proliferate under the stimulation of cytokines like interleukin 1 and 2 (49). These cells will subsequently migrate back to the area of skin that had the original contact with the allergen and cause a typical dermatitis to form. This initial sensitization may take several days to complete and will last a lifetime.

Once sensitized, subsequent challenges from the same allergen will result in a rapid elaboration and amplification of lymphocytes into the dermis and result in a dermatitis within 24–96 hours (49). The ability to become sensitized appears to be strongly mediated by genetics. Hence, at conception, the predisposition to allergic reactions from exogenous chemicals is determined. Recent work has linked specific HLA to specific sensitivity to allergens like nickel. The processes of sensitization and ultimate elicitation of an allergic contact dermatitis during a lifetime will depend on whether future exposure is sufficient to trigger the immune cascade (11).

Allergic and irritant contact dermatitis may not be clinically distinguishable. History may be useful in differentiating these eruptions since; allergic reactions usually require a longer induction period than occurs with primary irritation effects. Also cutaneous sensitizers generally do not affect large numbers of workers simultaneously except when dealing with very potent sensitizers such as epoxy resin systems, phenol–formaldehyde plastics, or plant resins such as poison ivy. Some other well-known sensitizers associated with occupation exposures are potassium dichromate by itself or contained in cement, nickel sulfate, hexamethylenetetramine, mercaptobenzothiazole, and tetramethylthiuram disulfide, among several other agents (1, 6, 7, 9, 10, 11, 13, 15, 20, 21, 50). Table 6.3 lists common allergens and their likely sources of exposure (37).

3.2 Plants and Woods

Many plants and woods cause injury to the skin through direct irritation or allergic sensitization by their chemical nature. Additionally, irritation can result from contact with sharp edges of leaves, spines, thorns, and so on, which are appendages of the plants. Photosensitivity may also be a factor.

Although the chemical identity of many plant toxins remains undetermined, it is well known that the irritant or allergic principle can be present in the leaves, stems, roots, flower, and bark (51, 52). High-risk jobs include agricultural workers, construction workers, electric and telephone linemen, florists, gardeners, lumberjacks, pipeline installers, road builders, and others who work outdoors (53).

Poison ivy and poison oak are major offenders. In California several thousand cases of poison oak occupational dermatitis are reported each year. Poison ivy and oak and sumac are members of the *Anacardiaceae*, which also includes a number of chemically related allergens as cashew nut shell oil, Indian marking nut oil, and mango. The chemical toxicant common to this family is a phenolic (catechol), and sensitization to one family member generally confers sensitivity or cross-reactivity to the others (1, 9, 11–14, 51–53).

Plants known to cause dermatitis are carrots, castor beans, celery, chrysanthemum, hyacinth and tulip bulbs, oleander, primrose, ragweed, and wild parsnip. Other plants including vegetables have been reported as causal in contact dermatitis (9, 11, 15; 51–53). One study reported the most common plant species to cause allergic contact dermatitis are *Grevillea, Compositae, Rhus*, and *Alstromeria* (54). Some plants such as dandelion and tansy and pesticides such as folpet and captafol may also produce photoallergic contact dermatitis (55).

Allergic contact dermatitis is rare among woodworkers, and occurs more commonly with hardwoods than softwoods. Procedures that produce dust or shavings, such as sanding and shaping are most likely to promote contact dermatitis. These scenarios produce airborne contact dermatitis distributions with involvement primarily on face, neck, chest, armpits, waistband, and groin areas (56).

Contact with wood may cause a skin reaction through chemical irritation, sensitization, or both Pao ferro and Brazilian rosewood are potent sensitizers, and Brazilian box tree wood (*Aspidosperma* spp.) and teak are both irritants and sensitizers. The sensitizers found

Table 6.3. Common Allergens and Sources of Exposure

Topical Medications

- Antibiotics
  - Aminoglycosides
  - Neomycin
  - Bacitracin
  - Polymyxin
  - Sulfonamides
- Therapeutics
  - α-Tocopherol (vit. E)
  - Benzocaine
  - Corticosteroids

Personal Hygiene Products/Cosmetics

- Benzophenones
- Cinnamic aldehyde
- Ethylenediamine
- Fragrances
- Lanolin
- *p*-Phenylenediamine
- Propylene glycol
- Thioglycolates

Antiseptics

- Chlorhexidine
- Chloroxylenol
- Glutaraldehyde
- Hexachlorophene
- Mercurials
- Phenylmercuric acetate
- Thimerosal (Merthiolate)

Leather

- Formaldehyde
- Glutaraldehyde
- Potassium dichromate

Glues and Bonding Agents

- Acrylic monomers
- Bisphenol A
- Cyanocrylates
- Epichlorohydrin
- Epoxy resin
- Formaldehyde
- Metharcrylates
- *p*-(*t*-Butyl)formaldehyde resin
- Toulene sulfonamide resins
- Urea–formaldehde resins

Cement

- Chromium

Preservatives

- Benzalkonium chloride
- Methylchloroisothiazolinone
- Methyldibromoglutaronitrile
- Formaldehyde
- Formaldehyde releasers
  - Diazolidinyl urea
  - DMDM Hydantoin
  - Imidazolidinyl urea
  - Quarternium 15

Plants and Trees

- Abietic acid
- Balsam of Peru
- Pentadecylcatechols
- Sesquiterpene lactone
- Rosin (colophony)
- Tuliposide A
- α Methylene gamma butyrolactone

Rubber Products

- Benzothiazoles
- Carbamates
- Dithiocarbamates
- diphenylguanidine
- hydroquinone
- Mercaptobenzothizole
- *p*-Phenylenediamine
- Sulfenamides
- Thioureas
- Thiurams

Paper Products

- Abietic acid
- Dyes
- Formaldehyde
- Rosin (colophony)

Metals

- Beryllium
- Chromium
- Cobalt
- Gold
- Mercury
- Nickel
- Palladium

in different woods have largely been members of a few categories: quinones, terpenes, phenols, stilbenes, and other miscellaneous allergens. Other agents capable of causing cutaneous injury are the chemicals used for wood preservation purposes as arsenicals, chlorophenols, creosote, and copper compounds (9, 15, 52, 57).

3.3 Photosensitivity

Dermatitis resulting from photoreactivity is an untoward cutaneous reaction usually to the ultraviolet (UV) band of the electromagnetic spectrum. This band of light spans from 200 to 400 nm. It is further stratified into three sub-bands of light; UV-A, UV-B, UV-C.

UV-C, spanning 200–280 nm, does not penetrate the upper atmosphere and has no clinical significance from natural light sources. In artificial settings, UV-C can cause marked sunburn within hours of an exposure, faster than other bands of UV light. It has been exploited for its antimicrobial potential in air handling systems. UV-B, 280-320 nm, has potent effects on keratinocytes and is capable of causing substantial acute damage to the epidermis. It can penetrate the majority of the epidermis, but cannot reach the upper portions of the dermis. It is the principal UV sub-band responsible for sunburn. UV-A, 320-400 nm, reaches the earth's surface in greater quantity but has less potency in causing acute epidermal damage than UV-B. It is a potent stimulator of melanin production, is capable of penetrating the skin to the upper dermis, and can cause DNA damage to keratinocytes. These combined effects result in damage to elastic tissue and dermal supporting structure, which causes photoaging changes such as wrinkling. It is the likely band most responsible for carcinogenesis in the skin (58).

3.3.1 Phototoxicity

The effect of exposure to UV light may be phototoxic, which is analogous to primary irritation or it may be allergic. While sunburn represents a purely phototoxic reaction, chemical phototoxicity represents the interaction of the skin with a combination of a photoactive chemical and ultraviolet light to produce a photoproduct. Photoproducts often produce free radicals capable of inducing cell death. In that regard, the epidermal damage is similar to irritant dermatitis.

Thousands of outdoor workers in construction, road building, fishing, forestry, gardening, farming, and electric and phone line erection are potentially exposed to sunlight and photosensitizing chemicals. Additionally, electric furnace and foundry operators, glassblowers, photoengravers, steelworkers, welders, and printers in contact with photocured inks experience exposure to artificial ultraviolet light. Phototoxic reactions due to certain plants, pesticides, medications, and fragrances have been well documented (11, 20, 55, 59, 60).

In the coal and tar industry, distillation can offer exposure to anthracene, phenanthrene, and acridine, all of which are well-known phototoxic chemical agents. Related products such as creosote, pitch roof paint, road tar, and pipeline coatings have caused hyperpigmentation from the interaction of tar vapors or dusts with sunlight (2, 15, 59).

Occupational photosensitivity is complicated by a number of topically applied and ingested drugs that can interact with specific wavelengths of light to produce a phototoxic or photoallergic reaction. Among such agents known to produce these effects are drugs related to sulfonamides, certain antibiotics, tranquilizers of the phenothiazine group, and a number of phototoxic oils that are used in fragrances (11, 59–61). Among the plants known to cause photosensitivity reaction are members of the Umbellifera. They include celery that has been infected with pink rot fungus, cow parsnip, dill, fennel, carrot, and wild parsnip (15, 18, 62).

The development of classic signs of allergic contact dermatitis resulting from the combined interaction of a purported plant photoallergen and ultraviolet light is termed phytophotodermatitis (Table 6.4). The photoactive chemicals in these plants are psoralens or furocoumarins and have been used for decades as therapeutic agents in the treatment of photoresponsive dermatoses like psoriasis and eczematous dermatitis.

3.3.2 Photoallergy

Photoallergic contact dermatitis may occur in the setting of ultraviolet light exposure. In allergic photocontact dermatitis, prior sensitization to the allergen is required, in contrast to phototoxic reaction, which may occur on initial introduction to a chemical. The photoallergens, in the absence of ultraviolet light, are not inherently allergenic and are incapable of causing a dermatitis in the sensitized individual. The introduction of ultraviolet light can cause changes in the molecule that render it allergenic and hence capable of inducing a rash.

3.3.2.1 Phototesting. It is often necessary to test for sensitivity to potentially photoactive chemicals since the cause of a dermatitis may not be obvious. In such a setting photopatch

Table 6.4. Photosensitizing Plants

Umbelliferae	*Rutaceae*
Anthriscus sylvestris (cow parsley)	*Citrus aurantifolia* (lime)
Apium graveolens (celery, pink rot)	*Dictamnus* (gas plant)
Daucus carota var., *savita* (carrot)	*Ruta graveolens* (rue)
Pastinaca sativa (garden parsnip)	*Citrus aurantium* (bitter orange)
Foeniculum vulgare (fennel)	*Citrus limon* (lemon)
Anethum graveolens (dill)	*Citrus bergamia* (bergamot)
Peucedanum ostruthium (masterwort)	*Compositae*
Heracleum spp. (cow parsnip)	*Anthemis cotula* (stinking mayweed)
Moracae	*Cruciferae*
Ficus carica (fig)	*Brassica* spp. (mustard)
Ranunculaceae	
Ranunculus (buttercup)	
Hypericaceae	
Psoralea corylifolia (scurfy pea, bavchi)	
Hypericum perforatum (St. John's wort)	

testing may be performed where suspected photoallergens are placed on a patient's back in duplicate. After 24 hours one test set is exposed to ultraviolet A light. (See patch test section for greater details.) Reactions on the irradiated side with negative reactions on the nonirradiated side confirm the diagnosis of photocontact allergic dermatitis. A majority of the chemicals listed in Table 6.5 are capable of causing photoallergic contact dermatitis and may be tested by photopatch testing methods. Those that typically cause phototoxic reaction such as tar derivative are generally not tested in clinical settings since history and

Table 6.5. New York University Photoallergen Series

1-(4-Isopropylphenyl)-3-phenyl-1,3-proandione	Achillea millefolium
3-(4-Methylbenzylidene) camphor	Alantolactone
6-Methylcoumarin	*p*-Aminobenzoic acid
Benzophenone-4 (sulisobenzone)	Arnica montana
Bithionol	Benomyl
Chlorhexidene	Captafol
Chlorpromazine	Captan
Cinoxate	Chamomilla romana
Dichlorophen	Chrysanthemum cinerariaefolium
Diphenhydramine	Diallydisulfide
Fentichlor	Lichen acid mix
Folpet	α-Methylene-β-butyrolactone
Hexachlorophene	Musk ambrette
Maneb	Oxybenzone (BZP-3)
Methyl anthranilate	Petrolatum control
Musk Ambrette	Propolis
Octyl dimethyl PABA	Pyrethrum
Octyl methoxycinnamate	Tanacetum vulgare
Permethrin	Taraxacum officinale
Promethazine	Zineb
Sandalwood oil	Ziram
Sesquiterpene lactone mix	
Sulfanilamide	
Thiourea	
Tribromosalicylanilide	
Trichlorocarbanilide	
Triclosan	
Other Photosensitizing Chemicals (not routinely tested)	
Acridine	
Anthracene	
Certain chlorinated hydrocarbons	
Coal tar	
Creosote	
Phenanthrene	
Pitch	

physical exam are generally revealing. Table 6.5 outlines potentially photosensitizing chemicals and those used for photopatch testing at New York University Medical Center as of September, 1999 (Occupational and Environmental Dermatology Unit, Department of Dermatology, New York University Medical Center, 1999).

3.4 Mechanical

Work-incurred cutaneous injury may be mild, moderate, or severe. The injuries include cuts, lacerations, punctures, abrasions, and burns, and these account for about 35% of occupational injuries for which worker's compensation claims are filed (16, 63). Almost 1.5 million occupational skin injuries occur annually (30).

Contact with spicules of fiberglass, copra, hemp, and so on induce irritation and stimulate itching and scratching. Skin can react to friction by forming a blister or a callus; to pressure by changing color or becoming thickened; and to shearing or sharp force by denudation or a puncture wound. Any break in the skin may become the site of a secondary infection (1, 2, 6, 9, 13–15).

Thousands of workmen use air-powered and electric tools that operate at variable frequencies. Exposure to vibration in a certain frequency range can produce painful fingers, a Raynaud-like disorder resulting from spasm of the blood vessels in the tool-holding hand. Slower-frequency tools such as jackhammers can cause bony, muscular, and tendon injury (64, 65). Jackhammer drillers and mine blasters may show soft tissue wasting, nerve impairment, and Dupuytren's contracture (66).

3.5 Physical

Heat, cold, electricity, and electromagnetic radiation sources can induce cutaneous injury and sometimes systemic effects.

3.5.1 Temperature

Thermal burns are common among welders, lead burners, metal cutters, roofers, molten metalworkers, and glass blowers. Miliaria (prickly heat) often follows overexposure to increased temperature and humidity. Increase in sweating causes over-hydration of sweat ducts and their skin exits. Usually symptoms are confined to the affected areas and involve itching. Dysregulation of core temperature from failure of integumentary thermoregulatory processes can result in serious illness. Heat exhaustion results in muscle cramping, nausea, vomiting, and fainting. Treatment involves cooling and rehydrating. Untreated heat exhaustion can progress to heat stroke. Elevated core temperature, neurologic symptoms, and lack of sweating are hallmark symptoms of heat stroke. Death occurs if aggressive electrolyte replacement and core temperature cooling is not provided (9, 67–69).

Frostbite is a common injury caused by water crystal formation within cells. These crystals mechanically disrupt cell membranes and directly cause cell death. Fingers, toes, ears, and the nose are the usual sites of injury in policemen, firemen, postal workers, farmers, construction workers, military personnel, and frozen food storage employees. The hallmark of therapy rests in the preservation of viable tissue through rapid rewarming.

Adjuvant approaches such as medication, surgery, and hemoperfusion continue to be investigated (1, 69–70).

3.5.2 *Electricity*

Severe local or widespread burns can occur from electricity. Direct contact with live wires or devices with short circuit will cause erythema, pain, or necrosis at and near the site of contact. If a circuit is made using the skin, exit wounds may also be seen. Lightning and other high-intensity electrical injuries produce a ramifying fern-shaped red lesions emulating the track of the electricity (71).

3.5.3 *Electromagnetic Radiation*

The electromagnetic spectrum spans from radio wave to gamma rays. Only some of the sub-bands in this spectrum are capable of producing dermatologic disease. Ultraviolet light has already been discussed above.

3.5.3.1 Microwaves. Thermal burn is the major hazardous potential from microwave radiation sources (68).

3.5.3.2 Lasers. Laser radiation has enjoyed exponentially expanded utility in medicine, but particularly in dermatology, ophthalmology, and surgery. Dozens of lasers are now available for therapeutic use in the treatment of cutaneous disease. They work by selectively destroying pathological tissue while sparing normal surrounding tissue. Lasers with light outputs in the visible spectrum are capable of destroying tissue containing complementary colors. For example, a laser producing yellow light will cause destruction of red tissue. This may be useful in eliminating disfiguring vascular birthmarks without the necessity of surgery and the resultant scar. Normal tissue is spared because the tissue without overt redness does not absorb the energy of the light. This protective phenomenon is augmented by an extremely short pulse time that allows normal surrounding tissues to cool before heat damage occurs. This is termed thermal relaxation time. Other lasers like the carbon dioxide laser produce nonvisible light that causes destruction to any tissue through the production of intense heat. Adverse effects of inadvertent exposure of the skin to lasers primarily rest in the thermal damage cause by absorption of the light (72).

3.5.3.3 Ionizing Radiation. Ionizing radiation exposures can occur in occupations were fissionable materials, radioisotopes, X-ray diffraction machines, electron beam operations, industrial X-rays for detecting metal flaws, diagnostic and therapeutic radiology chemicals are utilized. Erythema, scaling, depilitation and poorly healing ulcers can occur after acute large dose radiation exposures (1, 68). Lower-level exposures to ionizing radiation may cause chronic radiation dermatitis that manifests as skin thinning, scarring, and ulceration occurring in sites of previous radiation exposure years earlier (18).

3.6 Biological

Biological agents may cause a variety of primary and secondary dermatologic infections. The skin is often a portal of entry for infectious agents that can systemically affect the worker. Infections at work may occur through direct integumentary trauma. A preexisting skin condition may substantially increase the risk of acquiring a primary cutaneous or systemic infection by compromising barrier function. Personal protective equipment, particularly gloves, may result in either contact dermatitis from rubber constituents or maceration and breakdown resulting in suboptimal protection against infectious agents. For example, health care workers are continuously in contact with infectious fluids and therefore utilize gloves frequently.

3.6.1 Bacteria

Common human pathogenic bacteria can gain entrance into the skin through routine trauma in almost any occupation. The most common agents responsible for routine, trauma-induced cutaneous infections are staphylococcus and streptococcus. Office workers may be susceptible to these banal bacterial infections through paper cuts, or they may suffer from folliculitis due to over heated offices resulting in increased perspiration. Veterinarians exposed to bacteria such as *Listeria* and *Salmonella* during delivery of cattle or sheep have been reported to develop pustular dermatitis (73).

Anthrax is caused by the cocci-bacillus *Bacillus anthracis* which produces a highly toxic endotoxin resulting in death of infected animals. Anthrax causes cutaneous injury in the area of contact with infected animals and their hides. Workers at risk include those handling goat hair, wool, and hides from animals in endemic areas. A small, erythematous papule enlarges into an edematous mass surmounted by a hemorrhagic bulla, followed by necrosis and the formation of satellite pustules, most commonly on hands, arms, face or neck following minor abrasions. Systemic symptoms may include fever, malaise, weakness, lymphadenopathy and tachycardia. Anthrax is typically treated with systemic antibiotics including penicillin and aminoglycosides. Recent advances in vaccine development may result in immunity to spores as well as the toxin (74, 75).

Brucellosis (Malta fever, undulant fever) is caused by three species of a nonmotile gram negative rod, induced by handling contaminated animals or drinking infected milk. Veterinarians, meat packers, and animal laboratory workers are at risk (83). Ulcers may form at the site of inoculation, leading to systemic symptoms including fever, headache, and weakness. Chronic infection may lead to splenic absess formation. Effective treatment includes aminoglycoside and tetracycline antibiotics (74).

Erysipeloid is caused by the gram-positive rod *Erysipelothrix rhusiopathiae* that inhabits the surfaces of fresh and saltwater fish, crustacea, and poultry (especially turkey and quail) (78). Fisherman and butchers are at risk following abrasions and puncture wounds in contact with infected animals (76). A painful, progressively enlarging, well demarcated purple to erythematous plaque forms at the site of inoculation. Lymphangitis, lymphadenitis with fever joint pain, and rarely endocarditis may occur. Spontaneous resolution occurs in three to four weeks, and penicillin can hasten recovery.

Tuberculosis and atypical mycobacterium can cause primary and secondary skin infections. Cutaneous tuberculosis infections may result via minor trauma from infected tissue in pathologists, surgeons, veterinarians, farmers and butchers. The description of the various primary tubercular lesion is beyond the scope of this text. The most frequent atypical mycobacterium to occupationally infect humans is the water-borne *Mycobacterium mar-*

inum. Papules and nodules form and may eventually ulcerate, propagating along lymphatics resembling sporotrichosis. Fisherman, pet store attendants, and swimmers are at risk (79). *Mycobacterium marinum* may result in positive tuberculin tests and thus affect the predictive value of this test in areas of high prevalence (140). Treatment includes Minocycline, Rifampin and Ethambutol and Sulfa drugs (74).

Tularemia, caused by the gram-negative cocci *Bacillus francisella tularensis*, is transmitted to humans by fleas, ticks, and deerflies. Host animals include deer, squirrels, skunks, muskrats, and most commonly in the United States, cottontail rabbits. Hunters, farmers, butchers, and forest workers are at greatest risk. The inoculation site ulcerates, and regional lymph nodes become markedly enlarged (81). A generalized morbilliform and hemorrhagic eruption may accompany fever, headache, weakness, arthralgia, and myalgia. Treatment includes aminoglycoside antibiotics (74).

3.6.2 Fungi

Fungi can produce localized cutaneous disease. Yeast infections (*Candida*) can occur among those employees engaged in wet work (e.g., bartenders, cannery workers, and fruit processors). A breakdown of barrier function precedes an entrenched infection. Treatment with topical imidazole antifungals is generally effective, although oral imidazoles may be necessary in recalcitrant cases.

Superficial fungal infections are often caused by dermatophytic fungi with some occupational specificity. Table 6.6 lists potential exposure sources and the corresponding animal species. An expanding, erythematous, scaling plaque with a raised, advancing border and a clearing center is seen. Treatments with imidazole antifungals and allylamines are effective. Systemic imidazoles and allylamines may be necessary in widespread infections or those involving hair follicles.

Sporotrichosis, is seen among garden and landscape workers, florists, farmers, and miners who regularly contact soil and foliage following skin trauma. Painful erythematous nodules propagate through lymphatic channels, creating a semilinear row of nodules. Treatment with systemic imidazole antifungals is effective (82).

Coccidioidomycosis, caused by *Coccidioides immitis* fungal spore inhalation, is endemic in the semiarid portions of the United States and South America. A self limiting asymptomatic to flulike illness is the usual presentation. During the acute infection, cutaneous hypersensitivity reactions such as erythema nodosum, erythema multiforme, and urticaria can occur. In disseminated cases, cutaneous abscesses that are minimally inflammatory can occur (141).

Table 6.6. Dermatologic Fungal Infections

Fungus	Source
Microsporum canis	Dog, cat
Microsporum equinum	Horses
Trichophyton gallinae	Fowl
Trichophyton mentagrophytes	Dogs, cats, birds, cattle
Trichophyton verrucosum	Cattle

3.6.3 *Viruses*

Several occupational diseases are associated with virus infections, such as Q fever, Newcastle disease, and ornithosis. In fact, viral diseases are becoming the most important class of biological agents to cause severe illness. Pox viruses from sheep and cows cause Orf and milkers nodules, characteristic nodules with some lymphadenopathy that undergo spontaneous resolution in a few weeks. Herpetic whitlow is a herpes simplex infection usually occurring on the hands of health care workers that touch grossly infected human skin. Recurrences of blistering red skin in the area of original infection are common. Antiherpes virus medications are effective to treat recurrences but cannot eradicate human infection.

3.6.4 *Parasites/Rickettsiae*

Bakers, grain harvesters, grocers, and longshoremen are targets for certain parasitic mites that inhabit cheese, grain, and other foods. Other mites which may attack humans live on animals and fowl. *Ancylostoma caninum* and *Ancylostoma brasiliense*, are zoonotic hookworm larvae which are deposited in sandy soils of the southeastern United States and lead to infection among construction workers, farmers, plumbers, and others by entering the skin through fissures or abrasions usually in the foot. Cutaneous larval migrans, the characteristic eruption, reveals a serpiginous, slow moving, worm-like form under the skin. Early lesions may be treated with topical thiabendazole.

Seabathers eruption may occur in divers, professional swimmers, lifeguards, dockworkers, and other workers in water-based industries. A water-borne trematode cercariae, derived from birds produces this extremely itchy, papular eruption. Treatment is symptomatic (84).

Ticks, fleas, and insects can produce troublesome skin reactions and in certain instances, systemic disease such as Rocky Mountain Spotted Fever, Lyme disease, Yellow fever, and Malaria, among other vector-borne diseases with dermatologic manifestations (1, 13, 63, 85).

Outdoor workers may be susceptible to bites from animals such as snakes, sharks, dogs and cats which result in aggressive, necrotizing infections of the skin or rabies. Insects (particularly brown recluse spiders) can cause very painful and life threatening necrotic bites in victims. Hornet, wasp, and bee stings can cause painful local reactions and are life threatening if individuals are sensitized to their venoms. Such severe allergic reactions can result in compromised breathing and blood pressure with potential shock or death (1, 13, 15, 23).

4 CLINICAL FEATURES OF OCCUPATIONAL AND ENVIRONMENTAL SKIN DISEASE

The hazardous potential of the work environment is unlimited. Animate and inanimate agents can produce a wide variety of clinical displays that differ in appearance and in histopathological pattern. The nature of the lesions and the sites of involvement may provide a clue as to a certain class of materials involved, but only in rare instances does clinical appearance indicate the precise cause. Except for a few strange and unusual effects, the majority of occupational dermatoses can be placed in one of the following reaction patterns. Several materials known to be causal for each clinical type are included (1, 2, 6, 7, 9, 13, 15, 40).

4.1 Acute Eczematous Contact Dermatitis

Most of the occupational dermatoses can be classified as acute eczematous contact dermatitis. Heat, redness, swelling, vesiculation, and oozing are the clinical signs; itch, burning, and general discomfort are the major symptoms experienced. The backs of the hands, the inner wrists, and the forearms are the usual sites of attack, but acute contact dermatitis can occur anywhere on the skin. When forehead, eyelids, ears, face, and neck are involved, dust and vapors are suspected. Generalized contact dermatitis comes from massive exposure, the wearing of contaminated clothing, or autosensitization from a preexisting dermatitis, or from systemic exposure.

Usually a contact dermatitis is recognizable but whether the eruption has resulted from contact with a primary irritant or a cutaneous sensitizer can be ascertained only through a detailed history, a working knowledge of the materials being handled, their behavior on the skin, and a proper application and evaluation of diagnostic tests. Severe blistering or destruction of tissue generally indicates the action of an absolute or strong irritant; however, the history is what reveals the precise agent.

Acute contact dermatitis can be caused by hundreds of irritant and sensitizing chemicals, plants, and photoreactive agents. Some examples are noted in Table 6.7.

4.2 Chronic Eczematous Contact Dermatitis

Hands, fingers, wrists, and forearms are the favored sites affected by chronic eczematous lesions. The skin is dry, thickened, and scaly with cracking and fissuring of the affected areas. The process of skin thickening with accentuation of skin markings is called lichenification. Chronic contact dermatitis may occur from repeated exposure to marginal irritant or less commonly to low-grade exposure to sensitizers. Occasionally acute flares of dermatitis can occur when allergic contact dermatitis complicates an existing irritant dermatitis. Onychodystrophy, or abnormal nail plates may accompany chronic contact dermatitis when the distal fingers or toes are involved. A large number of materials sustain the marked dryness that accompanies this chronic recurrent skin problem. Some examples are noted in Table 6.8.

Table 6.7. Causes of Acute Contact Dermatitis[a]

Acids, dilute	Herbicides	Resin systems
Alkalies, dilute	Insecticides	Rubber accelerators
Anhydrides	Liquid fuels	Rubber antioxidants
Detergents	Metal salts	Soluble emulsions
Germicides	Plants and woods	Solvents

[a]Refs. 1, 2, 6, 7, 9, 11–15, 86, 87.

Table 6.8. Causes of Chronic Dry Appearing Skin[a]

Abrasive dusts (pumice, sand, fiber glass)	Chronic fungal infections
Alkalies	Oils
Cement	Resin systems
Cleaners (industrial)	Solvents
Cutting fluids (soluble)	Wet work

[a]Refs. 1, 2, 6, 7, 9, 11–15, 88.

4.3 Follicular Diseases

4.3.1 Acne and Folliculitis

Hair follicles on the face, neck, forearms, backs of hands, fingers, lower abdomen, buttocks, and thighs can be affected in any kind of work entailing heavy soilage. Comedones (blackheads/whiteheads) and follicular infection are common among garage mechanics, certain machine tool operators, oil drillers, tar workers, roofers, and trades-men engaged in generally dusty and dirty work.

Acne caused by industrial agents usually is seen on the face, arms, upper back, and chest; however, when exposure is severe, lesions may be seen on the abdominal wall, buttocks, and thighs. Machinists, mechanics, oil field and oil refinery workers, road builders, and roofers exposed to tar are at risk. Such effects are far less prevalent than was noted in the past.

The term folliculitis implies inflammation of the follicles caused either by an irritant chemical or infection. Acne is a term reserved for a skin eruption affecting follicles on the skin. The hallmark signs include comedones, papules (bumps less than 1 cm), cysts, pustules, and subsequent scars (89). Often folliculitis and acne are coexistent in the same person, since chemicals known to cause one can cause the other. Site of exposure will also determine the type of follicular disease that is manifested such that the face will usually manifest with acne, and the arms and legs with folliculitis. Some known causes of folliculitis and acne are listed below.

1. Asphalt
2. Creosote
3. Crude oil
4. Greases
5. Insoluble cutting oil
6. Lubricating oil
7. General purpose petroleum oils
8. Heavy plant oils
9. Pitch
10. Tar

4.3.2 Chloracne

Chloracne is a disease with similarities to acne that are limited to follicular involvement and similar-sounding names. Chloracne is not merely severe acne but differs from regular acne entirely. Clinically, yellow, straw-colored cysts first appear on the sides of the forehead, around the lateral aspects of the eyelids, and behind the ears. Comedones, pustules, pigmentary disturbances, and subsequent scarring abound. Typically the head and neck, chest, back, groin and buttocks are involved. Atypical areas may be involved if these areas are heavily or chronically exposed (86).

Chloracne differs from acne in several ways. First, the histopathology of chloracne is distinct from acne. In chloracne sebaceous glands are conspicuously destroyed rather than being large and overactive (89). Second, the exposure patterns of chloracne are distinct and for the most part include exposure to halogenated aromatic hydrocarbons. Chloracnegens are:

Hexachlorodibenzo-*p*-dioxin
Polybrominated biphenyls
Polybrominated dibenzofurans
Polychlorinated biphenyls
Polychlorinated dibenzofurans
Polychloronaphthalenes
Tetrachloroazobenzene
Tetrachloroazoxybenzene
Tetrachlorodibenzo-*p*-dioxin

Third, the natural history of chloracne and acne are divergent. Chemically induced acne and folliculitis will resolve without treatment within a week or two, and sooner if treatment is instituted. Chloracne is classically resistant to treatment, including systemic retinoids, and may last for decades after exposure.

Polychlorinated biphenyls (PCB) and dioxins are epidemiologically the most frequent cause of chloracne. During the Vietnam War, chloracne occurred from exposure to a 2,3,7,8-tetrachlorodibenzo-*p*-dioxin contaminated chlorophenoxyacetic acid herbicide called Agent Orange. These herbicides are capable of causing a variety of neurologic and irritant dermatological findings. The chloracne was likely produced by the dioxin contaminate, a potent chloracnegen.

Domestically, chloracne occurs in the occupational setting from polychlorinated biphenyls. These dielectric compounds were commonly utilized in electrical transformers. Although production of PCBs has been discontinued since the 1970s, hundreds of millions of pounds have escaped into the environment or are still present in some transformers in relatively high concentration. PCB exposure has been reported to cause illness in practically every organ system and has been described as a potential carcinogen. Despite this, rigorous epidemiological studies in humans have failed to link PCB exposure with any specific illness other than chloracne (87, 88). Hence chloracne remains the only reliable indicator of PCB exposure in humans (88). Further studies are necessary before conclusions can be drawn about other disease linkages.

4.3.3 *Miliaria*

Miliaria (prickly heat) results from waterlogged keratin that eventually blocks the sweat duct and surface opening. Workers confined to extremely warm environments may develop miliaria rubra (90). Intertrigo occurs at sites where the skin opposition allows sweat and warmth to macerate the tissue. Favored locations are the armpits, the groins, between the buttocks, and under the breasts (6, 9).

4.4 Pigmentary Abnormalities

Color changes in skin can result from percutaneous absorption, inhalation, or a combination of both entry routes. The color change may represent chemical fixation of a dye to keratin or an increase or decrease in epidermal pigment through stimulation or destruction of melanocytes, respectively (melanin).

Hyperpigmentation from excessive melanin production may follow inflammatory dermatoses, exposure to sunlight alone, or the combined action of sunlight plus a number of photoactive chemicals or plants. The opposite (loss of pigment or leukoderma) results from direct injury to the epidermis and melanin-producing cells by burns, cold injury, chronic dermatitis, trauma, or chemical interference with the enzyme system that produces melanin. Antioxidant chemicals used in adhesives, cutting fluids and lubricants, sanitizing agents, and rubber have caused complete loss of pigment (leukoderma) (91).

Inhalation or percutaneous absorption of certain toxicants such as aniline or other aromatic nitro and amino compounds causes methemoglobinemia. Jaundice can result from hepatic injury by carbon tetrachloride or trinitrotoluene, among other hepatotoxins. Pigmentary abnormalities are caused chemical agents are listed in Table 6.9 (9, 10, 47).

Table 6.9. Chemicals Involved in Pigmentary Disturbances

Discoloration	Hyperpigmentation	Hypopigmentation
Arsenic	Burns	Antioxidants
Carbon dyes	Chloracnegens	Burns
Copper salts	Acute and chronic dermatitis	Butyl catechol
Disperse dyes	Coal tar products	Acute and chronic dermatitis
Nitric acid	Dihydroxyacetone	Hydroquinone
Organic amines	Petroleum oils	Monobenzyl ether of hydroquinone
Picric acid	Photoactive chemicals	Radiation (ionizing)
Rescorcinol	Photoactive plants	*tert.*-Butyl catechol
Silver nitrate	Ionizing and nonionizing radiation	*tert.*-Amyl phenol
Trinitrotoluene	Trauma	*tert.*-Butyl phenol
		Trauma

4.5 Neoplasms

Exposures to mutagens in the workplace may result in the development of benign or malignant neoplasms on the skin. Tumors, usually occur at sites of contact with a suspected carcinogen. Occasionally, however, distant sites can be involved from systemic exposures or idiopathic causes. Tar papillomas and pitch warts are benign neoplasms associated with recurrent topical exposures to known mutagens. Malignant lesions such as basal and squamous cell carcinomas are clearly associated with recurrent ultraviolet light exposure (18, 58). Controversy still exists whether malignant melanoma is associated with specific occupational exposures, but epidemiological evidence strongly suggests that ultraviolet light (particularly exposure during childhood) as a causal factor. Melanomas are an example of a neoplasm that may arise in sites ectopic to the purported mutagenic exposure. Hence, while a history would suggest recurrent, intermittent sun exposure with blistering sunburns to exposed areas in the past, lesions may arise in sun protected areas (15, 18, 58, 92–94).

Many chemical and physical agents are classified as occupational and environmental carcinogens. Few, however, are noted as causes of skin cancer. Some cutaneous carcinogens are

Anthracene
Arsenic
Benzo-[a]-pyrene
Coal tar pitch
Coal tar
Creosote oil
Crude oils
Dibenzanthracene
Dimethyl benzanthracene
Ionizing radiation
Methyl cholanthrene
Mineral oils containing various additives
Shale oil
Soot
Ultraviolet light

Since the probable leading cause of skin cancer, sunlight, is so ubiquitous, determining causation of skin cancer can be very difficult. Sunlight is probably the major cause of occupational basal and squamous cell carcinoma, particularly among those engaged in agriculture, construction, fishing, forestry, gardening and landscaping, oil drilling, road building, roofing, and telephone and electric line installations (1, 2, 15, 43, 62, 94–98). Melanoma does not seem to be elevated in high sun exposure occupations. Rather, high income, white-collar jobs possess a higher risk of melanoma than workers routinely exposed to the sun occupationally. Intermittent intense sun exposure may prove to be more mutagenic for melanocytes than regular ultraviolet light exposures (92).

The International Agency for Research on Cancer determined that mineral oils containing various additives and impurities used in mule spinning, metal machining, and jute processing were carcinogenic to humans. Oils formerly in use that were responsible for cutaneous cancers including those affecting the scrotum in mule spinner and pressman were not as well refined as the lubricating oils being used today (43).

Systemic exposures to carcinogens that result in cutaneous neoplasms are rare. Arsenic, however, is well known to cause a variety of benign and premalignant growths as well as basal and squamous cell carcinomas. They are clinically and histopathologically unique such that they have distinct names such as arsenical keratoses and arsenical carcinomas (48).

4.6 Ulcerations

Cutaneous ulcers were the earliest documented skin changes observed among miners and allied craftsmen. In 1827, Cumin reported on skin ulcers produced by chromium (99). Today the chrome ulcer (hole) caused by chromic acid or concentrated alkaline dichromate is a less familiar lesion among chrome platers and chrome reduction plant operators (100). Perforation of the nasal septum also occurs among these employees, though in smaller numbers than occurred 20 years ago because many of the operations are now well enclosed (101). Punched out ulcers on the skin can result from contact with arsenic trioxide, calcium arsenate, calcium nitrate, and slaked lime (102). Nonchemical ulcerations may be associated with trauma and ulcers of the lower extremities in diabetics, pyogenic infections, vascular insufficiency, and sickle cell anemia (1, 2, 13, 15, 103).

4.7 Granulomas

Many agents of animate and inanimate nature cause cutaneous granulomas. Such lesions are characterized by chronic, indolent inflammatory reactions that can be localized or systematized and result in severe scar formation. Granulomatous lesions can be the result of bacterial, fungal, viral, or parasitic elements such as atypical mycobacterium, Sporotrichosis, milker's nodules, and tick bite, respectively. Additionally, minerals such as silica, zirconium, and beryllium and substances such as bone, chitin, coral, thorns, and grease have produced chronic granulomatous change in the skin (1, 15, 40, 104).

4.8 Other Clinical Patterns

The clinical patterns described above represent well-known forms of occupational skin diseases. However, there are a number of other disorders affecting skin, hair, and nails that do not fit into these patterns. Some examples follow.

4.9 Contact Urticaria

Contact urticaria (hives) results from skin contact with an allergen or nonallergenic chemical urticariant. Allergic contact urticaria represents immediate-type (type I) hypersensitivity and requires initial sensitization to occur. A recognized syndrome manifesting principally as contact urticaria with other associated symptoms such as rhinitis, conjunctivitis, asthma, and rarely anaphylaxis and death has been associated with latex proteins found in rubber products.

In the mid-1980s during the emergence and recognition of the human immunodeficiency virus (HIV) and the escalating incidence of hepatitis, standards of personal protection developed. Ultimately the term universal precautions arose and charged health-care personnel to handle all blood and body fluids as if they were infected with a serious transmissible agent. As such, the appropriate use of protective gloves made of latex became widespread. This resulted in a tremendous increase in the number of gloves being used by over five million health care workers (105, 106). For health care workers with occupational exposures to rubber gloves, the recognition of the syndrome proved useful in determining an etiology for their symptoms. The exact source of initial sensitization, however, is extremely difficult to ascertain since natural rubber is so ubiquitous in society. In fact while earlier dogma indicated a substantially higher incidence of latex allergy in health care workers, recent epidemiological studies of serum anti-Latex IgE indicates this group may have similar rates of sensitization as the general public (107–111). Post sensitization exposures at work make result in greater recognition of the allergy in affected workers.

Glove powders are known to bind latex protein allergen and become airborne when gloves are removed. Inhaled latex protein can trigger severe allergic reactions. A recent study of occupational exposures to powdered and nonpowdered latex gloves revealed no difference in the incidence of latex allergy (112). Nevertheless, latex-safe environments should be utilized to avoid symptoms in those already sensitized.

Risk factors for Latex allergy are enumerated below (113–116).

1. Atopic patients: history of eczema, hayfever, asthma.
2. Spina bifida: *carry the highest risk factor*.
3. Multiple surgical procedures.
4. History of hand eczema.
5. Frequent exposures to latex products.

Frequently contact urticaria and contact dermatitis may coexist in the same patient. Moreover, chronic urticaria of the hands may be difficult to differentiate clinically from contact dermatitis. Hence it is essential to use diagnostic tests to elucidate that exact cause of a possible rubber induced allergic or irritant phenomenon. Misdiagnosis often leads to inappropriate treatment and failure to improve. Health care institutions are now recognizing this problem and are developing strategies to detect latex allergy in employees and patients and cope with them. At New York University Medical Center an algorithm is used to screen potentially allergic employees (Fig. 6.1). Here, employees are first screened with a latex radioallergosorbent assay (RAST) test to detect latex-specific immunoglobulin E. These tests are now available through most major laboratories and are approved by the Food and drug administration. If negative, a "use" test utilizing a latex glove under supervised setting is performed, first with one finger, than an entire hand. If these tests are negative, eluted latex protein in solution is used for prick or scratch testing. The presence

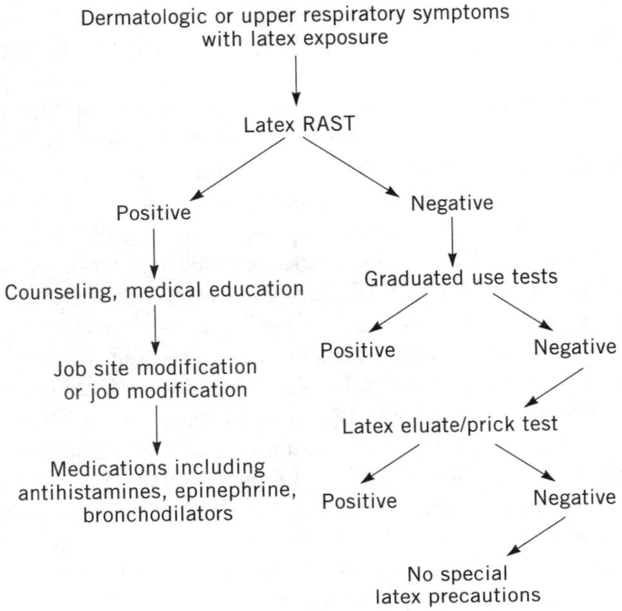

Figure 6.1. New York University Medical Center Latex Allergy Evaluation.

of a hive at the test site with appropriate controls indicates latex hypersensitivity. If found to be positive, employees use alternative gloves, and the latex in the immediate worksite is minimized. To date all new latex-sensitive employees at NYU have successfully had job-site or job-modification protocols enacted.

Occupational contact urticaria may also occur from a variety of other allergenic sources such as foods like apple, carrot, egg, fish, beef, chicken, lamb, pork, and turkey. Other sources have arisen from animal viscera and products handled by veterinarians and food dressers; from contact with formaldehyde, rat tail, guinea pig, and streptomycin. It is likely that many of these cases go unreported and unrecognized as such (10, 11, 80, 117, 118).

4.10 Nail Discoloration and Dystrophy

Chemicals such as alkaline bichromate induce an ochre color in nails; tetryl and trinitrotoluene induce yellow coloring; dyes of various colors may change the nail color; carpenters may have wood stains on their nails. Systemic medications and chemotherapeutic agents may cause specific colors and patterns on nail plates (18). Dystrophy can follow chronic contact from acids and corrosive salts, alkaline agents, moisture exposures, sugars, trauma, and infectious agents such as bacteria and fungi (1, 9, 11, 40, 119, 120). Nail cosmetics have also been shown to induce contact dermatitis and subsequent onychodystrophy (121, 122).

4.11 Facial Flush

This peculiar phenomenon has been reported from the combination of certain chemicals such as tetramethylthiuram disulfide, trichloroethylene, or butyraldoxime following the ingestion of alcohol. Because trichloroethylene is known to cause liver damage, the facial flush may be related to the intolerance associated with alcohol ingestion (1, 15, 123–125).

4.12 Acroosteolysis

Several years ago, a number of workers involved in cleaning vinyl chloride polymerization reaction tanks incurred a peculiar vascular and bony abnormality involving the digits, hands, and forearms. Raynaud's symptoms and sclerodermatous-like changes of the hands and forearms accompanied bone resorption of the digital tufts. Removal from the tank cleaning duties led to vascular and bone improvement (126, 127). Stricter permissible exposure limits for vinyl chloride monomer over the past 25 years has led to a virtual elimination of this problem in the United States.

5 CLINICAL FEATURES OF PERCUTANEOUS ABSORPTION

Physiologic issues of percutaneous absorption have been discussed above. A number of chemicals with or without direct toxic effect on the skin can cause systemic intoxication following percutaneous entry. To do so the toxicant must traverse the stratum corneum keratinocytes, and then pass through the basement membrane. At this point the toxicant has ready access to the vascular system and conveyance to certain target organs. Examples of chemicals commonly associated with systemic toxicity from percutaneous absorption are listed below (40, 128, 129).

1. Aniline
2. Benzidene
3. Carbon disulfide
4. Carbon tetrachloride
5. Chlorinated naphthalenes, biphenyls, and dioxins
6. Ethylene glycol ethers
7. Methyl butyl ketone
8. Organophosphate pesticides
9. Tetrachlorethane
10. Toluene

Over a period of years, industrial and contract laboratories have developed and used tests for predicting the toxicological behavior of a new product or one in use whose toxic action becomes questioned. Such tests on animals and humans are designed to demonstrate signs of systemic and cutaneous and toxic effects, for example, primary irritation, allergic hypersensitivity, phototoxicity, photoallergenicity, interference with pigment formation, sweat and sebaceous gland activity, dermal absorption routes, metabolic markers, and cellular aberrations indicating malformation or frank carcinoma. Conducting these tests leads to prediction of toxicological potential and diminishment thereby of the number of untoward reactions that might appear if such tests had not been done (40, 128).

6 DIAGNOSIS

It is a common assumption among employees that any skin disease they incur has something to do with their work. At times the supposition is correct, but often there is no true relationship to the work situation. Arriving at the correct diagnosis may be quite easy, but such is not a routine occurrence. The industrial health practitioner has a distinct advantage in being familiar with agents within the work environment and the conditions associated with contacting them. The dermatologist may find the diagnosis difficult if unfamiliar with contact agents in the work environment. The practitioner with little or no interest in dermatologic problems associated with work and also lacking dermatologic skill will find it most difficult to make a correct diagnosis.

6.1 History

Only through detailed questioning can the proper relationship between cause and effect be established. Taking a thorough history is time consuming because it should cover the past and present health and work status of the employee. Relevant personal and family history, particularly of allergies can be important in generating a differential diagnosis. Work history including nature of the work performed, materials handled, lengths of time on the current and past jobs will help narrow the scope of the investigation. History of the present illness including timing and location of the rash, response of the rash to medications and time away from work, and whether other employees were affected will help discern an occupational relatedness. The presence of concurrent skin disease is critical in the total evaluation of an occupational skin disease (1, 6, 7, 9, 10, 13, 15, 40).

6.2 Appearance of the Lesions

The eruption should fit into one of the clinical types in its appearance. Although the majority of the occupational dermatoses are acute or chronic eczematous contact dermatitides, other clinical types such as follicular/acneiform, pigmentary, neoplastic, ulcerative, and granulomatous can occur. Further, one must be on the lookout for the oddities that show up in unpredictable fashion, for example, Raynaud's disease and contact urticaria (1, 6, 9).

6.3 Sites Affected

A large number of cases of occupational skin diseases affect the exposed surfaces on the upper extremities such as the hands, fingers, wrists, and forearms, since they are most frequently contacted by external adverse agents. However, the forehead, face, "V of the neck", and ears may also display active lesions, particularly when employees are exposed to dusts, vapors, fumes or ultraviolet light. Although most of the work dermatoses are usually seen in the above sites, generalization can occur from massive exposure, contaminated work clothing, and also from autosensitization (id reactions) of an already existent rash. Id reactions are incompletely characterized immune phenomena that result in eczematous like lesion distinct from the site of another dermatologic process (1, 6, 7, 9, 13, 15, 18, 40).

6.4 Diagnostic Tests

Laboratory tests should be employed when necessary for the detection of bacteria, fungi, viruses, and parasites. Such tests include direct microscopic examination of surface specimens, culture of bacterial or fungal elements and biopsy of one or more lesions for histologic evaluation. When allergic reactions are suspect, diagnostic patch tests can be used to ascertain occupational, as well as nonoccupational allergies, including photosensitization. At times, useful information can be obtained through the use of analytical chemical examination of blood, urine, or tissue (skin, hair, nails) (1, 6, 7, 9, 10–15, 40).

6.5 The Patch Test

Diagnostic patch testing, properly performed and interpreted, is an invaluable procedure. The test is based on the theory that when a given sensitizing agent causes an acute or chronic eczematous dermatitis, application of the suspected material to an area of unaffected skin for 48 hours will cause an inflammatory reaction at the site of application. A positive test usually indicates that the individual has an allergic sensitivity to the test material. When the employee is working with primary irritants and fellow employees also are affected with dermatitis, the cause is self-evident and patch testing is neither necessary nor indicated. Exception to this rule occurs when the employees are working with irritant agents that can also sensitize or when an agent is highly sensitizing and many employees have become allergic. There are approximately 65,000 known irritant chemicals and 3,000 described allergic chemicals, and so the patch test is necessary in identifying a causal agent or distinguishing an irritant from allergic contact dermatitis (11, 15, 20).

If the test is to have relevance and reliability, it must be performed by one with a clear understanding of the difference between a primary irritant and a sensitizer. When patch tests are conducted with strong or even marginal irritants, a skin reaction is inevitable and usually not relevant or interpretable. However, this does not mean that a patch test cannot be performed with a diluted primary irritant. There is an abundance of published material indicating proper patch test concentrations and appropriate vehicles considered safe for skin tests (9, 10, 11, 20). The goal of testing is to use concentrations of chemicals known to cause reactions in sensitized individuals and be negative in those not allergic. Hence, test concentrations are ideally below the threshold necessary to cause an irritant reaction, but above that necessary to elicit an allergic reaction. For the most part, these concentrations do not overlap for most chemicals. Further, the one performing the tests should have a working knowledge of environmental contactants, particularly those well known as potential sensitizers (14, 15, 77, 130, 131).

The technique of the test is simple. Liquids, powders, or solids are applied under occlusive conditions under stainless steel or aluminum discs or in a hydrogel suspension to the back. The test panel should include relevant chemicals based on the history and distribution of the rash. A standard allergen panel is not useful since exposure patterns across various occupations is so divergent (132).

The North American Contact Dermatitis Group and the International Contact Dermatitis Group advocate standardized test concentrations applied in vertical rows on the back and covered by hypoallergenic tape. Contact with the test material is maintained on the skin for 48 h, and readings are made 30 min after removal, and an additional time at the 72nd to 168th hour. Reading the tests and interpreting them for the degree of reaction requires experience. True allergic reactions tend toward increased intensity for 24 to 48 h after test removal, whereas irritant reactions usually subside within 24 to 48 h after removal (11, 14).

Interpreting the significance of the test reactions is of paramount importance. A positive test can result from exposure to an irritant or a sensitizer. When specific sensitization is the case, it means that the patient is reactive to the allergens at the time of the test. When the positive test coincides with a positive history of contact, it is considered strong evidence of an allergic etiology. Conversely, the examiner must be aware that clinically irrelevant positive tests can occur if the patient is tested (*1*) during an active dermatitis phase leading to one or several nonspecific reactions; (*2*) with a marginal irritant; or (*3*) with a sensitizer to which the patient had developed an early sensitization, for example, thimerosal, but which is not relevant to the present occupational dermatitis. The patch test is incapable of testing the irritancy potential of a suspect chemical. Irritant reactions on a patch test should never be correlated with any workplace exposures. *No conclusions regarding the cause of a dermatitis can be drawn from an irritant reaction on a patch test.*

A negative test indicates the absence of an irritant or an allergic reaction. However, a negative reaction can also mean (*1*) testing omitted an important allergen; (*2*) insufficient concentration and quantity of the test allergen; (*3*) poor test condition; (*4*) hyporeactivity by the patient at the time of the test; (*5*) insufficient contact time of the allergen on the skin (133).

Performing the patch test with unknown substances the employee has brought to the physician's office can be most misleading and potentially hazardous. The material could be a strong irritant and a chemical burn or perhaps no reaction will occur. In either case, this misuse of the test will provide no help in diagnosing the cause. Useful information concerning unknown materials can be obtained by contacting the plant manager, physician, nurse, industrial hygienist, or safety supervisor (9, 11, 20, 20–22, 131, 134, 135).

Photopatch testing utilizes the same techniques described above and add ultraviolet. A light to the process. Table 6.10 lists potentially photosensitizing chemicals. This is discussed in the photosensitivity section of the chapter. Most of the photodermatoses incurred in the work area are phototoxic and thus do not require photopatch tests for diagnosis. Most extensive phototesting is performed in medical centers equipped with light boxes capable of delivering specific quantities of light (9, 11, 20, 64, 134, 135).

7 TREATMENT

Immediate treatment of an occupational dermatosis does not differ essentially from that used for a similar eruption of nonoccupational nature. In either case, treatment should be

Table 6.10. Photosensitizing Chemicals[a]

Acridine	Biothionol	Octyl methoxycinnamate
Anthracene	Chlorhexidene	Promethazine
Certain chlorinated hydrocarbons	Chlorpromazine	Sandalwood oil
Coal tar	Cinoxate	Selected plant and pesticides
Creosote	Dichlorophen	Sulfanilamide
Phenanthrene	Diphenhydramine	Thiourea
Tar pitch	Fentichlor	Tribromosalicylanilide
1-(4-Isopropylphenyl)-3-phenyl-1,3 proandione	Hexachlorophene	Trichlorocarbanilide
3-(4-Methylbenzylidene) camphor	Methyl anthranilate	Triclosan
6 Methylcoumarin	Musk ambrette	
Benzophenone-4	Octyl dimethyl PABA	

[a]Italics indicate chemicals not tested under usual phototesting conditions.

directed toward providing fairly rapid relief of symptoms. The choice of treatment agents depends upon the nature and severity of the dermatitis. Most of the cases are either an acute or a chronic contact eczematous dermatitis, and most of these can be managed with ambulatory care. However, hospitalization is indicated when the severity of the eruption warrants in-patient care.

Acute eczematous dermatoses caused by a contactant generally respond promptly to wet dressings and topical steroid preparations, but systemic therapy with corticosteroids should be used when deemed necessary. Corticosteroids have definitely lessened the morbidity in the acute and chronic eczematous dermatoses caused by work. Once the dermatitis is under good control, clinical management must be directed toward:

1. Ascertaining the cause.
2. Returning the patient to the job when the skin condition warrants, but not before.
3. Instructing the patient in the means necessary to minimize or prevent contact at work with the offending material.

In any contact dermatitis it is essential to establish the causal agents or situations that contributed to the induction of the disease. Follicular or acneiform skin lesions, notably chloracne, are notoriously slow in responding to treatment. Pigmentary change similarly may resist the run of therapeutic agents and remain active for months. New growths can be removed by an appropriate method and studied histopathologically. Ulcerations inevitably lead to the formation of scar tissue. Similarly, granulomatous lesions generally scar.

Almost all cases of occupational skin disease respond to appropriate therapy; however, when chloracne or pigmentary changes or chronic dermatitis due to chrome or nickel are the problem, therapeutic response may take months or years. It cannot be overemphasized that contact with the causative agents must be minimized, if not eliminated; otherwise return to work is accompanied by the return of the rash (1, 6–9, 13, 15).

7.1 Prolonged and Recurrent Dermatoses

As a rule, an occupational dermatosis can be expected to disappear or to be considerably improved within a period of two to eight weeks after initiating treatment and removing the cause. Yet there are cases that refuse to respond to appropriate treatment and continue to plague the patient with chronic recurrent episodes. This situation is commonly noted when cement, chromium, nickel, mercury, or certain plastics caused the dermatosis. However, all cases of recurrent disease are not necessarily associated with the above materials. The following situations may be operable in prolonged and recurrent disease (43, 142):

1. Incorrect clinical diagnosis.
2. Failure to establish cause.
3. Failure to eliminate the cause even when direct cause has been established.
4. Improper treatment.
5. Poor hygiene habits at work.
6. Supervening secondary infections.
7. Cross-reactions with related chemicals.
8. Self-perpetuation for gain.

8 PREVENTION

The key to preventing occupational skin disease is to eliminate or at least minimize skin contact with potential irritants and sensitizers present in the workplace. To do so requires:

1. Recognition of the hazardous exposure potentials.
2. Assessment of the workplace exposures.
3. Establishment of necessary controls.

Achieving these steps is more likely to occur in large industrial establishments with trained personnel responsible for the maintenance of health and safety practices. In contrast, many small plants or workplaces that employ the largest percentage of the work force have neither the money nor personnel to initiate and monitor effective prevention programs. Nonetheless, any work establishment has the responsibility of providing those preventive measures that at least minimize, if not entirely eliminate, contact with hazardous exposures.

8.1 Direct Measures

Time-tested control measures known to prevent occupational diseases are classed as primary (immediate) and indirect. The primary categories are

1. Substitution.
2. Process change.

3. Isolation/enclosure.
4. Ventilation.
5. Good housekeeping.
6. Personal protection.

8.2 Indirect Measures

The indirect measures include:

1. Education and training of management, supervisory force, and employees.
2. Medical programs.
3. Environmental monitoring.

Although small plants generally lack in-house medical and industrial hygiene services, they do have access to such services through state health departments or through private consultants knowledgeable in health and safety measures.

8.2.1 Substitution and Process Change

When a particular agent or process is recognized as a trouble source, substituting a less hazardous agent or process can minimize or eliminate the problem. This has been done with a number of toxic agents, for example, substituting toluene for benzene or tetrachloroethylene for carbon tetrachloride. Substitution has potential value in allergen replacement of known offenders such as with chromium, nickel, certain antioxidants and accelerators in rubber manufacture, and certain biocides in metalworking fluids. When feasible, substituting a nonallergen for an allergenic agent is a recommended procedure.

8.2.2 Isolation and Enclosure

Isolation of an agent or a process can be used to minimize hours of exposure or the number of people exposed. Isolation can mean creation of a barrier, or distance, or time, as the means to lessen exposure. Enclosure of processes provides a high level of safety when hazardous agents are involved. Local enclosures against oil spray and splash from metal-working fluid lessen the amount of exposure to the machine operators.

Radiation exposures can be shielded with proper barriers and remote control systems. Bagging operations can be enclosed to lessen, if not entirely eliminate, exposure to the operators involved.

8.2.3 Ventilation

Movement of air can mean general dilution and/or local exhaust ventilation used to reduce exposures to harmful airborne agents. Local exhaust ventilation is effective in controlling vapors of degreasing tanks and in mixing, lay-up, curing, and tooling of epoxy, polyester, and phenol–formaldehyde resin systems. Air handling systems should also effective maintain normal humidity conditions that balance between over dryness leading to dry skin eczema and preventing growth of biohazards.

8.2.4 Good Housekeeping

A clean shop or plant is essential in controlling exposure to hazardous materials. This means keeping the workplace ceilings, windows, walls, floors, workbenches, and tools clean. It means providing adequate storage space, properly placed warning signs, sanitary facilities adequate in number, cleanup of spills, and emergency showers for use after accidental heavy exposure to harmful chemicals.

8.2.5 Personal Protection

8.2.5.1 Clothing. It is not necessary for all workers to wear protective clothing, but for those jobs in which it is required, good-quality clothing should be issued as a plant responsibility. Protective clothing against cold, heat, and biological and chemical injury to the skin is available. Depending upon the need, equipment such as hairnets, caps, helmets, shirts, trousers, coveralls, aprons, gloves, boots, safety glasses, and face shields are available. Similarly, clothing to protect against ultraviolet light and ionizing, microwave, and laser radiation is readily available.

Once protective clothing has been issued, its laundering and maintenance should be the responsibility of the plant. When work clothing is laundered at home it becomes a ready means of contaminating family members apparel with chemicals, fiber glass, or harmful dusts. At times workers may become allergic to their protective clothing since treatment of fabrics is necessary to impart specific safety qualities such as fire protection. Formaldehyde resins are often used to improve the quality and appearance of protective cloths and uniforms and the commons ones are listed below.

Dimethyloldihydroxyethylene urea
Dimethylolpropylene urea
Tetramethylolacetylene diurea
Ethylene urea
Urea–formaldehyde
Melamine–formaldehyde

Specific information concerned with protective equipment can be obtained from the National Institute for Occupational Safety and Health and from any of the manufacturers listed in the safety and hygiene journals. These organizations can be accesses through the World Wide Web.

8.2.5.2 Gloves. Gloves are an important part of protective gear because the hands are valuable instruments at work. Leather gloves, though expensive, offer fairly good protection against mechanical trauma (friction, abrasion, etc.). Cotton gloves suffice for light work, but they wear out sometimes in a matter of hours. Neoprene and vinyl-dipped cotton gloves are useful in protecting against mechanical trauma, chemicals, solvents, and dusts. Unlined rubber gloves and plastic gloves can cause maceration and sometimes contact dermatitis from the chemical accelerators or antioxidants that leach-out by sweat in the wearer's glove. Of no small importance in choosing gloves is the reason for use. Much time and money can be saved by reviewing the catalogs and tables provided by the manufacturers.

8.2.5.3 Hand Cleansers. Of all the measures advocated for preventing occupational skin disease, personal cleanliness is paramount. Although ventilating systems and monitoring are important in controlling the workplace exposures, there remains no substitute for washing the hands, forearms, and face and keeping clean. To do this the plant must provide conveniently located wash stations with hot and cold running water, good-quality cleansers and disposable towels.

Several varieties of acceptable cleansers are available on the market and these include conventional soaps of liquid, cake, or powdered variety. Liquid varieties including "cream" soaps are satisfactory for light soil removal. Powdered soaps are designed for light frictional removal of soil and may contain pumice, wood fiber, or corn meal.

Waterless cleansers are popular among those who contact heavy tenacious soilage such as tar, grease, and paint. They should not serve as a substitute for conventional removal of soilage. Daily use of waterless cleansers leads to dryness of the skin and, at times, eczematous dermatitis from the solvent action of the cleanser.

In choosing an industrial cleanser the following are suggested:

1. It should have good cleansing quality.
2. It should not dry out the skin through normal usage.
3. It should not harmfully abrade the skin.
4. It should not contain known sensitizers.
5. It should flow readily through dispensers.
6. It should resist insect invasion.
7. It should resist easy spoilage and rancidity.
8. It should not clog the plumbing.

8.2.5.4 Protective Creams. Covering the skin with a barrier cream, lotion, ointment is a common practice in and out of industry. Easy application and removal plus the psychological aspect of protection account for the popularity of these materials. Obviously, a thin layer of barrier cream is not the same as good environmental control or an appropriate protective sleeve or glove. There are currently no protective creams available that can provide adequate defense against irritants and allergens outside of laboratory testing conditions. Contact dermatitis may only be controlled with avoidance of offending allergens and good hygiene. The use of sunscreens offer incontrovertible protection against sunburn from ultraviolet light exposure. Their use is encouraged for all workers routinely exposed to sunlight or artificial ultraviolet light sources. DEET and citronella oil containing lotions can offer protection for outdoor worker against arthropod assault; however, they may on occasion cause allergic contact dermatitis. A newer protective cream containing an organo-clay may provide some protection against the active allergen in poison ivy (136).

The development of protective creams has clearly been an effort to reduce the necessity for wearing gloves that may reduce manual dexterity. Current glove technology has sufficiently advanced to allow excellent dexterity for workers performing intricate duties. For decades now neuro-, cardiac, and microsurgeons have successfully performed surgery requiring the highest degree of adroitness while having their hands completely enveloped in latex. Few if any occupations require higher degrees of deftness that make it impossible to wear protective gloves. However, workers exposed to strong solvent chemicals may find it difficult to locate gloves that are resistant to breakdown or provide adequate barrier protection under these harsh circumstances (1, 6, 7, 9, 13, 20, 23, 134, 137–139).

9 CONTROL MEASURES

9.1 Education

An effective prevention and control program against occupational disease in general, including diseases of the skin, must begin with education. Education has been shown to decrease the incidence of occupational dermatitis. A joint commitment by management, supervisory personnel, workers, and worker representatives is required. The purpose is to acquaint managerial personnel and the workers with the hazards inherent in the workplace and the measures available to control the hazards. The training should be in the hands of well qualified instructors capable of instructing the involved people with:

1. Identification of the agents involved in the plant.
2. Potential risks.
3. Symptoms and signs of unwanted effects.
4. Results of environmental and biological monitoring in the plant.
5. Management plans for hazard control.
6. Instructions for emergencies.
7. Safe job procedures.

Worker education cannot be static. It must be periodic through the medical and hygiene personnel, during job training, and periodically thereafter through health and safety meetings. Special training courses are available at several universities with departments specializing in occupational and environmental health and hygiene.

9.3 Environmental Monitoring

Periodic sampling of the work environment detects the nature and extent of potential difficulties and also the effectiveness of the control measures being used. Monitoring for skin hazards can include wipe samples from the skin as well as the work sites and use of a black light for detecting the presence of tar product fluorescence on the skin before and after washing. Monitoring is particularly required when new compounds are introduced into plant processes.

9.4 Medical Controls

Sound medical programs contribute greatly to preventing illness and injury among the plant employees. Large establishments have used in-house medical and hygiene personnel quite effectively. Daily surveillance of this type is not generally available to small plants; however, small plants do have access to well trained occupational health and hygiene specialists through contractual agreements. At any rate, medical programs are designed to prevent occupational illness and injury, and this begins with a thorough preplacement physical examination, including the condition of the skin. When the preplacement examination detects the presence of or personal history of specific skin disease, extreme care must be exerted in placement to avoid worsening a preexisting disease.

Plant medical personnel, full time or otherwise, should make periodic inspections of the plant operations to note the presence of skin disease, the use or misuse of protective gear, and hygiene breaches that predispose to skin injury. When toxic agents are being handled, periodic biological monitoring of urine and blood for specific indicators or metabolites should be regularly performed.

Plant medical and industrial hygiene personnel should have constant surveillance over the introduction of new materials into the operations within the plant. Failure to do so can lead to the unwitting use of toxic agents capable of producing serious problems (139).

Of great importance in the medical control program is the maintenance of good medical records indicating occupational and non-occupational conditions affecting the skin, as well as other organ systems. Medical records are vital in compensation cases, particularly those of litigious character (1, 6, 7, 9, 13, 15).

A well-detailed coverage of prevention of occupational skin diseases is present in the publications, "Proposed National Strategies for the Prevention of Leading Work-Related Diseases and Injuries, Part 2" (92) and the "Report of the Advisory Committee on Cutaneous Hazards to the Assistant Secretary of Labor, OSHA, 1978" (16).

ACKNOWLEDGMENTS

This chapter is adapted from the earlier chapter by Donald J. Birmingham, M.D., F.A.C.P.

BIBLIOGRAPHY

1. L. Schwartz, L. Tulipan, and D. J. Birmingham, *Occupational Diseases of the Skin*, 3rd ed., Lea & Febiger, Philadelphia, 1957.
2. R. P. White, *The Dermatoses or Occupational Affections of the Skin*, 4th ed., H. K. Lewis & Company, London, 1934.
3. B. Ramazzini, *Diseases of Workers* (translated from the Latin text, *De Morbis Artificum*, 1713, by W. C. Wright), Hafner, New York-London, 1964.

4. P. Pott, *Cancer Scroti*, Chirurgical Works, London, 1775, pp. 734; 1790 ed., pp. 257–261.
5. "History of Technology," in *The New Encyclopedia Brittanica Macropedia*, 15th ed., Vol. 18, 1978.
6. R. R. Suskind, "Occupational Skin Problems. I. Mechanisms of Dermatologic Response. II. Methods of Evaluation for Cutaneous Hazards. III. Case Study and Diagnostic Appraisal," *J. Occup. Med.* 1 (1959).
7. M. H. Samitz and S. R. Cohen, "Occupational Skin Disease," in L. Moschella and H. J. Hurley, eds., *Dermatology,* Vol. II, S. W. B. Saunders Company, Philadelphia, 1985.
8. K. E. Malten and R. L. Zielhius, "Industrial Toxicology and Dermatology," in *Production and Processing of Plastics*, Elsevier, New York, 1964.
9. R. M. Adams, *Occupational Dermatology*, Grune & Stratton, New York, 1983.
10. H. I. Maibach, *Occupational and Industrial Dermatology*, 2nd ed., Year Book Medical Publishers, Chicago, 1987.
11. R. L. Rietschel and J. F. Fowler, *Fisher's Contact Dermatitis*, 4th ed., Williams & Wilkins, Baltimore, 1995.
12. E. Cronin, *Contact Dermatitis*, Churchill Livingston, London, 1980.
13. G. A. Gellin, *Occupational Dermatoses*, Department of Environmental, Public and Occupational Health, American Medical Association, 1972.
14. R. J. G. Rycroft, "Occupational Dermatoses," Chapter 16 in A. Rook, D. S. Wilkinson, F. J. G. Ebling, and J. L. Burton, eds., *Textbook of Dermatology*, 4th ed., Vol. I, Blackwell Scientific Publications, Oxford, 1986.
15. D. J. Birmingham, "Occupational Dermatoses," in G. D. Clayton and F. E. Clayton, eds., *Patty's Industrial Hygiene and Toxicology,* 3rd ed., Vol. I, John Wiley & Sons, Inc., New York, 1978.
16. "Occupational Illness by Type," Bureau of Labor Standards Annual Survey 1984, Fig. 1 in A Proposed National Strategy for the Prevention of Dermatological Conditions, Part 2, Association of Schools of Public Health Under a Cooperative Agreement with NIOSH, U.S. Government Printing Office, Washington, DC, 1988.
17. American Academy of Dermatology, *National Conference on Environmental Hazards to the Skin*, Comprehensive Position Statement 1992.
18. I. M. Freedberg, ed., *Fitzpatrick's Dermatology in General Medicine*, 5th ed, McGraw-Hill, New York, 1999.
19. A. M. Kligman, "The Biology of the Stratum Corneum," in W. Montagna and W. C. Lobitz Jr., eds., *The Epidermis*, Academic Press, New York, 1964, pp. 387–433.
20. E. Cronin, *Contact Dermatitis*, Churchill Livingston, London, 1980.
21. F. Henry, V. Goffin, H. I. Maibach, G. E. Pierard, "Regional differences in stratum corneum reactivity to surfactants. Quantitative assessment using the corneosurfametry bioassay," *Contact Dermatitis* **37**(6), 271–275 (1997).
22. G. A. Gellin, P. A. Possick, and V. B. Perone, "Depigmentation from 4-*tert*-Butyl Catechol. An Experimental Study," *J. Invest. Dermatol.* **55**, 190–197 (1970).
23. Report of Advisory Committee on Cutaneous Hazards to Assistant Secretary of Labor, OSHA, U.S. Department of Labor, 1978.
24. G. Kahn, "Depigmentation Caused by Phenolic Detergent Germicides," *Arch. Dermatol.* **102**, 177–187 (1979).
25. K. Malten, E. Seutter, and I. Hara, "Occupational Vitiligo due to *p-tert*-Butyl Phenol and Homologues," *Trans. St. John Hosp. Dermatolol. Soc.* **57**, 115–134 (1971).

26. K. Jimbow, T. Iwashina, F. Alena, K. Yamada, J. Pankovich, and T. Umemura, "Exploitation of Pigment Biosynthesis Pathway as a Selective Chemotherapeutic Approach for Malignant Melanoma," *J. Invest. Dermatol.* **100**(2 Suppl.), 231s–238s (1993).

27. K. Jimbow, "N-Acetyl-4-*S*-Cysteaminylphenol as a New Type of Depigmenting Agent for the Melanoderma of Patients with Melasma," *Arch. Dermatol.* **127**(10), 1528–34 (1991).

28. National Institute for Occupational Safety and Health (NIOSH): National Occupational Research Agenda (NORA). Cincinnati, US Department of Health and Human Services, DHHS Publication (NIOSH) 96-115, 1996.

29. C. D. Klaassen ed., *Casarett and Doull's Toxicology: The Basic Science of Poisons*, 5th ed., McGraw-Hill Health Professions Division, New York, 1995.

30. W. N. Rom, ed., *Environmental & Occupational Medicine* 3rd ed., Lippincott-Raven Publishers, Philadelphia, 1998.

31. P. M. Elias, "Role of Lipids in Barrier Function Of the Skin," in H. Muktar, ed., *Pharmacology of the Skin*, CRC Press, Boca Raton, FL, 1992, pp. 389–416.

32. R. O. Potts, R. H. Guy, "Predicting Skin Permeability," *Pharmacol. Res.* **9**, 663–669 (1992).

33. M. E. Johnson, D. Blankschtein, and R. Langer, "Evaluation of Solute Permeation through the Stratum Corneum: Lateral Bilayer Diffusion as the Primary Transport Mechanism," *J. Pharm. Sci.* **86**(10), 1162–1172 (1997).

34. V. B. Keeble, L. Correll, and M. Ehrich, "Evaluation of Knit Glove Fabrics as Barriers to Dermal Absorption of Organophosphorus Insecticides Using an *in vitro* Test System," *Toxicology* **81**(3), 195–203 (1993).

35. J. L. Stauber, T. M. Florence, B. L. Gulson, and L. S. Dale, "Percutaneous Absorption of Inorganic Lead Compounds," *Science of the Total Environment* **145**(2), 55–70 (1994).

36. R. J. Scheuplein and I. H. Blank, "Permeability of the Skin," *Physiol. Rev.* **51**, 702–747 (1971).

37. R. H. Rice and D. E. Cohen, "Toxic Responses of the Skin," *Casarett and Doull's Toxicology, The Basic Science of Poisons*, 5th ed., Pergamon Press, Jan. 1996.

38. W. P. McNulty, "Toxic and Fetotoxicity of TCDD, TCDF and PCB Isomers in Rhesus Macaques," *Environ. Health Perspectives* **60**, 77–88 (1985).

39. H. Mukthar and D. R. Bickers, "Comparative Activity of the Mixed Function Oxidases, Epoxide Hydratase, and Glutathione-*S*-Transferase in Liver and Skin of the Neonatal Rat," *Drug Metab. Dispos.* **9** 311–314 (1981).

40. Report of Advisory Committee on Cutaneous Hazards to Assistant Secretary of Labor, OSHA, U.S. Department of Labor, 1978.

41. J. P. McFadden, S. H. Wakelin, and D. A. Basketter, "Acute Irritation Thresholds in Subjects with Type I—Type VI Skin," *Contact Dermatitis* **38**(3), 147–149 (1998).

42. C. Kwangsukstith and H. I. Maibach, "Effect of Age and Sex on the Induction and Elicitation of Allergic Contact Dermatitis," *Contact Dermatitis* **33**(5), 289–298 (1995).

43. *Occupational Medicine State of Art Review*, Vol. 1–3, Hanley & Belfus, Philadelphia, 1986, pp. 219–228

44. P. Clarys, I. Manou, and A. O. Barel, "Influence of Temperature on Irritation in the Hand/Forearm Immersion Test," *Contact Dermatitis* **35**(6), 240–243 (1997).

45. R. A. Tupker, P. J. Coenraads, V. Fidler, M. C. De Jong, J. B. Van der Meer, and J. G. De Monchy, "Irritant Susceptibility and Weal and Flare Reactions to Bioactive Agents in Atopic Dermatitis: II. Influence of Season," *Br. J. Dermatolol.* **133**(3), 365–370 (1995).

46. N. I. Denig, A. Q. Hoke, and H. I. Maibach, "Irritant Contact Dermatitis. Clues to Causes, Clinical Characteristics, and Control," *Postgraduate Medicine* **103**(5), 199–200, 207–208, 212–213 (1998).
47. D. Elder, ed., *Lever's Histopathology of the Skin*, Lippincott-Raven, Philadelphia, 1997.
48. A. C. deGroot, J. W. Wheeland, and J. P. Nater, "Unwanted Effects of Cosmetics and Drugs Used in Dermatology," 3rd ed., Elsevier, Amsterdam, New York, 1994.
49. A. H. Enk, "Allergic Contact Dermatitis: Understanding the Immune Response and Potential for Targeted Therapy Using Cytokines," *Molecular Medicine Today* **3**(10) 423–428 (1997).
50. D. V. Belsito, "The Rise and Fall of Allergic Contact Dermatitis," *Am. J. Contact Dermat.* **8**(4), 193–201 (1997).
51. K. F. Lampke and R. Fagerstrom, *Plant Toxicity and Dermatitis* (*A Manual for Physicians*), Williams & Wilkins, Baltimore, 1968.
52. T. Barber and E. Husting, "Plant and Wood Hazards," in M. M. Key et al, eds., *Occupational Diseases, A Guide to Their Recognition*, rev. ed., U.S. Department of Health, Education and Welfare, PHS, CDC, NIOSH, DHEW-NIOSH Publication 77-181, U.S. Government Printing Office, Washington, DC, 1977.
53. G. A. Gellin, C. R. Wolf, and T. H. Milby, "Poison Ivy, Poison Oak and Poison Sumac Common Causes of Occupational Dermatitis," *Arch. Environ. Health* **22**, 280 (1971).
54. D. K. Cook and S. Freeman, "Allergic Contact Dermatitis to Plants: and Analysis of 68 Patients Tested at the Skin and Cancer Foundation," *Australasian J. Dermatol.* **38**(3), 129–131 (1997).
55. K. A. Mark, R. R. Brancaccio, N. A. Sotor, and D. E. Cohen, "Allergic Contact and Photoallergic Contact Dermatitis to Plant and Pesticide Allergens," *Arch. Dermatol.* **135**(1), 67–70 (1999).
56. K. L. Watsky, "Airborne Allergic Contact Dermatitis from Pine Dust," *Am. J. Contact Dermatol.* **8**(2), 118–120 (1997).
57. *Wood Preservation Around the Home and Farm*. Forest Products Laboratory Publication No. 1117, Canadian Department of Forestry, Ottawa, 1966.
58. H. Lim and N. Soter, eds., *Clinical Photomedicine*, Marcel Dekker, Inc., New York, 1993.
59. V. De Leo and L. C. Harber, "Contact Photodermatitis," in A. A. Fisher, ed., *Contact Dermatitis*, Williams and Wilkins, Baltimore, 1995, chapt. 23.
60. J. Ferguson and R. Dawe, "Phototoxicity in Quinolones: Comparison of Ciprofloxacin and Grepafloxacin," *J. Antimicrobial Chemother.* **40**(suppl A), 93–98 (1997).
61. E. Selvaag, "Studies on the Phototoxic Effects of Oral Antidiabetics and Diuretics," *Arzneimittel-Forschung* **47**(1), 97–100 (1997).
62. J. Epstein, "Adverse Cutaneous Reactions to the Sun," in F. D. Malkinson and R. W. Pearson, eds., *Year Book of Dermatology*, Year Book Medical Publishers, Chicago, 1971.
63. National Electric Injury Surveillance System, U.S. Consumer Product Safety Commission, 1984; U.S. Bureau of Labor Statistics Supplementary Data Systems [SDS], 1983 data.
64. N. Williams, "Biological Effects of Segmental Vibration," *J. Occup. Med.* **17**, 37–39 (1975).
65. G. A. Suvorov and I. K. Razumov, "Vibration," in *Encyclopedia of Occupational Safety and Health*, 3rd ed., Vol. II, International Labor Office, Geneva, 1983.
66. A. K. Dasgupta and J. Harrison, "Effects of Vibration on the Hand–Arm System of Miners in India," *Occupat. Med.* **46**(1), 71–78 (1996).
67. B. E. De Galan and J. B. Hoekstra. "Extremely Elevated Body Temperature: A Case Report and Review of Classical Heat Stroke," *Netherlands J. Med.* **47**(6), 281–287 (Dec. 1995).

68. E. Meso, W. Murray, W. Parr, and J. Conover, "Physical Hazards: Radiation," in *Occupational Diseases A Guide to Their Recognition*, rev. ed., U.S. Department of Health, Education and Welfare, NIOSH, U.S. Government Printing Office, Washington, DC, 1977.
69. F. N. Dukes-Dobos and D. W. Badger, "Physical Hazards Atmospheric Variations," in *Occupational Diseases Guide to Their Recognition*, rev. ed., U.S. Department of HEW, NIOSH, U.S. Government Printing Office, Washington, DC, 1977.
70. J. Foray, "Mountain Frostbite. Current Trends in Prognosis and Treatment (from Results Concerning 1261 Cases)," *Internat. J. Sports Med.* **13**(Suppl 1), S193–S196, (Oct. 1992).
71. O. Braun-Falco, G. Plewig, H. H. Wolff, and R. K. Winkelman, *Dermatology*, Springer-Verlag, New York, 1991, p. 376.
72. R. G. Wheeland, *Cutaneous Surgery,* W. B. Saunders Co., Philadelphia, 1994.
73. I. J. Visser, "Pustular Dermatitis in Veternarians following Delivery in Domestic Animals: An Occupational Disease," *Nederlands Tijdschrift voor Geneeskunde* **140**(22), 1186–1190 (1996).
74. J. G. Barnett, *Pocket Book of Infectious Disease Therapy*, Williams and Wilkins, Baltimore, 1996, pp. 20–40.
75. A. V. Stepanov, L. I. Marinin, A. P. Pomerantsev, and N. A. Staritsin, "Development of Novel Vaccines against Anthrax in Man," *J. Biotech.* **44**(1–3), 155–160 (Jan. 26, 1996).
76. V. M. Frolov and A. N. Baklanov, "Occupational Causes of Erysipeloid Morbidity among the Workers Engaged in Meat-processing Industry in the Lugansk Region," {Russian}. *Meditsina Truda I Promyshlennaia Ekologiia* (8) 18–20 (1995).
77. D. E. Cohen and R. R. Brancaccio, "What's New in Clinical Research in Contact Dermatitis," *Dermatologic Clinics* **15**, 137–148 (1997).
78. A. Mutalib, R. Keirs and F. Austin, "Erysipelas in Quail and Suspected Erysipeloid in Processing Plant Employees," *Avian Diseases* **39**(1) 191–193 (Jan.–Mar. 1995).
79. S. J. Gluckman, "Mycobacterium marinum." {Review} {40 refs}. *Clinics in Dermatology,* **13**(3), 273–276 (May–June 1995).
80. F. el Saved, D. Seite-Bellezza, B. Sans, P. Vavie-Lebev, M. C. Marguery, and J. Bazex, "Contact Urticaria from Formaldehyde in a Root-Canal Dental Paste," *Contact Dermatitis* **33**(5), 353, (Nov. 1995).
81. B. F. Kodama, J. E. Fitzpatrick, and R. H. Gentry. Tularemia. *Cutis* **54**(4), 279–280 (Oct. 1994).
82. T. Carrada-Bravo "Update on Sporotrichosis," *Australian Family Physician* **24**(6), 1070–1071, 1074 (June 1995).
83. G. F. Araj and R. A. Azzam, "Seroprevalence of Brucella Antibodies among Persons in High-risk Occupation in Lebanon," *Epidemiology Infection* **117**(2), 281–288 (Oct. 1996).
84. F. Allerberger, G. Wotzer, M. P. Dierich, C. Moritz, P. Fritsch and W. Haas, "Occurrence of swimmer's itch in Tyrol." *Immunitat und Infektion* **22**(1), 30–32 (Feb. 1994).
85. D. S. Wilkinson, "Biological Causes of Occupational Dermatoses," in H. I. Maibach and G. A. Gellin, eds., *Occupational and Industrial Dermatology*, Year Book Medical Publishers, Chicago, 1982.
86. H. Urabe and M. Asahi, "Past and Current Dermatological Status of Yusho Patients," *Environmental Health Perspectives*, **59**, 11–15 (1985).
87. M. H. Sweeny, G. M. Calvert, G. A. Egeland, M. A. Fingerhut, W. E. Halperin, "Review and Update of the Results of the NIOSH Medical Study of Workers Exposed to Chemicals Contaminated with 2,3,7,8-Tetrachlorodibenzodioxin," *Teratogenesis, Carcinogenesis, Mutagenesis* **17**(4–5), 241–247 (1997–1998).

88. R. C. James, H. Busch, C. H. Tamburro, et al., "Polychlorinated Biphenyl Exposure and Human Disease," *J. Occupat. Med.* **35** 136–148 (1993)
89. G. Plewig and A. M. Kligman, *Acne and Rosacea*, 2nd ed., Springer-Verlag Publishers, New York, 1993.
90. E. Feng and C. K. Janniger, "Miliaria," *Cutis.* **55**(4), 213–216 (1995).
91. D. Iliev and P. Elsner, "An Unusual Hypopigmentation in Occupational Dermatology: Presentation of a Case and Review of the Literature," *Dermatology* **196**(2) 249–250 (1998).
92. I. A. Pion, D. S. Rigel, L. Garfinkel, M. K. Silverman, and A. W. Kopf, "Occupation and the Risk of Malignant Melanoma," *Cancer* **75**(2 suppl.), 637–644 (1995).
93. G. S. Rogers, A. W. Kopf, D. S. Rigel, M. L. Levenstein, R. J. Friedman, M. N. Harris, F. M. Golomb, P. Hennessey, S. L. Bumport, D. F. Roses et al., "Influence of Anatomic Location on Prognosis of Malignant Melanoma: Attempt to Verify the BANS Model," *J. Am Acad. Dermatol.* **15**(2 Pt 1) 231–237 (1986).
94. F. C. Combs, *Coal Tar and Cutaneous Carcinogenesis in Industry*, Charles C Thomas, Springfield, IL, 1954.
95. E. A. Emmett, "Occupational Skin Cancer-A Review," *J. Occup. Med.* **17**, 44–49 (1975).
96. W. D. Buchanan, *Toxicity of Arsenic Compounds*, Elsevier, Amsterdam, 1962.
97. I. Berenblum and R. Schoental, "Carcinogenic Constituents of Shale Oil," *Br. J. Exp. Pathol.* **24**, 232–239 (1943).
98. E. Bingham, A. V. Horton, and R. Tye, "The Carcinogenic Potential of Certain Oils," *Arch. Environ. Health* **10**, 449–451 (1965).
99. W. Cumin, "Remarks on the Medicinal Properties of Madar and on the Effects of Bichromate of Potassium on the Human Body," *Edinburgh Med. Surg. J.* **28**, 295–302 (1827).
100. N. Williams, "Nasal Septal Ulceration and Perforation in Jiggers," *Occupat. Med.* **48**(2), 135–137 (1998).
101. S. Cohen, D. Davis, and R. Kozamkowski, "Clinical Manifestations of Chromic Acid Toxicity Nasal Lesions in Electoplate Workers," *Cutis.* **13,** 558–568 (1974).
102. D. J. Birmingham, M. M. Key, D. A. Holaday, and V. B. Perone, "An Outbreak of Arsenical Dermatoses in a Mining Community," *Arch. Dermatol.* **91,** 457–465 (1964).
103. M. H. Samitz and A. S. Dana, *Cutaneous Lesions of the Lower Extremities,* J. B. Lippincott, Philadelphia, 1971.
104. H. Pinkus and A. H. Mehregan, "Granulomatous Inflammation and Proliferation," Section IV in H. Pinkus and A. H. Mehregan, eds., *A Guide to Dermatopathology,* 2nd ed., Appleton-Century-Crofts, New York, 1976.
105. Centers for Disease Control, "Recommendations for Prevention of HIV transmission in Health-Care Settings," *MMWR* 36(suppl. 2), 1S–18S (1987).
106. K. Turjanmaa. "Incidence of Immediate Allergy to Latex Gloves in Hospital Personnel," *Contact Dermatitis* **17** 270–275 (1987).
107. T. Kibby and M. Akl, "Prevalence of Latex Sensitization in a Hospital Employee Population," *Ann. Allergy, Asthma, Immunol.* **78** 41–44, (1997).
108. R. G. Kaczmarek, B. G. Silverman, T. P. Gross, R. G. Hamilton, E. Kessler, J. T. Arrowsmith-Lowe, R. M. Moore, Jr. "Prevalence of Latex-Specific IgE Antibodies in Hospital Personnel," *Ann. Allergy, Asthma, Immunol.* **76** 51–56 (1996).
109. D. R. Ownby, H. E. Ownby, J. McCullough, and A. W. Shafer, "The Prevalence of Anti-Latex IgE Antibodies in 1000 Volunteer Blood Donors," *J. Allergy Clin. Immunol.* **97** 1188–1192 (1996).

110. M. H. Lebenbom-Mansour, J. R. Oesterle, D. R. Ownby, M. K. Jennett, S. K. Post, and K. Zaglaniczy, "The Incidence of Latex Sensitivity in Ambulatory Surgical Patients: A Correlation of Historical Factors with Positive Serum Immunoglobin E Levels," *Anesth. Analg.* **85** 44–49, (1997).

111. F. Porri, C. Lemiere, J. Birnbaum, L. Guilloux, A. Lanteaume, R. Didelot, D. Vervloet, and D. Charpin, "Prevalence of Latex Sensitization in Subjects Attending Health Screening: Implications for a Perioperative Screening," *Clin. Exper. Allergy,* **27** 413–417, (1997).

112. G. L. Sussman and G. M. Liss et al., "Incidence of Latex Sensitization among Latex Glove Users," *J. Allergy Clin. Immunol.* **101** 171–178, (1998).

113. D. Jaeger, D. Kleinhaus, A. B. Czuppon, and X. Baur, "Latex-Specific Proteins causing Immediate-Type Cutaneous, Nasal, Bronchial, and Systemic Reactions," *J. Allergy Clin. Immunol.* **89** 759–768 (1992).

114. V. J. Tomazic, E. L. Shampaine, A. Lamanna, T. J. Withrow, N. F. Adkinson Jr, and R. G. Hamilton "Cornstarch Powder on Latex Products is an Allergen Carrier," *J. Allergy Clin. Immunol.* **93** 751–758 (1994).

115. B. L. Charous, R. G. Hamilton and J. W. Yungunger "Occupational Latex Exposure: Characteristics of Contact and Systemic Reactions in 47 Workers," *J. Allergy Clin. Immunol.* **94** 12–18 (1994).

116. R. Asa, "Allergens Spur Hospitals to Offer Latex-free Care," *Mater. Manag. Health Care* 28–34 (June 1994).

117. R. B. Odom and H. I. Maibach, "Contact Urticaria: A Different Contact Dermatitis in Dermatotoxicology and Pharmacology," in F. N. Marzulli and H. I. Maibach, eds., *Advances in Modern Toxicology,* Vol. 4, Hemisphere Publishing Corp., Washington, DC, 1977.

118. H. I. Maibach, "Contact Urticaria Syndrome from Mold on Salami Casing," *Contact Dermatitis* **32**(2), 120–121 (1995).

119. F. Ronchese, "Occupational Nails," *Cutis* **5**, 164 (1965).

120. F. Ronchese, *Occupational Marks and Other Physical Signs,* Grune & Stratton, New York, 1948.

121. L. Kanerva, A. Lauerma, T. Estlander, K. Alanko and M. L. Henriks-Eckerman. "Occupational Allergic Contact Dermatitis caused by Photobonded Sculptured Nails and a Review of (Meth) Acrylates in Nail Cosmetics," *Am. J. Contact Dermatitis* **7**(2), 109–115 (1996).

122. S. Freeman, M. S. Lee, and K. Gudmundsen, "Adverse Contact Reactions to Sculptured Acrylic Nails: 4 Case Reports and a Literature Review," *Contact Dermatitis* **33**(6), 381–385 (1995).

123. W. Lewis and L. Schwartz, "An Occupational Agent (*N*-Butyraldoxime) Causing Reaction to Alcohol," Med. Ann. D. C., 25, 485–490 (1956).

124. R. D. Stewart, C. L. Hake, and I. E. Peterson, "Degreaser's Flush, Dermal Response to Trichloroethylene and Ethanol," *Arch. Environ. Health* **29,** 1–5 (1974).

125. T. Chittasobhaktra, W. Wannanukul, P. Wattanakrai, C. Pramoolsinsap, A. Sohonslitdsuk, and P. Nitiyanant, "Fever, Skin Rash, Jaundice and Lymphadenopathy after Trichloroethylene Exposure: A Case Report," *J. Med. Assoc. Thailand* **80** (Suppl 1), s144–148 (1997).

126. R. H. Wilson, W. G. McCormick, C. F. Tatum, and J. L. Creech, "Occupational Acroosteolysis: Report of 31 Cases," *J. Am. Med. Assoc.* **201,** 577–581 (1967).

127. D. K. Harris and W. G. Adams, "Acroosteolysis Occurring in Men Engaged in the Polymerization of Vinyl Chloride," *Br. Med. J.* **3,** 712–714 (1967).

128. I. R. Tabershaw, H. M. D. Utidjian, and B. L. Kawahara, "Chemical Hazards," Section VII in *Occupational Diseases A Guide to Their Recognition,* rev. ed., U.S. Department of HEW-NIOSH Publication No. 77–181, 1977.
129. N. H. Proctor and J. P. Hughes, *Chemical Hazards in the Workplace,* J. B. Lippincott, Philadelphia, 1978.
130. M. B. Sulzberger and F. Wise, "The Patch Test in Contact Dermatitis," *Arch. Dermatol. Syphilol.* **23**, 519 (1931).
131. H. I. Maibach, "Patch Testing: An Objective Tool," *Cutis* **13,** 4 (1974).
132. D. E. Cohen, R. Brancaccio, D. Andersen, and D. V. Belsito, "Utility of a Standard Allergen Series Alone in the Evaluation of Allergic Contact Dermatitis: A Retrospective Study of 732 Patients," *J. Am. Acad. Dermatol.* **36**(6 Pt 1), 914–918 (1997).
133. M. K. Kosann, R. R. Brancaccio, J. L. Shupack, A. G. Franks Jr, and D. E. Cohen, "Six-hour Versus 48-Hour Patch Testing with Varying Concentrations of Potassium Dichromate," *Am. J. Contact Dermat.* 9(2):92–5 (1998).
134. S. Fregert, *Manual of Contact Dermatitis,* Munksgaard, Copenhagen, 1974.
135. K. E. Malten, J. P. Nater, and W. G. Von Ketel, *Patch Test Guidelines,* Dekker & Van de Vegt, Nigmegen, 1976.
136. J. G. Marks Jr, J. F. Fowler Jr, E. F. Sheretz, R. L. Rietschel, "Prevention of Poison Ivy and Poison Oak Allergic Contact Dermatitis by Quaternium-18 Bentonite," *J. Am. Acad. Dermatol.* **33**(2 Pt 1), 212–216 (1995).
137. C. D. Calnan, "Studies in Contact Dermatitis XXIII Allergen Replacement," *Trans. St. John's Hosp. Dermatol. Soc.* **56,** 131–138 (1970).
138. D. J. Birmingham, *The Prevention of Occupational Skin Disease, Soap & Detergent* Association, New York, 1975.
138. Prevention Planning, Implementation, Evaluation and Recommendations in Proposed National Strategy for the Prevention of Dermatological Conditions in Proposed National Strategies for the Prevention of Leading Work-Related Diseases and Injuries, Part 2, The Association of Schools of Public Health under a Cooperative Agreement with NIOSH, 1988.
139. D. Hood, "Practical and Theoretical Considerations in Evaluating Dermal Safety," in V. Drill and P. Lazar, eds., *Cutaneous Toxicity,* Academic Press, Inc., New York, 1977.
140. L. Joe and E. Hall, "*Mycobacterium macinum* Disease in Anne Arundel County: 1995 Update," *Md. Med.* **44,** 1043–1046 (1995).
141. R. D. Feigin, P. G. Shackelford, R. D. Lins, and D. K. Fey, "Subcutaneous Abscess Due to *Coccidioides immitis,*" *Am. J. Dis. Child.* **124,** 734–735 (1972).
142. G. E. Morris, "Why Doesn't the Worker's Skin Clear Up? An Analysis of Factors Complicating Industrial Dermatoses," *Arch. Ind. Hyg. Occup. Med.* **10,** 43–49 (1954).

CHAPTER SEVEN

Theory and Rationale of Exposure Measurement

Jeremiah R. Lynch, CIH

1 INTRODUCTION

This chapter explains why workplace measurements of air contaminants are made, discusses the options available in terms of number, time, and location, and relates these options to the criteria that govern their selection and the consequences of various choices. In addition, this chapter discusses industrial hygiene exposure assessment methods in the broader context of exposure assessment as it is used outside the workplace (1).

A person at work may be exposed to many potentially harmful agents for as long as a working lifetime, upward of 40 years in some cases. These agents occur singly and in mixtures, and their concentration varies with time. Exposure may occur continuously or at regular intervals or in altogether irregular spurts. The worker may inhale the agent or be exposed by skin contact or ingestion. As a result of these exposures and depending on the magnitude of the dose, some harmful effect may occur. All measurements in industrial hygiene ultimately relate to the dose received by the worker and the harm it might do.

Early investigators of the exposure of workers to toxic chemicals encountered obviously unhealthy conditions as evidenced by the existence of frank disease. Quantitative measurements of the work environment to estimate the dose received by the afflicted were not needed to establish cause-and-effect relationships and the need for exposure remediation measures. At the same time, the ability of these early industrial hygienists to make measurements was severely limited because convenient sampling equipment did not exist and analytical methods were insensitive. Pumps were driven by hand, equipment was large and heavy, filters changed weight with humidity, gases were collected in fragile glass

Patty's Industrial Hygiene, Fifth Edition, Volume 1. Edited by Robert L. Harris.
ISBN 0-471-29756-9

vessels, absorbing solutions spilled or were sucked into pumps, and laboratory instrument sophistication was bounded by the optical spectrometer. To collect and analyze only a few short-period samples required several days of work. The probability of failure due to one of many possible equipment defects or other mishaps was high. Consequently, few measurements were made, and much judgment was applied to maximize the representativeness of the measurements or even as a substitute for measurement.

Changes in working conditions, in technology, and in society have changed the old methods of measurement.

- With few exceptions, workplace exposure to toxic chemicals is much below what is commonly accepted as a safe level.
- As a consequence of the reduction of exposure, frank occupational disease is rarely seen. Much of the disease now present results from multiple factors of which occupation is only one.
- Workers have the right to know how much toxic chemical exposure they receive, and this often results in a need to document the absence of exposure.
- Technology provides enormously improved sampling equipment that is rugged and flexible. This equipment, used with analytical instruments of great specificity and sensitivity, has largely replaced the old "wet" chemical methods.

As a consequence of these changes in the workplace and advances in technology, it is now both necessary and possible to examine in far more detail the way in which workers are exposed to harmful chemicals. Personal sampling pumps permit collection of contaminants in the breathing zone of a mobile worker. Pump-collector combinations are available for long and short sampling periods. Passive dosimeters, which do not require pumps, are available for a wide range of gases and vapors. Systems that do not require the continual attention of the sample taker permit the simultaneous collection of multiple samples. Data loggers can continuously record instrument readings in a form easily transferable to a computer. Automated sampling and analytical systems can collect data continuously. Sorbent-gas chromatograph techniques permit the simultaneous sampling and analyses of mixtures and, when coupled with mass spectrometers, identify obscure unknowns. Sensitivities have improved to the degree that tens and hundreds of ubiquitous trace materials begin to be noticeable.

At the same time the demands placed on our information gathering systems are greater. Now we must not only answer the question, "Is exposure to this agent likely to harm anyone?" but provide data for many other purposes. Worker exposure must be documented to comply with the law (2). Employees are demanding to be told what they are breathing, even in the absence of hazard. Epidemiologists need data on substances not thought to be hazardous to relate to possible future outbreaks of disease (3). Design engineers need contaminant release data to relate to control options (4). Process operators want continuous assurance that contaminant levels are within normal bounds. Management information systems that issue status reports when queried require monitoring of data inputs. Data needs are so pervasive that there is a tendency to monitor without a clear idea of what the data will be used for or whether it will meet the need. An overall purpose of this chapter is to suggest the objectives that need to be considered in an analysis of the value of exposure measurement.

2 OBJECTIVES OF EXPOSURE MEASUREMENT

The central question that must be asked before measuring exposure is, "What use will be made of the data?" That is, what questions will the data answer or what external information need is to be satisfied? The collection of data should be looked on as part of a decision-making process. If no decision is to be made, or if nothing is to be done differently either in the short or long run as a result of the data collected, regardless of the result, then why collect the data? In some situations a correct control decision may be perfectly clear without any measurements, although measurements may serve to reinforce the decision or to convince others. In other cases it is difficult to see how any obtainable data will aid in making decisions, or it may be clear that the cost of obtaining the data needed exceeds the cost of making the wrong decision. To make decisions under uncertainty, as is usually required in industrial hygiene, the techniques of decision analysis (5) are useful. These techniques also permit the calculation of the value of information, which can then be compared with the cost. While the cost of information for an identified decision is important, it is also useful to consider what other questions will need to be answered or what other information will be needed. The resources available for the measurement of exposure are usually limited, so data must serve several purposes. Some of those purposes are discussed in the following sections.

2.1 Hazard Recognition

As a starting point for a complete assessment of the risk to health posed by an occupational environment, it is necessary to know the substances to which workers are exposed. Systematic recognition of all possible hazards requires inventories of materials brought into the workplace, descriptions of production processes, and identification of any new substances, by-products, or wastes. However, these sources of information may not be enough to identify all substances, particularly those present as trace contaminants or substances generated by production process, either inadvertently or as unknown by-products. To complete the identification of all substances present, before going to the next step of evaluating exposure and risk, it may be necessary to make some substance recognition measurements. Since these measurements, which are typically made by such techniques as gas chromatography–mass spectrometry (GC-MS), are not intended to evaluate exposure, they may be area samples rather than personal samples and may be large-volume samples for maximum sensitivity.

2.2 Exposure Evaluation

The most common reason for measuring worker exposure to a toxic chemical is to evaluate the health significance of that exposure. These evaluations are usually made by comparing the result with some reference level. Traditionally, the threshold limit values (TLVs) for airborne contaminants of the American Conference of Governmental Industrial Hygienists (ACGIH) (b) have been used to represent safe levels as accepted in the United States.

The introduction to the TLV list states that the TLVs are "conditions under which it is believed that nearly all workers may be repeatedly exposed day-after-day without adverse effect" and goes on to discuss the classes of workers who may not be protected by the limits. It notes that "these limits are not fine lines between safe and dangerous concentrations nor are they a relative index of toxicity. They should not be used by anyone untrained in the discipline of industrial hygiene." The reason for this last stipulation is that industrial hygiene training covers the caveats that apply to the TLVs. Appropriate judgment needs to be used in their application to take into account their uncertainty and to properly describe to workers and management what results are to be expected from using the TLVs.

In 1970 the U.S. Occupational Safety and Health Administration (OSHA) was given the responsibility for setting legally enforceable permissible exposure limits (PELs) for U.S. workplaces (7, 8). The ACGIH TLV list of 1968 was the source of the original PELs, and a later TLV list was one of the sources of a later update. Other countries have also used the TLV list as a basis for their standards, but more recently they have established their own standard-setting mechanisms (9).

Unfortunately, established limits such as the OSHA PELs, ACGIH TLVs, and AIHA Workplace Environmental Exposure Limits (WEELs) cover only a small fraction of the chemicals that are found in industrial workplaces, albeit they include the most common ones. Where there is exposure to a substance for which there is no established or recommended limit, it may be necessary to develop a supplemental standard for use in a particular plant or company. Standards for substances whose toxicology is not well known are generally set to avoid acute effects in humans or animals and, often by analogy to other better-documented substances, at a level low enough to make chronic effects unlikely. When very little is known about a substance, it may be possible only to estimate a lower level at which it is reasonably certain that no adverse effects occur and an upper level at which adverse effects are likely. The width of the gap between these two levels is a zone of uncertainty that needs to be considered in evaluating the results of exposure measurements.

2.3 Control Effectiveness

When changes in equipment or processes are made that affect the release of substances that are contributing to worker exposure, measurements of the magnitude of that change may be needed. These measurements provide empirical data on control effectiveness to confirm design expectations or are used as a basis for the design of other modifications. In the simplest case, before-and-after measurements are made when a new control, such as a local exhaust hood, is installed at a contaminant release point. From the results of these measurements, it is possible to predict the reduction in worker exposure, which can be confirmed by subsequent exposure measurement.

Unfortunately, the situation is rarely that simple. Most worker exposures are caused by multiple release points, creating a work environment of complex spatial and temporal concentration variations through which the worker moves in a not-altogether-predictable manner. Furthermore, interaction between several release points or other factors in the environment may confound the results. A needed improvement such as a new exhaust hood may seem to be without effect because the building is air starved, or a poor hood may seem to function well because of exceptional general ventilation. The time of contaminant release may depend on obscure and uncontrollable process operation factors. As with measurements made for other purposes, control evaluation studies must be carefully designed and are likely to consist of a series of factorial measurements analyzed by statistical methods.

2.4 Model Validation

Various physical models have been developed to predict contaminant concentrations in a workplace (10, 11, 12, 13, 14). These models may describe near-field dispersion from a source such as a valve leak or an open tank, or the mixing of a contaminant in an enclosed space. The models can be used to predict exposure to a chemical that has not yet been manufactured, to estimate past exposure, and to extend the range of exposure measurement to situations not measured. As physical models, they are based on first principles (heat and mass transfer) and empirical observations. The models are likely to be accurate when the assumptions made in developing the models are true. However, these assumptions are rarely perfectly true, and so it becomes important to know how sensitive the model is to deviations from the assumptions. To learn this, it is necessary to validate the model by comparing the model with actual exposure measurements over a wide range of conditions. The number of physical models in use has increased more rapidly than validation studies, so there are a number of unvalidated models.

Statistical models based on exposure data coupled with factors, such as job tasks, believed to be associated with exposure are also used to estimate worker exposure (15). Some statistical technique such as regression analysis is used to combine the data so that the factors contributing to exposure may be used as independent variables to predict unknown exposures. Since the model is developed from the data, there is no need for validation as long as the model is applied within the range of the measurements used to create it. Additional measurements may be used to extend the model, increase confidence in the output, or to replicate the model in apparently similar circumstances.

2.5 Methods Research

Industrial hygiene methods development research hypotheses often take the following form: "Will sampling and analytical method A give the same result as method B?" If we are unable to reject this hypothesis in a carefully designed experiment, then we accept that methods A and B are equivalent within our limits of error and given the bounds of the experimental conditions (16, 17, 18, 19).

Sampling and analytical methods development research usually requires extensive laboratory work, but in most cases field testing is necessary because completely realistic environments with interferences usually cannot be generated in laboratory chambers, and the difficulties of making field measurements introduce errors that may affect one method more than another. For these reasons, most practicing industrial hygienists tend to distrust assertions of equivalence that are not backed up by field data. To be credible, experiments of this kind should clearly define the range of concentrations and conditions over which the equivalence has been tested. Personal versus area equivalence of coal mine dust measurements made in long-wall mines should not be assumed to hold in room-and-pillar mines. Manual versus automated asbestos counting relationships based on chrysotile do not apply when counting amosite fibers.

Enough data should be collected not only to determine whether the methods are correlated but also to determine the ability of a measurement by one method to define the confidence limits on a prediction of the result that would have been obtained by the other method (Figure 7.1). Often a high correlation coefficient is obtained when many pairs of measurements have been made, indicating that the two methods are certainly related; yet the scatter is such that one method may be used only as a predictor of the other method to within an order of magnitude. The design of experiments for the purpose of measuring method equivalence is discussed elsewhere in this series. Considerations such as the environmental variability related to location, time, and numbers, which are described later, should be taken into account to maximize the range of conditions over which the equivalence is evaluated without introducing so much error that the relationship has no predictive ability.

2.6 Source Evaluation

Measurements can be made to evaluate the magnitude or trends in the generation of air contaminants at the source. These measurements are made to detect leaks, inadvertent loss

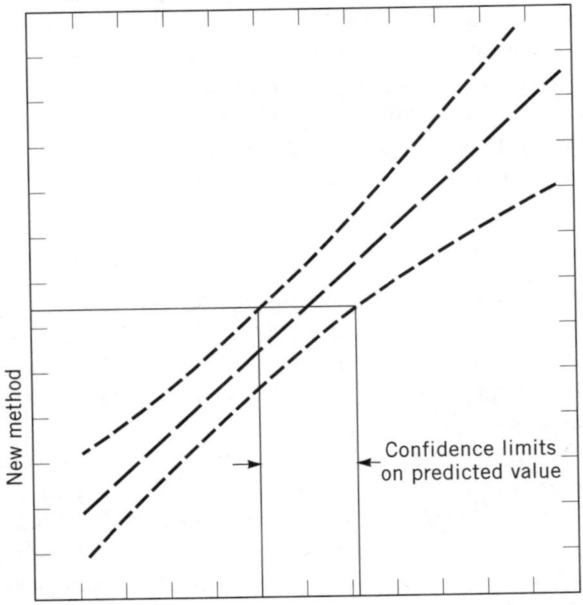

Figure 7.1. Methods comparison.

of control, or other events that may cause a change in the amount of contaminant released into the workplace. Patterns of contaminant generation may change because of loss of temperature control in a vessel or tank, dullness of a chisel in a mine or quarry, or bacterial contamination of a cutting oil. These events cannot always be detected by changes in process parameters. It may be unacceptable to wait until they show up in exposure measurements because the margin for error in a control system designed for every strict standard, such as vinyl chloride, may be too small, or as in the case of hydrogen sulfide, because the consequences of overexposure, even for a short period, are too serious.

Automatic leak detection systems may be installed where it is important to instantly detect any leak, or leaks may be detected by periodic manual surveys that check spot concentrations near pump and valve seals, flanges, and so forth. Various environmental laws administered by the U.S. Environmental Protection Agency (USEPA) require the monitoring of equipment leaks for reporting purposes under the Superfund Amendments and Reauthorization Act (SARA) Toxic Release Inventory and for the control of fugitive emissions of volatile organic compounds (VOCs) or volatile hazardous air pollutants (VHAPs). These emissions are the same emissions that are the sources of most worker exposure in chemical plants and oil refineries. The control of these emissions will reduce worker exposure so that leak detection and repair (LDAR) programs required for environmental reasons also serve occupational health objectives. Substantial work has been done to develop and evaluate methods for this purpose (20, 21, 22).

2.7 Epidemiology

Occupational epidemiology is concerned with the relationships between occupational factors and disease trends. The kinds of conclusions the epidemiologist may draw from a study depend on the kind of data used in the study. If the epidemiologist knows only the industry or place of employment of the individuals in the study group, then the conclusions can only be in the form that a disease excess may be associated with work in that industry or establishment. This is a scientifically sound conclusion, but it does not lead the managers and industrial hygienists responsible for worker health to the chemical agent, if any, which may be causing the disease. To come closer to useful causal relationships, the epidemiologist needs to know to what substances each worker was exposed. With data on the degree of exposure it is possible for the epidemiologist to detect dose-response relationships that can aid in confirming a causal relationship between an agent and a disease. If the degree of exposure is accurately known over time, a dose-response relationship, which can be used to estimate safe levels of exposure, may be calculated.

Ideally, the epidemiologist would like to have measurements of the exposure of all workers to all substances over time from the beginning of employment. In the National Coal Board study of pneumoconiosis in coal workers in the United Kingdom, exposure and health status were measured over a long enough period so that the results could be used as a basis for the present coal dust standard in the United States. Major prospective studies such as this, however, are rare. Historically, studies of causality and dose-response for chronic diseases of occupational origin used employment within an industry and/or job as a surrogate for exposure. This resulted in major misclassification of employees and/or of exposure categories, with the result that causal relationships were obscured and significant associations not found (23). As a result, most modern occupational epidemiologists conducting retrospective studies enlist the aid of industrial hygienists to reconstruct exposure by making use of whatever exposure measurements are available. In addition, they will use data on plant, process, and maintenance events that influence exposure, and employee recollection of exposure conditions. Typically, this assembly of information is converted into a job time-exposure matrix by one or a panel of several experienced industrial hygienists. Validation studies may be used to compare judgments with known exposures. These methods are being continuously improved so that the degree of reduction in misclassification now depends largely on the effort put into retrospective exposure assessment (24, 25, 26).

Prospective exposure assessment for epidemiology is hindered by the fact that most industrial hygiene measurements are made for purposes other than epidemiology (i.e., hazard evaluation, problem solving) and do not represent the exposure of the whole population. Harris (27) has addressed the question of what the plant industrial hygienist, who has day-to-day responsibility for worker protection, can do to provide for future epidemiology studies, without displacing more immediate and urgent work. These actions include taking more samples chosen to represent the exposure of all groups of workers, including those not at risk of overexposure by current standards, and to represent exposure to substances not presently known to present a special risk. Obviously, the development of such a program is complicated by our inability to predict what workers or substances will be of interest in future research. It may be that a few "fingerprint" samples, analyzed by such detailed methods as capillary GC with mass spectroscopy, will be the best choice to generally characterize the kinds of exposures that are occurring. In addition to more sampling, the collection and preservation of process, maintenance, and job history data will greatly benefit future exposure reconstruction. When the resources are available and the need for future epidemiology is evident, exposure estimation schemes may be used to fill in the gaps between what is measured and what is not. These schemes rely on experience based on observation to place workers in exposure categories over time.

In deciding what data to collect for use in future epidemiological studies, it should be known if the data will meet the need. Close early cooperation between industrial hygienists and epidemiologists is needed to avoid expensive data collection programs that fulfill no need.

2.8 Illness Investigations

When an employee has a frank occupational disease, such as lead poisoning, confirmed by both physical findings and analysis of biological materials, or is known to have been overexposed based on analysis of biological materials, the industrial hygienist should determine the cause (source) of the overexposure. When the conditions that led to the overexposure still exist, they may be evaluated by measurements made after the event. It is also possible to evaluate overexposure that resulted from past episodes that were not observed and evaluated when they occurred. A history of past exposure opportunities can be constructed and used to estimate those that led to the present case. In some instances, it may be necessary to reenact or simulate an event to measure what may have happened—

being careful, of course, to ensure that all participants are protected. Exposure, obviously, need not always be by inhalation and may include off-the-job activities.

A much more difficult investigation is the search, in an occupational setting, for the cause of an outbreak of illness or complaints of illness that may or may not be of occupational origin, or even if related to occupation may result from factors other than exposure to toxic substances. Marbury and Woods (28, p. 306) relate that "since the early 1970s, outbreaks of work-related health complaints have occurred in large numbers in a wide variety of nonindustrial workplaces such as hospitals, schools, and office buildings. In some cases, careful evaluation of the building or the affected population has revealed an agent responsible for the outbreak. In most cases, however, no specific etiologic agent can be identified as its cause."

The assessment of indoor air quality and its relation to work-related health complaints involves environmental and personal monitoring for such known agents as environmental tobacco smoke, carbon monoxide, nitrogen dioxide, formaldehyde, and volatile organic compounds (VOCs) In addition, carbon dioxide can be used as a surrogate for the absence of fresh (outdoor) air. Indoor air can also be a vector for the transmission of infectious disease organisms such as Legionella. Beyond the measurement of air contaminants, the resolution of indoor air problems involves detailed knowledge of human responses, building system performance, and factors affecting the service of the building air-handling plant. Indoor air quality assessment is discussed in detail in Volume IV of this series and by Samet and Spengler (29) and Nagda et al. (30).

2.9 Legal Requirements

Section 6b7 of the Occupational Safety and Health Act provides for "monitoring or measuring employee exposure at such locations and intervals, and in such a manner as may be necessary for the protection of the employees" (7). Under this act, OSHA has responsibility for establishing exposure limits (PELs) in the working environment. Elsewhere in the act OSHA is required to set standards that require monitoring to be performed. Responsible and effective implementation of the congressional intent behind these provisions require that OSHA devise a scheme that requires monitoring where it will be of value in the protection of worker health and not elsewhere. The ideal regulation should not apply to the majority of establishments, which have no conceivable hazard resulting from the substance being regulated and should apply requirements of increasing strictness to other establishments as the significance of the hazard in the establishment increases, ultimately calling for measurements of sufficient frequency to ensure that the potential for harm is fully assessed in the few establishments where exposures are great enough to create significant risk. Further, this should be done by a regulatory scheme that is easy to understand and implement.

This sorting of workplaces by level of risk can be done by prescribing a series of thresholds or triggers that lead to increasingly stringent requirements for a decreasing number of employers. First, all employers who do not have the substance present in the workplace should not be required to monitor. Although the "presence" of a material seems a simple enough criterion that everyone would interpret in the same way, the extreme bounds of interpretation, which are of concern in legal arguments, include the presence of as little as a few molecules of a gas or a single asbestos fiber. As analytical techniques become more sensitive, almost everything is to be found almost everywhere, at least at the level of a few molecules. What is needed in a regulation is an exclusion, such as a percentage concentration in a liquid, below which the presence of a substance is of no health significance.

For employees who work in a place where a substance is present above the excluded level, the next step should be to determine whether there is any possibility that the substance is released into the workplace such that workers may be exposed. The setting of this threshold must reflect consideration of the conditions under which the substance is present and the consequences of release. Thus, nuclear reactor decay products are continuously monitored against the possibility of leaks even when they are hermetically sealed. Such high toxicity materials are not released into the workplace except under very rare emergency circumstances. On the other hand, cadmium released into the workplace as a result of silver soldering should be monitored, but it is not necessary to measure exposure to cadmium where cadmium-coated auto parts are stored. Since no simple "potential for release" trigger has yet been devised, there is some regulatory error (employers included who should not have been, and vice versa) at this decision point. A further step is needed, therefore, before a full monitoring program with its consequent expense is mandatory. One step is to use a small number of measurements of the exposure of the maximum risk employee under conditions when the exposure is likely to be the greatest. If the results of these measurements are sufficiently below the PEL so that it is possible to be confident that the PEL is not likely to be exceeded, then further monitoring to demonstrate compliance with the PEL is not necessary. However, any change in circumstance that may increase the risk is cause for a reevaluation. This initial measurement scheme, however, is not appropriate with a highly toxic substance, for which continued vigilance must be maintained against the possibility of leaks or other inadvertent releases.

In cases where it has been established, by means of an initial measurement or data from other sources, that significant exposure is occurring, possibly at a level that is over the PEL on occasion, a regular program of periodic monitoring should be required. The frequency of monitoring should relate to the level of exposure and should consider trends between measurements that might lead to conditions with unacceptable consequences.

Monitoring programs undertaken to meet legal regulatory requirements also meet the needs of worker protection if the PEL is appropriate and if unregulated substances and risk situations are evaluated by other means, but it should be understood that compliance testing is not identical to risk assessment (31).

2.10 Routine Monitoring

Many employers attempt to meet most of the purposes of exposure measurement through a simple routine monitoring program. Schemes for the logical stepwise analysis of data to arrive at decisions regarding monitoring have been developed by the U.S. National Institute for Occupational Safety and Health (NIOSH) (32), European Council of Chemical Manufacturers' Federations (CEFIC) (33), the West German Federal Ministry of Labor (BMA) (34), and the American Industrial Hygiene Association (35). The CEFIC occupational exposure analysis flowchart is shown in Figure 7.2. At the start, a chemical inventory of

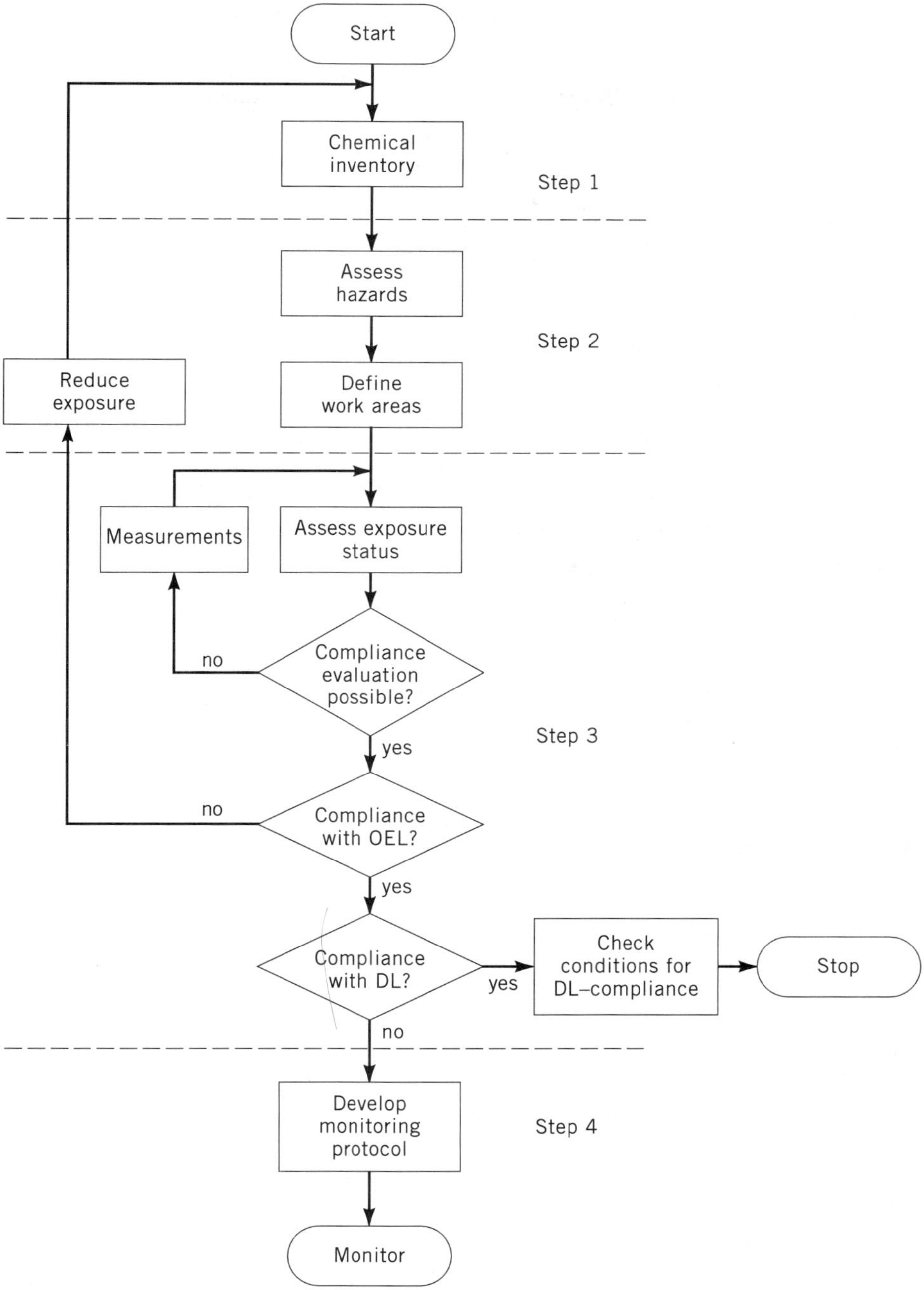

Figure 7.2. CEFIC occupational exposure analysis chart. The occupational exposure level (OEL) and decision level (DL) used depend on the substance being evaluated.

products, by-products, intermediates, and impurities is assembled and annotated with data about hazards, limit values, regulations, and standards. The hazard (potential for exposure) is assessed based on the process, equipment, material volume, temperature, pressure, ventilation, work practices, and precautions. This information is analyzed to determine where and when substances may be released into the workplace and what exposures are possible as a result. The exposure status of workers in work areas identified by this analysis is then initially assessed using such a priori information as earlier measurements or computed concentrations based on comparable installations, work processes, materials, and working conditions. A compliance evaluation based on this computation is now made, if possible; if not, exposure measurements are made of the maximum-risk (most exposed) employee for the job function under study. If the results are out of compliance, exposure reduction measures are taken and the process repeated. If conditions are in compliance but exposure levels are greater than a decision level (DL), then an occupational exposure monitoring protocol is developed and implemented. If exposures are below the DL, the process stops. The DL, which is expressed as a fraction of the occupational exposure limit, is based on judgment. In general, it would not be greater than 0.5, would usually be 0.25, and may be as low as 0.1 of the occupational exposure limit (OEL) for special circumstances such as carcinogens. Unlike the OSHA action level, the DL is used only for monitoring decisions, not for decisions involving training, medical examinations, and so forth.

This scheme is one approach to the general problem of designing employee exposure monitoring programs that fit the need. Several investigators (36, 37, 38, 39, 40, 41) have commented on the limitations of various other schemes and have proposed alternatives. In general, exposure assessment strategies fit the pattern shown in Fig. 7.3.

3 SOURCES OF WORKER EXPOSURE

The core concern of industrial hygiene is the prevention of disease arising out of the workplace. Toxic substances cause disease when some amount, or dose, enters the body or comes in contact with it. Workers are exposed to toxic substances by inhalation, ingestion, skin contact, and even, under rare industrial circumstances, injection. To accurately assess the total exposure of a worker, it is necessary to understand the sources and characteristics of exposure.

3.1 Production Operations

Industry is generally thought of in terms of continuous repetitive operations that generate air contaminants to which workers are more or less continuously exposed. Paint is sprayed on parts passing continuously in front of a worker. Dust is generated every few seconds by a foundry shakeout on a continuous casting line. Fumes seep steadily from cracks in aluminum smelting pot enclosures. Welders join structural members on a production basis, with only short breaks between welds. Operators watch controls in the midst of chemical plant air contaminated by fugitive releases. In some cases, such as a grinder cleaning sand from a casting, the concentration of the contaminant is closely related to the work performed and is probably higher in the workers' breathing zone than it would be several feet

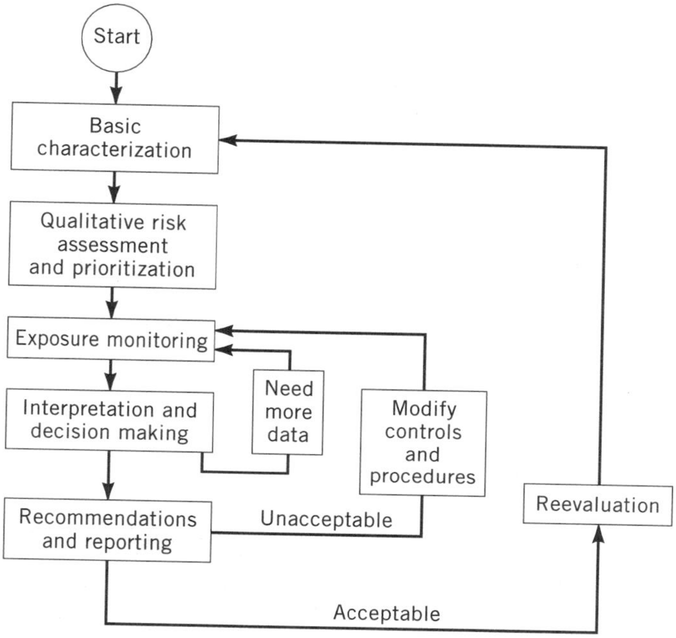

Figure 7.3. Overall flow diagram of an exposure assessment strategy.

away. Other workers, such as dorfers and creelers in a cotton spinning mill, are exposed to dust released from hundreds of bobbins, and their own activities, short of leaving the workplace, have little effect on their exposure. All these continuous exposures are actual rather than potential and present the least difficulty in evaluation.

3.2 Episodes

Much of the exposure workers receive occurs as a result of events or episodes that occur intermittently. Glue is mixed on Wednesday. The shaker mechanism breaks and a mechanic must enter the baghouse. Coke strainers preceding a pump need to be dumped when the pressure drop becomes excessive. A drum falls off a pallet and ruptures on the floor. Samples of product are taken every two hours. A pressure relief valve opens. A packing gland bursts. A reactor vessel cover is driven from its hinges by overpressure in the vessel.

These exposure events can be periodic, or they can occur at irregular intervals. They may be planned and predictable or altogether unanticipated. Some events are frequent and result in small exposures, whereas others may be catastrophic events causing massive exposures and even fatalities. As a class, episodic exposure events result in a significant fraction of the exposure burden of many workers and may not be ignored. While their evaluation is extremely complex, and often only broad estimates of the probability of an unlikely event and the consequent risk may be available, it is useful to consider episodes separately from continuous operations because episodes may be missed altogether by routine monitoring programs. Some possible ways to "catch" unpredictable events include continuous, instantaneous real-time monitors (usually area monitors), discussions with workers and supervisors on the details of processes and events, and a nonthreatening incident reporting system. The analysis of episode data will be discussed later.

3.3 Dermal Exposure

When relating exposure to biological effects, either in an epidemiological study or because of an outbreak of illness that might be related to occupation, the total dose is the relevant quantity, not merely the dose received by inhalation. Amounts of a substance entering the body by any route may contribute to the total dose.

Many substances, especially fat-soluble hydrocarbons and other solvents, can enter the body and cause systemic damage directly through the skin when the skin has become wet with the substance by splashing, immersion of hands or limbs, or exposure to a mist or liquid aerosol. Some substances, such as amines and nitriles, pass through the skin so rapidly that the rate at which they enter the body is like that of substances inhaled or ingested. The prevention of skin contact to phenol is as important as preventing inhalation of airborne concentrations. A few drops of dimethyl formamide on the skin can contribute a body burden similar to inhaling air at the TLV for 8 hours.

The "skin" notation in the TLV list identifies substances for which skin absorption is potentially a significant contribution to the overall exposure (42, 43). While the TLV list offers no precise definition of a significant contribution, the following two criteria for the skin notation have been proposed (44):

- "Dermal absorption potential, which relates to dermal absorption raising the dose of nonvolatile chemicals or biological levels of volatile chemicals 30% above those observed during inhalation exposure to TLV-TWA only."
- "Dermal toxicity potential, which relates to dermal absorption that triples biological levels as compared with levels observed during inhalation exposure to TLV-TWA only. (This toxicity criterion may not be valid for TLVs based on irritation or discomfort.)"

Note that the skin notation relates to absorption or toxicity via a skin route and not to the potential to cause skin damage and dermatitis. These later effects are important, however, and should be indicated on such hazard data sources as material safety data sheets (MSDSs).

For some substances with low vapor pressure, such as benzidine, skin absorption is the most important risk. Benzene, on the other hand, though absorbed through the skin, is absorbed at such a low rate that skin contact probably contributes little to the body burden. To judge the degree to which skin absorption is contributing to exposure, it is necessary to consider both the rate of absorption and the degree of contact. Clothing wet with a substance that remains on the worker for prolonged periods provides the maximum contact short of immersion. On the other hand, the poultice effect does not occur on wet unclothed skin; thus, evaporation can take place and the result is less severe. Contact with mist that does not fully wet the clothes or body but merely dampens them is not as severe as being splashed with the bulk liquid. Some substances, which by themselves do not significantly penetrate the skin, can have their absorption significantly enhanced by the presence of vehicles in solutions or mixtures. The TLV list notes that direct skin contact with the liquid substance is probably of greater importance than contact with the vapor. This follows from the fact that the gas in contact with the skin contains mostly (>99%) air for typical occupational vapor concentrations. Human and animal studies have estimated that dermal absorption of vapor typically accounts for between 0.1 and 10% of total uptake including inhalation for nonpolar vapors such as benzene, xylene, and others. In a study of dermal uptake of solvents from the vapor phase (45) it was found that dermal uptake of MEK vapor contributed about 3% of the body burden. These facts lead to the general conclusion that the absorption of vapor through the skin is not usually a problem. An exception appears to be 2-butoxyethanol, for which dermal uptake from ambient air was found to be appreciably higher than respiratory uptake, particularly in hot, humid conditions (46). This is important because clothing that protects against liquid contact will not necessarily prevent vapor contact.

Protective clothing that is impervious to the substance will reduce skin absorption to nil on protected areas (47). However, a leaky glove that has become filled with a solvent is providing contact with the hand equivalent to immersion. Barrier creams are often used to prevent dermatitis but may not always prevent skin absorption (48). Although skin contact and absorption must be considered as contributing to the dose for many materials, few quantitative data on rate of absorption are available, and these are in a form that is difficult to apply in an industrial setting (49, 50). Furthermore, such factors as the part of the skin exposed, sweating, and the presence of abrasions or cuts can cause order of magnitude differences (51).

Dermal exposure assessment is a complex matter that has not received much attention in the practice of industrial hygiene. Estimation of dermal exposure involves exposure scenarios and pathways, contact duration and frequency, body surface area in contact, and substance adherence (52, 53). The amount of contaminant that crosses the skin barrier and enters the body is influenced by the properties of the substance and the properties and condition of the skin at the exposed site. Models of dermal absorption have been developed (54, 55, 56). These models start with the partition coefficients of the substance, its molecular weight and solubility, and the diffusivity of the compound in the lipid and protein phases of the skin. Complex theoretical models have been simplified and combined with empirical models to yield an estimate of the chemical specific dermal permeability constant. The dermally absorbed dose (mg/kg day) can then be calculated from this constant and the skin contact area, exposure time, frequency and duration, and body weight. Since these quantities are rarely known, they must be assumed. When conservative assumptions are used, the degree of conservatism may compound to yield a very conservative estimate of the dermal dose.

3.4 Ingestion

Although ingestion is an uncommon mode of exposure for most gases and vapors (57), it cannot be ignored in the case of certain metals such as lead. Indeed, for workers exposed in pigment manufacture and use, ingested lead, either coughed up and swallowed or taken on food, though less well adsorbed than inhaled lead, constitutes a significant fraction of the total burden. Spot tests (58) and tests with tracers have shown that materials present in a workplace tend to be widely dispersed over surfaces, thus supporting the essential rule prohibiting eating and smoking where toxic substances are present.

3.5 Injection

Toxic substances may pass through the skin by intentional or unintentional injection. Opportunities for all kinds of materials to enter the body by injection connected with drug abuse or therapeutic accident are obvious. Bulk liquids may also break through the skin and enter the bloodstream without the aid of a needle when driven into the body as high velocity projectiles released from high pressure sources. Airless paint spray and hydraulic systems (59) often use pressures in this range, and such pressures often occur inside pipes and vessels in chemical plants. Inadvertent cracking of a flange under pressure can cause the traumatic introduction of a toxic substance into the body. Solid particles may also enter this way; and if soluble or radioactive, they may cause damage beyond the initial injury.

3.6 Nonoccupational Exposure

In addition to the exposure to toxic substances a worker receives at work, an increment of dose may also be delivered during nonworking hours. Air (60), water (61), and food all contain small amounts of toxic substances that may also be found in the occupational environment. As a rule, the nonoccupational dose is at least an order of magnitude below the occupational dose, just as community air standards are much lower than TLVs, and many substances present in industry are very rare in the community. Yet, in certain cases, these pollutants have an impact. The consequence of arsenic exposure resulting from copper smelting in northern Chile is difficult to assess since arsenic poisoning is endemic there as a result of naturally occurring drinking water contamination. Chronic bronchitis resulting from air pollution in the industrial midlands of England is confounded with the lung diseases of coal miners. The cardiovascular consequences of urban carbon monoxide exposure may be not unlike those of marginal industrial exposures. The benzene exposure of smokers probably exceeds typical present-day occupational exposure.

Wallace et al. (62) have shown that indoor exposure to chemicals is often higher than outdoor exposure, although the compounds involved may be different. This fact, together with the observation that people spend 80–90% of time indoors (63, 64) makes indoor exposure important. The sources of indoor exposure in homes are passive smoking, cooking, off-gassing of furniture and building materials, perchloroethylene from dry-cleaned fabrics, household chemicals and pesticides, release of VOCs in water from showers and washing machines, and infiltration of outdoor air pollutants. Office buildings have similar sources plus office machines (65).

Many workers have hobbies or other leisure activities that result in exposure to chemicals and physical agents (65, 67, 68). Some of the specific hazards are lead frits in ceramic glazing, solvents in coatings, benzidine compounds in fabric dyes, cadmium in silver solder use for making jewelry, plastic hardeners and curing agents, silica flour used in lost wax casting, welding fumes in metalworking and auto repair, and noise from loud music and target practice. Lead exposure on police and presumably private firing ranges is significant. Epoxies are used in home workshops, and garden chemicals contain a variety of poisons. Although data are scarce, it seems likely that leisure activity exposure to toxic substances rivaling high but permissible work exposure is rare. Yet, significant overexposure does occur as evidenced by the fact that cases of illness have been reported as a result of such exposures.

Intentional exposure to solvents for their narcotic effects is well documented (69, 70, 71, 72, 73, 74). Glue sniffing, gasoline sniffing, ingestion of denatured alcohol, methanol, and even turpentine have been reported. This category of chemical abuse results in exposure and often damage far beyond that normally encountered in industry. The industrial hygienist must be alert to the possibility that a case of disease may be related to intentional addictive exposure.

4 CHARACTERISTICS OF EXPOSURE AGENTS

The physical and chemical properties of a substance are important because of the effect they have on exposure measurement quite apart from their effect on the magnitude of the exposure and its consequences. The most important properties of any substance that affect sampling are those that determine whether it can be collected on a sampling medium and treated or removed in a manner that permits analysis. Many vapors are rapidly adsorbed and desorbed from one or another of a variety of solid sorbents, but some high boiling materials are very difficult to desorb and some gases do not adsorb well. Similarly, most dust particles are collected efficiently by membrane filters with 0.8-mm and even larger pore sizes; however, fresh fumes may pass through and hot particles can burn holes. Beyond these apparent physical and chemical dependencies of exposure measurement methods, there are some less obvious complications that frequently occur.

4.1 Vapor Pressure

Since liquids with high vapor pressure tend to evaporate completely when they are aerosolized, it is rarely necessary to measure them as mists. Liquids with very low vapor pressure may be present only as mists and must be measured with methods appropriate for aerosols. The situation can be very complicated for a liquid of intermediate volatility, high enough to produce a saturated vapor 10% or more of the TLV, yet low enough that mists will not quickly evaporate. If samples are collected on filters, only the mist will be caught, and part of this liquid may be lost by evaporation into the air flowing through the filter during sampling, particularly if samples are collected over long periods. This effect undoubtedly results in a significant loss of the lower molecular weight two- and three-ring aromatic compounds present in coke oven effluents, when sampled by the usual filter methods.

When mixed mist-vapor atmospheres are sampled with a solid sorbent, the vapor is likely to be caught and retained efficiently, but the charcoal granules and associated support plugs do not constitute an efficient particulate filter (75). As a consequence, compound devices with a filter preceding a sorbent tube have been developed. Despite severe sampling rate limitations, these systems are satisfactory when the sum of aerosol and vapor concentrations is to be related to a standard or effect. If, however, the aerosol and vapor mist are to be considered separately, perhaps because of different deposition sites or rates of adsorption, the single sum of the two concentrations is not enough. Furthermore, it cannot be assumed that the material collected on the filter is the entire aerosol, since part of the liquid evaporates and is recollected on the charcoal. Since the reverse cannot be true, in cases when the vapor is more hazardous, this assumption is safe (i.e., protective). Size-selective presamplers that collect part of the aerosol and none of the vapor may approximate the respiratory deposition-adsorption differential and so deliver a sample weighted toward the vapor to a biologically appropriate degree. However, even here the possibility of evaporation from the cyclone, elutriator, impactor, or other component must be acknowledged. Several investigators have studied this problem and suggested ways of dealing with it for airborne styrene (76, 77) and paint spray (78).

4.2 Reactivity

Most toxic substances are stable in air at the concentrations of interest under the usual ranges of temperature and pressure encountered in inhabited places to the degree that they may be sampled, transported to the laboratory, and analyzed without significant change or loss. Some few substances are in a transient state of reaction during the critical period when the worker is exposed. In the spraying of polyurethane foam (79), isocyanates present in aerosol droplets are still reacting with other components of the polymer while the aerosol is being inhaled, thus collection must be in a reagent that halts the reaction and yields a product that can be related to the amount of isocyanate that was there. Similarly solid sorbents such as charcoal can be coated with a reagent to react with the substance collected to yield a new compound, which will be retained and can be analyzed.

Unwanted reactions can also occur on the collecting media. For example, a substance that will hydrolyze may do so if brought in contact with water on the solid sorbent, particularly if a sorbent such as silica gel, which takes up water well, is used. Substances that may coexist in air and not react or react only slowly because of dilution may react rapidly when concentrated on the surface of a collection medium.

4.3 Particle Size

The route of entry, site of deposition, and mode of action of an aerosol all depend to some degree on the size of particle, usually but not always the aerodynamic equivalent diameter. Exposure measurement methods must discriminate among different size particles to sort out those of greater or lesser biological effect. The most common instance of such size selection is the exclusion of particles not capable of penetrating into the terminal alveoli when measuring exposure to pneumoconiosis-producing dust. Such "respirable mass sampling" by cyclone, horizontal elutriator, or various impactors is discussed at length elsewhere in this series and is not covered in detail here. However, penetration into the smallest parts of the lung is not the only division point in size selective sampling. Cotton dust particles probably do not produce the chest tightness response by penetration deep into the lung. Rather, histamine may be released in large airways (bronchioles), which can be reached by larger particles; thus a different size selective criterion (50% at 15 μm) and device (vertical elutriator) are used (80). Even larger particles may enter the body by ingestion, if they are caught in the ciliated portion of the bronchus, transported to the epiglottis, and swallowed.

At some size particles are so large they cease to be inhalable and thus should be excluded from exposure samples (81). It is difficult to select a criterion on which to base size selectors for inhalable dust samplers. Particles with falling (setting) speeds greater than the up-flow velocity into the nose could be said to be noninhalable, except that some individuals breathe through their mouths. These particles turn out to be quite large and thus have falling speeds that cause them to be removed from all but the most turbulent or recently generated dust clouds. Because of their size, of course, they represent a mass far out of proportion to their number. By using the general tendency of open face filter samples to undersample large particles when pointed down (82, 83), it is possible to use this vertical elutriation effect to discriminate against noninhalable particles in most cases.

To respond to the dependence of biological effect on particle size, the ACGIH (6, 84) has established three classes of TLV for airborne particulate matter:

- *Inspirable Particulate Mass.* This is the quantity of particulate matter that should be measured for those materials that are hazardous when deposited anywhere in the respiratory tract. This class includes some fraction of particles as large as 100 μm.
- *Thoracic Particulate Mass.* These are the particles that are deposited anywhere within the lung airways and the gas exchange region. The sample for thoracic mass collects particles with a median aerodynamic diameter of 10 μm, which is similar to the cotton dust sampler.
- *Respirable Particulate Mass.* These particles are hazardous when deposited in the gas exchange region of the deep lung. They include the traditional pneumoconiosis-producing dusts such as silica and are collected with a particle size selector that passes 50% at 3.5 μm.

4.4 Exposure Indices

Most toxic substances encountered in the workplace are clearly defined chemical compounds that can be measured with as much specificity as we please. Benzene need not be confused with other compounds, and other closely related compounds (e.g., ethyl benzene) have different toxicology. For some toxic substances we tend to think of an element (lead) in any one of a number of possible compounds. Thus specificity is defined in terms of the element rather than the compound. Often, however, the situation becomes more complex because the compound containing the element of interest has a significant effect on its uptake, metabolism, toxicity, or excretion (85). Although the influence of the chemical structure containing the element probably varies even among similar compounds, for simplicity the compounds are usually grouped as organic/inorganic (lead, mercury), soluble/insoluble (nickel, silver), or by valence (chromium).

The problem of differentiating the several classes of compounds of a toxic element in a mixed atmosphere adds complexity to sampling method selection, and it is sometimes necessary to make, and clearly state alongside the results, certain simplifying assumptions. It is commonly assumed when measuring lead exposure in gasoline blending, for example, that all the lead measured is organic. Similarly when measuring the more toxic soluble form of an element, the safer assumption may be made that the entire element present was soluble.

The greatest complexity occurs when toxicity is based on the effects of a class of compounds or of a material of a certain physical description. Some polynuclear aromatic hydrocarbons (PNAs) are carcinogens of varying potency, and they usually exist in mixtures with other PNAs and with compounds (activators, promoters, inhibitors) that modify their activity. Analysis of each individual compound is very difficult and when done does not yield a clear answer, since given the complexity of the mixture of biologically active agents and their interactions, a calculated equivalent dose would have little accuracy. In these instances it is common to measure some quantity related to the active agents and to base the TLV on that index. For PNAs a TLV has been based on the total weight of benzene- or hexane-soluble airborne material. While this limit may be appropriate for the coal tar pitch volatiles for which it was developed, it may not work for other PNA-containing materials. Crude oil and cracked petroleum stocks may contain PNAs; but, whereas the coal dust particles mixed in with coal tar pitch volatiles are not soluble in benzene, almost all of the petroleum-derived materials admixed with PNAs are soluble in benzene. For example, heavy aromatic naphtha (HAN) may or may not contain PNAs, depending on the manufacturing process, but is completely soluble in benzene. Thus a measurement of the benzene-soluble fraction of a HAN aerosol will reveal nothing about the PNA content. Alternate indices include the single carcinogen B(a)P, the sum of a subset of six carcinogenic PNAs (Table 7.1) or 14 or more individual PNAs (86).

Asbestos is another toxic substance for which the parameter of greatest biological relevance is difficult to define (87, 88, 89). In the early studies when measurements were made with an impinger, few fibers were seen and consequently a count of all particles present was used as an index of overall dustness. More recently the TLV has been based on counts of fibers longer than 5 mm as seen with a light microscope. Since most fibers present are usually shorter than 5 mm and too thin to be visible under a light microscope, it has been suggested that counts of all fibers seen by an electron microscope would provide the most meaningful estimate of risk. Long fibers may be more dangerous; however, short fibers will dominate the count; thus some adjustment may be necessary.

Lippmann (90) has observed that the several diseases caused by asbestos fiber inhalation (asbestosis, mesothelioma, lung cancer) are associated with differing parameters of fiber

Table 7.1. Carcinogenic PNA Subset

Benz[*a*]anthracene
Benzo[*b*]fluoranthene
Benzo[*j*]fluoranthene
Benzo[*a*]pyrene
Benzo[*e*]pyrene
Benzo[*k*]fluoranthene

exposure. Based on this, the asbestos exposure indices shown in Table 7.2 are recommended.

Byssinosis appears to be caused not by cotton itself but by inhalation of cotton plant debris dust baled with the cotton. The total dust airborne in a cotton textile mill is mostly lint (Lynch, 1970a) (80), and indices have aimed at excluding these "inert" cellulose fibers by collecting a "lintfree" fraction by use of a screen or vertical elutriator. More relevant indices could include plant debris only or the specific biologically active agents, if known.

Indices are used where a group of compounds interact to produce a biological effect, or where the active agent is unknown or unmeasurable. In choosing an index we try to maximize biological relevance with a method of measurement that is practical to use. On the one hand very simple parameters like gross dust or total count are easy to use but include much irrelevant material. On the other hand counting fibers by scanning electron microscope or detailed analyses of individual PNAs is difficult, expensive, and not likely to be undertaken frequently. In making the choice, it is important to remember that the primary objective is the protection of the worker. Very exact and highly relevant methods, though scientifically satisfying, may be so tedious that very few samples are taken, and because of the larger variability of worker exposure, the true accuracy of the exposure estimate is lower than it would have been if many samples had been taken by a less specific method.

When the overall contaminant level has been reduced and with it the level of the biologically active agent, the level of all correlated indices will be lower. The danger of less relevant, more indirect indices is that serious systematic bias may occur, particularly when an index from the workplace where the health effect relationship was estimated is used in other quite different workplaces. The carcinogenic risk of roofers using asphalt is far lower than for coke oven workers at the same level of exposure as measured by benzene solubles. Byssinosis patterns may be different in mills that garner old rags or process linters than in raw cotton card rooms. In using a measurement that is not perfectly specific, it must be remembered that the result obtained has a less than direct connection with the biological process, and the stronger the effect of extraneous factors, the more care is needed in interpreting the result.

4.5 Mixtures

Industrial workplaces rarely contain only one airborne contaminant, although it is uncommon for there to be several toxic substances each at or near its TLV. The measurement

Table 7.2. Summary of Recommendation on Asbestos Exposure Indices

Disease	Relevant Exposure Index
Asbestosis	Surface area of fibers with length >2 μm; diameter <0.15 μm
Mesothelioma	Number of fibers with length >5 μm; diameter <0.1 μm
Lung cancer	Number of fibers with length >10 μm; diameter >0.15 μm

Source: From Lippmann (90).

problems caused by the presence of gases, vapors, or dust—some in even higher concentrations than those of primary interest—are discussed later. The question of what to measure and how to interpret the result when a worker is simultaneously exposed to several agents involves their biological mechanism and can be answered only by considering the mode of action of the substances in the body. Possible effects of mixtures can be divided into three broad categories (6):

- *Additive*. When two or more hazardous substances act on the same organ system, their combined effect, rather than that of any one by itself, should be the criterion for overexposure. For additive substances the ratios of the concentration, C_n, of each substance to its limit, T_n, are summed:

$$\frac{C_1}{T_1} + \frac{C_2}{T_2} + \ . \ . \ . + \frac{C_n}{T_n}$$

  When this summation exceeds one, then the TLV for the mixture as a whole is exceeded.
- *Independent*. In some cases there is good reason to believe that the major effects of the components of the mixture are not additive but act on altogether different organ systems; for example, an eye irritant and a pneumoconiosis-producing dust, or a neurotoxic substance with a liver poison. In these cases the TLV is exceeded when one of the components exceeds its TLV.
- *Synergistic*. It is possible that components of a mixture, or a workplace exposure plus a lifestyle risk (tobacco, alcohol), may act together synergistically, or potentiation may occur. This action usually occurs in high rather than low doses. The interpretation of exposure data where synergism or potentiation may occur needs to be evaluated on a case-by-case basis using the scientific evidence that such action may happen and at what levels it happens.

Although possible interactions of substances are extremely complex, it has been the custom to accept the simplifying assumption that in the absence of information to the contrary, the effects of the different hazards should be considered as additive.

Hydrocarbons used as fuels or solvents are usually a mixture of a large number of individual aliphatic compounds and their isomers, often so numerous that analyzing for each individual compound and comparing the result with an individual limit is impractical. Indeed, since TLVs exist for only a few of the compounds present, limits must be stated for the mixture by considering the compounds as a class. Gasoline (91), for example, may contain aliphatic and aromatic hydrocarbons and such oxygenate additives as methyl *tert*-butyl ether (MTBE). In view of the differences in toxicity of the several substances and variations in content, calculation of a TLV for gasoline is a complex matter (92, 93, 94). One difficulty in expressing a TLV for any vapor mixture is that it has been customary to state TLVs for gases or vapors in parts per million. Analytical results, typically from gas chromatographs, emerge initially as the weight in milligrams of each fraction present, from which the concentration, in milligrams per cubic meter, can be calculated. To convert this to parts per million, it is necessary to know the molecular weight, which will not be known for unidentified homologues or for the mixture as a whole. For this reason, it is preferable to state TLVs for vapor mixtures as a weight concentration (mg/m^3) rather than as a volume fraction (ppm).

Certain combinations of substances present a far more complex situation than can be described by either independent or simple interaction. Benzo[a]pyrene and particulates, with and without sulfur dioxide, carbon monoxide and hydrogen cyanide, ozone, and oxides of nitrogen, and some other mixtures result in complex interactions that may cause effects beyond those predicted by the merely additive case.

4.6 Period of Standard

The concentration of industrial air contaminants varies with time, and recordings such as that in Figure 7.4 are typical. In general, where the concentration varies with time, the height of the maximum and depths of the minimum are greater as the period of the measurement is shortened. If it were possible to make truly instantaneous or zero-time measurements, the peaks and troughs would be very great indeed. Real measurements using continuous-reading instruments do not show quite such wide extremes because the response time of the instrument causes some averaging, thus preventing true zero-time measurements. Even so, instruments with short response times, such as those with solid-state sensors, show wide variation from the average, and even those with relatively long response times such as a beta adsorption type of dust monitor, still reveal peaks and valleys that are more than double or less than half the average. As longer and longer period measurements are made, the extremes regress toward the average, and obviously, if we define our average over an 8-h period, a single integrated sample over that period would show no extremes, but only the average. However, even daily averages have highs and lows compared with monthly or yearly averages in all but perfectly nonvarying environments, which do not occur in the real world.

Given the variance of the universe of instantaneous concentrations and the probability distribution of the concentration over one averaging time, it should be possible to determine the probability distribution of the concentration over any other averaging time. The mathematics of this relationship has been addressed (95, 96), and as expected, the arithmetic mean of the exposure distribution is constant and independent of averaging time, and the variance decreases with increasing averaging time. For lognormal distributions, however, while the arithmetic mean is independent of averaging time, the geometric mean decreases with averaging time as the geometric standard deviation increases (25). These changes depend in part on the degree of autocorrelation of instantaneous concentrations.

The relationship of averaging time and environmental variance has the consequence that measurements made over different integrating times have a relationship to each other that is a function of the variance. For example, if a given value of the concentration had a 50% (or 90%) chance of occurring over one averaging time, there would be a unique higher and lower pair of values that had an equal chance of occurring over one averaging time, and there would be a unique higher and lower pair of values that had an equal chance of occurring over a shorter averaging time. Since not all values of environmental variance are equally likely but rather tend to be in the range of geometric standard deviations of

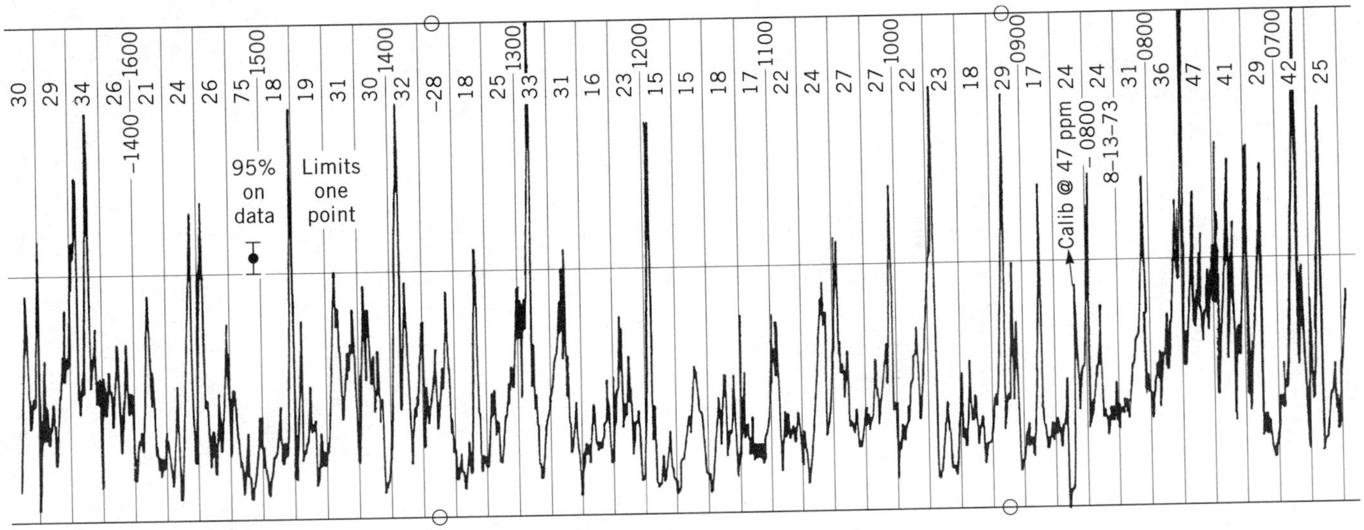

Figure 7.4. Actual industrial hygiene data showing intraday environmental fluctuations. Range of carbon monoxide data on chart is 0 to 100 ppm.

1.5 to 3 or 4, it is possible to say in some cases that certain values for different averaging times are inconsistent with each other. Thus, it is very unlikely that an industrial environment would be so constant that the 100-ppm, 8-h limit for ethyl benzene could be exceeded without exceeding the 125-ppm, 15-min short-term exposure limit (STEL). The effect of setting short-term limits close to long-term limits is to force the effective long-term limit down. Thus to avoid exceeding a 125-ppm, 15-min STEL, it will probably be necessary to achieve an 8-h average value of much less than 100 ppm of ethyl benzene, perhaps even lower than 50 ppm. There may be valid toxicological reasons (97, 98) for setting a short-term limit based on, for example, acute irritation, and a time-weighted average (TWA) that is aimed at preventing some chronic effect; however, it should be recognized that the two are not independent, and when they are set outside the range of approximately equal likelihood, holding concentrations below the limit for one averaging time means holding them far below the limit for the other averaging time; thus one limit is in effect forcing the other.

The TLV committee of ACGIH is aware of this issue and has revised and removed a number of STELs that had no independent toxicological basis and were close to the 8-h TWA. Based on typical distribution of exposures, the TLV committee now makes the following recommendation for substances that do not have STELs: "Excursions in worker exposure levels may exceed three times the TLV-TWA for no more than a total of 30 minutes during a work day and under no circumstances should they exceed five times the TLV-TWA provided that the TLV-TWA is not exceeded" (6) (p. 6).

In terms of sampling strategy, the significance of limits for different averaging times that are statistically inconsistent is that since one has a relatively greater likelihood of being exceeded than the other, regardless of the absolute likelihood of either, there is an opportunity to devise schemes that emphasize measurements to detect the likely event and use these measurements and knowledge of variance derived from them to draw inferences about the less likely event.

5 SAMPLING STRATEGY

The term *sampling strategy* as used here means the assembly of decisions about how to make a set of measurements to represent exposure for a particular purpose. The measurements should yield a data set that is logically and statistically adequate to satisfy the objective of the measurements as discussed in Section 2. An optimum strategy is selection of elements under the control of the exposure assessor that most efficiently achieves the objective, given the physical circumstances and environmental variability (99). Sampling strategy issues have been addressed by the American Industrial Hygiene Association (AIHA) (100), Hewett (101), and Tielemans et al. (102), among others.

5.1 Environmental Variability

An important factor in the design of any measurement scheme is the degree of variability in the system being observed. This variability has a primary effect on the number of samples to be taken and the accuracy of the results that can be expected. When the system being observed is the exposure of a worker to a toxic substance in a workplace, variability tends to be quite high. During the course of a day there are minute-to-minute variations, and daily averages vary from day to day (103).

A typical recording of actual intraday environmental fluctuations appears in Figure 7.4. Highly variable environmental data of this kind, which are truncated at zero, generally have been found to be best described by the log-normal rather than the normal distribution (Figure 7.5). This two-parameter distribution (104, 105, 106) is described by the geometric mean (GM) and the geometric standard deviation (GSD), which is the antilog of the standard deviation of the log-transformed data (107). A rough equivalence between GSDs and the more familiar coefficient of variation (Table 7.3) is valid up to a GSD of about 1.4. Figure 7.6 illustrates the consequence of various values in terms of spread of data. Models

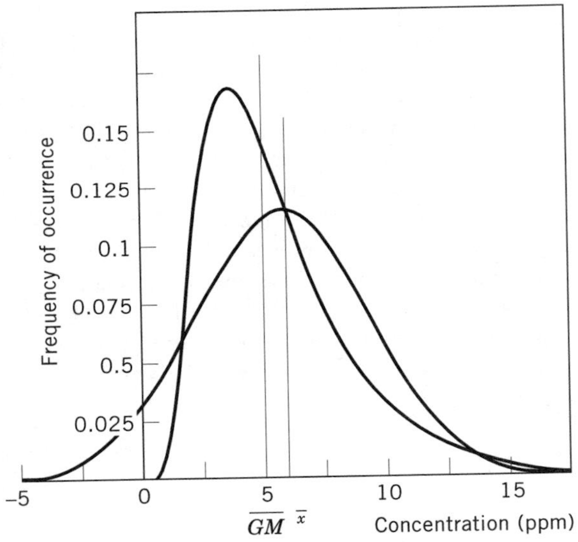

Figure 7.5. Log-normal and normal distributions with the same arithmetic (mean) and standard deviation. GM, geometric mean.

Table 7.3. Log and Arithmetic Standard Equivalence

Geometric Standard Deviation (GSD)	Coefficient of Variation (CV)
1.05	0.049
1.10	0.096
1.20	0.18
1.30	0.37
1.40	0.35

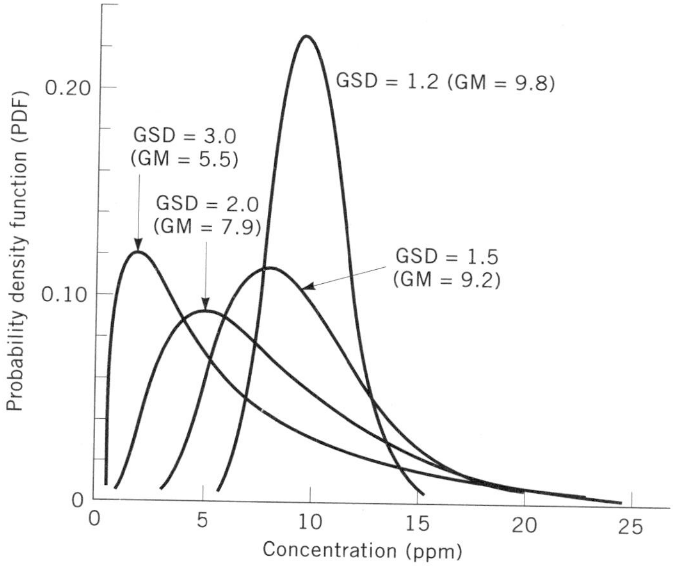

Figure 7.6. Log-normal distributions for arithmetic mean concentration of 10 ppm.

and data derived from community air pollution measurements (108, 109, 110, 111) have been useful in studying the in-plant microenvironment. Some studies of occupational environmental variability (112) have found log-normal distributions of data with GSDs in the range of 1.5–1.7, with few less than 1.1 and as many as 10% exceeding 2.3. Rappaport (25) summarized various studies with GSDs ranging from 1.2 to 9.3, with most under 3.0. While the GM and GSD are the parameters that fully describe the lognormal distribution, the mean may be the more appropriate parameter to estimate dose. This issue is discussed by Hewitt (113). Various methods for estimating the mean of a log-normal distribution have been examined (114, 115).

One can speculate on the probable causes for this variability in worker exposure. Fugitive emissions, which are like frequent small accidents rather than main consequences of the production process, occur almost randomly. Production rates change. Patterns of overlapping multiple operations shift irregularly. Distribution of contaminants by bulk flow, random turbulence, and diffusion is uneven in both time and space. Through all this our target system, the worker, moves in a manner that is not altogether predictable. These and other uncertainties are the probable causes of the variability typically observed, but the situation is far too complex to pinpoint each individual source of variation and its consequences.

Variance can be divided into within-worker variance and between-worker variance (116, 117). While within-worker variance is a result of changes in the exposure of a worker from day to nominally similar day, between-worker variation is at least theoretically controllable by the way workers are grouped for sampling purposes. Since it is not usually practical to measure the exposure of all workers all the time, it is customary to group workers who are relatively homogenous with respect to exposure. In some cases, this can be done by observation, but when the between-worker variance is seen to be large with respect to the within-worker variance, it may be necessary to use sampling data to group workers.

Sampling schemes may deal with the variability from all sources as a single pool and derive whatever accuracy is required by increasing the number of samples. Alternatively one can postulate that a large part of the variability is due to some observable factor or factors and by means of a factorial design account for this portion of the variance, leaving only the residual to be dealt with as error. No hard and fast rules can be made regarding the choice except that it seems logical to expect the factors identified to account for a statistically significant fraction of the variance (F test) if it is to be worthwhile to sample and analyze the data in this manner. Shift, season of the year, wind velocity, and rate of production are some of the factors affecting worker exposure that may be worth singling out to ensure that part of the gross variance can be assigned to an accountable cause. Even if successful, it is likely that the residual or error variance will still be large, much larger than the variance due to measurement method inaccuracy, and will exert a major influence on our decisions. Monte Carlo simulation modeling is a means of analyzing uncertainty as an aid in decision making (118).

5.2 Purpose of Measurement

The development of a sampling strategy requires a clear understanding of the purpose of the measurement. Rarely are data collected purely for their own sake. Even when data are collected because of a demand by others, such as the government or employees, the use of the data should be considered.

What questions will be answered by the data? What decisions could depend on those answers? For example, do we want to know whether these workers are overexposed? And if so, will a decision to take control action follow? Is the control likely to be a minor change in a work practice or an expensive engineering modification? Or is it intended to combine the data with those of health effects to answer the question of what level of exposure is safe (119)? Will it then be decided to modify the TLV or establish a new PEL by regulation? Or are the numbers to be assembled to answer the question of what level of control is presently being achieved in industry? And may this answer lead to new decision regarding what control is feasible? Last, will workers use the result to find out whether their health is at risk and, as a consequence, decide to change jobs or seek changes in the conditions of work? Often there are many questions that need answers, and thus data are collected for multiple purposes. As the discussion that follows indicates, the purpose of the data determines the design of the measurement scheme. All too often data intended for multiple purposes turn out not to be suitable for any purpose. Thus it is usually necessary to focus on the prime need to be sure that the strategy will meet this requirement; then, if possible, minor adjustment or additions can be made to meet other needs, if this can be done without losing the main purpose. The optimum sampling strategy is that which combines the choice of method and sampling scheme with respect to sampling location, time, and frequency so that we are confident that the data are adequate for the decisions that follow.

5.3 Location

The most common purposes of measurement of exposure of workers is to estimate the dose so as to prevent or predict adverse health effects. These health effects result from a substance entering the body by some route, and it is possible to estimate the dose by measurement of the substance or a by-product on the way in or in the way out. The use of biomonitoring methods and biomarkers for the reconstruction of internal dose is discussed later. However, due to the limitation in these methods and the need to know the source/time pattern of the substance intake, industrial hygienists most often estimate inhaled dose by measurement of the concentration of a substance in inhaled air. Although air samples are sometimes collected from inside respirator facepieces, it is generally not possible to directly sample the air being inhaled. Therefore, the location of the sample collector inlet in relation to the subject's nose and mouth is important. We categorize sampling methods in terms of their closeness to the subject and the point of inhalation as personal samples, breathing zone or vicinity samples, and area or general air samples. A personal sample is one that is collected by a sampling device worn on the person of the worker, which travels with the worker. Breathing zone samples are those collected in the envelope or breathing zone around the worker's head, which is thought, based on observation and the nature of the operation, to have approximately the same concentration of the contaminant being measured as the air breathed by the worker. Area or general air samples are the most remote and are collected in fixed locations in the workplace.

Obviously, personal samples are the preferred method of estimating dose of the inhalation route since they most closely simulate inhaled air. OSHA (120) enforcement operations reflect a longstanding belief that personal sampling generally provides the most accurate measure of an employee's exposure. However, even personal samples may not sample exactly the air being breathed. A few inches difference in the placement of the filter head of a personal sampler has been reported to make a significant difference in the concentration measured (121, 122), particularly when dust comes from point sources or is resuspended from clothing. For uniformly dispersed aerosols, however, there appears to be no bias between forehead or lapel versus nose locations (123).

If personal sampling cannot be used, some other means of estimating exposure must be accepted. Breathing zone measurements, made by a sample collector who follows the worker, can come close to measuring exposure. However, this intrusive measurement method may influence worker behavior, and the inconvenience of the measurement will limit the number of measurements and therefore reduce accuracy, as discussed later.

When fixed station samplers are used, knowledge of the quality of the relation between their measurements and the exposure of the workers is necessary if worker exposure is to be estimated. An experimental design that collects large numbers of pairs of measurements of quantities that are in any way related will yield a significant correlation coefficient. The important question in the use of general air measurements is what confidence can be placed in the estimate of worker exposure. Not only is the regression line important, but also the width of the bounds on the confidence limits of a predicted exposure value from some set of fixed station measurement, as shown in Figure 7.1.

In studying the relation between area and personal data with respect to asbestos, the British Occupational Hygiene Society concluded the following (124, pp. 26–27):

> The relationship between static and personal sampling results varies according to the characteristics of the dust emission sources and the general and individual work practices adopted in a particular work area.
>
> (i) When identical sampling instruments are deployed simultaneously at personal and static sampling points and the distances between them are reasonably small, at least two-thirds of the personal sampling results obtained in a given working location are higher than those obtained from static sampling.
>
> (ii) The differences found between the two types of result tend to be particularly great where the static sampling points are relatively remote from dust emission points, as, for example, when "background" static testing is adopted.
>
> (iii) In certain cases, results from personal sampling may be lower than those from static sampling. For example, the positioning of the sampling point with respect to air extraction systems could give this result.
>
> (iv) The correlation coefficient between the personal and static measurements is statistically significant but, even so, no consistent relationship of great practical utility could be found in the limited data available.

Many attempts have been made to estimate worker exposure from fixed station air sampling schemes (125, 126, 127, 128). The static sampling arrangements ranged from manually operated fixed stations in some studies to computer-based automated monitoring systems in others. In some studies the time a worker spent in an area was taken from observations of work patterns. Leidel et al. (32) analyzed these studies and concluded that exposure estimates based on general air (area) monitoring should be used only where it can be demonstrated that general air methods can measure exposure with appropriate accuracy. Linch and Pfaff (129) on the other hand conclude that "only by personal monitoring could a true exposure be determined."

Although worker exposure measurements are most often used in relation to health hazards, not all measurements made for the protection of health need be measurements of exposure. When it has been established that an industrial operation does not produce unsafe conditions when it is operating within specified control limits, fixed station measurements that can detect loss of control may be the most appropriate monitoring system for workers' protection. Local increases in contaminant concentration caused by leaks, loss of cooling in a degreaser, or fan failure in a local exhaust system can be detected before important worker exposure occurs. Continuous air monitoring (CAM) equipment, which detects leaks or monitors area concentrations, is often used in this way. All such systems should be validated for their intended purpose.

5.4 Period

Free of all other constraints, the most biologically relevant time period over which to measure or average worker exposure should be derived from the time constants of the uptake, action, and elimination of the toxic substance in the body (130, 131, 132). These periods range from minutes in the case of fast-acting poisons such as chlorine or hydrogen sulfide, to days or months for slow systemic poisons such as lead or quartz. In the adoption of guides and standards, this broad range has been narrowed, and the periods have not always been selected based on speed of effect of half-life. It has been observed (119, p. A4) however, that "transient exposures of a shift or less are unlikely to affect the risk of chronic disease provided that the following conditions are met: (*1*) the damage induced by the agent is reversible; (*2*) elimination of the toxicant is first-order; (*3*) the mean exposure is less than one-quarter to one-eighth of that which corresponds to the threshold burden; and (*4*) the exposure distribution is stationary and adequately described by a log-normal distribution."

When these conditions are met, measurement of the long-term average (multishift) exposure is much more efficient than measuring single-shift peaks. However, where standards have been developed or have been interpreted as single-shift limits, an increase in the averaging time is in effect an increase in the standard, and so it may be necessary to lower the numerical value of the standard to remain risk neutral. For most substances, a time-weighted average over the usual work shift of 8 hours has been accepted since it is long enough to average out extremes and short enough to be measured in one work day. Several systems have been proposed for adjusting limits to novel work shifts (133, 134, 135).

Once the time period over which exposure is to be averaged has been decided for either biological or other reasons (128, 136), there are available several alternate sampling schemes to yield an estimate of the exposure over the averaging time. A single sample could be taken for the full period over which exposure is to be averaged (Figure 7.7). If such a long sample is not practical, several shorter samples can be strung together to make up a set of full-period, consecutive samples. In both cases, since the full period is being measured, the only error in the estimate of the exposure for that period is the error of sampling and the analytical method itself. However, when these full-period measurements are used to estimate exposure over other periods not measured, the interperiod variance will contribute to the total error.

It is often difficult to begin sample collection at the beginning of a work shift, or an interruption may be necessary during the period to change sample collection device. Several assumptions may be made with respect to the unsampled period. It may be assumed that exposure was zero during this period, in which case the estimate for the full period could be regarded as a minimum. Alternatively, it could be assumed that the exposure during the unmeasured period was the same as the average over the measured period. This is the most likely assumption in the absence of information that the unsampled period was different. However, it is difficult to calculate confidence limits on the overall exposure estimate since the validity of the assumption is a factor, there is no internal estimate of environmental variance, and the statistical situation is complex.

When only very short period or grab samples can be collected, a set of such samples can be used to calculate an exposure estimate for the full period. Such samples are usually collected at random; thus, each interval in the period has the same chance of being included as any other, and the samples are independent. This sampling scheme of discrete measurements within a day is analogous to a set of full-period samples used to draw inferences

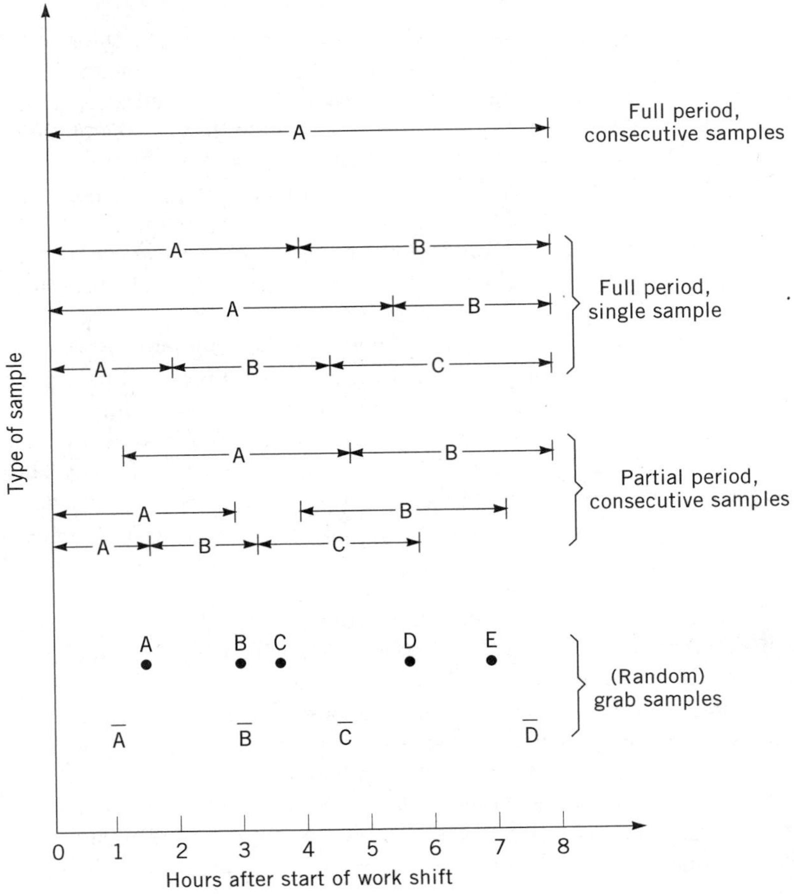

Figure 7.7. Types of exposure measurements that could be taken for an 8-h average exposure standard.

about what is happening over a large number of days. In both cases the environmental variance, which is usually large, has a major influence on the accuracy of the results.

Short-period sampling schemes can be useful with dual standards. For example, if a toxic substance has both a short-term, say 15-min limit and an 8-h limit, 15-min samples taken during the 8-h period could be used to evaluate exposure against both standards. This involves some compromise, however, since samples taken to evaluate short-period exposure are likely to be taken when exposure is likely to be at a maximum rather than a random level. Statistical techniques for evaluating exposure with respect to dual standards are also available (137, 138). As discussed earlier, when a dual standard is inconsistent, so that one limit is more likely to be exceeded than the other, sampling schemes that evaluate exposure with respect to the limit more likely to be exceeded can be used to provide some confidence about the other limit.

The traditional method of estimating full-period exposure is by the calculation of the time-weighted average. In this method the workday is divided into phases based on observable changes in the process or worker location. It is assumed that concentration patterns are varying with these changes and are homogeneous with each phase. A measurement or measurements, usually shorter than the length of the phase, are made in each phase, and the exposure estimate E is calculated according to the following formula:

$$E = \frac{C_1T_1 + C_2T_2 + \ldots + C_nT_n}{8}$$

where C_n = concentration measured in phase n
T_n = duration (h) of phase n ($ST = 8$)

Figure 7.8 represents this procedure graphically. Although the exposure estimate itself is simple to make, calculation of the confidence limits on this estimate can be very complex. Each phase must be treated separately. A set of samples must be collected to determine the mean and standard deviation for the phase. These data must then be combined in a manner that weights the variance to obtain an error estimate for the whole. This complex calculation does not include, however, any consideration of the imprecision in selection of phase boundaries. Given the number of samples required in each phase to provide adequate error estimates and the lack of confidence in the end results due to the several layers of assumptions, an equal number of grab samples collected at random over the whole period or stratified by task may be more efficient.

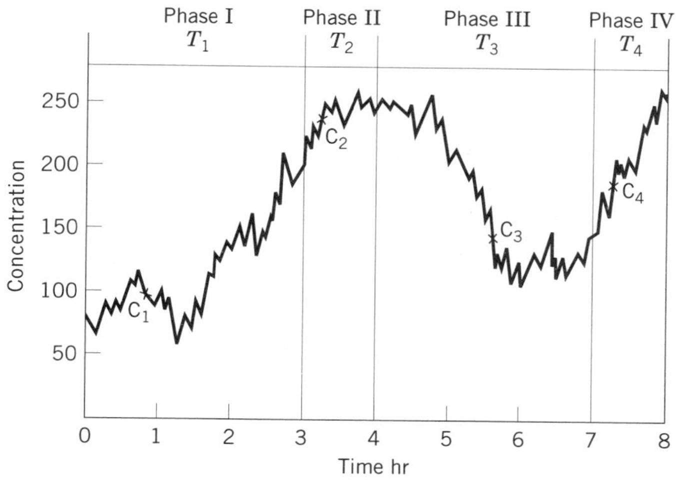

Figure 7.8. Time-weighted average.

The recommendation that samples should be taken over a full shift to determine employee exposure to toxic substances has been questioned (139). It is maintained that it is possible in some instances to characterize a worker's exposure with a few short-period samples. There are arguments to support both sides; ultimately, however, the issue can be decided in each individual case based on the answer to two questions: Can the risk of error in the decision to be based on these measurements be calculated from the data? Is this risk acceptable, given the consequences of the decision?

When an averaging time longer than a full shift is needed, that long-term average (LTA) is usually calculated from some number of full-shift samples. For example, the coal mine dust standard is based on the average of five full-shift measurements. Single-sample means of measuring multiple shifts, while excluding nonshift periods, are not widely used.

When workplace measurements are made for purposes other than the estimation of worker exposure, different considerations apply. While a single 8-h sample may be an accurate measure of a worker's average exposure during that period, the exposure was probably not uniform, and the single sample gives no information on the time history of contaminant concentration. To find out when and where peaks occur, with the aim of knowing what to control, short-period samples or even continuous recordings are useful. Similarly, when a control system is evaluated or sampling methods are compared, measurements need be only long enough to average out system fluctuations and provide an adequate sample for accurate analysis. As in the case of the decision on location, the purpose of the measurement is a primary consideration in the selection of a time period of a measurement.

5.5 Frequency

By increasing the number of measurements made over a period of time or in a sampling session, the magnitude of the confidence limits on the mean result or on the fraction of periods that may exceed the limit can be reduced. With smaller (tighter) confidence limits it becomes easier to arrive at a decision at a chosen level of confidence or to be more confident that a decision is correct. In Figure 7.9 decisions are possible in cases A and C, but not in case B. By collecting more samples, it might be possible to tighten the lower confidence limit (LCL) in case B1, for example, permitting the conclusion that these data do in fact represent an overexposure. The choice of the number of samples to be collected rests on three factors: the magnitude of the error variance associated with the measurement, the size of a difference in results that would be considered important, and the consequence of the decision based on the result.

The error variance associated with the measurement depends in most cases on the environmental variance. An exception is the rather limited instance of evaluating the exposure of a worker over a single day by means of a full-period measurement. In that case the error variance is determined by only the sampling and analytical error, and confidence limits tend to be quite narrow. Usually, however, our concern is with the totality of a worker's exposure, and we wish to use the data collected to make inferences about other times not sampled. There is little choice; unless the universe of all exposure occasions is measured; we must sample, that is, make statements about, the whole based on measurement of some parts.

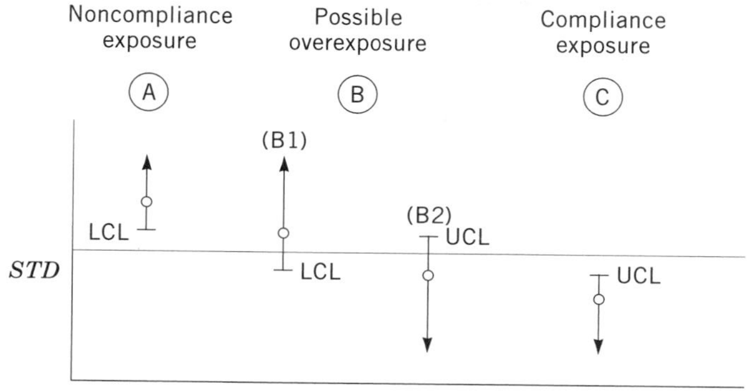

Figure 7.9. One-sided confidence limits.

As discussed earlier, the universe has a large variance, quite apart from the error of the sampling and analytical method. In terms of our decision-making ability, the error of the sampling and analytical method may have very little impact. In Figure 7.10 the inner pair of curves define the decision zone for an environmental coefficient of variation equal to .60 with no sampling or analytical error; the outer zone includes a typical detector tube error having a coefficient of variation of .25 (140). Even the relatively large error of one of the less accurate methods results in only a slight contribution of analytical variability to the total variability of measurements, as shown in Figure 7.11 (141).

The American Industrial Hygiene Association has addressed the issue of appropriate sample size (35) and recommends in the range of 6–10 random samples per homogeneous exposure group. Fewer than 6 leaves a lot of uncertainty, and more than 10 results in only

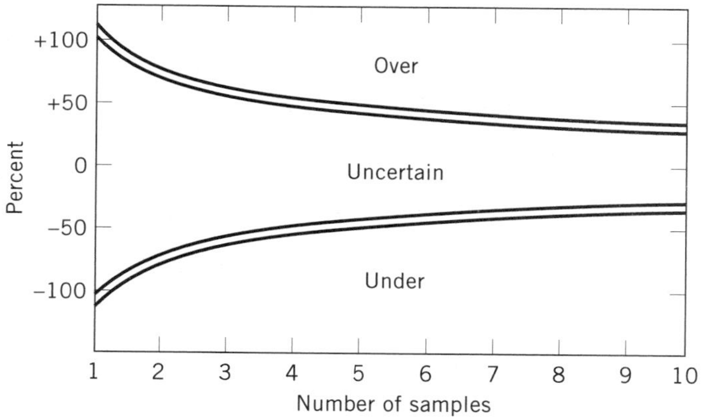

Figure 7.10. Difference between mean measured concentration and TLV required for a decision at the 95% level versus number of samples averaged.

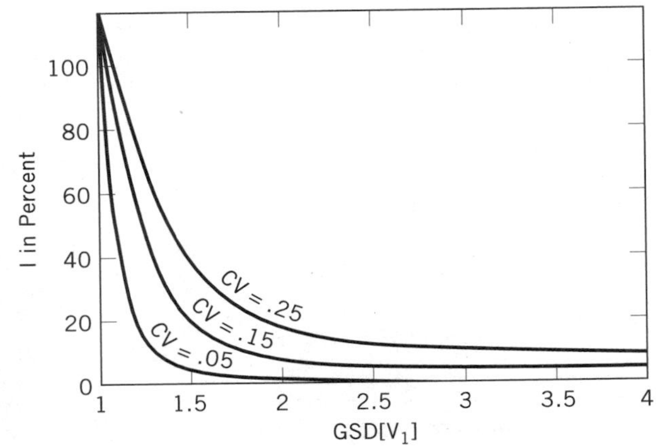

Figure 7.11. The percent contribution of analytical variability to total variability in measurements of 8-h TWAs (141).

marginal improvement in accuracy. Also, it is usually possible to make a reasonable approximation of the exposure distribution with 10 samples, although a rigorous goodness-of-fit test often requires 30 or more. Since the confidence interval is very sensitive to the sample estimate of the GSD, Buringh and Lanting (142) have recommended using an assumed GSD (of approximately 2.7) with small sample sets.

Figure 7.12 also illustrates the effect of sample size or our ability to arrive at a conclusion. These curves give the number of grab samples required in order to be confident that an overexposure did not occur for several typical levels of environmental variance. As can be seen, the difference between the mean and the standard necessary to achieve confidence in the conclusion decreases sharply as the number of samples used increases from 3 to 11. An important conclusion is that for a fixed sampling cost and level of effort, many samples by an easy but less accurate method may yield a more accurate overall result than a few samples by a difficult but more accurate method due to the effect of increasing sample numbers on the error of the mean in highly variable environments.

In selecting a sample size, note that it is possible to make a difference statistically significant by increasing the number of samples even though the difference may be of small importance. Thus given enough samples, it may be possible to show that a mean of 1.02 ppm is significantly different from a TLV of 1.0 ppm, even though the difference has no importance in terms of biological consequence. Such a statistical significance difference is not useful. Therefore, in planning our sampling strategies, we should first decide how small a difference we would consider important in terms of our use of the data and then select a frequency of sampling that could prove this difference significant, if it existed.

The consequences of the decision made on the basis of the data collected should be the deciding factor in selecting the level of confidence at which the results will be tested. Although the common 95% (1 in 20) confidence level is convenient because its bounds are two standard deviations from the mean, it is arbitrary, and other levels of confidence

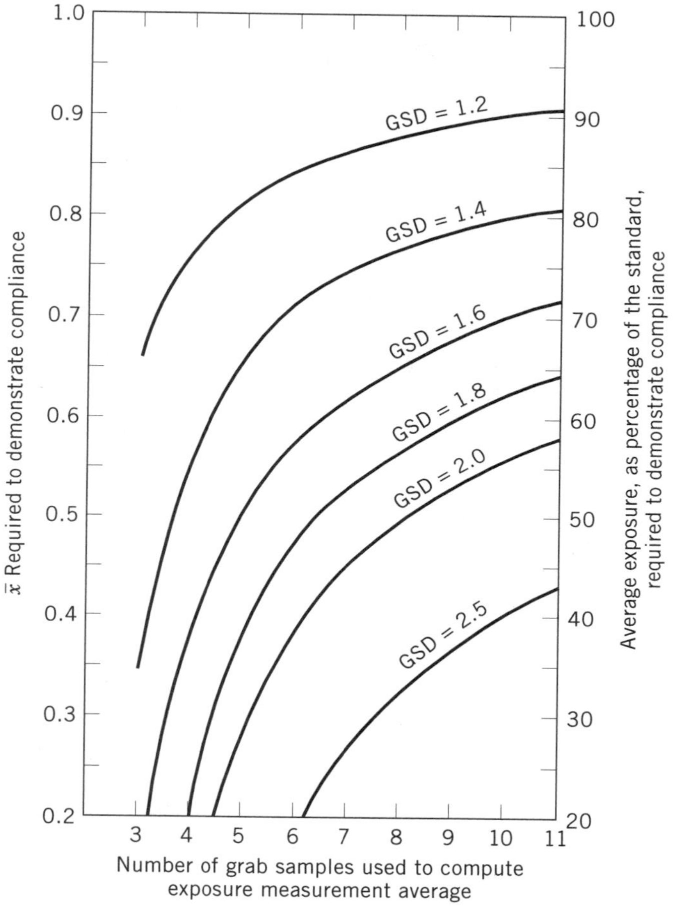

Figure 7.12. Effect of grab sample on compliance demonstration. GSD, variation of grab samples.

may be more appropriate in some situations. When measurements are made in a screening study to decide on the design of a larger study, it may be appropriate to be only 50% confident that an exposure is over some low trigger level. On the other hand, when a threat to life or a large amount of money may hang on the decision, confidence levels even beyond the three standard deviations common in quality control may be appropriate. To choose a confidence limit, first consider the consequence of being wrong and then decide on an acceptable level of risk.

Since sampling and analysis can be expensive, some thought should be given to ways of improving efficiency. Sequential sampling schemes in which the collection of a second or later group of samples is dependent on the results of some earlier set are a possibility. This common quality control approach results in infrequent sampling when far from decision points but increases the sampling as a critical region is neared. Another means of economizing is to use a nonspecific, direct-reading screening method, such as a total hydrocarbon meter, to obtain information on limiting maximum concentrations that will help to reduce the field of concern of exposure to a specific agent.

Few firm rules can be provided to aid in the selection of a sampling strategy because data can be put to such a wide variety of uses. However, the steps to be followed to arrive at a strategy can be listed:

1. Decide on the purpose of the measurements in terms of what decisions are to be made. When there are multiple purposes, select the one or several most important for design.
2. Consider the ways in which the nature of the environmental exposure and the agent relates to measurement options.
3. Identify the methods available to measure the toxic substance as it occurs in the workplace.
4. Select an interrelated combination of sampling method, location, time, and frequency that will allow a confident decision in the event of an important difference with a minimum of effort.

5.6 Biological Monitoring

Exposure can be measured after it has occurred by reconstruction of internal dose (143). If the internal dose can be estimated and the connection between the applied dose and the internal dose (pharmacokinetics) is known, it is possible to work back to the applied, external dose. The internal dose can be estimated by means of biomonitoring or biomarkers, provided the relationship with the dose can be established and interfering substances or metabolites are absent.

Biological monitoring can sometimes be used to estimate dose by measuring the substance itself in blood or urine, a metabolite of the substance, a biological effect of the substance, or the amount of the substance or its metabolite bound to target molecules. Various methods and their uses have been discussed (144, 98, 145, 146, 147). Biological monitoring methods are available for only some chemicals. While biological monitoring often can provide very direct information about past exposure, the relationship with exposure is weak or nonexistent for most chemicals.

6 MEASUREMENT METHOD SELECTION

The assortment of tools available to the industrial hygienist for measuring worker exposure to toxic substances is much larger now than a decade ago, although there are still many gaps. This section describes the choices available with respect to the location at which the measurement is made, time period or averaging time of the measurement, ability to select certain size aerosols and reject others, and the degree to which human involvement can be lessened by automation and computer analysis. The attributes of the various sampling methods, which are important in deciding if they are appropriate for a particular sampling strategy, are summarized. The detailed techniques for sample collection and analysis are covered elsewhere in this text and in other references (148).

6.1 Personal, Breathing Zone, and General Air Samples

Section 5 described the locations at which samples could be collected with relation to the air actually inhaled by the worker and defined the terms *personal, breathing zone*, and *general air samples*. Personal samplers are devices that collect a sample while being worn on the worker's clothing with the sample inlet positioned as close to the mouth as practicable, usually on the lapel. One kind of personal sampler consists of a pump on the worker's belt or pocket and a sampling head containing the sorbent tube, filter, or other collection medium, clipped to the lapel close to the nose and mouth. These pump-type active samplers have been largely replaced for vapor sampling by passive personal samplers that collect the air contaminants by diffusion onto a solid sorbent. When initially introduced, many questions were raised about the accuracy and reliability of passive dosimetery, and many studies have been performed. These studies have been reported by Berlin et al. (149) and reviewed by Harper and Purnell (150). In general, passive samplers, when properly designed and used, are as accurate as active samplers. Since passive sampling is less expensive and more convenient than powered sampling, more samples can be collected, so overall accuracy, including environmental variance, is better.

Personal samplers of both types are worn by the worker, so they must be lightweight, portable, and not affected by motion or position. These restrictions tend to limit the size of sample that can be collected. In the case of passive dosimeters, the geometry of the diffusion channel and the diffusion constant of the substance result in an equivalent sampling rate that is usually quite low, thus requiring long sample periods. Personal sampling with wet collectors, such as impingers or bubblers, is difficult due to the danger of spillage and glasswear breakage. Spill-proof impingers have, however, been used. Gravitational size selectors such as horizontal elutriators, which are affected by position, have not made successful personal samplers, although inertial devices such as cyclones or impactors are satisfactory.

When limitations of weight, complexity, wet collecting media, or position prevent successful personal sampling, it is still possible to make an approximate measurement of worker exposure by collecting the sample in the breathing zone of the worker. This vaguely defined zone is the envelope of air surrounding the worker's head that is thought to have approximately the same concentration of the contaminant being measured as the air breathed by the worker. Breathing zone samples may be collected by a fixed sampler with the inlet near the nose of a stationary worker or, in the case of a mobile worker, by carrying the sample-collecting equipment and holding the sample inlet near the worker's head, while moving around the work site with the worker. The obviously awkward and time-consuming nature of this kind of sampling limits its usefulness.

When measurements of workers' exposures are not needed or indirect estimates are adequate, the concentration of a contaminant in the general air of a workplace may be measured at some fixed station. Many of the equipment limitations imposed by personal and breathing zone sampling systems do not apply to general air samplers. Portability is not critical, so electrical components may be either battery or line powered. When line power is available, powerful pumps may be used to provide enough vacuum for critical orifices to obtain precise flow control or to operate high volume vacuum sources capable of collecting very large samples. Wet devices and both horizontal and vertical elutriators are practical. Very large samples, which may be needed to obtain sufficient sensitivity for trace analyses, may be collected on heavy or bulky collecting media. Multiple samplers of different types may be arrayed close to each other to provide sampler comparison data. Most new methods of measurement of worker exposure were first tested in fixed station arrays, where they were compared with older methods before being adapted to personal sampling.

6.2 Short- and Long-Period Samplers

Available sampling methods have limited flexibility in the period of time over which the sample can or must be collected. Some methods are inherently grab samplers, although the increased interest in long-period samplers has led to their adaptation. Most detector tubes (151) were originally intended to produce a result after a few pump strokes of up to several hundred milliliters. The interval between strokes could be lengthened, but instead of increasing the sampling period, this produces an average of several short samples taken over a longer time. Automatic systems have been developed as fixed station samplers that can extend the low range of some tubes by repeated pump strokes spread out over a long period. In the continuous flow mode, short-period detector tubes have been recalibrated for use at very low flow rates over long periods, and special tubes have been developed (152, 153) for long-term samples for up to 8-h flow rates of 5–50 cm^3/min. Other inherently short-period methods are those that use a liquid, particularly a volatile liquid, in a bubbler or impinger. As the air passes through the sampler, the collecting medium evaporates, eventually to dryness, with loss of sample unless terminated in time. Usually, without the addition of liquid, sampling periods of in excess of 30 min are impractical with wet devices.

Vessels that collect a whole air sample are convenient grab samples, and some can be adapted for longer-period sampling. Canister-based methods (154) used for measuring pollutants can maintain constant flow rates for time-integrated continuous sampling over 24 h or more. Many low flow rate personal sampler pumps have an air outlet fitting that can be used to fill a bag. Allowing for possible contamination due to the air passing through the pump, long-period samples may be collected. When the bag is carried in a sling on the worker's back, personal samples may be collected.

Greater time flexibility is available with solid sorbents. Even though there is a fixed volume of air that can be sampled at a concentration before breakthrough occurs, the freedom to use a wide range of low flow rates permits personal and fixed sampling over periods of 8 h or more. Other air contaminants, competing for active sorbent sites and particulates adding to the resistance flow, limit the maximum volume of air that can be sampled in much the same way as breakthrough.

Systems that tend to be suitable for long-period samples only are usually those where the sampler can barely collect the minimum amount of material required by the sensitivity of the analytical method. A personal sample for respirable quartz using a 10-mm cyclone-size selective presampler, operated at 2 L/min or less, will sample less than a cubic meter of air in 8 h and thus collect less than 100 mg of quartz at the TLV. Shorter samples will confront the serious sensitivity limitations of most methods for quartz (148). The same problem occurs with personal samples for beryllium and for detailed analyses for multiple compounds such as PNAs.

High volume fixed station or even personal particulate samplers using large-size selectors and filters, where necessary, allow shorter sampling periods that are still longer than grab samples. Even when analysis is not necessary and only gross or respirable weight is being measured, analytical balance limitations prevent very short sample times for particulates. There are, however, various more nearly instantaneous mass monitors based on beta absorption (155) or the piezoelectric effect (156) that are capable of making a measurement in as little as one minute.

6.3 Size Selection

When it is necessary in sampling for particulates to include or exclude certain size particles, limitations are created by the nature of the size selecting devices available (157). The unsuitability of elutriators as personal samplers due to their orientation requirements has already been mentioned. All size selectors make their stated cut only at a predetermined flow rate, which must be held constant over the period of the sample against changing filter resistance and battery conditions (158). Whereas cyclones and impactors tend to compensate for flow rate changes and elutriators compound the error, all need pumps that not only sample a known volume over the sampling period but do so at a known constant flow rate. Such pumps are usually larger and heavier than low flow pumps, which need only sample a reliably known volume, and approach the limit of practicality as personal samplers. In addition to the requirement for constant flow, pump pulsation must be damped out if it upsets the size selector.

Isokinetic conditions usually must be established when sampling for particulates in high velocity streams (159). In stacks, particles with high kinetic energy due to their weight and velocity can be improperly included or excluded from the sample under nonisokinetic conditions. However, workplace sampling is generally done at low ambient air velocities of less than 300 ft/min. Further, air velocity and direction is usually continually changing. Under these conditions, isokinetic sampling of the kind used in stack sampling is not necessary or practical. However, consideration must be given to the effect of the velocity and direction of the air at the filter inlet. Open-face filters may oversample when face up or undersample when face down. Some undersampling may be desirable to avoid collection of large, noninhalable particles. Davies (82) has developed theoretical relationships for filter inlet performance. For asbestos fiber sampling in accordance with NIOSH methods 7400 or 7402 (148), a conductive plastic cowl is attached to the filter inlet to improve the distribution of fibers on the filter surface and to reduce charge effects.

6.4 Continuous Air Monitors

Continuous air monitors (CAMs) with fixed sensors in critical locations connected to alarms and remote indicators are commonly used for carbon monoxide, hydrogen sulfide, chlorine hydrocarbons, and other acutely toxic or explosive gases. These systems may use passive sensors, they may pump the air through a detection cell located at the point of collection, or they may pump contaminated air to a remote analyzer. Sequential valving arrangements allow one remote analyzer to be coupled to many sample lines. Provision for automatic calibration and zeroing may be included.

Systems that make measurements automatically at a number of locations and gather the data at a central readout point have been installed in a number of plants. Adaptation of these systems to the estimation of worker exposure has led to the development of complex computer-based monitoring systems (Figure 7.13). Since the contaminant sensors do not detect the presence of a worker, some data on worker location must be added to estimate exposure. Time and motion studies that yield percentage of time in various measured locations could be used with the daily average fixed station measurement for that location to calculate a weighted average exposure. The drawbacks are that time/activity distributions exhibit considerable variation, even under routine conditions, and the most significant exposures often occur during nonroutine periods. Also an assumption is made that the concentration at a time is independent of the worker's activity.

Alternative schemes provide a device that reads a card carried by each worker to signal the computer of each entry and departure from a monitored area. Time in the area can be multiplied by the general or weighted average concentration from the sensors in that area as measured while the worker was present. This situation is analogous to estimating exposure from fixed station sampler measurements. An even more elaborate system could place small transmitters on each worker that are tracked automatically by a sensing network, and these detailed worker location data would be combined with fixed station measurements for various locations to estimate exposure.

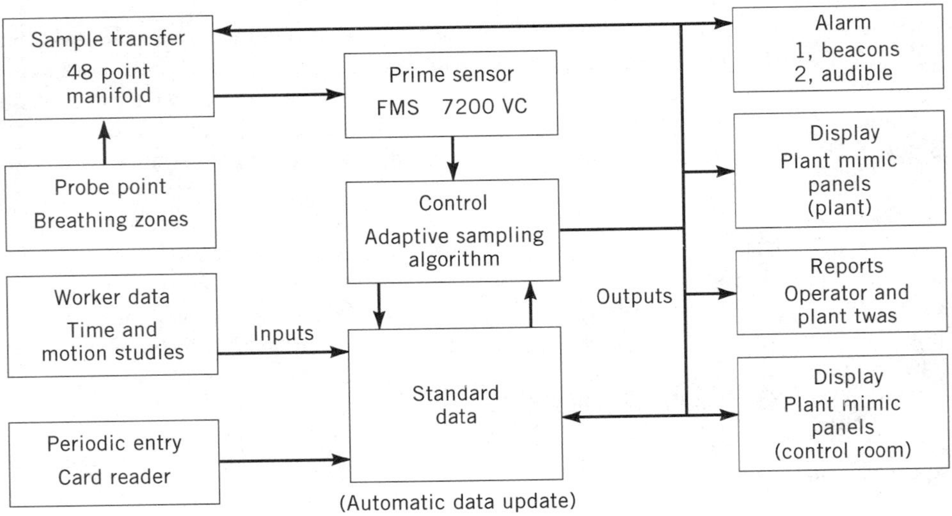

Figure 7.13. Vinyl chloride monomer (VCM) monitoring system. (Courtesy of Eocom Corporation).

Continuous monitoring may also be done with remote sensing systems such as those that use Fourier transform infrared (FTIR). These systems project a beam of electromagnetic radiation in the infrared region over a path through the potentially contaminated air volume to a receiving telescope and spectral analyses either directly or via a mirror. These systems can sense and measure a large number of gases or path lengths of hundreds of meters and give real-time readout. While not directly applicable to worker exposure measurement, the use of remote sensing devices as fugitive emission detectors is a useful addition to closed process chemical plant control. Systems that combine a video record of worker activity with continuous measurement of exposure have been developed (160, 161, 162). These techniques allow the study of determinants of exposure by correlating activity with concentration.

6.5 Mixtures

Toxic substances seldom occur in isolation. Mixtures not only cause difficulties in estimating the biological consequences of exposure, but also complicate exposure measurement. The other substances present in an air mixture need to be considered, even when only one component of the mixture is being measured and the other components are far below toxic effect levels, because of the effects of the other substances on the sampling systems. Charcoal is a very useful sampling medium because it will absorb and retain so many substances. However, because of this property, it is possible for all the active sorption sites to be occupied by other materials. The substance being measured may then break through long before the recommended sample volume, based on collection of the pure substance, has been passed through the tube.

An analogous case is the measurement of a low concentration of asbestos fibers in an environment containing a great deal of other, nonfibrous dust. To collect enough fibers to give the fiber density necessary for an adequate count without counting an unreasonable number of fields would result in the collection of so many grains that the fibers would be obscured. Thus the total dust present limits the sample volume, and counting a large number of fields compensates for low fiber density. Overload from other airborne substances can also result in plugging of filter samplers, particularly when liquid accumulates on the surface of membrane filters and blinds the pores. The use of thick depth filters of glass or cellulose fibers provides greater capacity, although at the possible sacrifice of some efficiency.

Where mixtures of substances present in the workplace are additive, an exposure index E is calculated as discussed earlier in this chapter. A difficulty with this approach is how E is distributed and where the variance in E comes from. Kumagai and Matsunaga (163) have proposed a model that splits the variance into specific factors that cause one component of a mixture to vary, and common factors that effect all components. This model, which has been empirically validated, offers the generality that E is log-normally distributed, and so all previously proposed methods for evaluating occupational exposure can be applied.

6.6 Method Selection

The purpose of this chapter is to explain the considerations that go into the choice of a sampling strategy (location, period, and frequency) and the selection of a method that will

Table 7.4. Sampling and Analysis Method Attributes

Method	Sampling Period	Ability to Concentrate Contaminant	Ability to Measure Mixtures	Time to Result	Intrusiveness	Proximity to Nose and Mouth
Personal sampler/solid sorbent						
Sorption only	Medium to long	Yes	Yes—gases	After analysis	Medium	Very close
Sorption plus reaction	Medium to long	Yes	No	After analysis	Medium	Very close
Personal sampler/filter						
Gross gravimetric	Medium to long	Yes	Yes—particulate	After weighing or analysis	Medium	Very close
Respirable gravimetric	Long	Yes	Yes—particulate	After weighing or analysis	Medium	Very close
Count	Medium to long	Yes	Yes—particulate	After counting	Medium	Very close
Combination filter and sorbent	Medium to long limited	Yes	Yes	After analysis	Medium	Very close
Passive dosimeter	Long	Yes	Yes—gases	After analysis	Low	Very close
Breathing zone impinger/bubbler						
Analysis	Medium-limited	Yes	Yes	After analysis	High	Close
Count	Medium-limited	Yes	Yes—particulate	After counting	High	Close
Detector tubes						
Grab	Short	MA	No	Immediate	High	Close
Long period	Medium to long	NA	No	Immediate	Medium	Very close
Gas vessels						
Rigid vessel	Short to long	No	Yes—gases	After analysis	High	Medium
Gas bag	Short to long	No	Yes—gases	After analysis	High	Close
Evacuated/critical orifice	Medium to long	No	Yes—gases	After analysis	Medium	Distant

Direct-reading portable meters						
Nonspecific	Instantaneous or recorder	NA	Yes	Immediate	High	Slightly distant
Specific	Instantaneous or recorder	NA	No	Immediate	Medium	Slightly distant
Multiple compound	Instantaneous or recorder	Some	Yes	Almost immediate	High	Slightly distant
Mass monitor	Short	Yes	No	Almost immediate	High	Slightly distant
Particle counters	Short	No	No	Almost immediate	High	Slightly distant
Sensor with dataloger	Short or long	No	No	Hours	Medium	Close
Fixed station						
High volume	Medium to long	Yes	Yes—particulate	After analysis	Low	Remote
Horizontal or vertical elutriator	Long to short	Yes	Yes—particulate	After analysis	Low	Remote
Installed monitor	Short to long	Some	No	Almost immediate	Low	Remote
Freeze trap	Medium	Yes	Yes—vapors	After analysis	Low	Remote
FTIR	Instantaneous	No	Yes—gases	Immediate	Low	Remote

Method	Specificity	Convenience Rating	Sample Transportability	Recheck of Analysis Possible	Accuracy
Personal sampler/solid sorbent					
Sorption only	High by analysis	High	Good	Elution; yes. Thermal des.: no.	Good
Sorption plus reaction	High by analysis	High	Good	Yes	Good
Personal sampler/filter					
Gross gravimetric	None for weight only—high by analysis	High	Fair	Yes	Good
Respirable gravimetric	High by analysis	Medium	Fair	Yes	Fair
Count	Fair—depends on particle identification	High	Good	Yes	Poor
Combination filter/sorbent	High by analysis	Medium	Good	Yes	Fair
Passive dosimeter	High by analysis	Very high	Good	Yes	Fair
Breathing zone impinger/bubbler					
Analysis	High by analysis	Low	Poor	Yes	Fair
Count	Fair—depends on particle analysis	Low	Poor	Yes	Poor

Table 7.4. (continued)

Method	Specificity	Convenience Rating	Sample Transportability	Recheck of Analysis Possible	Accuracy
Dector tubes					
Grab	Medium—some interference	High	No sample	No	Fair
Long period	Medium—some interference	High	No sample	No	Fair
Gas vessels					
Rigid vessel	High by analysis	Low	Fair	Yes	Good
Gas bag	High by analysis	Low	Fair	Yes	Good
Evacuated/critical orifice	High by analysis	Low	Good	Yes	Good
Direct reading portable meters					
Nonspecific	None—total of measured class	High	No sample	No	Good
Specific	Medium—some interference	High	No sample	No	Good
Multiple compound	Medium—frequent overlap	Medium	No sample	No	Fair
Mass monitor	Mass only	High	No sample	Not usually	Fair
Particle counters	Count/size only	High	No sample	No	Fair
Sensor with dataloger	Medium—some interference	High	No sample	No	Good
Fixed station					
High volume	High by analysis	Low	Fair	Yes	Good
Horizontal or vertical elutriator	High by analysis	Low	Fair	Yes	Good
Installed monitor	Medium—may be interferences	High	No sample	No	Good
Freeze trap	High by analysis	Very low	Poor	Yes	Fair
FTIR	Medium—may be interferences	High	No sample	No	Fair

permit the accomplishment of that strategy. Table 7.4 summarizes the degree to which the most common methods possess the attributes of importance in selecting various sampling strategies. The column headings are explained as follows:

Method. The methods listed are sampling methods, but the ratings of attributes that follow assume the usual range of analytical methods that can be applied to the size and type of sample collected.

Sampling Period. The term *short* means essentially grab samples, whereas the term *long* means 8 h or longer for a single sample.

Ability to Concentrate Contaminant. Sampling methods that extract contaminant from the air and collect it in a reduced area or volume are potentially able to improve analytical sensitivity by several orders of magnitude. However, the concentrating mechanism (filtration, sorption) may introduce errors.

Ability to Measure Mixtures. Most sampling methods provide a sample that can be analyzed for more than one gas or vapor, but usually not for both gases and vapors or particulates.

Time to Result. Certain decisions (vessel entry) must be made immediately while others can wait until after the sample is transferred to a laboratory and analyzed.

Intrusiveness. When the method requires the presence of a person to collect the sample, or the wearing of a heavy or awkward sampling apparatus, this intrusion of the sampling system into the work situation may affect worker behavior and exposure (164).

Proximity to Nose and Mouth. As discussed earlier, locating a sampler inlet even a small distance from a worker's mouth may bias the exposure measurement. Samplers remote from the worker may not be measuring the air inhaled at all.

Specificity. Some methods give only nonspecific information such as total weight of all dust particles or concentration of all combustible gases while others provide a sample that can be analyzed for any species or element.

Convenience Rating. These are estimates of the amount of work or difficulty involved in collecting samples.

Sample Transportability. If the sample must be transported to a distant laboratory for analysis, the ability to withstand shock, vibration, storage, and temperature and pressure changes without being altered or destroyed is important.

Recheck of Analysis Possible. Some samples may be analyzed only once while others are in a form such that rechecks, reanalysis at different conditions, or analysis for other substances is possible.

Accuracy. Given all the possibilities for error from sampler calibration, sample collection, transport, and analysis, an overall coefficient of variation of 10% is considered good. Some count methods are subject to such counter variability that poor accuracy is usual. Method inaccuracy should not be judged alone but should be seen in combination with the inaccuracy caused by environmental variability, which is usually larger, when making decisions about whether a method is sufficiently accurate for a purpose.

Table 7.4 shows that not all sampling strategies are possible because for some strategies the sampling and analytical method with the necessary combination of attributes may not exist. The industrial hygiene technology gaps thus revealed are fruitful areas for future research and development.

BIBLIOGRAPHY

1. V. G. Zartarian, W. R. Ott, and N. Duan, "Quantitative Definition of Exposure and Related Concepts," *J. Expos. Anal. Environ. Epidemiol.*, **7**, 411–437 (1997).
2. M. Corn, *Am. Ind. Hyg. Assoc. J.*, **37**, 353–356 (1976).
3. S. M. Rappaport and T. J. Smith, Eds. *Exposure Assessment for Epidemiology and Hazard Control*, Lewis, Chelsea, MI, 1991.
4. S. Lipton and J. Lynch, *Health Hazard Control in the Chemical Process Industry*, Wiley, New York, 1994.
5. H. Raiffa, *Decision Analysis: Introductory Lectures on Choices under Uncertainty*, Addison-Wesley, Reading, MA, 1970.
6. ACGIH (American Conference of Governmental Industrial Hygienists), *Threshold Limit Values for Chemical Substances and Physical Agents: Biological Exposure Indices*, ACGIH, Cincinnati, OH, 1999.
7. OSHA, *Occupational Safety and Health Act of 1970*, PL 91-596, 1970.
8. OSHA, *Occupational Safety and Health Standards*, 29 CFR 1910, 1992.
9. J. H. Vincent, "International Occupational Exposure Standards: A Review and Commentary," *Am. Ind. Hyg. Assoc. J.*, **59**, 729–742 (1998).
10. M. A. Jayjock, *Am. Ind. Hyg. Assoc. J.*, **49**, 380–385 (1988).
11. P. B. Ryan, *J. Expos. Anal. Environ. Epidemiol.*, **1**, 453–474 (1991).
12. USEPA, *Chemical Engineering Branch Manual for the Preparation of Engineering Assessments*, PN 3786-64, U.S. Environmental Protection Agency, 1991.
13. P. H. Reinke and L. M. Brosseau, "Development of a Model to Predict Air Contaminant Concentrations Following Indoor Spills of Volatile Liquids," *Ann. Occup. Hyg.*, **41**, 415–435 (1997).
14. M. R. Flynn, B. L. Gatano, J. L. McKernan, K. H. Dunn, B. A. Blazicko, and G. N. Carlton, "Modeling Breathing-Zone Concentrations of Airborne Contaminants Generated During Compressed Air Spray Painting," *Ann. Occup. Hyg.*, **43**, 67–76 (1999).
15. M. Nicas and R. C. Spear, *Am. Ind. Hyg. Assoc. J.*, **54**, 211–227 (1993).
16. E. M. Thompson, H. N. Treaftis, T. F. Tomb, and A. J. Beckert, *Am. Ind. Hyg. Assoc. J.*, **38**, 523–535 (1977).
17. H. M. Donaldson and W. T. Stringer, *Beryllium Sampling Methods*, Publication (NIOSH) 76-201, Department of Health, Education, and Welfare, Cincinnati, Ohio, 1976, p. 21.
18. C. S. McCammon, Jr., and J. W. Woodfin, *Am. Ind. Hyg. Assoc. J.*, **38**, 378–386 (1977).
19. L. D. Horowitz, *Am. Ind. Hyg. Assoc. J.*, **37**, 227–233 (1976).
20. CMA (Chemical Manufacturers Association), *Improving Air Quality: Guidance for Estimating Fugitive Emissions from Equipment*, 2nd ed., CMA, Washington, DC, 1989.
21. USEPA, *Protocol for Generating Unit-Specific Emission Estimates for Equipment Leaks of VOC and VHAP*, Pub. No., EPA 450/3-88-010, Research Triangle Park, NJ, 1988.

22. USEPA, *Determination of Volatile Organic Compound Leaks*, Code of Federal Regulations, Title 40 Part 60 Appendix A, Reference Method 21, 1990.
23. P. A. Stewart and R. F. Herrick, *Appl. Occup. Environ. Hyg.*, **6**, 421–427 (1991).
24. P. A. Stewart, R. F. Herrick, A. Blair, H. Checkoway, P. Dray, L. Fine, L. Fischer, R. Harris, T. Kauppinens, and R. Saracci, *Scand. J. Work Environ. Health*, **17**, 281–285 (1991).
25. S. M. Rappaport, *Ann. Occup. Hyg.*, **15**, 61–121 (1991).
26. R. W. Smith, J. D. Sahl, M. A. Kelsh, and J. Zalinski, "Task-Based Exposure Assessment: Analytical Strategies for Summarizing Data by Occupational Groups," *Am. Ind. Hyg. Assoc. J.*, **58**, 402–412 (1997).
27. R. L. Harris, *Guidelines for Collection of Industrial Hygiene Exposure Assessment Data for Epidemiologic Use*, Chemical Manufacturers Association, Washington, DC, 1993.
28. M. C. Marbury and J. E. Woods, in *Indoor Air Pollution*, J. M. Samet and J. D. Spengler, Eds., Johns Hopkins University Press, Baltimore, MD, 1991, pp. 306–332.
29. J. M. Samit and J. D. Spengler, *Indoor Air Pollution*, Johns Hopkins University Press, Baltimore, 1991.
30. N. L. Nagda, H. E. Rector, and M. D. Koontz, *Guidelines for Monitoring Indoor Air Quality*, Hemisphere, New York, 1987.
31. R. Tornero-Valez, E. Symanski, H. Kromhout, R. C. Yu, and S. M. Rappaport, "Compliance versus Risk in Assessing Occupational Exposures," *Risk Anal.* **17**, 279–292 (1997).
32. N. A. Leidel, K. A. Busch, and J. R. Lynch, *Occupational Exposure Sampling Strategy Manual*, Publication (NIOSH) 77-173, Department of Health, Education and Welfare, Cincinnati, OH, 1977.
33. CEFIC, *Report on Occupational Exposure Limits and Monitoring Strategy*, European Council of Chemical Manufacturers Association, Brussels, 1982.
34. BMA, *Measurement and Evaluation of Concentrations of Airborne Toxic or Health Hazardous Work-Related Substances*, TRgA401 Sheet 1, West German Federal Ministry of Labor (BMA), 1979.
35. N. C. Hawkins, S. K. Norwood, and J. C. Rock, Eds., *A Strategy for Occupational Exposure Assessment*, American Industrial Hygiene Association, Akron, OH, 1991.
36. R. P. Harvey, "Statistical Aspects and Air Sampling Strategies," in C. F. Culles and J. G. Firth, eds., *Detection and Measurement of Hazardous Gases*, Heinemann, London–New York, 1981, p. 147.
37. S. M. Rappaport, S. Selvin, R. C. Speer, and C. Keil, *Am. Ind. Hyg. Assoc. J.*, **42**, 831–838 (1981).
38. S. M. Rappaport, S. Selvin, R. C. Speer, and C. Keil, *ACS Symp. Series*, **149**, 431 (1981).
39. R. M. Tuggle, *Am. Ind. Hyg. Assoc. J.*, **42**, 493–498 (1981).
40. R. M. Tuggle, *Am. Ind. Hyg. Assoc. J.*, **43**, 338–346 (1982).
41. J. C. Rock, *Am. Ind. Hyg. Assoc. J.*, **43**, 297–313 (1982).
42. V. Fiserova-Bergerova, "Relevance of Occupational Skin Exposure," *Ann. Occup. Hyg.*, **37**, 673–685 (1993).
43. J. D. Cock, D. Heederik, H. Kromhout, and J. S. M. Boleij, "Strategy for Assigning a 'Skin Notation': A Comment," *Ann. Occup. Hyg.*, **40**, 611–614 (1996).
44. V. Fiserova-Bergerova, J. T. Pierce, and P. O. Dray, *Am. J. Ind. Med.*, **17**, 617–635 (1990).

45. I. Brooke, J. Cocker, J. I. Delic, M. Payne, K. Jones, N. C. Gregg, and D. Dyne, "Dermal Uptake of Solvents from the Vapour Phase: an Experimental Study in Humans," *Ann. Occup. Hyg.* **42,** 531–540 (1998).
46. G. Johanson and A. Boman, *Br. J. Ind. Med.*, **48**, 788–792 (1991).
47. G. R. Oxley, *Ann. Occup. Hyg.*, **19**, 163–167 (1976).
48. R. R. Lauwerys, T. Dath, J. M. Lachapelle, J. P. Buchet, and H. Roels, *J. Occup. Med.*, **20**, 17–20 (1978).
49. S. Fukabori and N. Nakaaki, *J. Sci. Labour*, **53**, 89–95 (1976).
50. S. Fukabori and N. Nakaaki, *J. Sci. Labour*, **52**, 67–81 (1976).
51. E. Cronen and R. B. Stoughton, *Arch. Dermatol.*, **36**, 265, (1962).
52. J. W. Cherrie and A. Robertson, "Biologically Relevant Assessment of Dermal Exposure," *Ann. Occup. Hyg.*, **39**, 387–392 (1995).
53. D. H. Brouwer, R. Kroese, and J. J. Van Hemmen, "Transfer of Contaminants from Surface to Hands: Experimental Assessment of Linearity of the Exposure Process, Adherence to the Skin, and Area Exposed During Fixed Pressure and Repeated Contact with Surfaces Contaminated with a Powder," *Appl. Occup. Environ. Hyg.* **14**, 231–239 (1999).
54. B. Berner and E. R. Cooper, in A. F. Kydoniew and B. Berner, eds., *Models of Skin Permeability in Transdermal Delivery of Drugs*, Vol. II, CRC Press, Boca Raton, FL, 1987, pp. 41– 55.
55. G. L. Flynn, "Physiochemical Determinants of Skin Absorption," in T. R. Geritz and C. J. Henry, eds., *Principles of Route-to-Route Extrapolation for Risk Assessment*, Elsevier Science, 1990, pp. 93–127.
56. H.-W. Leung and D. J. Paustenbach, "Techniques for Estimating the Percutaneous Absorption of Chemicals Due to Occupational and Environmental Exposure," *Appl. Occup. Environ. Hyg.*, **9**, 187–197 (1994).
57. M. M. Key, Ed., *Occupational Diseases: A Guide to Their Recognition*, U.S. Government Printing Office, Washington, DC, 1977.
58. R. W. Weeks, Jr., B. J. Dean, and S. K. Yasuda, *Occup. Health Safety*, **46**, 19–23 (1977).
59. J. V. LeBlanc, *J. Occup. Med.*, **19**, 276–277 (1977).
60. T. D. Sterling and D. M. Kobayashi, *Environ. Res.*, **3**, 1–35 (1977).
61. D. T. Wigle, *Arch. Environ. Health*, **32**, 185–190 (1977).
62. L. A. Wallace, E. D. Pellizzar, T. D. Hartwell, R. Whitmore, H. Zelon, R. Perritt, and L. Sheldon, *Atmos. Environ.*, **22**, 2141–2163 (1988).
63. W. R. Ott, "Human Activity Patterns: A Review of the Literature for Estimating Time Spent Indoors, Outdoors and in Transit," in *Proceedings of the Research Planning Conference on Human Activity Patterns*, USEPA 600, 04.89.004, 1989.
64. P. J. Lioy, *Environ. Sci. Technol.*, **24**, 938–945 (1990).
65. A. T. Hodgson, J. M. Daisey, and R. A. Grot, *J. Air Waste Mgt. Assoc.*, **41**, 1461–1468 (1991).
66. M. McCann, *Artists Beware*, Watson-Guptill, New York, 1979.
67. E. W. Helper, N. K. Napier, M. E. D. Hillman, and S. Davidian, *Product/Industry Profile on Art Materials and Selected Craft Materials*, CPSC-C-78-0091, U.S. Consumer Product Safety Commission, 1979.
68. C. Hart, *J. Environ. Health*, **49**, 282–287 (1987).
69. A. Poklis and C. D. Burkett, *Clin. Toxicol.*, **11**, 35–41 (1977).
70. J. S. Oliver, *Lancet*, **1**, 84–86 (1977).

71. R. Korobkin, A. K. Asbury, A. J. Sohner, and S. L. Nielsen, *Arch. Neurol.*, **32**, 158–162 (1975).
72. J. W. Hayden, E. G. Comstock, and B. S. Comstock, *Clin. Toxicol.*, **9**, 169–184 (1976).
73. R. A. III Warriner, A. S. Nies, and W. J. Hayes, *Arch. Environ. Health*, **32**, 203–205 (1977).
74. B. L. Weisenberger, *J. Occup. Med.*, **19**, 569–570 (1977).
75. C. I. Fairchild and M. I. Tillery, *Am. Ind. Hyg. Assoc. J.*, **38**, 277–283 (1977).
76. R. F. Malek, J. M. Daisey, and B. S. Cohen, *Am. Ind. Hyg. Assoc. J.*, **47**, 524–529 (1986).
77. R. B. M. Geuskens, M. J. M. Jongen, J. C. Ravensberg, J. Vander Tuin, L. H. Leenheers, and J. F. Vander Wal, *Appl. Occup. Environ. Hyg.*, **6**, 364–369 (1990).
78. T. L. Chan, J. B. D'Arey, and R. M. Schreck, *Am. Ind. Hyg. Assoc. J.*, **47**, 411–417 (1986).
79. J. E. Peterson, R. A. Copeland, and H. P. Hayles, *Am. Ind. Hyg. Assoc. J.*, **23**, 345–352 (1962).
80. J. R. Lynch, "Air Sampling for Cotton Dust," in *Transactions of the National Conference on Cotton Dust and Health*, University of North Carolina, Chapel Hill, NC, 1970, p. 33.
81. J. H. Vincent, and D. Mark, *Ann. Occup. Hyg.*, **24**, 375–390 (1981).
82. C. N. Davies, *J. Appl. Phys., Ser. 2*, **1**, 921–932 (1968).
83. J. K. Agarual and B. Y. H. Liu, *Am. Ind. Hyg. Assoc. J.*, **41**, 191–197 (1980).
84. ACGIH (American Conference of Governmental Industrial Hygiene), *Particle Size-Selective Sampling for Health-Related Aerosols*, ACGIH, Cincinnati, OH, 1998.
85. B. R. Roy, *J. Am. Ind. Hyg. Assoc.*, **38**, 327–332 (1977).
86. K. A. Schulte, D. J. Larsen, R. W. Hornung, and J. V. Crable, (1974). *Report on Analytical Methods Used in a Coke Oven Effluent Study*, National Institute for Occupational Safety and Health, 1974.
87. R. K. Zumwalde and J. M. Dement, *Review and Evaluation of Analytical Methods for Environmental Studies of Fibrous Particulate Exposures*, Department of Health, Education, and Welfare, Publication (NIOSH) 77-204, NIOSH, Cincinnati, OH, 1977, p. 66.
88. J. A. Merchant, *Environ. Health Perspect.*, **88**, 287–293 (1990).
89. N. A. Esmen and S. Erdal, *Environ. Health Perspect.*, **88**, 277–286 (1990).
90. M. Lippmann, *Environ. Res.*, **46**, 86–105 (1988).
91. C. A. Halder, G. S. Van Gorp, N. S. Hatowm, and T. M. Warne, *Am. Ind. Hyg. Assoc. J.*, **47**, 164–172 (1986).
92. H. E. Runion, *Am. Ind. Hyg. Assoc. J.*, **38**, 391–393 (1977).
93. H. J. McDermott and S. E. Killiany, *Am. Ind. Hyg. Assoc. J.*, **39**, 110–117 (1978).
94. WGD, *Health-Based Recommended Occupational Exposure Limit for Gasoline*, Report WGO 91-318-5, Directorate General of Labour, Den Haag, Netherlands, 1991.
95. R. C. Spear, S. Selvin, and M. Francis, *Am. Ind. Hyg. Assoc. J.*, **48**, 365–368 (1986).
96. B. Preat, *Am. Ind. Hyg. Assoc. J.*, **48**, 877–884 (1987).
97. D. Turner, *Ann. Occup. Hyg.*, **19**, 147–152 (1976).
98. R. R. Lauwerys, *Industrial Chemical Exposure: Guidelines for Biological Monitoring*, Biomedical Publications, Davis, CA, 1983.
99. HSE (Health and Safety Executive), *Monitoring Strategies for Toxic Substances*, Guidance Note, EH 42, HSE, Bootle, Meseyside, UK, 1989.
100. AIHA (American Industrial Hygiene Association), *A Strategy for Assessing and Managing Occupational Exposures*, AIHA, Fairfax, VA, 1998.

101. P. Hewett, "Interpretation and Use of Occupational Exposure Limits for Chronic Disease Agents," *Occup. Med. State Art Rev.*, **11**, 561–589 (1996).
102. E. Tielemans, L. L. Kupper, H. Kromhout, and R. Houba, "Individual-Based and Group-Based Occupational Exposure Assessment: Some Equations to Evaluate Different Strategies," *Ann. Occup. Hyg.*, **42**, 115–119 (1998).
103. S. Kumagai and I. Matsunaga, "Within-Shift Variability of Short-Term Exposure to Organic Solvents in Indoor Workplaces," *Am. Ind. Hyg. Assoc. J.*, **60**, 16–21 (1999).
104. J. Aitchinson and J. A. C. Brown, *The Lognormal Distribution*, University Press, Cambridge, England, 1963.
105. B. E. Saltzman, "Health Risk Assessment of Fluctuating Concentrations Using Lognormal Models," *Air Waste Manage. Assoc.*, **47**, 1152–1160 (1997).
106. R. H. Lyles, L. L. Kupper, and S. M. Rappaport, "A Lognormal Distribution-Based Exposure Assessment Method for Unbalanced Data," *Ann. Occup. Hyg.*, **41**, 63–76 (1997).
107. Lynch, J. R. *Nat. Safety News*, **113**(5), 67–72 (1976).
108. K. E. Bencala and J. H. Seinfeld, *Atmos. Environ.*, **10**(11), 941–950 (1976).
109. R. I. Larsen, *A Mathematical Model for Relating Air Quality Measurements to Air Quality Standards*, Publication AP-89, U.S. Environmental Protection Agency, Research Triangle Park, NC, 1971.
110. Y. Kalpasanor and G. Kurchatora, *J. Air. Pollut. Control Assoc.*, **26**, 981 (1976).
111. A. C. Stern, Ed., *Air Pollution*, 3rd ed., Vol. 3, *Measuring Monitoring, and Surveillance of Air Pollution*, Academic Press, New York, 1976, p. 799.
112. N. A. Leidel, K. A. Busch, and W. E. Crouse, *Exposure Measurement Action Level and Occupational Environmental Variability*, U.S. Government Printing Office, Washington, DC, 1975.
113. P. Hewitt, "Mean Testing, I: Advantages and Disadvantages," *Appl. Occup. Environ. Hyg.*, **12**, 339–346 (1997).
114. P. Hewitt, "Mean Testing, II: Comparison of Several Alternative Procedures," *Appl. Occup. Environ. Hyg.*, **12**, 339–346 (1997).
115. X.-H. Zhou, "Estimating the Mean Value of Occupational Exposures," *Am. Ind. Hyg. Assoc. J.*, **59**, 785–788 (1998).
116. H. Kromhout, E. Symanski, and S. M. Rappaport, "A Comprehensive Evaluation of Within- and Between-worker Components of Occupational Exposure to Chemical Agents," *Ann. Occup. Hyg.*, **37**, 243–270 (1993).
117. S. M. Rappaport, R. H. Lyles, and L. L. Kupper, "An Exposure-Assessment Strategy Accounting for Within- and Between-Worker Sources of Variability," *Ann. Occup. Hyg.*, **39**, 469–495 (1995).
118. M. A. Jayjoch, "Uncertainty Analysis in the Estimation of Exposure," *Am. Ind. Hyg. Assoc. J.*, **58**, 380–382 (1997).
119. E. E. Campbell, *Am. Ind. Hyg. Assoc. J.*, **37**, 6, A-4, (1976).
120. OSHA, *OSHA Technical Manual*, OSHA, Washington, DC, 1999.
121. B. B. Chatterjee, M. K. Williams, J. Walford, and E. King, *Am. Ind. Hyg. Assoc. J.*, **30**, 643–645 (1969).
122. R. Butterworth and J. K. Donoghue, *Health Phys.*, **18**, 319 (1970).
123. B. S. Cohen, A. E. Chang, N. H. Harley, and M. Lippmann, *Am. Ind. Hyg. Assoc. J.*, **43**, 239–243 (1982).

124. S. A. Roach, S. Holmes, W. H. A. Beverley, J. L. Bonsall, L. H. Capel, R. D. Hunt, M. Jacobsen, J. G. Morris, W. H. Smither, J. Steel, R. Sykes, and S. J. Silk, *Ann. Occup. Hyg.*, **1**, 1–55 (1983).

125. A. J. Breslin, L. Ong, H. Glauberman, A. C. George, and P. LeClare, *Am. Ind. Hyg. Assoc. J.*, **28**, 56–61 (1967).

126. R. J. Sherwood, *Am. Ind. Hyg. Assoc. J.*, **27**, 98–109 (1966).

127. D. B. Baretta, R. D. Stewart, and J. E. Mutcheler, *Am. Ind. Hyg. Assoc. J.*, **30**, 537–544 (1969).

128. E. J. Calabrease, *Am. Ind. Hyg. Assoc. J.*, **38**, 443–446 (1977).

129. A. L. Linch and H. V. Pfaff, *Am. Ind. Hyg. Assoc. J.*, **32**, 745–752 (1971).

130. S. A. Roach, *Am. Ind. Hyg. Assoc. J.*, **27**, 1–12 (1966).

131. S. A. Roach, *Ann. Occup. Hyg.*, **20**, 65–84 (1977).

132. P. Droz and M. Yu, "Biological Monitoring Strategies," in *Exposure Assessment for Occupational Epidemiology and Hazard Control*, S. M. Rappaport and T. J. Smith, Eds., Lewis Publishing, Chelsea, MI, 1990.

133. R. S. Brief and R. A. Scala, *Am. Ind. Hyg. Assoc. J.*, **36**, 467–469 (1975).

134. J. W. Mason and H. Dershin, *J. Occup. Med.*, **18**, 603–606 (1976).

135. M. E. Anderson, M. G. MacNaughton, H. J. Clewell, and D. P. Paustenbach, *Am. Ind. Hyg. Assoc. J.*, **48**, 335–343 (1987).

136. J. L. S. Hickey and P. C. Reist, *Am. Ind. Hyg. Assoc. J.*, **38**, 613–621 (1977).

137. R. S. Brief and A. R. Jones, *Am. Ind. Hyg. Assoc. J.*, **37**, 474–478 (1976).

138. S. M. Rappaport, S. Selvin, and S. Roach, *Appl. Ind. Hyg.*, **3**, 310–315 (1988).

139. D. D. Douglas, *Am. Ind. Hyg. Assoc. J.*, **38**, A-6 (1977).

140. J. R. Lynch, "Uses and Misuses of Detector Tubes," in *Transactions of the 32nd Meeting of the American Conference of Governmental and Industrial Hygienists*, ACGIH, Cincinnati, OH, 1970.

141. M. Nicas, B. P. Simmons, and R. C. Spear, *Am. Ind. Hyg. Assoc. J.*, **52**, 553–557 (1991).

142. E. Buringh and R. Lanting, *Am. Ind. Hyg. Assoc. J.*, **52**, 6–13 (1991).

143. USEPA, *Guidelines for Exposure Assessment*, FR 57, 104:22888–22939, May 29, 1992.

144. R. C. Baselt, *Biological Monitoring Methods for Industrial Chemicals*, Biomedical Publications, Davis, CA, 1980.

145. R. L. Zielhuis, *Scand. J. Work Environ. Health*, **4**, 1–18 (1978).

146. T. Kneip and J. V. Crable, Eds., *Methods for Biological Monitoring*, APHA, New York, 1988.

147. P. J. Hewitt and J. T. Sanderson, "The Role of Occupational Hygiene in Exposure Assessment by Biological Indices," *Ann. Occup. Hyg.*, **37**, 579–581 (1993).

148. NIOSH, *Manual of Analytical Methods*, 4th ed. with updates through 1998, NIOSH, Cincinnati, OH, 1998.

149. A. Berlin, R. H. Brown, and K. J. Saunders, Eds, *Diffusive Sampling, An Alternative Approach to Workplace Air Monitoring*, Royal Society of Chemistry, London, 1987.

150. M. Harper and C. J. Purnell, *Am. Ind. Hyg. Assoc. J.*, **48**, 214–218 (1987).

151. AIHA (American Industrial Hygiene Association), *Direct Reading Calorimetric Indicator Tubes Manual*, 2nd ed., AIHA, Fairfax, VA, 1993.

152. D. Jentysch and D. A. Fraser, *Am. Ind. Hyg. Assoc. J.*, **42**, 810–823 (1981).

153. M. J. Keane, *Ann. Am. Conf. Govern. Ind. Hyg.*, **1**, 241–245 (1981).

154. W. A. McClenny, J. D. Pleil, G. F. Evans, K. D. Oliver, M. W. Holdren, and W. T. Winberry, *J. Air Waste Manage. Assoc.*, **41**, 1308–1318 (1991).
155. P. Lilienfeld and J. Dulchunos, *J. Am. Ind. Hyg. Assoc.*, **33**, 136–145 (1972).
156. G. J. Sem, K. Tsurubayashi, and K. Homma, *Am. Ind. Hyg. Assoc. J.*, **38**, 580–588 (1977).
157. P. A. Baron, "Personal Aerosol Sampler Design: A Review," *Appl. Occup. Environ. Hyg.* **13**, 313–320 (1998).
158. D. L. Bartley and G. M. Brewer, *Am. Ind. Hyg. Assoc. J.*, **43**, 520–528 (1982).
159. N. A. Fuchs, *Atmos. Environ.*, **9**, 697–707 (1975).
160. M. G. Gressel, W. A. Heitbrink, and P. A. Jensen, "Video Exposure Monitoring—a Means of Studying Sources of Occupational Air Contaminants Exposure, Part I: Video Exposure Monitoring Techniques," *Appl. Occup. Environ. Hyg.*, **8**, 334–338 (1993).
161. M. G. Gressel, W. A. Heitbrink, and P. A. Jensen, "Video Exposure Monitoring—a Means of Studying Sources of Occupational Air Contaminants Exposure, Part II: Data Interpretation," *Appl. Occup. Environ. Hyg.*, **8**, 339–343 (1993).
162. P. Martin, F. Brand, and M. Servais, "Correlation of the Exposure to a Pollutant with a Task-Related Action or Workplace: The CAPTIV System," *Ann. Occup. Hyg.*, **43**, 221–233 (1999).
163. S. Kumagi and I. Matsunaga, *Ann. Occup. Hyg.*, **36**, 131–143 (1992).
164. J. W. Cherrie, G. Lynch, B. S. Bord, P. Heathfield, H. Cowie, and A. Robertson, "Does the Wearing of Sampling Pumps Affect Exposure?" *Ann. Occup. Hyg.*, **38**, 827–838 (1994).

CHAPTER EIGHT

Workplace Sampling and Analysis

Robert D. Soule, CIH, CSP

1 INTRODUCTION

Industrial hygiene sampling is done to identify and quantify specific contaminants present in the environment, determine exposures of workers in response to complaints, assess compliance with respect to various occupational health standards, evaluate the effectiveness of engineering controls installed to minimize workers' exposures, and for other purposes. The reasons for sampling dictate to some extent the sampling strategy that should be used. Sampling can be conducted in a strategic pattern throughout an area to document the environmental characteristics of the workplace. Personal breathing zone samples can be obtained to document actual exposure conditions. The substances being evaluated determine the type of sampling devices to be used, and the analytical requirements specify time and perhaps flow rate of sampling. The occupational health standard or reference indicates whether continuous or instantaneous sampling is appropriate. In short, consideration must be given to a number of variables pertaining to the fundamental purpose of the sampling.

Many analytical methods available to the industrial hygienist have become so standardized and simplified that they require relatively little experience to apply. On the other hand, many seemingly simple tests call for a basic understanding of solubility and gas laws, partial pressures, and chemical reactions. In many instances, questions arising from such considerations can be answered only by qualified specialists. The ultimate method of analysis to be used depends on the problem at hand rather than mere application of a standard method. The trend in recent years has been toward development of methods that

Patty's Industrial Hygiene, Fifth Edition, Volume 1. Edited by Robert L. Harris.
ISBN 0-471-29756-9

give relatively prompt results with a high degree of accuracy. The latter aspect has increased in importance because of legal significance given to occupational health standards, particularly as promulgated under authority of the Occupational Safety and Health Act of 1970. The National Institute for Occupational Safety and Health (NIOSH) and the Occupational Safety and Health Administration (OSHA) have developed specific methods for sampling and analyzing many atmospheric contaminants in the workplace. The objective of these procedures typically is an accuracy of at least ±25% with 95% confidence at the permissible exposure limit.

This chapter presents a discussion of overall strategy for sampling in the workplace, statistical bases for industrial hygiene sampling, sampling techniques for gaseous and particulate contaminants, summaries of techniques available for analyzing atmospheric samples, and a discussion of biological monitoring as it relates to industrial hygiene sampling. Relatively complete discussions of detailed methods of analysis for specific contaminants are presented elsewhere in this series (1). Many of the early descriptions of industrial hygiene sampling/analysis (2, 3) still serve as foundations for modern, more sophisticated methods.

2 GENERAL CONSIDERATIONS

Chemical and physical stresses can be evaluated in various ways. One form of evaluation is qualitative, that is, using one or more of the human senses without taking any actual measurements. This kind of inspection and evaluation of a work situation is very beneficial, particularly when done by an experienced industrial hygienist. Another form of evaluation is quantitative, that is, involving collection and analysis of samples representative of actual workers' exposures. Generally, this type of evaluation is most desirable and necessary in many cases, particularly when the purpose of the sampling is to determine compliance with occupational health standards or to form the basis for designing engineering controls.

2.1 Preliminary Survey

An experienced, professional industrial hygienist often can estimate, quite accurately and in some detail, the magnitude of chemical and physical stresses without benefit of any instrumentation. In fact, professionals use this qualitative evaluation every time they make a survey, whether it is intended to be the total effort of their work or a preliminary survey prior to actual sampling and analysis. Qualitative evaluation can be applied by anyone familiar with an operation, from the worker to the professional investigator, to ascertain some of the potential problems associated with work activities.

The first step in evaluating the occupational environment is to become as familiar as possible with particular operations. The person evaluating the operation should be aware of the types of industrial process and the chemical raw materials, by-products, and contaminants encountered. The evaluator should also know what protective measures are provided, how engineering controls are being used, and how workers are exposed to contaminants generated by specific job activities.

It is important that responsible industrial hygienists establish and maintain lists of the chemical and physical agents encountered in their particular areas of jurisdiction. In fact, OSHA's hazard communication standard, and many state right-to-know laws, make such inventorying a legal obligation. The number of chemical and physical agents capable of producing occupational illnesses continues to increase. New products that require the use of new raw materials or new combinations of familiar substances are continually being introduced. This is particularly true in the chemical industries, where new chemical products and operations for their processing are being developed. The composition of the products and by-products and as many as possible of the associated contaminants and undesirables should be known. This means that the industrial hygienist must obtain complete information on the composition of various commercial products. In most instances the desired information can be obtained from descriptive material provided by the suppliers in the form of a Material Safety Data Sheet. Similarly, labels on containers of the material should be read carefully. Although there are explicit requirements under hazard communication and right-to-know regulations, labels still do not always give complete information, and further investigation of the composition of the materials is necessary.

After the inventory is obtained, it is necessary to determine the toxicity and other hazardous properties of the chemicals. Information of this type can be found in several excellent reference texts on toxicology and industrial hygiene (4–9).

During a qualitative walk-through evaluation, many potentially hazardous operations can be detected visually. Operations that produce large amounts of dusts and fumes can be spotted. However, *visible* does not necessarily mean *hazardous*; airborne dust particles that cannot be seen by the unaided eye normally are more hazardous, because they are more likely to be inhaled into the lungs. Concentrations of dust of respirable size usually must reach extremely high levels before they are visible. Thus the absence of a visible cloud of dust is not a guarantee that a dust-free atmosphere exists. However, operations where activities generate dust that can be spotted visually are likely to warrant implementation of additional controls.

In addition to sight, the sense of smell can be used to detect the presence of many vapors and gases. Trained observers are able to estimate rather accurately the concentrations of various gases and solvent vapors present in the workroom air. For many substances the odor threshold concentration, that is, the lowest concentration that can be detected by smell, is greater than the permissible exposure level. In these cases, if the substances can be detected by their odors, excessive levels are indicated. However, many substances, notably hydrogen sulfide, can cause olfactory fatigue, that is, anesthesia of the olfactory nerve endings, to the extent that even dangerously high concentrations cannot be detected by odor.

Although it is usually possible to determine the presence or absence of potentially hazardous physical agents at the time of the qualitative evaluation, rarely can the potential hazard be evaluated without the aid of special instruments. As a minimum, however, the sources of physical agents such as radiant heat, abnormal temperatures and humidities, excessive noise, improper or inadequate illumination, ultraviolet radiation, microwaves, and various other forms of radiation can be noted.

Inspection of the types of control measure in use at a particular operation is an important aspect of the qualitative evaluation. In general, the control measures include such features as shielding from radiant or ultraviolet energy, local exhaust and general ventilation provision, respiratory protection devices, and other personal protective measures. General indicators of the relative effectiveness of these controls are (*1*) presence or absence of accumulated dust on floors, ledges, and other work surfaces; (*2*) condition of ductwork for the ventilation systems, that is, whether there are holes or badly damaged sections of ductwork, whether the fans for ventilation systems appear to provide adequate control of contaminants generated by the process; and (*3*) the manner in which personal protective measures are accepted and used by the workers.

2.2 Representative Quantitative Surveys

Although the information obtained during a qualitative evaluation or walk-through inspection of a facility is important and always useful, only by measurement can the hygienist document the actual level of chemical or physical agents associated with a given operation. The strategy used for any given air sampling program depends to a great extent on the purpose of the study. Specific objectives of any sampling program may include one or more of the following:

- Provide a basis on which unsatisfactory or unsafe conditions can be detected and the sources identified
- Assist in designing controls
- Provide a chronicle of changes in operational conditions
- Provide a basis for correlating disease or injury with exposure to specific stresses
- Verify and assess the suppression of contaminants by methods designed to do so
- Document compliance with health and safety regulations

These objectives can be condensed into two major categories:

- Sampling for industrial health engineering surveillance, testing, or control
- Sampling for health research or epidemiological purposes

A sampling program for engineering purposes should be designed to yield the specific information desired. For example, one might need only single samples before and after a change in ventilation to determine whether the change has had the desired effect. On the other hand, industrial hygiene is directed primarily at predicting the health effects of an exposure by comparing sampling results with hygienic guides, determining compliance with health codes or regulations, or defining, as precisely as possible, environmental factors for comparison with observed medical efforts.

Regardless of the objective or objectives of the sampling program, the investigating industrial hygienist must answer the following questions to be able to implement the correct strategy:

- Where should samples be obtained?
- Whose work area should be sampled?

- For how long should the samples be taken?
- How many samples are needed?
- Over what period of work activity should the samples be taken?
- How should the samples be obtained?

In answering these questions, the importance of adequate field notes is emphasized. While the sample is being obtained, notes should be made of the time, duration, location, operations underway, and all factors pertinent to the sample and the exposure or condition it is intended to define. Printed forms with labeled spaces for essential data help to avoid the common failure to record needed information. Industrial hygiene record keeping is discussed in detail elsewhere.

2.2.1 *Where to Sample*

Three general locations are used for collection of air samples: at a specific operation, in the general workroom air, and in a worker's breathing zone. The choice of sampling location is dictated by the type of information desired, and combination of the three types of sampling may be necessary. Most frequently, sampling is intended to determine the level of exposure of a worker or group of workers to a given contaminant throughout a work day. To obtain this type of information it is necessary to collect samples at the worker's breathing zone as well as in the areas adjacent to his particular activities. When the purpose of the survey is to determine sources of contamination or to evaluate engineering controls, a strategic network of area sampling would be more appropriate.

2.2.2 *Whom to Sample*

Logically, samples should be collected in the vicinity of workers directly exposed to contaminants generated by their own activities. In addition, however, samples should be taken from the breathing zone of workers in nearby work areas not directly involved in the activities that generate the contaminant, and from those of workers remote from the exposure who have either complained or have reason to suspect that the contaminants have been drawn into their work areas.

2.2.3 *Sample Duration*

In most cases, minimum sampling time is dictated by the time necessary to obtain an amount of the material sufficient for accurate analysis. The duration of the sampling period therefore is based on the sensitivity of the analytical procedure, the acceptable concentration of the particular contaminant in air, and the anticipated concentration of the contaminant in the air being sampled.

Preferably, the sampling period should represent some identifiable period of time of the worker's exposure, usually a minimum of one complete cycle of activity. This is particularly important in studying nonroutine or batch-type activities, which are characteristic of many industrial operations. Exceptions include operations that are highly automated and enclosed operations where the processing is done automatically and the operator's exposure is relatively uniform throughout the workday. In many cases it is desirable to sample the worker's breathing zone for the duration of the full shift. This is particularly important if sampling is being done to determine compliance status relative to occupational health standards.

Evaluation of a worker's daily time-weighted average exposure is best accomplished, when analytical methods permit, by allowing the person to work a full shift with a personal breathing zone sampler attached. The concept of full-shift integrated personal sampling is much preferred to that of short-term or general area sampling, if the results are to be compared to standards based on time-weighted average concentrations. When methods that permit full-shift integrated sampling are not applicable, time-weighted average exposures can be calculated from alternative short-term or general area sampling methods.

The first step in calculating the daily, time-weighted average exposure of a worker or a group of workers is to study the job descriptions obtained for the people under consideration and to ascertain how much time during the day they spend at various tasks. Such information usually is available from the plant personnel office or foreman on the job. In many situations, the investigator must make time studies to obtain the correct information. Information obtained from plant personnel should be checked by the investigator because in many situations job activities as observed by the investigator do not fit official job descriptions. From this information and the results of the environmental survey, a daily, 8-h, time-weighted average exposure can be calculated, assuming that a sufficient number of samples have been collected, or measurements obtained with direct-reading instruments under various plant operating conditions to represent accurately the exposure profiles.

When sampling is done for the purpose of comparing results with airborne contaminants whose toxicologic properties warrant short-term and ceiling limit values, it is necessary to use short-term or grab sampling techniques to define peak concentrations and estimate peak excursion durations. For purposes of further comparison, the 8-h, time-weighted average exposures can be calculated using the values obtained by short-term sampling.

2.2.4 Number of Samples

The number of samples needed depends to a great extent on the purpose of sampling. Two samples may be sufficient to estimate the relative efficiency of control methods, one sample being taken while the control method is in operation and the other while it is off. On the other hand, several dozen samples may be necessary to define accurately the average daily exposure of a worker who performs a variety of tasks. The number of samples also depends to some extent on the concentrations encountered. If the concentration is quite high, a single sample may be sufficient to warrant further action. If the concentration is somewhat near the acceptable level, a minimum of three to five samples usually is desirable for each operation being studied. There are no set rules regarding the duration of sampling or the number of samples to be collected. These decisions usually can be reached quickly and reliably only after much experience in conducting such studies.

2.2.5 When to Sample

The type of information desired and the particular operations under study determine when sampling should be done. If, for example, the operation continues for more than one shift, it usually is desirable to collect air samples during each shift. The airborne concentrations of chemicals or exposure to physical agents may be quite different for each shift. It usually is desirable to obtain samples during both summer and winter months, particularly in plants located in areas where large temperature variations occur during different seasons of the year. For one thing, there generally is more natural ventilation provided to dilute the airborne contaminants during the summer months than during the winter.

2.2.6 Choosing a Method

In general, the choice of instrumentation when sampling for a particular substance depends on a number of factors including the following:

- Type of analysis or information desired
- Efficiency of the instrument or method
- Reliability of the equipment under various conditions of field use
- Portability and ease of use
- Availability of the instrument
- Personal choice of the industrial hygienist based on past experience and other factors

No single, universal air sampling instrument is available today, and it is doubtful that such an instrument will ever be developed. In fact, the present trend in the profession is toward a greater number of specialized instruments. The sampling instruments used in the field of industrial hygiene generally can be classified by type as follows: (*1*) direct-reading, (*2*) those that remove the contaminant from a measured quantity of air, and (*3*) those that collect a fixed volume of air. The three methods are listed in order of their general application and preference in use today.

Industrial hygienists must consider a proposed sampling program in relation to their own familiarity with the sampling and analytical method. As a rule, a method should never be relied on unless and until the hygienist has personally evaluated it under controlled conditions, such as by the following:

- Sampling a synthetic atmosphere from a proportioning apparatus or gastight impervious chamber of sufficient size to permit creating and sampling mixtures without introducing significant errors
- Introducing measured amounts of contaminants into a device attached to the sampling arrangement in a manner that utilizes the entire amount deposited
- Comparing performance of the method with a device of proven performance by sampling from a common manifold over the same period of time

In other words, the sampling program must be preceded by appropriate calibration of all equipment to be used. Regardless of the sampling instrumentation selected for use in conducting industrial hygiene surveys, it is critical and imperative that the actual performance characteristics be known. This requires that various types of calibration be done periodically, or at any time that the performance of the device is questioned.

3 STATISTICAL BASES FOR SAMPLING

Statistical tests for noncompliance are applied when environmental data are used to make a decision concerning a worker's exposure to a specific contaminant. The decision options concerning compliance and noncompliance are given in Table 8.1. In statistical terms, these options correspond to the null hypothesis that the worker is in compliance with the industrial hygiene standard. The detailed discussion of procedures is presented from a compliance officer's viewpoint. In this approach, samples are collected to see whether it is possible to reject the hypothesis that compliance exists with appropriate certainty. The type I error is the probability of declaring noncompliance given that the true state is compliance. This probability is a measure of any uncertainty that noncompliance does exist, and it should be kept small to ensure that noncompliance decisions are correct.

The employer's responsibility, on the other hand, is the protection of the employees. The employer's goal is to keep the type II error, which is the probability of declaring compliance when the true state is noncompliance, as small as possible. In statistical decision terms, therefore, the employer wants to assume that a given worker is in a state of noncompliance (null hypothesis). Then data are collected to show that the hypothesis can be rejected, with a goal of keeping as small as possible the probability of wrongly rejecting the null hypothesis.

Three questions are of primary importance when considering a statistical basis for industrial hygiene sampling. Over what time span should each sample be taken? How many samples should be obtained? At what times during the workday should the samples be obtained?

3.1 Duration of Sampling

For some types of sampling unit, such as the common colorimetric detector tubes, the sampling period is predetermined. In other cases, such as sampling for asbestos, the sampling period is defined fairly closely by the requirements of the analytical procedure. However, in most cases, the industrial hygienist has a choice over a wide range of sampling times from a few seconds to a complete work shift. The tendency might be to maximize the sampling period, expecting that this would increase significantly the reliability of the data, that is, that it would ultimately provide a better answer. This is not true for short-term, or grab, samples, however. In such cases the primary consideration in arriving at sampling duration should be the requirements of the analytical method. Each analytical procedure requires some minimum amount of material for reliable analysis. This should

Table 8.1. Decision Options for Compliance versus Noncompliance

Action \ True State	Compliance with Standard	Noncompliance with Standard
Declare Compliance	No Error	Type II Error
Declare Noncompliance	Type I Error	No Error

be known in advance by the industrial hygienist, and the sampling period should be selected accordingly. Any increase in the duration of the sampling past this minimum time required to collect an adequate amount of material is both unnecessary and unproductive (10, 11).

When attempting to make a decision on a possible noncompliance situation, as the statistical discussion in this chapter is directed, it is better to take shorter samples because this allows more samples to be taken in a given time period. It is much more important to collect several samples of short duration than to collect one medium-length (partial-shift) sample covering the same total sampling period. The random sampling and analytical errors can be averaged out, along with the longer-term environmental fluctuations during the sampling period, by taking a mean of random independent short-term samples. If sampling can be conducted over essentially 100% of the work period, either by a single sample or several consecutive samples, a better estimate of the true average exposure of the employee can be obtained. The sampling and analytical error must be contended with, and these are typically much smaller than the environmental fluctuations that affect short-term samples.

Thus there is a marked advantage in using a single, full-period sample (or several consecutive samples over the full work period) when attempting to demonstrate noncompliance with an occupational health standard. It is much more difficult to demonstrate noncompliance using the mean of several grab samples because the additional variability due to environmental fluctuations lowers the lower confidence limit (LCL) of the sampling result. For much industrial hygiene data the geometric standard deviations range between 1.5 and 2.5. Within this range, and for sample sizes of 3 to 10, it generally is necessary to obtain a measured mean exposure of 150 to 250% of the occupational health standard to demonstrate noncompliance.

If a full-period sample, or series of several consecutive samples, is used in attempting to demonstrate noncompliance, the degree to which the observed mean must exceed the occupational health standard is much lower. For sampling and analytical schemes having coefficients of variation (CVs) ranging from 0.05 to 0.25, sample sizes between one and four result in data for which the average concentration must exceed the occupational health standard by 5 to 14% to demonstrate noncompliance. This range of CVs includes methods for sampling and analysis of many of the common contaminants: total dust (CV = 0.05), respirable dust (CV = 0.09), organic vapor on charcoal (CV = 0.10), colorimetric indicator tubes (CV = 0.14), and asbestos fiber counting (CV = 0.22).

3.2 Number of Samples

The second question of importance to the industrial hygienist is how many samples should be obtained. This question is vital because it relates directly to the confidence that can be placed in the resulting estimate of the airborne concentration of the contaminant in question, and subsequently the employee exposure. Statistical assessment of the effects of sample size on requirements for demonstrating noncompliance demonstrates that relatively little is to be gained beyond sample sizes of seven or eight (12).

For full-period consecutive samples, these studies show that an appropriate number of samples is between four and seven. Practical considerations include costs of sampling and analyses, and the impossibility of running some long-duration sampling methods for longer than about 4 h per sample. Thus, most full-period consecutive sampling strategies result in at least two samples when an 8-h, time-weighted average standard is being applied. For a sampling and analytical technique with a coefficient of variation of 10%, the degree to which the estimate of the mean concentration must exceed the standard drops from 12% to 6% with an increase from two to seven samples. The relatively small decrease in percentage, with sample size exceeding seven, normally cannot be justified when compared with the time and effort required to obtain the additional samples. Thus on a cost-benefit basis, it can be concluded that two consecutive samples covering a full work period, for example, two consecutive 4-h measurements, is the best number of samples to be obtained when comparison is made with an 8-h, time-weighted average standard.

For grab samples, the estimate of the mean exposure concentration must exceed the standard by unreasonably large amounts to demonstrate noncompliance when less than four grab samples are obtained. As discussed earlier, there is a point beyond which little is gained in attempting to reduce errors in the mean by taking more than seven grab samples. Because the level of variability in the mean of grab samples usually is much higher than for the same number of full-period samples, however, it might be necessary to take more than seven grab samples to attain the same level of precision afforded by even four or fewer full-period samples. Thus on the basis of the statistical criterion that can lead to economies in sampling by permitting reduction in the sampling effort with a calculable degree of confidence, it can be concluded that the optimal number of grab samples to be taken over the time period appropriate to the standard is between four and seven.

When consecutive samples are obtained in a series over only a portion of the total period for which the standard applies, that is, the full work period is not included in the consecutive sampling period the relationships of sample size, total time covered by all samples and demonstration of noncompliance are much more complex. Obviously, the taking of partial-period consecutive samples is a compromise between the preferred full-period sampling and the less desirable approach with grab samples. Comparative statistics suggest that, in general, if it is not possible to sample for at least 70% of the time period appropriate to the standard, for example, 5.5 h for an 8-h standard, it is better to use the grab sampling strategy.

3.3 Grab Sampling

The last of the three statistical questions to be answered concerns the periods of exposure during which grab sampling should be conducted. The accuracy of the probability level for the test depends on assumptions of the log-normality and independence of sample results that are averaged. These assumptions are not highly restrictive if precautions are taken to avoid any bias when selecting the sampling times over the period for which the standard is defined. It usually is preferable to choose the sampling period in a statistically random manner. For a standard that is defined as a time-weighted average concentration over a period significantly longer than the sampling interval, an unbiased estimate of the true mean can be assured by taking samples at random. On the other hand, it is valid to sample at regular intervals if the contaminant level varies randomly about a constant mean, and any fluctuations are of short duration relative to the length of the sampling interval. If means and their confidence limits are to be calculated from samples taken at regularly spaced intervals, however, biased results could occur if the industrial operation being monitored is cyclic and in phase with the sampling periods. Results from random sampling would be valid nevertheless, even when cycles or trends occur during the period of the standard. In this context, *random* refers to the manner of selecting the sample, and any particular sample could be the outcome of a random sampling procedure. A practical definition of random sampling is a strategy by which any portion of the work shift has the same chance of being sampled as any other.

Strictly speaking, sampling results are valid only for the portion of the work period during which measurements were obtained. However, if it is not possible to sample during the entire workday or the entire length of a particular operation, professional judgment may permit inferences to be made about concentrations during other unsampled portions of the day. Reliable knowledge concerning the operation obviously is required to make these types of extrapolation.

3.4 Statistical Procedures

For those interested in, or required to use, statistical procedures, the following should be used to compare sampling results with the applicable occupational health standard. As is the case with other material in this section, the statistics have been oriented toward determining whether noncompliance with the time-weighted average, ceiling, or excursion standard exists.

3.4.1 Full-Period, Single Sample

The following procedure can be used to determine noncompliance when a single sample is being tested and compared with either a time-weighted average or a ceiling standard. For a time-weighted average standard, the sample must have been taken for the entire period for which the standard is defined (usually 8 h). The variability of the sampling, expressed either as a standard deviation or as a coefficient of variation, and the analytical methods used to collect and analyze the sample must be well known from previous measurements. The statistical test given is the one-sided comparison-of-means test using the normal distribution at the 95% confidence level.

Only if the lower confidence limit of the sample exceeds the standard is there 95% confidence that the true average concentration exceeds the standard, and thus that a condition of noncompliance exists. The lower confidence limit can be expressed as the following:

$$\text{LCL} = x - 1.645\sigma$$

where x = measurement being tested
1.645 = critical standard normal deviate for 95% confidence
σ = standard deviation of sampling/analytical method

If the coefficient of variation (CV) is known, the LCL can be computed:

$$LCL = x - [(1.645)(CV)(standard)]$$

The nomogram in Figure 8.1 can be used to aid this calculation

3.4.2 Full-Period, Consecutive Samples

The procedure involving consecutive samples for a full period should be used to determine noncompliance with either a time-weighted average or a ceiling standard. That is, several consecutive samples are taken for the entire time period for which the standard is defined. If the samples do not cover the entire time period of the standard, refer to the procedure described in Section 3.4.3. The variability (standard deviation or coefficient of variation) of the sampling and analytical methods used to collect and analyze the samples must be

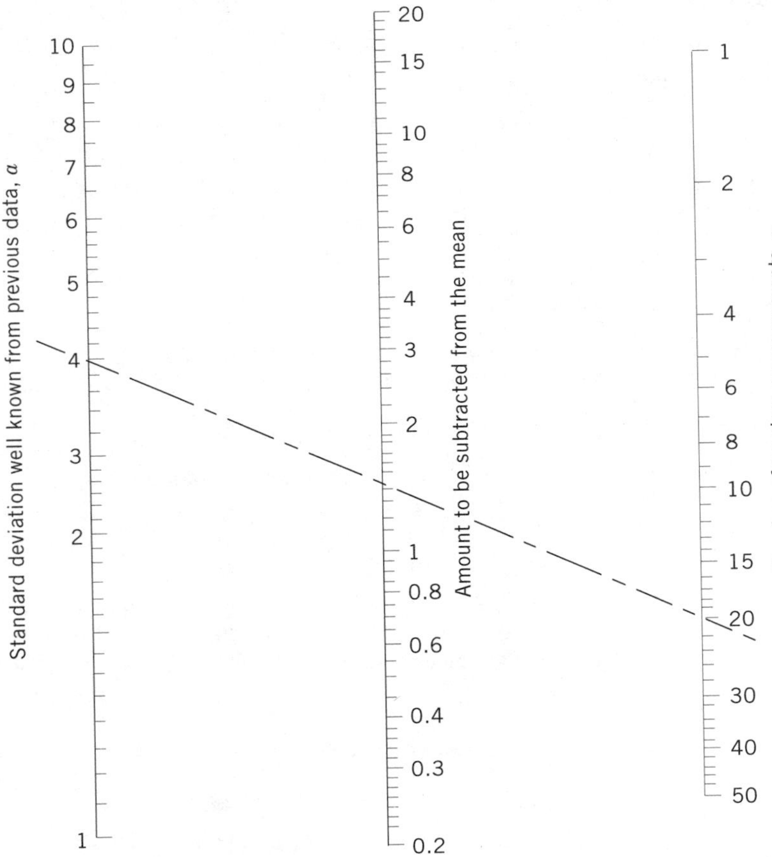

Figure 8.1. Nomogram for full-period or partial-period procedures using a well-known standard deviation; both σ and the amount can be multiplied by the same power of 10 (7).

well known from previous measurements. The statistical test given is the one-sided comparison-of-means test using the normal distribution at the 95% confidence level.

Only if the lower confidence limit of the mean of the consecutive samples exceeds the standard is there 95% confidence that the true average concentration exceeds the standard and that a condition of noncompliance exists.

$$\text{LCL} = x - 1.645\sigma_x$$

where x = time-weighted average of n samples = $(T_1x_1 + T_2x_2 + \ldots + T_nx_n)/T$
T_i = duration of ith sample
x_i = measurement of concentration in ith sample
$T = T_1 + T_2 + \ldots + T_n$ = total duration for n consecutive samples
$\sigma_x = (\sigma/T)(T_1^2 + T_2^2 + \ldots + T_n^2)/2 = \sigma/[(n)^{1/2}]$ if $T_1 = T_2 = \ldots = T_n$

If the coefficient of variation is known, σ can be computed:

$$\sigma = (\text{CV})(\text{standard})$$

Again, Figure 8.1 can be used to aid this calculation for the case of n equal-duration samples.

3.4.3 Partial-Period, Consecutive Samples

One sample or a series of consecutive samples collected over less than the period for which a standard is defined is referred to as a partial-period sample. Because the concentration during the period not covered by the sample could not be less than zero, the full-period standard can be multiplied by a factor to obtain a conservative partial-period standard. The lower confidence limit then can be calculated as in the preceding section and compared with the partial-period standard.

$$\text{Factor} = \text{actual time of sample/time period of standard}$$

Thus for an 8-h, time-weighted average standard, typical factors would be as follows:

Total Time of Sample(s) (h)	Factor
8.00	1.000
7.75	1.032
7.50	1.067
7.25	1.103
7.00	1.143
6.75	1.185
6.50	1.231
6.25	1.280
6.00	1.333

3.4.4 *Grab Sampling Data*

If full-period samples of industrial contaminant concentrations are available, the best method of modeling the uncertainties of the result is with a normal distribution. When only a set of grab samples is available, however, the log-normal distribution best describes the uncertainties of the process. Procedures for analyzing grab samples by estimating the average concentration of a contaminant and making a decision on the level of the contaminant have been developed. Advantages offered by this technique are listed here:

- It is contaminant independent. Thus it becomes possible to use only a single decision chart for any contaminant.
- It is capable of both the noncompliance and the no-action decision. Each of these decisions is subject to a predetermined probability of type I or type II error. A type I error is said to occur if the noncompliance decision is wrongly asserted. A type II error is said to occur if the no-action decision is wrongly asserted.
- The estimation and decision procedures are implemented by a single, straightforward nomographic method. For estimation, the procedure yields the best estimate of the actual average contaminant level.

One disadvantage of the procedure is that the lower confidence limit is not directly computed. However, the simplicity of the calculations required and the plotting of a single point on the decision chart far outweigh any advantages that direct calculation of a lower confidence limit would yield. Modifications of this basic approach are being used by industrial hygienists today in a variety of employment settings. (13–16).

4 SAMPLING FOR GASES AND VAPORS

For purposes of definition, a substance is considered to be a gas if, at 70°F and atmospheric pressure, the normal physical state is gaseous. A vapor is the gaseous state of a substance in equilibrium with the liquid or solid state of the substance at the given environmental conditions. This equilibrium results from the vapor pressure of the substance causing volatilization or sublimation into the atmosphere. The sampling techniques discussed in this section are applicable to a substance in gaseous form regardless of whether it is technically a gas or a vapor.

4.1 General Requirements

Particulate substances can be readily scrubbed or filtered from sampled airstreams because of the larger relative physical dimension of the contaminant and the operation and interaction of agglomerative, gravitational, and inertial effects. Gases and vapors, however, form true solutions in the atmosphere, thus requiring either sampling of the total atmosphere using a gas collector or the use of a more vigorous scrubbing mechanism to separate the gas or vapor from the surrounding air. Sampling reagents can be chosen to react chemically with the contaminants in the airstream, thus enhancing the collection efficiency of the sampling procedure. In the development of an integrated sampling scheme it is necessary to consider the following basic requirements:

- The method must have an acceptably high efficiency of collecting the contaminant of interest.
- A rate of airflow that can provide a sufficient sample for the required analytical procedure, maintain the acceptable collection efficiency, and be accomplished in a reasonable time period must be available.
- The collected gas or vapor must be kept in the chemical form in which it exists in the atmosphere under conditions that maintain the stability of the sample before analysis.
- The sample must be submitted for analysis in a suitable form and medium.
- A very minimal amount of analytical procedure in the field must be associated with the overall method.
- To the extent practicable, the use of corrosive (e.g., acidic or alkaline) or relatively toxic (e.g., benzene) sampling media should be avoided.

Of these general requirements, perhaps the most important is the first, that is, knowing the collection efficiency of the sampling system chosen or anticipated. This efficiency information can be obtained either from published data describing the method or as a result of independent evaluations as an essential part of planning the industrial hygiene survey. In making such evaluations, known concentrations of the gases must be prepared by means of either dynamic or static test systems. Once the known atmosphere is generated, the sampling device should be used as anticipated or intended and the efficiency defined in terms of such variables as characteristic of the sampler, rate of airflow, stability of the sampling during collection, and apparent losses by adsorption on the walls of the sampling device.

The collection of sufficient sample for the subsequent analytical procedure is a matter that must be clearly understood not only by the laboratory analyst but by the field investigator as well. Understanding can best be promoted by discussions between the two professionals before the industrial hygiene field survey. The field investigator must discuss as fully as possible with the chemist the nature of the process involved in the survey so that together they may select the best combination of sampling and analytical methods to satisfy the sensitivity requirements of the analytical method, minimize effects of potential interferences, and complete the sampling within a time frame consistent with process conditions or potential exposure.

4.2 Collection Techniques

There are two basic methods for collecting samples of gaseous contaminants: instantaneous or grab sampling, and integrated or long-term sampling. The first involves the use of a gas-collecting device, such as an evacuated flask or bottle, to obtain a fixed volume of a contaminant-in-air mixture at known temperature and pressure. This is called grab sampling because the contaminant is collected almost instantaneously, that is, within a few seconds or minutes at most. Thus the sample is representative of atmospheric conditions at the sampling site at a given point in time. This method commonly is used when atmospheric analyses are limited to such gross contaminants as mine gases, sewer gases, carbon dioxide, or carbon monoxide, or when the concentrations of contaminants likely to be found are sufficiently high to permit analysis of a relatively small sample. However, with the increased sensitivity of modern instrumental techniques such as gas chromatography and infrared spectrometry, instantaneous sampling of relatively low concentrations of atmospheric contaminants is becoming feasible.

The second method for collection of gaseous samples involves passage of a known volume of air through an absorbing or adsorbing medium to remove the contaminants of interest from the sample airstream. This technique makes it possible to sample the atmosphere over an extended period of time, thus integrating the sample. The contaminant that is removed from the airstream becomes concentrated in or on the collection medium; therefore it is important to establish a sampling period long enough to permit collection of a quantity of contaminant sufficient for subsequent analysis but not so long that the medium's capacity is exceeded.

4.2.1 *Instantaneous Sampling*

Various devices can be used to obtain instantaneous or grab samples. These include vacuum flasks or bottles, gas or liquid displacement collectors, metallic collectors, glass bottles, syringes, and plastic bags. The temperature and pressure at which the samples are collected must be known, to permit reporting of the analyzed components in terms of standard conditions, normally 25°C and 760 mm Hg for industrial hygiene purposes.

Grab samples usually are collected when analysis is to be performed on gross amounts of gases in air (e.g., methane, carbon monoxide, oxygen, and carbon dioxide). The samplers should not be used for collecting reactive gases such as hydrogen sulfide, oxides of nitrogen, and sulfur dioxide unless the analyses can be made directly in the field. Such gases may react with dust particles, moisture, wax sealing compound, or glass, altering the composition of the sample.

The introduction of highly sensitive and sophisticated instrumentation has extended the application of grab sampling to low levels of contaminants. In areas where the atmosphere remains constant, the grab sample may be representative of the average as well as the momentary concentration of the components; thus it may truly represent an integrated equivalent. Where the atmospheric composition varies, numerous samples must be taken to determine the average concentration of a specific component. The chief advantage of grab sampling methods is that collection efficiency is essentially 100%, assuming no losses due to leakage or chemical reaction preceding analysis. The more common types of grab sampling devices are discussed here in detail.

4.2.1.1 Evacuated Containers. Evacuated flasks are heavy-walled containers of glass or other suitable material, usually of 250- or 300-mL capacity, but frequently holding as much as 500 or 1000 mL, from which essentially all (99.97% or more) of the air has been removed by a heavy-duty vacuum pump. The internal pressure after the evacuation is practically zero. The container is closed; for example, for glass units, the neck is sealed by heating and drawing during the final stages of evacuation. These units are simple to use because no metering devices or pressure measurements are required. The pressure of the sample is taken as the barometric pressure reading at the site. After the sample has been collected by breaking the heat-sealed end of the glass unit or opening the container, the flask is resealed and transported to the laboratory for analysis. A variation of this procedure with evacuated flasks is to add a liquid absorbent to the flask before it is evacuated and sealed to preserve the sample in a desirable form following collection.

Partially evacuated containers or vacuum bottles are prepared with a suction pump just before sampling is performed, although frequently they are evacuated in the laboratory the day before a field visit. No attempt is made to bring the internal pressure to zero, but temperature readings and pressure measurements with a manometer are recorded after the evacuation, and again after the sample has been collected. Examples of this type of collector are heavy-walled glass bottles and metal or heavy plastic containers with tubing connectors closed with screw clamps or stopcocks.

4.2.1.2 Displacement Collectors. Gas or liquid displacement collectors include 250- to 300-mL glass bulbs fitted with end tubes that can be closed with greased stopcocks or with rubber tubing and screw clamps. They are used widely in collecting samples containing oxygen, carbon dioxide, carbon monoxide, nitrogen, hydrogen, or other combustible gases for analysis by an Orsat or similar analyzer. Another device operating on the liquid displacement principle is the aspirator bottle, which has exit openings at the bottom through which the liquid is drained during sampling.

In applying the gas displacement technique, the samplers are purged conveniently with a bulb aspirator, hand pump, small vacuum pump, or other suitable source of suction. Usually satisfactory purging can be achieved by drawing a minimum of 10 air changes of the test atmosphere through the gas collector.

The devices used for gas displacement collectors can be filled by liquid displacement; the most frequently employed liquid is water. In sampling, liquid in the container is drained or poured out slowly in the test area and replaced by air to be sampled. Of course, this method is limited to gases that are insoluble in and nonreactive with the displaced liquid. The solubility problem can be minimized by using mercury or water conditioned with the gas to be collected. Mercury, however, must be used with caution, because it may create an exposure problem if handled carelessly.

4.2.1.3 Flexible Plastic Bags. Evacuated flasks and displacement containers have become much less common since the introduction of flexible containers. Flexible plastic bags can be used to collect air and breath samples containing organic and inorganic vapors and gases in concentrations ranging from parts per billion to more than 10% volume in air. They also are convenient for preparing known concentrations of gases and vapors for equipment calibration. The bags are available commercially in a variety of sizes, up to 9 ft^3, and can be made easily in the laboratory.

Sampling bags are manufactured from various plastic materials, most of which can be purchased in rolls or sheets cut to the desired size. Some materials, such as Mylar, may be sealed with a hot iron using a Mylar tape around the edges. Others, such as Teflon, require high temperature and controlled pressure in sealing. Certain plastics, including

Mylar and Scotchpak, may be laminated with aluminum to seal the pores and reduce the permeability of the inner walls to sample gases and the outer walls to sample moisture. Sampling ports may consist of a sampling tube molded into the fabricated bag and provided with a closing device or a clamp-on air valve.

Plastic bags have the advantages of being light, unbreakable, and inexpensive, and they permit the entire sample to be withdrawn without the difficulty associated with dilution by replacement air, as is the case with rigid containers. However, they must be used with caution because generalization of recovery characteristics of a given plastic cannot be extended to a broad range of gases and vapors. Important factors to be considered in using these collectors are absorption and diffusion characteristics of the plastic material, concentration of the gas or vapor, and reactive characteristics of the gas or vapor with moisture and with other constituents in the sample. Information on the storage properties of gases and vapors in plastic containers has been published (17, 18).

The bags must be leak tested and should be preconditioned for 24 h before they are used for sampling. Preconditioning consists of flushing the bag three to six times with the test gas, the number of times depending on the nature of the bag material and the gas. In some cases it is recommended that the final refill remain in the bag overnight before the bag is used for sampling. Such preconditioning usually is helpful in minimizing the rate of decay of a collected gas. At the sampling site the air to be sampled is allowed to stand in the bag for several minutes, if possible, before removal and subsequent refilling of the bag with a sample. Once collected, the interval between sampling and analysis should be as short as practicable.

4.2.2 Integrated Sampling

Integrated sampling of the workroom atmosphere is necessary when the composition of the air is not uniform, when the sensitivity requirements of the method of analysis necessitate sampling over an extended period, or when compliance or noncompliance with an 8-h, time-weighted average standard must be established. Thus the professional observations and judgment of the industrial hygienist are called on in devising the strategy for the procurement of representative samples to meet the requirements of an environmental survey of the workplace. This deliberation is discussed earlier in this chapter.

4.2.2.1 Sampling Pumps. Integrated air sampling requires a known and controllable source of suction as an air-moving device. A vacuum line, if available, may be satisfactory. However, the most practical source for prolonged periods of sampling is a pump powered by electricity. These pumps come in various sizes and types and must be chosen for the sampling devices with which they will be used.

If electricity is not available or if flammable vapors present a fire hazard, aspirator bulbs, hand pumps, portable units operated by compressed gas, or battery-operated pumps are suitable for sampling at rates up to 3 L/min. The latter have become the workhorses of the industrial hygiene profession, particularly in judging compliance with health standards. For higher sampling requirements, ejectors using compressed air or a water aspirator may be employed.

When compressed air or batteries are to be used as the driving force for a pump, the duration of the sampling period is important in relation to the supply of compressed air or the life of the rechargeable battery. These units must be capable of running unattended, and periodic checks on the airflow must be made.

The common practice in the field is to sample for a measured period of time at a constant, known rate of airflow. Direct measurements are made with rate meters such as rotameters and orifice or capillary flowmeters. These units are small and convenient to use, and have become quite accurate even at the very low flow rates common with modern industrial hygiene sampling, that is, 10 to 20 cm^3/min. The sampling period must be timed carefully with a stopwatch or with the timing device incorporated into the most recent models of pumps.

Many pumps have inlet vacuum gauges or outlet pressure gauges attached. These gauges, when properly calibrated with a wet or dry gas meter, can be used to determine the flow rate through the pump. The gauge may be calibrated in terms of cubic feet per minute or liters per minute. If the sample absorber does not have enough resistance to produce a pressure drop, a simple procedure is to introduce a capillary tube or other resistance into the train behind the sampling unit.

Samplers are always used in assembly with an air-moving device (source of suction) and an air-metering unit. Frequently, however, the sampling train consists of filter, probe, absorber (or adsorber), flowmeter, flow regulator, and air mover. The filter serves to remove any particulate matter that may interfere in the analysis. It should be ascertained that the filter does not also remove the gaseous contaminant of interest. The probe or sampling line is extended beyond the sampler to reach a desired location. It also must be checked to determine that it does not collect a portion of the sample. The meter that follows the sampler indicates the flow rate of air passing through the system. The flow regulator controls the airflow. Finally at the end of the train the air mover provides the driving force.

4.2.2.2 Absorbers. Four basic types of absorber are employed for collecting gases and vapors: simple gas-washing bottles, spiral and helical absorbers, fritted bubblers, and glass-bead columns. Use of such units has become quite rare in routine exposure monitoring, finding application primarily in research applications. The absorbers provide sufficient contact between the sampled air and the liquid surface to ensure complete absorption of the gaseous contaminant. In general, the lower the sampling rate, the more complete the absorption.

Simple gas-washing bottles include Drechsel types, standard Greenburg-Smith devices, and midget impingers. The air is bubbled through the liquid absorber, without special effort, to secure intimate mixing of air and liquid, and the length of travel of the gas through the collecting medium is equivalent to the height of the absorbing liquid. These scrubbers are suitable for gases and vapors that are readily soluble in the absorbing liquid or react with it. One or two units may be enough for efficient collection, but in some cases several may be required to attain the efficiency of a single fritted-glass bubbler. Advantages of these devices are simplicity in construction, ease of rinsing, and the small volume of liquid required.

Spiral and helical absorbers provide longer contact between the sampled air and the absorbing solution. The sample is forced to travel a spiral or helical path through the liquid 5 to 10 times longer than that in the simpler units. In fritted-glass bubblers air passes through porous glass plates and enters the liquid in the form of small bubbles. The size of the air bubbles depends on the liquid and on the diameter of the orifices from which the bubbles emerge. Frits are classified as fine, coarse, or extra coarse, depending on the number of openings per unit area. The extra-coarse frit is used when a more rapid flow is desired. The heavier froth generated by some liquids increases the time of contact of gas and liquid. These devices are more efficient collectors than the simple gas-washing bottles and can be used for the majority of gases and vapors that are soluble in the reagent or react rapidly with it. Flow rates between 0.5 and 1.0 L/min are used commonly. These absorbers are relatively sturdy, but the fritted glass is difficult when used for contaminants that form a precipitate with the reagent, in which cases a simple gas-washing bottle would be preferable.

Packed columns are used for special situations calling for a concentrated solution. Packing material, such as glass pearl beads, wetted with the absorbing solution, provides a large surface area for the collection of a sample and is especially useful when a viscous liquid is required. The rate of sampling is low, usually 0.25 to 0.5 L of air/min.

4.2.2.3 Adsorption Media. When it is desired to collect insoluble or nonreactive vapors, an adsorption technique frequently is the method of choice. Activated charcoal and silica gel are common adsorbents (Figures 8.2 and 8.3). Solid adsorbents require less manipulative care than do liquid adsorbents; they can provide high collection efficiencies, and with improved adsorption tube design and a better definition of desorption requirements,

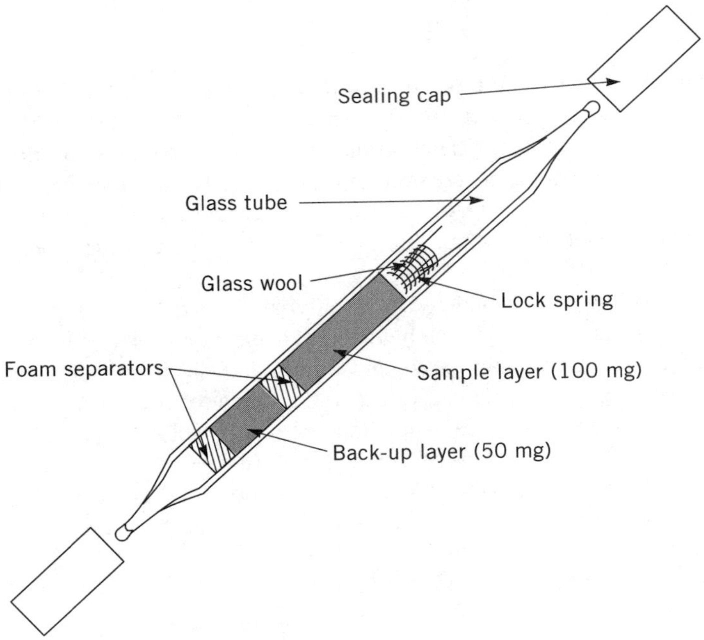

Figure 8.2. Activated charcoal sampling tube.

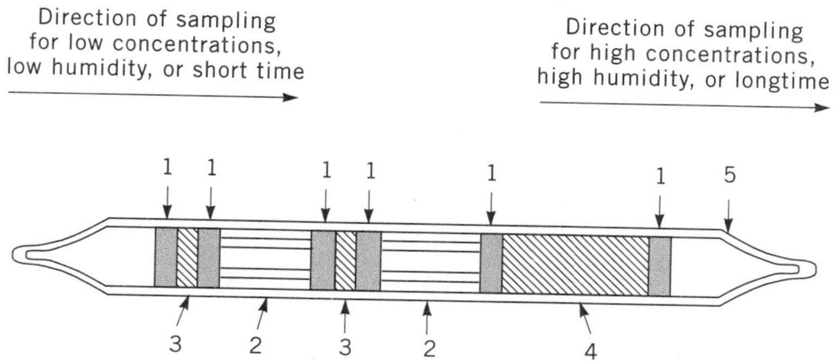

Figure 8.3. Silica gel sampling tube for aromatic machines: 1, 100-mesh stainless steel screen plugs; 2, 12-mm glass tube separator; 3, 150-mg silica gel section, 45/60 mesh; 4, 700-mg silica gel section, 45/60 mesh; 5, 8-mm ID glass tube.

they are becoming increasingly popular as indicated by surveys of industrial hygiene professionals.

Activated charcoal is an excellent adsorbent for most vapors that boil above 0°C. It is moderately effective for gaseous substances that boil at low temperatures (between −100 and 0°C), particularly if the carbon bed is refrigerated, but it is a poor collector of gases having boiling points below −150°C. Its retentivity for adsorbed vapor is several times that of silica gel. Because of their nonpolar characteristics, organic gases and vapors are adsorbed in preference to moisture, and sampling can be performed for long periods of time.

Silica gel has been used widely as an adsorbent for gaseous contaminants in air samples. Because of its polar character it tends to attract polar or readily polarizable substances preferentially. The general order of decreasing polarizability or attraction is: water, alcohols, aldehydes, ketones, esters, aromatic compounds, olefins, and paraffins. Organic solvents are relatively nonpolar in comparison with water, which is strongly adsorbed onto silica gel; such compounds are displaced by water in the entering airstream. Consequently, the volume of air sampled under humid conditions may have to be restricted. Despite this limitation, silica gel is a very useful adsorbent.

4.2.2.4 Condensation. In condensation methods, vapors or gases are separated from sampled air by passing the air through a coil immersed in a cooling medium such as dry ice and acetone, liquid air, or liquid nitrogen. The device is not ordinarily considered to be a portable field technique. It may be necessary to use this method when the gas or vapor might be altered by collecting in liquid or when it is difficult to collect by other techniques. A feature of this method is that the contaminating material is obtained in concentrated form. The partial pressure of the vapor can be measured when the system is brought back to room temperature.

4.2.2.5 Collection Efficiency. The collection efficiency (the ratio of the amount of contaminant retained by the absorbing or adsorbing medium to that entering it) need not be

100% as long as it is known, constant, and reproducible. The minimum acceptable collection performance in a sampling system is usually 90%, but higher efficiency is certainly desirable. When the efficiency falls below the acceptable minimum, sampling may be carried out at a lower rate, or in the case of liquid absorbers, at a reduced temperature by immersing the absorber in a cold bath to reduce the volatility of both the solute and solvent.

Frequently, the relative efficiency of a single absorber can be estimated by placing another in series with it. Any leakage is carried over into the second collector. The absence of any carryover is not in itself an absolute indication of the efficiency of the test absorber, because the contaminant may be stopped effectively by either absorber. Analysis of the various sections of silica gel or activated charcoal tubes used in sampling a contaminant is a useful check on the collection efficiency of the first section of the tube. Another valuable technique is the operation of the test absorber in parallel or in series with a different type of collector having a known high collection efficiency (an absolute collector if one is available) for the contaminant of interest. By running the test absorber at different flow rates, the maximum permissible rate of flow for the device can be ascertained.

4.3 Direct-Reading Techniques

Various direct-reading techniques that can be used to evaluate airborne concentrations of gases and vapors are available to the industrial hygienist. These include instruments capable of direct response to airborne contaminants, various reagent kits that can be used for certain substances, colorimetric indicator (detector) tubes, and passive dosimeters.

4.3.1 Instruments

A direct-reading instrument is an integrated system capable of sampling a volume of air, making a quantitative analysis and displaying the result. Direct-reading instruments can be portable devices or fixed-site monitors. Generally these devices are characterized by disadvantages that limit their application for measuring the low concentrations of significance to the industrial hygienist.

Direct-reading instruments are used commonly for on-site evaluations for a variety of reasons, depending primarily on the understood purpose of the survey. Direct-reading instruments are useful in the following applications:

- To find sources of emission of hazardous substances "on the spot"
- To determine the performance characteristics of specific operations or control devices, usually by comparing results of before-and-after surveys
- As a qualitative industrial hygiene monitoring instrument to ascertain whether specific air quality standards are being complied with
- As continuous monitoring devices, by establishment of a network of sensors at fixed locations throughout a plant; readout from such a system can be used to activate either an alarm or an auxiliary control system in the event of process upsets, or to obtain permanent recorded documentation of concentrations of contaminants in the workroom atmosphere

The advantages of having direct-reading instruments available for industrial hygiene surveys are obvious. Such on-site evaluations of atmospheric concentrations of hazardous substances make possible the immediate assessment of unacceptable conditions and enable industrial hygienists to initiate immediate corrective action in accordance with their judgment of the seriousness of the situation without causing further risk of injury to the workers. It cannot be overemphasized that great caution must be employed in the use of direct-reading instruments and, more important, in the interpretation of their results. Most of these instruments are nonspecific, and before recommending any action industrial hygienists often must verify their on-site findings by supplemental sampling and laboratory analyses to characterize adequately the chemical composition of the contaminants in a workroom area and to develop the supporting quantitative data with more specific methods of greater accuracy. Such precautions become mandatory if the industrial hygienist or other professional investigator conducting the sampling has not had extensive experience with the process in question, or when a change in the process or substitution of chemical substances may have occurred.

The calibration of any direct-reading instrument is an absolute necessity if the data are to have any meaning. Considering this to be axiomatic, it must also be recognized that the frequency of calibration depends on the type of instrument. Certain classes of instruments, because of their design and complexity, require more frequent calibration than others. It is also recognized that peculiar quirks in an individual instrument produce greater variations in its response and general performance, thus requiring a greater amount of attention and more frequent calibration than other instruments of the same design. Direct personal experience with a given instrument serves as the best guide in this matter.

Another unknown factor that can be evaluated only by experience is the environmental variability of sampling sites. For example, when locating a particular fixed-station monitor at a specific site, consideration must be given to the presence of interfering chemical substances, the corrosive nature of contaminants, vibration, voltage fluctuations, and other disturbing influences that may affect the response of the instrument.

Finally, the required accuracy of the measurements must be determined initially. If an accuracy of $\pm 3\%$ is needed, more frequent calibration must be made than if $\pm 25\%$ accuracy is adequate in the solution of a particular problem.

As indicated earlier, direct-reading instruments for atmospheric contaminants are classified as devices that provide an immediate indication of the concentration of contaminants by a dial reading, a strip-chart recording, or a tape printout. When properly calibrated and used with full cognizance of their performance characteristics and limitations, these services can be extremely helpful to industrial hygienists who are engaged in on-site evaluations of potentially hazardous conditions. Many types of instruments depend on certain physical or chemical principles for their operation. They are discussed briefly here and are described in detail in other publications (19, 20). In general, the advantages of direct-reading instruments include the following:

- Rapid estimation of the concentration of a contaminant, permitting on-site evaluations and implementation of appropriate measures
- Provision of permanent 24-h records of contaminant concentrations, when used as continuous monitors

- Easy incorporation of alarm systems into continuous monitoring instruments to warn workers of buildup of hazardous conditions
- Reduction of the number of manual tests needed to accomplish an equivalent amount of sampling
- Similar reduction of the number of laboratory analyses
- Provision of evidence of the monitoring of environmental conditions for presentation in litigation proceedings
- Reduced cost per sample of obtaining data

The disadvantages of direct-reading instruments usually include at least one of the following:

- High initial cost of instrumentation and, if used as a continuous monitor, installation of the sensing network
- Need for frequent calibration
- General lack of adequate calibration facilities
- Lack of portability
- Lack of specificity, the most critical negative factor

Several common types of direct-reading instrument are described in the following section. These instruments can be used if necessary precautions are taken and the limitations of the devices are understood. In the following discussions, when the lower end of the working range is indicated as zero, the reader should recognize that it is a relative term related to the sensitivity of the instrument.

4.3.1.1 Colorimetry. Colorimetry is the measurement of the relative power of a beam of radiant energy (colorimetry) in the visible, ultraviolet, or infrared region of the electromagnetic spectrum, which has been attenuated as a result of passing a suspension of solid or liquid particulates in air or other gaseous medium, or a photographic image of a spectral line or an X-ray diffraction pattern on a photographic film or plate. Such photometers contain a lamp or other source of energy generation, an optical filter arrangement to limit the bandwidth of the incident beam of radiation, and an optical system to collimate the filtered beam. The beam then is passed through the sample system contained in a cuvette or gas cell to a photocell, bolometer, thermocouple, or pressure-sensor type of detector, where the signal is amplified and fed to a readout meter or to a strip-chart recorder. The more sophisticated technique, termed *spectrophotometry*, employs prisms made of glass (visible region), quartz (ultraviolet), and sodium chloride or potassium bromide (infrared), or diffraction gratings, instead of optical filters, to provide essentially monochromatic radiation as the source of energy.

Spectrophotometers are used mostly in laboratories for highly specific and precise analytical determinations, although field colorimetric analyzers have been designed to function primarily as fixed-station monitors for active gases such as oxides of nitrogen, sulfur dioxide, total oxidant content, ammonia, aldehydes, chlorine, hydrogen fluoride, and hydrogen sulfide. These instruments require frequent calibration with zero and span gases at the sampling site to assure generation of reliable data. However, built-in automated calibration systems, which regularly standardize zero and span controls against pure air and a calibrated optical filter, are now available for several gases including nitrogen oxides, sulfur dioxide, and aldehydes.

4.3.1.2 ***Heat of Combustion.*** In heat-of-combustion instruments, a combustible gas or vapor mixture is passed over a filament heated above the ignition temperature of the substance of analytical interest. If the filament is part of a Wheatstone bridge circuit, the resulting heat of combustion changes the resistance of the filament, and the measurement of the imbalance is related to the concentration of the gas or vapor in the sample mixture. The method is nonspecific, since it responds to a property of the material, but it may be made more selective by choosing appropriate filament temperatures for individual gases or vapors or by using an oxidation catalyst for a desired reaction, such as Hopcalite for carbon monoxide.

Like all direct-reading instruments, combustible gas indicators must be calibrated for their response to the anticipated individual test gases and vapors. These instruments are definitely portable and they are valuable survey meters in the industrial hygienist's collection of field instruments. Readings are in terms of 0 to 1000 ppm or 0 to 100% of the lower explosive limit (LEL). However, industrial atmospheres rarely contain a single gaseous contaminant, and these indicators will respond to all combustible gases present. Hence supplementary sampling and analytical techniques should be used for a complete definition of environmental conditions.

4.3.1.3 ***Electrical Conductivity.*** Electrical conductivity instruments function by drawing a gas-air mixture through an aqueous solution. Gases that form electrolytes, such as vinyl chloride, produce a change in the electroconductivity as a summation of the effects of all ions thus produced. Hence the method is nonspecific. If the concentrations of all other ionizable gases are either constant or insignificant, the resulting changes in conductivity may be related to the gaseous substances of interest. Temperature control is extremely critical in conductance measurements; if thermostated units are not used, electrical compensation must enter into the measurements to allow for the 2%/°C conductivity temperature coefficient average for many gases.

One of the most common applications of electrical conductivity instruments has been in the continuous monitoring of sulfur dioxide in ambient atmospheres. Lightweight portable analyzers that use a peroxide absorber to convert SO_2 to H_2SO_4 are available; these battery-operated instruments can provide an integrated reading of the SO_2 concentration over the 0 to 1 ppm range within 1 min. A larger portable model that may be operated from a 12-V automobile battery is also available for the high concentration ranges of SO_2 encountered in field sampling.

4.3.1.4 ***Thermal Conductivity.*** Specific heat of conductance is a property of a gas or vapor that can be used to measure its concentration in a carrier gas such as air, argon, helium, hydrogen, or nitrogen. However, thermal conductivity measurements are nonspecific, and the method has had only limited application as a primary detector. It has found its greatest usefulness in estimating the concentrations of separately eluted components from a gas chromatographic column. This method operates by virtue of the loss of heat from a hot filament to a single component of a flowing gas stream, the loss being registered as a decrease in electrical resistance measured by a Wheatstone bridge circuit. This method is applied mainly to binary gas mixtures, and uses are based on the electrical imbalance produced in the bridge circuit by the difference in the filament resistances of the sample and reference gases passed through separate cavities in the thermal conductivity cell.

4.3.1.5 Flame Ionization. The hydrogen flame ionization detector (FID) typically consists of a stainless steel burner in which hydrogen is mixed with the sample gas stream in the base of the unit. Combustion air or oxygen is fed axially and diffused around the jet through which the hydrogen-gas mixture flows to the cathode tip, where ignition occurs. A loop of platinum, set about 6 mm above the tip of the burner, serves as the collector electrode. The current carried across the electrode gap is proportional to the number of ions generated during the burning of the sample. The detector responds to essentially all organic compounds, but its response is greatest with simple hydrocarbons and diminishes with increasing substitution of other elements, notably oxygen, sulfur, and chlorine. A low noise level of 10–12 A provides a high sensitivity of detection, and the detector is capable of the wide linear dynamic range of 107. Its usefulness is enhanced by its insensitivity to water, the permanent gases, and most inorganic compounds, thus simplifying the analysis of aqueous solutions and atmospheric samples. It serves to great advantage in both laboratory and field models of gas chromatographs (an application discussed in detail subsequently), as well as in hydrocarbon analyzers that are set up as fixed-station monitors of ambient atmospheres in the laboratory or field.

Hydrocarbon analyzers, operating with an FID, are literally carbon counters; their response to a given quantity of a typical C_6 hydrocarbon is six times that to methane, at a fixed flow rate of the sample stream. Thus the instrument's characteristics, such as sensitivity, usually are given as a methane equivalent. In addition to hydrocarbons, these analyzers respond to alcohols, aldehydes, amines, and other compounds, including vinyl chloride, which produces an ionized carbon atom in the hydrogen flame.

4.3.1.6 Gas Chromatography. Gas chromatography, a physical process for separating components of complex mixtures, is used routinely as a portable technique for in-plant studies. A gas chromatograph has the following components:

- Carrier gas supply complete with a pressure regulator and flowmeter
- Injection system for the introduction of a gas or vaporizable sample into a port at the front end of the separation column
- Stainless steel, copper, or glass separation column containing a stationary phase consisting of an inert material such as diatomaceous earth, used alone as in gas-solid chromatography or as a support for a thin layer of a liquid substrate, such as silicone oils, in gas-liquid chromatography
- Heater and oven assembly to control the temperature of the column(s) injection port and detector unit

- Detector
- Recorder for the chromatograms produced during the separations

Separations are based on the varied affinities of the sample components for the packing materials of a particular column, the rate of carrier gas flow, and the operating temperature of the column. Improved separations are made possible by the use of temperature programming. The sample components, as a consequence of their varied affinities for a given column, are eluted sequentially; thus they evoke separate responses by the detection system, from which the signal is amplified to produce a peak on the strip chart or other output. The height and area of the peak are proportional to the concentration of the eluted sample components. Calibrations can be made using known mixtures of the pure substance in a gas-air mixture prepared in a flexible plastic bag or other suitable container. The time of retention on the column and supporting analytical techniques (e.g., infrared spectrophotometry) can be used in qualitative analysis of the individual peaks of a chromatogram. The method is capable of providing extremely clean-cut separations and is one of the most useful techniques in the field of organic analysis. Commonly used detectors include flame ionization, thermal conductivity, and electron capture.

Rugged, battery-operated, portable gas chromatographs have been refined sufficiently to be practical for many field study applications. These instruments may be obtained with a choice of thermistor, thermal conductivity, flame ionization, and electron capture detectors. Complete with gas sampling valve, rechargeable batteries, appropriate columns, and self-contained supplies of gas, these chromatographs have much to offer the industrial hygienist engaged in on-site analyses of trace quantities of organic compounds and the permanent gases. Gas lecture bottles provide 8 to 20 h of operation, depending on the flow rates, and they must be recharged using high pressure gas regulators. The retention times of the compounds of analytical interest must be determined in the laboratory for a given type of column, as is true for the laboratory-type chromatographs.

4.3.1.7 Other Principles of Operation. Various additional types of direct-reading instrument are available, although their range of applicability is more limited. The principles of operation include chemiluminescence, coulometry, polarography, potentiometry, and radioactivity. Table 8.2 contains a summary of the types of direct-reading instruments available, along with typical operating characteristics and examples of gases and vapors for which the instruments have been used successfully.

4.3.2 Reagent Kits

Direct-reading colorimetric techniques, which utilize the chemical reaction of an atmospheric contaminant with a color-producing reagent, are available in a variety of forms. Detector kit reagents may be in either liquid or solid phase or supplied in the form of chemically treated papers. The liquid and solid reagents are generally supported in sampling devices through which a measured amount of contaminated air is drawn. Chemically treated papers are usually exposed to the atmosphere, and the reaction time for a color change to occur is noted.

Table 8.2. Direct-Reading Physical Instruments

Principle of Operation	Applications and Remarks	Code[a]	Range	Repeatability (Precision)	Sensitivity	Response Time
Aerosol photometry	Measures, records, and controls particulates continuously in areas requiring sensitive detection of aerosol levels; detection of 0.05–40 μm diameter particles. Computer interface equipment is available	A and B	10^{-3} to 10^{2} μg/L	Not given	10^{-3} μg/L (for 0.3 μm DOP)	Not given
Chemiluminescence	Measurement of NO in ambient air selectively and NO_2 after conversion to NO by hot catalyst. Specific measurement of O_3. No atmospheric interferences	B	0–10,000 ppm	±0.5–3%	Varies: 0.1 ppb to 0.1 ppm	ca 0.7 sec, NO mode and 1 sec, NO_x mode; longer period when switching ranges
Colorimetry	Measurement and separate recording of NO_2-NO_x, SO_2, total oxidants, H_2S, HF, NH_3, Cl_2 and aldehydes in ambient air	A and B	ppb and ppm	±1–5%	0.01 ppm (NO_2, SO_2)	30 sec to 90% of full scale
Combustion	Detects and analyzes combustible gases in terms of percent LEL on graduated scale. Available with alarm set at ⅓ LEL	A	ppm to 100%	—	ppm	<30 sec
Conductivity, electrical	Records SO_2 concentrations in ambient air. Some operate off a 12-V car battery. Operates unattended for periods up to 30 days	A and B	0–2 ppm	<±1–10%	0.01 ppm	1–15 sec (lag)

Coulometry	Continuous monitoring of NO, NO_2, O_3 and SO_2 in ambient air. Provided with stripchart recorders. Some require attention only once a month	A and B	Selective: 0–1.0 ppm overall, or scale to 100 ppm (optional)	±4% of full scale	varies: 4–100 ppb dependent on instrument range setting	<10 min to 90% of full scale
Flame ionization (with gas chromatograph)	Continuous determination and recording of methane, total hydrocarbons, and carbon monoxide in air. Catalytic conversion of CO to CH_4. Operates up to 3 days unattended	B	Selective: 0–1 ppm: 0–100 ppm	±1% of full scale	Not given	5 min (cycle time)
	Separate model for continuous monitoring of SO_2, H_2S, and total sulfur in air. Unattended operation up to 3 days	B	0–20 ppm	±4% of full scale	0.005 ppm (H_2S); 0.01 ppm (SO_2)	5 min (cycle time)
Flame ionization (hydrocarbon analyzer)	Continuous monitoring of total hydrocarbons in ambient air; potentiometric or optional current outputs compatible with any recorder. Electronic stability from 32 to 110°F	B	0–1 ppm as CH_4; ×1, ×10, ×100, ×1000 with continuous span adjustment	±1% of full scale	1 ppm to 2% full scale as CH_4; 4 ppm to 10% as mixed fuel	<0.5 sec to 90% of full scale
Gas chromatograph, portable	On-site determination of fixed gases, solvent vapors, nitro and halogenated compounds, and light hydrocarbons. Instruments available with choice of flame ionization, electron capture, or thermal conductivity detectors and appropriate columns for desired analyses. Rechargeable batteries	A	Depends on detector	Not given	<1 ppb (SF_6) with electron capture detector; <1 ppm (HCs)	—

Table 8.2. (continued)

Principle of Operation	Applications and Remarks	Code[a]	Range	Repeatability (Precision)	Sensitivity	Response Time
Infrared analyzer (photometry)	Continuous determination of a given component in a gaseous or liquid stream by measuring amount of infrared energy absorbed by component of interest using pressure sensor technique. Wide variety of applications include CO, CO_2, Freons, hydrocarbons, nitrous oxide, NH_3, SO_2, and water vapor	B	From ppm to 100% depending on application	±1% of full scale	0.5% of full scale	0.5 sec to 90% of full scale
Photometry, ultraviolet (tuned to 253.7 nm)	Direct readout of mercury vapor; calibration filter is built into the meter. Other gases or vapors that interfere include acetone, aniline, benzene, ozone, and others that absorb radiation at 253.7 mμ	A	0.005–0.1 and 0.03–1 mg/m^3	±10% of meter reading or ± minimum scale division, whichever is larger	0.005 mg/m^3	Not given
Photometry, visible (narrow-centered 394 nm band pass)	Continuous monitoring of SO_2, SO_3, H_2S, mercaptans, and total sulfur compounds in ambient air. Operates more than 3 days unattended	B	1–3000 ppm (with airflow dilution)	±2%	0.01–10 ppm	<30 sec to 90% of full scale
Particle counting (near forward scattering)	Reads and prints directly particle concentrations at 1 of 3 preset time intervals of 100, 1000 or 10,000 sec, corresponding to 0.01, 0.1, and 1 ft^3 of sampled air	B	Preset (by selector switch); particle size ranges:	±0.05% (probability of coincidence) 0.3, 0.5, 1.0, 2.0, 3.0, 5.0, and 10.0, μm; counts up to 10^7 particles per ft^3 (35×10^3/L)	—	Not given

Polarography	Monitor gaseous oxygen in flue gases, auto exhausts, hazardous environments, and in food storage atmospheres and dissolved oxygen in wastewater samples. Battery operated, portable, sample temperature 32 to 110°F, up to 95% relative humidity. Potentiometric recorder output. Maximum distance between sensor and amplifier is 1000 ft	A	0–5 and 0–25%	±1% of reading at constant sample temperature	Not given	20 sec to 90% of full scale
Radioactivity	Continuous monitoring of ambient gamma- and X-radiation by measurement of ion chamber currents, averaging or integrating over a constant recycling time interval, sample temperature limits 32 to 120°F; 0 to 95% relative humidity (weatherproof detector); up to 1000 ft remote sensing capability. Recorder and computer outputs. Complete with alert, scram, and failure alarm systems. All solid-state circuitry	B	0.1–10^7 mR/h	±10% (decade accuracy)	—	<1 sec

Table 8.2. (continued)

Principle of Operation	Applications and Remarks	Code[a]	Range	Repeatability (Precision)	Sensitivity	Response Time
Radioactivity	Continuous monitoring of beta- or gamma-emitting radioactive materials within gaseous or liquid effluents; either a thin-wall Geiger-Müller tube or a gamma scintillation crystal detector is selected depending on the isotope of interest; gaseous effluent flow, 4 cfm; effluent sample temperature limits 32 to 120°F using scintillation detector and 65 to 165°F using G-M detector. Complete with high radiation, alert and failure alarms	B	$10–10^4$ counts/min	±2% full scale (rate meter accuracy)	$<10^{-7}$ μCi of ^{131}I per cm^3 of air and 10^{-7} μCi of ^{137}Cs per cm^3 of water using a scintillation detector	0.2 sec at 10^6 counts/min (rate meter)
Radioactivity	Continuous monitoring of radioactive airborne particulates collected on a filter tape transport system; rate of airflow, 10 scfm; scintillation and G-M detectors, optional but a beta-sensitive plastic scintillator is provided to reduce shielding requirements and offer greater sensitivity. Air sample temperature limits 32 to 120°F; weight 550 lb. Complete with high and low flow alarm and a filter failure alarm	B	10 to 10^6 counts/min	±2% of full-scale (rate-meter accuracy)	10^{-12} μCi of ^{137}Cs per cm^3 of air using a scintillation detector	0.2 sec at 10^4 counts/min (rate meter)

[a]Codes: A, portable instruments; B, fixed monitor "transportable" instruments.

Liquid reagents come in sealed ampules or in tubes for field use. Such preparations are provided in concentrated or solid form for easy dilution or dissolution at the sampling site. Representative of this type of reagent are the *o*-tolidine and the Griess-Ilsovay kits for chlorine and nitrogen dioxide, respectively. Although the glassware needed for these applications may be somewhat inconvenient to transport to the field, methods based on the use of liquid reagents are more accurate than those that use solid reactants. This is due to the inherently greater reproducibility and accuracy of color measurements made in a liquid system.

Chemical reagents impregnated into special papers have found wide applications for many years for the detection of toxic substances in air. Examples include the use of mercuric bromide papers for the detection of arsine, and lead acetate for hydrogen cyanide. When a specific paper is exposed to an atmosphere containing the contaminant in question, the observed time of reaction provides an indication of the concentration of that substance. For example, a 5-sec response time by the *o*-tolidine–cupric acetate paper indicates a concentration of 10 ppm HCN in the tested atmosphere.

Similarly, sensitive detector crayons have been devised for the preparation of a reagent smear on a test paper for which response to a specific toxic substance in a suspect atmosphere may then be timed to obtain an estimation of the atmospheric concentration of a contaminant. Crayons for phosgene, hydrogen cyanide, cyanogen chloride, and lewisite (ethyl dichloroarsine) have been developed.

4.3.3 Colorimetric Indicators

Colorimetric indicating tubes containing solid reagent chemicals provide compact direct-reading devices that are convenient to use for the detection and semiquantitative estimation of gases and vapors in atmospheric environments. Presently there are tubes for more than 300 atmospheric contaminants on the market; seven companies manufacture these devices.

Whereas it is true that the operating procedures for colorimetric indicator tubes are simple, rapid, and convenient, there are distinct limitations and potential errors inherent in this method of assessing atmospheric concentrations of toxic gases and vapors. Therefore, dangerously misleading results may be obtained with these devices unless they are used under the supervision of an adequately trained industrial hygienist who does the following:

- Enforces rigidly the periodic (as required) calibration of individual batches of each specific type of tube for its response to known concentrations of the contaminant, as well as the refrigerated storage of all tubes to minimize their rate of deterioration
- Informs the staff of the physical and chemical nature and extent of interferences to which a given type of tube is subject and limits the tube's usage accordingly
- Stipulates how and when other independent sampling and analytical procedures will be employed to derive needed quantitative data.

A manual describing recommended practice for colorimetric indication tubes (21) discusses in detail the principles of operation, applications, and limitations of these devices. A brief summary is given here.

Detector tubes are filled with a solid granular material, such as silica gel or aluminum oxide, that has been impregnated with an appropriate chemical reagent. The ends of the glass tubes are sealed during manufacture. When a tube is to be used, its end tips are broken off, the tube is placed in the manufacturer's holder, and the recommended volume of air is drawn through the tube by means of the air-moving device provided by the manufacturer. This device may be one of several types, such as a positive displacement pump, a simple squeeze bulb, or a small electrically operated pump with an attached flowmeter. Each air-moving device must be calibrated frequently, for example, after sampling 100 tubes as an arbitrary rule, or more often if there are reasons to suspect changes due to effects of corrosive action from contaminants in the tested atmospheres. An acceptable pump should be correct to within approximately 5% volume; with use, its flow characteristics may change. It should also be checked for leakage and plugging of the inlet after every 10 samples.

Usually, a fixed volume of air is drawn through the detector tube, although in some systems varied amounts of air may be sampled. The operator compares either an absolute length of stain produced in the column of the indicator gel or a ratio of the length of stain to the total gel length against a calibration chart, to obtain an indication of the atmospheric concentration of the contaminant that reacted with the reagent. To make estimates using another type of tube, a progressive change in color intensity is compared with a chart of color tints. In a third type of detector, the volume of sampled air required to produce an immediate color change is noted; it is intended that this air volume be inversely proportional to the concentration of the atmospheric contaminant. Basic mathematical analyses of the relationships among the variables affecting the length of stain (i.e., the concentration of test gas, volume of air sample, sampling flow rate, grain size of gel, tube diameter, and other variables) have been published (22–24). These sources should be consulted for a full appreciation of the complex interrelationships among the factors affecting the kinetics of indicator tube reactions. It is sufficient to point out here that the length of stain is proportional to the logarithm of the product of gas concentration and air sample volume. It is important to control the flow rate, which may produce a greater effect on the length of stain than does the concentration of the test gas. Ideally, the reaction rate between contaminant and reagent will be rapid enough to permit the establishment of equilibrium between the two, thus producing a stoichiometric relationship between the volume of stained indicating gel and the quantity of the absorbed test gas. Such equilibrium conditions may be assumed to exist when stain lengths are directly proportional to the volume of sampled air and are not affected by the sampling flow rate. With this situation a plot (log-log) of stain length versus concentrations for a fixed sample volume may be prepared in the calibration of a given batch of tubes.

Because of the complexity of the heterogeneous phase kinetics of indicator tube reactions, the quality-control problems associated with their manufacture and storage, and the difficulties posed by interfering substances, it is obvious that frequent, periodic calibration of these devices should be made by the user. Dynamic dilution systems for the reliable preparation of low concentrations of a test gas or vapor are recommended for this purpose.

4.3.4 *Passive Monitors*

Passive dosimetry has become popular for some personal monitoring in recent years, taking advantage of gaseous diffusion. The dosimeters, or monitors, utilize Brownian motion to control the sampling process into a collection medium. This technology is particularly well suited for personal monitoring devices, resulting in lightweight, low-cost monitors that require no power source. The monitors rely on a concentration gradient across a static or placid layer of air to induce a mass transfer. By choosing an effective collection surface, the mass transfer or collection rate is proportional to the ambient vapor concentration.

The accuracy and precision of the sampling process are functions of the measured exposure time, velocity effects, and temperature effects. The accuracy and precision of the reported concentration are functions of the calibration standards, collection media, and analytical method used. Of these factors, the potential velocity and temperature effects distinguish this type of monitoring device from the conventional dynamic or flow monitor. All other factors are common to both methods.

The thickness of the attached boundary layer on the surface of the barrier film is a function of the velocity of air movement over the face of the monitor. Sampling that is independent of this air velocity can be achieved when the length or depth of the static layer is large compared with the average boundary layer thickness. For temperatures between 50 and 90°F, the temperature factor is constrained to less than 2%. For use at higher or lower temperatures, the temperature data may be rerecorded to allow corrections to be made.

In use, this type of personal monitor is attached to the worker in his or her breathing zone. The total exposure time is noted, and analysis results give the amount of vapor collected. These data provide an average mass collection rate, which can then be used to calculate the time-weighted average concentration. The physical parameters of the sampler design are chosen according to desired exposure time and the substance to be monitored.

Corroborative testing of passive monitors for mercury, nitrogen dioxide and other contaminants and comparison of results to reference methods have shown excellent agreement. With the increasing emphasis on development of specific methods for monitoring workers' exposures to contaminants, and the extension of the concept to colorimetric indicator tubes, it is likely that the concept of passive monitoring will become a vital basis for a new generation of industrial hygiene monitoring equipment.

5 SAMPLING FOR PARTICULATES

Particulate, or *aerosol*, normally refers to any system of liquid droplets or solid particles dispersed in a stable aerial suspension. This requires that the particles or droplets possess physical characteristics that allow them to remain suspended for significant periods of time.

Liquid particulates usually are classified into two subgroups, mists and fogs, depending on particle size. The larger particles generally are referred to as mists, whereas small particle sizes result in fogs. Liquid droplets normally are produced by such processes as condensation, atomization, and entrainment of liquid by gases.

Solid particulates usually are subdivided into three categories: dusts, fumes, and smoke; the distinction among them is primarily related to the processes by which they are produced. Dusts are formed from solid organic or inorganic materials by reducing their size through some mechanical process such as crushing, drilling, or grinding. Dusts vary in size from the visible to the submicroscopic, but their composition is the same as the material from which they were formed. Fumes are formed by such processes as combustion, sublimation, and condensation. The term is generally applied to the oxides of zinc, magnesium, iron, lead, and other metals, although solid organic materials such as waxes and some polymers may form fumes by the same methods. These particles are very small, ranging in size from 1 mm to as small as 0.001 mm in diameter. *Smoke* is a term generally used to refer to airborne particulate resulting from the combustion of organic materials (e.g., wood, coal, or tobacco). The resulting smoke particles are all usually less then 0.5 mm in diameter.

The nature of the airborne particulate dictates to a great extent the manner in which sampling of the environment must be accomplished. Sampling is performed by drawing a measured volume of air through a filter, impingement device, electrostatic or thermal precipitator, cyclone, or other instrument for collecting particulates. The concentration of particulate matter in air is denoted by the weight or the number of particles collected per unit volume of sampled air. The weight of collected material is determined by direct weighing or by appropriate chemical analysis. The number of particles collected is determined by counting the particles in a known portion or aliquot of the sample and extrapolating to the whole sample.

5.1 General Requirements

Most of the general requirements discussed earlier for gases and vapors apply to particulate contaminants as well. However, because of the wide range of particle size of airborne particulates confronting industrial hygienists, there are several aspects of sampling that apply only to particulates.

The sampling train for particulates consists of the following components: air inlet, particle collection medium (and preselector, if classification of the total particulate is being done), flowmeter, flow rate control device, and air mover or pump. Of these, the most important is the particle collection medium, which is used to separate the particles from the sampled air stream. Both the efficiency of the device and its reliability must be high. The pressure drop across the medium should be low to keep the size of the required pump to a minimum. The medium may consist of a single element, such as a filter or impinger, or there may be two or more elements in series, to classify the particulate into two different size ranges. Proper selection, care, and calibration of the other components of the sampling train, particularly the flowmeter and flow control mechanism, are discussed elsewhere in this book.

The sampling method should not alter chemical or physical characteristics of the particles collected. For example, if the material is soluble, it cannot be collected in a medium capable of dissolving it. If the particles have a tendency to agglomerate, and it is important to be able to distinguish individual particles, deep-section collection on a filter should not be used.

An additional consideration, of greater concern with particulates than with gases and vapors, is the variation of concentration in space. Many cases have been reported of significant differences in concentrations being documented with sampling units placed equidistant from a source of contaminant generation. Similarly, with personal breathing zone sampling, it is not uncommon for simultaneous samples obtained on both shoulders of a worker to indicate substantially different concentrations. The importance of these observations lies in the understanding that particulate sampling results by themselves indicate conditions within a short distance of the sampling unit and should be augmented with additional information, such as studies of airflow patterns within the workroom.

5.2 Collection Techniques

The concept of grab sampling for particulates is not as valuable as for gases or vapors, although there are methods for collecting instantaneous samples of airborne particles. One such device is the konimeter, which although of perhaps only historical interest today, served a very useful purpose in industrial hygiene studies. This device draws a small, measured volume of air into the instrument and literally blasts it at high velocity against a glass plate on which the particles deposit. The particles then can be examined microscopically and counted or defined in other terms. Because there are millions of particles per cubic foot in most industrial workrooms, a very small volume of air is needed for this technique.

Settling chambers also were used to obtain an instantaneous sample of particulates. With this device, a chamber was opened in the atmosphere being tested and closed rapidly to trap the sample. The particles in the air then were allowed to settle by gravitational forces and collected on a glass plate for subsequent microscopic analysis.

Today, the most meaningful sampling for particulates is done over extended periods of time with various collection techniques, depending on the material being collected and the availability of sampling equipment. The most common collection techniques are filtration, impingement, impaction, elutriation, electrostatic precipitation, and thermal precipitation.

5.2.1 Filtration

Filtration is the most common method of collecting airborne particulate. The fibrous type of filter matrices consists of irregular meshes of fibers, usually about 20 μm or less in diameter. Air passing through the filter changes direction around the fibers, and the particles impinge against the filter, where they are retained. The largest particles (30 μm and greater) deposit to some extent by sieving action; the smaller particles (submicrometer sizes) also deposit through their Brownian motion, which carries them into the filter material. Efficiency of collection generally increases with airstream velocity, density, and particle size for particles greater than 0.5 μm in diameter. Deposition by diffusion dominates for the smallest particle sizes and decreases as the diameter of the particle increases. Thus, there is a size at which the combined efficiency by impingement and diffusion is at a minimum; this is usually 0.1 μm diameter. Because the total weight of particles less than 0.1 μm diameter usually is less than 2% of the total collected dust, deposition by diffusion can practically be ignored. Of course, there are exceptions, such as sampling for freshly formed metal fumes, where diffusion deposition is very significant.

Filters are available in a wide variety of matrices including cellulose, glass, asbestos, ceramic, carbon, metallic, polystyrene, and other polymeric materials (Table 8.3). Filters made of these fibrous materials consist of thickly matted fine fibers and are small in mass

Table 8.3. Common Applications of Filters

Filter Matrix	Common Applications
Cellulose ester	Asbestos counting, particle sizing, metallic fumes, acid mists
Fibrous gloss	Total particulate, oil mists, coal for pitch volatiles
Paper	Total particulate, metois, pesticides
Polycarbonate	Total particulate, crystalline silica
Polyvinyl chloride	Total particulate, crystalline silica, oil mists
Silver	Total particulate, coal for pitch volatiles, crystalline silica
Teflon	Special applications (high temperature)

per unit face area, making them useful for gravimetric determinations. Of these, cellulose fiber filters are the least expensive, are available in a wide range of sizes, have high tensile strengths, and are relatively low in ash content. Their greatest disadvantage is their hygroscopicity, which can present problems during weighing procedures. The filters made of synthetic fibers, particularly glass and polyvinyl chloride, are more common, partly because stable tare weights can be determined easily.

Membrane filters, microporous plastic films made by precipitation of a resin under controlled conditions, are used to collect samples that are to be examined microscopically, although they can also serve for gravimetric sampling and for specific determinations using instrumentation. Thus the cellulose ester membrane filters are the most commonly used filters for sampling for such substances as asbestos (analyzed microscopically for fiber count) and metal fumes (analyzed for specific elements by atomic absorption techniques).

5.2.2 *Impaction and Impingement*

These devices take advantage of a sudden change in direction in airflow and the momentum of the dust particles to cause the particles to impact against a flat surface. Usually impactors are constructed in a series of several stages, to separate dust by size fractions. The particles adhere to the plate, which may be dry or coated with an adhesive, and the material on each plate is weighed or analyzed at the conclusion of sampling. It is imperative that the impactors be calibrated for the particular material of interest, because the manufacturer's calibration typically is based on a uniformly sized particle and may not give accurate results for the substance of interest.

Impingers also utilize inertial properties of particles to collect samples. The impinger consists of a glass nozzle or jet submerged in a liquid, frequently water. Air is drawn through the nozzle at high velocity, and the particles impinge on a flat plate, lose their velocity, are wetted by the liquid, and become trapped. Gases that are soluble in the liquid also are collected. Usually the contents of the impinger samples are analyzed microscopically, gravimetrically, or in a few cases, by specific methods.

The principles of collection for impaction and impingement are quite similar. The primary distinction between them is that with impaction, the particles are directed against a dry or coated surface, whereas a liquid collecting medium is used in impingement.

5.2.3 Elutriation

Elutriators have been essential elements in the sampling trains used to characterize dust levels in many mineral dust surveys, usually as preselectors at the front of the sampling train. Elutriators can have either a horizontal or a vertical orientation. They are quite similar to inertial separators in the theoretical basis of operation. The primary difference is that elutriators operate at normal gravitational conditions, whereas the inertial collectors induce very high momentum forces to achieve collection of particles. Horizontal elutriators make use of the fact that as air moves across a horizontal channel with laminar flow, particles of greatest mass tend to cross the streamlines and settle because of gravity. The smaller particles remain airborne by resistance forces of the air for longer times and distances, depending on their size and mass. It is imperative that elutriators be used and operated exactly as described by the manufacturer to avoid disturbances at the inlet, to ensure laminar flow along the elutriator, and to avoid risk of redispersion of settled particles.

Vertical elutriators utilize the same dependence on gravitational forces to separate the dust into fractions, except that with the vertical elutriator the natural force works in a direction opposite to the induced airflow instead of normal to it. The vertical elutriator is recognized as a practical device to sample for cotton dust when it is desired to avoid collecting the lint or fly.

5.2.4 Electrostatic Precipitation

Instruments using electrostatic precipitation have been used for many decades for industrial air analysis in workrooms. These systems have the advantage of negligible flow resistance, no clogging, and precipitation of the dust onto a metal cylinder or foil liner whose weight is unaffected by humidity. In most units, the wire-and-tube arrangement is used; a stiff wire, supported at one end, is aligned along the center of the tube and serves as the charged electrode. The tip of the wire is sharpened to a point, and a high voltage (10 to 25 kV dc) is applied to the electrode. The corona discharge from the tip charges particles suspended in the air that is drawn through the tube. The electrical gradient between the wire and the wall of the cylinder (or foil) causes the charged particles to migrate to the inside surface of the tube. The migration velocity of the charged particles greater than 1 mm in diameter increases in proportion to particle diameter. This velocity is almost independent of particle diameter for particles smaller than about 1 mm. Therefore very high separation efficiencies are attainable with electrostatic precipitators, and they have become particularly well suited for particles of submicrometer size, such as those in metal fumes.

5.2.5 Thermal Precipitation

Particles are directed away from a high temperature source by the differential bombardment from gas molecules in the thermal gradient maintained in the air around the hot source, the basis for design of thermal precipitation units. Air is drawn past a hot wire or plate, and the dust collects on a cold glass or metal surface opposite the hot element. Because a high thermal gradient is needed, the gap between the wire or plate and the deposition surface is kept very small, typically less than 2 mm. The migration velocity induced by the thermal gradient is small and is very nearly independent of particle diameter. Because of the severe limitations on maximum flow rate possible with high deposition efficiency, however, these units have been used only for collecting sufficient particulate matter for examination under a microscope. An additional limitation is the inappropriateness of these devices for sampling mists or other liquid particulates, unless their boiling points are high enough to ensure that the liquid is not volatilized by the operating temperatures of the instrument.

5.2.6 Centrifugal Collection

When industrial hygiene sampling is done to document concentrations of respirable dust, it is necessary to separate the total particulate into respirable and nonrespirable portions. This is most commonly done with centrifugal separators, such as the cyclone. Air enters the cyclone tangentially through an opening in the side of a cylindrical or inverted cone-shaped unit. The larger particles are thrown against the side of the cyclone and fall into the base of the assembly. The smaller particles are drawn toward the center of the unit, where they swirl upward along the axis of a tube extending down from the top. The air in the cyclone rotates several times before leaving, and consequently the dust deposits as it would in a horizontal elutriator having an area several times that of the cyclone's outer surface. Thus the volume of a cyclone is much smaller than a horizontal elutriator or other inertial collector with the same flow rate and efficiency. Cyclones used to sample for respirable dust, such as in determining compliance with the respirable mass standard for silica-containing dusts, should have performance characteristics meeting the criteria of the American Conference of Governmental Industrial Hygienists (ACGIH) (8). The orientation of the cyclone is not as critical as for the elutriators, and small 10-mm diameter cyclones have become commonplace for personal breathing zone sampling.

5.3 Direct-Reading Techniques

The development of methods for continuous monitoring of particles in air has been more substantial in the field of air pollution than in industrial hygiene. In general, instruments for continuous monitor have limitations on their sensitivity from two primary sources: the random property fluctuations of the accompanying gas molecules, and the noise level of the electronic circuitry, which converts the physical change to a measurable signal. Instruments that read out particle sizes are often calibrated using well-characterized aerosols not necessarily representative of the particles sampled; thus the accuracy of these instruments may be highly variable. The response of the instruments is nonspecific; that is, the devices respond to a property of the substance (size, shape, or mass), rather than to the material itself. However, there are needs of the industrial hygienist that can be met by direct-reading instruments for particulates, and continuing application of these devices is anticipated.

5.3.1 Aerosol Photometry (Light Scattering)

Instruments using light scattering techniques are based on generation of an electrical pulse by a photocell that detects the light scattered by a particle. A pulse height analyzer estimates the effective diameter. The number of pulses is related to the number of particles counted per unit flow rate of the sampled medium. Instruments that give a size analysis based on the measurement of total particle concentration in a large illuminated volume are used in monitoring particulate concentrations in experimental rooms or exposure chambers.

Aerosol photometry can usually provide only an approximate analysis of particulates, classified according to particle size, in industrial surveys. Calibrating the instrument with each type of particulate to be measured is not practicable. The great variations in shape, size, agglomerative effects, and refractive indexes of the various components in a given dust suspension make such calibrations virtually impossible. Whereas aerosol photometry can indicate the particulate concentration in the different size ranges of interest, it usually is necessary to perform size distribution analyses by microsieving or microscopic procedures.

5.3.2 Respirable Mass Monitors

The first direct-reading respirable mass monitor to be introduced to the industrial hygiene field was based on the attenuation of beta radiation resulting from collection of a sample of dust by taking comparative readings of a beta-radiation source. The instrument uses a two-stage collection system. The first stage consists of a cyclone precollector for the retention of the nonrespirable fraction of the dust. The precollector retains the larger particles and allows those of respirable size to pass to the second stage, which consists of a circular nozzle impactor and beta absorption assembly with an impaction disk. The dust collected by impaction on the thin plastic film increasingly absorbs the beta radiation from a ^{14}C source being monitored by a Geiger-Mueller detector. The penetration of low-energy beta radiation depends almost exclusively on the mass per unit area of the absorber and the maximum beta energy of the impinging electrons; it is independent of the chemical composition or the physical characteristics of the collected, absorbing matter. This unit, and modifications thereof, incorporate a digital readout of the respirable mass concentration.

A second type of unit is based on the change in resonant frequency of a piezoelectric quartz crystal accompanying deposition of particulate on the face of the crystal. This unit uses an impactor inlet to separate the dust into respirable and nonrespirable fractions. Modifications have utilized electrostatic precipitation for particle collection, as well as multistage impactors. The mass concentration-to-frequency relationship is linear, and the instrument has sufficient sensitivity to measure ambient particulate mass concentration in a few seconds, sampling at 1 to 3 L/min.

5.3.3 Limitations

Direct-reading instruments are not personal monitoring devices. Because the readings are instantaneous, they cannot be used directly to determine compliance with 8-h, time-weighted average standards. Thus such units are not be used as compliance instruments. However, the usefulness of direct-reading equipment for analyzing environmental conditions "on the spot" has been great, and continued development and refinement is likely.

6 ANALYSIS OF SAMPLES

It has become commonplace to expect that advances and improvements in analytical capabilities make it possible to measure extremely small quantities of specific compounds, ions, or elements. The industrial hygienist can process a very small sample of air and accurately determine the presence of suspected contaminants. As emphasized earlier in this chapter, the field industrial hygienist must work closely with the industrial hygiene analyst to become familiar with the limitations of the analytical equipment of interest, thus making it possible to plan a sampling strategy with maximum efficiency.

It is beyond the intent of this chapter to outline specific procedures for industrial contaminants. Instead, brief descriptions of the various analytical methods and techniques that have been applied to industrial hygiene samples are presented, with the expectation that the reader will consult more detailed sources for complete understanding of analytical requirements for particular substances of interest. Of particular note are the descriptions of analytical methods published by NIOSH and OSHA (14, 15).

6.1 Gravimetric Techniques

The most frequently employed analytical method for industrial hygiene samples is gravimetric analysis of filters or other collection media to determine the weight gain. This requires careful handling and processing of the media before collecting the sample, as well as the conditioning of the media after collection of the sample in the exact manner as used to obtain tare weights. In so doing, any necessary correction for the "blank" can be incorporated into the analysis. Often gravimetric analysis is done as a gross analysis, or general indicator of conditions, with subsequent analyses performed for specific constituents of the sample.

Another type of gravimetric technique involves the formation of a precipitate by combining a sample solution with a precipitating agent, with subsequent weighing of the solid precipitate formed.

6.2 Titrimetric Methods

Acid-base and oxidation-reduction volumetric procedures are outstanding examples of simple but useful analytical methods still employed in the analysis of industrial hygiene samples. Hydrogen chloride gas and sulfuric acid mist can be collected in an impinger or bubbler containing a standard sodium hydroxide solution and quantified by back-titration with a standard acid. Ammonia and caustic particulate matter can be collected in acid solution with similar apparatus, and the airborne concentration determined by titration with a standard base. Oxidation-reduction titrations, principally iodometric, are useful for measuring sulfur dioxide, hydrogen sulfide, and ozone. Improved volumetric methods utilize electrodes to indicate acid-base null points, and amperometric methods are available for oxidation-reduction titrations. These electrical techniques increase analytical precision and speed up the analyses but do not affect the sensitivity appreciably.

6.3 Optical Methods

Much of the sampling for dust requires analysis by microscopic techniques. The use of light microscopy for dust counting is decreasing and is being replaced by more specific, and more reproducible, mass sampling techniques. However, it is frequently necessary to determine the particle size distribution of airborne particulate, and optical methods offer an effective way of doing this.

In addition to the more classic counting applications of microscopy, the present sampling and analysis method for asbestos is based on actual fiber counting at 400 to 450× magnification using phase contrast illumination. As with all optical methods, the analytical results (i.e., actual counts) are somewhat analyst dependent, because much of the technique requires subjective analysis by the microscopist.

6.4 Colorimetric Procedures

Changes of color intensity or tone have been the bases of many useful industrial hygiene analytical methods. For example, the use of Saltzman's reagent in a fritted glass bubbler to determine the airborne concentrations of nitrogen dioxide is a classic application of such methods (22). Under controlled conditions of sampling, the concentration of NO_2 in the air is inversely proportional to the time required to produce the color change. Titrations employing acid-base and iodometric reactions with color indicators are conducted in similar fashion.

Usually such titrations of air samples lack the accuracy and precision obtainable with careful laboratory procedures, but they are adequate for most field studies and have the great advantage of giving a direct and immediate indication of the environmental concentrations. Relatively sensitive and specific analyses of many contaminants can be made, using the spectrophotometers available as both laboratory and field instruments.

Colorimetric methods involve analytical reactions to produce a color in proportion to the quantity of the contaminant of interest in the sample. For example, in the determination of metals, the dithizone extraction method is able to determine selectively the various metallic elements, depending on the pH of the solution.

6.5 Spectrophotometric Methods

In addition to the colorimetric procedures, which take advantage of spectrophotometry operating in the visible range, infrared and ultraviolet spectrophotometers have considerable application in the industrial hygiene area. The interaction of electromagnetic radiation with matter is the basis for such analytical techniques. Principles of operation extend from the infrared radiation spectra, to the ultraviolet and, in fact, to the X-ray region. The latter can be used to provide information on elemental composition (fluorescence) and crystal structure (diffraction).

In most cases the sample, whether gas, liquid, or solid, is exposed to radiation of known characteristics and specific wavelengths (fluorescence), and the fractions transmitted or scattered are determined and quantified. Color production, turbidity, and fluorescence are examples of properties determined by electromagnetic radiations that are widely used for quantifying industrial hygiene air samples.

6.6 Spectrographic Techniques

Because the smallest trace of materials can be detected by the spectrograph, spectrographic procedures may be employed for small amounts of metallic ions and elements when other procedures cannot be used. The chief limitations are the high cost and the need for a highly trained technician who has access to a rather complete spectrographic laboratory to do quantitative work. Generally the degree of sensitivity afforded by these units is not required by the industrial hygienist, although it frequently is desirable to obtain a complete elemental analysis of a sample of unknown composition as a starting point for an elaborate analytical program.

In applying emission spectroscopy, a solid sample is vaporized in a carbon arc, causing the formation of characteristic radiation, which is dispersed by a grating or a prism, and the resulting spectrum is photographed. Each metallic or metallike element can be identified from the spectra that are formed. Elemental analyses of body tissues, dust, ash, and air samples can be qualitatively analyzed by this technique.

With mass spectroscopy, the gases, liquids, or solids are ionized by passage through an electron beam. The ions thus formed are projected through the analyzer by means of an electromagnetic or electrostatic field, or simply by the time necessary for the ions to travel from the gun to the collector. Each compound has a characteristic ionization pattern that can be used to identify the substance. This analytical tool, in conjunction with gas chromatography, has become a powerful technique for separating and identifying a wide range of trace contaminants in industrial hygiene and ambient air samples.

6.7 Chromatographic Methods

The development of chromatographic methods of analysis has given the industrial hygienist an extremely versatile means of quantifying low concentrations of airborne contaminants, particularly organic compounds. Gas chromatography utilizes the selective absorption and elution provided by appropriately chosen packings for the columns to separate mixtures of substances in an air sample or in a desorption solution. The various compounds in the air sample have different affinities for the material in the column, thus slowing some of the constituents more than others, with the result that as the individual compounds reach the detector associated with the chromatograph, they can be quantified by running standards of known concentration of the various substances along with the unknowns. This separation is achieved without any appreciable change in the entities; thus the chromatograph can serve as an analytical technique in its own right by attaching an appropriate detector. Thermal conductivity, flame ionization, and electron capture detectors are commonly used for this purpose. The chromatograph can also serve to purify a sample by separating the constituents and selecting a narrow portion of the eluted sample. This portion then can be subjected to other more sophisticated types of analysis, such as mass spectroscopy.

6.8 Atomic Absorption Spectrophotometry

With atomic absorption spectrophotometry (flame photometry), monochromatic radiation from a discharge lamp containing the vapor of a specific element, such as lead, passes through a flame into which the sample is aspirated. The absorption of the monochromatic radiation is measured by a double-beam method, and the concentration is determined. This technique permits rapid determination of almost all metallic elements. Solutions of the metals are aspirated into the high-temperature flame, where they are reduced to free atoms. The absorption generally obeys Beer's law in the parts per million range, where quantitative determinations can be made. The characteristic absorption gives this technique high selectivity. Most interferences can be overcome by proper pretreatment of the samples. Atomic absorption methods have found substantial use in industrial hygiene in the determination of both major and trace metals in industrial hygiene samples, as well as in blood, urine, and other body fluids and tissues.

6.9 Other Techniques

Many additional, specialized analytical methods are available to the industrial hygienist and analytical chemist for application to specific qualitative and quantitative needs. For detailed information on particular methods or procedures, the reader is advised to consult the references listed at the end of this chapter and, more important, to keep abreast of current developments in the industrial hygiene analytical field by subscribing to journals or routinely reviewing the wealth of new information constantly coming forth.

As a general guide to the industrial hygienist, a summary of sampling and analytical techniques appropriate for a variety of contaminants commonly encountered in industry is presented in Table 8.4.

7 BIOLOGICAL MONITORING

The degree of success associated with attempts to provide a safe and healthful work environment must, in the final analysis, be determined by assessing the amount of the contaminant actually absorbed by the workers. Regardless of the degree of sophistication applied to environmental and personnel monitoring, the extent to which workers have absorbed the contaminant should be determined by some clinical measurement on the individual. In the industrial hygiene profession, this can be derived from direct quantitative analysis of body fluids, tissue, or expired air for the presence of the substance or metabolite. An indirect determination of the effect of the substance on the body can be made by measurements on the functioning of the target organ or tissue. With the possible exception of carcinogenic substances, even the most hazardous materials have some no-effect level below which exposure can be tolerated by most workers for a working lifetime without incurring any significant physiological injury.

An ideal approach to establishment of no-effect or tolerance levels would require a classical chemical engineering materials balance applied to humans and their environment. Simply stated, the amount of any substance entering the body must equal the products and by-products leaving the system, plus any accumulation within the body. A material balance such as this was established for lead by and through the efforts of Kehoe and his coworkers (25). The study involved quite elaborate test facilities for analyzing food, beverages, air intake, and urinary, fecal, and expired breath outputs, as well as a closely controlled environment in which volunteer subjects were willing to put in 40-h weeks under conditions that closely simulated actual work environments. The studies indicated that lead did not undergo metabolism to other forms within the body. This made the determination

Table 8.4. Sampling and Analytical Methods

	Sampling Medium					Analytical Method				
Substance	Charcoal/ Silica Gel	Direct-Reading Unit	Filter	Solvent/ Reagent	Std. Acid/ Base	Atomic Absorption	Gas Chromatography	Gravimetric/ Colorim.	Ion-selective Electrode	X-ray Diffraction
Acid mists					X				X	
Alcohols	X						X			
Ammonia					X				X	
Asbestos			X							X
Carbon monoxide		X								
Coal tar pitch volatiles			X					X		
Hydrocarbon	X						X			
Hydrogen sulfide				X				X		
Metals			X			X				
Oil mists			X					X		
Organic vapors	X						X			
Pesticides			X	X			X			
Phenol					X		X			
Silica			X							X
Solvents	X						X			

of the overall material balance relatively straightforward. Unfortunately, such is not the case with most occupational contaminants of interest today, particularly the organic materials. Material balances are much more difficult to establish for compounds that undergo metabolic change to other chemical structures within, and during passage through, the human body.

7.1 Urine Analysis

Historically, urine is perhaps the most common biological fluid analyzed in attempts to determine the extent to which individuals have been exposed to contaminants. As illustrative of urine analysis, for urinary phenol samples, spot urine specimens of about 10 mL are collected as close to the end of the workday as possible. For other contaminants, it might be preferable to collect the specimen at the beginning of the work shift or perhaps composite a 24-h sampling. If the level of the analyte in any worker's urine exceeds the biologic exposure index, procedures should be instituted immediately to determine the cause of the elevated levels and to reduce the exposure of the worker. Weekly, or other appropriate interval, specimens are collected as described until three consecutive determinations indicate that the urinary phenol, or other analyte, levels are below the prescribed limit.

To collect the sample, workers should be instructed to wash their hands thoroughly with soap and water and to provide a urine sample from a single voiding into a clean, dry specimen container having a tight closure and at least 120-mL capacity. The containers may be of glass or of wax-coated paper or other disposable materials. After collection of the specimens, 1 mL of a 10% copper sulfate solution, or other appropriate preservative, should be added to each sample. The samples should be stored immediately under refrigeration, preferably below 4°C. The stability of specimens for various metabolites or contaminants will vary; this factor should be ascertained before establishing a biological monitoring program. If shipment of samples is necessary, the most rapid method available should be employed, using acceptable packing procedures specified by the carrier. Each specimen must include proper identification, including the worker's name, the date, and time of collection.

Although the specific requirements of methods for specific analytes may differ significantly from those outlined above, it should be emphasized that the specific concentrations indicating excessive buildup of contaminants or metabolites within the body are not known for most compounds with any certainty. The biological threshold limit values, that is, concentrations indicative of excessive exposure for some compounds, have been determined; biological exposure indexes are published annually by the ACGIH (8).

7.2 Blood Analysis

Collection of samples of workers' blood and subsequent analysis for specific indications of exposure, either for the contaminant in question directly or for key metabolites, is another useful biological monitoring technique. A typical procedure for collecting and analyzing blood samples, based on lead-in-blood determinations, is presented as an illustration of the technique for this method.

A 10-mL sample of whole blood is collected using a vacutainer and a sterilized, stainless steel needle. It is important that the vacutainers have been determined to be free of the contaminant of interest, because any leaching of the material from the vacutainers into the blood samples would distort the analytical result. The possibility of such distortion has been of particular concern in the analysis of blood samples for lead. In the laboratory, the sample is transferred to a tared, lead-free beaker; no aliquoting of the sample is permissible, because most of the lead is present in the clotted portion. The weight of the blood sample is determined to the nearest 0.01 g, weighing rapidly to minimize evaporation of the sample. A system for ashing the sample is employed to permit the analyst to handle the sample easily, for the blood clot will break up readily and smoothly. The sample then is evaporated just to dryness, and after cooling, additional acid is added and ashing continued until the residue fails to darken upon additional heating. The residue then is put into solution and analyzed directly by atomic absorption spectrophotometry or other suitable method.

Another useful application of blood analysis as an index of exposure to contaminants is the determination of carboxyhemoglobin in the blood as an indication of exposure to carbon monoxide. Extensive studies of the concentration of carbon monoxide in the air and consequent level of carboxyhemoglobin in the blood have been made (24). Figure 8.4 presents a family of curves giving the length of time of exposure to various concentrations of carbon monoxide in air necessary to achieve 5% carboxyhemoglobin levels in the blood as a function of the relative activity of the workers, that is, sedentary, light work, or heavy work.

7.3 Expired Breath Analysis

A biological monitoring technique that is increasing in application is analysis of expired air for contaminants for which equilibrium between the body and respired air can be used as an indication of the concentration of the contaminant in the workroom air to which the individual has been exposed. For the most part, this type of analysis has been limited to the chlorinated hydrocarbon solvents such as methylene chloride, carbon tetrachloride, vinyl chloride, 1,1,1-trichloroethane, trichloroethylene, tetrachloroethylene, and some of the freons (26).

In collecting the sample, the worker is instructed to take several deep breaths and direct the expired air into a flexible bag or to pass it through a glass tube after flushing the contents several times with expired air. In either case, the container is tightly sealed and the contents subsequently analyzed directly for the substance, usually using gas chromatographic techniques. The analytical result is compared to breath decay curves, such as in Figure 8.5, to determine the concentration of the contaminant to which the worker had been exposed recently (25).

7.4 Future Considerations

With the continuing interest of the environmental/safety/health professional in providing safe work conditions for persons exposed to a tremendously wide range of contaminants,

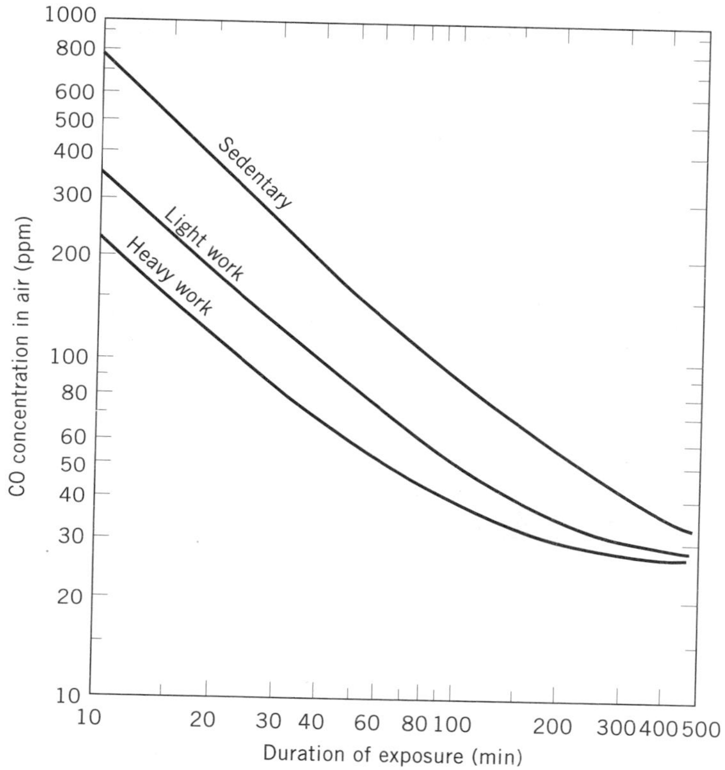

Figure 8.4. Length of time to achieve 5% carboxyhemoglobin at various concentrations of carbon monoxide (25).

development of cause-and-effect relationships between the level of exposure to a particular contaminant and the amount of the substance remaining in the body will become a more integral part of the total occupational health monitoring procedure. Accordingly, greater emphasis is being placed on the medical, and even epidemiological, aspects of the total monitoring effort. The ultimate objective of this cooperative effort is to document, as the end result, that workers have not been exposed to excessive levels of contaminants. When environmental and biological data are inconsistent, that is, where results of biological monitoring indicate excessive exposures, and the environmental monitoring indicates concentrations are within acceptable limits, analysis of individuals' work practices, or at least specific analysis of an individual's work activity, may be required to ascertain the cause of the elevated readings in a given individual. As such, the biological monitoring program is a viable and extremely useful supplement to the ongoing environmental and medical surveillance programs for contaminants for which such coordinated efforts are possible (27, 28).

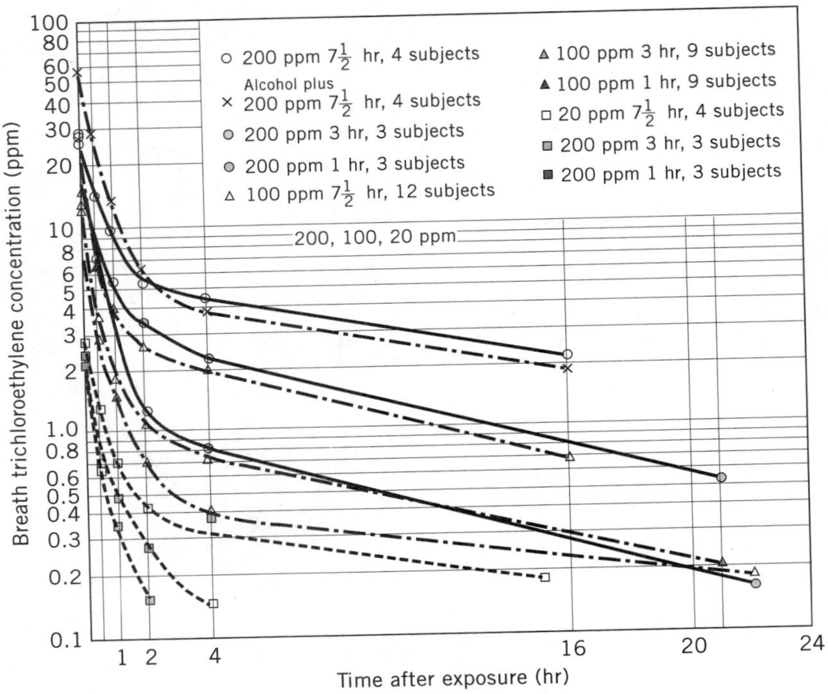

Figure 8.5. Trichloroethylene postexposure breath decay curves.

BIBLIOGRAPHY

1. G. D. Clayton and F. E. Clayton, Eds., *Patty's Industrial Hygiene and Toxicology*, Vol. 2, 4th rev. ed., Wiley Interscience, New York, 1991.
2. M. B. Jacobs, *The Analytical Toxicology of Industrial Inorganic Poisons*, Wiley Interscience, New York, 1967.
3. H. B. Elkins, *The Chemistry of Industrial Toxicology*, 2nd ed., Wiley, New York, 1959.
4. A. Hamilton and H. Hardy, *Industrial Toxicology*, 3rd ed., Publishing Sciences Group, Acton, MA, 1974.
5. C. D. Klaassen, Ed., *Casarett and Doull's Toxicology: The Basic Science of Poisons*, 5th ed., McGraw-Hill Health Professions Division, New York, 1996.
6. P. L. Williams and J. L. Burson, *Industrial Toxicology: Safety and Health Applications in the Workplace*, Van Nostrand Reinhold, New York, 1985.
7. E. Hodgson, R. B. Mailman, and J. E. Chambers, *Dictionary of Toxicology*, Van Nostrand Reinhold, New York, 1988.
8. American Conference of Governmental Industrial Hygienists, *Documentation of the Threshold Limit Values and Biologic Exposure Indices*, 6th ed., ACGIH, Cincinnati, OH, 1991.
9. International Labour Office (ILO), *Occupational Exposure Limits for Airborne Toxic Substances*, 3rd ed., Geneva, Switzerland, 1991.

10. Y. Bar-Shalom, A. Segall, D. Budenaers, and R. B. Shainker, *Statistical Theory for Sampling of Time-Varying Industrial Atmospheric Contaminant Levels*, Systems Control, Inc., Report to NIOSH Contract HSM-99-73-78, Palo Alto, CA, 1974.
11. D. Budenaers, Y. Bar-Shalom, A. Segall, and R. B. Shainker, *Handbook for Decisions on Industrial Atmospheric Contaminant Exposure Levels*, U.S. Department of Health, Education and Welfare, Cincinnati, OH, 1975.
12. N. A. Lediel, K. A. Busch, and J. R. Lynch, *Occupational Exposure Sampling Strategy Manual*, National Institute for Occupational Safety and Health (NIOSH), Cincinnati, OH, 1977.
13. C. A. Esche, J. H. Groff, P. C. Schlecht, and S. A. Shulman, *Laboratory Evaluations and Performance Reports for the Proficiency Analytical Testing (PAT) and Environmental Lead Proficiency Analytical Testing (ELPAT) Programs*, DHHS/NIOSH Pub. No. 95-104, National Institute for Occupational Safety and Health, Cincinnati, OH, 1994.
14. Occupational Safety and Health Administration, *OSHA Technical Manual*, 4th ed., U.S. Government Printing Office, Washington, DC, 1996.
15. Occupational Safety and Health Administration, *OSHA Analytical Methods Manual*, Vols. 1–4, OSHA Technical Center, Salt Lake City, 1993.
16. N. C. Hawkins, S. K. Norwood and J. C. Rock, Eds., *A Strategy for Occupational Exposure Assessment*, AIHA, Akron, OH, 1991.
17. F. J. Schuette, "Plastic Bags for Collection of Gas Samples," *Atmos. Environ.*, **1**, 515 (1967).
18. G. O. Nelson, *Controlled Test Atmospheres, Principles and Techniques*, Ann Arbor Sciences Publishers, Ann Arbor, MI, 1971.
19. B. S. Cohen and S. V. Herings, Eds., *Air Sampling Instruments for Evaluation of Atmospheric Contaminants*, 8th ed., ACGIH, Cincinnati, OH, 1995.
20. B. A. Plog. J. Niland, and P. J. Quinlan, Eds., *Fundamentals of Industrial Hygiene*, 4th ed., National Safety Council, Chicago, IL, 1992.
21. American Industrial Hygiene Association, *Direct Reading Colorimetric Indicator Tubes Manual*, AIHA, Akron, OH, 1986.
22. B. E. Saltzman, "Direct Reading Colorimetric Indicators, Section S," in *Air Sampling Instruments for Evaluation of Atmospheric Contaminants*, 4th ed., American Conference of Governmental Industrial Hygienists, Cincinnati, OH, 1972.
23. P. M. Eller and M. Cassinelli, Eds., *NIOSH Manual of Analytical Methods*, 4th ed., DHHS/NIOSH Pub. No. 94-113, U.S. Department of Health and Human Services, Cincinnati, OH, 1994.
24. OSHA Analytical Methods Manual, *OSHA Analytical Laboratory*, Salt Lake City, UT, 1985.
25. A. Linch, *Biological Monitoring for Industrial Chemical Exposure Control*, CRC Press, Cleveland, OH, 1974.
26. R. D. Stewart et al., *Biological Standards for the Industrial Worker by Breath Analysis: Trichloroethylene*, U.S. Department of Health, Education and Welfare, Cincinnati, OH, 1974.
27. Que Hee, S. S., Ed., *Biological Monitoring: An Introduction*, Van Nostrand Reinhold, New York, 1993.
28. R. R. Lauwerys and P. Hoet, *Industrial Chemical Exposure: Guidelines for Biological Monitoring*, 2nd ed., Lewis Publishers, Boca Raton, FL, 1993.

CHAPTER NINE

Assessment of Exposures to Pneumoconiosis-Producing Mineral Dusts

Howard E. Ayer, CIH and Carol H. Rice, Ph.D., CIH

1 INTRODUCTION

Pneumoconiosis is produced by significant long-term exposure to certain mineral dusts of particle sizes that can penetrate to the gas exchange region of the lungs. Thus, assessment of exposure may include size separation as well as measurement of concentration and composition of the dusts. When the dust is fibrous, as the various forms of asbestos, the assessment is by collection of air samples, light microscopic fiber counts of the samples, and possibly electron microscopic identification of the fiber types. For all other minerals with particles of regular shapes, current methods of assessment use air sampling devices that discriminate against particles too large to penetrate to the sensitive regions of the lung. The dust collected is then identified as the "respirable" fraction, using an international definition of the fraction. Chemical or physical analysis of the samples may then be required.

Assessment also includes the length and pattern of exposure. For estimation of pneumoconiosis hazard the cumulative exposure over a period of days, weeks, months and years is considered. Air sampling for hazard estimation is typically done over full working shifts. For design of efficient control measures those tasks or operations that produce the major portions of the exposures must be identified. To accomplish this identification, the minute by minute exposure pattern may be required.

Patty's Industrial Hygiene, Fifth Edition, Volume 1. Edited by Robert L. Harris.
ISBN 0-471-29756-9

The principal pneumoconiosis-producing dusts as they are now recognized in the United States are

1. Those containing free crystalline silica (quartz or cristobalite).
2. Coal dust (both bituminous and anthracite).
3. Asbestos (chrysotile, amosite, crocidolite, anthophyllite and tremolite).

Other minerals may also produce lung fibrosis, but usually only at very much higher concentrations.

As with other physical and chemical hazards, it is impossible to predict with certainty the precise hazard to a particular individual. The degree of pneumoconiosis hazard from respirable coal dust depends not only on the cumulative exposure, but in a not well understood manner to the chemical and mineralogical composition of the seam being mined. The pneumoconiosis hazard from respirable quartz is greater from freshly fractured particles than from "aged" dust. Furthermore, the hazard from quartz-containing dust is modified upwards or downwards by the other components of the dust. To these differences in collective hazard is added the differences in individual susceptibility of the people exposed. For this reason, the concentration limits and measurement methods are intended to protect the more susceptible individuals against the agents with greater potential to cause disease. Further changes in the limits and measurement methods may be expected, but if experience is a guide, they will be decade by decade rather than year by year.

This chapter describes the methods in use at the beginning of the twenty-first century, and the development of these methods over a major part of the twentieth century.

2 REGULATORY REQUIREMENTS

The largest numbers of samples for assessment of hazard are taken to satisfy governmental requirements. In such cases, usually the type, location and frequency of sampling are specified. In both the United States and United Kingdom, for example, the estimation of underground coal miner exposure has been specified in explicit detail for many years. Although mining methods have caused the instruments and sampling locations selected in the two countries to be different, both now use the international standard for respirable dust. Both countries use full shift measurements, typically for eight hours per sample.

For potentially silicosis-producing exposures to dusts other than coal dust, governmentally specified measurement methods may be less explicit. However, if exposure measurement is required it will generally be for most, if not all, of a full shift, using instruments that meet the international size-selective criteria for respirable dust. In many cases, personal sampling will be required. It is neither feasible nor desirable to take a full-shift personal sample on every potentially exposed worker every day. Where there are governmental requirements, these may specify what proportion of workers in which operations at what frequency must be sampled. Lacking governmental requirements, an industrial hygienist has the task of balancing the degree of potential health hazard, economic feasibility, and statistical reliability in choosing where to sample, who to sample, and how often to take samples.

Measuring the dust exposure of workers involves consideration of the temporal and spatial differences between exposures, affecting the intra- and inter-worker differences in daily and average exposures. The exposure assessment strategies have been outlined by Lippmann (1). Likewise, measuring the effectiveness of control measures requires quantitation of dustiness in the air of the workplace. The degree of health hazard, if any, will depend upon the composition and size distribution of the airborne dust. For design of control measures information on size distribution may be necessary. For concentration, composition and size distribution the measurement methods should, ideally, be accurate, precise, rapid, simple and inexpensive. At no time have the measurement methods available met all these requirements. There are methods for estimating the size distribution of the dust to which the workers may be exposed. There are methods for measuring the temporal variation of the dust exposures. Given an unlimited budget and adequate time, virtually any aspect of dust exposure may be measured. The task of the occupational hygienist is to use those methods at his/her disposal to get the best information available that answers the question at hand within the constraints of time and money.

Regulatory agencies in the United States include the Occupational Safety and Health Administration (OSHA), the Mine Safety and Health Administration (MSHA), a State agency acting under federal guidelines, or a successor to one of these agencies. The United States Environmental Protection Agency (EPA) may also have procedures in place for measurements of pneumoconiosis-producing dusts, but in general the concentrations involved are far less than those that are capable of causing pneumoconiosis. Where the measurements are made to satisfy a particular agency's requirements, the measurement method may be specified in some detail, and must be followed strictly. Fortunately, international standards are resulting in the coming together of many national methods. Unfortunately, there may be a considerable time lag between consensus on an appropriate measurement method and the adoption of that method by the State agency involved. Particularly in the United States, the possibility of civil litigation far in the future means that not only must the current regulatory agency method be considered, but the method generally believed to be the best currently available must be in place.

3 ASSESSMENT FOR CONTROL MEASURE DESIGN

Regulatory compliance measurements are generally made to estimate the pneumoconiosis hazard to particular individuals if the dust concentrations observed over one or more working shifts are added to those previously accumulated. Measurements may also be made to determine progress toward goals in control. These control progress measurements may use methods that are an index of relative dustiness for the particular operation or circumstance and are selected for convenience, cost, compatibility with previous measurements, real time measurement, or the ability to keep a continuous record.

4 RESPIRABLE DUST

The respirable fraction of dust is one of three fractions that have been internationally accepted for particulate air sampling. It is the fraction used in assessing pneumoconiosis hazard. The fractional criteria, initially proposed by Soderholm (2), are now accepted by the American Conference of Governmental Industrial Hygienists (ACGIH), the European Community or Comite Europeen de Normalisation (CEN) and the International Standards Organization (ISO) The definitions of "inspirable", "thoracic" and "respirable" dust fractions are as follows: For the inspirable fraction,

$$SI(d) = 0.5(1 - e^{-0.06d})$$

For the thoracic fraction,

$$ST(d) = SI(d)[1 - F(x)]$$

where $x = [\ln(d/\Gamma)]/[\ln(\Sigma)]$
$\Gamma = 11.64\ \mu m$
$\Sigma = 1.5.$
$F(x) =$ the cumulative probability of a standardized normal random variable

For the respirable fraction $SR(d) = SI(d)[1 - F(x)]$ where $\Gamma = 4.25 \mu m$, $\Sigma = 1.5$. For the above d is particle aerodynamic diameter, $SI(d)$ is the sampling efficiency of an ideal collector of the inspirable fraction (also called the inhalable fraction), $ST(d)$ is the sampling efficiency of an ideal collector of the thoracic fraction, $SR(d)$ is the sampling efficiency of an ideal collector of the respirable fraction, Γ is a parameter with units of length that is sometimes used when calculating the thoracic or respirable fraction, and Σ is a unitless parameter that is sometimes used when calculating the thoracic or respirable fraction. A particle's aerodynamic diameter is the diameter of a unit density sphere with the same settling velocity in still air as the particle.

4 RESPIRABLE MASS SAMPLING

4.1 Horizontal Elutriation

For monitoring hazards from pneumoconiosis-producing dusts, a relatively simple method of monitoring is required to assure wide use. Instruments such as the cascade impactor (3) separate the dust into size fractions. The various stages can be then be analyzed, and that fraction likely to penetrate to susceptible areas of the lung may be calculated. A method requiring a number of microscopic or chemical analyses for each sample is mainly useful for research purposes. For field use, a relatively simple method for collecting that fraction of the dust that could penetrate to the lung is required. The National Coal Board, responsible for the production of coal in Great Britain from shortly after World War II until privatization under the Tory government, conducted research into the evaluation and control of coal workers' pneumoconiosis. The scientists in the National Coal Board worked in close collaboration with the Pneumoconiosis Research Unit of the (British) Medical Research Council. They were engaged in developing a sampling device for measuring respirable mass of workplace dust. Mass sampling of respirable dust requires use of a two-stage sampling device, the first stage of which is a separator for removing from the sampled air those particles larger than those defined as respirable (i.e. remove particles too large to penetrate to the gas exchange portions of the lungs), and a second stage, usually a filter, which collects for weighing the respirable sized particles which penetrate the first stage. The theory of elutriation applicable to respirable mass sampling of industrial dust clouds developed by W. H. Walton of the National Coal Board was communicated by him to the Medical Research Council in 1950, and published in 1954 (4). He outlined advantages of horizontal elutriation as the first steps of a respirable mass samples, demonstrating that it was capable of giving as sharp a cutoff as other elutriation processes. He demonstrated that for a flow tube in a horizontal elutriator,

$$P = 1 - fA/Q$$

where P = the fractional penetration f = the settlement speed of the particle size in question A = the total floor area Q = the total flow. His theory also covered circular ducts, ducts of nonuniform cross section, and multiple ducts of different shapes. In practice, the overriding advantages of multiple horizontal ducts of uniform cross section, as illustrated in his paper, led to their being the only type constructed. A major advantage of elutriation is that the size selecting device can be designed using known physical principles for the desired cutoff. The cutoff selected, primarily on the basis of analysis of data such as that from Brown, et al. (5), was 50% at 5 μm aerodynamic diameter. The theory, closely followed by the samplers, gives zero penetration of 7.07 μm aerodynamic diameter. The research of these groups led to the recommendation by Panels of the (British) Medical Research Council Industrial Pulmonary Diseases Committee (6) that

> for purposes of estimating airborne dust in relation to pneumoconiosis, samples for . . . assessment of concentration for a bulk measurement such as that of mass . . . should represent only the 'respirable' fraction of the cloud and that "the 'respirable fraction' is to be defined in terms of the free falling speed of the particles, by the equation $C/C_0 = 1 - f/f_c$ where C and C_0 are the concentrations of particles of falling speed f in the 'respirable' fraction and in the whole cloud, respectively, and f_c is a constant equal to twice the falling speed in air of a sphere of unit density 5μ dia.

Also recommended by the Pneumoconiosis Conference in Johannesburg, South Africa in 1959, this definition of respirability became known as either the Johannesburg convention or the British Medical Research Council (BMRC) criteria. The first commercial instrument designed according to these criteria was reported in 1954 by Wright (7). His device used 118 channels 0.8 cm deep, formed by two sets of 58 aluminum plates in a rectangular box separated by steel wires. The size separating section was 25.1 cm long, and each set of plates was 3.55 cm wide, giving a total floor area of 10,900 square cm. According to theory, with this floor area and 100 liters per minute total air flow, only 2% of 1 μm aerodynamic diameter dust would settle, but 50% of the 5 μm size and 100% of the 7.07 μm aerodynamic diameter dust would be removed. The "respirable" dust passing the elutriating section was collected in a Soxhlet thimble, with the airflow developed by a compressed air ejector. The horizontal elutriator with the Soxhlet filter led to naming the device the Hexhlet. With a total length of less than 35 cm and a cross section of about 10 cm by 10 cm, the Hexhlet was readily portable, though requiring a source of compressed air for its use. It was soon used for obtaining comparisons with thermal precipitator sample results in British collieries participating in the National Coal Board's Pneumoconiosis Field Research (8). For various reasons, including the unavailability of compressed air at desired sampling sites, the Hexhlet was determined to be unsuitable for routine sampling in the British coal mines. However, as the only unit readily available for respirable dust sampling, it became the *de facto* standard for use in other industries and with which other respirable dust sampling devices were compared, particularly in the United Kingdom (9–12).

As noted above, the Hexhlet did not meet the needs of sampling in the British coal mines. The coal mines in Great Britain use the long wall system of mining. In this system, the coal is mined along a face that commonly exceeds 100 meters in length, with tunnels at either end for access. Fresh air enters from one end and the contaminated air is withdrawn from the other end. As the face advances, the miners move along its length, so sampling in the outlet air can result in an estimate of their average dust exposure. For research and dust control, sampling by the thermal precipitator instruments for dust count had been done in this outlet air, and an instrument was needed that could replace these devices. This led to the development by the Mining Research Establishment of a unit about the size and weight of an electric flat iron (about 23 cm long, 11 cm wide and 18 cm high including handle, weighing 3.7 kg) (13). It used a four channel horizontal elutriator for size selection, sampled onto a 5.5 cm glass fiber filter, with airflow provided by a diaphragm pump driven by an intrinsically safe motor, powered by a 6 V, 900 ma-h rechargeable nickel–cadmium battery. It sampled at a constant rate of 2.5 L/min. for a period of up to 10 hours. Still in use, it is currently known as the MRE 113 A sampler. The instrument was used as a standard for dust measurement by the United States Congress in the 1969 Coal Mine Act. The system of dust measurement 70 m downwind from the face in British long wall mines was revaluated in 1993 by the Institute of Occupational Medicine, with a recommendation that it be continued (14).

Horizontal elutriators, with their inherent ability to design for any aerodynamic size and airflow, have been used for many research purposes. The conversion from an impinger count to a respirable mass threshold limit value (TLV), for example, relied on data partly produced by a custom-built 10 L/min. horizontal elutriator (15). Knight has described a stepped horizontal elutriator that may be used as a research standard to match the current ISO/CEN/ACGIH criterion for respirable dust sampling (16).

4.1.2 Cyclone Size-Selectors

In discussing the advantages of horizontal elutriation, Walton had mentioned the potential use of cyclone separators to remove that fraction of dust that was not respirable. He rejected them because the performance of a given design would not be predictable. However, in 1953, Harris and Eisenbud, of the Health and Safety Laboratory, New York Operations Office, Atomic Energy Commission (AEC), described the potential use of a commercially available double inlet, 2 inch diameter cyclone, the Aerotec 2, to achieve a respirable dust sample (17). As calibrated by the Harvard Air Cleaning Laboratory, the Aerotec 2 closely matched the lung deposition curve of Brown, et al (5) when operated at 0.019 m^3/s

(40 cfm). The AEC was interested in such a device for isolating those radioactive isotopes that could reach the gas exchange region of the lung. Further work at the Health and Safety Laboratory resulted in the development of a family of cyclones that were more applicable to field sampling. These were reported by Lippmann and Harris (18) along with a curve for respirable dust sampling presented by Theodore Hatch at a 1961 Los Alamos meeting. The percent passing "the Los Alamos curve" was described by five points. Percentage of aerodynamic size particles passing a first stage collector was 100% at 2 μm, 75% at 2.5 μm, 50% at 3.5 μm, 25% at 5 μm and 0% at ≥10 μm. As initially calibrated by microscopic particle sizing, a ½-in. cyclone modeled on the Aerotec design met the Hatch criteria at 18 L/min., a 1-in. diameter version at 240 L/min., and a commercially available Dorr-Oliver 10-mm nylon hydroclone at 2.8 L/min. Data were presented on sampling in uranium refining and fuel element fabrication and in beryllium refining. Respirable dust sampling appeared to have advantages for prediction of possible health effects of both uranium and beryllium. However, probably for administrative reasons, respirable dust standards for beryllium and uranium were never adopted. A commercial version of the ½ inch cyclone was used by Bloor et al. to do personal sampling with a helmet mounted sampler and an external pump connected by a hose (11). The commercial version of the ½-in. cyclone was also used by Reno et al., along with horizontal elutriators in the first attempt to measure a respirable mass equivalent to the American Conference of Industrial Hygienists count TLV for silica dust (19).

An advantage of cyclones over horizontal elutriators is that they are, to some extent, self-compensating for flow rate. If the flow rate of an horizontal elutriator is increased above the rated flow rate for the cutoff desired, not only is the amount of air sampled increased, but a coarser fraction is obtained, thus magnifying the error. For a cyclone separator, on the other hand, as the flow rate increases a finer fraction is obtained. In chamber sampling of a polydisperse size dust with a cyclone size separator, for example, one may observe virtually the same weight passing as the flow rate is increased or decreased by twenty percent or more. Although this does not mean that flow rate is unimportant, it does mean that a significant difference in a respirable mass obtained from dust passing a cyclone is unlikely to have been caused by a flow rate error.

In 1960, shortly before Lippmann and Harris reported the family of cyclones, Sherwood and Greenhalgh described a personal sampler for radioactive dust (20). This used a battery-powered diaphragm pump that could be worn on the belt and connected to a sampling head on the worker's shoulder. Personal sampling pumps of similar design soon became commercially available in the United Kingdom and the United States.

4.2 Personal Respirable Dust Sampling Using Cyclones

4.2.1 The 10-mm Nylon Cyclone

Following a missionary visit by the British scientist W. H. Walton, research on size-selective sampling for respirable silica dust was conducted in the early 1960s by engineers and chemists at the Cincinnati facility of the National Institute for Occupational Safety and Health (NIOSH). [It was not known as NIOSH until christened by the U.S. Congress, but it had existed in the U.S. Public Health Service as the Office of Industrial Hygiene,

Division of Industrial Hygiene, Division of Occupational Health, Occupational Health Program, or Bureau of Occupational Safety and Health since 1912 (21)]. Rather than attempt to keep track of the name and organizational changes, it will generally be referred to in this chapter as NIOSH, although that acronym meant nothing until 1970. In this laboratory, Talvitie and Hyslop (22) had developed a colorimetric method for free silica analysis that took analytical sensitivity from the milligram to the microgram range. The 10-mm Dorr-Oliver hydroclone suggested for air sampling by Lippmann was mated to 37-mm membrane filter cassettes, using a commercially available pump and Talvitie's colorimetric silica analysis method. This system was first used for personal size-selective silica dust sampling in measurements on heavy equipment operators during construction of the California aqueduct. The results demonstrated that although the soil (bulk sample) was high in quartz, the respirable dust was not, leading to a significant difference in the perceived potential silicosis hazard. The first published reports were of comparative impinger-respirable mass comparisons in Michigan foundries (23) as part of the development of the respirable mass TLV (24) for quartz.

These earliest NIOSH personal respirable mass samplers were used without pulsation damping of the diaphragm pump airflow. The intermittent flow resulted in a sharper size cutoff than steady flow. Before general adoption of the system the pulsation problem was recognized (25–27) and commercial sampling instruments built to meet coal mine dust sampling regulations included flow smoothing devices.

Initially reported air flows by Lippmann (18) for respirable dust sampling with cyclones were found to be too high. Recalibration by Lippmann and others (28–31) gave results varying from 1.4 L/min. to 2.0 L/min. for the 10-mm nylon cyclone. A guide produced by the Aerosol Technology Committee of the American Industrial Hygiene Association recommended a flow rate of 1.7 L/min. for the 10-mm cyclone (32). The rationale for the guide, including both air sampling principles and physiological relevance was presented by Lippmann (33). For quartz-containing dusts, the 1.7 L/min. flow rate with the 10-mm nylon cyclone as size separator was used by the Occupational Safety and Health Administration (OSHA), NIOSH and the general U.S. industrial hygiene community until the adoption of the current international standard for size selection. Based on experiments in the 1990s, Bartley, et al. concluded that the 10-mm nylon cyclone best met the international standard at 1.7 L/min. (34).

For coal dust sampling in United States mines, the Coal Mine Health and Safety Act of 1969 mandated an exposure limit in terms of the MRE horizontal elutriator sampler, an instrument impractical for use with room and pillar mining methods in the United States. The Bureau of Mines, therefore, performed extensive comparison sampling in coal mines and determined that the appropriate flow in the 10-mm nylon cyclone was 2.0 L/min. (35). For coal mining samples the 2.0 L/min. flow rate was used.

4.2.2 British Cyclones

In the United Kingdom, a personal sampling cyclone developed under the auspices of the British Cast Iron Research Association was reported in 1968 by Higgins and Dewell (12). They calibrated their cyclone to meet the BMRC criterion of 50% passage of 5 μm aerodynamic diameter particles.

Also in 1968, Harris and McGuire described a personal sampler, the SIMPEDS, built into a miner's cap lamp and battery assembly (36). Their device included a cyclone and filter holder and used the miner's cap-lamp battery as its source of power. Their preliminary results suggested that the device, sampling at 1.85 L/min., oversampled by about 10% as compared with the MRE horizontal elutriating instrument. The following year, however, they reported that preliminary underground mine trials gave SIMPEDS results not significantly different, or less than those estimated by MRE samples at the routine sampling point in the return airway (37). As a more compact sampler than the BCIRA device, the SIMPEDS became commercially available. Verma et al., compared the BCIRA, SIMPEDS, and 10-mm nylon cyclone samplers and the MRE instrument in an underground uranium mine and found the three cyclone results to be highly correlated with one another (R^2 = 89% to 98%), although the BCIRA sampler gave results 16% higher than the SIMPEDS, a statistically significant difference (38).

4.3 Personal Respirable Dust Sampling with Porous Foam

Porous foam was suggested by Brown as an alternative to the miniature cyclone as a size selector (39). Gibson and Vincent tested the use of porous foams as a prefilter on a respirator so that a backup membrane filter could be used to collect the respirable dust to which a worker might be exposed (40). Chen et al., by using parallel foams, were successful in producing a penetration curve that "nearly matched the new international standard for respirable fraction" when using dioctylphthalate. They were unsuccessful in matching the curve with their solid particulate, potassium sodium tartrate (41).

4.3.1 The CIP 10

A porous foam personal sampler was developed at the Centre d'Etudes et Recherches de Charbonnages de France (CERCHAR), and became commercially available. This battery powered sampler, the CIP 10, using a rotating foam cup as the air mover and filter, drawing air through an impinger and a static porous foam was described in 1984 (42). In a later publication data were presented from tests with coal dust and aluminum oxide, concluding that the CIP 10 instrument is a very compact air sampling instrument, incorporating its own source of power, suitable for measuring the respirable (alveolar) fraction of ambient solid aerosol at places of work. The combination of static polyurethane foam of grade 45 pores for the main particle size selection stage, and rotating polyurethane foam of grade 60 makes the instrument match the conventional definitions of respirable (alveolar) particles. The long running time and the relatively large flow rate for its size, 101. $min.^{-1}$, increase its capabilities. The upper impaction cup, trapping most of the coarse particles, makes possible a long working time without significant change of the selection properties of the static foam (43).

Gero and Tomb evaluated the CIP 10 in the laboratory with Arizona road dust, coal and silica aerosols (44). Their final paragraph was:

> Although the CIP 10 provides the user with several advantages by collecting a larger sample mass, eliminating connecting tubing which could become caught or kinked, and reducing

orientation effects on particulate collection, it fails to sample according to any of the adopted respirable dust criteria. If measurements made with the CIP 10 are to be compared with standards based on the accepted respirable dust criteria, the relationships between measurements made with this instrument and measurements which meet the criteria must be determined.

5 CRYSTALLINE FREE SILICA ANALYSIS

Very few of the worker exposures to quartz or cristobalite are to dusts of almost pure silica. Most are to dusts containing from 5–75% free silica. At very low percentages of crystalline free silica in a dust any health effects will be from other constituents. Though the threshold for silica effect is indeterminate, a common practice has been to consider the other constituents for limits when the dust contains less than 5% quartz. Assessing silicosis hazard in terms of the respirable mass concentration of silica involves analysis of the respirable quartz and/or cristobalite on the membrane filter in a cyclone/filter personal sampler. Sampling rates for the cyclones principally used at this writing are at, slightly above or slightly below 2 L/min. At these rates a nominally full-shift sample will result in less than 100 μg of quartz/cristobalite when the concentration is at a limit of 0.1 mg/m^3; less than 50 μg if at a limit of 0.05 mg/m^3. When the sampling time is much less than a full shift the quantity of crystalline silica representing the limit is correspondingly less. In the mid-1960s, the analytical method readily available in most analytical laboratories was the colorimetric differential solubility method of Talvitie (22). That method, requiring extraordinary analytical skill, is still practiced by some of the laboratories participating in the Proficiency Analytical Testing program now operated by the American Industrial Hygiene Association, although most use either an infrared spectrometry or x-ray diffractometry method. Limit of detection for each of the methods is about 5 μg, but accuracy is poor, especially at filter loadings of 30 μg or less (45). The colorimetric method has, over the years, tended to give somewhat lower results than the other two methods, particularly at higher loadings, where results average 20 μg lower. However, silica analysis results have improved in recent years for all three methods (46). Personal sampling pumps have improved over the years, and alternatives to the 10-mm nylon or the Higgins-Dewell cyclone are increasingly considered. With cyclones designed for higher flow rates personal respirable dust sampling can produce larger air samples, so that the difficulties resulting from operating so close to the limit of detection will decrease. One advantage noted for the CIP 10 porous foam sampler is the higher flow rate and thus greater mass of silica collected.

6 THE MEMBRANE FILTER METHOD FOR FIBER COUNT

As part of a NIOSH restudy of asbestos processing factories in the 1960s, a fiber counting method was developed (47). The samples were collected on 37 mm diameter mixed cellulose ester membrane filters in plastic cassettes, the filters were rendered transparent with a mixture of dimethylphthalate and diethyloxalate, and fibers longer than 5 μm were counted by microscope with a 4 mm objective and phase contrast illumination. A Porton eyepiece graticule was used to outline a counting field and estimate a fiber length of 5 μm. Fiber concentrations of less than 2 fibers/cc were seldom encountered, and a sampling time of one to one and one half hour at 2 L/min. generally resulted in countable samples. It was found that, on average, a total airbourne particle concentration of one mppcf was equivalent to 5.9 fibers per cubic centimeter (f/cc) in textile mills, 2.2 f/cc in friction product manufacture, and 1.9 f/cc in asbestos-cement pipe manufacture (48). Based on these data from all plants studied, the 5 mppcf limit would thus have been equivalent to roughly 30 chrysotile fibers per cubic centimeter, and this was the conversion used by the ACGIH TLV Committee in 1968. A more extensive study of one of the mills by Dement (49) indicated a ratio of approximately 3 f/cc per mppcf for most operations in that mill, indicating that even in asbestos textile manufacture the ratios could vary by facility.

As counts decreased with controls in asbestos product manufacture and the substitution of nonasbestos materials in most products following the promulgation of the emergency asbestos standard in 1972, the method was improved. A change to a British system of rendering the filter transparent with acetone vapor and fixing it with triacetin gave slides that were semipermanent, rather than lasting only hours as with the previous method. A change from 37 mm to 25 mm diameter cassettes more than doubled the sensitivity of the method. The Walton-Beckett microscope eyepiece graticule (50) was designed specifically for asbestos counting and ordered to microscope specifications so that all fields counted were the same size, with lengths and diameters shown for comparison. NIOSH method 7400 "ASBESTOS and OTHER FIBERS by PCM" (51) details the sampling method, filter preparation, microscopic equipment, calibration procedures and counting rules. A comparable procedure has been published in Great Britain (52).

7 OTHER ASSESSMENT METHODS

In the United States from the mid-1920s on until the late 1960s measurements of airborne particulate concentrations were made almost exclusively by the impinger method. The history of dust measurements and their relationships to the perception of "safe" levels of exposure is important. Most of the data for setting current limits relies at least partially on measurements made before the present respirable mass methods were developed. This is true in other nations as well; the difference is in the previous dust evaluation methods used. The brief history of dust measurement methods that follows gives information that can be useful as methods are developed for unusual situations.

8 DUST SAMPLING METHODS, GENERAL

In describing the various principles, sampling instruments and measuring methods that have been used over the years to sample and enumerate dust, the 1925 arrangement of Greenburg (53) is a useful outline. Most of the principles now used in sampling atmospheric dust were included in his review. These included condensation, filtration, washing, sedimentation, impinging, electrostatic precipitation, and resistance. Other than thermal precipitation and light scattering as a direct dust measurement, he included most of the principles used by industrial hygienists in the three-fourths of a century since, though increasing sophistication may have made the methods unrecognizable. Measurements of atmospheric particulate by filtration go back in the U.S. to at least 1894 (54) and in England to 1897 (55). Filtration media reported by Greenburg included cotton wool, cotton cloth, asbestos cloth, shredded asbestos, filter paper, paper thimbles, and soluble chemicals. Chemicals used were collodion wool, nitrocellulose and, most commonly, refined sugar. Although not a problem at the high concentrations noted in the Joplin mines, at lower concentrations the variability of the insoluble content of refined sugar became a problem in sugar tube measurements. The efficiency for smaller particles, and compensation for moisture content of cellulosic media were of particular interest for the filtration methods. Some of the media, cheese cloth for example, were very inefficient when tested with smoke clouds. With filter papers, the variation of weight with humidity was a problem. If the samples were ashed for chemical analysis, a low ash filter paper was required.

8.1 Total (Unselected) Mass Concentration

The impetus for measurement and control of dust grew in the 1920s and 1930s, with emphasis by governmental industrial hygiene units in several States, and joint studies by the States, the U.S. Public Health Service (USPHS) and/or the U.S. Bureau of Mines.

Many of the early studies of dustiness used mass concentration, measured by weighing particulate material collected in a known volume of air sampled. It is now recognized that the mass concentration found in a sample of coarse dust is dependent upon the sampler entrance conditions. Inhalable dust, by the international standard definition includes 50% of the 100 μm aerodynamic diameter particles. Most of the air sampling for total dust, continuing up to the present time, has included much less of the dust larger than 10 μm aerodynamic diameter than defined by the inhalable criterion, although there was no intentional size selection. Drawing air through sugar tubes to sample for insoluble particles had reached the status of a standard method early in the 20th century. This method allowed both mass and count to be measured. A 1912 committee of the Laboratory Section of the American Public Health Association (56) mentioned the 1909 recommendation of "the sugar tube and microscopic counting method". In the 1912 report they stated that "What is needed for sanitary purposes is a reasonabiy accurate determination of the comparatively few, but rather large, particles which lacerate the epithelial tissue and favor the development of disease. The filtration method gives this information, accurately measuring the number of particles large enough to be seen under a ⅔ in. objective." The method was too slow because of high resistance in the small sugar filter, so instead of using a small glass tube, as recommended in the 1909 report, the committee in its recent studies has used a chemical adapter (a conical piece of glassware seven inches long having openings of ⅞ in. at the smaller end and 1½ in. at the larger end). The small end is closed with a perforated rubber stopper over which is placed a piece of bolting cloth to support the filtering layer made up of about 2.5 g of granulated sugar forming a layer 1 cm deep. Five cubic feet of air can be drawn through this filter in eighteen minutes. The sugar is dissolved and the contained dust suspended in 10 cc of distilled water and 1 cc of the suspension is placed in a Sedgwick-Rafter counting cell and counted under the ⅔-in. objective.

A sugar tube method was used by Lanza and Higgins in their studies of silicosis in the Joplin, Missouri lead–zinc mines in 1914 and 1916 (57, 58). This early study of pneumoconiosis in the United States may serve as an example of the way in which exposure estimation was used in conjunction with medical findings to provide information to be used in reducing hazards of pneumoconiosis, in this case silicosis and silico-tuberculosis in a particular group of workplaces. The sampling method in these studies, which used a miner being sampled as the source of suction, was described in their reports as follows:

> The device consists essentially of the following parts:
>
> 1. A glass bulb or container for the filtering medium (4½ in. long, 1¼ in. inside diameter at the large end, and ½ in. inside diameter at the stem). This bulb is partly filled with granulated sugar, which has been found in tests made in South Africa and in tests by the Bureau of Mines, to be superior to any other medium as a dust filter. The granulated sugar is prevented from passing into the sampling device by means of a perforated glass partition inserted just above the stem of the bulb, but at a point where the diameter is not less than 1 in.
> 2. That part of the oxygen-breathing apparatus known as the mouthpiece, with inhalation and exhalation valves attached. The mouthpiece may be held securely in position, when the device is in use, by means of a cap to which straps are attached. It is not necessary, however, to use a cap when one has become familiar with the device.
> 3. A nose clip to prevent air from passing in or out of the nostrils.
> 4. What is known as a Draeger liter bag, having a capacity of 35–45 L.
>
> In order to collect a sample the apparatus is adjusted in the mouth with a dust bulb and liter bag attached. The nose clip is then placed on the nose and the person taking the sample breathes naturally. When air is inhaled the valve in the inhalation tube opens, allowing the air to pass to the lungs; at the same time the valve in the exhalation valve closes, making it impossible to draw air from the liter bag. When air is exhaled from the lungs the valve in the inhalation tube closes and that in the exhalation tube opens. In taking a sample, the air passes through the bulb (where the dust is intercepted by the sugar) into the lungs, and thence through the exhalation tube into the liter bag where it is measured. Ordinarily, sufficient accuracy may be obtained by inhaling and exhaling a total of 35 L of air. However, if the place in which the sample is taken is only slightly dusty the bag may be filled a second time, thus making a total of 70 L of air for the sample.
>
> In order to determine the amount of dust collected, the contents of the bulb are washed out into a weighed Gooch crucible and dried at 105°C. After successive washings with water, to dissolve and remove all of the sugar, the crucible is again dried at 105°C and weighed. The increase in weight represents the weight of rock dust in the sample.

This apparatus was bulky. The illustration in the manuscript shows it to be, with the 35-L bag, about the height of the person using it. The analysis in the laboratory, though simple in description, was obviously time-consuming and required the care of an analytical chemist. The sample was said to take about two minutes, with the sample-taker breathing at a rate similar to that of the miner and holding the bulb near his face. Results in the paper were given in mg/100 L. In the more common units of today, 39 samples of the

1914 preliminary study, taken near the nostrils of a drill man or shoveler, ranged from 5.7 mg/m^3 to 77.1 mg/m^3, with an average of 50.4 mg/m^3. The more extensive study of 1916 took a further 183 samples, covering a number of operations, some conducted with the use of water sprays, but most without water. The investigators concluded that with the use of water, dust concentrations could generally be kept below 10 mg/m^3, and this was suggested as a tentative standard for the Joplin District.

The 1914 and 1916 Joplin District mining investigations also included microscopic dust sizing, both of dust collected in air sampling and dust from sputum samples collected by Dr. Lanza. The sputum dust was described by Reinhardt Thiessen as follows: "The rock dust is of the same nature and appearance as that gathered in tests of the suspensions of dust particles. . . . There are comparatively few particles measuring more than 10 microns in diameter, but a large number measure between 2 and 5 microns, with rapidly increasing numbers of smaller sizes, down to colloidal dimensions." He further wrote that the dust resembled that collected from air samples. The size distributions also demonstrated the reduction in large particles as distance from the dust source increased. The authors discussed the differences in potential hazard for the same weight of dust depending upon the particle size. Their experience, and the similar experience of others led to the virtual abandonment of gross weight concentration as an index of hazard for pneumoconiosis. Instead, the number concentration of dust as measured by microscopic examination of samples became the index of hazard. However, given a specific size distribution and composition, total mass serves now, as in the early twentieth century, to be a useful index of relative hazard.

8.2 Condensation Methods

Condensation nuclei counting, although useful in aerosol research, air pollution applications and respirator fit testing, has little current application in sampling for pneumoconiosis producing dusts. However, condensation methods were used from 1889 when John Aitken described his koniscope, with an improved version, the Aitken dust counter slightly later. As described, the Aitken dust counter was very large, and by no means portable. To quote Greenburg (53).

> It consisted of a receiver, an air pump, an air measuring device, an illuminometer, and a gasometer. The air is drawn into the receiver by means of the gasometer. In passing into the receiver it is measured and mixed with a known quantity of dust-free air and saturated with water. The air is then rarefied by means of the pump. The rain, which is produced by the condensation of water on the dust present, falls on the ruled polished silver plate constituting the bottom of the receiver. The number of droplets on the counting plate multiplied by the proper factors for the amount of air sampled and the dilution with dust-free air give a count of the number of particles present in the original sample.

The large, nonportable apparatus required intricate manipulation promoting large experimental and personal error.

A modified form of the Aitken dust counter was reported by Cohen (59). It was significantly simpler, smaller, thus portable, and supported on a tripod in use. To saturate the air, strips of damp blotting paper were used. The principal disadvantage, according to Greenburg, was that all particles of dust receive equal significance irrespective of size, since this method takes account of particles of an ultramicroscopic size. For example, a churchyard in Leeds had a count of 59.6 particles per cubic centimeter whereas a flour mill in the same city had a count of 51.4 particles per cubic centimeter. Obviously, any observer would have judged the flour mill to be far dustier. Similar results are obtained with current condensation nuclei counters, where a foundry may show lower counts than outdoor air.

8.3 Obtaining Dust for Microscopic Counting

Virtually any dust evaluation method used consistently in similar dust production situations can monitor dust exposure as related to pneumoconiosis hazard. For example, Tyndallometer light scattering readings in German coal mines were closely correlated with pneumoconiosis incidence. But because microscopic counts measured primarily dust of less than 10 μm, like that found by pathologists in the lungs of silicotics at autopsy, dust counting became the preferred method throughout the world. If the samples were obtained on glass slides or cover slips, they could be counted directly. If the dust was collected, dissolved or washed into a liquid, a portion of the liquid could be placed in a glass cell for settling and then counted.

In South Africa the hazard was measured by konimeter counts, in the United Kingdom by thermal precipitator counts, and in the United States primarily by impinger counts. Translating a set of counts by one method to that of another, however, is difficult. The efficiency of the sampling devices for various dust sizes differed. Some, like the impinger, were suggested to disaggregate agglomerates, and others such as the konimeter were prone to particle bounce of larger particles. Furthermore, the microscopic counting methods could differ in both magnification and illumination. Higher magnification objectives and dark field illumination both greatly increased counts as compared to the commonly used 16 mm, 10X objective and bright field illumination. It was soon observed that methods promoting high counts of particles much less than 0.5 μm in diameter caused an undue influence of ambient air pollution. In mining and manufacturing the pneumoconiosis-producing dust was primarily produced by such operations as drilling, blasting, grinding, crushing and screening. These dust-producing processes resulting in silicosis or coal workers' pneumoconiosis had in common the fact that they were breaking up solids. The resulting dust larger than 10 μm that did not settle was removed in the nose and throat so that it did not reach the lung. The dust smaller in diameter than 0.5 μm was not necessarily in higher concentration than that outdoors. Thus the methods most desired were those that produced high efficiencies of particles between 0.5 μm and 10 μm in diameter. Other than the early Aitken instrument that collected droplets by settling, the instruments for collecting dust on glass slides or cover slips in the field relied either on impaction or thermal precipitation. The konimeter was by far the most commonly used impaction instrument, but the Owens jet and its modification, the Bausch and Lomb dust sampler, deserve mention. The thermal precipitator was primarily used in Great Britain, but in its initial and long-running version gave the historical background for the British and American coal dust standards.

8.3.1 *The Konimeter*

Developed in South Africa to measure dustiness in their deep, high silica, gold mines,the konimeter was one of the earliest devices for collecting samples for count. The model most used in the United States (60) used a small cylinder with a spring-loaded piston to draw a 2.5 or 5 cc air sample through a jet at high speed. The jet of air impinged upon a glass slide lightly coated with adhesive, leaving a dust spot for microscopic counting. The slide consisted of a round disc with 30 sampling positions. Various models of konimeter were produced over the years for use in these South African gold mines. Comparisons of some of these, such as the Kotze konimeter, circular konimeter, Witwatersrand konimeter and Gathercole konimeter were performed both in laboratory and in mines. Commercial models of the konimeter were widely used elsewhere, with published reports of sampling in Germany, England, Canada, Australia and the United States. Some instruments had a self-contained microscope so that samples could be readily counted on site.

8.3.2 *The Owens Jet Dust Counter*

The use of moist blotting paper to saturate the air, as in Cohen's modified Aitken dust counter was also used by Owens in his jet dust counting apparatus (61). Dr. Owens used a tube lined with blotting paper as a saturating chamber. A hand operated piston in a cylinder at right angles drew air through a slit shaped orifice, with the air impinging on a microscope cover slip held in place by a clamp. In his experiments a $\frac{1}{12}$ in. (2.1 mm) oil immersion objective was used for counting the dust particles collected. With well saturated air, as the air rapidly expanded coming through the slit, water condensed on dust nuclei, the droplets were impinged on the cover slip, and the oil immersion objective permitted counting of particles down to 0.3 μm diameter. In its commercial version, the apparatus became widely used. The Owens Jet sampler was used to take some of the earliest samples in the Australian gold mines before being replaced by the konimeter.

8.3.3 *The Bausch and Lomb Dust Counter*

A modification of the Owens Jet dust counter that included a dark-field microscope was sold as the Bausch and Lomb dust counter (62) prior to World War II. Built for the American market, it had a hand pump volume of 28.3 cc or 0.001 cubic foot. For a one stroke sample the average count across the dust ribbon within the limits of the ocular grid, divided by 10, equaled the dust concentration in the traditional American dust count unit, millions of particles per cubic foot (mppcf).

8.3.4 *Thermal Precipitator*

Difficulties in obtaining an unbiased sample of fine dust by impaction methods were recognized. Efficiencies approaching the lower desirable limit of 0.5 μm tended to be low, and particle bounce was a problem only partially solved by coating the impaction surface with adhesive. The hot-wire thermal precipitator was designed to be efficient for fine particle collection (63). In this device, electrically heated nichrome wires were stretched across a narrow channel between microscope cover slips. The cover slips rested on brass blocks separated by mica strips on which the cover slips rested. As air streamed slowly past the heated wires, the suspended particles were precipitated on the cover slips by thermal repulsion. With a sampling time of about 15 minutes a deposit of suitable density for microscopic counting was usually achieved. A readily portable commercial version was developed. It contained a small tank with an inward airflow of a few cc per minute created by outward flowing water. A battery was included to heat the wire. The dust was counted using a 4 mm oil immersion objective (dry). An eyepiece reticule was used to enable the persons counting to record only those particles between 0.5 and 5.0 μm in diameter.

The sampling time of 15 minutes for the unit meant that a number of samples, ideally covering most of the time, were necessary to estimate the exposure for an entire working shift. To alleviate this problem the "long-running thermal precipitator" was developed (64). The instrument combined gravitational settling of the coarser particles with hot-wire thermal precipitation of the smaller particles. The sampling rate of the cam-driven, battery-powered diaphragm pump was 2 cc/min. The collection area on the slide or cover slip was much greater than that of the conventional thermal precipitator, i.e., 0.25 in.^{2} (1.6 cm^2) vs. 0.02 in.^{2} (0.13 cm^2), allowing the sampling time to increase to 8 hours for an acceptable count density. The long running thermal precipitator replaced the standard thermal precipitator as the major dust sampling instrument in the long term study of coal workers' pneumoconiosis in British coal mines, until replaced in its turn by the gravimetric measuring instrument (65).

8.3.5 The Palmer Apparatus

With the obvious fact that particulate air pollution was reduced by rainfall, the removal of particles by washing seemed a promising technique. Among the techniques described by Greenburg (63) were washing or scrubbing. He listed papers from 1874 to 1923 describing various sampling devices for collecting dust in water. The majority of these used relatively high air flows, collecting enough dust so that it could either be placed in a cell for microscopic counting, or filtered, dried and weighed. One method, however, drew only 100 cc of air into a syringe with 20 cc of water. The user shook it to wet the dust and then placed a drop in a blood counting cell for a dust count. Dr. Greenburg dismissed this method as giving "results which are, to say the least, somewhat erroneous."

The Palmer method was reported in 1916 (66). It used a small 110 V centrifugal fan to draw air at 0.14 m^3/s (5 cubic feet per minute (cfm)) through a sample of water in a specially designed bulb. The air was drawn through a fountain spray curtain of water which retains by actual test 98% or more of finely divided dust added at the inlet (67). It was contained in a leather dress suit case. A later model from Wallace and Tiernan was more compact and more rigid in its wooden case. This later model was apparently the most widely used industrial dust sampling method between its introduction and the development of the Greenburg-Smith impinger. Efficiency tests of the commercially available Palmer apparatus were reported in 1920 by Katz (68) and as part of the joint Public Health Service–Bureau of Mines methods comparison in 1925 (69, 70). Katz found an efficiency of 30% against finely divided silica dust at an air flow of 0.11 m^3/s (4 cfm), the maximum sampling rate that could be obtained in the newer compact unit without liquid carryover. In his paper on the electrostatic method of dust collection Bill reported a Palmer apparatus efficiency of 61.6% by weight and 59.9% by count with the compact unit (71). Smyth and Izzard, however, reported that when the original unit of Palmer was run at the original recommendation of 0.14 m^3/s (5 cfm) the efficiency would have been significantly higher in either the Katz et al or Bill tests (72).

8.3.6 *The Drinker, Thompson, Fitchett Scrubber*

An improved scrubber was reported by Drinker, et al. in 1923 (73). It drew 0.14 m^3/s (6 cfm) of air through two identical stages of dust removal. In each a thimble with perforations was used to create the bubbles, and a spiral in the column was used to prevent carry-over of liquid. Its development was contemporaneous with that of the Greenburg-Smith impinger, a much simpler and equally efficient apparatus, so it never was widely used.

8.3.7 *The Impinger*

In 1922, Leonard Greenburg of the U.S. Public Health Service and G. W. Smith of the U.S. Bureau of Mines developed the impinger (74). It sampled by drawing air at 28.3 L (1 cubic foot) per minute through a submerged orifice with an area of 4.2 mm^2. The impingers as used by the Bureau of Mines over the years are shown in Figure 9.1 (75).

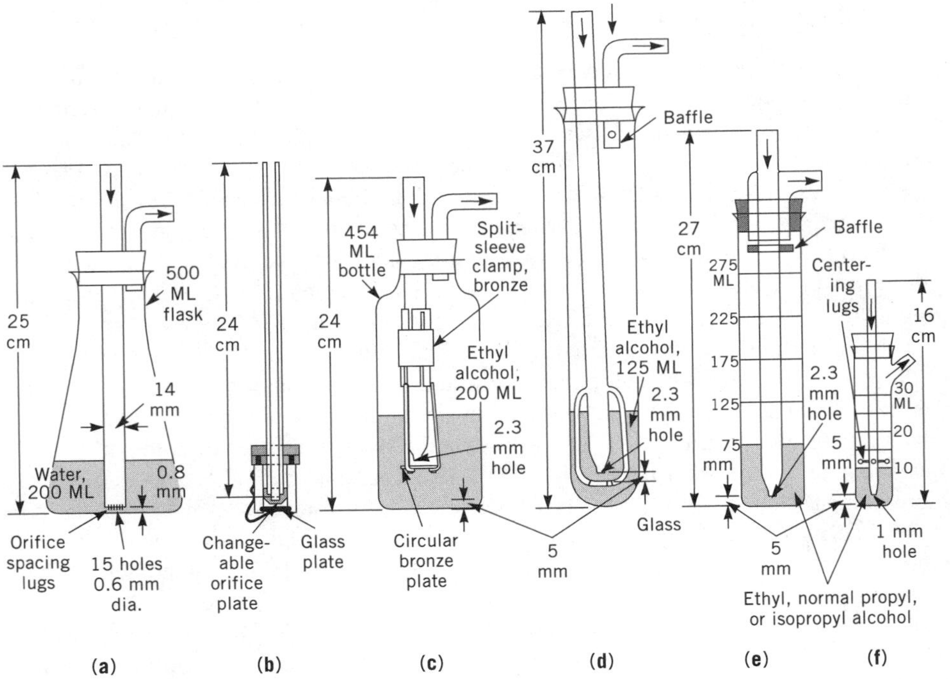

Figure 9.1. Impingers used by the U. S. Bureau of Mines. See text for explanation.

Impinger ***a*** was used in the comparative dust-collection efficiency study in progress when the impinger was devised. Impinger tube ***b*** was designed to study such features as orifice shape and size, impinging distance or distance between orifice and impinging surface, depth of immersion of orifice in the impinging liquid, and nature (smooth or rough) of impinging surface.

Impinger ***c*** was the first to be used in the field and the one used until the first part of 1932. The 454-mL or 16-oz bottle was selected partly because of its convenience for shipping samples to the Pittsburgh Experiment Station of the Bureau of Mines for determination of the amount of collected dust.

Impinger ***d*** was designed to be attached to the worker for collection of samples from his immediate breathing zone. The impinger was made long so that the liquid would not run out when the worker bent over. Design ***d*** differed from the modification described by Greenburg and Bloomfield (76) in having the glass impinging plate attached to the tube by two glass legs instead of three.

The next model used, ***e*** was described by Hatch, Warren, and Drinker (77). Impinger ***e*** uses the floor of the container as the impinging surface, the inlet and outlet tube are combined in one piece, and the container is graduated. This was the first impinger made widely available commercially.

The recently developed midget impinger, ***f*** (78) is the one now used mainly by the Bureau of Mines. It has an orifice 1 mm in diameter instead of 2.3 mm, uses a sampling rate of 0.1 cubic foot (2.83 L) per minute instead of 1 cubic foot, and at this sampling rate has a resistance to air flow of 30 cm (12 in.) of water column height instead of about 90 cm. The investigation of comparative samples by this impinger and the preceding one, ***e*** is still in progress, and the results to date have been satisfactory.

The midget impinger (78) gradually replaced the Greenburg-Smith impinger in general industrial hygiene use. A commercially produced, hand cranked, four cylinder pump with a spring loaded diaphragm for adjustment of suction was used for field sampling. The pump, vacuum gauge and a wooden block holding nine midget impingers flasks were contained in a case with a shoulder strap. In use, the midget impinger, connected to the pump by rubber tubing through a hole in the case, was held by the user's left hand (also holding a stop watch) in the worker's breathing zone. The crank was turned by the right hand with little effort and maintained a steady vacuum at any rate over about 30 revolutions per minute. (No left hand models were provided.) By the 1950s, even the NIOSH industrial hygienists were using the midget impinger exclusively for field studies.

9 COUNTING DUST SAMPLES IN LIQUID

When dust was collected in a liquid, it could either be counted, or the liquid could be evaporated and the residue weighed. When counted, the liquid samples had the major advantage over devices collecting directly on the counting slide, e.g., the konimeter and the thermal precipitator, in that the sample could be diluted for optimum particle density in the counting area. Thus no overlap correction, as used with the thermal precipitator, was necessary. As noted in the report of the Standard Methods Committee in 1912 (56), in describing the sugar tube method, the samples were typically counted in a Sedgwick-

Rafter cell after settling. The cell, developed in 1889 for examination of microorganisms in the Boston water supply was described by Rafter in 1892 (79). To count the microorganisms after concentration filtering through sand and washing, method states that 1 cubic centimeter is taken with a 1 cubic centimeter pipette, and transferred to a cell 50 by 20 millimeter area and exactly 1 millimeter in depth. Such a cell, of course, contains 1,000 cubic millimeters, or 1 cubic centimeter. The top of the cell is ground perfectly smooth, and with a little practice one can float a thick cover glass without losing a drop. The work for the Boston Water Works was assisted by Mr. Geo. C. Whipple, who further described the use of the cell (80).

> This consists of a brass rim cemented with Canada balsam to an ordinary glass slip. A thick cover glass (No. 3) having dimensions equal to those of the brass rim (55 mm by 25 mm), prevents vibration of the liquid in the cell. The cell is then filled in such a manner as to distribute the organisms over the entire area. This is done by placing the cover glass diagonally over the cell so that an opening is left at either end, allowing the water to flow in at one end while the air escapes at the other (see Figure 9.2).
>
> When the cell is full the cover slip automatically slides into place. The cell is placed upon the stage of a microscope and subjected to examination. In order to obtain quantitative estimates of the number of organisms present in each field examined, the microscope must be equipped with an ocular micrometer that defines the area of the field. The Whipple micrometer ordinarily used in the Sedgwick-Rafter method consists of a square ruled upon a thin glass disk which is placed upon the diaphragm of the ocular. The square is so dimensioned that with a certain combination of objective, ocular and tube length of the microscope, the area on the stage covered by the ocular micrometer is exactly one square millimeter. Hence, with a cell one millimeter deep, the volume within the outline of the ruled square is one cubic millimeter. For convenience in determining the size of the organisms found, the cell is further subdivided as shown in Figure 9.3. The best micrometers are made by engraving, but a serviceable micrometer for occasional use may be made by photography. The square ruled for the Whipple micrometer is 7 mm. on a side.

Although the Sedgwick-Rafter cell was initially used for dust counting because of its availability, it was recognized by Whipple (or his revisers) that even for microorganisms,

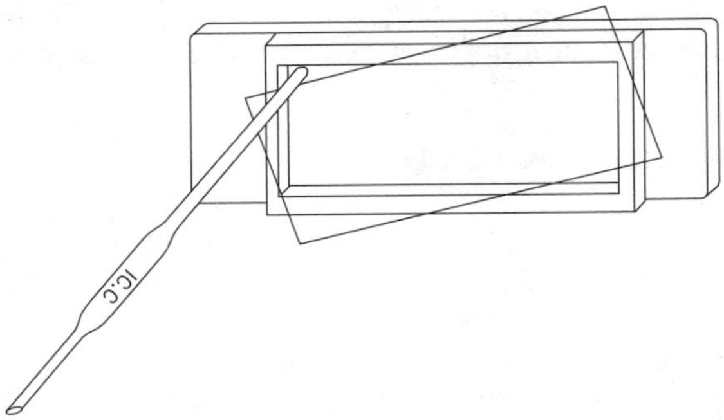

Figure 9.2. Sedgewick-Rafter cell.

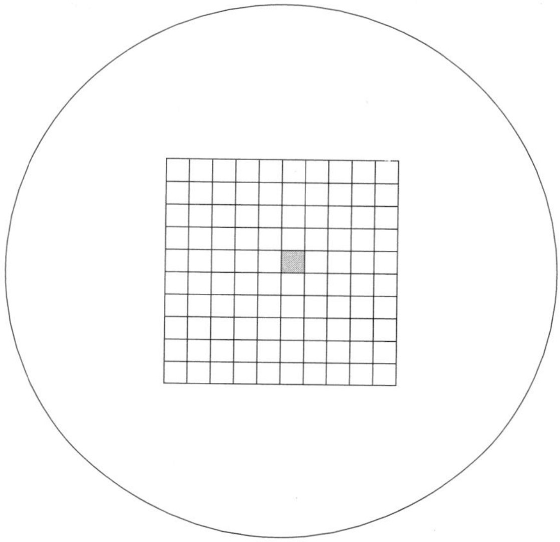

Figure 9.3. Whipple disc.

Whipple states that it is not necessary to use a rectangular cell. A circular cell is equally satisfactory, much cheaper, and more easily cleaned. The capacity of the cell is immaterial. The depth of the cell must be exactly established. A depth of 1 mm is commonly used.

The disadvantages of a rectangular cell were further described by Couchman and Schulze in 1938 (81).

1. The edges and corners formed by the cell walls make it difficult to remove all dust particles during cleaning.
2. Through usage the bottom of the cell may become scratched to the extent that it interferes with the accurate counting of dust particles.
3. A water–alcohol mixture, used for the collection of types of dust which tend to clump in water alone, acts as a solvent on the cement which holds the cell walls in place, thus necessitating recementing or replacement.
4. Recementing the cell walls may entail an appreciable error in the depth of the cell, hence a resultant error in the dust count. Although this discrepancy may be of little significance in comparison with others introduced in the sampling technique, its elimination does improve the accuracy of the procedure.
5. The solvent effect on the cement of oils used as a media for comparing the refractive indices of dust particles makes it impracticable to use the Sedgwick-Rafter cell for this purpose.
6. The amount of breakage of cells is increased by its all-glass construction. (Note the replacement of brass by glass for the cell rim since the 1920s)

They suggested a circular glass disk held between two metal frames, using a 1 mm depth machined as the upper part of the frame, on which was placed the cover slip. Although their design may never have been put into production, by the 1950s commercially available all-glass circular cells were in virtually exclusive use. The Dunn cell (82), perhaps the most commonly used, was a three-piece unit consisting of a wide glass slide, a square ground glass piece the same width as the slide, 1 mm thick with a round opening, and a cover glass the size of the ground glass piece. All pieces could be washed, and the less expensive top and bottom pieces could be replaced when scratched.

The commonly used cell depth of one millimeter was also not universal. A Hatch cell of 0.25 mm depth was sometimes used (83), and Williams proposed the use of a Spencer Bright-Line Hemacytometer (84). The data provided by Williams (84) and Pool et al. (83) suggest that the difference between the standard light field count and the hemacytometer count is small relative to the differences in exposure normally encountered. Perhaps more importantly, this author's experience in training individuals to count dust is that inexperienced persons using the hemacytometer had counts within the range of experienced dust counters using the "standard" method. The same was not true when both used the standard method, with the neophytes tending to grossly undercount. Given the fact that there are few (if any) persons experienced in impinger dust counting still in the field for training and for comparison of counts, anyone currently intending to perform dust counts on impinger samples would be more likely to reproduce counts of previous eras if the counts were done using a commercially available bright-line hemacytometer.

9.1 The Standard Method of 1942

The counting of impinger samples continued the techniques used for sugar tubes and Palmer apparatus samples. It was done much as described for the biological samples by Whipple, using a 16-mm (⅔-in.) 10X objective. The Standard Methods Subcommittee, Committee on Technical Standards of the American Conference of Governmental Industrial Hygienists proposed a standard method in 1942 (85). As will be noted in the method, some options as to counting cell type and depth were given. The method follows:

Determination of Atmospheric Concentration of Fibrosis-Producing and Similar Dusts (Proposed standard impinger sampling and counting technic)

I. *Application.* For the determination of the atmospheric concentration of an insoluble inorganic dust such as that associated with the production of lung fibrosis, the particle size of which is generally below 10 μ and, from the hygienic standpoint, is not of interest below 0.5 μ. This method is not recommended for the determination of organic dusts, metallic fumes, vapors or gases.

II. *Sampling Instrument.* The sampling instrument shall be the standard impinger, operating at a rate of 1.0 cfm at 3 in. Hg negative pressure (U.S. Public Health Service or modified form), or the micro-impinger operating at 0.1 cfm at 12 in. H_2O negative pressure (U.S. Bureau of Mines)

III. *Flow-Producing Apparatus and Flow Meter.* Any type of suction apparatus may be employed, provided it is capable of maintaining the rate of flow within ± 3% of the rated capacity, i.e., 1.0 cfm for standard and 0.1 cfm for the micro-impinger, and is provided with a calibrated flow meter and indicating gauge. The use of a constant-flow orifice is recommended with the compressed air ejector and an automatic negative pressure regulator with the motor or hand driven pump. If an indirect indicating gauge is used, it shall be checked against a U tube or gas meter at regular intervals.

IV. *Counting Cell.* The standard counting cell shall have a liquid depth of 1.0 mm. Counting cells less than 1.0 mm deep down to 0.1 mm may be employed, provided the liquid depth is reported with dust concentration data thus obtained.

V. *Microscope and Accessories*:

A. The microscope shall be a standard biological instrument or its equivalent preferably with an adjustable tube length. It shall be provided with an Abbe' condenser with iris diaphragm, 16 mm objective 7.5 or 10 $\times$ ocular.

B. The counting area shall be defined by ocular grid or by rulings on the counting cell. It shall have an area equivalent to 1/4 square mm in the counting cell and shall be divided into 25 subsquares.

C. The microscope tube length shall be adjusted so that the counting area corresponds to 1/4 mm^2 in the counting cell or, in the case of a fixed tube length, the ruled area shall be equivalent to ¼ $mm^2 \pm 3\%$. The counting area shall be accurately measured by means of a stage micrometer and recorded with dust concentration data.

D. Illumination shall be provided by means of a 15-W substage lamp, placed directly under the condenser or a 75-W microscope lamp, placed approximately 10 in. from the plain mirror. A "Daylight" filter shall be used.

E. A microprojector of the type described by the Bureau of Mines (86) may be used instead of the preceding system.

VI. *Collecting Liquid.* Distilled water with or without alcohol (ethyl or isopropyl) shall be used as the collecting liquid. The collecting and subsequent washing and diluting liquids shall show not more than six (6) countable particles in 1/4 mm^2 counting area in the cell. The "blank" count shall be determined under the microscope for every series of samples.

VII. *Sampling Technic*:

A. All glassware shall be clean and the equipment protected against dust contamination in the field. One impinger flask shall be treated exactly like the others except that no air shall be drawn through it. The liquid in it shall be used to determine the blank count which must be subtracted from the sample counts, as described below.

B. The number and spacing of the samples at a given operation, the location of the sampling point, and the length of each sampling period shall be so chosen as to yield a representative measure of the exposure throughout the cycle of operation and also to indicate the peak concentrations. The minimum number of samples at each sampling point shall be two.

C. In addition to samples collected in the breathing zone of the exposed workers, samples should be taken to determine the dust concentration in the general workroom atmosphere and, in special studies, to determine the sources of the dust dispersion or distribution.

VIII. *Treatment of Collected Sample*:

A. The sample shall be counted within 36 hours after collection except that in the case of flocculating dusts, the permitted holding time shall not exceed twelve hours.

Special collecting liquids may be employed to minimize the degree of flocculation for certain dusts.

B. The sample shall be made up to a known volume, after washing down and removing the impinging nozzle. The concentration of the suspension shall be such as to give a dust count at least four (4) times the "blank" count and not greater than 150 particles in the 1/4 mm^2 counting area. When further dilution from the initial volume is required to reduce the field count to the limits specified above, the dilution shall be made in steps not exceeding 1:10. Complete mixing of the suspension by vigorous agitation is required before removing a portion for dilution and volume removal shall not be not less than 10% of the total volume of the suspension to be diluted, except that the minimum volume to be removed from the original sample for dilution shall be not less than 5 cc. Excessive dilutions are to be avoided by proper gauging of the sampling time in relation to the dust concentration.

IX. *Preparation of the Sample for Counting*:

A. Cells shall be cleaned before use and the effectiveness of cleaning shall be checked by microscopic examination of each cell before being filled.

B. The suspension to be counted shall be shaken vigorously and a portion transferred immediately to the cell by means of a clean pipette, taking care to prevent the inclusion of air bubbles.

C. One or more cells shall be filled from each sample

D. Cell contents shall settle before counting, at least 30 minutes for the 1.0 mm cell and proportionately less for the cells of less depth.

E. One cell shall be made up from the "blank" flask for every series of samples.

F. Cells shall not be filled in advance of the time required for settling before counting.

X. *Counting*:

A. Counting shall be done in a darkened room, using artificial illumination in the microscope, as described above.

B. The light and mirror shall be centered so that no shift in position of the particles is seen when focusing up and down. The condenser and diaphragm shall be adjusted to give an even moderate illumination. Each observer should determine and employ the optimum lighting for his eye.

C. Before counting, the field of the microscope shall be examined and, if necessary, the ocular grid cleaned to remove dust particles.

D. The count shall be made in a plane just above the surface of the cell rather than throughout the depth of the liquid. The microscope may be focused up and down by means of the slow-motion adjustment in order to bring individual particles into focus. Before making the count, the cell contents shall be carefully examined throughout the entire depth to make certain that all countable particles have settled.

E. Five fields of 1/4 mm^2 in area shall be counted, one in the central part and the other four toward the edges of the cell.

F. Except in the case of abnormal samples, no distinction shall be made between the number of particles above and below 10 μ.

G. The use of a hand tally is a useful aid during counting.

XI. *Computation of Results*. The concentration of dust shall be expressed by millions of particles per cubic foot of air (mppcf), reported to two (2) significant figures only.

Concentration is determined as follows:

$$Conc.\ (mppcf) = \frac{(N_s - N_c) \times CF \times V_w}{V_a}$$

where N_s = Mean count per field for the sample (five or more fields)
N_c = Mean count per field for the blank (five or more fields)
CF = Cell factor = 1000/ ((area of counting field in mm^2) $\times$ (depth of cell in mm.))
V_w = Total water volume (equivalent total, when portion diluted, i.e., initial sample volume times dilution factor)
V_a = Air volume sampled in $ft.^3$.

The standard method represented the way in which most counts were being done. It may be noted that universally, only a quarter of the Whipple disc was counted, i.e., 0.25 mm^2. Rather than counting one corner of the grid, it was suggested that a disc be made with only the 25 squares outlining the 0.25 mm^2, or 3.5 mm on a side of the ocular micrometer placed in the center of the disc (87).

The impinger dust count was recognized virtually at the outset to be a relative index of dustiness rather than an absolute measure (88), and differences between individuals and laboratories were recognized (89). The limit of visibility for the microscopic method employed was approximately 1 μm. Typical industrial dusts have a median diameter around 1 μm. Small changes in the observer's visual discrimination or minor differences in the microscopic setup thus could result in significant differences in the count.

9.2 The Standard Method of 1969

NIOSH scientists and engineers, with extensive outside collaboration, examined the precision and accuracy of the impinger dust count method in the 1950s and 1960s. Subjects of investigations of impinger counting, published and unpublished, included reliability (90), variability (91), cell settling time in water-alcohol mixtures (92), intra- and inter-observer variance (93–95), interaction of individual counter and cell type with repeated counts of the same sample. These experiments led to the conclusions that: "In carefully controlled experiments, systematic differences in dust counts may be detected. These include differences which may be attributed to (*1*) interobserver difference, (*2*) (sampling) media difference, (*3*) illumination difference, (*4*) microscope difference, (and/or) (*5*) cell difference (93)." The conclusions were that the method might, for one or two samples, produce disagreements between two observers of as much as a factor of two to four. However, the same judgment on the hazard from dust at a given operation, or the effectiveness of control measures, would generally be reached as long as each based the judgment on an average of several samples. For compliance with a recommended or mandatory limit at an operation to be demonstrated by a small number of samples, a considerable margin below the limit would be necessary (96). Further research has demonstrated that the same statement about the margin below the limit may be made from exposure variability alone (97).

An ACGIH-AIHA *ad hoc* joint committee on uniform methods in dust counting formed in the mid-1960s presented the following method in their 1969 report (98). Although little used, in that impinger counting was generally abandoned shortly thereafter, it may be of use to anyone attempting to reproduce former counts. It represents the improvement in microscopic equipment available to industrial hygienists in the intervening 27 years, and presents the method in some detail. As noted previously, this author would suggest that those not having access to practiced observers by this method would suggest the use of the bright line hemocytometer as a counting cell.

I. *Application.* For the determination of the number concentration of airborne mineral and coal dusts insoluble in, and wetted by, the collection liquid used in the impinger.

II. *Sampling Instrument.* The sampling instrument shall be the standard impinger, operated at a rate of 1.0 efm ± 5% at 3″ Hg negative pressure, or the midget impinger operated at 0.1 cfm ± 5% at 12″ H_2O negative pressure. The sampling instrument and the indicating gauge on the flow producing instrument shall be calibrated at regular intervals.

III. *Counting Cell.* The standard counting cell shall be 1.0 mm in depth with an allowable variation of ± 5%.

IV. *Optical System.*

A. The microscope shall be equipped with the following:

Objective	10× (16 mm) 0.25 N.A.
Ocular (eyepiece)	10×
Condenser	0.25 N.A. or greater

B. The counting area shall be defined by an ocular grid such as a Whipple disc and shall be accurately measured by means of a stage micrometer. The counting area shall be recorded with the count data.

C. Kohler illumination shall be used except that, to improve contrast, the eyepiece shall be removed and the iris diaphragm of the microscope condenser shall be closed until the diameter of the disc of light seen in the back lens of the objective is about one-half that of the lens. Further reduction of brightness may be accomplished, if desired, with neutral density filters.

V. *Collection Liquid.* The collection liquid shall be ethyl alcohol containing no more than 5% water; mixtures with other liquids shall not be used.

VI. *Treatment of Collection Equipment and Samples*

A. All glassware shall be clean and the equipment protected against dust contamination in the field.

B. As a control, one impinger (a "blank") shall be treated exactly like the others except that no air shall be drawn through it.

C. The dilution liquid shall be ethyl alcohol containing no more than 5% water.

D. The impinger nozzle shall be rinsed down inside and out with diluting liquid as the sample is made up to a known volume. Samples having low concentrations of dust shall be diluted as little as practicable. Heavier samples shall be diluted so no more than 2000 particles/square millimeter will appear in the counting area of the cell. Not less than 5 mL of original or dilute sample shall be taken for further dilution, and dilutions shall be made in steps not exceeding 10 parts of dilution liquid to I part of original or diluted sample. The dust suspension shall be shaken vigorously by hand for a minimum of 30 seconds before a portion is removed for counting or dilution.

VII. *Preparation for Counting*

A. The sample to be counted shall be shaken to insure a uniform suspension and a portion transferred immediately to a clean cell by means of the impinger nozzle or a clean pipette, taking care to prevent the inclusion of air bubbles.

B. Two cells shall be filled from each sample, and from the "blank".

C. Counting of each cell shall start at the end of the following listed settling times and shall be completed no later than 10 minutes after starting. For mineral dust, a settling time = 30 min; for coal dust, settling time = 120 min.

VIII. *Counting*

A. Before counting, the field of view of the microscope shall be examined and, if necessary, the ocular grid cleaned to remove dust particles.

B. The counting plane shall be the bottom liquid–glass interface of the cell. The microscope shall be focused up and down slightly with the fine focus adjustment in order to bring individual particles in and out of focus for more positive detection and counting.

C. Fields selected for counting shall be uniformly distributed over the counting plane of the cell. Observation shall not be made through the microscope while fields are being selected.

D. At least five fields of equal area shall be counted in each of two cells.

E. For any sample, when the first five fields of the first cell yield a total count of less than 100 particles, additional fields of known area shall be counted until at least 100 particles have been counted; the total area counted shall be recorded and used in calculation of concentration. The same total area shall be counted in the second cell as is counted in the first.

F. Total counts from the two cells of the same sample shall be compared, and when the ratio of the greater to the lesser count is larger than 1.2, additional pairs of cells shall be counted until a pair yields counts which satisfy this criterion. The count of this pair shall be used for calculating the concentration of the sample.

G. Five fields of the same area as that used for dust sample counting shall be counted in each of two cells from a "blank." The blank count should not exceed 30 particles per square millimeter of counted area. The average blank count shall be used in calculation of net count.

H. Observers are cautioned that their counting effectiveness probably improves during the first few minutes of counting as their eyes become accustomed to the task. A brief period of counting is suggested prior to recording data.

Fatigue can cause deterioration in counting effectiveness; good judgment should be exercised on when to discontinue counting because of fatigue.

IX. *Reporting of Results*. The concentration of dust shall be expressed as millions of particles per cubic foot of air (mppcf) reported to two significant figures only. The following formula may be used for calculation of concentration:

$$\text{Mppcf} = \frac{(N_{\text{avg}} - N_{\text{blank}})\left(\dfrac{1000}{\text{single field area, sq. mm}}\right)(\text{diluted impinger volume, ml.})}{(\text{Air volume of sample, ft}^3) \times 10^6}$$

where N_{avg} = average count per single field
N_{blank} = average count per blank field

Note: If further dilution is carried out, the concentration must be multiplied by the dilution factor

X. *Reliability*. When this method is used by practiced counters, average counts may be expected to be reasonably consistent within and between laboratories. It has been observed that unpracticed counters tend to give results below the general average.

10 IMPINGER DUST STUDIES FOR STANDARDS DEVELOPMENT

10.1 Silicosis

Silica, chemically SiO_2, may be either crystalline or amorphous. Crystalline silica exists as either combined silica, where silica is a part of the crystalline structure of a compound, or as free silica. It is the crystalline free silica that, given sufficient exposure, can result in silicosis. The predominant form of crystalline free silica is quartz, the commonest material in the earth's crust, occurring either relatively pure in many sands, or in minerals such as granite, clays, and shales. Amorphous silica, for example naturally occurring diatomaceous earth, is much less harmful, although simple fibrosis may occur with excessive exposures. Amorphous silica or quartz may be transformed into another crystalline form, cristobalite, at high temperatures, or in the production of flux-calcined diatomaceous earth. A third form, tridimite, is not normally encountered in occupational exposures. Most silicosis has been produced by quartz dust, encountered in metal mines, quarries, and the processing or use of sand and quartz-containing minerals. The fact that silicosis from cristobalite has been much less frequent is only because exposure to its dust is so much less common.

The development of the dust counting methods for evaluation of silicosis hazard has been detailed earlier in the chapter. In the United States the impinger, in its Greenburg-Smith or midget form was almost the exclusive dust exposure sampling device from 1925 until 1968. The relationship of silicosis to dust count and composition was examined in a series of studies by NIOSH in cooperation with State Health Departments from the 1920s through the 1950s. These included the granite monument manufacturing industry in Barre, Vermont (99–101) where successful prevention of silicosis by keeping dust counts below

10 mppcf led to the derivative respirable mass limit for quartz of 0.1 mg/m^3 (15). A similar medical-environmental study demonstrated the silicosis problem in the pottery industry (102). The hazards of cristobalite in flux-calcined diatomaceous earth (103) and successful control by dust reduction (104) were reported. Dust concentrations of silica-containing dusts in general industry were monitored by more progressive manufacturing firms as well as by occupational health units in State Health or Labor Departments. Dust counts in mines were made by the larger mining companies, State mine inspectors and the U.S. Bureau of Mines. Where dust counts were kept within accepted limits, new silicosis cases seldom developed.

10.2 Coal Dust

10.2.1 Anthracite

The anthracite mines in Pennsylvania were a source of smokeless coal primarily for home heating until replaced by inexpensive natural gas in the late 1940s. The mines were recognized as a potential source of chronic incapacitating miner's asthma by a Pennsylvania Governor's Commission in 1932. In 1933, at the Governor's request, and with cooperation from the anthracite coal operators and the United Mine Workers of America, a study was conducted by a team from the Office of Industrial Hygiene and Sanitation, U. S. Public Health Service (105). On the basis of 300 impinger samples and examinations of 2,711 mine employees, it was indicated that where quartz content was less than 5 percent, "less than 50 million dust particles per cubic foot would produce a negligible number of cases of anthraco-silicosis" . . . where the quartz content of dust was about 13 percent a safe limit appeared to be 10 to 15 million particles per cubic foot. The limit of toleration for rock workers was set tentatively at 5 to 10 million particles per cubic foot of air.

10.2.2 Soft Coal

A study of three Utah soft coal mines was made by the U.S. Public Health Service's Division of Industrial Hygiene in cooperation with the Utah State Board of Health in 1939 (106). A total of 168 impinger samples were taken to evaluate the dust exposure of the miners, with medical examinations of 545 workers employed in the three mines. Excluding workers with other dusty trade exposures and surface workers, none of whom were diagnosed with anthraco-silicosis left 348 underground workers with 16 cases of anthraco-silicosis. For dust concentration groups above 20 mppcf, incidence of anthraco-silicosis rose from 3.9% for those with 10–19 years exposure to 50% for the 12 persons employed 30 or more years.

Another study, of 774 bituminous coal workers of the Southern Appalachian region indicated a tendency of bituminous miners to develop presilicotic and silicotic lung changes after prolonged employment (107). At the mine where the study was made 2% of the workers showed presilicotic changes and 1% showed silicotic nodulation in their chest roentgenograms. Dust counts in the mines studied confirmed the (U.S. Bureau of Mines) ranking of dust concentrations at different occupations in the industry, but dust counts were not listed.

The introduction to a Pennsylvania Department of Health study reported in 1961 stated that the concept that there is little or no pneumoconiosis problem in the bituminous coal industry has prevailed throughout the United States for many years (108). The study included 4,182 active and 796 retired miners presenting themselves at 14 central Pennsylvania coal mines for medical examination. Suspected or definite pneumoconiosis found ranged from 5% of the active miners under age 35 to 47% of those age 55 and over. In retired miners under age 55 the prevalence was 37%, 46% in those from age 55 to 64, and 61% in those age 65 and over. The authors stated that the only conclusion that can be drawn at this time is that there is a sizable amount of pathology resulting from exposure to dust in the central Pennsylvania bituminous mining population. A companion paper reported dust counts in the 14 mines (109). At locations or operations in the mines, median dust counts by the impinger method varied from 1.2 mppcf at the mine entrance to 40.4 mppcf at undercutting. There was no way to estimate time-weighted average exposures for the miners examined.

The prevalence of coal worker's pneumoconiosis in the United States was confirmed by a NIOSH study from 1963 to 1965 (110). By a controlled probability sampling method it was found that the prevalence of definite pneumoconiosis among working underground coal miners varied from 4.8 percent in Utah, to 7.5% in Illinois-Indiana, to 11.1% in Appalachia (111). It was noted that the prevalence of pneumoconiosis in the Utah mines as measured by the ILO/UC criteria varied little from the anthraco-silicosis reported by Flinn et al. (106) 28 years earlier. Although there were no obvious major differences in either coal quality or mining methods between Utah and Appalachia, the earlier Utah study had resulted in a significant underestimate of the prevalence of pneumoconiosis in bituminous coal mining as noted by Lieben et al. (108). The NIOSH study included no dust measurements. The studies, with coal workers pneumoconiosis popularized as "black lung" resulted in federal legislation, the Coal Mine Health and Safety Act of 1969. Convinced that respirable mass concentration was a superior method of estimating health hazard, but with no measurements in U.S. coal mines, preliminary British data were used by Congress to set the U.S. standard at 2 mg/m^3 of respirable dust.

10.3 Asbestosis

The studies of the dusty trades by (NIOSH) in the 1930s were all conducted with the Greenburg-Smith impinger as the dust collection method and light field counting of samples using the 10× objective, the Whipple disc, and a 7.5× or 10× eyepiece. One of these studies was a joint NIOSH–North Carolina Health Department study of asbestos textile workers in 1938 (112). It was recognized that it was the asbestos fibers breathed by the workers that was the cause of asbestosis. The four asbestos textile mills studied used only chrysotile, and few fibers were seen in impinger samples counted in the usual manner. Because the dust seen was from chrysotile asbestos, the only mineral present, the investigators counted the usual specks plus the roughly 3% fibers for their index of exposure. Examination of former workers showed a high prevalence of asbestosis, but based upon dust measurements during the study the investigators concluded that if the dust concentrations in asbestos factories could be kept below 5 million particles per cubic foot (mppcf), new cases of asbestosis probably would not appear. Although the study could not directly address lung cancer, references in the Bulletin noted it as a complication of asbestosis. From this recommendation in Public Health Bulletin 241, a threshold limit value of 5 mppcf of asbestos dust was used by States with asbestos processing and adopted in the first ACGIH list in 1946. The limit for asbestos did not distinguish between chrysotile and the less common amosite, crocidolite and anthophyllite. Unfortunately, the 5 mppcf limit, so successful in preventing silicosis where it was observed, proved grossly inadequate even for chrysotile. A misinterpretation of the relation between prevalence and cumulative dust counts, and the general belief that 5 mppcf was a reasonable limit, caused the investigators to err in their recommendation. In retrospect, no impinger limit would have been adequate; even for prevention of asbestosis the limit for chrysotile dust would have had to be less than 0.5 mppcf, a background level for many industrial areas.

Furthermore, unlike the TLV for quartz, the TLV for asbestos gave no guidance as to what limit to use for mixed dusts, or at what percentage asbestos the limit applied. Many chrysotile asbestos products, for example, asbestos cement pipe and thermal insulation, were predominantly minerals other than asbestos. Not surprisingly, industrial hygienists had varying opinions as to the dust count corresponding to the TLV for such mixed dusts. For the same material containing 15% asbestos, for example, the TLV has been said to be 5, 10, and 20 mppcf.

When amosite was the asbestos mineral involved, the proportion of fibers to grains in impinger or impactor samples was much greater. In this author's only experience of counting dust from an amosite processing operation in an impinger sample, the fibers and grains in the count were roughly equivalent. Fleischer et al. (113) in their konimeter study of asbestos in World War II ship construction, where the asbestos in thermal insulation was predominantly or exclusively amosite, listed fiber count as asbestos dust in their tables. Presumably impinger counts of crocidolite or anthophyllite dust would also have shown a larger proportion of fibers in the counts.

10.3.1 Fiber Counts and Asbestosis

The 1960s NIOSH study included medical examinations of textile workers. The study was abandoned before final reports could be written, and there were no diagnoses of asbestosis by the medical officers, but data showed both x-ray findings and pulmonary function decreases consistent with asbestosis among many of the asbestos textile workers with more than 10 years experience. The pulmonary function data suggested that for these workers, decreases did not occur below 50 fiber per cc-years, with the low exposures in workers with less than 15 years experience (114). These data suggest that average chrysotile exposures below 1 f/cc over a working lifetime probably would not result in effects on the cardiopulmonary system that lead to a significant decrease in the ability to perform previous activities or a decrease in life expectancy. Statistical data demonstrate that for an operation to comply with an OSHA permissible exposure limit (PEL) of 2 f/cc, concentrations must average less than 1 f/cc. For chrysotile, this means that the 1976 PEL of 2 f/cc would probably have been adequate to prevent asbestosis. However, to prevent lung cancer from asbestos exposures, as well as mesothelioma hazards from the amphibole varieties, the PEL was reduced to 0.1 f/cc in 1996.

The OSHA and EPA regulations on asbestos in the United States, and the regulations in other nations are all designed to prevent lung cancer and mesothelioma. Exposures are kept far below those found to cause asbestosis in the past. Membrane filter sampling is now practiced throughout the world. The associated fiber counting method, standardized over the years, is now international as well. Current asbestos regulations of governmental bodies for controlling and monitoring asbestos exposures have been designed for prevention of cancer. The maximum worker exposures permitted are far below those that can result in asbestosis.

10.4 The Replacement of Impinger Dust Sampling

It was not the variability of dust counting (88–95) that caused impinger dust sampling to be abandoned as the principal method of evaluating the pneumoconiosis hazard from dust exposures. The number of impinger samples suggested by statistical analysis to be necessary was seldom achieved. By taking the 5- to 20-minute samples over several shifts and various workers, a time-weighted average dust concentration for comparison with the threshold limit could be obtained. This was only practical where there was a person on site to take samples over a period of weeks or months. Instead, the typical procedure was to sample the operations with the apparently highest concentrations. For those operations where a sample exceeded the limit, further sampling and/or controls were suggested.

It was the development of superior methods of estimating long term exposures to dust of a size range that could result in coal workers pneumoconiosis or silicosis that resulted in the abandonment of the traditional dust counting methods for coal dust and silica-containing dusts. Likewise, the development of the membrane filter and methods for making it transparent made possible the quantitative collection of airborne asbestos fibers and their counting at higher magnification and resolution. The United Kingdom was ahead of the United States in both respirable mass sampling for coal dust and fiber counting for asbestos. However, within a few years comparable methods were developed in the United States for coal dust, and the abandonment of counting for silica-containing dusts was even earlier in the United States. International agreement on asbestos fiber counting was in place in the 1980s, and international criteria for size selection were adopted in the 1990s. A spirited defense of impinger dust counting, with a recommendation that OSHA give a dust counting option was presented by Horowitz (115), and a dual limit was, in fact, promulgated. The enthusiasm of Horowitz for impinger sampling and microscopic dust counting was not widely shared, and few impinger samples were taken after the equipment for respirable mass sampling became widely available. The fiber counting method for asbestos completely replaced impinger counting as soon as it became generally available.

11 RECONSTRUCTING IMPINGER COUNT CONCENTRATIONS

For various reasons it may become necessary now to estimate dust counts at a particular operation or occupation as it was in the 1960s and before. In general, recollections of visual dustiness are so subject to vagaries of memory and subjectivity as to be virtually worthless in terms of even order of magnitude estimates. What one remembers as "so dusty you could only see a few feet" another will remember as "relatively clear". A potential solution is to set up a reproduction and take impinger samples.

The first problem is reasonably reproducing the situation. There are a number of operations that are still carried out in much the same way as they were in that time period. There are others where the change has been recent enough so that there are still workers available who performed the operations and the equipment of the type they used may still be obtained. There still remains the potential problem of reasonably reproducing the conditions, i.e., ventilation, room size, etc. under which they worked. Unfortunately, many of the situations that might call for such reconstruction involve litigation, and it becomes difficult to get the necessary parties to agree on an appropriate protocol.

If all these conditions can be met, the samples should be taken in accordance with the 1969 recommendations of the joint ACGIH-AIHA committee given herein. The one exception, if no one who actually counted dust in that period is available, is the possible substitution of the bright line hemacytometer as a counting cell.

11.1 Comparisons of Count and Respirable Mass Data

For epidemiological studies it was necessary to consider pre-1970 exposures measured by count methods. In the United States, these counts had been by impinger sampling with light field count using a 10X objective. In the United Kingdom, the counts had been by either standard or long running thermal precipitator. In either case, the ratio of count to respirable mass varied by operation, so such ratios have been determined, for example, in studies of coal mining (116), granite monument manufacture, (15, 19, 117), the taconite industry (118), and diatomaceous earth mining and milling (119).

12 INTERNATIONAL STANDARDIZATION

Much had been made since the 1960s of the difference between the BMRC criteria for respirable dust, and that of the ACGIH. The curves for definition of respirability have different shapes, but in practice any differences in performance are trivial in comparison to sampling and environmental variables. It was in an effort to rationalize the standards, that Soderholm, then chairman of the ACGIH Air Sampling Procedures Committee, with the endorsement of the full committee, proposed the respirable mass sampling criteria that in effect split the theoretical difference between the ACGIH and BMRC, with a 50% passing point of 4 μm. These criteria also defined inhalable and thoracic fractions of dust for purposes of dust sampling.

The predominant methods of removing the fraction of dust that is not respirable are the horizontal elutriator and the miniature cyclone. Although neither the horizontal elutriator nor any of the various cyclones that have been used follow the international criteria exactly, both types can come close enough so that the differences are not of practical significance. The same has been reported for the CIP 10 porous foam personal sampler. The degree to which various devices follow the published criteria have been extensively investigated. These investigations continue.

13 REAL TIME MONITORING

Even with the maximum sample size that may reasonably be obtained with higher flow rate personal sampling pumps and redesigned cyclones, collecting a respirable mass large enough for analysis will require the major portion of a working shift. One of the advantages of the impinger method cited by Horowitz (115) was the ability to take short samples for measurement of the high concentrations that may contribute most of a worker's exposure. The count methods, however, whether impinger or the more rapid konimeter, require numbers of samples and extensive microscopic counting time to characterize exposures over a shift. Real time aerosol monitors, with data loggers, can give concentrations by the minute or second throughout the shift. The instruments measure light scatter, and the absolute value of mass, or respirable mass, of a given dust cloud may require calibration for the situation. However, the devices can readily indicate the operations and times that are contributing most of the exposure. NIOSH has demonstrated simultaneous video taping of the operation to indicate the specific parts of the operation involved, for example Gressel (120) and O'Brien, et al (121). Control of the peak exposures revealed by such sampling may be adequate to protect the worker. Identification of these situations can also allow the worker to take appropriate action to reduce dust generation or, where this is infeasable, to don personal protective equipment. The real time monitors and associated equipment are at present, and may remain, too costly for purchase by many who might use them, but equipment rental or use by consultants could make them more widely available.

BIBLIOGRAPHY

1. M. Lippmann, *Ann. Occup. Environ. Hyg.* **10**, 981–990 (1995).
2. S. C. Soderholm, *Ann. Occup. Hyg.* **33**, 301–320 (1989).
3. K. R. May, *J. Sci. Instr.* **22**, 187–195 (1945).
4. W. H. Walton, *Brit. J. Appl. Phys.* (Supp. 3), S29–S40 (1954).
5. J. H. Brown, K. M. Cook, F. G. Ney, and T. Hatch, *Am. J. Public Health* **40**, 450–459 (1950).
6. R. J. Hamilton and W. H. Walton in *Inhaled Particles and Vapours*, Pergamon Press, London, 1961, p. 465.
7. B. M. Wright, *Brit. J. Industr. Med.* 284–288 (1954).
8. J. W. J. Fay, *Ann. Occup. Hyg.* **1**, 314–343 (1960).
9. J. R. Ashford and C. O. Jones, *Ann. Occup. Hyg.* **7**, 85–113 (1960).
10. W. A. Bloor and A. Dinsdale, *Ann. Occup. Hyg.* **9**, 29–39 (1966).
11. W. A. Bloor, R. E. Eardley, and A. Dinsdale, *Ann. Occup. Hyg.* **11**, 81–86 (1968).
12. R. I. Higgins and P. Dewell, *A Gravimetric Size-Selecting Personal Dust Sampler*, British Cast Iron Research Association, Alvechurch, Birmingham, BCIRA report 908, 1968, pp. 112–119.
13. J. H. Dunmore, R. J. Hamilton and D. S. G. Smith, *J. Sci. Instrum.* **41**, 669–672 (1964).
14. P. L. Bodsworth, J. F. Hurley, C. L. Tran, P. Weston, and G. Z. Wetherill, *Technical Memorandum Series, IOM Report TM/93/06*, Institute of Occupational Medicine, Edinburgh, 1993, 197 pp.
15. H. E. Ayer, *Appl. Occup. Environ. Hyg.* **10**, 1027–1030 (1995).

16. G. Knight, *Am. Industr. Hyg. Assoc. J.* **49**, 248–254 (1988).
17. W. B. Harris and M. Eisenbud, *Arch. Industr. Hyg. Occup. Med.* **8**, 446–452 (1953).
18. M. Lippmann and W. B. Harris, *Health Physics* **8**, 155–163 (1962).
19. S. J. Reno, B. T. H. Levadie and H. B. Ashe, "A Comparison of Count and Respirable Mass Sampling Techniques in the Granite Industry," Presented at the 1966 Annual Meeting of the American Industrial Hygiene Association, Pittsburgh, 1966
20. R. J. Sherwood and D. M. S. Greenhalgh, *Ann. Occup. Hyg.* **2**, 127–132 (1960).
21. H. E. Mock, *J. Ind. Hyg.* **1**, 1–8 (1919).
22. N. A. Talvitie and F. Hyslop, *Am. Ind. Hyg. Assoc. J.* **19**, 54–58 (1958).
23. H. E. Ayer, G. W. Sutton and I. H. Davis, *Am. Ind. Hyg. Assoc. J.* **29**, 336–342 (1968).
24. H. E. Ayer, *Am. Ind. Hyg. Assoc. J.* **30**, 117–125 (1969).
25. J. A. Lamonica and H. N. Treaftis, *Report of Investigations 7545*, Bureau of Mines, Pittsburgh, 1971.
26. D. P. Anderson, J. A. Seta and J. F. Vining, III, *The Effect of Pulsation Dampening on the Collection Efficiency of Personal sampling Pumps*, TR-70, NIOSH, Cincinnati, 1971, 15 pp.
27. P. A. LaViolette and P. C. Reist, *Am. Ind. Hyg. Assoc. J.* **33**, 279–282 (1972).
28. R. H. Knuth, *Am. Ind. Hyg. Assoc. J.* **30**, 379–385 (1969).
29. H. J. Ettinger, J. E. Partridge and G. W. Royer, *Am. Industr. Hyg. Assoc. J.* **31**, 537–545 1970.
30. G. Knight and L. Licht, *Am. Ind. Hyg. Assoc. J.* **31**, 437–441 (1970).
31. R. W. Weidner, *Personal Respirable Mass Procedure*, Research Report 7, 1968, NIOSH, Cincinnati, 1968, 13 pp.
32. Aerosol Technology Committee, H. J. Ettinger, Chair, *Am. Ind. Hyg. Assoc. J.* **31**, 133–137 (1970).
33. M. Lippmann, *Am. Ind. Hyg. Assoc. J.* **31**, 138–159 (1970).
34. D. Bartley, C.-C. Chen, and T. J. Fischbach, *Am. Ind. Hyg. Assoc. J.* **55**, 1036–1046 (1994).
35. M. Jacobson and J. A. Lamonica, *Personal Respirable Dust Sampler*, Technical Progress Report No. 17, Mineral Industry Program Bureau of Mines, Pittsburgh, 1969, 8 pp.
36. G. W. Harris and B. A. McGuire, *Ann. Occup. Hyg.* **11**, 195–201 (1968).
37. B. A. Maguire and D. Barker, *Ann. Occup. Hyg.* **12**, 197–201 (1969).
38. D. K. Verma, A. Sebestyen, J. A. Julian, and D. C. F. Muir, *Ann. Occup. Hyg.* **36**, 23–24 (1992).
39. R. C. Brown, *J. Aerosol Sci.* **11**, 151–179 (1979).
40. H. Gibson and J. H. Vincent, *Ann. Occup. Hyg.* **24**, 205–215 (1981).
41. C-C. Chen, C-Y. Lai, T-S. Shih, and W-Y. Yeh, *Am. Ind. Hyg. Assoc. J.* **59**, 766–773 (1998).
42. P. Courbon, J. F. Fabries, and R. Wrobel, "Dust Measurements at Worksites: the CIP 10 Dust Sampler" in *Proceedings of the First International Aerosol Conference, Aerosols-Science, Technology and Industrial Applications of Airborne Particles*, Minneapolis, 1984.
43. P. Courbon, R. Wrobel, and J. F. Fabries, *Ann. Occup. Hyg.* **32**, 129–143 (1988).
44. A. Gero and T. Tomb, *Am. Ind. Hyg. Assoc. J.* **49**, 286–291 (1988).
45. C. D. Lorberau and M. T. Abell, *Scand. J. Work Environ. Health.* **21**, (Supple 2), 35–38 (1995).
46. S. A. Shulman, J. H. Groff, and M. T. Abell, *Am. Ind. Hyg. Assoc. J.* **53**, 49–56 (1992).
47. G. H. Edwards and J. R. Lynch, *Ann. Occup. Hyg.*, **11**, 1–6 (1968).
48. J. R. Lynch, H. E. Ayer, and D. L. Johnson, *Am. Ind. Hyg. Assoc. J.* **31**, 598–604 (1970).

49. J. M. Dement, R. L. Harris, Jr., M. J. Symons, and C. M. Shy, *Am. J. Ind. Med.* **4**, 399–419 (1983).
50. W. H. Walton and S. T. Beckett, *Ann. Occup. Hyg.* **20**, 19–23 (1977).
51. *NIOSH Manual of Analytical Methods (NMAM)*, 4th ed., Cincinnati, 1994.
52. Asbestos International Association, AIA Health and Safety Recommended Technical Method #1 (RAMIE), *Airborne Asbestos Fiber Concentrations at Workplaces by Light Microscopy (Membrane Filter Method)*, London, 1979.
53. L. Greenburg, *Public Health Reports* **40**, 765–786 (April 17, 1925).
54. C. Arens, *Arch. f., Hyg.* **21**, 325 (1894).
55. J. B. Cohen, *J. Soc. Chem. Indust.* **16**, 411–412 (May 31, 1897).
56. Committee on Standard Methods for the Examination of Air, *Am. J. Public Health* **2**, 76–86, (1912).
57. A. J. Lanza and E. Higgins, *Technical Paper 105*, Dept. of the Interior, Bureau of Mines, Washington, 1915.
58. E. Higgins, A. J. Lanza, F. B. Laney, and G. S. Rice, *Bull. 132*, Dept. of the Interior, Bureau of Mines, Washington, 1917.
59. J. B. Cohen, *Smithsonian Misc. Collections*, No. 1073 (1896).
60. P. Drinker and T. Hatch, *Industrial Dust*, McGraw Hill Book Co., Inc., New York, 1954 pp. 135–140.
61. J. S. Owens, *J. Ind. Hyg.* **4**, 522–534 (1923).
62. P. Drinker and T. Hatch, *Industrial Dust*, 2nd ed., McGraw-Hill Book Co., Inc., New York, 1954, pp. 142–144.
63. R. Whytlaw-Gray and R. Lomax, His Majesty's Stationery Office, London, *Special Report Series No. 199*, pp. 23–38 1935.
64. R. J. Hamilton, *J. Sci. Instrum.* **33**, 395–399 (1956).
65. Walton, W. H., "Progress of the 25-Pit Scheme" in H. A. Shapiro, ed., *Pneumoconiosis: Proceedings of the International Conference, Johannesburg*, Oxford University Press, Cape Town, 1970, pp. 292–294.
66. G. T. Palmer, *Am. J. Public Health* **6**, 54–55 (1916).
67. H. F. Smyth, *J. Ind. Hyg.* **1**, 140–149 (1919).
68. S. H. Katz, E. S. Longfellow, and A. C. Fieldner, *J. Ind. Hyg.* **2**, 167–177 (1920).
69. S. H. Katz, G. W. Smith, W. M. Myers, L. J. Trostel, M. Ingels, and L. Greenburg, *Public Health Bulletin 144*, 1925, 82 pp.
70. L. Greenburg, *Public Health Reports*, 591–1603 (July 31, 1925).
71. L. P. Bill, *J. Ind. Hyg.* **1**, 323–342 (1919).
72. H. F. Smyth and M. Iszard, *J. Ind. Hyg.* **2**, 159–167 (1921).
73. P. Drinker, R. M. Thomson and S. M. Fitchett, *J. Ind. Hyg.* **4**, 62–78 (1923).
74. L. Greenburg and G. W. Smith, *A New Instrument for Sampling Aerial Dust*, U. S. Dept. of Interior, Bureau of Mines, Serial No. 2392, 1922, 3 pp.
75. C. E. Brown and H. H. Schrenk, *A Technique for Use of the Impinger Method*, Bureau of Mines Information Circular 7026, Washington, DC, 1938, 24 pp.
76. L. Greenburg and J. J. Bloomfield, *U. S. Public Health Repts.* **47**, 654–675 (1932).
77. T. Hatch, H. Warren, and P. Drinker, *J. Ind. Hyg.* **14**, 301–311 (1932).

78. J. B. Littlefield, F. L. Fecht and H. H. Schrenk, *Bureau of Mines Midget Impinger for Dust Sampling*, U. S. Dept. of Interior, Bureau of Mines, Report of Investigations 330, 1937, 7 pp.
79. G. W. Rafter, *The Microscopical Examination of Potable Water*, D. Van Nostrand Company, 1892, pp. 65–79.
80. G. C. Whipple, *The Microscopy of Drinking Water*, Revised by G. M. Fair and M. C. Whipple, 4th ed. John Wiley & Sons, Inc., New York 1927, pp. 95–97.
81. C. E. Couchman and W. H. Schulze, *Public Health Reports* **53**, 348–350 (1938).
82. K. L. Dunn, *J. Ind. Hyg. Toxicol.*, 202–203 (1939).
83. C. L. Pool, J. Wuraftic and R. J. Kelly, *Industrial Hygiene Supplement, Industr. Med.* **2**, 39–45 (1941).
84. C. R. Williams, *J. Ind. Hyg. Toxicology* **21**, 226–230 (1939).
85. Subcommittee on Standard Methods, *Transactions of the American Conference of Governmental Hygienists*, Washington DC. 1942, pp. 171–175.
86. C. E. Brown, L. A. H. Brum, W. P. Yant, and H. H. Schrenk, *Microprojector Method for Counting Impinger Dust Samples*, Bureau of Mines Report of Investigations 3373, 1938, 9 pp.
87. R. T. Page, *Public Health Reports* **52**, 1315–1316 (Sept. 17, 1937).
88. J. M. DallaValle, *Public Health Reports* **54**, 25 (June 23, 1939).
89. P. O. Halley, *Am. Ind. Hyg. Assoc. Quart.* **7**, 15 (1946).
90. H. M. Chapman and R. C. Ruhf, *Am. Ind. Hyg. Assoc. Quart.* **16**, 201 (1955).
91. R. G. Edwards, C. H. Powell and M. A. Kendrick, *Am. Ind. Hyg. Assoc. J.* **27**, 546–554 (1966).
92. F. E. Hall, R. E. Kupel, and R. L. Harris, Jr., *Am. Ind. Hyg. Assoc. J.* **26**, 537–543 (1965).
93. H. E. Ayer, *Transactions of the Amer. Conf. Industr. Hygienists*, 10–21 (1966).
94. H. J. Paulus, unpublished data, quoted in Ref. 93, 1955.
95. J. R. Lynch, unpublished data, quoted in Ref. 93, 1966.
96. S. A. Roach, E. J. Baier, H. E. Ayer and R. L. Harris, *Am. Ind. Hyg. Assoc. J.* **28**, 543–553 (1967).
97. N. A. Leidel and K. A. Busch, *Amr. Ind. Hyg. Assoc. J.* **36**, 839–840 (1975).
98. ACGIH-AIHA Ad Hoc Committee on Uniform Methods in Dust Counting, *Trans. ACGIH*, 204–209 (1969).
99. A. E. Russell, R. H. Britten, L. R. Thompson, and J. J. Bloomfield, *The Health of Workers in Dusty Trades. II. Exposure to Siliceous Dust (Granite Industry)*, Pub. Health Bull. 187, U. S. Public Health Service, Washington, D.C., 1929, 206 pp.
100. A. E. Russell, *The Health of Workers in Dusty Trades. VII. Restudy of a Group of Granite Workers*, Pub. Health Bull. 269, U. S. Public Health Service, Washington, 1941, 71 pp.
101. A. D. Hosey, V. M. Trasko, and H. B. Ashe, *Control of Silicosis in Vermont Granite Industry*, Occupational Health Program, U. S. Public Health Service, Rep. No. 557, 1957, 79 pp.
102. R. H. Flinn, W. C. Dreessen, T. I. Edwards, E. C. Riley, J. J. Bloomfield, R. R. Sayers, J. F. Cadden, and S. C. Rothman, *Silicosis and Lead Poisoning among Pottery Workers*, Pub. Health Bull. 244, U. S. Public Health Service, Washington, 1939, 185 pp.
103. W. C. Cooper and L. J. Cralley, *Pneumoconiosis in Diatomite Mining and Processing*, Public Health Service Publication 601, Washington, 1958, 96 pp.
104. W. C. Cooper and E. N. Sargent, *J. Occup. Med.* **26**, 456–460 (1984).
105. R. R. Sayers, J. J. Bloomfield, J. M. DallaValle, R. R. Jones, W. C. Dreessen, D. K. Brundage and R. H. Britten *Anthraco-Silicosis among Hard Coal Miners*, Public Health Bulletin No. 221, Government Printing Office, Washington, 1936.

106. R. H. Flinn, H. F. Seifert, H. P. Brinton, J. L. Jones, R. W. Franks, *Soft Coal Miners Health and Working Environment*, Public Health Bulletin No. 270, Government Printing Office, Washington, 1941.

107. B. G. Clarke and C. G. Moffet, *J. Ind. Hyg. Tox.* **23**, 177–186 (1941).

108. J. Lieben, E. Pendergrass, and W. W. McBride, *J. Occup. Med.* **3**, 493–506 (1961).

109. E. J. Baier & R. Diakun, *J. Occup. Med.* **3**, 507–521 (1961).

110. W. S. Lainhart, H. N. Doyle, P. E. Enterline, A. Henschel, and M. A. Kendrick, *Pneumoconiosis in Appalachian Bituminous Coal Miners*, Public Health Service Publication No. 2000, Government Printing Office, Washington, DC, 1969, 148 pp.

111. W. S. Lainhart, *J. Occup. Med.* **11**, 399–408 (1969).

112. W. C. Dreessen, J. M. DallaValle, T. I. Edwards, J. R. Miller, and R. R. Sayers, *A Study of Asbestosis in the Asbestos Textile Industry*, Public Health Bulletin No. 241, U. S. Government Printing Office, Washington, DC, 1938, 133 pp.

113. W. E. Fleischer, F. J. Viles, Jr., R. L. Gade, and P. Drinker, *J. Ind. Hyg. Tox.* **28**, 9–16 (1946).

114. H. E. Ayer and J. R. Burg, "Cumulative Asbestos Exposure and Forced Vital Capacity" Presented at the XVIII International Congress on Occupational Health, Brighton, England, 1975; summarized in W. H. Walton, ed., *Inhaled Particles V*, Pergamon Press, Oxford, 1977, pp. 797–798.

115. L. D. Horowitz, *Am. Ind. Hyg. Assoc. J.* **37**, 227–232 (1976).

116. W. H. Walton, "Conversion Factors Between Particle Numbers, Surface Area and Mass Concentrations for Coal Dust", in H. A. Shapiro, ed, *Pneumoconiosis: Proceedings of the International Conference, Johannesburg*, Oxford University Press, Cape Town, 1970, pp. 584–586.

117. H. E. Ayer, J. M. Dement, K. A. Busch, H. B. Ashe, B. T. H. Levadie, W. A. Burgess, and L. Deberardinis, *Am. Ind. Hyg. Assoc. J.* **34**, 206–211 (1973).

118. J. W. Sheehy and C. E. McJilton, *Am. Industr. Hyg. Assoc. J.* **48** 914–918 (1987)

119. N. S. Seixas, N. J. Heyer, E. A. E. Welk, and H. Checkoway, *Ann. Occup. Hyg.* **41**, 591–601 (1997).

120. M. G. Gressel, W. A. Heitbrink, J. D. McGlothin, and T. J. Fischbach, *Appl. Ind. Hyg.* **3**, 316–320 (1988).

121. D. O'Brien, P. A. Froehlich, M. G. Gressel, R. M. Hall, N. J. Clark, P. Bost and T. Fischbach, *Am. Ind. Hyg. Assoc. J.* **53**, 42–48 (1992).

CHAPTER TEN

Basic Aerosol Science

Parker C. Reist, Sc.D., PE

1 INTRODUCTION

Knowledge of aerosol properties and behavior are important to the industrial hygienist. In many instances sampling an aerosol contaminant can be quite different from sampling a gaseous contaminant. Health effects caused by aerosols can vary depending on particle size and size distribution as well as on the composition of the particles. Respirator selection may depend greatly on aerosol size, shape, and composition. Other aerosol behavior from light scattering to coagulation can also vary markedly with particle size. This chapter discusses fundamental concepts of aerosol science, illustrating some of the calculations that can be made to predict aerosol behavior.

1.1 Definitions

There are a number of definitions in aerosol studies that are used in a wide variety of contexts. The more important are listed below:

1.1.1 Aerosol

An aerosol is a suspension of solid or liquid particles in a gas, a colloid. Included in the definition of aerosol would be a number of other definitions that describe one or several specific properties of the aerosol

1.1.1.1 Dust. Solids formed by disintegration processes like crushing, grinding, blasting and drilling, e.g., house dust, silica dust, saw dust. Particles are small replicas of the parent material. Size can be from submicroscopic to macroscopic.

1.1.1.2 Fume. Solids produced by physiochemical reactions such as combustion, sublimation, and distillation. Lead oxide, ZnO, and FeO are typical metallurgical fumes. These

Patty's Industrial Hygiene, Fifth Edition, Volume 1. Edited by Robert L. Harris.
ISBN 0-471-29756-9

particles are much smaller than dust particles (below 1 μm in diameter). They appear to flocculate readily.

1.1.1.3 Smoke. In this case the optical density is presupposed. The particles are produced from systems of organic origin such as smoke from coal, oil, wood, or other carbonaceous fuels. Smoke particles have the same sizes as metallurgical fumes.

1.1.1.4 Mists and Fogs. Particles produced by the disintegration of liquid or the condensation of vapor. These are spherical droplets small enough to float in moderate air currents. When these droplets coalesce to drops greater than about 0.4 mm, they may appear as rain.

1.1.1.5 Haze. Fairly small particles with some water vapor, as observed in the atmosphere.

1.1.1.6 Smog. Particles which are usually photochemical reaction products combined with water vapor (smoke + fog = smog).

1.1.2 Monodisperse Aerosol

An aerosol in which all of the particles have the same size.

1.1.3 Polydisperse Aerosol

An aerosol made up of particles of various sizes.

1.1.4 Aerodynamic Diameter

The diameter of a unit density sphere (density = 1 g/cm^3) which has the same settling velocity as the particle in question.

1.1.5 Stokes or Sedimentation Diameter

The diameter of a sphere having the same density and the same settling velocity as the particle in question.

1.1.6 Agglomerates or Flocs

Irregularly shaped aerosol particles made up of a number of smaller particles (often spherical), the result of agglomeration of highly charged or highly concentrated small particles (dense smokes or metal fumes, are examples of agglomerates).

1.2 Units

Up until about 1960 most aerosol data were reported in the English system of units, using such concepts as millions of particles per cubic foot (mppcf), grains per cubic foot, feet per minute (fpm), or similar English or Imperial units. This proved to be cumbersome. At present, units are usually either all in the centimeter–gram–second system of units or Le

Systeme International d'units (SI units). In this chapter cgs units will be used for discussion; conversion to other sets of units can be made after a problem is solved (this includes conversion to SI units). Table 10.1 shows the basic dimensions for the various systems of units.

The relationship between prefixes used on the various units is given in Table 10.2. For example, 1 nanometer (nm) $= 10^{-9}$ meters and one kilogram (kg) $= 10^3$ grams.

To determine the dimensions, and hence the units, for force, recall that $F = ma$, acceleration, $a = v/t$, and velocity, $v = s/t$. The symbol s represents a distance and t a time. Then the following can be written:

$$F = ma = \frac{mv}{t} = \frac{ms}{t^2}$$

In the cgs system of units, this is gram-cm/seconds^2 = dyne. Similarly, in the SI system of units, force is equal to kilogram-meter/seconds^2 = Newton.

To convert from one set of units to another it is only necessary to apply the appropriate conversion factors. To avoid confusion a consistent set of units must be used for problem solutions. For most aerosol problems, computations in cgs units are to be preferred. Then conversion to another set of units can be done in a single step after the problem is solved.

Example: Convert 100 cm/s to feet per minute. Notice in the example how the intermediate units cancel out.

Table 10.1. Systems of Units

Designation	Length	Mass	Time
English or fpm	Foot	Poundal[a]	Minute
SI	Meter	Kilogram	Second
cgs	Centimeter	Gram	Second
mks	Meter	Kilogram	Second

[a]In the English system, pound represents a force (mass × 32 ft/sec^2 and poundals a mass).

Table 10.2. Prefixes for Units

Prefix	Value	Abbreviation
pico	10^{-12}	p
nano	10^{-9}	n
micro	10^{-6}	μ
milli	10^{-3}	m
centi	10^{-2}	c
deci	10^{-1}	d
kilo	10^3	k

$$100\ \frac{\text{cm}}{\text{s}} \times 60\ \frac{\text{s}}{\text{min.}} \times \frac{1}{30.5}\ \frac{\text{ft}}{\text{cm}} = 196.7\ \frac{\text{ft}}{\text{min.}}$$

2 PHYSICAL PROPERTIES OF AEROSOLS

2.1 Particle Size

Aerosol sizes are usually referred to in terms of micrometers (μm). One micrometer is 10^{-6} meters or 10^{-4} centimeters. An older term is the "micron", equivalent in length to the micrometer. Its use is discouraged, but it is still easier to say than "micrometer." The sizes of aerosol particles range from about 10^{-2} μm diameter (smallest) to 10^{2} μm diameter (largest). The physiological aerodynamic diameters for aerosols are $>$ 10 μm—inspirable, 4–10 μm—thoracic, and $<$ 4 μm—respirable (1). Figure 10.1 illustrates the ranges of particle encountered in aerosol problems.

2.2 Shape

There are three general shapes of particles. Isometric particles are those for which all three dimensions are roughly the same. Spherical or regular polyhedral shapes and particles approximating these forms belong here. Platelets are particles having much greater lengths in two dimensions than in the third (leaf fragments, scales, etc.), and particles with great length in one direction and small lengths in the other two are known as fibers, prisms, needles, hairs, strings, or threads.

Particle shape varies with the formation method and the nature of the parent material. For example, liquid condensation aerosols are generally spherical in shape while particles formed by attrition are rarely spherical.

2.3 Surface Properties

The surface area per gram of an aerosol increases with decreasing particle size and decreases with increasing polydispersity. The boundary conditions between phases become more important as particle size decreases but also become more confused. It then is difficult to distinguish between small particles and large molecules. Some of the important surface properties of particles influenced by particle size are surface energy, adsorption, absorption, chemical reaction rates, wetability, and electrostatic effects. Table 10.3 shows surface area per gram of aerosol for aerosols of different mean diameters and degrees of dispersity.

3 PARTICLE SIZE DISTRIBUTIONS

Most frequently, in an aerosol the particles are present in a variety of sizes, that is, the aerosol is said to be polydisperse. Monodisperse aerosols are very rare in nature and when they do appear, generally do not last very long.

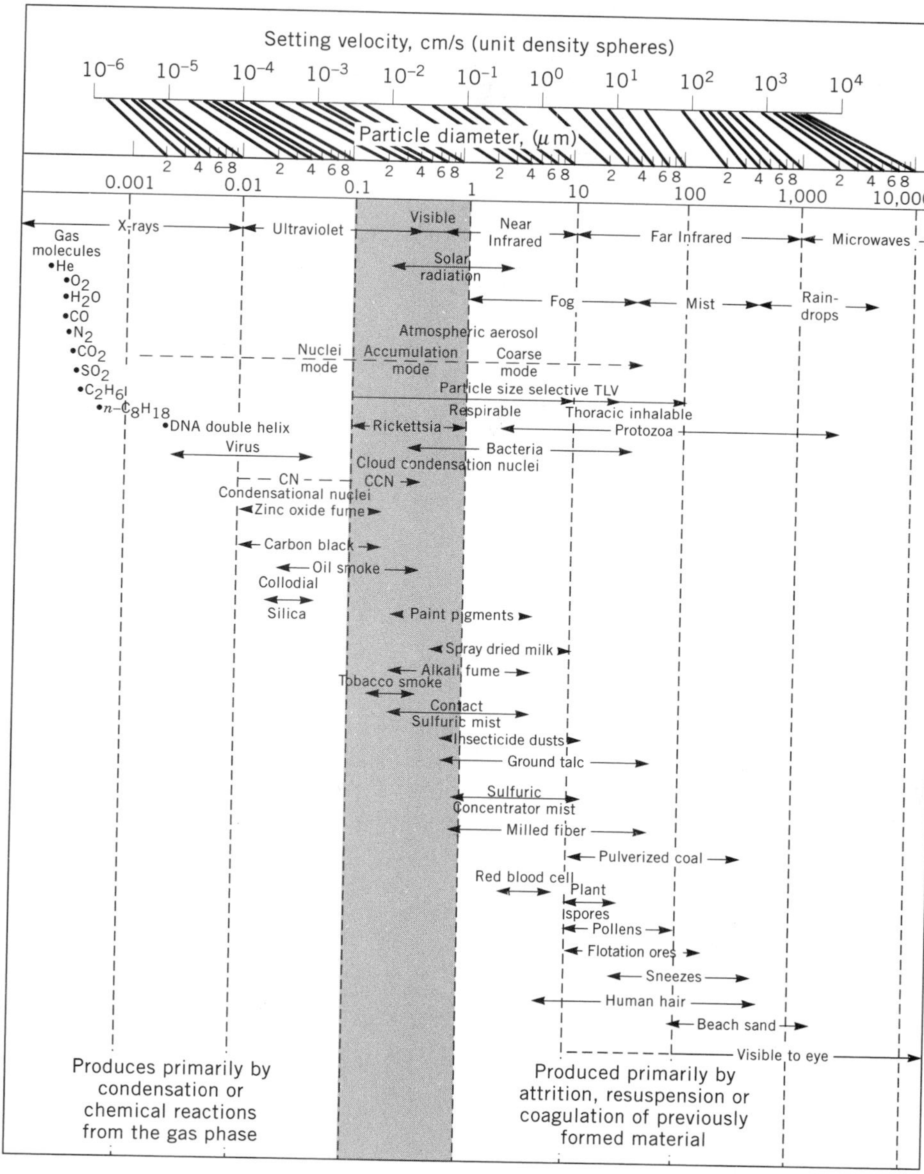

Figure 10.1. Molecular and aerosol particle diameters. Reist 1997 adapted from Ref. 2.

Table 10.3. Area per Gram of Aerosol Particles of Different Diameters and Geometric Standard Deviations[a]

d_g	σ_g	Area/g
10	2.5	735
10	2.0	1,805
10	1.5	3,978
10	1.0	6,000
5	2.5	1,471
5	2.0	3,610
5	1.5	7,956
5	1.0	12,000
1	2.5	7,355
1	2.0	18,051
1	1.5	39,779
1	1.0	60,000
0.5	2.5	14,710
0.5	2.0	36,102
0.5	1.5	79,558
0.5	1.0	120,000
0.1	2.5	73,549
0.1	2.0	180,512
0.1	1.5	397,790
0.1	1.0	600,000

[a]Density = 1 g/cm^3, log-normal distribution assumed

Sometimes it is satisfactory or sufficient to represent all particle sizes in a polydisperse aerosol by a single size, but usually more information is needed about size distribution. A simple plot of particle frequency versus size gives a picture of the sizes present in the aerosol, but this usually is not enough for a complete quantitative picture.

There are a number of ways polydisperse aerosols can be described, using mathematical or visual methods. Some of the more common methods will be discussed.

3.1 Mean and Median Diameter

The simplest way of treating a group of different particle diameters is to add together all the diameters and divide by the total number of particles. This gives the average diameter of the particles. Mathematically this can be expressed as

$$\bar{d} = \frac{\sum d_i \times n_i}{\sum n_i} \tag{1}$$

This quantity is known as the mean particle diameter.

The median particle diameter can be determined by listing all diameters in order from the smallest to the largest and then finding the particle diameter that splits the list into two equal halves.

Although simple in concept, neither the mean nor the median diameter alone conveys much information about the general range of particle diameters present. It is common practice to describe an aerosol solely by some mean or average value, completely ignoring considerations of particle size distribution. When this is done, estimates of aerosol properties are much less accurate than they would have been if all particle sizes had been taken into account.

3.2 Standard Deviation

Besides representing a set of data by some mean value, it is also useful to have some measure of dispersion or a way of viewing how the data is dispersed around the mean. This can be done graphically or by some mathematical technique.

The most important measure of dispersion is the standard deviation, a gauge of the amount each individual measurement varies from the mean. Standard deviation can be computed from the relationship

$$\sigma = \frac{\sum (d_i - \overline{d})^2 \times n_i}{\left(\sum n_i\right) - 1} \qquad (2)$$

where n_i is the number of particles with a diameter d_i and Σn_i the total number of particles. Usually the standard deviation is measured from a sample of finite magnitude, whereas the true value of the standard deviation should come from the population as a whole. The term σ represents the estimated standard deviation, based on a sample of the population, and is estimated by using $(\Sigma n_i) - 1$ in the denominator of equation 2 rather than Σn_i. When n is large there is essentially no difference between either value.

The standard deviation, σ, of a distribution can also be calculated using

$$\sigma^2 = \frac{\sum d_i^2 \times n_i}{\sum n_i - 1} - \overline{d}^2 \qquad (3)$$

As an example, Table 10.4 illustrates a method of computing the mean, median, and standard deviation of an aerosol with the listed size distribution characteristics.

The mean is calculated as follows:

$$\overline{d} = \frac{\sum d_i \times n_i}{\sum n_i} = \frac{75{,}627}{79{,}552} = 0.951\ \mu\text{m} \qquad (1)$$

Table 10.4. Illustrative Calculations for Determining Mean and Standard Deviation[a]

d_{j-1}, µm	d_j, µm	d_i µm	Log d_i	n_i	$n_i \times d_i$	$n_i \times d_i^2$
0.076	0.122	0.099	−1.004	46	5	0
0.122	0.195	0.159	−0.800	573	91	14
0.195	0.312	0.254	−0.596	3,768	957	243
0.312	0.500	0.406	−0.391	13,055	5,300	2,152
0.500	0.800	0.650	−0.187	23,853	15,504	10,078
0.800	1.280	1.040	0.0171	22,981	23,900	24,856
1.280	2.049	1.665	0.2213	11,674	19,437	32,363
2.049	3.279	2.664	0.4255	3,127	8,330	22,192
3.279	5.248	4.263	0.6297	442	1,884	8,033
5.248	8.398	6.823	0.8340	33	225	1,536
			Sums	79,552	75,633	101,467

[a] d_i is the mid-point size of the size increment d_{i-1} and n_i is the number of particles in this size increment.

And the standard deviation as

$$\sigma^2 = \frac{\sum d_i^2 \times n_i}{\left(\sum n_i\right) - 1} - \bar{d}^2$$

$$\sigma = \sqrt{0.372} = 0.610 \ \mu\text{m}$$

The median lies somewhere between 0.5 and 0.8 µm.

3.3 Histograms

Particle size distribution can also be graphically shown using a histogram or bar chart. The numbers of particles in various size intervals are plotted. Dividing the number of particles in an interval by the width of the interval normalizes the ordinate or height of each bar. Since the width of each bar represents the actual width of each size interval, the area of each block so normalized represents the relative frequency of particles in that particular size interval. Figure 10.2 shows a histogram of the data from Table 10.4.

Charts or graphs of this sort have the advantage of showing at a glance what the particle size distribution of an aerosol looks like and are perhaps the best way of visually representing complex size distribution data.

3.4 Log Normal Size Distribution

If the size interval of an aerosol is permitted to become very small, the resulting histogram begins to approximate a smooth curve. Then it is possible to represent the distribution by some mathematical function. One widely used form that is applicable to many industrial aerosols is the log-normal distribution.

It has been observed that although most particle size data will not fit a normal distribution, these same data will very often fit a normal distribution if frequency is plotted

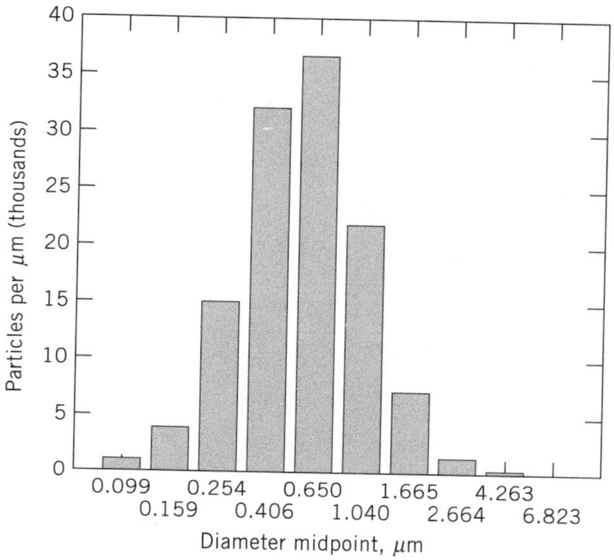

Figure 10.2. Histogram of data in Table 10.4.

against the logarithm of particle size. This tends to spread out the smaller size ranges and compress the larger ones. If the new plot then looks like a normal distribution, the particles are said to be log-normally distributed and the distribution is called a log-normal distribution. By analogy with a normal distribution, the mean and standard deviation are known as the geometric mean diameter, d_g, and geometric standard deviation, σ_g.

With a log-normal distribution, one geometric standard deviation represents a range of particle sizes within which lie 67% of all particles in the sample. In this case the range is from d_g/σ_g to $_{dg} \times \sigma_g$, unlike the simple additive case for a normal distribution. Ninety-five percent of all particles lie in a range d_g/σ_g^2 to $d_g \times \sigma_g^2$. For a monodisperse aerosol, σ_g is equal to 1 for a log normal distribution whereas σ is equal to 0 for a normal distribution.

3.5 Log Probability Paper

Because a log normal distribution can be expressed as a distinct mathematical function, graph paper can be constructed on which a cumulative lognormal distribution plots as a straight line. An example of such a plot using the data presented in Table 10.4 is shown in Figure 10.3. Data are plotted as the log of the upper size of particles in a size interval as a function of the cumulative percentage of particles up to and within that size interval. A straight line on such a plot implies a lognormal distribution.

If a straight line can be fitted to the plot, then the median particle diameter can be determined as being the 50% value on the plot (remember that when plotting number

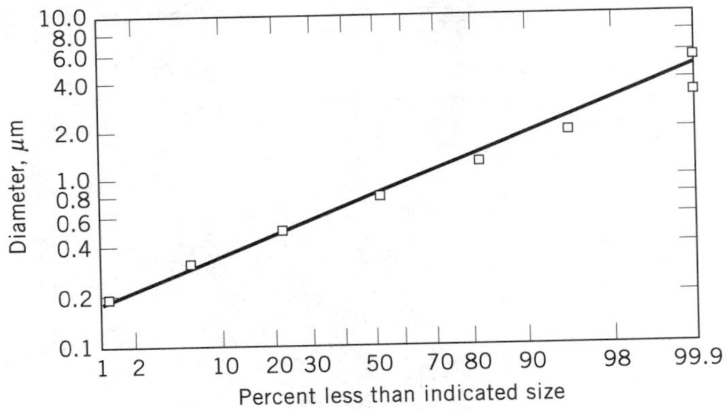

Figure 10.3. Near log-normal distribution plotted on a log-probability graph.

distribution, geometric mean and median for the number distribution are the same if there is a log-normal distribution). The geometric standard deviation is determined by the ratio

$$\frac{d_{84\%}}{d_{50\%}} = \frac{d_{50\%}}{d_{16\%}} = \sigma_g \tag{4}$$

Use of log probability paper is the simplest way to determine the mean and geometric standard deviation of an aerosol provided the distribution does indeed follow a log-normal shape, or at least approximates it.

3.6 Other Definitions of Means

There are a number of different mean or median values that can be defined for a particle size distribution. These means or medians are useful depending on where the data came from and how the data are to be used. For example, the diameter of average mass (volume) can be defined as representing the diameter of a particle whose mass (volume) times the number of particles gives the total mass (volume) of all the particles. Similarly, the diameter of average surface represents the diameter of a particle whose surface times the number of particles gives the total surface.

Choice of which average diameter to use in a given situation depends on how the diameter was measured or how it is to be used. For the case where aerosol mass is measured and the fractions collected are associated with specific particle diameters, the resulting average value is the mass mean or median diameter. If the diameters are expressed as aerodynamic diameters, this value is the mass median aerodynamic diameter (MMAD). In studying chemical reaction rates, the volume-surface mean diameter may be more important than just the arithmetic mean or geometric number mean.

Table 10.5 gives definitions for various "average" diameters (3). For a log-normally distributed aerosol the different diameters defined in Table 10.5 can be related by

Table 10.5. Definitions for Various "Average" Diameters

Indicated diameter	Symbol	Definition	Description
Mode	d_0 $p = -1$	d at maximum n_1	Diameter associated with the maximum number of particles in a distribution
Geometric mean	d_g $p = 0$	$\log^{-1} (\Sigma n_i \log d_i/\Sigma n_i)$	The Σnth root of the product of all particle diameters, also for a log-normal distribution the median diameter
Arithmetic mean	d $p = 0.5$	$\Sigma n_i d_i/\Sigma n_i$	The sum of all diameters divided by the total number of particles
d of average surface	d_s $p = 1.0$	$\sqrt{\Sigma n_i d_i^2/\Sigma n_i}$	The diameter of a hypothetical particle having average surface area
d of average volume (mass)	d_v $p = 1.5$	$\sqrt[3]{\Sigma n_i d_i^3/\Sigma n_i}$	The diameter of a hypothetical particle having average volume of mass
Surface median diameter	$d_{sm\,d}$ $p = 2.0$	$\log^{-1} (\Sigma n_i d_i^2 \log d_i/\Sigma n_i d_i^2)$	The geometric mean of the particle surface areas or for a log-normal distribution the area median diameter
Surface mean diameter (Sauter diameter)	d_{sm} $p = 2.5$	$\Sigma n_i d_i^3/\Sigma n_i d_i^2$	The average diameter based on unit surface area of a particle
Volume (mass) median diameter	$d_{mm\,d}$ $p = 3.0$	$\log^{-1} (\Sigma n_i d_i^3 \log d_i/\Sigma n_i d_i^3)$	The geometric mean of particle volumes (mass) or for a log-normal distribution the volume (mass) median diameter
Volume (mass) mean diameter	d_{vm} $p = 3.5$	$\Sigma n_i d_i^4/\Sigma n_i d_i^3$	The average diameter based on the unit volume (mass) of a particle

$$d_p = d_g \exp (p \ln 2\sigma_g) \quad (5)$$

The term d_p is the diameter associated with the value p in Table 10.5 is the geometric mean diameter, and p is a parameter in equation 5 which links the desired diameter in a log-normal distribution to the geometric mean diameter.

Eq. 5 is a more general form of the Hatch-Choate equation (4), a well-known relationship used for converting particle number measurements to mass measurements and vice versa. It is important to note that σ_g will be the same regardless of the definition of diameter used. That is, with a lognormal distribution, σ_g will be the same whether number, surface, or mass median diameters are being measured.

3.7 Effect of Polydispersity on Aerosol Concentration

Use of equation 5 can show how neglecting a particle size distribution can affect estimates of concentration. For example, suppose an aerosol is measured with a mean diameter of 2 μm and a concentration of 100 μg/m³. Assuming all particles are the same size and have a density of one, the particle concentration can be computed to be 2.39×10^7 p/cc. However, if the aerosol is actually log-normally distributed with a geometric standard deviation of 2.5, then to get the "true" number concentration we must find the diameter of average mass, $d_{1.5}$. Using equation 5 this is found to be 7.05 μm, giving a "true" particle concentration of 5.46×10^5 p/cc. From these calculations it can be seen that as the aerosol becomes more polydisperse, it takes fewer particles to give an equivalent mass concentration, or in other words, for the same particle number concentration d_g, the more monodisperse an aerosol is, the greater is its mass.

4 STOKES LAW AND NEWTON'S RESISTANCE LAW

An aerosol is a suspension of particles in a gaseous medium, usually air. Without the medium there would be no aerosol. The medium acts to restrain random particle motion, support the particles against the strong pull of gravity and in some cases act as a buffer between particles.

4.1 Knudsen Number

Although there is a tendency to always think of the medium as a continuum, like water, sometimes the medium needs to be considered as an ensemble of small particles in which the larger aerosol particles are suspended. How the medium is viewed depends to a great extent on the aerosol particle size and the mean free path of the gas molecules, that is, the mean distance a gas molecule travels before it hits another gas molecule. Whether the medium is thought of as a continuum or not depends on the ratio of the mean free path of the gas molecules to the aerosol particle diameter, known as the Knudsen number,

$$K_n = \frac{2\lambda}{d} \tag{6}$$

The term λ represents the mean free path of the gas molecules, and d the particle diameter. With normal temperature and pressure the molecular mean free path in air is about 0.0686 μm.

When K_n is small the medium must be treated as a continuum; when K_n is large, the molecular nature of the medium is most important. For air at normal temperature and pressure this change occurs with particles in the 0.1 to 1.0 μm range (see Fig. 10.1).

4.2 Stokes Law

With a continuous medium it is possible to derive an expression for the resistance of a sphere moving through a fluid, first done by G.G. Stokes in 1855 (5). This expression, known as Stokes Law, is written as

$$F = 3\pi\mu dv \tag{7}$$

where F is the force on the sphere, in dynes, μ is the viscosity of the medium in poises, v is the relative velocity between the air and the sphere in centimeters per second, and d is the diameter of the sphere in centimeters.

For Stokes' solution, he assumed an incompressible, continuous, viscous, and infinite medium with spherical, rigid particles. Stokes Law has been shown to give a reasonable approximation of the resisting force on moving spheres in many situations, and the only stipulation is that the assumptions listed above are not violated.

Although air is compressible, compression is not important for motion of very small particles. This assumption can be considered to be always valid.

Stokes Law becomes incorrect when the other assumptions used to derive it cannot be met. For most aerosol problems only three of the assumptions need be considered. Particle shape is important since most real-life particles are not spherical, but are irregularly shaped. However, if the particles are isometric, that is, if all three of the particle's major axes are approximately the same, then the particle approximates a sphere and Stokes Law can be used. There are corrections to Stokes Law that permit estimation of resistance for platelets (two long axes, one short one) and fibers (one long axis, two short ones) (see Ref. 3).

The Reynolds number is a dimensionless number describing the ratio of inertial to viscous forces acting on a particle. When the Reynolds number is low, viscous forces predominate. Hence, Stokes Law holds when Re is small, specifically when $\mathrm{Re} < 1$. In most cases in industrial hygiene problems this will be true. Reynolds number is defined as

$$\mathrm{Re} = \frac{dv\rho_m}{\mu} = \frac{dv}{\nu} \tag{8}$$

Where v is the relative velocity between the particle and its surrounding medium, ρ_m the air density and μ the air viscosity. Kinematic viscosity, defined as $\nu = \mu/\rho_m$, has a value of about 0.15 for air at normal temperature and pressure.

The other major assumption is that the aerosol particle is in a continuous medium. It was observed separately by R. A. Millikan (6) and E. Cunningham (7) in 1910 that as particle size gets smaller, a particle tends to settle faster than predicted by Stokes Law. This was explained as being caused by particles "slipping" between the gas molecules and thus experiencing less resistance per unit cross-sectional area.

To more accurately predict the aerodynamic behavior of sub-micron aerosol particles, a correction factor, C_c, known as the Cunningham, slip, or Millikan correction factor is used with Stokes Law. Then Stokes Law takes the form

$$F = \frac{3\pi\mu d v}{C_c} \tag{9}$$

with C_c defined as

$$C_c = 1 + K_n(1.257 + 0.4 \exp(-1.1/K_n)) \tag{10}$$

where K_n is the Knudsen number.

The term C_c is always equal to or greater than 1. Sometimes, when $d > 0.2\mu m$, C_c is approximated by the expression

$$C_c \approx 1 + 1.257K_n \tag{11}$$

The Cunningham correction factor represents the mechanism for transition from the continuum to the molecular case. For large values of d, the resisting force F is proportional to d whereas for very small values of d, F is proportional to d^2. In this case a large value of d is something greater than 1 μm.

The Cunningham correction factor, C_c, is an important correction to Stokes' law and should always be used when particles are less than 1-μm in diameter. The term C_c is usually neglected for particles $>$ 1 μm. However, the practice of neglecting C_c is only for convenience in calculations. With the advent of the programmable calculator or personal computer, it is now best to always include the factor C_c in computations wherever possible.

5 PARTICLE MOTION

5.1 Terminal Settling Velocity

As an aerosol particle moves in air, the resisting force of the air opposes the force causing the motion. As long as these two forces are unequal, the particle is either accelerating of decelerating. When the forces are equal, the particle will move at a constant velocity. For example, particles fall in air due the force of gravity and the air resists the fall. As mentioned above, for small particles Stokes Law gives this air resistance. When the two forces balance

$$F = mg = 3\pi\mu dv/C_c \tag{12}$$

so that for spherical particles of diameter d

$$v_T = \frac{1}{18}\frac{d^2(\rho_p - \rho_m)gC_c}{\mu} \tag{13}$$

where v_T is the particle equilibrium settling velocity, known as the terminal settling velocity, μ is the viscosity of the air (1.82×10^{-4} poises at normal temperature and pressure), g is the acceleration due to gravity (981 cm/s^2), d is the particle diameter in centimeters, C_c is Cunningham correction factor, and $(\rho_p - \rho_m)$ is the difference in density between the particle and the air, in g/cm^3. For most aerosol problems ρ_p is so much greater than ρ_m that equation 13 is often written as

$$v_T = \frac{1}{18}\frac{d^2\rho_p gC_c}{\mu} \tag{14}$$

5.2 The Relaxation Time, τ

By defining

$$\tau = \frac{1}{18}\frac{d^2(\rho_p - \rho_m)C_c}{\mu} \approx \frac{1}{18}\frac{d^2\rho_p C_c}{\mu} \tag{15}$$

an extremely useful parameter representing a relaxation time, τ, is created. Then terminal settling velocity can be written as

$$v_T = \tau g \tag{16}$$

The parameter τ is important in all of aerosol dynamics. It represents the relationship between the particle, its density, and the properties of the medium (viscosity and density). Use of τ generally simplifies equations involving aerosol dynamics.

By equating Stokes resistance with the gravitational force to determine the terminal settling velocity, important insights into the time-dependent cases of acceleration or deceleration to terminal velocity are not apparent. The rapidity with which the terminal settling velocity is reached is given by the factor exp $(-t/\tau)$. Thus the smaller the value of τ, the more quickly an aerosol particle reaches equilibrium or steady-state conditions. For example, for a 2-μm-diameter unit-density sphere $\tau = 1.35 \times 10^{-5}$ s. Since $\exp(-7)$ is about 0.001, equilibrium values are essentially reached when $t/\tau = 7$, or for the 2-μm sphere, within about 100 μs.

Table 10.6 gives values of C_c, τ and 7τ for unit-density spheres of a number of different diameters. It can be seen that particles smaller than several micrometers in diameter will rapidly accelerate or decelerate to equilibrium conditions.

Table 10.6. Cunningham Correction Factor and τ for Various Particle Diameters[a]

d	Kn	C_c	τ, s	$7 \times \tau$, s
0.01	13.72	23.311	7.12×10^{-9}	4.98×10^{-8}
0.02	6.86	11.960	1.46×10^{-8}	1.02×10^{-7}
0.05	2.744	5.184	3.96×10^{-8}	2.77×10^{-7}
0.08	1.715	3.517	6.87×10^{-8}	4.81×10^{-7}
0.1	1.372	2.971	9.07×10^{-8}	6.35×10^{-7}
0.2	0.686	1.918	2.34×10^{-7}	1.64×10^{-6}
0.5	0.2744	1.347	1.03×10^{-6}	7.20×10^{-6}
0.8	0.1715	1.216	2.37×10^{-6}	1.66×10^{-5}
1	0.1372	1.172	3.58×10^{-6}	2.51×10^{-5}
2	0.0686	1.086	1.33×10^{-5}	9.28×10^{-5}
5	0.02744	1.034	7.89×10^{-5}	5.53×10^{-4}
8	0.01715	1.022	2.00×10^{-4}	1.40×10^{-3}
10	0.01372	1.017	3.11×10^{-4}	2.17×10^{-3}
20	0.00686	1.009	1.23×10^{-3}	8.62×10^{-3}
50	0.002744	1.003	7.66×10^{-3}	5.36×10^{-2}
80	0.001715	1.002	1.96×10^{-2}	1.37×10^{-1}

[a]Particle density = 1 g/cm^3.
Viscosity = 0.000182 poises.
Mean free path = 0.0686 μm.

For particles injected into a moving air stream (similar to acceleration under the influence of gravity and similar to problems of particle deceleration), the difference between particle velocity and air stream velocity decreases by a factor of $e = 0.368$ for each time period $t = \tau$. Thus within 7τ for all practical purposes steady-state conditions are reached, and the particle has achieved the velocity of the moving airstream, regardless of its initial velocity.

5.3 Stop Distance

If a particle is given an initial velocity in one direction with no forces acting on it except air resistance, it will travel a certain distance before losing all its velocity. This is known as the stop distance, s, and is given by

$$s = \tau v_0 \tag{17}$$

where v_0 is the initial velocity. For example, a 1-μm-diameter particle projected into air at an initial velocity of 1000 cm/s will move a distance of only 0.0036 cm or 36 of its diameters before stopping.

5.4 Motion Outside of the Stokes Region

There are occasions when particle velocity is so great or particle diameter so large that Stokes' Law is no longer applicable (that is, when Re > 1). Then an empirical approach is taken using the "coefficient of drag" $= C_D$. The coefficient of drag is an experimentally determined parameter which is a function of Reynolds number. This is shown for spheres in Figure 10.4.

The resisting force of air for a spherical particle of diameter d can be defined as

$$F = C_D \pi d^2 v^2 / 8 \tag{18}$$

For sedimenting spherical aerosol particles, the drag force and gravitational force can be equated to give

$$v_T^2 = \frac{4d(\rho_p - \rho_m)g}{3C_D\rho_m} \tag{19}$$

Unfortunately C_D depends on Re which depends on v_T. This problem can be treated through the use of plots of $C_D\mathrm{Re}^2$ versus Re and C_D/Re versus Re.

If d is known and is desired to find v_T, then use

$$C_D\mathrm{Re}^2 = \frac{4}{3}\frac{d^3\rho_m(\rho_p - \rho_m)g}{\mu^2} \tag{20}$$

and solve for $C_D\mathrm{Re}^2$. This rearrangement eliminates the v term in equation 19. The result from equation 20 can be used to enter Figure 10.5, a plot of $C_D\mathrm{Re}^2$ vs. Re, and find Re, which can then be solved to give v_T.

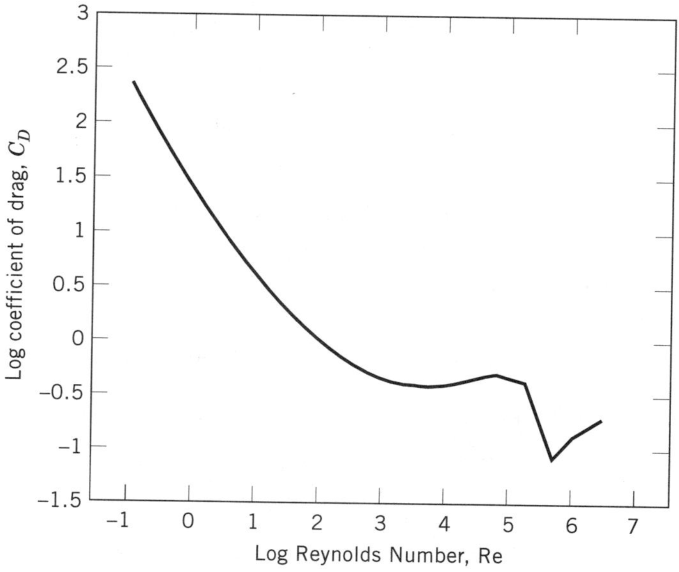

Figure 10.4. Coefficient of a drag for a sphere as a function of the Reynolds number.

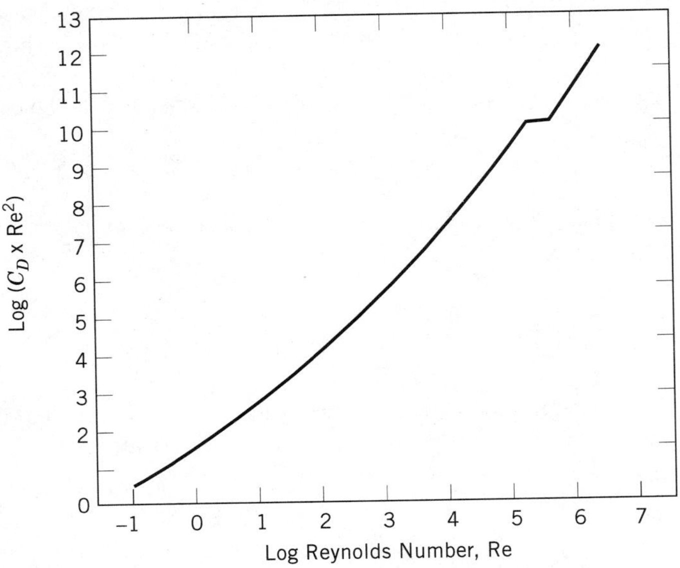

Figure 10.5. Coefficient of drag times the square of the Reynolds number as a function of the Reynolds number.

An approximation of the curve in Figure 10.5 is given by

$$\text{Re} = \log^{-1}(-5 + \sqrt{12.09 + 9.09 \log (C_D\text{Re}^2)}) \tag{20a}$$

If v_T is known and it is desired to find d, then use

$$\frac{C_D}{\text{Re}} = \frac{4\mu g(\rho_p - \rho_m)}{3v_T^3\rho_m^2} \tag{21}$$

and solve for C_D/Re. This rearrangement eliminates the d term in equation 19. The result from equation 21 can be used to enter Figure 10.6, a plot of C_D/Re vs. Re, and find Re, which can then be solved to give d.

An approximation for Re when C_D/Re is known is given by

$$\text{Re} = \log^{-1}(8.26 - \sqrt{56 + 8.7 \log (C_D/\text{Re})}) \tag{21a}$$

For very large particles the stop distance can be approximated by

$$s \approx \frac{4}{21}\frac{\rho_p}{\rho_m} d\sqrt{\text{Re}_i} \tag{22}$$

The utility of equation 22 lies in its generality. However, this solution is applicable only to one-dimensional flow or to those cases where one-dimensional flow can be approximated.

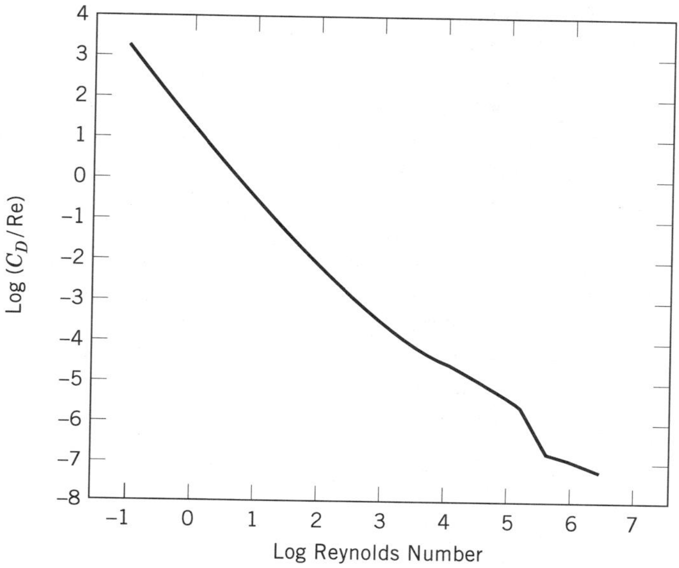

Figure 10.6. Coefficient of drag times the square of the Reynolds number as a function of the Reynolds number.

5.5 Ideal Stirred Settling

Although each particle making up an aerosol settles at its own terminal settling velocity, settling rarely takes place in absolutely still air. There is always some circulation and mixing of the air. This mixing has the effect of producing a uniform aerosol concentration throughout the space, which decreases with time because of sedimentation. Given a room of height H, the decrease is given by equation 23

$$c = c_0 \exp\left(-\frac{v_T t}{H}\right) \tag{23}$$

with c_0 the initial concentration and c the concentration after some time t. Equation 23 implies exponential decay of an aerosol concentration in a closed chamber. This is observed in practice.

5.6 Isokinetic Sampling

In aerosol sampling the measured concentration and size distribution of an aerosol sampled from a moving airstream should represent as closely as possible the concentration and size distribution of the original aerosol. There are several physical reasons why measured concentration can differ from the true concentration. Gravitational or inertial deposition of the sample as it flows into the sampling probe can result in loss of larger-sized particles. Also, deposition or selective collection at the mouth of the sampling probe could affect the proportion of various sizes of particles measured.

If the probe is not aligned with main flow some larger particles may be lost, giving a sample concentration which would be less than actual. Or, if the collection velocity is greater than stream velocity, some particles, because of their inertia, may fail to follow streamlines and therefore would not be collected, giving a sample concentration less than actual. Finally, the collection velocity could be less than the stream velocity and then the sample concentration could be greater than the actual.

If the probe is aligned with the flow, and the sample velocity is equal to the stream velocity, sampling is said to be isokinetic, and the concentration as collected should match the actual concentration. If sampling velocity differs from stream velocity, sampling is known as anisokinetic, which under some conditions can give rise to errors in the measured concentrations. With other conditions of sampling, depending on the Stokes number, $Stk = \tau u/r$, there may be no errors, even with anisokinetic sampling. Here r is the sampling probe radius. If c_s is the sample concentration, c the true concentration, u the stream velocity and u_s the sampling velocity, sampling results can be summarized as

$$\begin{aligned}
c_s/c &= 1 \text{ when } u = 0 \\
c_s/c &= 1 \text{ when } u > 0 \text{ but } Stk \ll 1 \\
c_s/c &> 1 \text{ when } u/u_s > 1 \\
c_s/c &< 1 \text{ when } u/u_s < 1 \\
c_s/c &= 1 \text{ when } u = u_s.
\end{aligned}$$

Figure 10.7 illustrates some isokinetic flow configurations.

6 IMPACTORS

6.1 Theory

When air that is carrying particles suddenly changes direction, the particles, because of their inertia, tend to continue along their original trajectories. If the change in air direction is caused by an object placed in the airstream, particles with sufficient inertia will strike the object. This process is known as impaction and the surface on which particles strike is the impaction surface. Impaction is the mechanism by which many large particles are removed from the atmosphere, it is one of the important mechanisms for removal of particles by the lungs, and it is important in air cleaning as well as aerosol sampling.

Impaction is characterized by the ratio of the stop distance of a particle divided by the half-width of the opening from which the particle is issuing. This ratio is known as the Stokes number,

$$Stk = \frac{\tau u}{W/2} \tag{24}$$

where u is the velocity of the airstream, $W/2$ the radius or half-width of the jet from which the airstream is issuing, and τ the relaxation time of the aerosol.

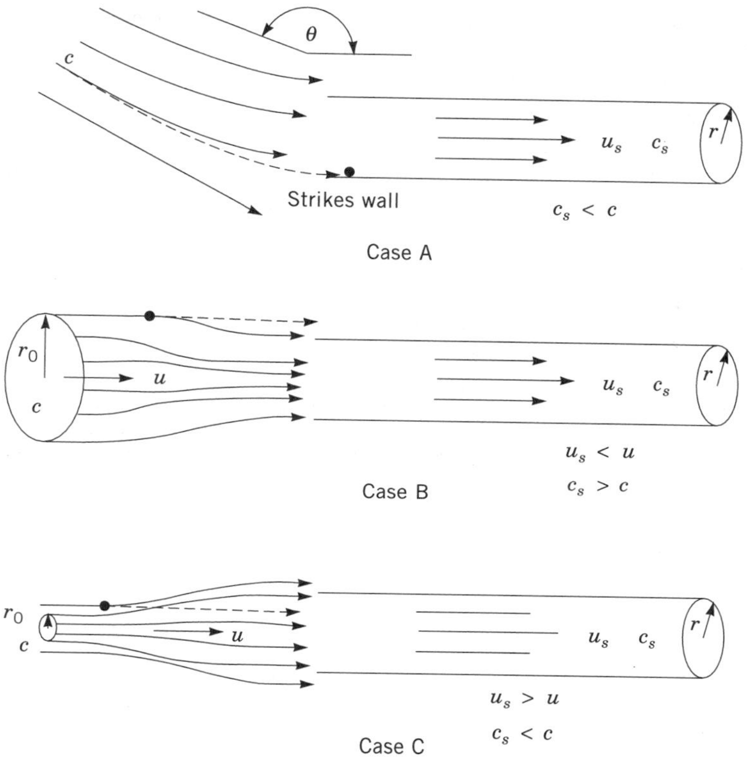

Figure 10.7. Types of anisokinetic sampling.

The Stokes number, a dimensionless parameter, is used to describe impactor behavior. For impactors with rectangular openings, W is the slit width; for circular openings W represents the diameter of the impactor opening. Some authors prefer to use the impaction parameter, ψ, rather than the Stokes number, to describe impactor properties. The impaction parameter is one-half the Stokes number.

It is common practice to plot impactor efficiency as a function of either $\sqrt{Stk}$ or $\sqrt{\psi}$. This is done because the particle diameter is present in either term as d^2, making the square root of the term proportional to particle diameter.

6.2 Impactor Operation

The characteristic behavior of impactors depends on factors such as the nozzle-to-plate distance, nozzle shape, flow direction and Reynolds-numbers for both the jet and the particle (3). Other factors of importance include the probability the particles will stick to the impaction surface and particle loss to the walls of the impactor. It is not surprising that with such a variety of possible variables it is quite difficult, if not impossible, to accurately predict impactor characteristics on purely theoretical grounds.

For a well-designed impactor, a typical plot of impactor efficiency vs. the square root of the Stokes number is shown in Figure 10.8. It can be seen that the efficiency curve may deviate from the ideal case. In the ideal case, there would be a single value of $\sqrt{Stk}$ for all efficiencies and hence a sharp size cut of the impactor. All particles larger than this size would be collected, and all smaller sizes would be passed. In actuality, this is not the case and a range of particle sizes is collected with varying efficiencies. To represent the stage collection characteristic, it is often the practice to choose the 50% efficiency point as being the representative cut point. The maximum slope at this point most nearly represents the ideal case. In Figure 10.8 both the actual and ideal cases would be considered to have the same characteristic cut point or cut size.

Impactor 50% cut points can be estimated from the equation

$$d_{50} = \sqrt{\frac{9Stk_{50}\mu LW^2}{C_c\rho_p Q}} \tag{25}$$

for rectangular jets of length L and width W or

$$d_{50} = \sqrt{\frac{9Stk_{50}\mu\pi W^3}{4C_c\rho_p Q}} \tag{26}$$

for round jets of diameter W. These are rearrangements of equation 24. Empirical estimates of Stk_{50} are 0.77 for rectangular jet impactors and 0.49 for round jet impactors (8).

The most common configuration for impactors used for aerosol sampling is to have a series of jets of decreasing size, arranged so that the air passes in series from the largest through the smallest slot. This cascade arrangement permits the aerosol to be fractionated into a number of size intervals, depending on the number of impactor stages used. Aerosol mass collected on the different impactor stages is then analyzed to provide size distribution information. The practical lower limit of collection for an impactor of this type is about 0.4 μm unit density sphere.

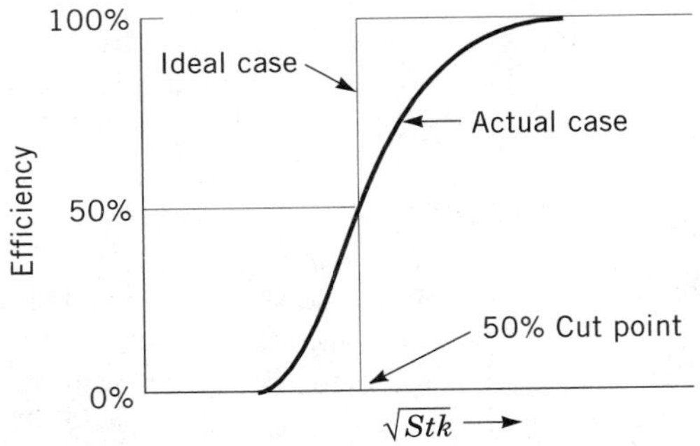

Figure 10.8. Typical impactor stage efficiency curve.

6.3 Particle Bounce

The surface on which particles impact is also an important factor in determining impactor efficiency. Particles which bounce off the impaction surface can be carried through the impactor and distort measurement data. Particle bounce will lower the collection efficiency of a given impactor stage and will lower the apparent mean diameter of the aerosol measured.

Internal deposition of material may take place within the impactor and not on the impactor stage. This has the effect of raising the value of Stk_{50} compared to the case of no wall deposition, and can increase the apparent mean diameter of the aerosol being sampled.

6.4 Impactors for Very Small Particle Sizes

Within the past few years there has been growing interest in measuring the diameters of very small particles, i.e., particles with diameters less than about 0.4 μm. Although traditional impactors are not adequate for this task, either low pressure or micro-orifice impactors are said to be able to collect particles with substantially smaller particle diameters than 0.4 μm.

With micro-orifice impactors W is made very small, in the order of 50 μm to 150 μm. Velocities are still kept somewhere below 100 m/s (Re = 500 to 3000), but the number of orifices is increased to provide a reasonable total flow so that an adequate amount of sample is collected.

Low-pressure impactors utilize the fact that C_c is a function not only of particle diameter but also pressure, through the gas mean free path. Therefore at low pressures C_c can be substantially larger than for the same size particle at atmospheric pressure, resulting in much smaller d_{50}'s for the impactor (9).

7 CYCLONES

In a cyclone, particle-laden air is introduced radially into the upper portion of a cylinder so that it makes several revolutions inside the cylinder before leaving axially along the cylinder centerline. While making these revolutions, particles in the air are accelerated outward to the cylinder wall where they either stick and are retained (low particle loading) or are swirled down to a collection port at the bottom of the cylinder (high particle loading). Figure 10.9 shows a sketch of a typical cyclone.

According to simple cyclone theory, collection efficiency for a given aerosol particle size is a function of τ as defined in equation 15 times the inlet velocity divided by some characteristic length, usually taken as the cyclone inlet width. There are problems with this simple theory which make it only approximate at best.

First, laminar flow is assumed, but turbulent flow is often observed in cyclones. The effect of turbulence will be to move particles away from the cyclone walls or resuspend deposited ones. Hence turbulence will decrease cyclone efficiency. Second, the width of the cyclone inlet is not as important a parameter as overall cyclone diameter, since it is

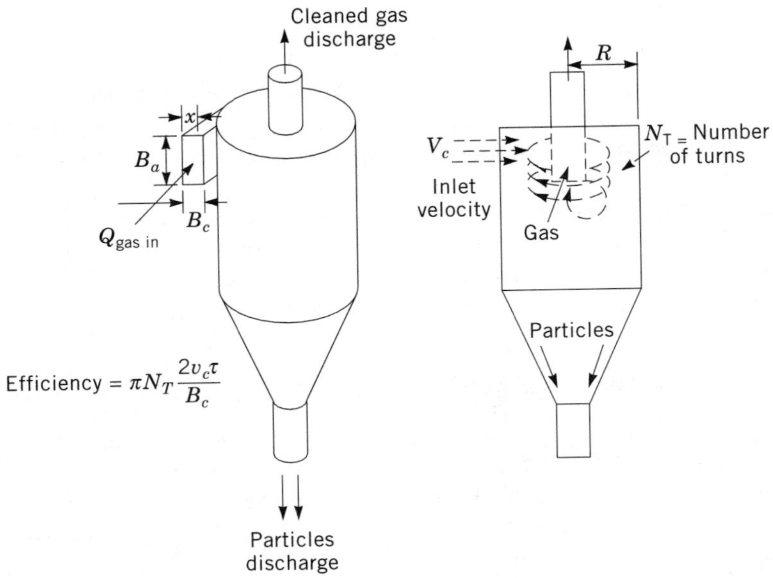

Figure 10.9. Sketch of a cyclone particle collector.

the width of an element of gas within the cyclone which determines particle deposition, and this width is not strongly controlled by inlet width. Finally, overall cyclone configuration will affect efficiency. This is not taken into account in simple theory. Aerosol concentrations also affect cyclone performance with increased concentrations increasing efficiency.

Simple cyclone theory does illustrate the general approach that has been used to refine cyclone theory, and it permits rough estimates of system performance to be made. As can be seen in Figure 10.10, however, the shape of the efficiency curve as predicted by simple theory is not consistent with experimental observations. It is true, however, that for any particular particle size, or τ, efficiency decreases as the flow through a given cyclone decreases. For a further discussion of cyclone theory, see references 3, 10, or 11

8 RESPIRABLE SAMPLING

As discussed elsewhere, for the purpose of estimating the toxic dose of an aerosol, aerosol particles that might enter the human body are considered to fall into one of three size ranges:

1. Inspirable particles: particles having aerodynamic diameters which permit them to be inhaled into the respiratory system.

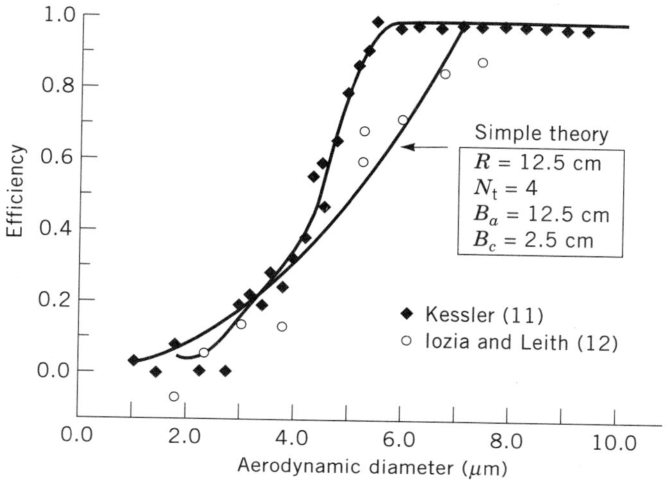

Figure 10.10. Comparison of simple theory with detailed theoretical efficiency and experimental data. Adapted from Kessler (12).

2. Thoracic particles: particles with aerodynamic diameters such that their deposition will take place mainly in the ciliated bronchial passages;
3. Respirable particles: particles with aerodynamic diameters such that deposition will take place mainly in the alveolar parts of the lung.

Because deposition is a function of particle size, shape, and particle density, particle size is usually given in terms of the aerodynamic diameter. Figure 10.11 shows the relative fraction of particles deposited in the various parts of the human airway as a function of particle aerodynamic diameter. In many cases the dose from airborne toxic materials is dependent on regional deposition in the lungs. A good estimate of this dose is possible if the size distribution of the aerosol is known. For this reason it is important to know mass concentrations within various size fractions. This information can be obtained by (*1*) carrying out a size distribution analysis of the airborne aerosol or (*2*) carrying out a size distribution analysis of the collected sample or (*3*) separating the aerosol into size fractions corresponding to anticipated regional deposition during the process of collection.

With mass respirable sampling, an attempt is made to separate the aerosol into the fraction representing the mass that would be deposited in the alveolar region. These particles are defined as "respirable". Presently respirable sampling is carried out through the use of a cyclone precollector followed by a filter sample. Collection efficiency of the cyclone mimics collection in the upper airways so that the particles passing through the cyclone represent the "respirable" fraction of the aerosol.

9 PM-10 AND PM-2.5 SAMPLING

Because of the importance attached to small particles which can be taken into the body by inhalation, in recent years general atmospheric sampling has centered around measuring

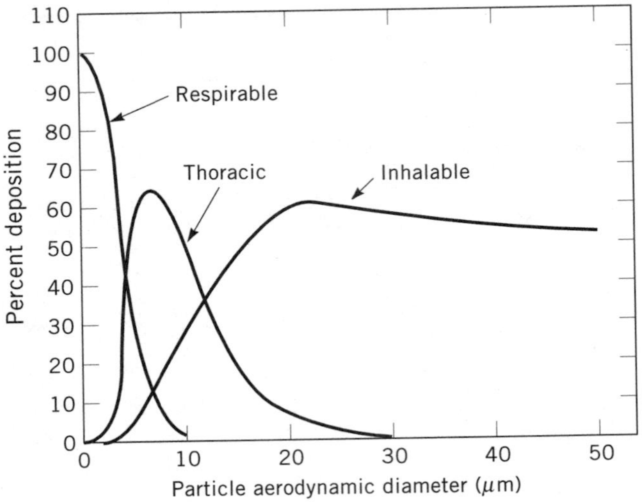

Figure 10.11. Theoretical percent deposition of inhalable particles by deposition site.

concentrations of particles with aerodynamic diameters of 10 μm or smaller. One type of sampler was designed to exclude for the most part all particles with aerodynamic diameters greater than 10 μm. Samplers of this sort are called PM-10 samplers, where PM stands for "particulate matter". More recently, samplers have been designed which primarily exclude all particles with aerodynamic diameters greater than 2.5 μm. These samplers are called PM-2.5 samplers.

Figure 10.12 shows the two peaks in sizes often found in typical size distribution data for atmospheric air. PM-2.5 sampling was chosen as the logical division between the coarse and fine mode portion of the aerosol. Some investigators have recently reported that health effects observed in high pollution areas have been primarily associated with the fine mode of the aerosol (14).

Fine and coarse mode particles have distinctly different origins and lifetimes. Fine mode particles are generally formed from chemical reactions of gases, or by nucleation or condensation of material on small nuclei, and have ambient air lifetimes of days to weeks. Coarse mode particles, on the other hand, come from crushing or grinding of larger material and get into the air by resuspension. Table 10.7 compares the two different modes of particles (15).

Both PM-10 and PM-2.5 samplers are filter samplers, with specially designed impactor inlets to exclude particles greater than stated size. Since the cut is by impaction, or a variation of it, the 10 μm and 2.5 μm exclusion points are 50% cut points as discussed above, with some larger particles being included and some smaller particles being excluded from the sample. As a result, different designs of PM-10 samplers, which are performance based, may not always agree with one another, while the PM-2.5 samplers, which are design based, should agree.

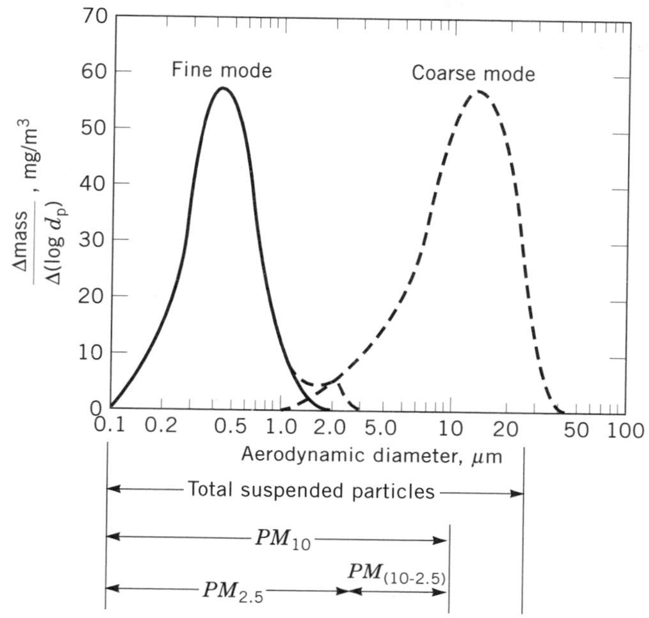

Figure 10.12. Typical size distribution curve for an atmospheric aerosol. Adapted from Wilson and Suh (15).

10 BROWNIAN MOTION AND DIFFUSION

Because of the constant motion of gas molecules, aerosol particles are also in a state of continuous motion. This so-called "Brownian motion" gives rise to mixing of particles in otherwise still air, and causes aerosol particles to move from regions of high concentration to regions of low concentration, a process known as "diffusion".

Although an accurate description of diffusion is quite complicated mathematically, simple theory can give good insights into many fundamental aspects of diffusion. For example, the distance a particle will travel on the average by diffusion in a particular interval of time is

$$s^2 = \frac{4}{\pi} Dt \tag{27}$$

Where s is the average distance traveled in cm, t is a time in seconds, and D the diffusion coefficient, defined as

$$D = \frac{C_c}{3\pi\mu d} kT \tag{28}$$

In equation 28, C_c is the Cunningham correction factor, μ the gas viscosity in poises, d the aerosol particle diameter in cm, k Boltzmann's constant, 1.38×10^{-16} erg/K, and T

Table 10.7. Comparison of Ambient Fine and Coarse Mode Particles[a]

Fine Mode	Coarse Mode
	Formed from
Gases	Large solids/droplets
	Formed by
Chemical reaction Nucleation Condensation Coagulation Evaporation of fog and cloud droplets in which gases have dissolved and reacted	Mechanical disruption (e.g., crushing and grinding abrasion of surfaces) Evaporation of sprays Suspension of dusts
	Composed of
Sulfate, $SO_4^=$ Nitrate, NO_3^- Ammonium, NH_4^+ Hydrogen ion, H^+ Elemental carbon Organic compounds (e.g., PAHs, PNAs) Metals (e.g., Pb, Cd, V, Ni, Cu, Zn, Mn, Fe) Particle-bound water	Resuspended dusts (e.g., soil dust, street dust) Coal and oil fly ash Metal oxides of crustal elements (Si, Al, Ti, Fe) $CaCO_3$, NaCl, sea salt Pollen, mold spores Plant/animal fragments Tire wear debris
	Solubility
Largely soluble hydroscopic and deliquescent	Largely insoluble and nonhydroscopic
	Sources
Combustion of coal, oil, gasoline, diesel, wood Atmospheric transformation products of NO_x, SO_2, and organic compounds including biogenic species (e.g., terpenes) High temperature processes Smelters, steel mills, etc.	Resuspension of industrial dust and soil tracked onto roads Suspension from disturbed soil (e.g., farming, mining unpaved roads) Biological sources Construction and demolition Coal and oil combustion Ocean spray
	Lifetimes
Days to weeks	Minutes to hours
	Travel Distance
100s to 1000s of kilometers	<1 to 10s of kilometers

[a]Adapted from Wilson and Suh (15).

the temperature in K. Table 10.8 gives values of D computed for various diameter particles, as well as the distance a particle will travel by diffusion in air in 100 seconds. Distance is given both in centimeters and the number of particle diameters. Not surprisingly, this distance is quite small, being larger for small particles and smaller for large particles. As might be expected, diffusion is important in aerosol studies only when particles are very small, or diffusion distances are very short.

For example, in Table 10.8 a 0.01 μm diameter particle will move a distance of over 250,000 diameters in 100 seconds, while a 10 μm diameter particle will only move about 1.75 diameters in the same time interval. This illustrates that for very small particles diffusion is an extremely important transport factor.

In the case of steady-state diffusion, particles diffuse from a constant concentration c across some distance x to a surface where the particles are then removed from the air and deposited. The diffusive flux, J, in particles per second per square centimeter, can be given as

$$J = -\frac{Dc_0}{x} \tag{29}$$

Suppose an aerosol has a concentration of 1000 p/cm^3 and is made up of 0.1 μm particles. The aerosol flows in a tube such that turbulent flow keeps the center of the tube

Table 10.8. Diffusion Coefficient for Various Particle Diameters[a]

Diameter, μm	C_c	D, cm/s^2	Dist. in 100 s, cm	Dist. in 100 s, μm	Number of Diameters
0.01	23.311	5.50×10^{-4}	0.265	2645.08	264,508
0.02	11.960	1.41×10^{-4}	0.134	1339.73	66,986
0.05	5.184	2.44×10^{-5}	0.056	557.85	11,157
0.08	3.517	1.04×10^{-5}	0.036	363.24	4,541
0.1	2.971	7.00×10^{-6}	0.030	298.60	2,986
0.2	1.918	2.26×10^{-6}	0.017	169.63	848.2
0.5	1.347	6.35×10^{-7}	0.009	89.92	179.8
0.8	1.216	3.58×10^{-7}	0.007	67.53	84.4
1	1.172	2.76×10^{-7}	0.006	59.32	59.3
2	1.086	1.28×10^{-7}	0.004	40.37	20.2
5	1.034	4.88×10^{-8}	0.002	24.92	4.98
8	1.022	3.01×10^{-8}	0.002	19.58	2.45
10	1.017	2.40×10^{-8}	0.002	17.47	1.75
20	1.009	1.19×10^{-8}	0.001	12.30	0.62
50	1.003	4.73×10^{-9}	0.001	7.76	0.16
80	1.002	2.95×10^{-9}	0.001	6.13	0.08

[a]Also showing diffusion distances.
viscosity = 0.000182 poises.
$k = 1.38 \times 10^{-16}$
mean free path = 0.0686 μm
T = 293 K

well-mixed, but there is laminar flow near the wall of the tube, where the air velocity goes to zero. If the distance x from the turbulent edge of the flow to the wall is 0.1 cm, then equation 29 can be used to predict the deposition rate of particles on the tube wall. From Table 10.8, D for 0.1 μm particles is 7×10^{-6} cm²/s. Substituting in equation 29 gives $J = 0.07$ particles per square centimeter per second deposited.

In actual practice as the aerosol is deposited, the main central stream concentration is reduced. Again simple theory predicts that if c_{in} is the incoming concentration to a sampling tube of radius R and length L, and c_{out} the outgoing concentration, with air moving in the tube at a velocity of u, then

$$\frac{c_{out}}{c_{in}} = 1 - \frac{4}{\sqrt{\pi}} \sqrt{\frac{DL}{uR^2}} \tag{30}$$

From this simple approach it appears that diffusive deposition in a tube is controlled by the dimensionless factor

$$\phi = \frac{DL}{uR^2} = \frac{\pi DL}{Q} \tag{31}$$

where Q is the volumetric flow rate of air through the tube.

In an exhaustive study by Cheng (16) empirical equations were developed for estimating tube deposition, based on φ as defined in equation 31. A plot of c/c_0 versus φ is given in Figure 10.13. When φ is greater than 0.5, c/c_0 is rapidly approaching zero. Thus large values of D, large values of L, or small values of Q can have a strong effect on wall deposition. One interesting point is that for a given volumetric flow rate, the choice of a large or small diameter tube for transporting an aerosol will have no effect on deposition.

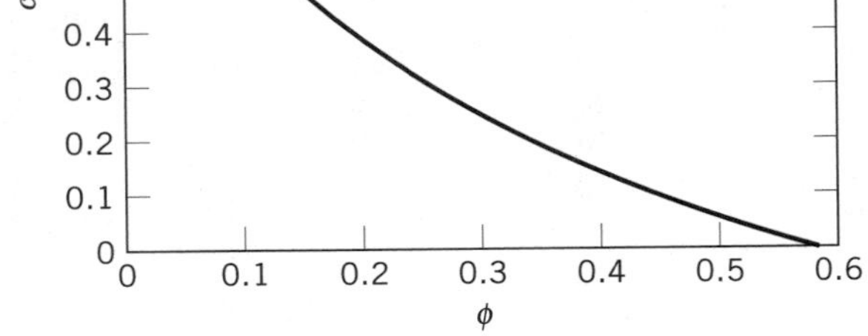

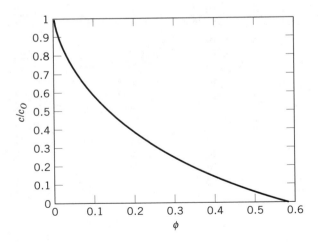

Figure 10.13. Deposition in a circular tube after Cheng (16).

11 CONDENSATION AND EVAPORATION

11.1 Introduction

When moist air is expanded adiabatically (no heat lost or gained) its saturation level can rise momentarily above 100% relative humidity causing clouds to form. In the 19th century Coullier (17) and Aitken (18) found that the presence of very small dust particles were also necessary to form clouds. Without the dust, no clouds would form. These small dust particles were called condensation nuclei. Later experiments showed that with supersaturations in the order of 500% to 600% relative humidity, clouds could form in the absence of condensation nuclei. This process is called homogeneous nucleation. Condensation on dust particles is called heterogeneous nucleation.

The saturation ratio, S, is defined as the ratio of the partial pressure of a condensable gas over its vapor pressure. For water vapor, $100 \times S$ is equal to the relative humidity. Under normal circumstances it is difficult to get values of S, even momentarily, much in excess of 1.00, although transient values slightly in excess of 1.00 are quite common. These are the transient values that produce clouds in the sky or fog on the ground.

Many vapors of industrial significance have relatively low vapor pressures at ambient temperature but quite high vapor pressures at elevated temperatures. These vapors will condense very rapidly when cooled whether there are condensation nuclei present or not. Hence it is important to understand the process of homogeneous nucleation in order to understand the conditions under which these vapors form aerosols.

11.2 Homogeneous Nucleation

If the energy balance in a nucleating drop is considered, it can be shown that for a given saturation ratio there is a critical drop size such that drops smaller than this size will evaporate and drops larger than this size will grow. Figure 10.14, Curve 1, shows the relationship of this critical drop size, d^*, to the saturation ratio for water, S. This graph is plotted from Kelvin's equation, equation 32,

$$\ln S = \frac{4\gamma M}{\rho R T d^*} \tag{32}$$

where γ is the surface tension of the liquid (dyn/cm), M its gram molecular weight (g), ρ the liquid density (g/cm^3), R the gas constant (8.1432×10^7 erg K^{-1} mol^{-1}) and T the temperature in K. If, for a given supersaturation and drop diameter, a point on Figure 10.14 falls to the left of the line, the drop will evaporate. On the other hand, if the point falls to the right of the line, the drop will grow by condensation.

11.3 Heterogeneous Nucleation

The presence of condensation nuclei in the atmosphere is strongly dependent on human activity, as well as to some extent on other life forms. Table 10.9 gives some example

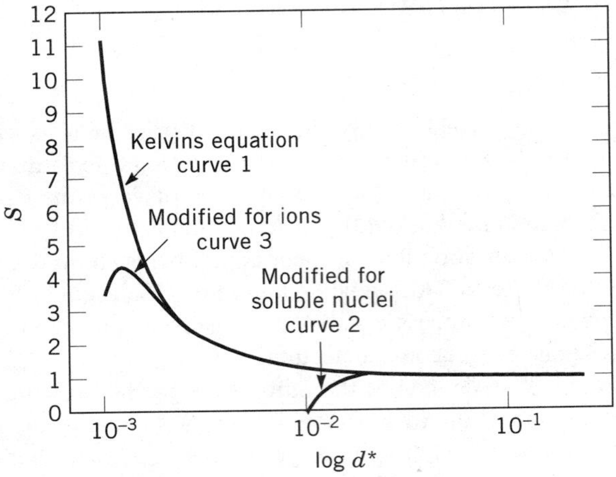

Figure 10.14. Kelvin's equation. Curve 1, homogeneous nucleation; Curve 2, nucleation on soluble nucleus; Curve 3, nucleation on an ion.

Table 10.9. Concentration Range for CN and CCN[a]

Region	Location	CN	CCN
Stratosphere	Hon-Fiji-Hon	200	30
South Pacific	Fiji Islands	200	50
North Atlantic	Azores-Greenland	300	60
Eastern Forest	Adirondack Mts., NY	1,500	150
Central Utah	Moonwater Point	2,000	250
Eastern New York	Mohawk Valley	12,500	300
New York City	East Side Piers	125,000	4,000

[a]Particles per cubic centimeter.
Adapted from Reist (3).

concentrations of condensation nuclei (CN, $d < 0.05$ μm) and cloud condensation nuclei (CCN, $d < 0.1$ μm) for various isolated or densely populated areas (19).

Besides being important for homogeneous nucleation, Kelvin's equation also forms the basis for understanding heterogeneous condensation. As a liquid condenses on a nucleus, the resulting droplet looks more and more like a drop of pure liquid, and so the picture for heterogeneous condensation evolves into the picture for homogeneous condensation. This can be seen in Figure 10.14, Curve 2, for condensation of water vapor on a soluble ammonium sulfate nucleus (3).

For a wettable insoluble nucleus Kelvin's equation can be used directly to predict whether there will be subsequent droplet growth or not. Supersaturations in excess of that predicted by the Kelvin equation for a given nucleus diameter, d^*, will cause droplet growth; those lower will bring about droplet evaporation. Particle growth will follow Kelvin's equation, Figure 10.14 Curve 1.

11.4 Nucleation on Ions

In 1897 Wilson (20) observed that condensation occurred at lower supersaturations when ions were present as nuclei. The shape of the S vs. d^* curve varied markedly at small values of d^* compared to that for homogeneous nucleation, but at larger values of d^* the two curves were essentially the same. This can be seen on Figure 10.14, Curve 3.

11.5 Aerosol Formation from the Gas Phase

Many gas phase reactions will produce small particles that go through a liquid phase before becoming solid reaction products. Because of their extremely low vapor pressures in the liquid phase these reaction products will experience extremely high supersaturations. Kelvin's equation applies also to the formation of these solid particles.

With vapor emissions, aerosol formation can be estimated by the Kelvin equation. Hot vapors that are cooled after emission can have partial pressures higher than their vapor pressure, making S greater than one and setting up conditions for either heterogeneous or homogeneous nucleation. Thus to prevent aerosol formation from emitted vapors it is necessary to take steps to keep $S < 1$. A more detailed presentation of condensation and evaporation phenomena is given in Reist (3).

11.6 Drying Times

The drying time for an airborne droplet having an initial diameter d_0 and a final diameter d, is given by Langmuir's equation,

$$t = \frac{\rho RT(d_0^2 - d^2)}{8DM_\infty(p_s - p_\ell)} \qquad (33)$$

where ρ is the drop density, D the vapor-air diffusion coefficient, p_s and p_ℓ the vapor pressure at the temperature of the drop and surrounding medium, respectively, and M the molecular weight of the vapor. This equation is also applicable to a condensing drop.

In equation 33 the temperature, T_s, represents the drop temperature. Drop temperature should not be approximated using ambient temperature. As the drop evaporates it rapidly cools to the wet-bulb or dew point temperature. This is the temperature that should be used in these equations. This temperature can be found from psychometric charts if the ambient temperature and relative humidity are known, or it can be estimated from

$$T_l - T_s = \frac{DML}{RK_t}\left(\frac{p_s}{T_s} - \frac{p_l}{T_l}\right) \qquad (34)$$

T_l and p_l are temperature and partial pressure of vapor, respectively, away from the drop, L is the latent heat of evaporation, K_t the thermal conductivity of air, and T_s the temperature within the drop. The term p_s is the vapor pressure at the temperature T_s. Figure 10.15 gives drop temperature as a function of ambient temperature for drops evaporating in 50% saturated air as calculated from equation 34.

When the drop is small (for water, $d \approx 0.5$ μm), Kelvin's equation (eq. 32) predicts that even under saturated conditions the drop will evaporate. The time for complete evaporation of a small drop at 100% relative humidity is given as

$$t = \frac{d_0^3}{3D\gamma p_\infty T}\left(\frac{\rho RT}{4M}\right)^2 \tag{35}$$

Here p_∞, the drop vapor pressure, is to be expressed in dynes/cm^2. A pressure of one millimeter of mercury equals 1.33×10^3 dynes/cm^2.

12 ELECTRICAL PROPERTIES OF AEROSOLS

The electrical force acting on an aerosol particle is the product of the electrical charge on the particle, q, and the strength of the electrical field in which the particle is placed, E. That is, $F = q \times E$. The charge on a particle is the product of the unit electrical charge, $e = 4.8 \times 10^{-10}$ esu, and the number of charges, n.

12.1 Maximum Charge on an Aerosol Particle

Since aerosol particles are constantly gaining or losing charge, and even the earth has an electric field associated with it, electrical forces are always acting on aerosol particles. In some cases these forces are small compared to other forces (gravity, inertial) so that electrical effects can be neglected. In other cases, electrical effects must be taken into account.

The maximum charge on a particle, q_m, depends on the ability of the surface of the particle to hold a charge, that is, the charge density. For a spherical particle of diameter d

$$q_m = e \times n_m = \frac{E_s d^2}{4} \tag{36}$$

where for charging with ions E_s is 6.67×10^5 statvolts/cm and for electron charging $E_s = 3.3 \times 10^4$ statvolts/cm. One statvolt in the cgs system of units equals 300 conventional volts.

With charged liquid drops the situation is somewhat different. The repelling force of the accumulated charge on the surface of the drop will eventually overcome the surface tension and the drop will literally explode, forming many smaller drops with more surface area. This charge limit, known as the Rayleigh limit, is given by

$$q_r = e \times n_r = \sqrt{2\pi\gamma d^3} \tag{37}$$

where γ is the liquid surface tension, which for water at normal temperature is 72.7 dynes/cm^2.

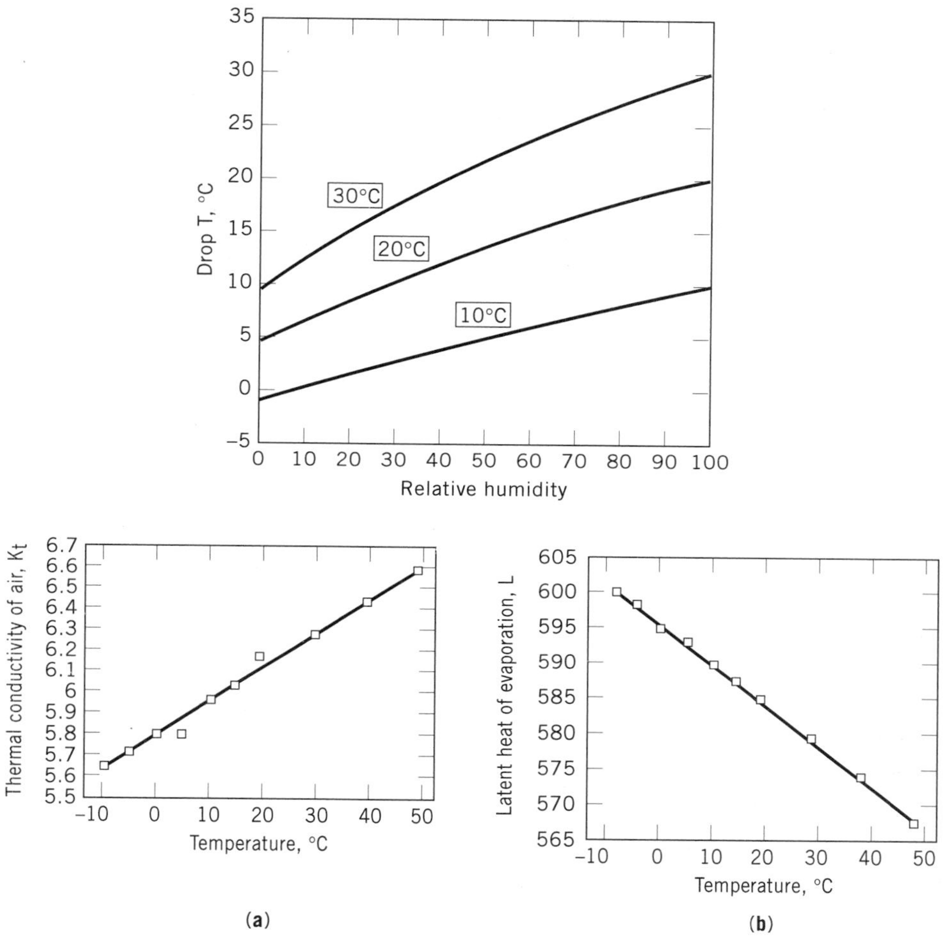

Figure 10.15. Drop temperature as a function of relative humidity.

12.2 Charge Equilibrium

Aerosol particles are constantly gaining and losing electrical charges from such diverse sources as atmospheric ions, gamma radiation from space and charges on other particles, to name a few. As a result there is a charge equilibrium on the aerosol particles, representing the average charge distribution expected to be found under normal circumstances. If it is assumed that the number of positive and negative charges available to the aerosol are equal, then Table 10.10 shows the relative number of various charges of either sign to be expected to be found on an aerosol particle of a given diameter at any time. The table also gives the average number of charges on all particles of a given diameter at equilibrium.

Table 10.10. Equilibrium Charge Distribution[a]

	Number of Charges on Particle									
d, μm	0	1	2	3	4	5	6	7	8	Average Charge
0.05	0.602	0.385	0.013	0.000	0.000	0.000	0.000	0.000	0.000	0.528
0.1	0.426	0.482	0.087	0.005	0.000	0.000	0.000	0.000	0.000	0.747
0.2	0.301	0.453	0.193	0.046	0.006	0.000	0.000	0.000	0.000	1.057
0.5	0.190	0.340	0.241	0.137	0.062	0.022	0.006	0.001	0.000	1.671
1.0	0.135	0.254	0.214	0.161	0.108	0.065	0.035	0.017	0.007	2.364
2.0	0.095	0.185	0.170	0.147	0.121	0.093	0.068	0.047	0.031	3.343
5.0	0.060	0.119	0.115	0.109	0.100	0.091	0.080	0.069	0.058	5.285
10	0.043	0.085	0.083	0.081	0.078	0.074	0.069	0.064	0.059	7.474

[a]Fraction of Particles Carrying Indicated Charge of Either Sign
Parameters
$e = 4.80 \times 10^{-10}$
$k = 1.38 \times 10^{-16}$
$T = 293$ K

Although Table 10.10 indicates that for particles greater than about 0.5 μm there will always be some charge present, this charge will generally be too small to be of any significance in most aerosol problems. This is shown by comparing the gravitational force on a unit density 1-μm diameter sphere to the electrical force on the same particle carrying the average number of equilibrium charges in an electric field with a strength equal to that of the earth's electrical field, about 2 volts/cm. In this case the gravitational force will be about 64 times greater than the electrical force.

The average number of charges on a monodisperse aerosol of a diameter *d* is

$$\bar{n} = \sqrt{\frac{dkT}{\pi e^2}} \tag{38}$$

Highly charged aerosols that are placed in strong bipolar ion fields will rapidly lose their charge and attain an equilibrium charge distribution. This is the principle behind all aerosol charge neutralizing devices. The resulting aerosol is not devoid of charge, it just has its overall charge reduced to charge equilibrium.

12.3 Unipolar Charging

12.3.1 Diffusion Charging

When an aerosol is immersed in a cloud of ions having the same sign, unipolar charging can take place. If the particles are small, diffusion charging predominates. With diffusion charging ions diffuse to the particle surface where they attach themselves, thus increasing the particle charge. For diffusion charging, a characteristic charge is defined as

$$n' = \frac{dkT}{2e^2} \tag{39}$$

and a characteristic time as

$$t' = \frac{2kT}{\pi d v_{avg} \bar{N} e^2} \tag{40}$$

where $\bar{N}$ is the average ion concentration and v_{avg} is the average ionic velocity. The number of characteristic charges attained by a particle of diameter d after a time of t seconds is, in a dimensionless form,

$$\frac{n}{n'} = \ln\left(1 + \frac{t}{t'}\right) \tag{41}$$

A plot of equation 41 is given in Figure 10.16. It can be seen that with diffusion charging the charge on the aerosol particles rises rapidly at first, but then goes at a much slower rate as time progresses.

12.3.2 Field Charging

Field charging takes place when the aerosol is placed in a moving field of ions all having the same sign, such as might be found in an electrostatic precipitator. The electric field acts to drive the ions onto the aerosol particle surfaces up to some maximum value de-

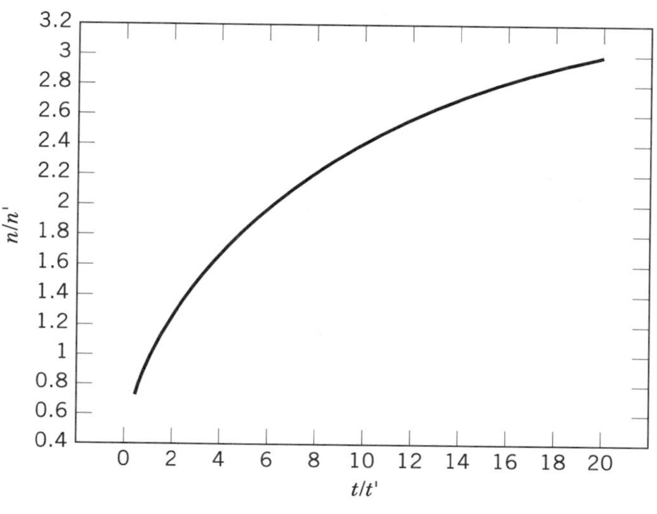

Figure 10.16. Dimensionless diffusion charging.

pending on the field strength. This maximum value, n_s, known as the saturation charge, is given by

$$n_s = \frac{\chi E_0 d^2}{4e} \tag{42}$$

where E_0 is the field strength, and χ a parameter relating to the dielectric properties of the particle. χ is usually taken as 3.

A characteristic time factor can also be defined as

$$t_0 = \frac{1}{\pi \bar{N} e Z} \tag{43}$$

Again $\bar{N}$ is the ion concentration and Z the ion mobility, that is, the ionic velocity when placed in a unit electrical field. In the case of positive ions, $Z \approx 1\ \text{cm}^2\ \text{s}^{-1}\ \text{V}^{-1}$, and for negative ions $Z \approx 2\ \text{cm}^2\ \text{s}^{-1}\ \text{V}^{-1}$. In the past an average value of $Z \approx 1.4\ \text{cm}^2\ \text{s}^{-1}\ \text{V}^{-1}$ has been used in field charging calculations (3). Remember that V in this case is in statvolts (1 statvolt = 300 volts).

The two dimensionless terms defined in equations 42 and 43 can be used in a dimensionless field charging equation, equation 44 which indicates how the saturation charge on a particle builds up as a function of time;

$$\frac{n}{n_s} = \frac{t}{t + t_0} = \frac{t/t_0}{1 + t/t_0} \tag{44}$$

A plot of equation 44 is given in Figure 10.17. It can be seen that with field charging the charge on the aerosol particles will initially rise rapidly and then reach an equilibrium value, n_s, as time progresses.

12.4 Aerosol Motion in an Electric Field

The force on a charged aerosol particle in an electric field is given by $F = qE$. If this force is balanced by the Stokes resistance, the particle velocity in an electric field, w, is

$$w = E\left(\frac{\chi E_0 d^2}{4}\right)\left(\frac{C_c}{3\pi\mu d}\right) = \frac{\chi E E_0 d C_c}{12\pi\mu}. \tag{45}$$

The term w is known as the electrical drift velocity. In equation 45, E_0 represents the charging field strength and E the strength of the electric field in which the charged aerosol particle is placed, or which is acting to move the aerosol particle. Sometimes the charging field and "moving field" are the same; often they are not. In either case, each is used as if it were a separate electric field.

The electric field between two parallel plates has a constant value, being the voltage across the plates divided by the distance between them, i.e., $E = V/x$. If a charged aerosol

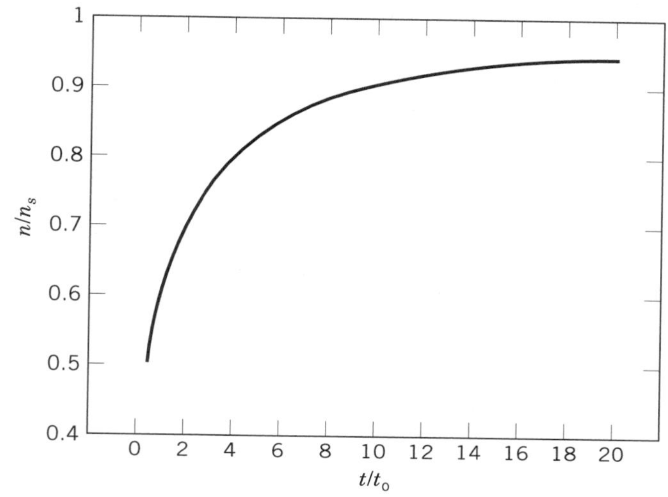

Figure 10.17. Dimensionless field charging.

particle is placed between two horizontal parallel plates the particle will have zero settling velocity when the electrical force on the particle balances the gravitational force. For a particle carrying n units of charge placed between plates x cm apart, the voltage required to levitate the particle is

$$V = \frac{\pi}{6} \frac{d^3 \rho_p x}{ne} \tag{46}$$

With a 10-μm diameter spherical, unit density particle carrying 5 units of charge, placed between horizontal plates spaced 2 cm apart, the voltage required to levitate the particle would be 0.44 statvolts, or $300 \times 0.44 = 130.9$ volts.

12.5 Efficiency of an Electrostatic Precipitator

The utility of the concept of drift velocity is used to estimate the theoretical efficiency of an electrostatic precipitator using the Deutsch Equation (eq. 47) (21), an equation which is applicable to both tube and plate types of precipitators,

$$\epsilon = 1 - \exp\left(\frac{-Aw}{Q}\right). \tag{47}$$

In this equation ϵ is the precipitator collection efficiency, w the electrical drift velocity, A the total collecting area of the precipitator, and Q the total flow through the precipitator.

The general form of this equation has been verified many times in practice. In using equation 47, the terms A, w, and Q must all have consistent units.

13 OPTICAL PROPERTIES OF AEROSOLS

Optical effects of aerosols represent the most spectacular and most important of all aerosol characteristics. Clouds, haze, and smokes all appear as they do because of the optical properties of the individual particles and the effects of these particles on each other.

All aerosol particles scatter light. Some, principally metallic or carbon particles, or very large particles, can also absorb light. The combination of scattering plus absorption is called extinction. Selective scattering and absorption are responsible for colors of things. Black smoke appears black because the smoke particles efficiently absorb all visible wavelengths of light. White smoke appears white because the smoke particles efficiently scatter all visible wavelengths (0.4 μm to 0.7 μm).

Scattering functions that describe the light scattered or absorbed by a particle can be computed for spherical or cylindrical shapes using a general mathematical theory known as Mie Theory after G. Mie (22). Although Mie's theory was formulated for spherical particles, later work by others indicate that angular scattering patterns and extinction predictions for isometric particles such as cubes or octrahedra differ very little from those for spherical particles of the same equivalent size (23).

For aerosol particles where there is appreciable absorption of radiation as well as scattering it is necessary to express the refractive index of a material as a complex number of the form

$$m = \nu - i\kappa \tag{48}$$

Here the parameter ν represents the real part of the index and κ the imaginary part. The real part represents scattering and the imaginary part absorption. Carbon particles, for example, have a refractive index of about 2-i. In the visible region, water has a refractive index of 1.33 (scattering only), and most atmospheric aerosols about 1.5–0.02i. Most aerosol materials will vary in their refractive index depending on the wavelength of light used, their chemical composition and, in some cases, their orientation with respect to the light source and receptor. Table 10.11 gives refractive indices for various materials.

13.1 Extinction, Scattering, and Absorption

It has been established theoretically and experimentally that a thin section of a medium such as air that contains particles will both scatter and absorb light in an amount proportional to the flux of light entering the section. This gives rise to Bouger's Law

$$\frac{I}{I_0} = \exp(-\gamma\ell) \tag{49}$$

where ℓ is the distance between the object and observer, and $\gamma = b + k$, the extinction coefficient, the sum of the scattering coefficient and absorption coefficient, respectively. The factor γ is also known as the turbidity or attenuation coefficient.

Table 10.11. Refractive Indices for Selected Materials[a]

Material	Density, g/cm^3	m	Wavelength
Vacuum	(−)	1.000	Visible light
Air	0.0012	1.0002918	Visible light
Alumina	3.9	1.67	Visible light
Ice	1.00	1.31	Visible light
Water	1.00	1.333	550 nm
Water	1.00	1.153-0.0968i	11 μm
H_2SO_4	1.841	1.430	at 550 nm
38% H_2SO_4, 62% H_2O	1.32	1.394	at 550 nm
38% H_2SO_4, 62% H_2O	1.32	1.46-0.38i	at 9.5 μm
Diamond	3.51	2.417	Visible light
Glass	2.45	1.51-2.00	Visible light
NaCl	1.33	1.5443	Visible light
PSL	1.05	1.5	Visible light
$(NH_4)_2SO_4$	1.769	1.528	at 550 nm
NH_4HSO_4	1.780	1.482	at 589 nm
$(NH_4)NO_4$	1.725	1.559	at 550 nm
$CaCO_3$	2.930	1.586	550 nm
SiO_2	2.17–2.66	1.478	550 nm
Atmospheric aerosol		1.5-0.02i	Visible light
Urban aerosol	1.60	1.5-0.1i	Visible light
Soot aggregates	1.00	1.56-0.47i	Visible light

[a]From Ref. 3.

With many particles of the same size,

$$\gamma = nQ_{ext}A \tag{50}$$

where n = the number of particles per unit volume of medium
A = the cross-section area of a particle
Q_{ext} = an extinction efficiency factor defined as

$$Q_{ext} = \frac{\text{Total energy flux extinguished by particle}}{\text{Total energy flux geometrically incident on particle}}$$

A scattering efficiency factor, Q_{scat}, and an absorption efficiency factor, Q_{abs}, can be similarly defined. Then $Q_{ext} = Q_{scat} + Q_{abs}$. For spherical particles equation 50 becomes

$$\gamma = nQ_{ext}\frac{\pi}{4}d^2 \tag{51}$$

With a polydisperse cloud having n_i particles per cubic centimeter of diameter d_i it is necessary to sum over all particle sizes, or,

$$\gamma = \sum_{i=1}^{\infty} n_i Q_{ext,i} \frac{\pi}{4} d_i^2 \quad (52)$$

When d is expressed in centimeters, γ has the units cm^{-1}. For a constant Q_e and mass of material, decreasing the particle size increases the extinction of light for an aerosol. That is, for the same amount of mass, small particles produce more haze in the atmosphere than large particles.

The term Q_e is an efficiency factor that depends on the particle size divided by wave length of light, times a factor *pi*, i.e., $\alpha = \pi d/\lambda$. When α is large, Q_{ext} has a constant value of 2.

For very small particles, $\alpha \ll 1$, with no adsorption

$$Q_{ext} \approx Q_{scat} \approx \frac{8}{3} \alpha^4 \left(\frac{m^2 - 1}{m^2 + 2}\right)^2 \quad (53)$$

In equation 53 m is the refractive index of the particle. This special case of light scattering is called "Rayleigh Scattering". One important example of Rayleigh scattering is the scattering of light by molecules making up the earth's atmosphere. Because of the strong dependence of Q_{ext} on α^4, i.e., $(1/\lambda)^4$, blue light is scattered by air molecules about nine times as efficiently as red light. Hence, the sky is blue!

For larger particles the relationship of Q_{ext} to α is more complex, rising from the fourth power relationship of Rayleigh scattering until it finally oscillates around and becomes asymptotic to a value of Q_{ext} 2. For typical aerosol particles, a plot of Q_{ext} versus α calculated from Mie theory is shown in Figure 10.18.

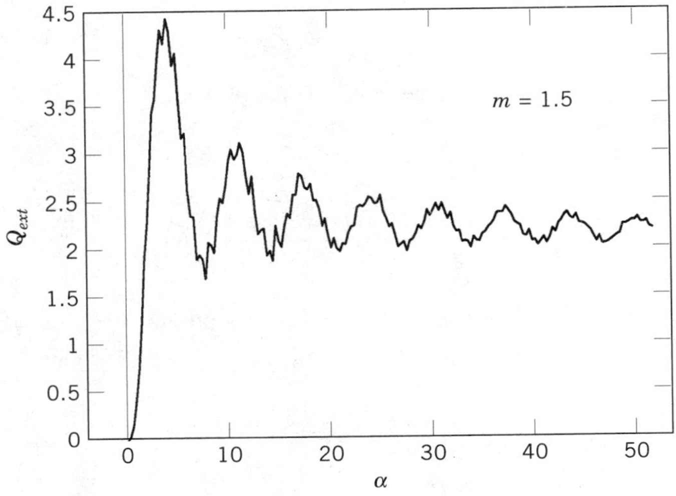

Figure 10.18. Extinction coefficient as a function of dimensionless size, α.

For visible light, aerosols are most optically active in the 0.1 to 1 μm-diameter range. This is shown in Figure 10.19 which is a plot of energy flux extinguished by the particle per unit volume versus particle diameter for light of wavelength 0.5 μm and a typical atmospheric aerosol. It can be seen that particles smaller than about 0.05 μm have little impact on light attenuation as do those larger than about 3 μm.

13.2 Contrast

In order for an object to be seen, there must be contrast between the object and its background. Aerosols in the atmosphere can lower this contrast, the amount of reduction being dependent on the aerosol particle size and concentration. If the contrast between an object and its background is defined as

$$C = \frac{B}{B_h} - 1 \tag{54}$$

where B the illuminance of the object and B_h the illuminance of the background, to see a black object against a white background the minimum value for contrast has been found to be -0.02.

A simple theory by Koschmieder (24) says that

$$C = -\exp(-\gamma\ell) \tag{55}$$

so that visual range can be found by rearranging equation 55 to give

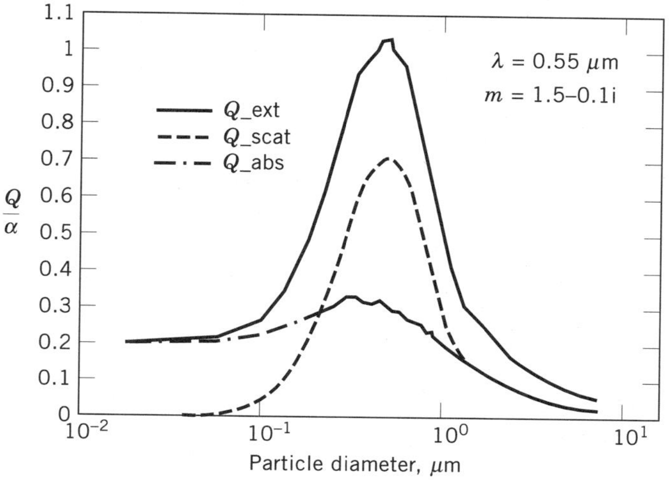

Figure 10.19. Energy flux extinguished versus particle diameter for an urban aerosol

$$\ell = -\frac{1}{\gamma} \ln 0.02 = \frac{3.91}{\gamma} \tag{56}$$

In this case aerosol concentration is averaged over the distance between the observer and the object.

It might be thought that the visual range equation could be used to measure aerosol mass concentration in the atmosphere since visual range is a fairly simple measurement. Indeed, some studies comparing predicted mass to measured mass concentrations tend to bear this assumption out. The difficulty lies in choosing a proper average particle diameter. The average size of atmospheric aerosol particles can vary markedly, depending on their moisture content, which will not be included in mass measurements. For soluble nuclei this can be further confounded by the hysteresis effect, by which the value of d will be determined by whether the nuclei are in an atmosphere of rising or falling humidity. Since very often this fact is difficult to ascertain, especially with a moving air mass measured at a stationary point, mass concentration measurements derived from extinction measurements should be considered to be valid only for cases where the atmospheric humidity is less than 40%.

13.3 Optical Properties—Angular Scattering

The terms Q_{scat}, Q_{abs} and Q_{ext} represent loss of radiation along the path from source to observer, that is, the extinction of light by an aerosol. But often interest centers more on the scattering of light in a single direction. For example, interest might lie in being able to estimate the quantity of light scattered from a single aerosol particle into a detector.

Light scattered along the direction of propagation is said to be scattered at a scattering angle of 0°, known as "forward scattering". Incident light is considered to be parallel and may or may not be polarized. If it is polarized, then the scattered light will also be polarized in the same plane. If the incident light is not polarized the scattered light may or may not be polarized.

Determination of the amount of light scattered by an aerosol particle in a particular direction for all values of alpha and the two polarization states is accomplished through the use of the Mie Theory (22). Unfortunately solutions of Mie's equations do not lend themselves readily to computation except through the use of computers. Manageable solutions other than Mie Theory are available for the cases where $\lambda \gg d$ (Rayleigh scattering) or when $\alpha \gg 1$ (geometric optics).

Mie scattering functions are generally presented in terms of the angular intensity functions, i_1 and i_2. The subscripts of these functions indicate perpendicular and plane polarization respectively. Besides being functions of the scattering angle theta, i_1 and i_2 are also functions of the particle properties m and α.

Plots of i_1 and i_2 as a function of alpha for $m = 1.55$ and three scattering angles, 0°, 90° and 180°, are given in Figure 10.20. Although representing only one refractive index, that of a typical aerosol, these Figures show the variations in i_1 and i_2 expected for many transparent aerosol materials. It can be seen that as alpha increases the i-values become more complex, especially for non-forward scattering.

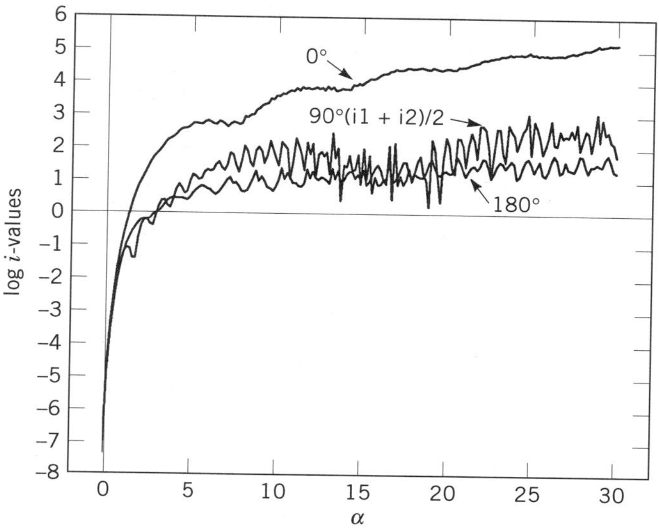

Figure 10.20. Plot of i-values versus α for three different scattering angles, $m = 1.5$.

With scattering from a volume containing a polydisperse aerosol both the total scattering cross-section and scattering functions show much less irregularity as maxima and minima are smoothed out by the variety of particle sizes present. According to computations by Reist (3) major irregularities in an i_1 versus theta curve are essentially gone for log-normally distributed aerosols when $\sigma_g > 1.8$ (Fig. 10.21).

For very small nonabsorbing spheres Rayleigh's theory gives

$$i_1 = \alpha^6 \left(\frac{m^2 - 1}{m^2 + 2}\right)^2 \tag{57}$$

and

$$i_2 = \alpha^6 \left(\frac{m^2 - 1}{m^2 + 2}\right)^2 \cos^2 \theta \tag{58}$$

Besides explaining why the sky is blue, Rayleigh's theory of scattering by very small spheres also helps explain why light is partially polarized by the atmosphere.

13.4 Optical Instruments

In recent years there have been a number of designs of aerosol sampling devices which rely on optical principles for their operation. Figure 10.22 shows sketches of the three

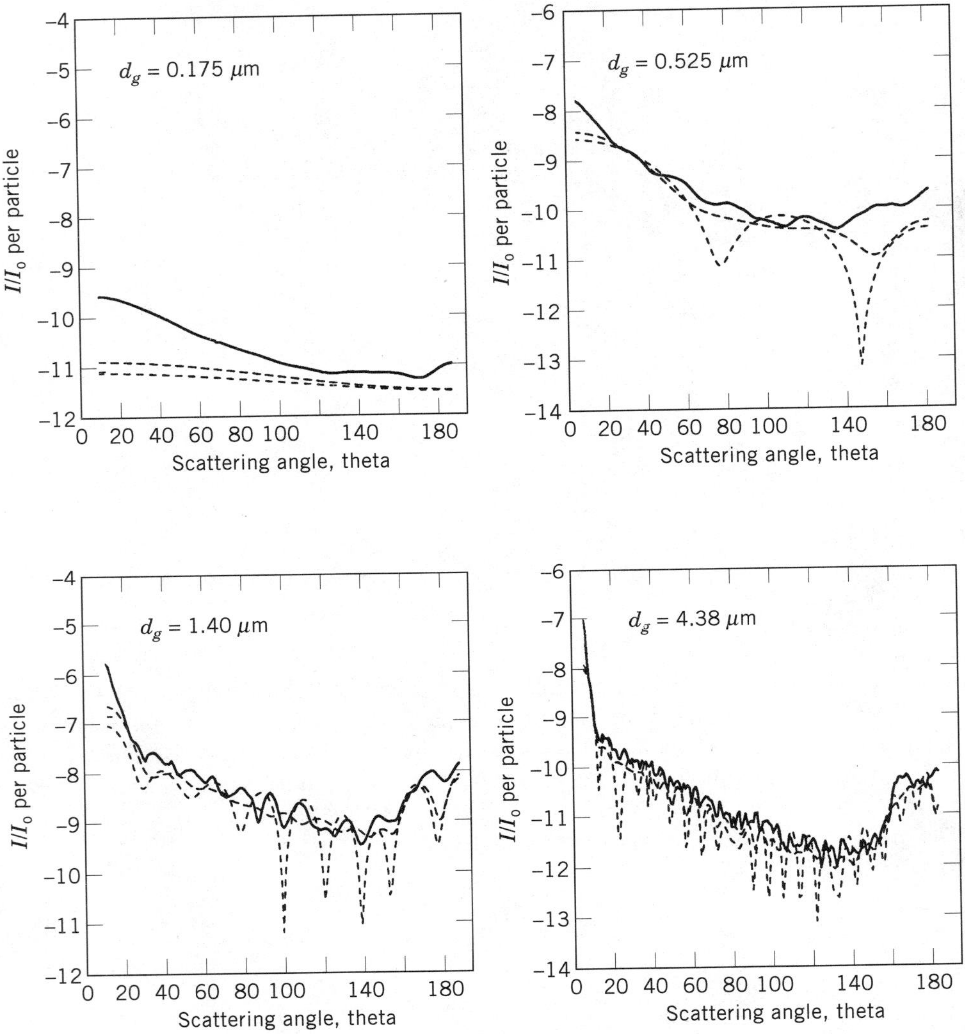

Figure 10.21. Angular intensity ratio, I/I_0, for log-normal aerosols with different degrees of polydispersity, $\lambda = 550$ nm. --- $\sigma_g = 1.0$, — — — $\sigma_g = 1.2$, ——— $\sigma g = 1.8$.

main types of instruments, light scattering photometers, single particle counters, and time-of-flight instruments.

In the first type, light scattering photometers, an aerosol is placed in front of a light source and light extinguished by the aerosol is measured. This gives a rough measure of aerosol concentration. Or light scattered by the aerosol is collected and measured by some sort of light detector. The amount of light scattered is approximately related to the mass

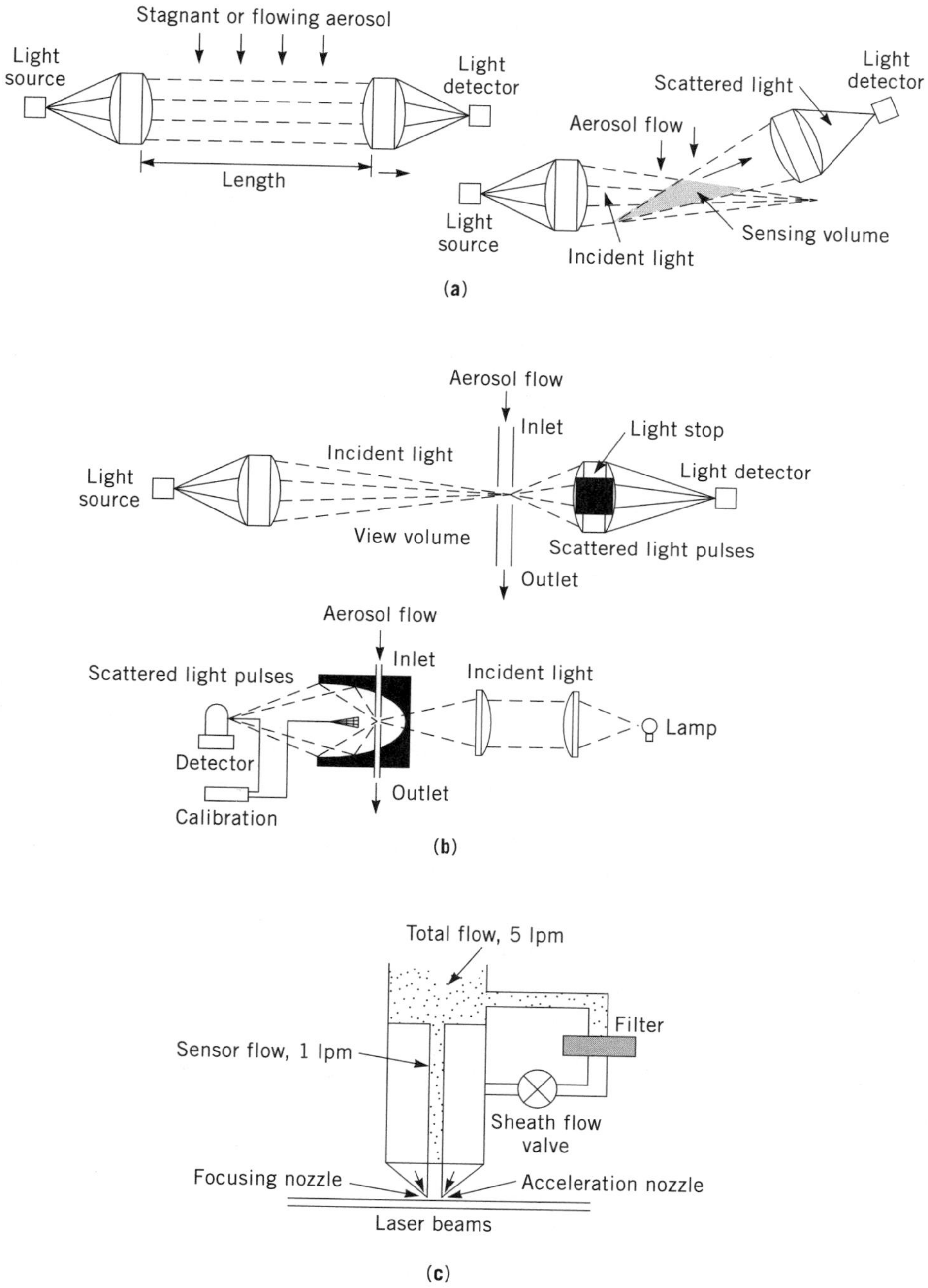

Figure 10.22. Optical instruments for measuring aerosol number and size distribution. (**a**) Total light scattering photometers. (**b**) Single particle counters. (**c**) Time-of-flight instruments.

of aerosol passing the detector. This is a fairly straightforward design used in a number of aerosol mass monitors.

Single particle counters, on the other hand, measure the light scattered from a single particle as it passes through a small, defined volume at a known rate of flow. Each particle that passes through is counted and the intensity of the scattered light pulse is measured and related in some way to the particle's size. Hence these counters give information on both particle concentration and size distribution. As might be expected, single particle counters are limited to measuring relatively low aerosol concentrations.

Time of flight instruments represent another class of optical measuring devices. These use optics to measure the time it takes for a decelerating aerosol particle to pass from one laser beam to the next. The deceleration of the particle is related to particle inertia, which in turn is related to the Stokes number. Hence these types of instruments can measure both concentration and size distribution as well.

14 COAGULATION

At high aerosol concentrations individual particles coalesce to form flocs made up of many particles. Coagulation can result from the random motion and subsequent collision of particles (often called thermal coagulation) or from external forces such as turbulence or electricity that act to increase the rate of coagulation. With coagulation, both particle number as well as particle size change with time. Simple models of coagulation are representative but inexact. In many cases results derived from these simplified models can be used to predict aerosol coagulation rates and number concentration with reasonable accuracy.

14.1 Coagulation of Monodisperse Spherical Particles

The simplest coagulation problem is thermal coagulation of monodisperse spherical particles. Since only the first several particle collisions are considered, the size of the resulting agglomerated particles will not be appreciably different from the size of the initial particles. This model is especially applicable to the coagulation of liquid drops since the size of the agglomerated drop increases only as the cube root of the number of drops making it up. The model has also been used for many years for solid particles and forms the basis for the definition of the coagulation coefficient. The approach, first presented by von Smoluchowski (24) gives

$$\frac{1}{c} - \frac{1}{c_0} = \frac{K_0}{2} t \tag{59}$$

The term t is in seconds and in the cgs system the units of K_0 are cubic centimeters per second, and K_0 is defined as:

$$K_0 = 8\pi dD = \frac{8kT}{3\mu} C_c \tag{60}$$

where d is the particle diameter in μm and D its diffusion coefficient in cm^2/sec (see eq. 28)

Equation 59 shows that the inverse of the concentration at any time is a linear function of the time, the slope of the line being determined by the coagulation constant. Experimental data from both monodisperse as well as polydisperse aerosols follow this general form, at least initially. For coagulation of very small particles, or two particles of unequal size, the coagulation constant may be appreciably larger than the theoretical value.

When c_0 is very large, approaching an infinitely large concentration, then equation 59 can be approximated by

$$c = \frac{2}{K_0 t} \tag{61}$$

Equation 61 indicates that for particles greater than 1 μm ($K_0 = 6 \times 10^{-10}$ cm^3/sec) the maximum possible aerosol concentration after one second of coagulation is somewhere in the order of $c \approx 2/(6 \times 10^{-10}) \approx 3.3 \times 10^9$ p/cm^3 since concentrations higher than this will rapidly coagulate to this value within one second.

14.2 Coagulation of Particles of Two Different Sizes

For the case of coagulation of particles of two different sizes, K_0 is computed by replacing d with $(d_1 + d_2)/2$ and D is replaced with $(D_1 + D_2)/2$ where the 1 and 2 refer to the two different sized particles. Then the coagulation coefficient becomes

$$K_{12} = 2\pi(D_1 + D_2)(d_1 + d_2) \tag{62}$$

The minimum coagulation constant occurs for coagulation of equal-sized particles. This can be seen in Figure 10.23 which shows a three-dimensional graph of the coagulation constant matrix. The valley indicated on the plot represents the constants for coagulation of equal-sized particles.

14.3 Coagulation of Many Sizes of Particles

It has been shown that if one is interested only in the change in the number of all the particles then it is possible to combine all coagulation coefficients into a single one that can be expressed in terms of mean values. For example, if particle sizes are initially distributed log-normally, an expression can be written for K' in terms of the geometric mean diameter d_g and geometric standard deviation, σ_g i.e.,

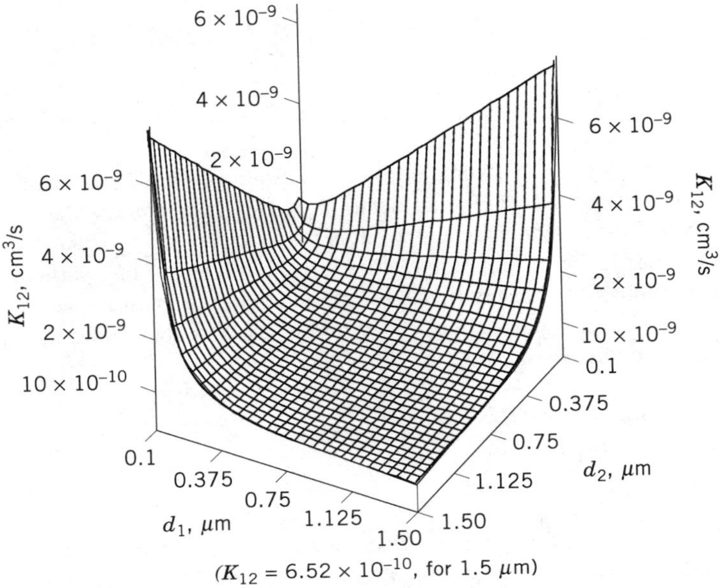

Figure 10.23. K_{12} for coagulation of two particles.

$$K' = \frac{4kT}{3\mu}\left[1 + \exp(\ln^2 \sigma_g) + \frac{2A\lambda}{d_g}\exp\left(\frac{1}{2}\ln^2 \sigma_g\right) + \frac{2A\lambda}{d_g}\exp\left(\frac{5}{2}\ln^2 \sigma_g\right)\right] \tag{63}$$

In equation 63 A is the Cunningham correction factor constant, 1.257, and λ is the gas mean free path. Table 10.12 lists values of K' for various d_g and σ_g computed from equation 62. With increasing polydispersity this coefficient can become quite large, implying that extremes in polydispersity are quickly reduced by agglomeration, particularly of the smaller particles. Also, it is clear that the coagulation rate for a polydisperse aerosol is greater than for a monodisperse one.

However, coagulation of a monodisperse aerosol initially increases polydispersity, so that for any coagulating aerosol the coagulation coefficient is not constant, but is itself a variable bound to the coagulation rate. There is little wonder that the interpretation of coagulation data is difficult and it is amazing that simple coagulation theory is as adequate as it is for many applications.

14.4 The "Self-Preserving" Size Distribution

It has been suggested by Friedlander (26), and Friedlander and Wang (27), that a coagulating aerosol should with time reach the same steady-state size distribution regardless of the aerosol's initial size distribution. This has been shown to be true by Lee (28), Lee and Chen (29), and the distribution is called a "self-preserving size distribution". When this

Table 10.12. Values of Coagulation Coefficient K' for Aerosols of Varying Degrees of Polydispersity (value $\times\ 10^{-10}$ cm^3/s)[a]

d_g	Geometric Standard Deviation 1.0	1.5	2.0	2.5
0.02	56.25	69.58	116.59	240.37
0.04	31.07	38.00	62.15	125.07
0.10	15.96	19.05	29.49	55.89
0.16	12.19	14.31	21.32	38.59
0.20	10.93	12.73	18.60	32.83
0.40	8.41	9.58	13.15	21.30
1.00	6.90	7.68	9.89	14.38
1.60	6.52	7.21	9.07	12.65
2.00	6.40	7.05	8.80	12.07
4.00	6.14	6.73	8.25	10.92
10.00	5.99	6.54	7.93	10.23
16.00	5.95	6.50	7.85	10.06

[a]Parameters:
$k = 1.38 \times 10^{-16}$ erg/K
$T = 293$ K
Viscosity $= 1.82 \times 10^{-4}$ poises
$A = 1.257$
Lambda $= 0.068$ μm

steady-state is reached, gains by coagulation in the number of particles of a given size are equaled by losses from that size either by coagulation or by sedimentation.

14.5 External Factors in Coagulation

There are a number of external physical factors which can act to enhance or retard the coagulation of aerosols. These include electrical effects such as the attraction or mutual repulsion of charged particles, polarization effects giving rise to induced forces, sonic agglomeration, gravitational coagulation and coagulation brought about by turbulence.

With electrically charged particles of the same or opposite sign coagulation may be enhanced or diminished depending on the signs of the two charges and their magnitude. For a unipolar aerosol it is necessary to consider electrostatic dispersion, that is, the tendency of charged particles of the same sign to move away from each other. This dispersion tends to reduce the concentration of an aerosol, decreasing the coagulation rate.

Coagulation can also take place in rapidly moving airstreams, and one might expect that by moving the particles coagulation would be enhanced. This motion could assume two forms. On one hand there could be an ordered flow of particles in one direction, with the particles moving at different velocities. This could occur, for example, in a polydisperse aerosol settling by gravity under quiescent conditions. On the other hand, motion could be disordered, as in turbulent mixing.

Particle agglomeration can occur in ordered flow such as that in a sonic field. Here agglomeration takes place by the different velocities imparted to particles of differing inertia. No complete and adequate theory for acoustic agglomeration as yet exists.

For turbulent agglomeration there are two cases which should be considered. First, if the inertia of the aerosol particles is approximately the same as the medium, the particles will move about with the same velocities as associated air parcels and can be characterized by a turbulence or eddy diffusion coefficient D_T. This coefficient can have a value 10^4 to 10^6 times greater than aerosol diffusion coefficients. Turbulent agglomeration processes are treated in a manner similar to conventional coagulation except that the larger diffusion coefficients are used.

A second method of aerosol coagulation in turbulent flows arises because of inertial differences between particles of different sizes. The particles accelerate to different velocities by the turbulence depending on their size and may then collide with each other. In both cases turbulence tends to increase agglomeration up to a certain point, where increased turbulence decreases agglomeration.

15 SUMMARY

Aerosols are both a bane and a blessing. For centuries man has suspected that dust could be harmful. At least, early writers indicated in their works a general connection between lung diseases and dust inhalation, even though they did not distinguish between the various types of respiratory diseases, only talking about inhalation of "fatal dust" or "the corrosive dust". With the industrial revolution in the 19th century and the advent of high speed machinery, dust exposure increased dramatically as did dust-caused diseases.

Many air pollutants originate in particulate form or become particulates soon after discharge and must be dealt with as such. Acid rain is an example of an aerosol problem where gas is transformed to a liquid—in this case sulfur dioxide is transformed in the air to a sulfuric acid aerosol.

But aerosols have a good side as well. They appear to play a major role in the removal of pollutant gases from the atmosphere either by adsorbing the pollutants on existing particles or through the creation of new particles. Aerosols play a major role in cloud formation and have a vitally important role in the hydrological cycle.

As might be surmised from the previous material, particles in air behave differently than the air in which they are suspended, and behave differently among themselves depending on their size, shape and composition. Collecting a representative sample of an aerosol for any purpose can be a frustrating and time consuming task, and a knowledge of aerosol properties and behavior is essential to maximize chances for adequate sample collection. This is especially true when using many of the automated sampling devices available today. The devices generate the numbers, whether they are accurate or not and it is up the investigator to interpret and understand what is being generated.

In this chapter a general introduction to aerosol properties and behavior has been presented. You, the reader, have available the tools that the ancients could only dream about. Specific applications involving aerosols are reserved for other chapters. The tools described here are for all to use. It is hoped that this material will be complementary to the requirements of these other chapters, and add to an understanding and comprehension of the role of aerosols in industrial hygiene, and how aerosol problems can be solved.

NOMENCLATURE

A	Cunningham correction factor constant
A_s	surface area of electrostatic precipitator
B_0	illuminance of an object
B_h	illuminance of horizon
C	contrast
Cc	Cunningham correction factor
C_D	coefficient of drag
D	diffusion coefficient
E_s	field strength
F	force
I	intensity of light at a distance ℓ
I_0	initial intensity of light (distance $= 0$)
J	diffusive flux
K'	average coagulation coefficient
K_0	coagulation coefficient for monodisperse particles
K_{12}	coagulation coefficient for polydisperse particles
Kn	Knudsen number
K_T	thermal conductivity of air
L	impactor slit length
L	tube length
L_h	latent heat of evaporation
M	molecular weight
$\bar{N}$	average ion concentration
Q	volumetric flow rate
Q_{abs}	absorption efficiency
Q_{ext}	extinction efficiency
Q_{scat}	scattering efficiency
R	gas constant, tube radius
Re	Reynolds number
Stk	Stokes number
T	temperature
V	voltage
W	impactor slit width or diameter

Z	ion mobility
$\bar{n}$	average number of charges
$\overline{d}$	average particle diameter
a	acceleration
b	scattering coefficient
c	concentration
c_0	initial concentration
c_s	sample concentration
d	particle diameter
e	unit charge
g	acceleration due to gravity
$i_{1,2}$	angular intensity functions
k	Boltzmann's constant
k_a	absorption coefficient
m	mass, particle mass, refractive index
n	number of charges, number of particles
n'	characteristic charge
n_i	number of particles in ith interval
p	diameter weighting factor
p_∞	saturation vapor pressure over infinite plane
p_1	partial pressure
p_s	vapor pressure
q	electrical charge
s	stop distance, a distance
t	time
t'	characteristic time
u	gas velocity
v	particle velocity
v_{avg}	average ionic velocity
V_T	terminal settling velocity
w	electric drift velocity
x	diffusion distance, a distance
α	dimensionless size parameter
χ	particle dielectric factor
λ	gas mean free path
γ	extinction coefficient, surface tension
μ	gas viscosity
κ	imaginary part of refractive index
ρ	density

σ	standard deviation
τ	relaxation time
ν	kinematic viscosity, real part of refractive index
Ψ	impaction parameter
ℓ	distance, visual range

BIBLIOGRAPHY

1. American Conference of Governmental Industrial Hygienists, *Particle Size-Selective Sampling in the Workplace*, Pub. No. 0830, ACGIH, Cincinnati OH, 1984.
2. P. C. Reist, 1997 from Lapple, *Stanford Research Institute Journal*, 3rd quarter; J. S. Eckerd and R. F. Strigle, *JAPCA* **24**, 961–966 (1974); Bird, Steward and Lightfoot, *Transport Phenomena*, Wiley, New York, 1960.
3. P. C. Reist, *Aerosol Science and Technology*, McGraw Hill, New York, 1993.
4. T. Hatch, T. and S. P. Choate, *J. Frank. Inst.* **207**, 369 (1929).
5. L. D. Landau and E. M. Lifshitz, *Fluid Mechanics*, Pergamon Press, New York, 1959.
6. R. A. Millikan, *Science* **32**, 436 (1910).
7. E. Cunningham, *Proc. Royal Soc. A* **83**, 357 (1910).
8. V. A. Marple and K. L. Rubow, "Theory and Design Guidelines," in J. P. Lodge, Jr., and T. L. Chan, eds., *Cascade Impactor, Sampling and Data Analysis*, AIHA, Akron, OH, 1986, p. 79.
9. S. V. Hering and V. A. Marple, "Low-Pressure and Micro-Orifice Impactors," in J. P. Lodge, Jr., and T. L. Chan, eds., *Cascade Impactor, Sampling and Data Analysis*, AIHA, Akron, OH, 1986, p. 103.
10. T. Chan and M. Lippmann, *Environ. Sci. Tech.* **11**, 372 (1977).
11. J. Abrahamson in, R. J. Wakeman, ed., *Progress in Filtration and Separation 2*, Elsevier Scientific Publishing Co., Amsterdam, 1981.
12. M. Kessler, *Flow Measurement and Efficiency Modeling of Cyclones for Particle Collection*, M.S. Thesis, University of North Carolina, Chapel Hill, NC, 1990.
13. D. L. Iozia and D. Leith, *Filtration and Separation*, **26**, 272 (1989).
14. D. W. Dockery, C. A. Pope III, X. Xu, J. D. Spengler, J. H. Ware, M. E. Fay, B. G. Ferris, and F. E. Speizer, *New Engl. J. Med.* **329**, 1753–1759 (1993).
15. W. Wilson and H. Suh, *Aerosol Sci. and Tech.* **16**, 200 (1996).
16. Y.-S. Cheng Cheng, in *Air Sampling Instruments for Evaluation of Atmospheric Contaminants*, 7th ed., p. 405, ACGIH, American Conference of Governmental Industrial Hygienists, Cincinnati, OH, 1989.
17. M. Coullier, *J. Pharm. Chim.* **22**, 165 (1875).
18. J. Aitken, *Trans. Royal Soc. Edinb.* **30**, 337 (1880).
19. V. J. Schaefer, and J. A. Day, *A Field Guide to the Atmosphere*, Houghton-Mifflin Co., Boston, 1981, p. 16.
20. C. T. R. Wilson, *Phil. Trans. Royal Soc.* **A189**, 265 (1897).
21. W. Deustch, *Ann. d. Phys.* **68**, 335 (1922).
22. G. Mie, *Ann. Physik.*, **25**, 377 (1908).

23. D. Deirmendjian, *Electromagnetic Scattering on Spherical Polydispersions*, American Elsevier, New York, 1969.
24. H. Koschmieder, *Beitr. Phys. Freien Atm.* **12**, 171 (1924).
25. M. von Smoluchowski, *Bull. Acad. Sci., Cracow* **1a**, 28 (1911).
26. S. K. Friedlander, *Smoke, Dust and Haze*, Wiley, New York, 1977, p. 43.
27. S. K. Friedlander and C. S. Wang, *J. Col. Int. Sci.* **22**, 126 (1966).
28. K. W. Lee, *J. Col. Int. Sci.* **92**, 315 (1983).
29. K. W. Lee and H. Chen, *Aerosol Sci. Tech.* **3**, 327 (1984).

CHAPTER ELEVEN

Computed Tomography in Industrial Hygiene

Lori A. Todd, Ph.D., CIH

1 INTRODUCTION

A variety of air monitoring methods are used by industrial hygienists to evaluate human exposures to contaminants, monitor process emissions and leaks, and determine the effectiveness of ventilation systems. Although methods may vary in the length of time over which the samples are obtained, essentially all of the methods use point samplers. Therefore, the results are spatially limited to the discrete locations of the sampling devices. In addition, when the concentrations are integrated over time, they are temporally limited to the length of the sample time. Limited spatial and temporal resolution is important because it reduces an industrial hygienist's ability to evaluate and control exposures to chemicals effectively.

When sampling devices are placed in the breathing zones of workers, the results relate only to the physical location of the workers or to the paths that they travel during the sampling period. Industrial hygienists take these spatially limited results and assume they are representative of the larger unsampled workforce. This assumption may not always be valid; in practice, it is difficult to select a representative subset of individuals to sample because the concentration distributions in a room are unknown. Before choosing a subset of workers to sample, a larger homogeneous group is usually created based upon similarities in the tasks they perform and in the local environments in the rooms where they work (1). The local environment is important because contaminant flow patterns are strong determinants of exposure; however, environmental similarity is difficult to predict. Data

Patty's Industrial Hygiene, Fifth Edition, Volume 1. Edited by Robert L. Harris.
ISBN 0-471-29756-9

on ventilation systems and airflow patterns are usually lacking and individuals selected for sampling may not be truly representative.

If a device is placed in a fixed location of a room to obtain an area sample, or is carried throughout an area to attempt to characterize flows, the results relate only to the physical location or path of the sampling device. Strategies for placing area samplers assume that locations with the highest or most relevant concentrations are represented and that the areas between the samplers are of less importance. In practice, this assumption is difficult to support; however, industrial hygienists use this information to guess the dispersion patterns of chemicals for the regions in the room between the area samplers. The assumptions and decisions currently made using spatially and temporally limited air sampling may not be valid; air concentrations are quite variable during the day due to changing ventilation patterns, the movement of people in a space, fluctuations in source emissions, and differences in work practices among people.

Temporal resolution is important for understanding and effectively controlling the generation and transport of contaminants in air, for identifying chemical intermediates produced during a process, and for monitoring chemicals that are acutely toxic. Ideally, sample results would enable us to construct a complete spatial and temporal profile of the flow of air contaminant concentrations for an entire day, week, or season. This would require an enormous number of real-time measurements to be concurrently obtained throughout an area. This is rarely performed. Real-time instruments can be used to provide good temporal resolution of concentration fluctuations, but the number of chemicals that an instrument can accurately measure at one time is usually limited.

In addition to the above limitations, air sampling is usually performed infrequently, and information about daily, weekly, or monthly variability is lost. Most industrial hygiene investigations involve short campaigns of one or two days. Thus, insufficient data are currently being used to evaluate the impact of industrial emissions and indoor air contaminants on the health of workers and residents in the community.

An entirely new air monitoring technique, for both the occupational and environmental field, may provide spatially and temporally resolved estimates of contaminant concentrations noninvasively (does not pump air out of a space through collection media), and in real-time, over large areas. This technique combines the real-time chemical detection methods of optical remote sensing, such as an open-path Fourier transform infrared (OP-FTIR) spectrometer, with the mapping capabilities of computed tomography (CT) (2–15). This environmental CT system generates near real-time spatially and temporally resolved two-dimensional concentration maps of multiple chemicals at low limits of detection (ppb–low ppm) for an entire area (see Fig. 11.1). Not just another nifty tool, this technology represents a major departure from conventional industrial hygiene air sampling methods and could allow researchers to understand and evaluate human exposures, source emissions, and chemical transport in ways that are unavailable using conventional methods. This technique provides a powerful tool for visualizing air contaminant species, concentrations, and flows in industry and outdoors in the community.

Each tomographic concentration map provides a snapshot, which represents a short time period (minutes), of the concentration and location of contaminant plumes in a slice or plane through the air. As measurements are obtained over the day, the reconstructed

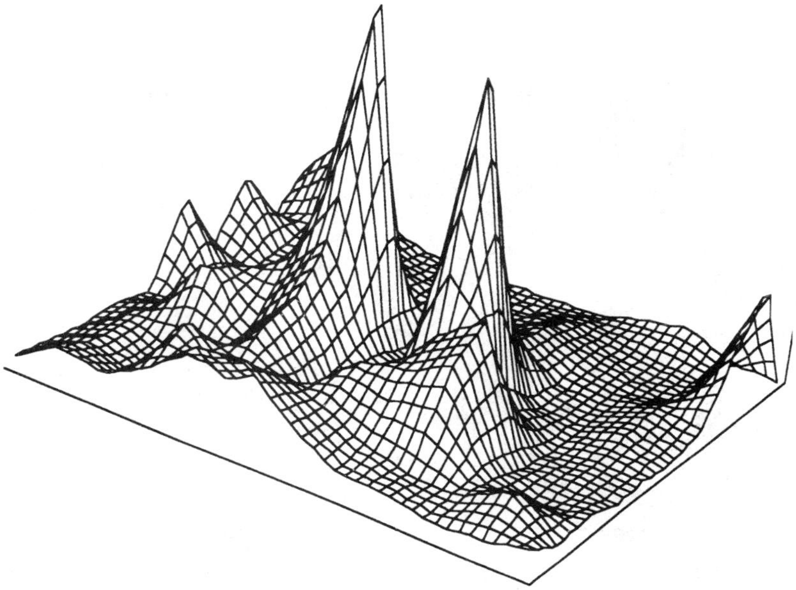

Figure 11.1. Example of a two-dimensional concentration map for a 10 foot by 12 foot room. The heights of the peaks represent the concentrations of the chemical at a specific location in the room. The diameters of the peaks represent the dispersion of the chemical.

concentration maps are linked together to provide a powerful tool for visualizing the flow of air contaminants over space and time. Thus, a video of concentration fluctuations, for single or multiple chemicals, can be created for a room or area. Using a CT system for industrial hygiene monitoring, multiple locations throughout a room are quantified simultaneously, reducing the need to first determine the few important hot spots to sample. For example, Figure 11.2 shows a series of tomographic concentration maps of a tracer gas released over time in one location, an exposure chamber (16). If this was a workplace and an integrated sampler was placed in the assumed highest concentration area in the chamber (A), the industrial hygienist would have missed significant concentrations accumulating in other areas (B). The concentrations maps generated with this CT system provide a powerful way to track the rapid movement of chemicals in the room. When maps are combined with ventilation measurements, tracer gas releases, information on process changes and workload, meteorological data, or information on the movement and location of workers in a room, the information can be used to quantify chemical emission rates, evaluate the effectiveness of ventilation systems, and track human exposures. With this environmental CT system, concentrations can be resolved in near real-time for an entire area with far fewer measurements than would be necessary to obtain the same level of detail using conventional point samplers.

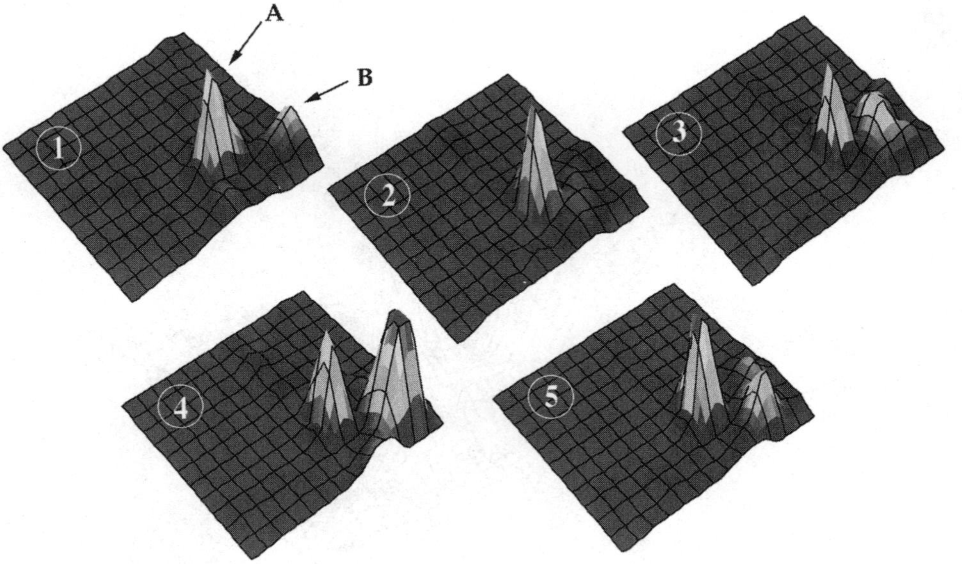

Figure 11.2. A series of five tomographic concentration maps reconstructed for sulfur hexafluoride in an exposure chamber. Each map represents a seven-minute snapshot.

1.1 History of Applying Tomography to Industrial Hygiene

The use of tomographic techniques dates to 1917 with a paper by mathematician J. Radon (17). He proved that a three-dimensional object could be reconstructed uniquely from the infinite set of all its projections (line integrals along the particular measurement path). CT has been used in astronomy for producing maps of microwave radiation emitted from the sun (18, 19), imaging of nuclear fuel pin bundles (20), optics (21, 22), and electron microscopy for complex biomolecules (23–26). However, its greatest achievement and progress has been in the field of medicine, where cross-sections of the human body are reconstructed from a large number of X-ray attenuation measurements through the section of the body of interest (27–30). In diagnostic radiology, a person is placed inside of a device similar to a large doughnutlike structure. Thousands of X-rays are shot from this structure through a body at many different angles along a plane (see Fig. 11.3). The first commercial clinical X-ray CT scanner was introduced in 1972 by EMI Ltd. (27). This noninvasive procedure used X-rays from many different angles to create images in three dimensions of the organs within the body. An X-ray projection consisted of 160 scans along a single direction; the unit was then rotated one degree for each of 180 projections. A total of 28,800 ray sums were processed by a tomographic reconstruction algorithm to reconstruct an image of the structures in the slice of the body that was sampled. To reconstruct a three-dimensional image, the ring of X-rays is sequentially moved to adjacent planes or slices of the body. Tomographic imaging overcomes the shortcoming of conventional radiography in which body structures are superimposed on a film making it

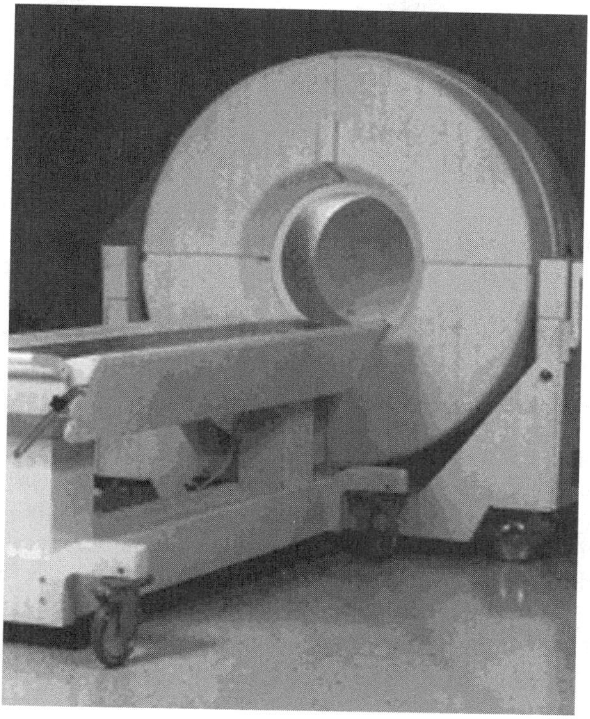

Figure 11.3. An example of a commercial medical CT device.

difficult to distinguish individual organs. In medicine, tomography is usually used to find small abnormal features, such as tumors, within the known features of the human anatomy. Therefore, to achieve high spatial resolution, a large number of rays must be used.

In industrial hygiene or environmental air sampling, CT reconstructs chemical concentrations and plume shapes. This could be an advantage because industrial hygienists are interested in the large structures (plumes) rather than small structures within the large structures. However, there are some obvious challenges when applying CT to air sampling: (*1*) instead of using X-rays, optical rays must be used that can safely, noninvasively, and in near-real time monitor plume concentrations; (*2*) in medical CT, a person can easily be placed inside a device that houses the X-rays; in industrial hygiene, a room cannot as easily be encircled by remote sensing instruments; (*3*) appropriate optical remote sensing (ORS) instrumentation must be available for indoor monitoring of chemicals across an open space; (*4*) people, equipment or building structures all restrict the number and symmetry of available line-of-sight measurements; (*5*) in contrast to most organs in the body, chemical concentrations are constantly fluctuating in both time and space; (*6*) the signal to noise ratio of available ORS instrumentation limits data acquisition speed; and (*7*) the current cost of ORS equipment limits the number of instruments and associated hardware that can be used. To be feasible, the use of CT in industrial hygiene must use minimal hardware and still achieve acceptable results.

The first investigations applying CT to air pollution were in 1979 and 1982, when Byer, Shepp, and Wolfe (31, 32), and Wolfe and Byer (33), described a theoretical laser-based system for measuring air pollution on an urban scale. Artificial concentration profiles were generated on a computer, and a simulated CT system was evaluated which used a circularly symmetrical set-up of laser fan beams. A circle of mirrors was used to reflect the laser fan beams across kilometer distances to detectors; thus, the mirrors created multiple virtual laser sources. These studies suggested rather complex virtual source beam set-ups (geometries) that required many mirrors and detectors. Consequently, the optical systems proposed in these studies would be difficult to align or maintain. The use of multiple detectors made the system costly, hardware intensive, and posed calibration and equipment maintenance difficulties. To date, these and other practical considerations have prevented the application of CT to large scale (>1 km) ambient air pollution monitoring.

The first investigation of the application of CT to industrial hygiene was in 1990 when Todd and Leith (8, 9) performed a theoretical study using computer generated static chemical profiles and a simulated CT system. This study showed the feasibility of applying CT to industrial hygiene by successfully reconstructing concentration profiles composed of overlapping bivariate Gaussians, using a simulated CT system with as few as four projection angles. An equal angle parallel projection geometry was used in these initial studies which had optical sources and detectors on three sides of a square room see Fig. 11.4. For a 40 by 40 foot room, this geometry placed one detector per foot along the side of the

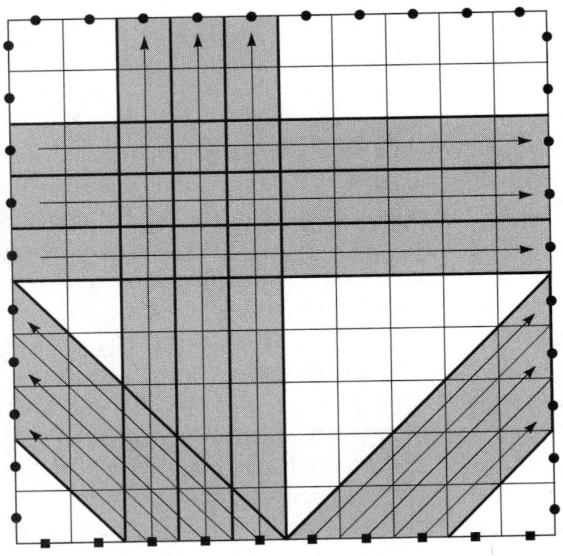

Figure 11.4. An equal-angle parallel-projection geometry. Representation of a 10 by 10 room with four projection angles, 0, 90, and ±45.

room. This geometry was chosen for its simplicity and applicability to a various workplace layouts. As shown in Figure 11.4, the acute angle that the light beam makes with the room width is called the projection angle. A projection is defined as all the parallel, equally spaced light beams with the same projection angle. For every positive acute angle, its negative counterpart is used. From four to twelve projection angles were evaluated using stationary test data. In these initial computer simulation studies, the mean percent errors for reconstructing the highest concentrations of plumes decreased (from 9 to 0.3%) as the number of projection angles increased (from four to twelve), respectively. In addition to the relatively low quantitative errors, the reconstructed concentration maps were visually similar to the original concentration maps. Although this geometry is not practical given the current remote sensing instrumentation, these studies were important for showing that CT is a feasible technology to apply to reconstructing chemical concentrations. The number of projections used was orders of magnitude fewer than required in other applications. Considerable progress has been made in this application of CT for measuring chemicals in indoor and outdoor air. There have been many theoretical and experimental studies, and a few field studies (2–15, 34–40). Outdoors, this technology is being applied to measuring emission rates of chemicals, such as ammonia, off of waste lagoons from intensive swine facilities (5, 16, 38). Progress in using CT to measuring aerosol distributions in air has been slower (41–43).

2 DESIGNING COMPUTED TOMOGRAPHY SYSTEMS

As applied to air monitoring for chemicals, computed tomography is the process of measuring a spatial concentration profile of a plane through a room using a network of line-integrated concentration data. The plane could be horizontal or vertical; the horizontal plane would primarily be used for the industrial hygiene application (see Fig. 11.5). In theory, one or more rotating ORS instruments, such as a tunable diode laser or an OP-FTIR spectrometer can be placed in an area, along with mirrors, detectors, and/or retroreflectors, to create a network of remote sensing optical beams. Retroreflectors return an optical beam directly back to its origin. As the ORS instrument scans the area, it measures the integrated concentrations of contaminants present in the volume of each scanned beam path. Using a tomographic algorithm, a computer takes all of the one-dimensional path-averaged measurements and produces a two-dimensional estimate (map) of the concentrations and locations of the chemicals in the sampled plane (see Fig. 11.6) (30). The tomographic map can be generated in a matter of minutes on a personal computer. The strategy is to obtain reconstructed concentration maps whose ray sums are similar to the measured ray sum data. The degree of spatial resolution (ability to differentiate between peaks) that can be reconstructed using the maps depends upon the number and orthogonality of the ORS beam paths for a given size area. The number and spatial placement of beams is partially determined by the type of ORS hardware. Constraints placed on the geometry of the CT system can translate into insufficient measurements, a limited number and range of angles and, ultimately, into artifacts in the reconstructed maps. To obtain a network of open-path measurements with the currently available instrumentation, each ORS instrument sequentially scans the air; thus, each open-path measurement is taken at a different point in time.

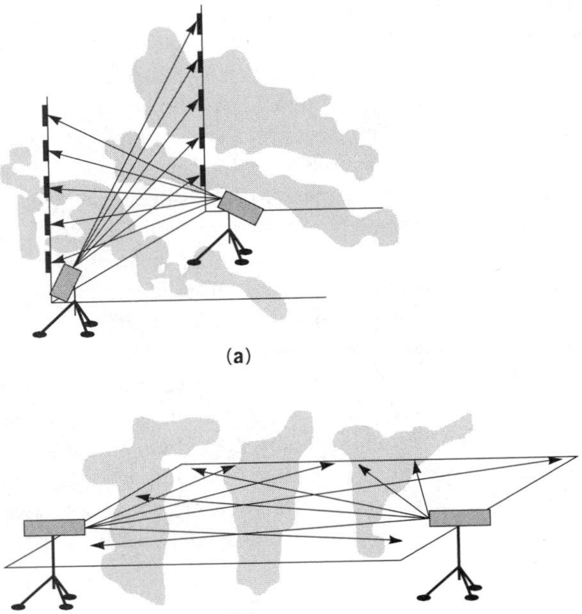
(a)

(b)

Figure 11.5. Horizontal and vertical planes for using CT to map chemicals in air. (**a**) Two ORS instruments monitoring a vertical plane to capture plumes coming off of an area. (**b**) Two ORS instruments monitoring a horizontal plane to capture plumes through a room.

Therefore, if geometries use a high density of rays, the overall time required to scan an area can be very long. Given that concentrations are in flux, this introduces inconsistencies in the data that adversely affect reconstruction accuracy. The degree of temporal resolution that can be obtained depends upon the ORS instrument measurement time for each individual beam, and the time required to completely sample an area. To map concentration profiles in flux accurately, the time required to sample the entire room (due to ray density and measurement time) must be balanced with the movement of air contaminants. Each complete set of ray measurements must be obtained very quickly to minimize data inconsistencies and to capture short-term fluctuations in concentrations and peak exposures. Using a given geometry, the choice of reconstruction algorithm can impact the spatial and temporal resolution of concentrations and the overall accuracy of the reconstructed maps.

Some of the decisions faced when using CT in industrial hygiene include: (*1*) selecting appropriate reconstruction algorithms; (*2*) designing and testing optimum remote sensing geometries; (*3*) generating test concentration data to evaluate the CT system; (*4*) selecting appropriate image quality measurements to evaluate the reconstructed concentration maps; and (*5*) selecting appropriate remote sensing instrumentation. The quality of the reconstructed concentration maps determines the ability of scientists to use the maps for exposure assessment, source monitoring, and leak detection. Scientists need to be able to use

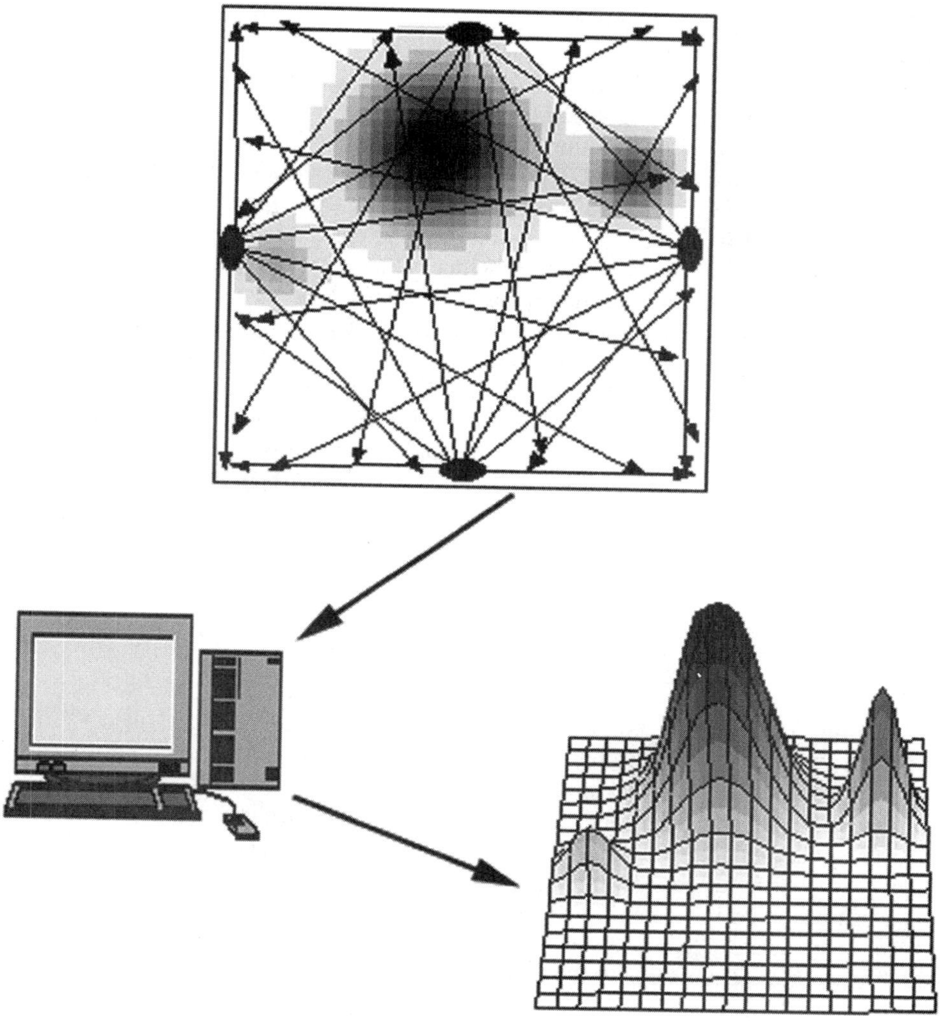

Figure 11.6. Sequence of events for creating concentration maps using a CT System. Top: Four ORS instruments scan a room. Middle: Computer using a tomographic algorithm to reconstruct concentration maps from open-path measurements. Bottom: Reconstructed concentration map.

the reconstructed maps to answer specific questions such as: (*1*) What are the quantities of chemicals leaking from an industrial process or emitted from an industry? (*2*) Where are the chemicals leaking from? (*3*) What quantities of chemicals do people potentially inhale? and (*4*) What are the chemical dispersion patterns in air?

In general, when developing a measurement system using CT, both the geometry of the ORS equipment and the reconstruction algorithm must be tailored to the specific application and evaluated using computer simulations and laboratory experiments before they are deployed in the field.

Computer simulation studies are an efficient way to design and evaluate potential ORS algorithms and geometries for a specific industrial hygiene application. This is accomplished by using a computer simulation program and a battery of appropriate test concentration maps to simulate the entire CT process from data acquisition to reconstruction of concentration maps. Test maps are high-resolution grids that represent a room or area; each cell in the grid is assigned a concentration. Test concentration data should model reality, test the limits of the system, and include as many potential sources of error as possible. Although static concentration maps can be used, time-varying images provide a more rigorous evaluation of an algorithm or geometry. Time-varying images are sets of multiple maps that represent spatial and temporal changes in concentrations of single or multiple chemicals in air. These test maps are usually computer generated because temporally and spatially resolved field data do not exist. It is important to use a variety of test concentration profiles to evaluate how well the reconstructed maps can perform specified tasks relevant to the application (44–46).

2.1 Test Concentration Maps

The test concentration profiles should be generated on a grid which represents the sample site; the resolution of the test map grid (number of cells) should be a minimum of two to four times the density of the resolution of the final reconstructed map. The level of spatial detail needed for the application determines the resolution of the final grid. Test maps are then fed into a simulation program and synthetic open-path data are created by back calculating the measurements from the maps. Test maps that have been used in simulation studies include: (*1*) krigged experimental point sample data from studies using an exposure chamber (static maps); (*2*) CT maps reconstructed from chamber studies using a prototype CT system (static or time-series maps), and; (*3*) static and spatially and temporally varying data generated using dispersion models (3, 4, 47, 48). Static maps can be used for preliminary tests; however, they should include a range of peak numbers, peak concentrations, and plume diameters. Static maps cannot reveal the impacts of temporally changing concentration profiles. Time-series maps can be created with varying source generation and decay rates, diffusion coefficients, air velocities, source locations, and source intensities for single and multiple contaminants. Figure 11.7 shows time-series maps, which model the movement and decay of a single contaminant source.

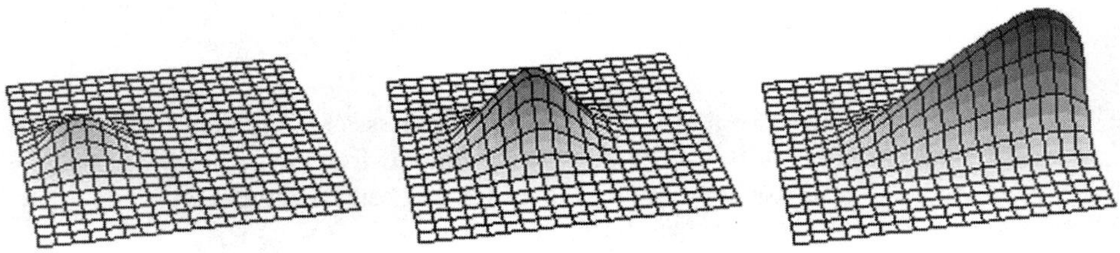

Figure 11.7. Example of time-series maps showing the movement of contaminants over time.

2.2 Image Quality Parameters

To evaluate CT geometries and algorithms, it is necessary to quantitatively and qualitatively (by visual assessment) compare the reconstructed concentration distributions with the true (test maps) concentration distributions. Quantitative measures that can be used in computer simulation studies include nearness, peak exposure error, peak height error, peak location error, and projection data distance. It is important to use a variety of quantitative measures because they evaluate different uses for the maps, and geometries and algorithms may not be ranked the same by all the measures. For example, nearness measures overall reconstruction accuracy and could reflect the presence or absence of artifacts. In contrast, exposure error measures how well the highest concentrations are reconstructed, and would not necessarily be related to how well the shape of the peaks are reconstructed.

2.2.1 *Nearness*

Nearness describes the discrepancy between the true and the reconstructed concentration map on a cell by cell basis, and is a global measure of the errors over all the cells in the map (49–52). Nearness is a dimensionless number; the smaller the nearness, the better the agreement (see Eq. 1). Nearness should only be compared between maps with the same number of grid cells. Nearness can only be used on experimental data where the true concentration distribution is known.

$$\text{Nearness} = \sqrt{\frac{\sum_{1}^{N^2} (c_i^* - c_i)^2}{\sum_{1}^{N^2} (c_i^* - \overline{c_i^*})}} \tag{1}$$

where c_i^* = concentration of the ith grid cells on the "true" map
c_i = concentration of the ith grid cells in the reconstructed map
$\overline{c_i^*}$ = mean concentration of all the grid cells in the "true" map
N^2 = total number of grid cells in the map (N by N grid)

2.2.2 *Peak Location Error*

"Peak location error" was developed to reflect how accurately a reconstructed map pinpoints the location of a chemical leak or emission source (9). It is the root mean square (RMS) difference in the location of the peaks in the original and reconstructed maps (see Eq. 2). For example, for a 40 by 40 meter area, reconstructed on a 40 by 40 grid, a location error of two would be equivalent to a distance of two meters. The significance of this error is related to the level of accuracy required to pinpoint the plume. For example, if one were only interested in locating the general area of a leak, a coarse approximation would be required.

$$\text{Peak Location Error} = \sqrt{(x - x^*)^2 + (y - y^*)^2} \tag{2}$$

where x = the x coordinate of the location of the peak in the reconstructed map
x^* = the x coordinate of the location of the peak in the test map
y = the y coordinate of the location of the peak in the reconstructed map
y^* = the y coordinate of the location of the peak in the test map

2.2.3 Peak Exposure and Peak Height Error

A measure of peak exposure error was developed to reflect the error in estimating the average concentration of a chemical inhaled by a person (2). Most occupational exposure limits for chronic toxicants are based upon averages of concentrations over time. Therefore, using time-series maps, concentrations can be averaged over an extended time frame (see Eq. 3). In addition, the spatial movement of the worker can be simulated and the appropriate size footprint can be factored into the equation.

$$\text{Peak Exposure Error} = -\frac{\sum\limits_{\text{time}}\sum\limits_{\text{space}} c_j^* - \sum\limits_{\text{time}}\sum\limits_{\text{space}} c_j}{\sum\limits_{\text{time}}\sum\limits_{\text{space}} c_j^*} \times 100 \tag{3}$$

where time varies from the 1st time series map to the last map in the series, space represents a window around the exact location of the peak or through the path that the worker travels
c_j^* = the concentration of the jth cell in the test map
c_j = the concentration in the jth cell in the reconstructed map

Peak height error uses Equation 3; however, space would represent a small window only around the location where the peak is at its highest. For static maps, the time factor would be removed.

2.2.4 Projection Data Distance

Projection data distance, Equation 4, is a measure of how closely the reconstructed ray sums match the original ray sums (51). A data distance value of zero implies a perfect match.

$$\text{Projection Data Distance} = \frac{\sqrt{\sum\limits_{i=1}^{M}\left(p_i - \sum\limits_{j=1}^{N^2} a_{ij}c_j\right)^2}}{M}, \tag{4}$$

where p_i = the true i^{th} ray sum
$\Sigma_{j=1}^{N^2} a_{ij}c_j$ = the calculated i^{th} ray sum
a_{ij} = the contribution of the jth cell to the ith ray
M = the total number of rays
N^2 = total number of grid cells in the map (N by N grid)

2.2.5 Visual Assessment

Visual evaluation of the reconstructed maps is the most useful method for evaluating CT geometries and algorithms. Although it is time consuming and subjective, it allows the finest distinctions to be made between geometries and algorithms, and it models what the industrial hygienist will be using in the field. Visually comparing surface plots of the reconstructed two-dimensional maps with the original maps is the most direct way to discern differences in peak location, peak shape, and peak resolution, and document the generation of artifacts. Artifacts can include unpredictable irregularities that look like noise, patterns such as streaking, and false concentration peaks. Artifacts and streaks can result in underestimation or overestimation of concentrations in the maps. By understanding artifact generation, industrial hygienists can determine the situations where geometries and algorithms might fail in the field, and can use the information to design more robust geometries and algorithms. False peaks and streaking could be due to an uneven placement of rays and sparse sampling. The video of reconstructed concentration maps can be evaluated as to how well dispersion, concentrations, and plume contours are reconstructed.

3 CT ALGORITHMS FOR THE INDUSTRIAL HYGIENE APPLICATION

Many different tomographic reconstruction algorithms have been proposed for CT applications: the three main types are analytic (29, 53), back-projection (54–56), and iterative methods (27, 19, 52, 57). Analytic techniques are used in most commercial medical CT scanners; however, they may not be the best algorithms to use for reconstructing chemical concentrations where data are sparse. Iterative methods, for example, may yield better performance in industrial hygiene applications where there are incomplete data, nonsymmetric ray geometries, and noise due to nonstationary chemical profiles. Iterative methods allow flexibility in the placement of remote sensing equipment in a room and allow constraints to be imposed on reconstructed measurements (56, 58). Some of the algorithms that have been used when applying CT to reconstructing concentrations include algebraic reconstruction (4, 9) and smooth basis function minimization (13, 14) techniques.

A single algorithm may not be appropriate for all industrial hygiene applications. An algorithm that can accurately pinpoint leaks might not provide the most accurate estimates of overall environmental concentrations. An algorithm used to reconstruct rapidly changing concentration distributions may not be applicable for a region with near steady-state concentrations. Therefore, it is important to evaluate and refine an algorithm for a given application. Some of the algorithms that have been used in industrial hygiene applications are described below.

3.1 Iterative Methods

Iterative methods can be considered as brute force methods to solve image equations. Using iterative methods, an idealized measurement space, such as a room, is broken into an N by N grid of cells. Figure 11.8 shows a 10 by 10 grid of 100 cells. Each cell, *j*, is assigned a concentration, *cj*, which is assumed to be uniform and non-negative. Ray sums are line integrals (path-integrated concentrations) of *cj* along various paths through the room. The rays have a finite width, and therefore, are approximated by strip sums. The relationship between cell concentrations and ray sums can be expressed as Equation 5.

$$p_i = \sum a_{ij} c_j, \quad i = 1 \ldots M, \quad j = 1 \ldots N^2 \tag{5}$$

where p_i = the ray sum at projection angle Θ
a_{ij} = the weighting factor representing the contribution of jth cell to the ith ray sum
c_j = the density function for the jth cell
M = the total number of rays
N^2 = the total number of cells in an N by N grid

Equation 5 represents an array of M equations and N^2 unknowns. For reconstructing chemical concentrations, typically a sparse geometry is used and the number of equations is far fewer than the number of unknowns. In addition, time constraints are placed on the sampling that generate inconsistent data. Therefore, iterative approaches have been used to solve these equations rather than using matrix inversions.

3.1.1 Algebraic Reconstruction Techniques (ART)

For ART1, an initial guess is made of the values in all the cells in the grid; based on this guess, the first path-integrated ray concentration is calculated. The values of all the cells

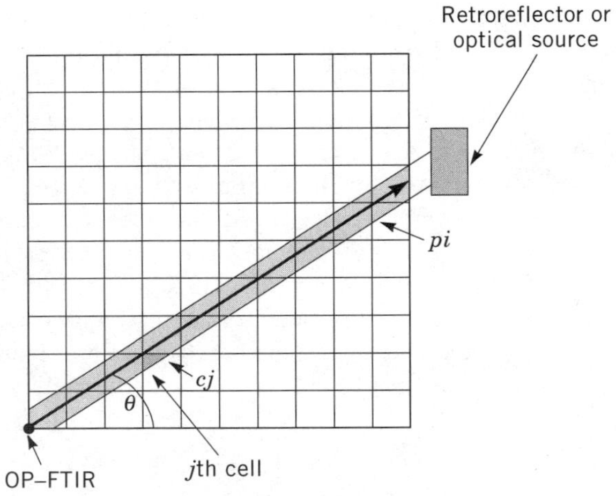

Figure 11.8. An idealized room is represented by 10 by 10 grid of 100 cells. The projected ray sum is p_i, with a projection angle of θ. The concentration of the contaminant in the jth cell is c_j. The ray only covers a portion of the cell.

that the ray passes through are then corrected so that the calculated ray sum equals the measured ray sum. The correction procedure is shown in Equation 6 and is repeated sequentially for each ray. With this algorithm, the weights of all cells through which the ray passes are assigned a value of one, and the weights of all cells through which the ray does not pass are assigned a value of zero. One iteration is completed when all the ray sums have been corrected. The ART1 algorithm is terminated after a fixed number of iterations, such as ten, or by using a stopping criteria, such as the fractional change in the variance of the reconstructed concentration map, from one iteration to the next (50, 51).

$$c_j^{(q+1)} = c_j^{(q)} + \frac{\left(p_i - \sum_{j=1}^{N^2} a_{ij} c_j\right) a_{ij}}{\sum_{j=1}^{N^2} aij^2} \tag{6}$$

where q = the iteration number

The ART3 algorithm is based on the premise that a precise solution to Equation 5 is generally impossible because the p_i's are physical measurements that contain measurements error (59). Therefore, ART3 finds a solution to the inequalities shown in Equation 7 and uses the same correction scheme as in Equation 6. However, the equations have to be satisfied only within a tolerance, which is estimated from prior knowledge of the experimental conditions and errors.

$$p_i - \epsilon_i \leq \sum_{j=1}^{N^2} a_{ij} c_j \geq p_i + \epsilon_i, \quad i = 1, 2, 3, \ldots M \tag{7}$$

where ϵ_i = an allowed tolerance

For ARTW, the initial guess and correction procedure is the same as ART1; however, the weights of each cell for a given ray, are proportional to the average length of the ray through the cell. The average length of a ray in a cell is calculated by estimating the area of the cell covered by a ray and dividing it by the width of the ray, which can assumed to be one-cell width wide.

SIRT (Simultaneous Iterative Reconstruction Technique) uses Equation 6 to determine the change in the jth cell; however, the value of the jth cell is not changed until all of the differences in the ray sums are calculated. At the end of each iteration, the cell concentrations are then changed once using an average value of all the corrections that would be made to that cell.

SART (Simultaneous ART) (60–62) makes cell corrections similar to SIRT; however, average corrections are applied to each cell after calculations are made for all ray sums in one projection. Therefore, when four projection angles are used, four averages are applied to each cell in an iteration.

MART (Multiplicative ART) is similar to ARTW, however, cell corrections are updated by a multiplicative adjustment scheme that maximizes an entropy function (63). The correction in the cell estimate is equal to a relaxation factor raised to the power of the ratio of the measured ray-sum and the back projected raysum. In MART, like ART, the cell estimates are corrected one ray at a time.

3.1.2 Maximum Likelihood Expectation Maximization (MLEM) Algorithm

MLEM is a simultaneous iteration technique that corrects all the cells at once after an entire set of ray sum data is obtained in each iteration (64, 65) (see Eq. 8). The correction factor for a cell is equal to the sum of the ratios of the reprojected values of the cells and the backprojected values of the cell. In MLEM, the cell estimates are corrected taking into consideration all the projections simultaneously. MLEM is usually allowed to iterate longer than the ART algorithms.

$$C_j(n + 1) = \left(c_j(n) \Big/ \sum t_{ij}\right) * \sum \left(t_{ij} p_i \Big/ \sum t_{ij} c_j n\right) \tag{8}$$

where $C_j(n)$ = concentration in j^{th} cell after the nth iteration
$\Sigma t_{ij} c_j n$ = reprojection of the image estimate $c_j n$
$\Sigma t_{ij} p_i$ = backprojection of the projection array

The concentration distribution is estimated by using the maximum likelihood criterion and is given by Equation 9,

$$\ln L(C) = \sum \left\{ -\sum t_{ij} c_j + p_i \ln \left(\sum t_{ij} c_j \right) - \ln (p_i) \right\} \tag{9}$$

where $L(C)$ = the likelihood of generating the image C
t_{ij} = the transfer matrix from image cell j to set of parallel projections i
c_j = the concentration at cell j
p_i = the raysum for the ith set of projections
i = the number of rays, 1 to M
j = the number of cell, 1 to $N2$

The likelihood strictly increases at each step unless it is at a maximum, therefore, all the cells have positive values of concentration.

3.2 Smooth Basic Function Minimization (SBFM)

The previous algorithms did not assume anything about the underlying concentration profile except that it is non-negative and the concentrations are homogeneous in discrete grid cells. SBFM does not artificially create a grid of cells and assumes *a priori* information about the underlying concentration profile by modeling the data with smooth basis functions (bivariate Gaussians defined in the x–y plane) (13, 14).

3.3 Simulation Studies of Reconstruction Algorithms

Rigorous testing should be performed when developing or modifying algorithms because they are based on different mathematical foundations which could make them sensitive to different types of ORS geometries and concentration profiles.

Todd (4, 9) has studied iterative algorithms using both simulated concentration data and data obtained from prototype CT systems placed in an exposure chamber. In general, MART and MLEM performed the best, followed by ARTW, ART1, ART3, and then SIRT and SART. For the ART algorithms, reconstructions were improved by using a more accurate weighting scheme to allocate the concentration corrections (ARTW). While MLEM and MART were similar, MART performed better when the concentration profiles had broad peaks, and MLEM performed better for rapidly changing concentration profiles. Regardless of the type of algorithm, most peaks were reconstructed within one foot of their true position using both static and time-series test maps. Peak exposure errors ranged from 25 to 50% using time-series maps and 8 to 15% using static maps. The time series maps not only provided a more rigorous test of the algorithms than static maps, they allowed greater distinctions to be made among the algorithms. When Drescher et al. (13, 14) evaluated SBFM using chamber generated and simulated data, they found it to be superior to ART in reconstructing concentrations and minimizing artifacts.

4 ORS GEOMETRIES FOR THE INDUSTRIAL HYGIENE APPLICATION

The geometry of a CT system involves the placement of the open-path instrumentation (optical sources, detectors, mirrors, or retroreflectors) in or around a room. In theory, because concentrations in a room are really not discretized into cells, an ideal geometry would be symmetrical and would completely flood the room with an infinite number of optical rays that span a wide range of angles. However, this is obviously unrealistic. In practice, limitations are placed on the geometry due to the layout of rooms, purpose of the space, ORS instrumentation, monetary resources, and the need to measure spatially changing concentration profiles quickly. Estimates of how many equally spaced angles are needed for good resolution vary and are specific to the application (23, 66).

Symmetry is important to ensure that all areas of the room are sampled equally; symmetrical geometries yield better quality reconstructions than asymmetric geometries. Unequal sampling can result in artifacts, streaking, and distortions of the images. Unlike medicine, reflected rays can be used in the application of CT to industrial hygiene. Reflected rays can improve reconstruction accuracy by adding orthogonal rays to the geometry and increasing coverage. Reflected rays are created by using mirrors to reflect rays across a space to a retroreflector that returns the beam directly back to its point of origin. However, mirrors are difficult to align quickly and to maintain alignment in the field. In contrast, retroreflectors are easy to align and maintain alignment, but they are currently very expensive.

As a rule of thumb, increasing the number of remote sensing instruments results in better reconstruction quality because the angular coverage of the beams is increased and multiple measurements can be taken within a short period of time. The most critical factor in designing geometries is the independence of the optical beams; by maximizing the angular separation between the different ray projections or rays, independence is maximized. With only a few remote sensing instruments, independence can be maximized by creating parallel reflections at many angles.

The higher the density of rays in a geometry, the higher the reconstruction resolution. However, when reconstructing nonsteady state plumes of contaminants, a trade-off exists between the ray density and the rate at which the rays are obtained. When using CT for industrial hygiene sampling, the air flow and contaminant generation conditions determines the point at which increasing the density of rays deteriorates reconstruction quality because of the increase in the overall sample rate.

4.1 Simulation Studies of Geometries

To use a traditional equal angle parallel projection geometry (ideal) in a room, hundreds of detectors would be placed along a wall. With the currently available ORS instrumentation, this set-up would be very expensive. A more feasible set-up involves the use of a scanning, fanlike beam geometry. For example, an entire OP-FTIR spectrometer, or a portion of the optics, is placed on a 360° rotating base. One or more of these spectrometers are placed in a room, and they sequentially scan the room shooting light at retroreflectors or receiving light from optical sources. When there is only one spectrometer, multiple mirrors are used to reflect light across the room to create a network of intersecting rays.

Todd (10, 67) has extensively studied many different CT geometries designed with from one to four scanning remote sensing instruments (see Fig. 11.9). Some of the geometries had optical beams with only straight lines of sight, others also used mirrors to reflect the beams across the room to retroflectors in order to increase spatial coverage. When two or more scanning instruments were used, they were placed along the walls of the rooms and scanned a range of 0 to 180°. When one instrument was used, it was placed in the center of the room and it scanned a range of 360°. Simulations were performed using both static and time-series test concentration maps; each time-series map represented changes in concentrations every 15 seconds over simulated three-hour time periods.

Geometries were designed on a 20 by 20 grid, and used a total of 120 rays. Therefore, one, two, three, and four ORS instruments used 120, 60, 40, and 30 rays per instrument, respectively. It was assumed that the time required to take each ray was equivalent regardless of the geometry. Therefore, the greater the number of instruments, the shorter the time required to sample an entire room.

Using the time-series maps, geometries with four instruments were superior and reconstructed maps that were quantitatively more accurate, visually smoother, and had fewer artifacts, than geometries with fewer instruments. The geometries using a single instrument in the center of the room with reflected rays gave very poor reconstructions with the time-series maps that simulated a high airflow. Large streaks were present in the reconstructions, radiating from the grid center where the instruments and the peaks were co-located. Peaks appeared jagged, there was streaking along some of the ray paths, peaks were poorly located and concentrations were significantly overestimated. One of the single instrument geometries both simulated parallel projections with reflected rays and used additional straight rays (see Fig. 11.9(**b**)). Even with this geometry, reconstructions were poor using maps that simulated high airflow in the room. However, for low airflow simulations, this geometry reconstructed the maps very well and equivalent to the four-instrument geometry. The importance of using a variety of time-series maps is illuminated by the results using a single instrument. When the static maps were used, several of the single instrument geometries had quantitative and qualitative results equivalent to the three and four instrument geometries. Park et al. (35) also tested similar single instrument geometries. In their studies, they used only static test maps generated from a bivariate normal distribution model and from experimental chamber data. Their results were similar to the Todd and Bhattacharyya (3) study using static maps.

When using time series maps, differences in the reconstruction quality can occur because of the difference in the times required to scan an area completely. For example, if 120 rays in total are used, and if it takes 15 seconds for each ray measurement, the four, three, two, and one instrument geometry could require 7.5, 10, 15, and 30 minutes to sample the room once, respectively. The difficulty that the single instrument geometries had with reconstructing the time series maps was due in part to the increased time required to obtain a complete sets of rays relative to the spatial and temporal changes in concentrations. If the overall sample times are made equivalent, then the geometries which have a higher density of orthogonal rays will produce better reconstructions. In practice, there is a limitation to the speed that each ray measurement can be taken. The faster the measurement of the chemicals, the lower the signal to noise ratios, and higher the chemical limits of detection.

An alternative to decreasing measurement time by reducing instrument measurement time is to use fewer optical rays. For a four-instrument geometry, when 30, 40, or 80 rays per instrument were used to scan a room, the smaller the number density of rays, the lower (better) the nearness, data distance, and exposure errors (2). When using 30 rays, exposure error improved over using 40 rays. The reconstructed maps using 30, 40, and 80 rays were visually very similar to each other, however, peak heights for reconstructed maps using 30 rays were closer to the original map than reconstructed maps using 40 or 80 rays. These simulations indicate that coarser sampling may be a better means of tracking changing concentration profiles over time than using finer sampling.

In general, most of the geometries tested could locate the highest concentration points in the peaks within a few grid cells. As the ventilation rates increased, nearness and peak location errors increased (worse). As the number of detectors increased, peak location and exposure errors decreased; the four instrument geometry had mean location errors of two grid cells and mean exposure errors of 12–15%. The number of concentration plumes had a greater adverse impact on reconstruction quality than the diameter and concentration of the peaks. The diameter of the concentration peaks is a limitation when the density of rays is so sparse that the rays miss the peaks.

4.2 Chamber Studies of Geometries

Several research groups have evaluated CT for industrial hygiene by using pilot-CT systems in exposure chambers and comparing tomographic map concentrations with concentrations obtained from point samples taken in the chamber.

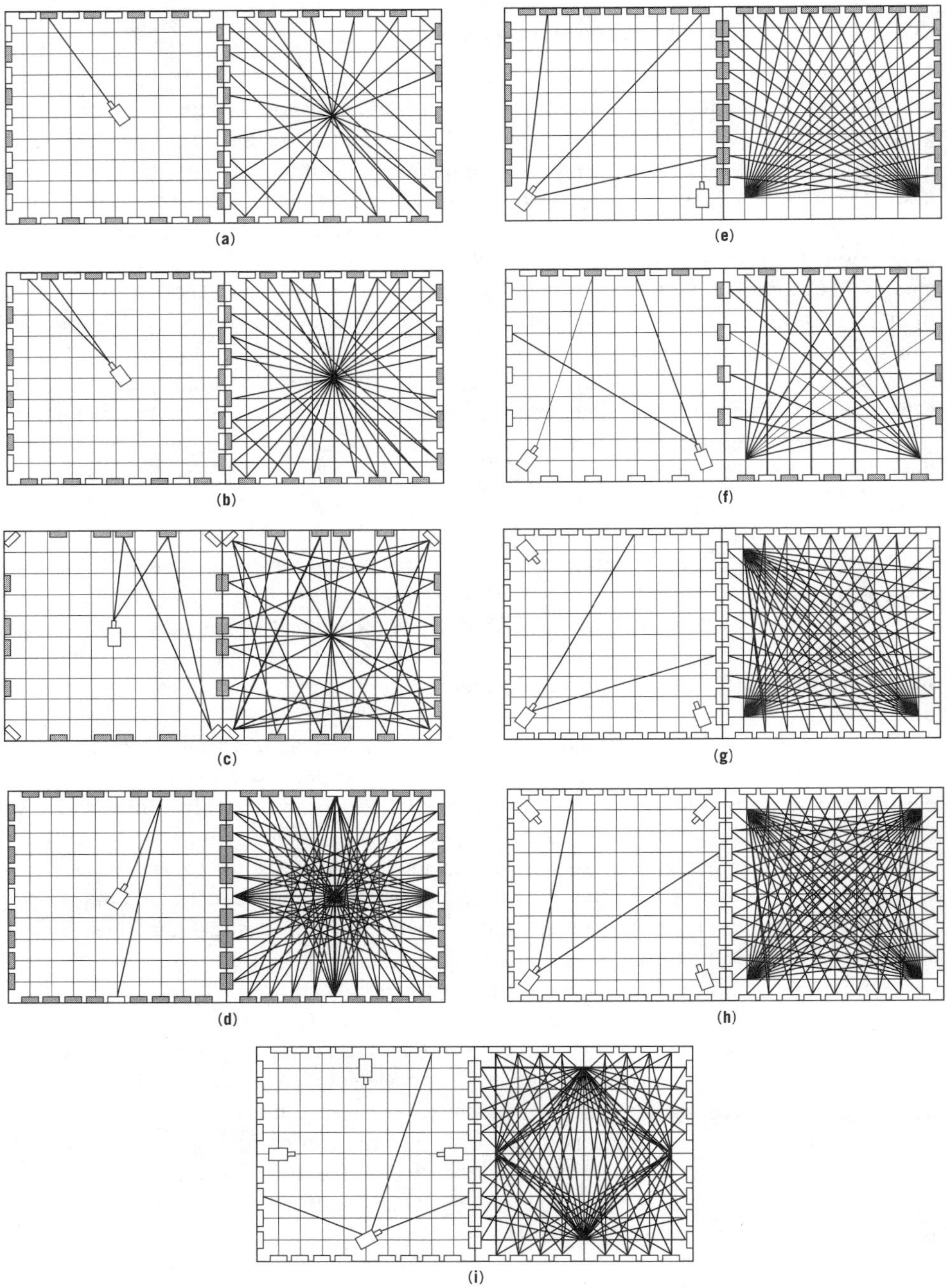

Figure 11.9. Each CT geometry is illustrated with a pair of diagrams. The left diagram shows a few rays and the diagram on the right shows the overall pattern of the rays. A dark box represents a mirror and a light box represents a retroreflect. (**a–d**), single instruments are placed in the center of a room. Mirrors are used to create virtual equal angle parallel projections (**a** and **b**). (**b**), uses additional straight rays. Mirrors are used to create fan-like beams (**c** and **d**). (**e–f**), two instruments, one with reflected rays. (**g**), three instruments. (**h–i**), four instruments.

Samanta and Todd (7, 36) tested a prototype CT system that used a single Midac open-path FTIR spectrometer (MIDAC Corporation, Irvine, CA) on a translatable table which was placed on a track along the periphery of the chamber. With this set-up, one instrument was used to simulate four instruments; a single OP-FTIR spectrometer scanned the chamber from many different angles to create a network of open-path measurements. Two configurations were tested, one with two instruments on each of two sides of the chamber (136 rays) and one with one instrument on each of four sides of the chamber (100 rays). The chamber was divided into a grid of 120 cells, with each cell being one square foot. The chamber was equipped with a point sampling system that collected tracer gas samples from specific points in the chamber directly above the beam.

Single and multiple plumes of sulfur hexafluoride were generated vertically up from the floor of the chamber. During the experiments, open-path measurements and point samples were obtained over approximately 1.5 hours. Recently, a rapid scanning system was developed that uses two rotating monostatic Midac spectrometers and two rotating mirrors to obtain a total of 48 rays. The rotating mirrors act as virtual spectrometers to create four OP-FTIR spectrometers. However, only two instruments are operating at a given time (see Fig. 11.10). Using this set-up, a complete set of open-path measurements is obtained every seven minutes.

For the chamber experiments, the tomographic reconstructed concentration maps were compared with the point sample maps created by krigging the point samples. Reconstructions were performed using MART and MLEM. In almost all of the experiments, regardless of the geometry, the CT system reconstructed the location of the plumes within one foot. The geometry using 136 rays reconstructed multiple peaks better than the geometry using 100 rays. This may be due to the fact that the 136-ray geometry has a greater angular range of coverage than the 100-ray geometry. The rapid scanning system was successful at mapping rapidly changing plume concentrations (see Fig. 11.2). Using the rapid scanning system, the peak concentrations in the tomographic maps were within <20% of the concentrations in the point sample maps.

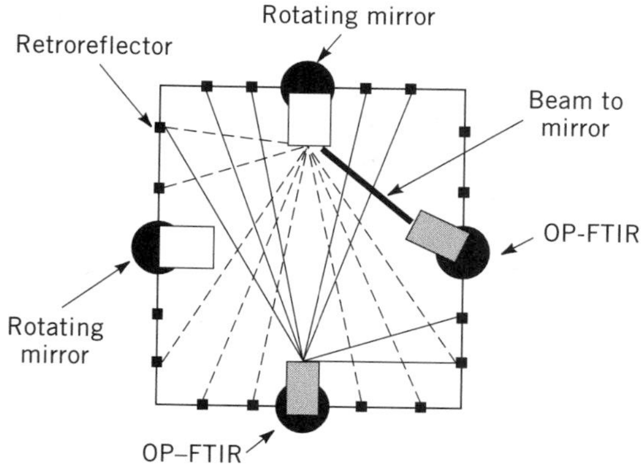

Figure 11.10. Rapid scanning set-up in the chamber studies using virtual spectrometers (Two rotating OP-FTIR spectrometers and two rotating mirrors; virtual spectrometers).

Yost et al. (12) performed chamber studies using a single OP-FTIR spectrometer (Nicolet Instrument Corp, Wisconsin) mounted on a translatable table, to simulate a parallel projection geometry with four projection angles (36 rays). Acetone was released in the chamber and three hours were required to complete the CT scans and obtain the point samples. Maps were reconstructed using an ART algorithm. While the peaks were located correctly, there was a large discrepancy between the concentration of the peaks reconstructed with the tomographic and point sample maps. This may have been due to fluctuating concentrations and airflow patterns in the chamber, and a very sparse geometry. More recently, Drescher et al. (14, 15) used a rapidly scanning system that had a single scanning OP-FTIR spectrometer in the center of a chamber, one retroreflector in each of the four corners the room, and 52 flat mirrors on the walls. This geometry had 56 optical rays with an overall sampling time of six minutes. Using this geometry and the SBFM algorithm, the tomographic maps were very similar to the point sample maps and quantitatively agreed to within 50%.

One of the problems encountered in laboratory validation studies is that a noninvasive, real-time, spatially resolved technique is compared to invasive point samples taken at a different height and time frame than the optical beams. In addition, the point samples must be processed, such as by krigging, to obtain spatial concentration maps. This can introduce errors. Thus, differences in concentrations between the methods may be due in part to technical problems associated with differences in the sampling methods.

5 FIELD STUDIES

To date, there have not been any field studies in workplaces using a CT system. However, small-scale CT systems have been used in outdoor studies to compare results with dispersion models and to calculate chemical emission rates (5).

A pilot environmental CT system with two scanning OP-FTIR spectrometers and eight retroreflectors was used on a 70 by 70 meter field. This highly underdetermined system was used to reconstruct concentration plumes of chemicals that were released from a simulated volume source. The reconstructed concentration maps were compared with maps created using the ISCST dispersion model. For many of the reconstructed time periods, the shapes, locations, and directions of the plumes in the reconstructed tomographic maps compared fairly well with the model predictions (see Fig. 11.11). For almost all of the time periods, the concentrations in the tomographic maps were higher than the concentrations in the model-generated maps. For the periods where the maps did not agree, the wind direction was found to be rapidly and repeatedly shifting. One of the reasons for this could be the fact that the model used average wind conditions to generate concentrations, whereas the CT system took measurements sequentially and in real-time. The CT system may have more accurately captured real-time variability.

The first large scale field implementation of a CT system was performed at an intensive swine confinement operation to measure ammonia and other contaminants generated from a waste lagoon. The CT system operated every day for up to 24 hr a day for several weeks during three seasons. Two monostatic OP-FTIR spectrometers were placed on 360° rotating platforms, on opposite sides of a six-acre swine waste lagoon, each within 1 m of the

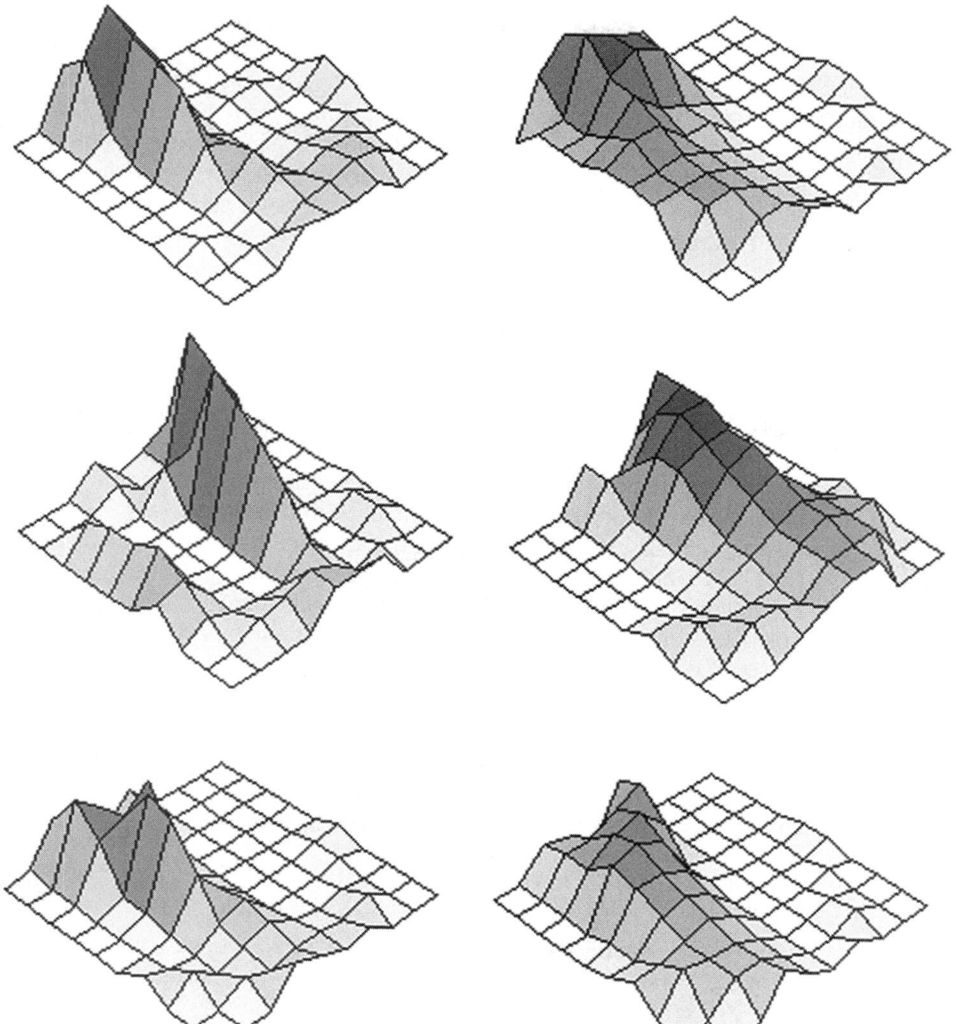

Figure 11.11. Comparison of model generated plumes and CT reconstructions of sulfur hexafluoride for three consecutive time periods. The first column has the CT maps and the second column has the model maps.

surface. This created a horizontal network of infrared rays. Twelve metal retroreflectors were positioned around the periphery of the lagoon. The OP-FTIR spectrometers sequentially scanned the retroreflectors, and both spectrometers operated simultaneously. Total path lengths ranged from 132 to 426 meters. A tracer gas was released from the center of the lagoon at a known emission rate and the network of rays simultaneously monitored the ammonia, other waste gases, and the tracer gas every two minutes over many days. Concentration maps of the gases were generated for the entire lagoon surface. The system was operated from computers that were located in a van; the entire system was connected using fiber optic cables. Except for replenishing the liquid nitrogen during the day to cool the detectors, the spectrometers were operated without supervision. The tomographic maps revealed that the concentration of ammonia varied widely over the surface of the lagoon, and was not homogeneous as is the assumption when using point sampling devices for these types of sources. This was an ideal application of CT. Using conventional point sampling equipment only a few locations can be easily sampled on a lagoon. In comparison, the CT system can measure a relatively large area. Figure 11.12 shows CT concentration maps for ammonia that represent three sequential time periods.

6 BACKGROUND OPTICAL REMOTE SENSING INSTRUMENTATION

In theory, any remote sensing instrument can be used to perform CT in industrial hygiene as long as it can quickly and safely measure path-integrated concentrations of chemicals in air along an open-path. Lasers, such as tunable diode lasers, have high scanning rates and offer the ability to sample the air over very long pathlengths (kilometers), however, they are usually limited to one or a few gases and may present an eye safety hazard (68–73). OP-FTIR spectrometers have primarily been used outdoors, in fenceline, air pollution, and hazardous waste applications, where one instrument with a long pathlength is needed (74–77). For indoor air monitoring, OP-FTIR spectrometers have advantages over lasers in that they can measure a wide range of multiple chemicals at low limits of detection (ppb–ppm) and they pose a low eye-safety hazard (78–87). OP-FTIR spectrometers are the only instruments that have been used thus far for applying CT to industrial hygiene and environmental (38, 16, 88) monitoring; therefore, they are described.

6.1 OP-FTIR Spectrometer Design

OP-FTIR spectrometers typically use a remotely placed optical source to transmit an infrared (IR) beam across the open air to a detector. Chemicals present in the optical path are measured in near-real time, and concentrations are integrated over the entire length of the optical path, to obtain path-averaged concentrations.

The OP-FTIR spectrometers are operated in two different configurations: bistatic and monostatic (see Fig. 11.13). These configurations differ in the transfer optics and the number of times that the IR beam passes through the air. In the bistatic mode, the IR source is placed at the opposite end of the optical path from the detector (can be greater than 500 meters apart), and the IR beam passes through the air once. In the monostatic mode, the IR source and detector are co-located at one end of the path, and the beam is sent out to a reflector placed at the other end of the path. In the monostatic mode, the reflector (flat mirror, corner cube or cat's eye retroreflector) sends the light directly back to the detector, and the IR light passes through the air twice. Telescopes can be used to collimate, send and receive the IR beam to and from the spectrometer to the retroreflector (monostatic systems), or to receive the IR light at the detector (bistatic systems). Telescopes may not

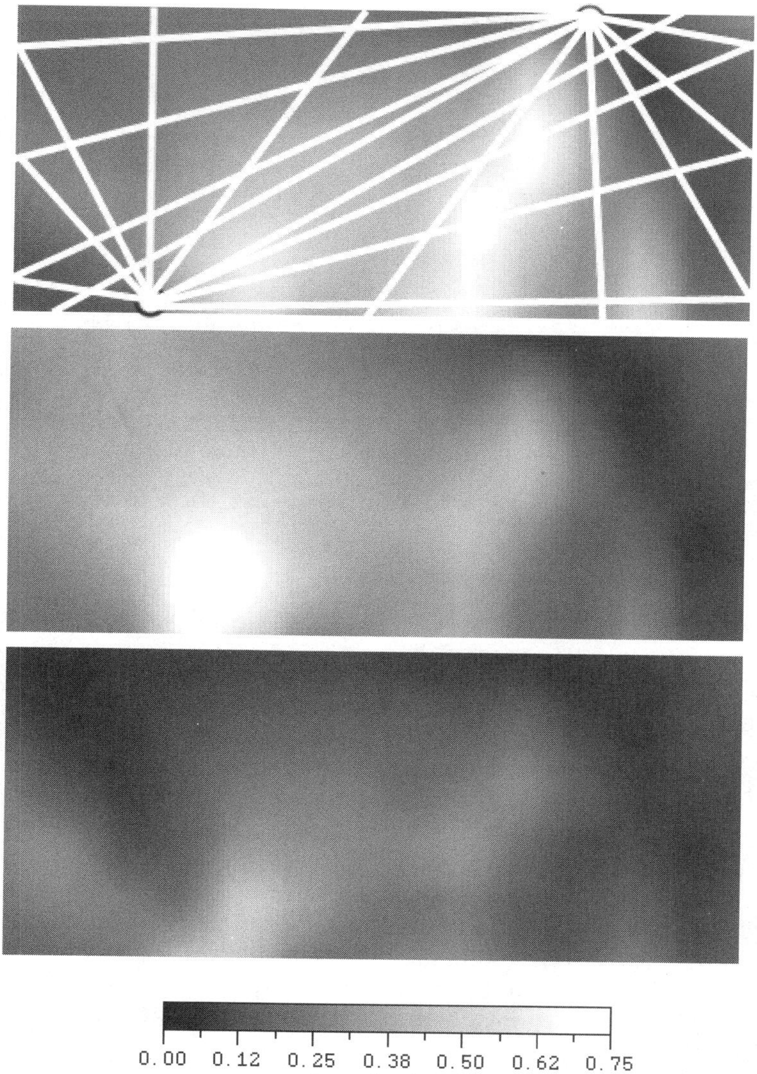

Figure 11.12. Three CT concentration maps of ammonia over a waste lagoon. The system used two scanning OP-FTIR spectrometers and twelve retroreflectors. Each map represents a fifteen-minute period. The lighter the picture, the higher the concentration.

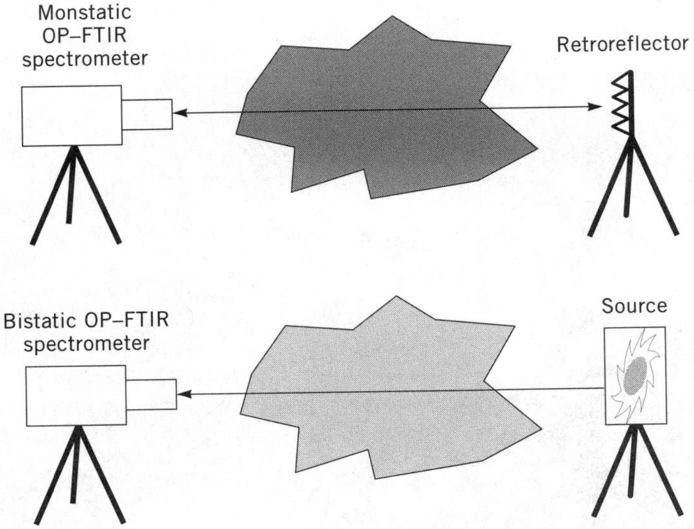

Figure 11.13. Two configurations of an OP-FTIR spectrometer. (**a**), monostatic spectrometer. (**b**), bistatic spectrometer.

be necessary when short path lengths are used for indoor workplace or exposure chamber applications.

6.2 Theory of Operation

OP-FTIR spectrometers use IR absorption to identify and quantify hundreds of chemicals in air. IR absorption techniques are based on the principle that compounds selectively absorb energy in the IR region of the electromagnetic spectrum. When molecules absorb incident electromagnetic radiation at specific wavelengths, the energy can boost electrons from their ground state to an excited state, or cause the molecules to stretch, bend, or rotate (89, 90). With the exception of elements, inorganic salts, and diatomic molecules (oxygen, nitrogen, chlorine), most substances absorb in the IR region.

In the mid-infrared region of the electromagnetic spectrum (4000 to 400 cm^{-1}), absorption usually occurs at several different wavelengths of IR light, which results in patterns or bands. This pattern of energy absorption, called the absorption spectrum, creates a unique fingerprint for each molecule and can be used to identify the chemical. Figure 11.14 shows an absorption spectrum for ammonia. Small differences in chemical structure can result in large differences in absorption spectra. To identify a compound, the number, location, and shape of absorption bands in a spectrum are compared to a reference spectrum for the same chemical. A reference spectrum is created using known concentrations of the chemical of interest under controlled conditions of temperature and pressure. The pattern of absorption is used for identification of compounds and the intensity of the spectral bands is used for quantification.

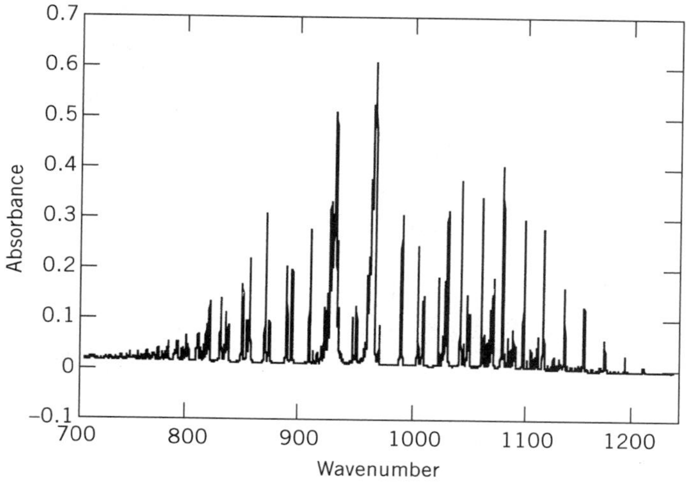

Figure 11.14. Absorption spectrum for ammonia.

Quantification is performed by selecting a wave number region in the absorption spectrum of the gas that ideally is free from interferences from other contaminants. Within constraints, there is a linear relationship between the intensity of the spectral bands and the concentration of the compound. This relationship is based on Beer's Law which states that at a constant path length, the intensity of the IR energy diminishes exponentially with the concentration of chemicals in the path (see Eq. 10)

$$I = I_0 \exp -aCL \tag{10}$$

$$\text{Transmission} = I/I_0 \tag{11}$$

$$\text{Absorption} = -\log (I/I_0) \tag{12}$$

$$\text{Absorption} = aCL \tag{13}$$

where I = the intensity of IR light when the absorbing chemical is present
Io = the intensity of IR light without the absorbing chemical present
a = the molecular absorption coefficient of the chemical
C = the concentration of the chemical
L = the path length of the IR radiation through the chemical

To measure concentration, the intensity of the IR light with and without contaminants must be obtained and converted to an absorption spectrum (see Eq 10–13). Background spectra (I_0) are generated either using ambient air that does not contain target contaminants, or are generated synthetically. The primary challenge facing users of open-path FTIR spectrometers is the acquisition of a clean background spectrum (82, 91).

An interferometer is the optical device used in OP-FTIR spectrometers to scan simultaneously and rapidly (seconds) the entire mid-IR region. An interferometer can achieve high spectral resolution (the ability to distinguish between absorption bands of compounds), high specificity, high signal-to-noise ratios, and low limits of detection (ppb) (92, 93). One type of interferometer is the Michelson interferometer (see Fig. 11.15) (90). An interferometer consists of a beam splitter, a stationary mirror and a moveable mirror. They are orthogonal to each other. A Nernst glower, which emits broad-band electromagnetic radiation, is used to generate IR light. The IR light travels to the beam splitter which divides the light into two beams of equal intensity; one half of the light is sent to the moveable mirror and one half is sent to the fixed mirror. After hitting the mirrors, the beams recombine at the beam splitter and exit to the detector. When the moving mirror and the fixed mirror are at the same distance from the beamsplitter, the IR light beams reflected off the mirrors are in phase with each other, and all the wavelengths of light constructively interfere in the recombined beam. This results in the highest intensity of light for all possible mirror positions. When the moving mirror moves along the optic axis, the half of the optical beam that hits the moving mirror travels a shorter distance than the optical beam which hits the fixed mirror. At any mirror position, the intensity of light is governed by constructive and destructive interferences in the recombined beam. When the light recombines, it hits the detector and a signal is recorded; the pattern of constructive and destructive interference is called the interferogram and depends upon the wavelengths of light present in the beam. Using an IR broad band radiation source, the incident IR radiation contains many wavelengths. Therefore, at each position of the moving mirror, the signal recorded by the detector is an integral of the intensities of all the wavelengths of light. The interferogram is a plot of light intensity vs. optical path difference (mirror position) and is used in all subsequent analyses.

To create absorption spectra, first the interferogram is converted to a single-beam spectrum by using appropriate software to perform a fast Fourier transform. Two single-beam

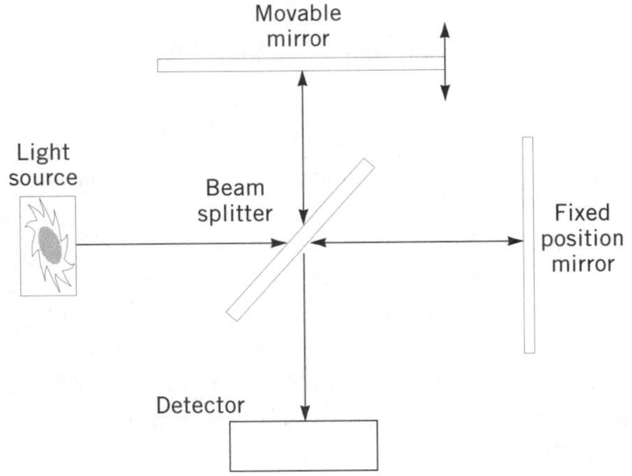

Figure 11.15. Schematic of a Michelson interferometer.

spectra (sample and background) are then ratioed to get a transmission spectrum (Eq. 11), and an absorption spectrum is generated by taking the negative logarithm of the transmission spectrum (Eq. 12). Absorption (Eq. 13) is proportional to the concentration of the chemical and the distance (path length) that the light travels through the chemical. Therefore, the longer the path length that the IR light travels through the air containing the compound of interest, before it reaches the detector, the greater the sensitivity and the lower the limits of detection.

6.3 Quantification

Quantitative analysis can be performed using a variety of methods including measuring peak height, measuring peak area, spectral subtraction, partial least squares, iterative least squares and classical least squares. Classical least squares (CLS) is most commonly used to match the measured spectra to a library of calibration reference spectra for the chemicals of interest (92–95). When the concentrations of chemicals are measured, they are integrated over the entire length of the optical path, and are reported as the product of concentration and path length (ppm-meters). The ppm-meter concentration is then divided by the path length to obtain a path-averaged concentration (ppm).

The resolution of an OP-FTIR spectrometer refers to the minimum separation in wave numbers that two absorption features can have and still be distinguished from one another. Typical OP-FTIR spectrometers used for environmental CAT scanning have minimum resolutions of 0.5 and 1.0 wave numbers; the higher the resolution, the lower the wave number. At any resolution, the OP-FTIR spectrometer can average individual measurements (scans) over time to decrease the SNR (signal to noise ratio). For tomographic imaging, spectrometer measurement rates should be minimized to reconstruct concentration profiles with multiple peaks or with rapidly changing concentration profiles. Therefore, measurements should be made at the lowest possible wavenumber resolution that adequately resolves spectral features of chemicals in the air because as resolution increases, scan time increases. In general, the SNR is directly proportional to the resolution, and to the square root of the measurement time; doubling the resolution doubles the noise.

7 CHALLENGES AND FUTURE DIRECTIONS

A new and innovative measurement technology for the field of industrial hygiene has emerged which may allow visualization of the spatial distribution of chemical concentrations in air over time. By applying computed tomography to industrial hygiene, spatially and temporally resolved two-dimensional concentration maps would be produced for a room or area. These maps could be used to evaluate pollutant dispersion in a workplace, characterize human exposures, support fundamental modeling research, and locate fugitive emissions.

There may always be limitations to this method due to sparse sampling in a room, and the available lines of sight, which would translate into concentration measurement errors.

For example, how can CT maps be used if the measurements are taken above the breathing zones of workers? However, the ability to use a tomographic system over hours, days and weeks in one work place, might result in more accurate estimates of human exposures than currently available using limited numbers of point samples taken sporadically in time. Once installed, a CT system could run continuously at low cost.

In order to implement and improve this technology, several challenges remain. These include: extensive laboratory and field validation studies; development of robust reconstruction algorithms; development of improved geometries for a variety of workplace designs; miniaturization and increased ruggedness of open-path instrumentation; and expansion of the technique to three-dimensional tomography. Ultimately, reconstructed concentration maps need to be evaluated in user studies in the field with industrial hygienists.

One improvement to current reconstruction algorithms is the use of nonsquare grid cells. The use of square grid cells is arbitrary and can actually decrease reconstruction accuracy. An adaptive multigrid reconstruction technique is being developed that alters the size of the grid cells in real-time. Various levels of refinement in the grid can be obtained to address peaks of different sizes and at different positions in a room. Therefore, in areas of low concentration, where information is not important or there is high uncertainty, larger cells can be used. A fine grid of cells can be used in areas near people, sources, or high concentration gradients.

To date, researchers have worked with fixed CT geometries. Another improvement is a dynamic geometry. A technique is being developed which, in real-time, pinpoints the regions of high concentrations and adapts the density of the rays to follow the movement of the plume. The use of adaptive ray sampling could improve reconstructions by optimizing the position of the rays. Researchers have used adaptive techniques in signal processing (96), image processing (97, 98) and CT (99), to achieve improved sampling of sharp spatial changes in objects. An adaptive sampling technique is being developed that selectively acquires ray measurements from the highest concentration regions in a room at finer intervals compared to the lower concentration regions. For example, retroreflectors could be evenly spaced along the walls of the room. The CT system would generate an initial tomographic map, and this map would be used to alter the geometry in real-time to reposition the rays, Thus, the density of rays would be increased where there is high information content, and would be reduced in places of low concentrations. Reconstruction accuracy can improve when only a few additional optical rays are placed through the areas of highest concentration (8).

With improved and less expensive ORS instrumentation, it will be possible to use three-dimensional tomography in indoor air. This will greatly enhance the usefulness of this technique for industrial hygiene. Currently, the optical beam in the sampled plane may have a diameter of a few inches up to a foot. Therefore, selection of the height of the beams through the sampled plane in a room is crucial for determining which exposures and concentrations are visualized. Three-dimensional maps could be reconstructed if instrumentation is developed that can generate and detect a wide fan beam, or if the cost of optical sources and retroreflectors dramatically drops so that complex geometries can be used. Ideally, sampling would then be possible over a several meter-diameter volume.

ACKNOWLEDGMENTS

Some of this work was supported by the U.S. Environmental Protection Agency cooperative agreement CR815152 with the University of North Carolina, and by the National Science Foundation's Presidential Faculty Fellows Award (94-53433).

BIBLIOGRAPHY

1. M. Corn, "Strategies of Air Sampling," *Scand. J. Work Environ. Health.* **11**, 173–180 (1985).
2. R. Bhattacharyya and L. A. Todd, "Spatial and Temporal Visualization of Gases and Vapours in Air Using Computed Tomography, Numerical Studies," *Ann. Occup. Hyg.* **41**(1), 105–122 (1997).
3. L. A. Todd and R. Bhattacharyya, "Tomographic Reconstruction of Air Pollutants: Evaluation of Measurement Geometries," *Applied Optics* **36** (30), 7678–7688. (1997).
4. L. A. Todd and G. Ramachandran, "Evaluation of Algorithms for Tomographic Reconstruction of Chemicals in Indoor Air," *Am. Ind. Hyg. Assoc. J.* **55**(5), 403–417 (1994).
5. P. R. Piper, L. A. Todd, and K. Mottus, "A Field Study Using Open-path FTIR Spectroscopy to Measure and Map Air Emissions from Volume Sources," *Field Analytical Chemistry and Technology* **3** (2), 69–79 (1999).
6. L. A. Todd, "Optical Remote Sensing/Computed Tomography Systems for Workplace Exposure Assessments," *Proceedings Optical Remote Sensing Applications to Environmental and Industrial Safety Problems, AWMA*, **SP81**, 356–360 (1992).
7. A. Samanta and L. Todd, "Mapping Air Contaminants Indoors using a Prototype Computed Tomography System," *Ann. Occup. Hyg.* **40** (6), 675–691 (1996).
8. L. Todd, "Optical Remote Sensing and Computed Tomography in Industrial Hygiene," Ph.D. Dissertation, University of North Carolina, Chapel Hill, NC, 1990.
9. L. Todd and D. Leith, "Remote Sensing and Computed Tomography in Industrial Hygiene," *Am. Ind. Hyg. Assoc. J.* **51**, (4), 224–233 (1990).
10. L. Todd and G. Ramachandran, "Evaluation of Optical Source-Detector Geometry for Tomographic Reconstruction of Chemical Concentrations in Indoor Air," *Amer. Ind. Hyg. Assoc. J.* **55**(12), 1133–1143 (1994).
11. L. Todd, "Optical Remote Sensing/Computed Tomography Systems for Workplace Exposure Assessments," *Proceedings Optical Remote Sensing Applications to Environmental and Industrial Safety Problems, AWMA*, **SP81**, 356–360 (1992).
12. M. G. Yost, A. J. Gagdil, A. C. Drescher, Y. Zhou, M. A. Simonds, S. P. Levine, W. W. Nazaroff, and P. A. Saisan, "Imaging Indoor Tracer Gas Concentration with Computed Tomography: Experimental Results with a Remote Sensing FTIR System," *Am. Ind. Hyg. Assoc. J.* **55**, 395–402 (1994).
13. A. C. Drescher; A. J. Gadgil, P. N. Price, and W. W. Nazaroff, "Novel Approach for Tomographic Reconstruction of Gas Concentration Distributions in Air: Use of Smooth Basis Functions and Simulated Annealing," *Atmospheric Environment* **30**(6), 929–940. (1996)
14. A. C. Drescher, D. Y. Park, M. G. Yost, A. J. Gadgil, S. P. Levine, and W. W. Nazaroff, "Stationary and Time-Dependent Indoor Tracer-Gas Concentration Profiles Measured by OP-FTIR Remote Sensing and SBFM Computed Tomography," *Atmospheric Environment* **31**(5), 727–740 (1997).

15. A. C. Drescher, "Computed Tomography and Optical Remote Sensing: Development For the Study of Indoor Air Pollutant Transport and Dispersion," PhD dissertation, University of California, Berkeley, 1995.
16. L. A. Todd, "Mapping the Air in Real-Time to Visualize the Flow of Gases and Vapors: Occupational and Environmental Applications," *Applied Occupational and Environmental Hygiene*, Accepted, 1999.
17. J. Radon, "On the Determination of Functions from Their Integrals Along Certain Manifolds," *Ber. Verh. Sachs. Akad. Wiss.* **69**, 262–277 (1917).
18. R. N. Bracewell and A. C. Riddle, "Inversion of Fan-beam Scans in Radio Astronomy," *Astrophysical J.* **150**, 427–434 (1967).
19. R. N. Bracewell, "Strip Integration in Radio Astronomy," *Aust. J. Phys.* **9**, 198–217 (1956).
20. J. G. Sanderson, "Reconstruction of Fuel Pin Bundles by a Maximum Entropy Method," *IEEE Trans. Nucl. Sci.* **NS-26**, 2685 (1979).
21. P. Rowley, "Quantitative Interpretation of Three-Dimensional Weakly Refractive Phase Objects Using Holographic Interferometry," *J. Opt. Soc. Amer.* **59**, 1496–1498 (1969).
22. M. V. Berry and D. F. Gibbs, "The Interpretation of Optical Projections," *Proc. R. Soc. Lond. A.* **314**, 143–152 (1970).
23. R. A. Crowther, D. J. DeRosier, and A. Klug, "The Reconstruction of a Three-Dimensional Structure from Projections and Its Application to Electron Microscopy," *Proc. R. Soc. Lond. A.* **317**, 319–340 (1970).
24. D. J. DeRosier and A. Klug, "Reconstruction of Three-Dimensional Structures from Electron Micrographs," *Nature* **217**, 130–134 (1968).
25. P. Gilbert, "Iterative Methods for the Three-dimensional Reconstruction of an Object from Projections," *J. Theor. Biol.* **36**, 105–117 (1972).
26. P. Gilbert, "The Reconstruction of a Three-Dimensional Structure from Projections and Its Application to Electron Microscopy II, Direct Methods," *Proc. R. Soc. Lond. B.* **182**, 89–102 (1972).
27. G. N. Hounsfield, "Computerized Transverse Axial Scanning (Tomography): Part I, Description of a System," *Br. J. Radiol.* **46**, 1016–1022 (1973).
28. A. M. Cormack, "Early 2-Dimensional Reconstruction (CT Scanning) and Recent Topics Stemming from It, Nobel Lecture, December 8, 1979," *J. Comput. Assist. Tomogr.* **4**, 658–664 (1980).
29. A. M. Cormack, "Representation of a Function by Its Line Integrals, with Some Radiological Applications, II," *J. Appl. Phys.* **35**, 2908–2913 (1964).
30. G. T. Herman, *Image Reconstruction from Projections: The Fundamentals of Computed Tomography*, Academic Press, Inc., New York, 1980, pp. 1–25.
31. R. L. Byer and L. A. Shepp, "Two Dimensional Remote Air Pollution via Tomography," *Appl. Optics Letters*, **4**(3), 75–79 (1979).
32. D. C. Wolfe and R. L. Byer, "Air Pollution Monitoring by Computed Tomography," in *Proceedings Computer Software and Applications Conference IEEE*, 867–870 (1979).
33. D. C. Wolfe and R. L. Byer, "Model Studies of Laser Absorption Computed Tomography for Remote Air Pollution Measurement," *Applied Optics* **21**(7), 1165–1177 (1982).
34. L. A. Todd, "Optical Remote Sensing/Computed Tomography Beam Geometries for Monitoring Workplace Gases and Vapors," *Proceedings Optical Remote Sensing Applications to Environmental and Industrial Safety Problems*, AWMA. **SP81**, 390–393 (1992).

35. D. Y. Park, M. G. Yost, and S. P. Levine, "Evaluation of Virtual Source Beam Configurations for Rapid Tomographic Reconstruction of Gas and Vapor Concentrations in Workplaces," *A&WMA J.* **47**(5), 582–591 (1977).
36. A. Samanta and L. Todd, "Mapping Chemical Concentrations Indoors Using Open-path FTIR Spectroscopy and Computed Tomography: Chamber Studies," in *Proceedings Optical Remote Sensing Applications to Environmental and Industrial Safety Problems*, AWMA, **SPIE 2365**, 187–194 (1994).
37. L. A. Todd, D. J. Norton, and R. Dishakjian, "CAT Scanning the Air to Map Concentrations: Field Study Results," *Optical Sensing for Environmental and Process Monitoring, an International Specialty Conference*, AWMA, McLean, VA, 1996, pp. 349–356.
38. L. A. Todd, M. G. Yost, and R. A. Hashmonay, "Trends and Future Applications of Optical Remote Sensing and Computed Tomography to Map Air Contaminants," *Proceedings of the Environmental Monitoring and Remediation Technologies*, SPIE, in press, 1999.
39. R. A. Hashmonay, M. G. Yost, Y. Mamane, and Y. Benayahu, "Emission Rate Apportionment from Fugitive Sources using Open-Path FTIR and Mathematical Inversion," accepted by *Atmospheric Environment.*
40. R. A. Hashmonay and M. G. Yost "Innovative Approach for Estimating Gaseous Fugitive Fluxes Using Computed Tomography and Remote Optical Sensing Techniques," accepted by *J. of A&WMA.*
41. G. Ramachandran, D. Leith, and L. Todd, "Extraction of Spatial Aerosol Distributions from Multispectral Light Extinction Measurements with Computed Tomography," *J. Opt. Soc. Amer. A.* **11**(1), 144–154 (1994).
42. G. Lorbeer, B. Siemund, and I. Willms, "Opto-Computer-Tomographic Methods as Aids for Characterizing Local Inhomogeneous Aerosol Distributions," *J. Aerosol Sci.* **15**, 287–293 (1984).
43. H. O. Luck, B. Siemund, and G. Lorbeer, "The Measurement of Spatial Aerosol Distributions in Enclosures By Means of Computed Tomography," *Part. Charact.* **2**, 137–142 (1985).
44. K. M. Hanson, "POPART-Performance Optimized Algebraic Reconstruction Technique," in *Visual Communications and Image Processing, 1001* SPIE, Cambridge, MA, SPIE, 1988, pp. 318–325.
45. K. M. Hanson and K. J. Myers, "Rayleigh Task Performance as a Method to Evaluate Image Reconstruction Algorithms," *Maximum Entropy and Bayesian Methods*, 303–312 (1991).
46. K. M. Hanson, "Method of Evaluating Image-Recovery Algorithms Based On Task Performance," *J. Opt. Soc. Am. A.* **7**, 1294–1304 (1990).
47. S. R. Hanna, G. A. Briggs, and R. P. Hoskar, *Handbook on Atmospheric Diffusion*, Prepared for the Office of Health and Environmental Research, U.S. Department of Energy, Technical Information Center, DOE/TIC/-11223, 1982.
48. S. R. Hanna, G. A. Briggs, and R. P. Hosker, Jr., "Gaussian Plume Model for Continuous Sources," *Handbook on Atmospheric Diffusion*, Technical Information Center, U.S. Department of Energy, 1992, pp. 25–35.
49. G. Frieder and G. T. Herman, "Resolution in Reconstructing Objects from Electron Micrographs," *J. Theor. Biol.* **33**, 198–211 (1971).
50. G. T. Herman and S. Rowland, "Three Methods for Reconstructing Pictures from X-Rays: A Comparative Study," *Computer Graphics and Image Processing*, **1**, 151–178 (1973).
51. G. T. Herman, A. Lent, and S. W. Rowland, "ART: Mathematics and Applications, A Report on the Mathematical Foundations and on the Applicability to Real Data of the Algebraic Reconstruction Techniques," *J. Theor. Biol.* **42**, 1–32 (1973).

52. R. Gordon, R. Bender, and G. T. Herman, "Algebraic Reconstruction Techniques (ART) for Three-Dimensional Electron Microscopy and X-ray Photography," *J. Theor. Biol.* **29**, 471–481 (1970).
53. A. M. Cormack, "Representation of a Function by Its Line Integrals, with Some Radiological Applications," *J. Appl. Phys.* **34**, 2722–2727 (1963).
54. D. E. Kuhl and R. Edwards, "Image Separation Radioisotope Scanning," *Radiology* **80**, 653–661 (1963).
55. R. A. Brooks and G. Di Chiro, "Principles of Computer Assisted Tomography (CAT) in Radiographic and Radioisotopic Imaging," *Phys. Med. Biol.* **21**, 689–732 (1976).
56. R. A. Brooks and G. Di Chiro, "Theory of Image Reconstruction in Computed Tomography," *Radiology* **117**, 561–572 (1975).
57. R. Gordon, "A Tutorial on ART," *IEEE Transactions on Nuclear Science*, **NS-21**, 78–93 (1974).
58. B. E. Oppenheim, "Reconstruction Tomography From Incomplete Projections. In: Reconstruction Tomography" in M. Ter-Pogossian et al, eds., *Diagnostic Radiology and Nuclear Medicine*, University Park Press, 1977.
59. G. T. Herman, "A Relaxation Method For Reconstructing Objects From Noisy X-Rays," *Mathematical Programming* **8**, 1–19 (1975).
60. A. H. Andersen and A. C. Kak, "Simultaneous Algebraic Reconstruction Technique (SART): A Superior Implementation of the ART Algorithm," *Ultrasonic Imaging* **6**, 81–94 (1984).
61. A. H. Andersen and A. C. Kak, "Digital Ray Tracing in Two-Dimensional Refractive Fields," *J. Acoust. Soc. Am.* **72**, 1593–1606 (1982).
62. A. H. Andersen, "Algebraic Reconstruction in CT from Limited Views," *IEEE Transactions on Medical Imaging* **8**(1), 50–55 (1989).
63. M. L. Reis and N. C. Roberty, "Maximum Entropy Algorithms for Image Reconstruction from Projections," *Inverse Problems* **8**, 623–644 (1992).
64. B. M. W. Tsui, X. Zhao, E. C. Frey, and G. T. Gulberg, "Comparison between ML-EM and WLS-CG Algorithms for SPECT Image Reconstruction," *IEEE Transactions on Nuclear Science* **38**, 1766–1772 (1991).
65. L. A. Shepp and Y. Vardi, "Maximum Likelihood Reconstruction for Emission Tomography," *IEEE Transactions on Medical Imaging* **MI-1**, 113–122 (1982).
66. G. T. Herman and S. Rowland, "Resolution in ART. An Experimental Investigation of the Resolving Power of an Algebraic Picture Reconstruction Technique," *J. Theor. Biol.* **33**, 213–223 (1971).
67. A. Samanta and L. A. Todd, "Mapping Chemical Concentrations Indoors Using Open Path FTIR Spectroscopy Computed Tomography: Chamber Studies," *International Symposium on Optical Sensing for Environmental Process Monitoring, SPIE* **2365**, 187–194 (1994).
68. R. M. Measures, *Laser Remote Sensing Fundamentals and Applications*, John Wiley & Sons, Inc., New York, 1984.
69. N. Menyuk, D. K. Killinger, and W. E. DeFeo, "Laser Remote Sensing of Hydrazine, MMH, and UDMH Using a Differential-Absorption CO_2 Lidar," *Appl. Opt.* **21**, 2275–2286 (1982).
70. E. Murray, "Remote Measurement of Gases Using Differential Absorption Lidar," *Opt. Eng.* **17**, 30–38 (1978).
71. E. R. Murray, "Remote Measurement of Gases Using Discretely Tunable Infrared Lasers," *Opt. Eng.* **16**, 284–290 (1977).
72. E. Uthe, "Airborne CO_2 DIAL Measurement of Atmospheric Tracer Gas Concentration Distributions," *Applied Optics*, **25**(15), 2492–2498 (1986).

73. W. B. Grant and R. T. Menzies, "A Survey of Laser and Selected Optical Systems for Remote Measurement of Pollutant Gas Concentrations," *J. Air Pollution Control Assoc.* **33**, 187–194 (1983).

74. W. B. Grant, R. H. Kagan, and W. A. McClenny, "Optical Remote Measurement of Toxic Gases," *J. Air Waste Management Assoc.* **42**(1), 18–30 (1992).

75. P. L. Hanst, N. W. Wong, and J. Bragin, "A Long-Path Infra-red Study of Los Angeles Smog," *Atmospheric Environment* **16**(5), 969–981 (1982).

76. W. Herget and J. Brasher, "Remote Fourier Transform Infrared Air Pollution Studies," *Opt. Eng.* **19**, 508–514 (1980).

77. S. P. Levine and G. M. Russwurm, "Fourier Transform Infrared Optical Remote Sensing for Monitoring Airborne Gas and Vapor Contaminants in the Field," *Trends in Analytical Chemistry* **13**(7), 258–262 (1994).

78. H. Xiao and S. P. Levine, "Application of Computerized Differentiation Technique to Remote-sensing Fourier Transform Infrared Spectrometry for Analysis of Toxic Vapors," *Anal Chem.* **65**(7), 2262–2269 (1993).

79. H. K. Xiao, S. P. Levine, W. F. Herget, J. B. D'arcy, R. Spear, and T. Pritchett, "A Transportable Remote Sensing, Infrared Air Monitoring System," *Am. Ind. Hyg. Assoc. J.* **52**, 449–457 (1991).

80. M. G. Yost, H. K. Xiao, R. C. Spear, and S. P. Levine, "Comparative Testing of an FTIR Remote Optical Sensor with Area Samplers in a Controlled Ventilation Chamber," *Am. Ind. Hyg. Assoc. J.* **53**(10), 611–616 (1992).

81. W. W. Herget and J. D. Brasher, "Remote Measurement of Gaseous Pollutant Concentrations Using a Mobile Fourier Transform Interferometer System," *Appl. Opt.* **18**(20), 3404–3420 (1979).

82. "Compendium Method TO-16. Long Path Open-Path Fourier Transform Infrared Monitoring of Atmospheric Gases," *Compendium of Methods for the Determination of Toxic Organic Compounds in Ambient Air*, 2nd ed. ORD, USEPA, EPA/625/R-96/010B, 1997.

83. Cone, S. K. Farhat, and L. Todd, "Development of QA/QC Performance Standards for Field Use of Open Path FTIR Spectrometers," *International Symposium on Optical Sensing for Environmental Process Monitoring, SPIE* **2365**, 334–338 (1994).

84. L. A. Todd, "Direct-Reading Instrumental Methods for Gases, Vapors, Aerosols," in S. DiNardi, ed., *The Occupational Environment—Its Evaluation and Control*, AIHA, Fairfax, VA, 1997.

85. M. Simonds, H. Xiao, and S. P. Levine, "Optical Remote Sensing for Air Pollutants-Review," *Am. Ind. Hyg. Assoc. J.* **55**(10), 953–965 (1994).

86. L. Todd, and G. Ramachandran, "Evaluation of an Infrared Open-Path Spectrometer using an Exposure Chamber and a Calibration Cell," *Am. Ind. Hyg. Assoc. J.* **56**(Feb.) 151–157 (1995).

87. L. A. Todd, "Evaluation of an Open-path Fourier Transform Infrared Spectrophotometer Using an Exposure Chamber," *Appl. Occupational Environ. Hyg.* **11**(11) 1327–1334 (1996).

88. R. A. Hashmonay, M. G. Yost, and Y. Mamane, "Mapping the Emission Rates over Fugitive Sources Using Open-Path FTIR and Inversion Techniques," *Proceedings of the A&WM International Conference on Optical Remote Sensing for Environmental and Process Monitoring*, Dallas, TX, 1996, pp. 227–236.

89. R. M. Silverstein, G. Clayton Bassler, T.C. Morrill, *Spectrometric Identification of Organic Compounds*, 5th ed., John Wiley & Sons, Inc., New York, Chapt. 3, pp. 91–164.

90. B. C. Smith, *Fundamentals of FTIR Spectroscopy*, CRC Press, Inc. Boca Raton, FL, 1996

91. G. M. Russwurm and J. W. Childers, *FT-IR Open-Path Monitoring Guidance Document*, 2nd ed., ManTech Environmental Technology, Inc. Research Triangle Park, NC, EPA/600/R-96/040, 1996.

92. P. R. Griffiths and J. A. de Haseth, *Fourier Transform Infrared Spectrometry*, John Wiley & Sons, Inc., New York, 1986.

93. G. Horlick, "Introduction to Fourier Transform Spectroscopy," *Appl. Spectros.* **22**(6), 617 (1968).

94. D. M. Haaland and R. G. Easterling, "Application of New Least Squares Methods for the Quantitative Infrared Analysis of Multicomponent Samples," *Appl. Spectros.* **36**(6), 665–673 (1982).

95. D. M. Haaland and R. G. Easterling, "Improved Sensitivity of Infrared Spectroscopy by the Application of Least Squares Methods," *Appl. Spectros.* **34**(5), 539–548 (1980).

96. D. M. Etter, "An Introduction to Adaptive Signal Processing Computers" *Elect. Eng.* **18**, 189–193 (1992).

97. H. T. Tanaka, "Accuracy Based Sampling and Reconstruction with Adaptive Meshes for Parallel Hierarchical Triangulation," *Computer Vision and Image Understanding* **61**, 335–350 (1995).

98. R. A. F. Belfor, M. P. A. Hesp, R. L. Lagendijk, and J. Biemond, "Spatially Adaptive Subsampling of Image Sequences," *IEEE Transactions on Image Processing* **3**, 492–500 (1994).

99. P. V. Shankar, O. Nalcioglu, and J. Sklansky, "Undersampling Errors in Region-of-Interest Tomography," *IEEE Transactions on Medical Imaging* **1**, 168–173 (1982).

CHAPTER TWELVE

Potential Endocrine Disruptors in the Workplace

Paige E. Tolbert, Ph.D.

1 INTRODUCTION

In recent years, increasing attention has focused on the possibility that some chemicals present in the general environment may disrupt endocrine function in humans and lead to serious health consequences. These chemicals have been labeled *environmental endocrine disruptors*. The science underlying this area of concern is rapidly evolving, and enormous gaps in our understanding remain to be addressed. While there is general agreement in the scientific community that the implications of such endocrine disruption could be far-reaching and profound, there is considerable controversy regarding the plausibility of the hypothesis and extent of the actual problem (e.g., 1, 2).

The goal of this chapter is to highlight some of the major findings to date in the field of environmental endocrine disruption that may be of interest to industrial hygienists and toxicologists. A number of reviews and consensus workshops regarding this topic have been conducted (e.g., 3–8), including a comprehensive assessment of the evidence recently completed by the National Academy of Sciences (9). In addition, an encyclopedic description of the major chemicals suspected to have endocrine-disrupting activity, their chemical properties, regulatory status, and summary of endocrine-related findings has been recently published (10). The reader is referred to these documents for a fuller treatment of this question.

This overview is limited to chemicals occurring in the environment; agents used therapeutically will not be covered, other than brief mention of diethylstilbestrol because of its

Patty's Industrial Hygiene, Fifth Edition, Volume 1. Edited by Robert L. Harris.
ISBN 0-471-29756-9

historical importance. Phytoestrogens, estrogenic compounds naturally present in certain plants such as soybeans, are also beyond the scope of this chapter.

Much of the literature on this topic has focused on exposures to the general population occurring through the food chain, such as DDT and its degradation products and polyhalogenated biphenyls. These chemicals were banned in the United States and many other countries in the 1970s but because of their persistence and bioaccumulation in the food chain, exposures through food continue to be of concern. While current exposure to these chemicals in occupational settings should generally be minimal or declining, evidence from these reports may be useful in (*1*) determining whether endocrine disruption by environmental agents is occurring, and (*2*) predicting activity of structurally similar chemicals still present in workplaces. Thus, while this chapter is intended to be useful to industrial hygienists and toxicologists dealing with occupational exposures, the evidence from nonoccupational exposures is also described.

A recent U.S. Environmental Protection Agency–sponsored workshop on research needs for the risk assessment of health and environmental effects of endocrine disruptors defined the term *environmental endocrine disruptor* to be "any exogenous agent that interferes with the production, release, transport, metabolism, binding, action, or elimination of natural hormones in the body responsible for the maintenance of homeostasis and the regulation of developmental processes" (6). This definition will apply for the purposes of this review.

2 THE ENDOCRINE SYSTEM AND MECHANISMS OF ENDOCRINE DISRUPTION

The human endocrine system is an exquisitely complex system that orchestrates key events in the body and maintains precise control over a number of central bodily functions. Through a variety of interconnected axes and feedback loops, it integrates developmental events and coordinates such basic physiological processes as metabolism, homeostasis, and reproduction. Hormones, the chemical messengers secreted by endocrine tissue into the bloodstream, exert their effects at distant sites by binding to specific receptors on cell surfaces or nuclei in target tissue and triggering a sequence of cellular events, generally through up- or down-regulation of specific genes. Hormones may be any of the following classes of compounds: glycoproteins, polypeptides, peptides, steroids, modified amino acids, catecholamines, prostaglandins, or retinoic acid. The classical endocrine glands include the hypothalamus, pituitary, pineal, thyroid, parathyroid, pancreas, adrenal, ovary, and testis, although endocrine activity has been more recently discovered in a number of additional organs such as the heart, gut, skin, thymus, and placenta. Target tissue under endocrine control includes the nervous system, bone, muscle, the immune system, reproductive organs, and the mammary glands (7, 11).

Exogenous chemicals may mimic a hormone by acting as an agonist binding to the hormone receptor and initiating the sequence of events normally elicited by the endogenous hormone. Alternatively, chemicals may act as antagonists, blocking the action of a hormone by preventing access of the endogenous hormone to the receptor. Some chemicals inhibit specific enzymatic steps in the synthesis of a hormone (e.g., inhibition of estrogen synthesis by aromatase inhibitors such as the fungicide fenarimol) (7). Other chemicals may operate by interfering with hormone storage, release, transport, or clearance. Another potential mechanism is interference with the cascade of cellular events initiated by binding of hormone to receptor. Finally, a chemical might alter the population of receptors (12). All of these mechanisms may ultimately interfere with normal endocrine function and thus may have profound impacts on the organism. A number of chemicals are known to have multiple mechanisms of endocrine-disrupting action.

3 SUBPOPULATIONS POTENTIALLY AT RISK

Concern regarding impacts of environmental endocrine disruption has focused on effects on the developing fetus and children. In addition to the obvious consideration that this is the phase during which growth and development are occurring and the individual is therefore vulnerable to disruption of these processes, other factors include the possibility that some negative feedback loops and other homeostatic mechanisms may not be operational until adulthood, and the fact that smaller body mass generally leads to higher effective concentration from a given dose (13).

Thus, in occupational environments, the primary concern is exposures to workers before or during their reproductively active years because of possible effects on fertility and offspring. However, whereas much of the literature has focused on general environmental exposures, in occupational settings direct impacts on the exposed adults as a result of the higher exposures experienced in the workplace may also warrant concern.

4 THE DES STORY

Diethylstilbestrol (DES) is a potent synthetic estrogen that was used clinically because it was thought to prevent miscarriages from 1948 to 1971. Two to three million women are estimated to have taken DES prior to its banning (14). Daughters of these women were found to be at elevated risk of developing vaginal clear-cell adenocarcinomas in their teens and twenties (15). These daughters were also reported to be at increased risk of the following: ectopic pregnancy, subfertility and infertility, anatomical masculinization, elevated serum testosterone levels, salpingitis isthmica nodosa of oviduct, structural abnormalities of the uterus, malformed cervical canal, cervical and vaginal hood and polyps, vaginal adenosis, immune dysfunction, and depression (1, 16, 17). Sons of treated women have been reported to have experienced reproductive tract problems such as subfertility, sperm abnormalities, decreased sperm counts, epididymal cysts, hypoplastic and cryptorchid testes, seminomas and rete tumors of the testes, anatomical feminization, microphallus, hypospadias, retention of Mullerian duct remnants, and prostatic inflammation (17). Experimental animals exposed prenatally or neonatally experienced many similar outcomes (17). Several mechanisms for the permanent induction of differentiation defects following DES exposure during development have been proposed, including premature or permanent induction of genes normally under steroid control in adults or the estrogen-receptor pathway (17).

These findings led to concern that exogenous estrogen exposure at critical stages of development could result in permanent cellular and molecular alterations that may become manifest as structural, functional or long-term pathological changes (17). Attention has since broadened beyond chemicals with estrogenic activity to include chemicals that may have activity interfering with other hormone axes.

5 WILDLIFE OBSERVATIONS

Very little direct assessment of endocrine disruption in humans has been undertaken. Evidence of possible endocrine disruption is largely indirect, either based on observations in wildlife or laboratory animals, or based on activity in in vitro assays.

The initial impetus that led to the banning of DDT was observations of impaired reproductive success in wildlife, particularly birds. Since then, further observations in a number of species in the wild have fueled concern about the potential impacts of environmental endocrine disruptors, both in terms of ecologic disturbance and human health effects. The EPA workshop on endocrine disruptor research needs (6) identified the following reports of effects in wildlife from either ecologic or laboratory studies: reproductive problems in wood ducks in Arkansas (18), wasting and embryonic deformities in Great Lakes fish-eating birds (19–25), feminization and demasculinization of gulls (26–28), developmental effects in Great Lakes snapping turtles (29), embryonic mortality and developmental dysfunction in lake trout and other salminiods in the Great Lakes (30–32), abnormalities of sexual development in Lake Apopka alligators (33–35), reproductive failure in mink from the Great Lakes area (36), and reproductive impairment of the Florida panther (37). Many of these studies lack substantiation of a link with specific putative endocrine disruptors. Nonetheless, they have triggered concern that analogous phenomena may be occurring in humans.

6 HEALTH OUTCOMES OF CONCERN

Selected evidence from human and laboratory studies relating to the major health endpoints that have been investigated in association with environmental endocrine disruptors are briefly summarized here.

6.1 Effects on the Male Reproductive System

6.1.1 Male Reproductive Tract Developmental Anomalies

Successful differentiation of the embryonal gonads into the male phenotype is dependent upon a coordinated sequence of events involving the hypothalamus, pituitary, and testes. In humans at age 8 weeks, the anti-Mullerian hormone (AMH) produced by the fetal Sertoli cells, acting in concert with testosterone produced by the fetal Leydig cells, causes regression of the Mullerian ducts. Complete regression is necessary for normal male reproductive tract development. In addition, testosterone controls other aspects of sexual differentiation, via interaction with androgen receptor in target tissue. The epididymus, vas deferens, and seminal vesicles are formed from the Wolffian ducts while the prostate gland, bladder and initial urethra are formed from the urogenital sinus following Mullerian duct regression. During the second and third trimester, development of the testes and penis occur, and the testes migrate and descend into the scrotum, events also under control of testosterone (7).

Interference with AMH action can lead to presence of components of the female reproductive tract in phenotypic males. Interference with testosterone action, depending on the extent and timing, can lead to the following outcomes: incomplete development of the penis (e.g., hypospadias or microphallus) or other genitalia; failure of the testes to descend into the scrotum (i.e., cryptorchidism), which is associated with impaired spermatogenesis and testicular cancer; incomplete proliferation or maturation of gonocytes or Sertoli cells, impacting sperm production; and incomplete proliferation of Leydig cells, possibly affecting androgen production, onset of puberty, and adult sexual behavior (7). Chemicals with antiandrogenic activity include the pesticides procymidone and vinclozolin, and the DDT degradation product *p,p'*-DDE (7). Certain chemicals with estrogenic activity have been found to also exhibit antiandrogenic activity (e.g., DES and estradiol), and thus effects that have been attributed to estrogenic mechanisms may in fact be a result of antiandrogenic effects; the antiandrogenicity of most estrogenic chemicals has not yet been assessed. Rats administered the antiandrogenic pesticide vinclozolin have been observed to have anomalies associated with interference with androgen receptor action, including reduced anogenital distance, impaired penis development, presence of vaginal pouches, delayed preputial separation, and seminiferous tubule atrophy (38). 2,3,7,8-Tetrachlorobenzo-*p*-dioxin (TCDD), an arylhydrocarbon (Ah) receptor agonist, impairs testosterone synthesis and appears to impair the feedback mechanism for leutinizing hormone (LH) synthesis and release (39). Rodents treated with TCDD exhibit reduced anogenital distance, delayed testis descent, impaired spermatogenic function, reduced accessory sex gland weights, and feminization of sexual behavior (40; review, 41).

There have been reports from several countries of an increasing secular trend in incidence of cryptorchidism (42, 43) and hypospadias (44–49; review, 50). These ecologic-level data are difficult to evaluate. There is extremely wide variation in the reported incidences, likely a result of differences in definitions of the anomalies, differences in completeness of reporting, and ethnic variation. Strongest evidence of an increasing secular trend comes from a pair of studies of cryptorchidism done in England. One study of approximately 3500 male infants born at a London hospital in the late 1950s found an incidence of cryptorchidism at age three months of 1.7% in boys less than 2500 g and 0.91% in boys over 2500 g, using carefully specified examination methods and definitions (51). A study of 7441 infants in Oxford in the 1980s using the same methods and definitions found rates of 5.2% and 1.6% for those less than and greater than 2500 g, respectively (52). Additional studies using comparable methods of assessment are needed to confirm an increasing trend in these two anomalies over this time period. A recent analysis of international monitoring data suggests that incidence of hypospadias and cryptorchidism in a number of countries may have leveled off since 1985 (53), suggesting that if there is an environmental etiology underlying the international trends the prevalence of exposure to the agent(s) has stabilized.

6.1.2 *Sperm Concentration*

Ecologic observations on time trends in sperm count data have led to concern that exposure to environmental endocrine disruptors may be impacting male fertility. A study in the early 1970s reported that sperm concentrations appeared to have fallen dramatically since the 1950s and proposed that an environmental exposure might be the cause (54). A subsequent study found an association of sperm density with seminal organochlorine level (55). Sharpe and Skakkebaek (56) have proposed that in utero exposure of the testis to estrogenic compounds decreases the multiplication of Sertoli cells, cells that support spermatogenesis. In support of this hypothesis, sons of women who took diethylstilbestrol during pregnancy have been reported to have decreased sperm counts (14).

In 1992, Carlsen published a meta-analysis of 61 studies of sperm density in "normal" men, concluding that there had been an appreciable decline in sperm density worldwide from 1938 to 1990 (57). Extensive debate regarding the validity of this conclusion ensued (e.g., 58–62). Swan (63) reanalyzed the data from 58 of these studies using several modeling methods and concluded that the apparent downward trend persisted. However, it is difficult to exclude the possibility that this trend is an artifact of confounding or selection bias. Concern about selection bias stems from the fact that the individuals studied are not random population samples. The meta-analyses excluded studies of infertile men, but selection criteria for the studies varied widely. A major potential confounder is abstinence time, as this is positively associated with sperm density. Secular trends in coital frequency, or differences in mean abstinence time in the men being compared (e.g., due to differences in mean age), could lead to confounding. Another serious potential confounder is geographic area. Swan found that differences across geographic areas were as large as the overall decline in sperm density. An additional possible confounder is the collection procedure. There is evidence that semen samples obtained via masturbation have lower counts than those collected during intercourse, a collection method used more commonly in the earlier studies (64).

In the absence of population-based studies with uniform protocols, it is difficult to draw conclusions regarding whether the observed decline in sperm density is real or artifactual. If the decline is indeed real, an etiologic relationship with putative endocrine disruptors has yet to be established. Finally, the impact of a real decline in sperm density on fertility is difficult to predict without an understanding of what is happening at the low end of the sperm density curve in the population. For a discussion of the evidence regarding the strong suppressive effect of occupation exposure to 1,2-dibromo-3-chloropropane on spermatogenesis, the reader is referred to Section 8.

6.2 Effects on the Female Reproductive System

6.2.1 *Disruption of Female Reproductive Function*

Few chemicals have been evaluated for effects on the full spectrum of female reproductive health endpoints. Certain chemicals are known to disrupt female reproductive function through disruption of normal sexual differentiation, ovarian function (follicular growth, ovulation, and corpus luteum formation and maintenance), fertilization, implantation, or pregnancy (7). Evidence of an endocrine-mediated mechanism has been developed for only a subset of these. Some findings largely from laboratory studies are mentioned here.

Kepone-treated rats have been reported to give birth to female pups with persistent vaginal estrus and anovulation (65). Perinatal exposure of rats to methoxychlor, a pesticide with estrogenic metabolites, has been associated with premature vaginal opening and persistent estrus in the adult, apparently as a result of masculinization of the developing brain (66). The rats were reportedly unable to achieve normal ovulatory LH surges, and their ovaries had a high frequency of polyfollicular or polycystic follicles and no corpora lutea. Prolonged exposure in adulthood leads to persistent estrus as a result of direct estrogenic action on the vagina and the adult female ovaries become atrophied due to suppression of gonadotropin secretion (7). Rats treated with methoxychlor in the first week of pregnancy exhibit reduced implantation as a result of increased speed of embryo transport through the oviduct, an estrogen-dependent process (67). Chlorotriazine herbicides have been reported to induce persistent estrus in Sprague-Dawley rats but not Fischer rats (68). Although results have been inconsistent across studies, it appears that atrazine and several other chlorotriazines disrupt ovarian function in the adult female rat via an endocrine pathway. The mechanism appears to be estrogen receptor independent, and the disturbance of estrus cycling appears to be due to disruption of hypothalamic-pituitary regulation of ovarian function (7). Treatment of female rats with heptachlor reduces implantations, increases resorptions, and decreases serum estrogen and progesterone levels (69). In a study of 47 women, levels of organochlorines such as DDT, PCBs, and hexachlorobenzene were associated with reduced oocyte recovery rates and embryo cleavage rates (70). Results of this small study should be replicated in further studies, and mechanisms underlying this association need to be investigated. Ovarian vacuolation was reported in a multigenerational rat study at high levels of dicofol exposure (7). Vinclozolin has been reported to bind to and inhibit the androgen receptor in rats, leading to increases in LH and testosterone levels (71), possibly impacting follicular growth and ovulation.

6.2.2 *Endometriosis*

Endometriosis is a condition in which uterine endometrial cells appear in aberrant locations, such as the ovaries, uterine ligaments, rectovaginal pouches, and pelvic peritoneum. It can cause pelvic pain, dysmenorrhea, and infertility. Proper functioning of the normal endometrium is dependent on well-organized intercellular interactions regulated locally by cytokines and growth factors under the direction of steroid hormones (72). The pathophysiology of endometriosis is not understood, but both immunological and endocrine disturbances appear to be important (72). In a small study of rhesus monkeys, TCDD exposure was reported to be associated with endometriosis, and severity appeared to be dose-dependent (73). In a study of nude mice injected with human endometrial tissue, normal endometrium treated with TCDD showed steroidal misregulation of matrix metalloproteinases (72). In contrast to the TCDD findings, rhesus monkeys treated with PCB Aroclor 1254 were not found to have elevated incidence or severity of endometriosis (74). In a comparison of 15 women with laparoscopically confirmed endometriosis and 15 age- and geographically matched controls, serum levels of dioxin, furan, and PCB congeners were similar (75). Larger studies will be needed to resolve the question of whether there is an association of any of these chemicals with endometriosis in women.

6.3 Thyroid Function

The thyroid gland secretes hormones critical for normal growth and differentiation and regulation of overall metabolism. In the fetal and neonatal period, thyroid hormone disturbances may lead to cognitive and motor disorders, while in children and adults, neuronal maintenance and metabolic function may be affected (76). Whereas the effects of thyroid disorders in adults are generally reversible, the effects in the developing fetus can be permanent (77). Thyroid hormones have profound effects on brain development through regulation of neuronal proliferation, migration, process outgrowth, synaptic development, and myelin formation (77). Too much thyroid hormone during brain development leads to earlier than normal differentiation and a deficit in the total number of neurons (77). Deficiency in thyroid hormone leads to delays in brain development and impairment of neuronal outgrowth, synaptic development, and establishment of specific neurotransmitters (77).

Studies in rats and monkeys have found decreased plasma T_4 levels and concomitant increased thyroid stimulating hormone (TSH) levels following exposure to PCDD, PCDF, or PCB (78–81). In a study of 105 infant-mother pairs, PCDD, PCDF, and PCB levels in breastmilk (combined using the toxic equivalence factor approach) were significantly correlated with reduced plasma levels of maternal total T_3 and total T_4 (78). These chemicals were also associated with higher plasma levels of TSH in the infants at 2 weeks and 3 months of age and with lower plasma free T_4 and total T_4 levels at 2 weeks. Another study found PCDD and PCDF levels in maternal milk to be associated with higher TSH levels in 38 breastfed infants, consistent with the study by Koopman-Esseboom, et al., but found higher, rather than lower, plasma total T_4 levels (82).

Mechanistic studies have indicated that polyhalogenated aromatic hydrocarbons interact with the thyroid hormone system at three levels: (*1*) thyroid gland function and morphology, (*2*) thyroid hormone metabolism, and (*3*) thyroid hormone transport binding proteins (83). Implications of disruption of the thyroid system on the developing fetus include effects on structure and function of the brain and sexual organs (83).

6.4 Neurobehavioral Effects

Endocrine disruption can clearly lead to profound effects on the nervous system and behavior since the endocrine and nervous systems are intimately related and interdependent. While some of the behavioral effects of the thyroid hormone system were mentioned in Section 6.3, the gonadal hormones also have major behavioral effects. During development, gonadal hormones direct organizational changes in the brain and behavior through permanent changes in wiring and sensitivity of the brain. Typically the behavioral effect of later gonadal hormone exposures is to activate neural systems that were organized early in life (84). In adults, circulating testosterone is associated with aggression and spatial ability, while estrogen has been reported to facilitate memory (84). Females exposed to elevated levels of testosterone in utero due to having a male twin exhibit tendencies toward more masculine behaviors: they have higher spatial ability, higher sensation seeking, and male-typical auditory function (84).

The epidemiologic findings regarding neurobehavioral assessments of children exposed prenatally to PCBs and several other polyhalogenated hydrocarbons are briefly summarized here. It should be emphasized, however, that while certain environmental exposures have been associated with nervous system effects, the specific mechanisms involved in these associations are not necessarily endocrine mediated (85). Moreover, sexually dimorphic behaviors have not been evaluated in these studies.

Two mass poisonings that occurred in Taiwan and Japan, the Yu-Cheng and Yusho episodes, respectively, involved contamination of rice oil with a mixture of PCBs, PCDFs, polychlorinated terphenyls, and polychlorinated quaterphenyls. Children born to exposed mothers exhibited a syndrome at birth involving a variety of physical and developmental problems including low birth weight, hyperpigmentation, and hyperbilirubinemia (86). Body weights of Yu-Cheng children were lower up to 10 years of age (87). Exposed children had higher rates of hyperactivity and disordered behavior, and reduced scores on intelligence tests (88, 89).

Neurologic and developmental status in relation to in utero exposure to environmental contaminants have been assessed in several birth cohorts. In the United States there have been two major cohorts: a North Carolina cohort that was essentially a general population sample, and a Michigan cohort comparing frequent consumers of Great Lakes fish to nonconsumers. In the North Carolina cohort, the top decile of prenatal PCB exposure was associated with hypotonicity and hyporeflexia at birth, and in subsequent examinations, slower motor development through 2 years of age (90). Measures of formal cognition at ages 3 to 5 (91) and school performance at ages 8 to 10 (92) did not differ by exposure status. In the Michigan cohort, children born to women who ate more Great Lakes fish also had higher rates of hyporeflexia (93, 94). This association was less pronounced when measured PCB values were used in the analysis rather than reported fish consumption. Poorer short-term memory was noted in the exposed children at age 7 months (95) and age 4 years (96, 97), and lower IQ at 11 years (98). Differences in congener-specific profiles or cognitive assessment methods may explain the inconsistencies between the Michigan and North Carolina cohorts in the follow-up investigations; another possible explanation is that the Michigan findings reflected other fish-associated contaminants such as methyl mercury (99). In a subsequent study that measured both methyl mercury and PCBs in children born in the Faroe Islands, Grandjean (100) reported that intrauterine PCB exposure was associated with poorer performance on verbal ability tests, while methyl mercury exposure was associated with measures of cognitive performance. A recent study of Dutch infants reported an association of intrauterine PCB exposure with hypotonia at birth (101) and reduced cognitive performance at age 42 months, particularly in those who had not been breastfed (102). Again, it must be emphasized that the mechanisms underlying these neurobehavioral effects have not been elucidated.

6.5 Cancer

6.5.1 Breast Cancer

Breast cancer risk appears to be related to cumulative lifetime exposure of breast tissue to circulating estrogens. This hypothesis derives support from the following observations: breast cancer risk increases with early age at menarche, late age at menopause, and late age at first pregnancy, and risk is inversely related to parity, history of lactation, and ovariectomy. How exposure to environmental chemicals that may have estrogenic activity relates to breast cancer risk is less well established. Evidence derives primarily from case-control studies in which serum organochlorine levels have been assessed. A couple of early positive studies have been followed up with some larger studies that have been largely null, but have not excluded modest excesses or elevated risk in subgroups.

In a study of 58 breast cancer cases and 171 matched controls from the NYU Women's Health Study cohort, the mean DDE levels in serum samples taken one to six months prior to diagnosis were significantly higher in the cases (103). Another case-control study of only 18 cases and 17 controls noted a positive association of DDE and PCBs with risk of estrogen receptor–positive breast cancer (104). In a larger study (120 case-control pairs) using banked serum collected an average of 14 years prior to diagnosis, Krieger et al. (105) reported overall null findings, but suggestive associations with DDE in racial subgroups. A case-control study nested in the Nurses Health Study using banked serum collected up to 2 years prior to diagnosis in 236 case-control pairs reported no association with DDE (106). Findings from several recent studies have been mixed (107–109). The Krieger study (105) remains arguably the most informative of the studies because the serum samples were collected many years prior to diagnosis, reducing the serious concern that the disease process altered the measurement of the analyte (e.g., through alterations in lipid metabolism).

6.5.2 *Testicular Cancer*

Incidence of testicular cancer, primarily germ cell tumors, has increased markedly over the past several decades (50, 110, 111). Incidence in men under 50 years of age has reportedly increased between 2 and 4% per year since the 1960s in Great Britain, Sweden, Finland, Norway, Denmark, Germany, Australia, New Zealand, and the United States (review: 50). Because testicular cancer is associated with conditions that involve in utero estrogenic exposure, such as cryptorchidism (112), the trend in testicular cancer has been hypothesized to be related to increased exposure to environmental estrogens (56). An association of suspect endocrine disruptors with testicular cancer risk has yet to be investigated extensively in epidemiologic studies. Leydig cell tumors are a common finding in animal studies, but the relevance of these tumors to human risk assessment is unclear (113). The apparent increasing trend in testicular cancer in men is cause for concern, and the factors underlying this trend warrant elucidation.

7 STUDIES OF OCCUPATIONALLY EXPOSED POPULATIONS

There are limited data on endocrine disruption in individuals exposed occupationally to chemicals suspected to be endocrine disruptors. While there is a substantial literature on reproductive, neurobehavioral, and immune system effects of occupational exposures, very little work has addressed endocrine mechanisms specifically. A sampling of results from occupational studies follows.

The adverse neurologic and reproductive effects of lead are well known and are not reviewed here. The endocrine aspects of these effects are less well understood. An assessment of 47 foundry workers with blood lead levels from 16 to 127 μg/dL noted a negative association of serum thyroxine and estimated free thyroxine levels with blood lead level, and no association with TSH and T_3 levels (114). The authors suggested that the data were consistent with several mechanisms: central depression of the thyroid axis, an alteration in thyroxine metabolism, or an effect on protein binding, but not a direct effect of lead on the thyroid. A recent study of lead smelter workers with blood lead levels less than 60 μg/dL found normal thyroxine, free thyroxine, and TSH (115). In another study, smelter workers employed for over 3 years were reported to have decreased serum testosterone, steroid-binding globulin, and free-testosterone index, and increased LH (116); mean blood lead concentrations were over 60 μg/dL. Ng et al. (117) found that workers exposed to lead for less than 10 years had elevated follicle-stimulating hormone (FSH) and LH and normal testosterone, while men exposed for over 10 years had normal FSH and LH and low testosterone levels. The authors suggest that their findings are consistent with a direct toxic effect of lead on the testes, leading to reduced testosterone production. They posit that in men with less chronic exposure, the hypothalmo-pituitary axis was capable of compensating the loss by increasing FSH and LH secretions, bringing testosterone back to normal, while in more chronically exposed men interference with this mechanism led to normal FSH and LH and low testosterone.

Egeland et al. reported an association of serum 2,3,7,8-TCDD levels with increased FSH and LH and decreased testosterone in 248 chemical production workers compared with neighborhood referents (118). These results are consistent with the animal literature describing dioxin-associated effects on the hypothalamic–pituitary–Leydig cell axis and on testosterone synthesis. While the reproductive significance of these findings is not clear, the findings warrant similar assessment in other exposed populations.

In a study of 38 transformer repairmen currently exposed to PCBs, 17 former repairmen, and 56 unexposed comparison workers, the PCB-exposed group was reported to have decreased serum T_4 levels (119). This result is consistent with results from epidemiologic studies of other populations and animal studies mentioned in Section 6.3.

In response to complaints about impotence and decreased sexual libido among male workers who manufactured 4,4′-diaminostilbene-2,2′-disulfonic acid (DAS), the National Institute for Occupational Safety and Health conducted a health hazard evaluation (120). In this cross-sectional survey, 30 current DAS-exposed workers and 20 former workers were compared to 35 plastics manufacture workers with respect to serum reproductive hormone levels. Mean total testosterone levels were significantly lower in the current and former DAS-exposed workers than in the plastics workers (458 and 442, respectively, versus 556 ng/dL). The plastics workers may have been exposed to phthalates so this may have been a conservative comparison (E. Whelan, personal communication, August 3, 1999). An association of duration of employment in DAS manufacture and testosterone levels was also noted. While exposure to this chemical is likely to be restricted to a relatively small group of workers, the structural similarity of the chemical to DES noted by the authors makes the findings intriguing.

Several investigators have pursued endocrine markers that would be useful in identifying workers affected by the well-known neurotoxicant styrene. Plasma prolactin levels, a measure of catecholaminergic dysfunction, have been reported to be elevated in styrene-exposed workers (121, 122). Bergamaschi et al. (122) suggest that the plasma prolactin

Table 12.1. Suspect Endocrine Disruptors, Uses, and Estimated Occupational Exposure

Chemical	Identifying Information	Description and Use	NOES Estimate of Number of U.S. Nonagricultural Workers Exposed
Acenaphthene	CAS #: [*83-32-9*] NIOSH reg. #: AB1000000 chem. form.: $C_{12}H_{10}$	Derived from coal tar and petroleum refining; used in synthesis of dye intermediates, plastics, pharmaceuticals, pesticides, and other chemical products	1,064
Alachlor	CAS #: [*15972-60-8*] chem. form.: $C_{14}H_{20}ClNO_2$	Herbicide for control of many broad-leaved weeds and grasses, used on corn, peas, soybeans, peanuts, beans, cotton, milo, and sunflowers; applied using ground or aerial methods, restricted use pesticide	—
Aldicarb	CAS #: [*116-06-3*] chem. form.: $C_7H_{14}N_2O_2S$	Extremely acutely toxic systemic carbamate insecticide; applied to soil as an insecticide, acaricide, and nematocide for use on cotton, sugar beets, potatoes, peanuts, ornamentals, yams, oranges, pecans, dry beans, soybeans, and sugarcane; also used on root crops against nematodes, millipedes, eelworms, and other insect pests; restricted use pesticide	—
Aldrin	CAS #: [*309-00-2*] NIOSH reg. #: I02100000 chem. form.: $C_{12}H_8Cl_6$	Chlorinated cyclodiene with high acute toxicity; manufacture and most uses in the U.S. have been discontinued due to adverse environmental effects; effective insecticide used primarily against soil insects that affect field, forage, vegetable, and fruit crops; used to kill cotton insects, turf pests, white grubs, and corn rootworms; also effective against termites and ants and was used as a wood preservative; in 1974, U.S.E.P.A. canceled all uses except subsurface ground insertion for termite control, dipping of nonfood roots and tops, moth-proofing by manufacturing processes in a closed system	—
Allethrin	CAS #: [*584-79-2*] chem. form.: $C_{19}H_{26}O_3$	A synthetic pyrethroid used in homes, food establishments and gardens for control of flies and mosquitoes and in combination with other pesticides to control flying or crawling insects, available in aerosol, coil, mat, dust, and oil formulations	1,366

Amitrole	CAS # [*61-82-5*] NIOSH reg. #: XZ3850000 chem. form.: $C_2H_4N_4$	A triazine used as defoliant, herbicide, a reagent in photography and a plant growth regulator, wide-spectrum herbicide that appears to act by inhibiting formation of chlorophyll; used on noncrop sites including rights-of-way, marshes, drainage ditches, ornamentals, and around commercial, industrial, agricultural, domestic, and recreational premises; applied as a spray using aerial or ground equipment; restricted use pesticide (for all uses except for homeowner uses); not produced in US; major importers (<800,000 lbs) are Union Carbide, American Cyanamid, Acelo Chemical	694
Anthracene	CAS #: [*120-12-7*] NIOSH reg. #: CA9350000 chem. form.: $C_{14}H_{10}$	Source of dyestuffs and used in the manufacture of anthraquinone, alizarin dyes, insecticides, wood preservatives and as scintillation counter crystals; product of incomplete combustion	2,303
Arsenic	CAS #: [*7440-38-2*] NIOSH reg. #: CG0525000	Used as an alloying agent for heavy metals, in special solders, in bronzing, pyrotechny, for hardening and improving the sphericity of shot, as a doping agent in solid-state devices, and in medicine; calcium and lead arsenates have been used as agricultural insecticides and poisons; gallium arsenide is used as a laser material	27,276
Atrazine	CAS #: [*1912-24-9*] NIOSH reg #: XY5600000 chem. form.: $C_8H_{14}ClN_5$	An s-triazine herbicide; used as a selective herbicide for weed control in agriculture; most widely used herbicide in world; production in U.S. is ~100 million lbs annually	1,002
Benomyl	CAS #: [*17804-35-2*] NIOSH reg #: DD6475000 chem form: $C_{14}H_{18}N_4O_3$	Fungicide used in control of a wide range of diseases of fruit, nuts, vegetables, mushrooms, field crops, ornamentals, turf, and trees; also used as acaricide against mites; used as a veterinary anthelmintic and as an oxidizer in sewage treatment	4,190
Benz(a)anthracene	CAS #: [*56-55-3*] NIOSH reg #: CV9275000 chem. form.: $C_{18}H_{12}$	PAH found in petroleum, wax, and smoke; universal product of combustion of organic matter	2,310
Benzo(a)pyrene	CAS #: [*50-32-8*] NIOSH reg #: DJ3675000 chem. form.: $C_{20}H_{12}$	PAH found in petroleum, wax, and smoke; ubiquitous product of incomplete combustion	896
Benzo(b)fluoranthene	CAS #: [*205-99-2*] NIOSH reg. #: CU1400000 chem. form.: $C_{20}H_{12}$	PAH found in petroleum, wax, and smoke; product of incomplete combustion of fuels including wood and fossil fuels	2,282

Table 12.1. (continued)

Chemical	Identifying Information	Description and Use	NOES Estimate of Number of U.S. Nonagricultural Workers Exposed
Benzo(k)fluoranthene	CAS #: [*207-08-9*] NIOSH reg. #: DF6350000 chem. form.: $C_{20}H_{12}$	PAH found in petroleum, wax, and smoke; ubiquitous product of incomplete combustion	2,282
β-BHC	CAS #: [*319-85-7*] NIOSH reg. #: GV4375000 chem. form.: $C_6H_6C_{16}$	Byproduct of the manufacture of chlorobenzenes, pentachlorophenol, vinyl chloride, tetrachloroethylene, many pesticides (e.g., atrazine and simazine) and other synthetic chemicals; also formed in electrolytic production of chlorine and in the incineration of chlorinated wastes	—
Bisphenol A	CAS #: [*80-05-7*] NIOSH reg #: SL63000000 chem. form.: $C_{15}H_{16}O_2$	Intermediate in manufacture of polymers, epoxy resins, polycarbonate, fungicides, antioxidants, dyes, phenoxy, polysulfon and polyester resins, flame retardant, and rubber chemicals; used in plastic dental fillings and teeth coatings; estimated U.S. production >1 billion lbs/yr	2,180
Butyl benzyl phthalate	CAS #: [*85-68-7*] chem. form.: $C_{19}H_{20}O_4$	Plasticizer for cellulosic resins, polyvinyl acetates, polyurethanes, and polysulfides and in regenerated cellulose films for packaging; in vinyl products such as synthetic leathers, floor tiles, acrylic caulking, adhesive for medical devices, and in the cosmetic industry as a dispersent and carrier for insecticides and repellents; as an organic intermediate, a solvent, and a fixative in perfume; estimated U.S. production >56 million lbs/yr	331,843
Cadmium	CAS #: [*7440-43-9*] NIOSH reg. #: EU9800000	Major use is electroplating; component of easily fusible alloys; soft solder and solder for aluminum; used as a deoxidizer in nickel plating, in process engraving, in electrodes for cadmium vapor lamps, in photoelectric cells, with photometry of ultraviolet sun-rays, and in nickel-cadmium storage batteries; powder used as an amalgam in dentistry	93,679

Carbaryl	CAS #: *[63-25-2]* NIOSH reg. #: FC5950000 chem. form.: $C_{12}H_{11}NO_2$	Widely used carbamate contact insecticide for control of many types of insects on fruits, vegetables, and ornamentals, and in flea powders for domestic animals; available as wettable powder, emulsifiable concentrates, suspension concentrates, powders, emulsifiable concentrates and oil-miscible concentrates; molluscicide	16,545
Chlordane	CAS #: *[57-74-9]* NIOSH reg. #: PB9800000 chem. form.: $C_{10}H_6C_{18}$	Chlorinated cyclodiene; nonsystemic contact and stomach insecticide with some fumigant activity; acaricide; wood preservative; used in termite control and as protective treatment for underground cables; banned from all use in U.S. in 1988	—
Chlorpyrifos	CAS #: *[2921-88-2]* chem. form.: $C_9H_{11}C_{13}NO_3PS$	Organophosphate; broad nonsystemic insecticide effective by contact, ingestion, and vapor action; acaricide; used to control flies, household pests, mosquito larvae and adults, various crop pests in soil, and on foliage, ectoparasites on livestock; used on poultry, dogs, livestock premises, domestic dwellings, terrestrial structures, stagnant water; one of largest use urban insecticides in U.S.; annual U.S. usage 9 million kg/yr	11,404
Chrysene	CAS #: *[218-01-9]* NIOSH reg. #: GC0700000 chem. form.: $C_{18}H_{12}$	Co-planar PAH; occurs in coal tar and formed during distillation of coal; formed in very small quantities during distillation or pyrolysis of many fats and oils; in particulates from smoke stacks	9,358
Cybermethrin	CAS #: *[5235-07-8]* chem. form.: $C_{22}H_{19}C_{12}O_3N$	Pyrethroid type of insecticide used on cotton, lettuce, and pecans; also applied into overhead sprinkler irrigation water and by pest control operators as crack, crevice, and spot spray treatment in and around buildings and vehicles; may be used in nonfood areas of schools, nursing homes, hospitals, restaurants, and hotels; food manufacturing, processing, and servicing establishments; insect repellent for horses	471
2,4-D	CAS #: *[94-75-7]* NIOSH reg. #: AG6825000 chem. form.: $C_8H_6Cl_2O_3$	Chlorinate phenoxy; defoliant and herbicide; used as to control broadleaf plants and as a plant growth regulator; applied to grasses, wheat, barley, oats, sorghum, corn, sugarcane, pasture and range land, lawns and turf; used on tomatoes to cause simultaneous ripening; used in forest management for brush control and conifer release; used to increase latex output of rubber trees; estimated U.S. production >49 million lbs/yr	—

Table 12.1. (continued)

Chemical	Identifying Information	Description and Use	NOES Estimate of Number of U.S. Nonagricultural Workers Exposed
p,p'-DDD	CAS #: [*72-54-8*] chem. form.: $C_{14}H_{10}C_{14}$	Degradation product of DDT; also used as an insecticide for contract control of leaf rollers and other insects on vegetables and tobacco; banned in U.S.	—
DDE	CAS #: [*72-55-9*] NIOSH reg. #: KV9450000 chem. form.: $C_{14}H_8C_{14}$	Impurity in, and degradation product of, DDT	—
DDT	CAS #: [*50-29-3*] NIOSH reg. #: KJ3325000 chem. form.: $C_{14}H_9C_{15}$	Used extensively as an insecticide until banned in U.S. in 1973; still manufactured in U.S. for export and is produced and used in developing countries; 96 tons exported from U.S. in 1991	—
1,2-Dibromo-3-chloropropane	CAS #: [*96-12-8*] NIOSH reg. #: TX8750000 chem. form.: $C_3H_5Br_2C_1$	Soil fumigant, nematocide, pesticide, intermediate in organic synthesis	—
2,4-Dichlorophenol	CAS #: [*120-83-2*] chem. form.: $C_6H_4C_{12}O$	Used in organic synthesis and in manufacture of 2,4-D; used as wood preservative, antiseptic, and seed disinfectant; estimated U.S. production >23 million lbs/yr	—
Dicofol	CAS #: [*115-32-2*] NIOSH reg. #: DC8400000 chem. form.: $C_{14}H_9C_{15}O$	Alcohol analog of DDT; synthetic nonsystemic organochlorine acaricide used primarily for control of mites on field crops, vegetables, citrus and noncitrus fruits, and in greenhouses	3,935
Dieldrin	CAS #: [*60-57-1*] NIOSH reg. #: IO1750000 chem. form.: $C_{12}H_8C_{16}O$	Nonsystemic, persistent insecticide with contact and stomach action; was used as a broadspectrum insecticide in U.S. until 1974 when EPA restricted its use to termite control by direct soil injection and nonfood seed and plant treatment; was used in tropical countries as a residual spray on inside walls and ceilings of homes for disease vector control, particularly malaria; industrial uses include timber preservation, termite-proofing of plastic and rubber coverings of electrical and telecommunication cables, of plywood and building boards, and as a termite barrier in building construction	—

Di(2-ethylhexyl)phthalate	CAS #: [*117-81-7*] NIOSH reg. #: TI0350000 chem. form.: $C_{24}H_{38}O_4$	Used as a plasticizer for polyvinyl chloride, particularly in manufacture of medical devices, and as a plasticizer for resins and elastomers; some PVCs contain 40% di(2-ethylhexyl)phthalate; also used in heat-seal coatings on metal foils for food packaging and in aluminum paper-foil laminates; used as a solvent in erasable ink and dielectric fluid; used in vacuum pumps; used as an acaricide in orchards; an inert ingredient in pesticides; a detector for leaks in respirators, testing of air filtration systems; and a component in cosmetic products; estimated U.S. production >214 million lbs/yr	34,097
di-*N*-butyl phthalate	CAS #: [*84-74-2*] NIOSH ref. #: TI0875000 chem. form.: $C_{16}H_{22}O_4$	In widespread use primarily as a plasticizer in plastics; used widely in PVC and nitrocellulose polyvinyl acetate; used in carpet backing hair spray, nail polish, and glue; plasticizer in coatings on cellophane; used in inks, cosmetics, safety glass, insecticides, paper coatings, adhesives, elastomers, and explosives; used as a solvent in polysulfide dental impression materials, solvent for perfume oils, perfume fixative, textile lubricating agent, and solid rocket propellent; estimated U.S. production >12 million lbs/yr	512,626
Endosulfan	CAS #: [*115-29-7*] NIOSH reg. #: RB9275000 chem. form.: $C_9H_6C_{16}O_3S$	Cyclodiene; a contact insecticide used widely on food crops; >20 million lbs/yr used worldwide, including ~2 million lbs in the U.S.; commonly sprayed on lettuce, tomatoes, artichokes, strawberries, pears, grapes, alfalfa, cotton, tea, tobacco, and nuts; available as emulsifiable concentrate, wettable powers, granules, dustable powder, and smoke tablets	3,205
Endrin	CAS #: [*72-20-8*] NIOSH reg. #: IO1575000 chem. form: $C_{12}H_8C_{16}O$	Was used as an insecticide, avicide, rodenticide, and pesticide; manufacture and use has been discontinued in U.S.; was used primarily on field crops such as cotton and grains; used to control the army cutworm, pale western cutworm, grasshoppers in noncropland, voles and mice in orchards; persistent	—
Heptachlor	CAS #: [*76-44-8*] NIOSH reg. #: PC0700000 chem. form.: $C_{10}H_5C_{17}$	Chlorinated cyclodiene; insecticide used for control of the cotton boll weevil, termites, ants, grasshoppers, cutworms, maggots, thrips, wireworms, flies, mosquitoes, soil insects, household insects, and field insects; some fumigant action; applied as a soil treatment, a seed treatment or directly to foliage; U.S. has canceled use except use through subsurface ground insertion for termite control and dipping of roots or tops of nonfood plants	1,033

Table 12.1. (continued)

Chemical	Identifying Information	Description and Use	NOES Estimate of Number of U.S. Nonagricultural Workers Exposed
Heptachlor epoxide	CAS #: [*1024-57-3*] NIOSH reg. #: PB9450000 chem. form.: $C_{10}H_5C_{17}O$	Degradation product of heptachlor, no commercial uses	—
Hexachlorobenzene	CAS #: [*118-74-1*] NIOSH reg. #: DA2975000 chem. form.: C_6Cl_6	Used in organic synthesis; as a fungicide for seeds (sunflower, safflower, seedling diseases), as a wood preservative, in the manufacture of pentachlorophenol in Europe, in the production of aromatic fluorocarbons and in the impregnation of paper; waste product in production of several chlorinated hydrocarbons and a contaminant in some pesticides	1,038
Indeno[1,2,3-c,d]pyrene	CAS #: [*193-39-5*] NIOSH reg. #: UR2625000 chem. form.: $C_{22}H_{12}$	PAH formed in most combustion of high-temperature processes involving hydrocarbons; known sources include coal, wood, and gasoline combustion, municipal incineration, coke ovens, and cigarette smoke; component of petroleum and coal products	—
Lead	CAS #: [*7439-92-1*] NIOSH reg. #: OF755000	Uses in paint and gasoline and lead salts as insecticides have been curtailed or eliminated, used as a container for corrosive liquids; alloys include solder, type metal, and antifriction metals; used in storage batteries, cable covering, plumbing, ammunition, and lead tetraethyl; used as a radiation shield, and to absorb vibration; lead oxide used in crystal and flint glass	773,328
Lindane	CAS #: [*58-89-9*] NIOSH reg. #: GV4900000 chem. form.: $C_6H_6Cl_6$	Chlorinated hydrocarbon insecticide and fumigant and a restricted use pesticide for treating soil; pediculicide, scabicide, ectoprasiticide; used as a foliar spray and soil application for insecticidal control of a broad spectrum of plant-eating and soil dwelling insects, animal ectoparasites, and public health pests; used on ornamentals, fruit trees, nut trees, vegetables, tobacco, and timber, used in baits and seed treatments for rodent control; stomach and contact poison and some fumigant action; used in control of malaria and other vector-borne diseases; used in pet shampoo to maintain luster and prevent ticks, lice, and sarcoptic mange mites; used on humans in treatment of head and crab lice	15,035

Malathion	CAS #: [*121-75-5*] NIOSH reg. #: WM8400000 chem. form.: $C_{10}H_{19}O_6PS_2$	Organophosphate used as an insecticide for fruits, vegetables, ornamentals, household and liverstock use; used as an acaricide, in control of flies and other insect pests in animal and poultry houses, in control of adult mosquitoes in public health programs, in control of human body and head lice and in flea and tick dips; used in veterinary medicine as an ectoparasiticide; U.S. usage 4 to 6 million lbs/yr	19,172
Mancozeb	CAS #: [*8018-01-7*] chem. form.: $(C_4H_6MnN_2S_4)x(Zn)y$	An ethylene bisdithiocarbamate registered as a general use pesticide in U.S.; combination of maneb and zineb; fungicide used with crops such as apples, potatoes, tomatoes, and onions; applied using ground or aerial equipment; U.S. usage 4 to 7 million lbs/yr	—
Maneb	CAS #: [*12427-38-2*] chem. form.: $(C_4H_6MnN_2S_4)x$	An ethylene bisdithiocarbamate registered as a general use pesticide in U.S. fungicide used on crops such as fruits, vegetables, seed crops, nuts, flax, and grains, and nonfood crops including ornamentals, lawns and turf, used primarily on apples, potatoes, tomatoes and corn; effective in control of foliar fungal diseases	593
Mercury	CAS #: [*7439-97-6*] NIOSH reg. #: OV4550000	Used for making thermometers, barometers, diffusion pumps, and other instruments; used in recovery of gold from ores; used in mercury-vapor lamps and advertising signs, in electrical apparatus, mercury cells for caustic soda and chlorine production, dental preparations, antifouling paint, batteries and catalysts; previous use in pesticides has been largely discontinued	71,933
Methomyl	CAS #: [*16752-77-5*] chem. form.: $C_5H_{10}N_2O_2S$	Carbamate insecticide applied to food and nonfood crops, aquatic food crops, forestry, and indoor human and animal premises; nematocide on tobacco, cotton, alfalfa, soybeans, and corn; used as a foliar treatment for control of insects such as aphids, armyworms, cabbage looper, tobacco budworm, tomato fruitworm, and cotton pests	2,367
Methoxychlor	CAS #: [*77-43-5*] NIOSH reg. #: KJ3675000 chem. form.: $C_{16}H_{15}C_{13}O_2$	Chlorinated hydrocarbon insecticide used to control a wide range of insects in field crops, forage crops, fruit, vines, flowers, vegetables, in forestry, in animal houses and dairies, and in household and industrial premises; used in veterinary medicine as an ectoparasiticide	3,418

Table 12.1. (continued)

Chemical	Identifying Information	Description and Use	NOES Estimate of Number of U.S. Nonagricultural Workers Exposed
Metiram	CAS #: [*9006-42-2*] chem. form.: $C_{16}H_{33}N_{11}S_{16}Zn_3$	An ethylene bisdithiocarbamate registered as a general use pesticide in U.S.; combination of maneb and zineb; used as fungicide on fruits and vegetables, tobacco, and roses	—
Metolachlor	CAS #: [*51218-45-2*] chem. form.: $C_{15}H_{22}ClNO_2$	Chloracetanilide; preemergence herbicide used to control broadleaf and grassy weeds with food crops, ornamental plants, and along railroad and highway rights of way	—
Metribuzin	CAS #: [*21097-64-9*] chem. form.: $C_8H_{14}N_4OS$	Pre- and postemergence triazone herbicide used to control broadleaf weeds and grasses in vegetable crops	—
Mirex	CAS #: [*385-85-5*] NIOSH reg. #: PC8225000 chem. form.: $C_{10}Cl_{12}$	Highly stable insecticide previously used for fire ant control in the southeastern U.S.; was also used as a flame retardant coating for plastics, rubber, paint, paper, and electrical goods; was used in antifouling paints, rodenticides, and additives for antioxidant and flame retardant mixtures for stabilized polymer, ablative, anthelmintic, and lubricant compositions; used in thermoplastic, thermosetting, and elastomeric resin systems; banned in U.S. in 1978	—
Nitrofen	CAS #: [*1836-75-5*] NIOSH reg. #: KN8400000 chem. form.: $C_{12}H_7Cl_2NO_3$	Contact herbicide for pre- and postemergence control of annual grasses and broadleaf weeds on a variety of food and ornamental crops, and on rights of way, and in nurseries on roses and chrysanthemums; no longer manufactured or sold in U.S. or Canada	—
Parathion	CAS #: [*56-38-2*] NIOSH reg. #: TF4550000 chem. form.: $C_{10}H_{14}NO_5PS$	Organophosphate used agriculturally as a broad spectrum insecticide; also used as an acaricide, fumigant, and nematocide	—
PCBs	CAS #: [*1336-36-3*] (other CAS #'s for specific Aroclors) NIOSH reg. #: TQ1360000 (Aroclor 1254) chem. form.: variable	Banned in the U.S. since 1977, PCBs were formerly used as dielectric fluids, fire retardants, heat transfer agents, hydraulic fluids, plasticizers, and other applications making use of their low flammability, low electrical conductivity, and stability to chemical and biological breakdown; PCBs have been used in railroad transformers, mining equipment, carbonless copy paper, pigments, electromagnets, microscopy immersion oil, optical liquids	>3,702

Pentachlorophenol	CAS #: [*87-86-5*] NIOSH reg. #: SM6300000 chem. form.: C_6Cl_5OH	Used in large quantities as a wood preservative for utility poles, crossarms, and fenceposts, and other wood, and as an insecticide for termite control, preharvest defoliant, general herbicide, molluscicide, fungicide, bactericide, antimildew agent, slimicide, and algaecide; also used in the synthesis of pentachlorophenyl ester, used in cooling towers; as additives to adhesives, in shingles, roof tiles, brick walls, concrete blocks, insulation, pipe scalant compounds, photographic solutions, textiles, and in drilling mud in the petroleum industry	26,805
Pentachloronitrobenzene	CAS #: [*82-68-8*] NIOSH reg. #: DA6650000 chem. form.: $C_6Cl_5NO_2$	Herbicide and fungicide for seed and soil treatment; used for damping off of cotton	757
Phenanthrene	CAS #: [*85-01-8*] NIOSH reg. #: SF7175000 chem. form.: $C_{14}H_{10}$	Component of coke oven emissions; product of incomplete combustion of wood and fossil fuels; used in synthesis of dyestuffs, explosives, and in many organic syntheses	2,241
Pyrene	CAS #: [*129-00-0*] NIOSH reg. #: UR2450000 chem. form.: $C_{16}H_{10}$	Present in coal tar, ubiquitous product of incomplete combustion; used in biochemical research and as an intermediate in the synthesis of 3,4-benzypyrene and other polyaromatic hydrocarbons	9,386
Simazine	CAS #: [*122-34-9*] NIOSH reg. #: XY5350000 chem. form.: $C_7H_{12}ClN_5$	Used for control of broadleaved and grass weeds in deep-rooted crops, as a preemergence herbicide and soil sterilant; also used to control algae in farm ponds and fish hatcheries; estimated U.S. production >700,000 lbs/yr	357
Styrene	CAS #: [*100-43-5*] NIOSH reg. #: WL3675000 chem. form.: C_8H_8	Used in manufacture of plastics, synthetic rubber, resins, and insulators; also used in some paints, acrylonitrile-butadiene-styrene, styrene-acrylonitrile polymer resins, styrenated polyesters, rubber-modified polystyrene, and copolymer resin systems, glass fibers in the construction of boats and in synthesis of styrene-divinylbenzene copolymers, as a matrix for ion exchange resins; flavoring agent for ice cream and candy	333,210
2,4,5-T	CAS #: [*93-76-5*] NIOSH reg. #: AJ8400000 chem. form.: $C_8H_5Cl_3O_3$	Widely used as a foliar translocated herbicide with residal action for control of woody weeds; selective weed killer used for control of shrubs and trees; used as a growth regulator to increase size of citrus fruits and reduce drop of deciduous fruits; used as defoliant in Vietnam war; may be contaminated with 2,3,7,8-TCDD or 2,4-D; use in U.S. cancelled in 1985	—

Table 12.1. (continued)

Chemical	Identifying Information	Description and Use	NOES Estimate of Number of U.S. Nonagricultural Workers Exposed
2,3,7,8-TCDD	CAS #: [*1746-01-6*] NIOSH reg. #: HP500000 chem. form.: $C_{12}H_4Cl_4O_2$	Contaminant in manufacture of Agent Orange, defoliant used in Vietnam, and present in certain herbicide and fungicide formulations such as 2,4,5-T and pentachlorophenol; present in emissions from incineration of municipal and hazardous wastes and in exhaust from automobiles using leaded gasoline; has been tested for use in flame proofing polyesters and against insects and wood-destroying fungi	—
Toxaphene	CAS #: [*8001-35-2*] NIOSH reg. #: XW5250000 chem. form.: $C_{10}H_{10}C_{18}$	Mixture of chlorinated terpenes produced by chlorination of camphene; used extensively as a pesticide and insecticide on cotton, soybeans, corn, wheat, peanuts, lettuce, tomatoes, grains, vegetables, fruit, and other crops, and on cattle and swine, controls livestock pests, grasshoppers, army-worms, mosquito larvae, bagworms, yellow jackets, and caterpillars; conditional and restricted use as an insecticide and miticide in foliar treatment of certain crops and animal treatment and noncrop areas	—
Trifluralin	CAS #: [*1582-09-8*] NIOSH reg. #: XU9275000 chem. form.: $C_{13}H_{16}F_3N_3O_4$	Selective herbicide for grasses and broadleaf weeds in field crops and pasture lands; contaminated with *N*-nitrosamine	
Vinclozolin	CAS #: [*50471-44-8*] chem. form.: $C_{12}H_9NO_3Cl_2$	Fungicide widely used on fruits	—
Zineb	CAS #: [*12122-67-7*] NIOSH reg. #: ZH3325000 chem. form.: $C_4H_6N_2S_4Zn$	Ethylene bisdithiocarbamate used as a fungicide and insecticide to protect fruit and vegetable crops from foliar and other diseases; home garden use canceled by EPA	3,205
Ziram	CAS #: [*137-30-4*] chem. form.: $C_6H_{12}N_2S_4Zn$	Carbamate applied to foliage and used in soil and seed treatment; used as a fungicide and a repellent to birds and rodents; also used as a rubber vulcanization accelerator and in adhesives used in food packaging, paper coats for nonfood contact, industrial cooling water, latex-coated articles, neoprene, paper and paperboard, plastics (polyethylene and polystyrene), and textiles	34,595

Source: Adapted from Keith (10), with additional data on number of exposed workers from NOES (127, 128) and production volume estimates from EPA (128, 129).

finding may be the result of impaired tubero-infundibular dopaminergic modulation of pituitary secretion.

In 1977, the dramatic suppressive effect of the nematocide 1,2-dibromo-3-chloropropane (DBCP) on spermatogenesis was noted in workers involved in production of the chemical (123, 124). In the severely affected individuals, a statistically significant increase in plasma FSH and LH and a nonsignificant decrease in testosterone were noted in a 17-year follow-up (125). The elevated FSH levels are thought to be the result of diffuse damage to the germinal epithelium and Sertoli cells, which secrete inhibin, a hormone that modulates FSH secretion at the pituitary level (125). Thus, DBCP appears to be an example of a reproductive toxin that has indirect endocrine effects accompanying direct toxic damage. Similarly, the altered sex ratio observed in offspring of exposed workers is thought to be a consequence of an adverse effect of DBCP on the fertility capacity of the Y-bearing sperm cells (125).

8 OCCUPATIONAL EXPOSURE TO SUSPECT ENDOCRINE DISRUPTORS

Table 12.1 lists the chemicals that appeared on a composite list of suspect environmental disruptors published by Keith (10), along with information on their use, production, and estimates of number of workers exposed. The chemicals included have varying degrees of evidence of endocrine activity, from extensive to almost none. There are also some chemicals with evidence of endocrine activity that do not appear on the list, such as diaminostilbene, perhaps because of limited opportunity for general population exposure. Nonetheless, this list includes the major chemicals that have been discussed in the literature on endocrine disruption to date. Where available, estimates of number of exposed U.S. workers from the National Occupational Exposure Survey (NOES) are included (126). The NOES survey provides estimates of the number of workers in nonagricultural, nonmining and nongovernmental workplaces exposed (>30 min/week) to specified chemicals, based on extrapolation of numbers from a sample of 4490 establishments surveyed in 1981–1983 (127). Some of these numbers may be out of date if there have been changes in exposures since the early 1980s. Because the survey excluded agricultural workplaces, it is not helpful for agricultural chemicals, which constitute a substantial proportion of the list of suspected endocrine disruptors, and no other sources of data on number of individuals exposed to the agricultural chemicals were located. U.S. production volume estimates were found for some of the agricultural chemicals from two EPA sources (128, 129).

9 CONCLUSION

While the potential impacts of disruption of endocrine function are profound, there is at present limited data on human health effects of exposure to suspect environmental endocrine disruptors, particularly in occupational settings. In the absence of data, it is always prudent to minimize exposure to chemicals that may have serious health consequences.

ACKNOWLEDGMENTS

The author would like to thank Heinz Ahlers for assistance in using the National Institute for Occupational Safety and Health's NOES survey data, and Michele Marcus for her review of the draft chapter. The chapter was drafted in part while the author was a consultant to CDC's National Center for Environmental Health, serving as a member of that agency's Endocrine Disruptor Leadership Panel.

BIBLIOGRAPHY

1. T. Colborn, F. S. vom Saal, and A. M. Soto, "Developmental Effects of Endocrine-Disrupting Chemicals in Wildlife and Humans," *Environ. Health Perspect.*, **101**, 378–384 (1993).
2. S. H. Safe, "Environmental and Dietary Estrogens and Human Health: is There a Problem?" *Environ. Health Perspect.*, **103**, 346–351 (1995).
3. P. T. C. Harrison, C. D. N. Humfrey, M. Litchfield, D. Peakall, and L. K. Shuker, *Assessment on Environmental Oestrogens: Consequences to Human Health and Wildlife Health*, JK: MRC Institute for Environment and Health, Leicester, 1995.
4. National Academy of Sciences, *Hormone-Related Toxicants in the Environment*, National Research Council, Washington, DC, 1995.
5. J. Toppari, J. C. Larsen, P. Christiansen, A. Giwercman, P. Grandjean, L. J. Guillette, Jr., B. Jegou, T. K. Jensen, P. Jouannet, N. Keiding, et al., *Male Reproductive and Environmental Chemicals with Estrogenic Effects*, Miljoprojekr nr. 290, Report of the Ministry of Environment and Energy, Danish Environmental Protection Agency, Copenhagen, 1995.
6. R. J. Kavlock, G. P. Daston, C. DeRosa, P. Fenner-Crisp, L. E. Gray, S. Kaattari, G. Lucier, M. Luster, M. J. Maczka, et al., "Research Needs for the Risk Assessment of Health and Environmental Effect of Endocrine Disruptors: A Report of the USEPA-Sponsored Workshop," *Environ. Health Perspect.*, **104**(Suppl. 4), 715–740 (1996).
7. T. M. Crisp, E. D. Clegg, R. L. Cooper, W. P. Wood, D. G. Anderson, K. P. Baetcke, J. L. Hoffmann, M. S. Morrow, D. J. Rodier, J. E. Schaeffer, L. W. Touart, M. G. Zeeman, and Y. M. Patel, "Environmental Endocrine Disruption: An Effects Assessment Analysis," *Environ. Health Perspect.*, **106**(Suppl. 1), 11–56 (1998).
8. Work Session on Environmental Endocrine-Disrupting Chemicals: Neural, Endocrine, and Behavioral Effects, "Statement from the Work Session on Environmental Endocrine-Disrupting Chemicals: Neural, Endocrine, and Behavioral Effects," *Toxicol. Ind. Health*, **30**(14), 1–8 (1998).
9. National Academy of Sciences, *Hormonally Active Agents in the Environment*, National Research Council, Washington, DC, 1999.
10. L. H. Keith, *Environmental Endocrine Disruptors: A Handbook of Property Data*, Wiley, New York, 1997.
11. L. J. DeGroot, M. Besser, H. G. Burger, J. L. Jameson, D. L. Loriaux, J. C. Marshall, W. D. Odell, J. T. Potts, Jr., and A. H. Rubenstein, *Endocrinology*, Vol. 1, 3rd ed., W. B. Saunders, Philadelphia, 1995.
12. C. DeRosa, P. Richter, H. Pohl, and D. E. Jones, "Environmental Exposures That Affect the Endocrine System: Public Health Implications," *J. Toxicol. Environ. Health*, Part B, **1**, 3–26 (1998).

13. L. S. Birnbaum, "Endocrine Effects of Prenatal Exposure to PCBs, Dioxins, and Other Xenobiotics: Implications for Policy and Future Research," *Environ. Health Perspect.*, **102**, 676–679 (1994).

14. R. J. Stillman, "In Utero to Diethylstilbestrol: Adverse Effects of the Reproductive Tract and Reproductive Performance and Male and Female Offspring," *Am. J. Obstet. Gynecol.*, **142**, 905–921 (1982).

15. A. L. Herbst, H. Ulfelder, and D. C. Posknzer, "Adenocarcinoma of the Vagina: Association of Maternal Stilbestrol Therapy with Tumor Appearance in Young Women," *N. Engl. J. Med.*, **284**, 878–881 (1971).

16. A. L. Herbst and H. A. Bern, *Development Effects of Diethylstilbestrol (DES) in Pregnancy*, Thieme-Stratton, New York, 1981.

17. R. Newbold, "Cellular and Molecular Effects of Developmental Exposure to Diethylstilbestrol: Implications for Other Environmental Estrogens," *Environ. Health Perspect.*, **103**(Suppl. 7), 83–87 (1995).

18. D. H. White and D. J. Hoffman, "Effects of Polychlorinated Dibenzo-*p*-dioxins and Dibenzofurans on Nesting Wood Ducks (*Aix sponsa*) at Bayou Meto, Arkansas," *Environ. Health Perspect.*, **103**, 37–39 (1995).

19. D. B. Peakall and G. A. Fox, "Toxicological Investigations of Pollutant-Related Effects in Great Lakes Gulls," *Environ. Health Perspect.*, **77**, 187–193 (1987).

20. D. J. Hoffman, B. A. Rattner, L. Sileo, D. Docherty, and T. J. Kubiak, "Embryotoxicity, Tetatogenicity, and Arylhydrocarbon Hydroxylase Activity in Foster's Terns on Green Bay, Lake Michigan," *Environ. Res.*, **42**, 176–184 (1987).

21. T. Colborn, "Epidemiology of Great Lakes Bald Eagles," *J. Toxicol. Environ. Health*, **33**, 395–453 (1991).

22. M. Gilbertson, T. Kubiak, J. Ludwig, and G. A. Fox, "Great Lakes Embryo Mortality, Edema, and Deformities Syndrome (Glemeds) in Colonial Fish-Eating Birds: Similarity to Chick-Edema Disease," *J. Toxicol. Environ. Health*, **33**, 455–520 (1991).

23. G. A. Fox, B. Collins, E. Hayakawa, D. V. Weseloh, J. P. Ludwig, T. J. Kubiak, and T. C. Erdman, "Reproductive Outcomes in Colonial Fish-Eating Birds: A Biomarker for Developmental Toxicants in Great Lakes Food Chains," *J. Great Lakes Res.*, **17**, 158–167 (1991).

24. J. P. Giesy, J. P. Ludwig, and D. E. Tillitt, "Deformities in Birds of the Great Lakes Region," *Environ. Sci. Technol.*, **28**, 128–135 (1994).

25. W. W. Bowerman, J. P. Giesy, D. A. Best, and V. J. Kramer, "A Review of Factors Affecting Productivity of Bald Eagles in the Great Lakes Region: Implications for Recovery," *Environ. Health Perspect.*, **103**, 51–59 (1995).

26. G. A. Fox, A. P. Gilman, D. B. Peakall, and F. W. Anderka, "Aberrant Behavior of Nesting Gulls," *J. Wildl. Manage.*, **42**, 477–483 (1978).

27. D. M. Fry, and C. K. Toone, "DDT-Induced Feminization of Gull Embryos," *Science*, **213**, 922–924 (1981).

28. D. M. Fry, C. K. Toone, S. M. Speich, and R. J. Peard, "Sex Ratio Skew and Breeding Patterns of Gulls: Demographic and Toxicological Considerations," *Stud. Avian Biol.*, **10**, 26–43 (1987).

29. C. A. Bishop, R. J. Brooks, J. H. Carey, P. Ng, R. J. Norstrom, and D. R. S. Lean, "The Case for a Cause-Effect Linkage between Environmental Contamination and Development in Eggs of the Common Snapping Turtle *(Chelydra S. serpentina)* from Ontario, Canada," *J. Toxicol. Environ. Health*, **33**, 521–547 (1991).

30. M. J. Mac and C. C. Edsall, "Environmental Contaminants and the Reproductive Success of Lake Trout in the Great Lakes: An Epidemiological Approach," *J. Toxicol. Environ. Health*, **33**, 75–394 (1991).

31. M. J. Mac, T. R. Schwartz, C. C. Edsall, and A. M. Frank, "Polychlorinated Biphenyls in Great Lakes Lake Trout and Their Eggs: Relations to Survival and Congener Composition 1979–1988," *J. Great Lakes Res.*, **19**, 752–765 (1993).

32. J. F. Leatherland, "Field Observations on Reproductive and Developmental Dysfunction in Introduced and Native Salmonids from the Great Lakes," *J. Great Lakes Res.*, **19**, 737–751 (1993).

33. L. J. Guillette, Jr., T. S. Gross, G. R. Masson, J. M. Matter, H. F. Percival, and A. R. Woodward, "Developmental Abnormalities of the Gonad and Abnormal Sex Hormone Concentrations in Juvenile Alligators from Contaminated and Control Lakes in Florida," *Environ. Health Perspect.*, **102**, 680–688 (1994).

34. L. J. Guillette, Jr., T. S. Gross, D. A. Gross, A. A. Rooney, and H. F. Percival, "Gonadal Steroidogenesis In Vitro from Juvenile Alligators Obtained from Contaminated or Control Lakes," *Environ. Health Perspect.*, **103**, 31–36 (1995).

35. J. C. Semenza, P. E. Tolbert, C. H. Rubin, L. J. Guilette, Jr., and R. J. Jackson, "Reproductive Toxins and Alligator Abnormalities at Lake Apopka, Florida," *Environ. Health Perspect.*, 105:1030–1032 (1997).

36. C. D. Wren, "Cause-Effect Linkages between Chemicals and Populations of Mink *(Mustela vison)* and otter *(Utra canadensis)* in the Great Lakes Basin," *J. Toxicol. Environ. Health*, **33**, 549–585 (1991).

37. C. F. Facemire, T. S. Gross, and L. J. Guillette, Jr., "Reproductive Impairment in the Florida Panther: Nature or Nurture?," *Environ. Health Perspect.*, **103**, 79–86 (1995).

38. L. E. Gray, Jr., J. S. Otsby, and W. R. Kelce, "Developmental Effects of an Environmental Antiandrogen: The Fungicide Vinclozolin Alters Sex Differentiation of the Male Rat," *Toxicol. Appl. Pharmacol.*, **129**, 46–52 (1994).

39. J. P. Whitlock, "The Aromatic Hydrocarbon Receptor, Dioxin Action and Endocrine Homeostasis," *Trends Endocrinol. Metab.*, **5**, 183–188 (1994).

40. L. E. Gray, Jr., W. R. Kelce, E. Monosson, J. S. Otsby, and L. S. Birnbaum, "Exposure to TCDD during Development Permanently Alters Reproductive Function in Male Long Evans Rats and Hamsters: Reduced Ejaculated and Epididymal Sperm Numbers and Sex Accessory Gland Weights in Offspring with Normal Androgenic Status," *Toxicol. Appl. Pharmacol.*, **1331**, 108–118 (1995).

41. R. E. Peterson, H. M. Theobald, and G. L. Kimmel, "Developmental and Reproductive Toxicity of Dioxins and Related Compounds: Cross-Species Comparisons," *Crit. Rev. Toxicol.*, **23**, 283–335 (1993).

42. C. Chilvers, M. C. Pike, D. Forman, K. Fogelman, and M. E. J. Wadsworth, "Apparent Doubling of Frequency of Undescended Testis in England and Wales in 1962–1981," *Lancet*, **ii**, 330–332 (1984).

43. D. M. Campbell, J. A. Webb, and T. B. Hargreave, "Cryptorchidism in Scotland," *Br. Med. J.*, **295**, 1237–1238 (1987).

44. T. Bjerkedal and L. S. Bakketeig, "Surveillance of Congenital Malformations and Other Conditions of the Newborn," *Int. J. Epidemiol.*, **4**, 31–36 (1975).

45. P. Matlai and V. Beral, "Trends in Congenital Malformations of External Genitalia," *Lancet*, **i**, 108 (1985).

46. A. Czeizel, "Increasing Trends in Congenital Malformations of Male External Genitalia," *Lancet*, **i**, 462–463 (1985).

47. B. Kallen, R. Bertollini, E. Castilla, A. Czeizel, L. B. Knudsen, M. L. Martinez-Frias, P. Mastroiacovo, and O. Mutchinick, "A Joint International Study on the Epidemiology of Hypospadias," *Acta Paediatr. Scand.*, **324**, 5–52 (1986).

48. World Health Organization, *Congenital Malformations Worldwide: A Report from the International Clearinghouse for Birth Defects Monitoring Systems*, Elsevier, Oxford, 1991.

49. L. J. Paulozzi, J. D. Erickson, and R. J. Jackson, "Hypospadias Trends in Two U.S. Surveillance Systems," *Pediatrics*, **100**, 831–834 (1997).

50. J. Toppari, J. C. Larsen, P. Christiansen, A. Giwercman, P. Grandjean, L. J. Guillette, Jr., B. Jegou, T. K. Jensen, P. Jouannet, N. Keiding, H. Leffers, J. A. McLachlan, O. Meyer, J. Muller, E. Rajpert-De Meyts, T. Scheike, R. Sharpe, J. Sumpter, and N. E. Skakkebaek, "Male Reproductive Health and Environmental Xenoestrogens," *Environ. Health Perspect.*, **104**(Suppl. 4), 741–803 (1996).

51. C. G. Scorer, "The Descent of the Testis," *Arch. Dis. Child.*, **39**, 605–609 (1964).

52. P. E. Ansell, V. Bennett, D. Bull, M. B. Jackson, L. A. Pike, M. C. Pike, C. E. D. Chilvers, N. E. Dudley, M. H. Gough, D. M. Griffiths, C. Redman, A. R. Wilkinson, A. MacFarlane, and C. A. C. Coupland, "Cryptorchidism: A Prospective Study of 7500 Consecutive Male Births: 1984–1988," *Arch. Dis. Child.*, **67**, 892–899 (1992).

53. L. J. Paulozzi, "International Trends in Rates of Hypospadias and Cryptorchidism," *Environ. Health Perspect.*, **107**, 297 (1999).

54. M. K. Nelson and R. G. Bunge, "Semen Analysis: Evidence for Changing Parameters of Male Fertility Potential," *Fertil. Steril.*, **25**, 503–507 (1974).

55. R. C. Dougherty, M. J. Whitaker, S. Y. Tang, R. Bottcher, M. Keller, and D. W. Kuehl, in J. D. McKinney, Ed., *Environmental Health Chemistry: The Chemistry of Environmental Agents as Potential Human Hazards*, Ann Arbor Science Publication 1981, Ann Arbor, MI, pp. 263–278.

56. R. M. Sharpe and N. E. Skakkebaek. "Are Oestrogens Involved in Falling Sperm Counts and Disorders of the Male Reproductive Tract?" *Lancet*, **341**, 392–395 (1993).

57. E. Carlsen, A. Giwercman, N. Keiding, and N. Skakkeback, "Evidence for Decreasing Quality of Semen during Past 50 Years," *Br. Med. J.*, **305**, 609–613 (1992).

58. A. Brake and W. Krause, "Decreasing Quality of Semen" (letter), *Br. Med. J.*, **305**, 1498 (1992).

59. G. W. Olsen, K. M. Bodner, and J. M. Ramlow, "Have Sperm Counts Been Reduced 50% in 50 Years?" *Fertil. Steril.*, **63**, 887–893 (1995).

60. A. J. Eccersley, "Declining Sperm Count: Data from Two Groups Should Not Have Been Combined in Analysis," *Br. Med. J.*, **313**, 43; discussion 44–45 (1996).

61. A. Lerchi and E. Nieschlag, "Decreasing Sperm Counts? A Critical (Re)view," *Exp. Clin. Endocrinol. Diabetes*, **104**, 301–307 (1996).

62. S. Becker and K. A. Berhane, "A Meta-analysis of 61 Sperm Count Studies Revisited," *Fertil. Steril.*, **67**, 103–108 (1997).

63. S. H. Swan, E. P. Elkin, and L. Fenster. "Have Sperm Densities Declined? A Reanalysis of Global Trend Data," *Environ. Health Perspect.*, **105**, 1228–1232 (1997).

64. P. M. Zavos and J. C. Goodpasture, "Clinical Improvements of Specific Seminal Deficiencies via Intercourse with a Seminal Collection Device versus Masturbation," *Fertil. Steril.*, **51**, 190–193 (1989).

65. R. J. Gellert and C. Wilson, "Reproductive Function in Rats Exposed Prenatally to Pesticides and Polychlorinated Byphenyls (PCBs)," *Environ. Res.*, **18**, 437–443 (1979).

66. L. E. Gray, Jr., J. Ferrell, J. Ostby, G. Rehnberg, R. Linder, R. Cooper, J. Goldman, V. Slott, and J. Laskey, "A Dose Response Analysis of Methoxychlor-Induced Alternations of Reproductive Development and Function in the Rat," *Fundam. Appl. Toxicol.*, **12**, 92–108 (1989).

67. A. M. Cummings and J. Laskey, "Effect of Methoxychlor on Ovarian Steroidogenesis: Role in Early Pregnancy Loss," *Reprod. Toxicol.*, **7**, 17–23 (1993).

68. J. C. Eldridge, D. G. Fleenor-Heyser, P. C. Extrom, L. T. Wetzel, C. B. Breckenridge, J. H. Gillis, L. G. Luempert, III, and J. T. Stevens, "Short-Term Effects of Chlorotriazines on Estrus in Female Sprague-Dawley and Fischer 344 rats," *J. Toxicol. Environ. Health*, **43**, 155–167 (1994).

69. B. E. Rani and M. K. Krishnakumari, "Prenatal Toxicity of Heptachlor in Albino Rats," *Pharmacol. Toxicol.*, **76**, 112–114 (1995).

70. M. Trapp, V. Baukloh, H. G. Bohnet, and W. Heeschen, "Pollutants in Human Follicular Fluid," *Fertil. Steril.*, **42**, 146–148 (1984).

71. W. R. Kelce, E. Monosson, M. P. Gamcsik, S. C. Laws, and L. E. Gray, Jr., "Environmental Hormone Disruptors: Evidence That Vinclozolin Developmental Toxicity Is Mediated by Antiandrogenic Metabolities," *Toxicol. Appl. Pharmacol.*, **126**, 276–285 (1994).

72. K. G. Osteen and E. Sierra-Rivera, "Does Disruption of Immune and Endocrine Systems by Environmental Toxins Contribute to Development of Endometriosis?" *Semin. Reprod. Endocrinol.*, **15**, 301–308 (1997).

73. S. E. Rier, D. C. Martin, R. E. Bowman, W. P. Dmowski, and J. L. Becker, "Endometriosis in Rhesus Monkeys *(Macaca mulatta)* Following Chronic Exposure to 2,3,7,8-Tetrachlorodibenzo-*p*-dioxin," *Fund. Appl. Toxicol.*, **21**, 433–441 (1993).

74. D. L. Arnold, E. A. Nera, R. Stapley, G. Rolnai, P. Claman, S. Hayward, H. Tryphonas, and F. Bryce, "Prevalence of Endometriosis in Rhesus *(Macaca mulatta)* Monkeys Ingesting PCB (Aroclor 1254): Review and Evaluation," *Fund. Appl. Toxicol.*, **31**, 42–55 (1996).

75. J. A. Boyd, G. C. Clark, D. K. Walmer, D. G. Patterson, L. L. Needham, and G. W. Lucier, "Endometriosis and the Environment: Biomarkers of Toxin Exposure" (abstract), Endometriosis 2000 Workshop, 15–17 May 1995, Bethesda, MD.

76. E. S. Sher, X. M. Xu, P. M. Adams, C. M. Craft, and S. A. Stein, "The Effects of Thyroid Hormone Level and Action in Developing Brain: Are These Targets for the Actions of Polychlorinated Biphenyls and Dioxins?" *Toxicol. Ind. Health*, **14**, 121–158 (1998).

77. S. P. Porterfield and L. B. Hendry, "Impact of PCBs on Thyroid Hormone Directed Brain Development," *Toxicol. Ind. Health*, **14**, 103–120 (1998).

78. C. Koopman-Esseboom, D. C. Morse, N. Weisglas-Kuperus, I. J. Lutkeschipholt, C. G. Van der Paauw, L. G. Tuinstra, A. Brouwer, and P. J. Sauer, "Effects of Dioxins and Polychlorinated Biphenyls on Thyroid Hormone Status of Pregnant Women and Their Infants," *Pediatr. Res.*, **36**, 468–473 (1994).

79. D. K. Ness, S. L. Schantz, J. Moshtaghian, and L. G. Hansen, "Effects of Perinatal Exposure to Specific PCB Congeners on Thyroid Hormone Concentrations and Thyroid Histology in the Rat," *Toxicol. Lett.*, **68**, 311–323 (1993).

80. C. H. Bastomsky, "Enhanced Thyroxine Metabolism and High Uptake Goiters in Rats after a Single Dose of 2,3,7,8-Tetrachlorodibenzo-*p*-dioxin," *Endocrinology*, **101**, 292–296 (1993).

81. D. W. Brewster, M. R. Elwell, and L. S. Birnbaum, "Toxicity and Disposition of 2,3,4,7,8-Pentachlorodibenzofuran (4PcCDF) in the Rhesus Money *(Macaca mulatta)*," *Toxicol. Appl. Pharmacol.*, **93**, 231–246 (1988).

82. H. J. Pluim, J. G. Koppe, K. Olie, J. W. Vd Slikke, J. H. Kok, T. Vulsma, D. Van Tijn, and J. J. De Vijlder, "Effects of Dioxins on Thyroid Function in Newborn Babies," *Lancet*, **339**, 1303 (1992).

83. A. Brouwer, D. C. Morse, M. C. Lans, A. G. Schuur, A. J. Murk, E. Klasson-Wehler, A. Bergman, and T. J. Visser, "Interactions of Persistent Environmental Organohalogens with the Thyroid Hormone System: Mechanisms and Possible Consequences for Animal and Human Health," *Toxicol. Ind. Health*, **14**, 59–84 (1998).

84. S. A. Berenbaum, "Early Hormonal Influences on Human Behavioral Development: Implications for Behavioral Assessment of Endocrine Disruptors," *Proceedings of the Prenatal Chemical Exposure Workshop, December 3–4, 1998*, National Center for Environmental Health, Atlanta, in press.

85. L. E. Gray, Jr., and J. Otsby, "Effects of Pesticides and Toxic Substances on Behavioral and Morphological Reproductive Development: Endocrine versus Nonendocrine Mechanisms," *Toxicol. Ind. Health*, **14**, 59–84 (1998).

86. W. J. Rogan, "PCBs and Cola Colored Babies," *Teratology*, **26**, 259, 261 (1982).

87. Y. L. Guo, G. H. Lambert, and C. C. Hsu, "Growth Abnormalities in the Population Exposed In Utero and Early Post-Natally to Polychlorinated Biphenyls and Dibenzofurans," *Environ. Health Perspect.*, **103**(Suppl. 6), 117–122 (1995).

88. W. J. Rogan, B. C. Gladen, K. L. Hung, S. L. Koong, L. Y. Shih, J. S. Taylor, Y. C. Wu, D. Yang, N. B. Ragan, and C. C. Hsu, "Congenital Poisoning by Polychlorinated Biphenyls and Their Contaminants in Taiwan," *Science*, **241**, 334–336 (1988).

89. Y. C. I. Chen, Y. L. Guo, C. C. Huse, and W. J. Rogan, "Cognitive Development of Yu-Cheng (Oil Disease) Children Prenatally Exposed to Heat-Degraded PCBs," *JAMA*, **268**, 3213–3218 (1992).

90. W. J. Rogan and B. C. Gladen, "PCBs, DDE, and Child Development at 18 and 24 Months," *Ann. Epidemiol.*, **1**, 407–413 (1991).

91. B. C. Gladen, W. J. Rogan, P. Hardy, J. Thullen, J. Tingelstad, and M. Tully, "Development after Exposure to Polychlorinated Biphenyls and Dichlorodiphenyl Dichloroethene Transplacently and through Human Milk," *J. Pediatr.*, **113**, 991–995 (1988).

92. B. C. Gladen and W. J. Rogan, "Effects of Perinatal Polychlorinated Biphenyls and Dichlorodiphenyl Dichloroethene on Later Development," *J. Pediatr.*, **119**, 58–63 (1991).

93. J. L. Jacobson, S. W. Jacobson, and H. E. B. Humphrey, "Effects of Exposures to PCBs and Related Compounds on Growth and Activity in Children," *Neurotoxicol. Teratol.*, **12**, 319–326 (1990).

94. J. L. Jacobson, S. W. Jacobson, and H. E. B. Humphrey, "Effects in Utero Exposure to Polychlorinated Biphenyls and Related Compounds on Cognitive Functioning in Young Children," *J. Pediatr.*, **116**, 38–45 (1990).

95. S. W. Jacobson, G. G. Fein, J. L. Jacobson, P. M. Schwartz, and J. K. Dowler, "The Effect of Intrauterine PCB Exposure on Visual Recognition Memory," *Child. Develop.*, **56**, 853–860 (1985).

96. J. L. Jacobson, S. W. Jacobson, R. J. Padgett, G. A. Brumitt, and R. L. Billings, "Effects of Prenatal PCB Exposure on Cognitive Processing Efficiency and Sustained Attention," *Develop. Psych.*, **8**, 297–306 (1992).

97. J. L. Jacobson and S. W. Jacobson, "Premeeting Comments for Workshop on Developmental Neurotoxic Effects Associated with Exposure to PCBs, in *EPA Workshop Report on Devel-*

opmental Neurotoxic Effects Associated with Exposure to PCBs, EPA/630/R-92/004:A75–A87, 1993.

98. J. L. Jacobson and S. W. Jacobson, "Intellectual Impairment in Children Exposed to Polychlorinated Biphenyls In Utero," *N. Engl. J. Med.*, **335**, 783–789 (1996).

99. S. L. Schantz, "Developmental Neurotoxicity of PCBs in Humans: What Do We Know and Where Do We Go from Here?" *Neurotoxicol. Teratol.*, **18**, 217–227 (1996).

100. P. Grandjean, P. Weihe, R. F. White, F. Debes, S. Araki, K. Yokoyama, K. Murata, N. Sorensen, R. Dahl, and P. J. Jorgensen, "Cognitive Deficit in 7-Year-Old Children with Prenatal Exposure to Methylmercury," *Neurotox. Teratol.*, **19**, 417–428 (1997).

101. M. Huisman, C. Koopman-Esseboom, V. Fidler, M. Hadders-Algra, C. G. van der Paauw, L. G. M. Th. Tuinstra, N. Weisglas-Kuperus, P. J. J. Sauer, B. C. L. Touwen, and E. R. Boersma, "Perinatal Exposure to Polychlorinated Biphenyls and Dioxins and Its Effect on Neonatal Neurological Development," *Early Hum. Dev.*, **41**, 111–172 (1995).

102. S. Patandin, C. J. Lanting, P. G. H. Mulder, E. R. Boersma, P. J. J. Sauer, and N. Weisglas-Kuperus, "Effects of Environmental Exposure to Polychlorinated Biphenyls and Dioxins on Cognitive Abilities in Dutch Children at 42 Months of Age," *J. Pediatr.*, **134**, 33–41 (1999).

103. M. S. Wolff, P. G. Toniolo, E. W. Lee, M. Rivera, and N. Dubine, "Blood Levels of Organochlorine Residues of Breast Cancer," *J. Natl. Cancer Inst.*, **85**, 648–652 (1993).

104. E. Dewailly, S. Dodin, R. Verreault, P. Ayotte, L. Sauve, J. Morin, and J. Brisson, "High Organochlorine Body Burden in Women with Estrogen Receptor–Positive Breast Cancer," *J. Natl. Cancer Inst.*, **86**, 232–234 (1994).

105. N. Kreiger, M. S. Wolff, R. A. Hiatt, M. Rivera, J. Vogelman, and N. Orentreich, "Breast Cancer and Serum Organochlorines: A Prospective Study Among White, Black, and Asian Women," *J. Natl. Cancer Inst.*, **86**, 589–599 (1994).

106. D. J. Hunter, S. E. Hankinson, F. Laden, G. A. Colditz, J. E. Manson, W. C. Willett, F. E. Speizer, and M. S. Wolff. "Plasma Organochlorine Levels and the Risk of Breast Cancer," *N. Engl. J. Med.*, **337**, 253–258 (1997).

107. L. Lopez-Carillo, A. Blair, M. Lopez-Cervantes, M. Cebrian, C. Rueda, R. Reyes, A. Mohar, and J. Bravo, "Dichlorodiphenyltrichloroethane Serum Levels and Breast Cancer Risk: A Case-Control Study from Mexico," *Cancer Res.*, **57**, 2738–2732 (1997).

108. P. van't Veer, I. E. Lobbezoo, J. M. Martin-Moreno, E. Guallar, J. Gomez-Aracena, A. F. M. Kardinaal, L. Kohlmeier, B. C. Martin, J. J. Strain, M. Thamm, P. van Zoonen, B. A. Baumann, J. K. Huttunen, and F. Kok, "DDT (Dicophane) and Postmenopausal Breast Cancer in Europe: Case-Control Study," *Br. Med. J.*, **31**, 81–85 (1997).

109. I. Romieu, M. Hernandez-Avila, E. Lazcano, J. P. Weber, and E. Dewailly, "Breast Cancer, Lactation History, and Serum Organochlorines," *Am. J. Epidemiol.*, in press.

110. D. Forman and H. Moller, "Testicular Cancer," *Cancer Surv.*, **19/20**, 323–341 (1994).

111. H. Adami, R. Bergstrom, M. Mohner, W. Zatonski, H. Storm, A. Ekbom, S. Tretli, L. Teppo, H. Ziegler, M. Rahu, R. Gurevicius, and A. Stengrevics, "Testicular Cancer in Nine Northern European Countries," *Int. J. Cancer*, **59**, 33–38 (1994).

112. M. A. Batata, F. C. Chu, B. S. Hilaris, W. F. Whitmore, and R. B. Golbey, "Testicular Cancer in Cryptorchids," *Cancer*, **49**, 1023–1030 (1982).

113. E. D. Clegg, J. C. Cook, R. E. Chapin, P. M. D. Foster, and G. P. Daston, "Leydig Cell Hyperplasia and Adenoma Formation: Mechanisms and Relevance to Humans," *Reprod. Toxicol.*, **11**, 107–121 (1997).

114. J. M. Robins, M. R. Cullen, B. B. Connors, and R. D. Kayne, "Depressed Thyroid Indexes Associated with Occupational Exposure to Inorganic Lead," *Arch. Intern. Med.*, **143**, 220–224 (1983).

115. C. Schumacher, C. A. Brodkin, B. Alexander, M. Cullen, P. M. Rainey, C. van Netten, E. Faustman, and H. Checkoway, "Thyroid Function in Lead Smelter Workers: Absence of Subacute or Cumulative Effects with Moderate Lead Burdens," *Int. Arch. Occup. Environ. Health*, **71**, 453–458 (1998).

116. M. Rodamilans, M. J. Martinez-Osaba, J. To-Figueras, F. Rivera Fillat, J. M. Marques, P. Perez, and J. Corbella, "Lead Toxicity on Endocrine Testicular Function in an Occupationally Exposed Population," *Hum. Toxicol.*, **7**, 125–128 (1988).

117. T. P. Ng, H. H. Goh, Y. L. Ng, H. Y. Ong, C. N. Ong, K. S. Chia, S. E. Chia, and J. Jeyaratnam, "Male Endocrine Functions in Workers with Moderate Exposure to Lead," *Br. J. Ind. Med.*, **48**, 485–491 (1991).

118. G. M. Egeland, M. H. Sweeney, M. A. Fingerhut, K. K. Wille, T. M. Schnorr, and W. E. Halperin, "Total Serum Testosterone and Gonadotropins in Workers Exposed to Dioxin," *Am. J. Epidemiol.*, **139**, 272–281 (1994).

119. E. A. Emmett, M. Maroni, J. Jefferys, J. Schmith, B. K. Levin, and A. Alvares, "Studies of Transformer Repair Workers Exposed to PCBs, II: Results of Clinical Laboratory Investigations," *Am. J. Ind. Med.*, **14**, 47–62 (1988).

120. B. Grajewski, E. A. Whelan, T. M. Schnorr, R. Mouradian, R. Alderfer, and D. K. Wild, "Evaluation of Reproductive Function among Men Occupationally Exposed to a Stilbene Derivative, I: Hormonal and Physical Status," *Am. J. Ind. Med.*, **29**, 49–57 (1996).

121. A. Mutti, P. P. Vescovi, M. Falzoe, G. Arfini, G. Valenti, and I. Franchini, "Neuroendocrine Effects of Styrene on Occupationally Exposed Workers," *Scand. J. Work Environ. Health*, **10**, 225–228 (1984).

122. E. Bergamaschi, A. Smargiassi, A. Mutti, S. Cavazzini, M. V. Vettori, R. Allinovi, I. Franchini, and D. Mergler, "Peripheral Markers of Catecholaminergic Dysfunction and Symptoms of Neurotoxicity among Styrene-Exposed Workers," *Int. Arch. Occup. Environ. Health*, **69**, 209–214 (1997).

123. M. D. Whorton, R. M. Drauss, S. Marshall, and T. H. Milby, Infertility in Male Pesticide Workers, *Lancet*, **2**, 1259–1261 (1977).

124. G. Potashnik, N. Ben-Aderet, R. Israeli, I. Yanai-Inbar, and I. Sober, "Suppressive Effect of 1,2-Dibromochloropropane on Human Spermatogenesis," *Fertil. Steril.*, **30**, 444 (1978).

125. G. Potashnik and A. Porath, "Dibromochloropropane (DBCP): A 17-Year Reassessment of Testicular Function and Reproductive Performance," *J. Occup. Environ. Med.*, **37**, 1287–1292 (1995).

126. National Institute for Occupational Safety and Health, *National Occupational Exposure Survey, 1981–1983*, unpublished electronic data as of 9/19/98, Cincinnati.

127. National Institute for Occupational Safety and Health, *National Occupational Exposure Survey*: Vol. 1, *Survey Manual*, National Institute for Occupational Safety and Health, Cincinnati, 1988.

128. U.S. Environmental Protection Agency, OPPT web site, available at ⟨http://www.epa.gov/opptintr/chemlist/hpv.htm⟩ as of 8/12/99.

129. A. Aspelin, *Pesticide Industry Sales and Usage: 1992 and 1993 Market Estimates*, Gov Pub # 733-K-94-001, U.S. Environmental Protection Agency, Washington, D.C., June 1994.

CHAPTER THIRTEEN

Atypical Human Responses to Low-Level Environmental Contaminants: The Problem of Multiple Chemical Sensitivities

Mark R. Cullen, MD

1 INTRODUCTION

Fundamental shifts in the nature of work, changing composition of the workforce, and changes in societal perceptions about health have brought to the practice of occupational medicine some new areas of focus. Where once one could be content with a clear understanding of the impact of chemical and physical hazards at the doses which caused clinically apparent illness or injury, primarily in manufacturing, mining, agriculture and construction, there has been a rapid increase in concern about the impacts of many of these same hazards as they arise in the nonindustrial setting, impacting far larger numbers of workers who often bring very different experiences and expectations regarding health in the workplace. Even in the industrial settings, the introduction of effective controls—which have markedly reduced, at least in developed countries, the well understood classic occupational diseases and risks—have had as an unexpected consequence a refocusing of attention from the devastating consequences of exposures above established Threshhold Limit Values (TLV's) or occupational exposure limits, onto the possibility of effects occurring at lower levels, confidently deemed safe by comparison to the "old" context of practice when the major hazards were still extant and uncontrolled.

Patty's Industrial Hygiene, Fifth Edition, Volume 1. Edited by Robert L. Harris.
ISBN 0-471-29756-9

Consider the classic model human dose–response curve for a direct acting cytotoxic chemical such as an irritant, solvent or pesticide depicted in Figure 13.1 (1). All theory from toxicologic research and reports from clinical practice of occupational medicine suggests that for each hazard there exists some "threshold" for each individual beneath which there is no demonstrable clinical effect, followed by responses of rising severity as the dose increases (Fig. 13.1**a**). From a population perspective (Fig. 13.1**b**) a composite threshold could be assumed (the level below which some arbitrarily small fraction of the exposed population does not react), and a curve of populational severity depicted, much like the classic toxicological dose–response curves in animal models. These thresholds in practice have been derived from syntheses of data from animal experiments, various short-term experimental challenges in human volunteers, reports of human clinical experience, and epidemiologic studies. Examples of this approach have been well summarized over the years in the ACGIH Documentation of TLVs publications (2). For industrial hygiene practice, such thresholds have been the basis for assigning targets for control, with some additional safety factor as a cushion, often supplemented by biologic monitoring and/or clinical testing where exposures could not uniformly be held low, or where uncertainty has existed about some health effects.

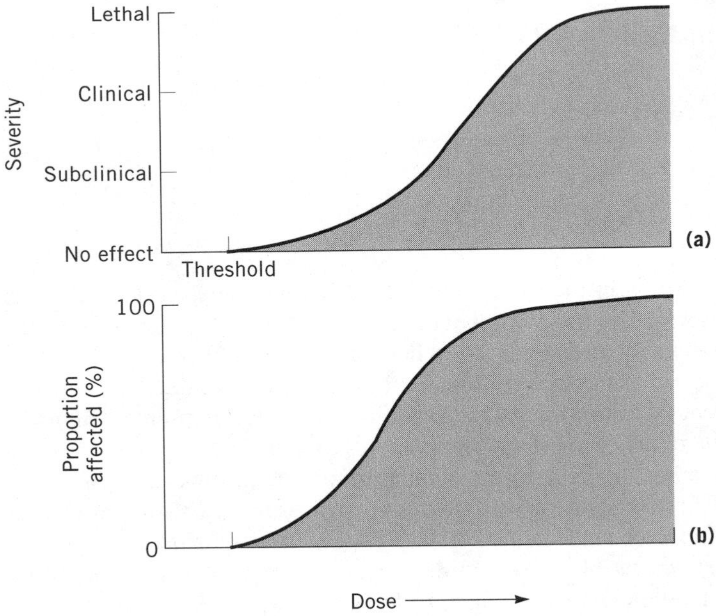

Figure 13.1. Theoretical dose–response curve for a cytotoxic chemical, demonstrating the hypothetical response for an individual (**a**), and for a population of individuals (**b**). The concept of "threshold" is different in the two cases: in the former, it represents the level beneath which there are no demonstrable effects; for the latter, a construct value beneath which some arbitrarily large proportion of the population in free of effects, or beneath which effects cannot be statistically distinguished. (Figure borrowed with permission from Ref 1.)

Although this paradigm has served well, and is probably defensible from a public health perspective when most workers were exposed to hazards in industrial, agricultural, mining or construction settings, the situation changes when the pattern of exposures changes, as one can appreciate from the demographic shift in work environments depicted in Figure 13.2. Extrapolating to the whole workforce one can envision that millions of people now have exposures well below any TLVs, with decreasingly few near or above. Whatever assumptions may have sufficed historically for defining the threshold (95% protective of even trivial effects, perhaps even 99% protective), it is a potentially huge population with exposures below these levels is now recognized. The scientific basis for the selection of these levels and their application to individual risk, once *relatively* unimportant, now becomes a crucial aspect of scientific inquiry, for which neither toxicology nor epidemiologic tools which have been thus far available entirely suffice.

There are a variety of reasons why existing estimates of thresholds may be inadequate for predicting what could be anticipated when large populations are exposed to various hazards at low levels:

1. End-points used for defining the Lowest and No Observable Effect Levels (NOELs and LOELs) in toxicologic studies have typically been gross clinical effects which are readily confirmable pathologically; subjective responses in human may occur at far lower levels.
2. Human experimental data, i.e., exposures to volunteers, may fail to give adequate results because of the testing conditions: the volunteers, usually healthy young stu-

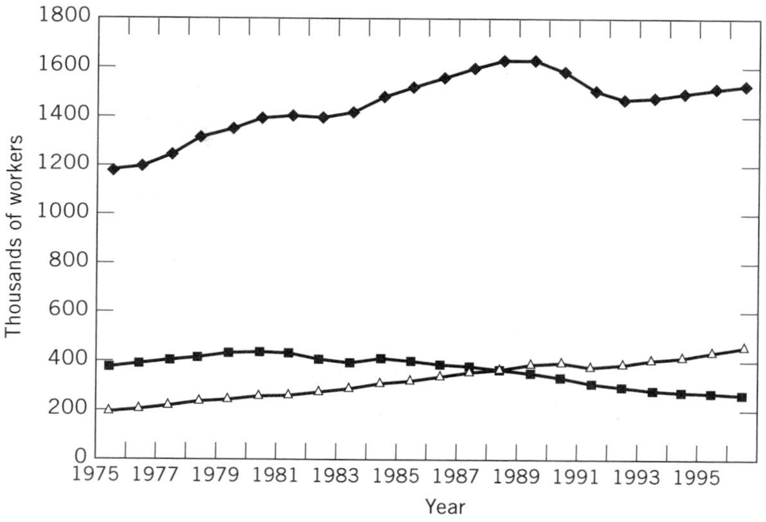

Figure 13.2. Employment curves for the State of Connecticut between 1975 and 1997, categorized by employment sectors. The dramatic reversal between manufacturing and service sector employment is illustrated. —◆— Total; —■— manufacturing; —△— service.

dents, may be different from those exposed; the duration of experimental exposure may be very short compared to the work day; the environmental background of the experiments may differ from the "mixed" environments of many work-places; and the size of chosen volunteers populations may be inadequate to include individuals with idiosyncratic responses. In fact such unusual subjects have often been intentionally excluded from such experimental studies for safety reasons.

3. Epidemiologic studies, though often designed to overcome the above shortcomings, must often rely on highly subjective end-points such as responses to symptom surveys, which have been heavily criticized as potentially biased and lacking in "objectivity", a serious scientific catch-22.

The conundrum is further amplified by remarkable changes in our social environment. The workers at potential risk for exposure are no longer the "rugged" men who comprised much of the workforce of previous eras, but are now men and women who have never worked in what most occupational health professionals would call a dangerous workplace, and have no particular reference to historic risks by which to be assuaged or reassured that things are now safe. People are far more medically aware and sophisticated as well, and no longer consider a little mucosal irritation or muskulo-skeletal pain as "normal". These "effects" are likely to be perceived, with or without toxicologic substantiation, as evidences of clinical illness or injury, possibly even serious, especially if they occur more noticeably in the workplace than without. Such complaints, which in a former world may never have received medical attention, are now the basis for frequent visits, inquiry, and not rarely request for the professional services of occupational health and safety specialists.

The burgeoning experience with these issues has become the basis for some important new scientific inquiry in the domain of indoor air quality, and has been extensively reviewed in this work (see Chapt. 69). Without intending to discourage very careful review of that chapter, as an almost necessary scientific preface to this one, the following key points are worthy of summary:

1. Workers in large numbers are developing clinical complaints in the nonindustrial environments is now indisputable.
2. Exhaustive characterization of environments where such complaints have arisen has demonstrated levels of VOCs, bioaerosols and other potential hazards, but at levels at least 1–2 orders of magnitude below established or putative TLVs for these materials.
3. The actual environmental "cause" of most complaints in these environments remains obscure, and may possibly be related to some additive or synergistic effect among hazards. Moreover, there is substantial host variability in response to these environments, which is also not well understood or predictable. But these major gaps in scientific knowledge notwithstanding, there is abundant experience from "sick" buildings that symptoms typically resolve when sources of hazards are minimized and ventilation is improved, strongly implicating a causal role for the physical/chemical work environment.

As if this were not perplexing enough, some individuals in these buildings as well as in other work-place and nonworkplace settings, are apparently responding even more unpredictably. Unlike workers in the outbreaks described by Morey (Chapt. 69), such individuals may "react" to chemicals in widely diverse settings, their symptoms may be more extensive and profound, and they do not necessarily respond to control measures which are typically effective despite our imperfect understanding of why. It is to this increasingly recognized problem, often referred to as Multiple Chemical Sensitivities (MCS), that this chapter is devoted.

Since this chapter is premised on the assessment which occurs as a result of the complaints of individual or groups of workers in setting where complaints would not traditionally be anticipated, I start with three case examples to illustrate the kinds of issues which arise. This is followed by a detailed delineation of the differential possibilities which must be considered when such cases arise. Current theories about the pathogenesis, and our present understanding of the clinical course and epidemiology of MCS will be followed by sections on management from the hygiene and medical perspectives.

2 CASE EXAMPLES

2.1 Case 1

A 36 year old library worker enjoyed good health until the onset of eye, nose and throat irritation, and recurrent headache occurring every day at work shortly after a renovation on her building had begun. She and many co-workers complained primarily of dust and paint-fume exposures, which were initially poorly controlled. After several weeks of discussion, the employer succeeded in establishing temporary ventilation for the work area; construction activities were deferred until evenings. Almost all of the patient's co-workers improved dramatically after these changes were instituted. The patient, however, felt no better and began experiencing recurrence of symptoms in her car, at various stores, and whenever she was around anything scented, but still suffering particularly in the office. She believed she was reacting to the small residual levels of construction material, but temporary transfer to another part of the library brought no relief. New symptoms, including difficulty breathing, muscle and joint aches and "confusion" now occurred both at work and at home, triggered by an increasing list of offensive odors, irritants and products. Efforts to clean her house of such materials, as well as a trial leave of absence from work resulted in minimal improvement.

On clinical evaluation the patient appeared well, without physical findings. Laboratory tests, including work-up for respiratory and neurologic lesions, were unrevealing, and consultants in pulmonary medicine, rheumatology, and neurology were not helpful, as were the various inhalers, pain relievers and migraine therapies these physicians prescribed. Because of the disparity between complaints and findings, the patient was referred to a psychiatrist who confirmed some depressive features, but could not explain the patient's complaints. A trial of antidepressants was not tolerated by the patient, who discontinued the drug after three days.

Finally, frustrated by what she perceived as unsympathetic physicians and employer, the patient took advice she got from the internet and sought an "environmental" physician who advised total avoidance of all chemical exposures (including her job) and a variety of nontraditional remedies based on results of blood and hair tests which purported to demonstrate organic chemicals and heavy metal "poisons," as well as immunologic "reactions" to a range of widely found chemicals such as formaldehyde. She remains highly symptomatic.

2.2 Case 2

A 52 year old machinist complained to the occupational health service about increasing headache and dizziness during degreasing operations in his tool shop. Symptoms occurred at work daily, persisting several hours after his shift before abating. Although he had been a tool and dye maker for over 25 years, symptoms had begun insidiously about three months earlier, and were getting progressively worse.

Evaluation of his work area revealed the presence of 1,1,1 trichloroethane as the major degreasing agent. Other exposures included semisynthetic and synthetic coolants, abrasives and metal dust, although most operations in the shop were "wet". Short and full shift samples revealed that exposures to solvent were generally well controlled, with even peak short term exposures less than 100 ppm, and TWA exposures less than 10 ppm. No other workers offered complaints on direct questioning.

Despite the outcome of the exposure assessment, the patient was given a job restriction precluding any degreasing activities. However, his complaints continued to intensify, and he began to notice that symptoms recurred away from the shop as well, especially after exposures to gasoline, diesel exhaust, and a variety of household products which were petroleum based. Central nervous system complaints persisted, but he also began to notice breathlessness and flushing after each contact with offending materials. He discontinued all consumption of wine or beer, and the use of any over-the-counter medications to which he became intolerant. His physician was unable to detect any abnormality on physical examination, EKG, routine blood chemistries, and a chest x-ray. He was referred to a psychiatrist who found no evidence of psychopathology and suspected a role for continued exposures to solvents at work. Based on this and his increasing symptomatology, he resigned from his job.

In the months after leaving work, his condition stabilized but persisted, with the patient developing symptoms in a wide range of circumstances when exposed to organic chemicals. He elected to move to a remote part of the country to "escape" these exposures. Although he enjoyed some periods free of symptoms in this isolated setting, all symptoms recurred severely with routine activities such as driving, shopping, visiting friends, etc.

2.3 Case 3

A 45-year old university departmental administrator was well until she developed severe upper and lower respiratory discomfort after an irritant gas was released in her office without warning as part of a "test" of the HVAC in her office building. Because she had responsibility for assuring the safe exit of other members of her department, she was the last person to leave the building, having spent about 20 minutes inhaling an aldehyde containing particulate irritant. Coughing and nauseated, she went home. Symptoms of cough and sore throat improved over 48 hours, but recurred when she reentered the office for return to work. Others who had been exposed were working comfortably, but the patient experienced recrudescence of cough and some chest discomfort, and left after a few minutes. A medical evaluation by her physician, including lung function tests and x-ray, were unrevealing. Nonetheless, upon return to work the following day, symptoms returned. More disturbingly, over the next several days the patient began to experience similar "flares" when exposed to tobacco smoke, vehicle exhaust or wood-burning stoves. Moreover, she began to experience more protracted effects from these exposures, as well as severe fatigue and migratory pain of her joints and muscles.

The patient underwent extensive medical evaluation for her complaints, but no cause was found. Attempts at treatment for suspected anxiety and depression were unsuccessful; the patient did not tolerate any of the prescribed medications. After weeks of attempts to return to her work environment (to which all co-workers had returned uneventfully) she decided to take an indefinite leave, but her overall condition did not improve. She remains largely house-confined.

3 CONSIDERATIONS WHEN WORKERS COMPLAIN IN APPARENTLY LOW-LEVEL EXPOSURE SETTINGS

Cases such as those listed above are coming with increasing frequency to the attention of managers, industrial hygiene, and safety personnel, and occupational health providers. Each of these cases, and the many variations of them, raise a host of differential possibilities which challenge a limited understanding of potential effects of chemical and biohazards at low level and the traditional assumptions of industrial health, namely that clinical effects are unlikely or impossible at dose levels far below established TLVs. The following conceptual possibilities are raised in each instance.

1. Complaints may be entirely unrelated to physical factors in the work environment, i.e., symptoms may reflect psychiatric disease such as claustrophobia or depression, or underlying medical disease, such as a cancer. Early medical evaluation by clinicians skilled both at considerations of work factors and alternative diagnostic explanations are essential for excluding this possibility.
2. Complaints may represent classic sensitization to an agent in the work, and possibly in the nonwork environment. Many chemical and biologic agents can trigger asthma, dermatitis, allergic alveolitis (an inflammatory lung disease), upper respiratory symptoms and often substantial accompanying systemic complaints such as fatigue. Some of these are relatively ubiquitous contaminants, such as latex protein, animal danders, bacterial and fungal proteins, and small molecules such as the diisocyanates. Since the occurrence of profound effects related to even trivial exposure in highly sensitized individuals is well recognized, the nature of both exposures and physiologic responses in the worker must be characterized to address this possibility adequately.
3. The occurrence of complaints from small exposures, developing after a more substantive single or recurrent exposure also raises the concern that initial exposures

may have been sufficient to cause some persistent injury, and that subsequent responses reflect the altered dose response relationship often ascribed to pre-existing host disease. For example, it is well known that people with chronic lung disease or asthma may develop symptoms from exposures to irritants such as SO_2 or ozone at levels far below those seen in healthy subjects (3–9). There is speculation, though little comparable physiologic evidence, that certain underlying disorders of central or peripheral nerves, such as that seen in diabetes or nutritional deficiencies, may predispose to neurotoxic effects at lower levels than are typically seen (10–12). So the question is raised whether the initial events may have caused an injury such as the reactive airway dysfunction syndrome (RADS) seen after acute airway injuries, or a CNS injury such as acute solvent or pesticide poisoning. Such a possibility requires adequate investigation of the circumstances under which symptoms first occurred (both the exposure and clinical aspects if they can be reconstructed) and the nature of the exposure-response situation ongoing.

4. Complaints may reflect persistent nonspecific building related illness (13), in which symptoms persist because the underlying problem has not been solved or the worker is not adequately removed from it. This can often be evaluated by history and the responses to change of co-workers when problems occur within an epidemic setting.
5. As most often happens, however, none of these explanations is entirely satisfactory, and consideration of MCS is raised. Although there remains vituperative debate about the underlying basis and nature of the problem, including concern about possible connotations of its most widely used name, Multiple Chemical Sensitivities is now generally accepted as a useful designation to describe individuals who develop the following (14–19):
   a. Symptoms occur after an occupational or environmental disorder such as an inhalation or toxic exposure. This precipitating event may be a single episode, as in third case above, or recurrent, as in the second case. Often the initial event or events are mild and may merge without clear demarcation into the syndrome which follows, as in the first case.
   b. Symptoms resembling those of the preceding exposure begin to occur after exposures to surprisingly lower levels of various materials, including chemicals, perfumes and other common work and household products, especially products that have a strong odor or are irritating.
   c. Symptoms appear referable to many organ-systems. Central nervous system problems, such as fatigue, confusion and headache, occur in almost every case.
   d. Complaints of chronic symptoms, such as fatigue, cognitive problems, gastrointestinal and musculoskeletal disturbances frequently occur, and render the patient persistently "ill," even in the absence of reactions to specific exposures. These more persistent symptoms may even predominate over acute "reactions" to chemicals in some cases.
   e. Objective impairment of the organs which would explain the pattern or intensity of complaints is typically absent.
   f. No other diagnostic consideration easily explains the range of responses or symptoms.

4 A BRIEF HISTORY OF MULTIPLE CHEMICAL SENSITIVITIES

The observation of individual reactions to chemicals and odors at ostensibly "non-toxic" levels is not new. Theron Randolph, an allergist, first published on this subject in 1952 (20), engaging a decade long debate with others in the allergy/immunology community who ultimately ostracized him. They rejected Randolph's expansive notion of allergy, which included diverse reactions to food additives, air-borne chemicals, and other contaminants of nature in favor of a narrower view of allergy restricted to events explicable by recently discovered immunologic mechanisms involving antibodies, lymphocytes, and related inflammatory responses. Randolph in turn founded the Society for Human Ecology, which over two decades attracted a range of practitioners, often known as clinical ecologists, who based patient care on a range of unconventional theories and therapies aimed at mitigating the effects of environmental toxins (21). Mainstream medical societies and specialty organizations publicly criticized these theories and practices (22–23). In 1985 the Society changed its name to the American Academy of Environmental Medicine, promulgating the following model as their underlying perspective (21):

> Environmentally triggered Illnesses result form a disruption of homeostasis by environmental stressors. This disruption may result from a wide range of possible exposures, ranging from a severe acute exposure to a single stressor to cumulative relatively low-grade exposures to many stressors over time. The disruption can effect any part of the body via dysfunctioning of any number of the body's biologic mechanisms and systems. The ongoing manifestations of Environmentally Triggered Illnesses are shaped by the nature of the stressors and the timing of exposures to them, by the biochemical individuality of the patient, and by the dynamic interactions over time resulting from various governing principles such as total load, the level of adaptation, the bipolarity of responses, the spreading phenomenon, the switch phenomenon and individual susceptibility (biochemical individuality).

In part because these theories were not well known to most physicians engaged in the practice of occupational medicine, and in part because many of these ideas lacked evidence based on usual scientific or medical practice, the appearance of patients such as the cases presented above in occupational clinics generated some considerable discussion. Starting in the mid-1980s this dialogue emerged for the first time in the mainstream medical and occupational health literature, including attempts both to describe and define the problem (24). It was at this time that the current designation Multiple Chemical Sensitivities was introduced for the first time, primarily for its phenomenologic utility (18). During the late 1980s and the 1990s numerous studies aimed at evaluating the theories and clinical practices of the clinical ecologists have been published (25–37). Nonetheless, controversy has persisted. Debate has centered around: (*1*) Whether such a discrete clinical syndrome actually exists; (*2*) How to define it (assuming it does); (*3*) What is its pathogenesis, in particular, how does it relate to the chemical exposures which appear to trigger symptoms; (*4*) What factors predispose to developing it; and (*5*) How should it be managed.

Despite this controversy, or perhaps because of it, there has been substantial government and legal activity involving MCS in North America, including development of a standard for MCS under social security disability (38); federal, state, and provincial government sponsored symposia and reports (19, 39–40); and targeted research funding initiatives. The emergence of poorly defined MCS-like complaints among a disturbingly high number of combatants and other participants in the Persian Gulf Conflict of 1990–1991 substantially increased focus and research efforts (41–45), with the addition of research programs under the auspices of the Department of Veterans Affairs and the Department of Defense. In 1997 an Interagency Workgroup on MCS was commissioned by eight federal agencies most directly impacted. Their report summarizes much of the relevant literature and governmental response (46). Internationally, the International Program on Chemical Safety, a body associated with WHO, has also sponsored a workshop (47), and similar initiatives appear to be emerging in Europe, Australia, and New Zealand.

4.1 Definitions of MCS

As has been noted there remains no consensus on a definition of MCS, reflecting in part the broader debate over the nature of the clinical entity, and MCS has not yet been included in the International Classification of Disease, the most widely applied system for medical nosology. Not surprisingly, the absence of a uniform definition or criteria for application of the term for clinical or epidemiologic purposes has, in turn, hampered research, as reports of findings in groups of patients and in populations are not easily comparable. Table 13.1 summarizes these definitions since 1985. Table 13.2 provides a side-by-side comparison of these definitions. Virtually all of the extant definitions include these common elements:

- Symptoms involve more than just one organ system, and invariably include some central nervous system complaints
- Symptoms are triggered by exposures to chemically dissimilar materials, and at doses which are usually considered at least an order of magnitude or more below typical threshold for human effects
- Symptoms are recurrent on a chronic basis for at least three months
- No laboratory test of physiologic function can explain the basis of the symptoms

Except for the last, all of these are clearly subjective issues (i.e., based on the report of the affected individual) and this has been the source of some skepticism. Even the final criterion, the failure to identify a physiologic basis is merely an expression of the *absence* of an established objective test for the disorder. In and of itself, of course, the absence of a specific objective sign on physical examination or laboratory testing does not preclude a biologic basis for disease, nor the usefulness of MCS as a diagnostic entity. Many clinical conditions were initially recognized in this way, and some, such as classic migraine or schizophrenia, both likely genetic disorders, remain without a "test". However, given the often contentious and litigious nature of occupational health practice, there is strong pressure to minimize reliance on self report. In order to "narrow" the definition and render it more discriminating, several authors including this author have proposed additional criteria, namely:

Table 13.1. Definition of MCS since 1985[a]

1985, Ontario Ministry of Health, ad hoc Committee (39)

More than three months duration.
Multisystem disorder.
Intolerance to foods, chemicals, environmental agents at levels generally tolerated by majority.
Symptoms diminish with avoidance; recur with exposure.

1987, Cullen (18)

Multiple Chemical Sensitivities is an acquired disorder characterized by recurrent symptoms, referable to multiple organ systems, occurring in response to demonstrable exposure to many chemically unrelated compounds at doses far below those established in the general population to cause harmful effects. No single widely accepted test of physiologic function can be shown to correlate with symptoms.

1991, Ashford and Miller (48)

The patient with multiple chemical sensitivities can be discovered by removal from the suspected offending agents and by rechallenge, after an appropriate interval, under strictly controlled environmental conditions. Causality is inferred by the clearing of symptoms with removal from the offending environment and recurrence of symptoms with specific challenge.

1992, American Academy of Environmental Medicine (21)

Ecologic illness is a chronic multisystem disorder, usually polysymptomatic, caused by adverse reactions to environmental incitants, modified by individual susceptibility and specific adaptation. The incitants are present in air, water, food, drugs, and our habitat.

1992, National Research Council (NRC), Workshop on Multiple Chemical Sensitivities, Working Group on Research Protocol for Clinical Evaluation (40)

Symptoms or signs related to chemical exposures at levels tolerated by the population at large that are distinct from such well recognized hypersensitivity phenomena IgE-mediated immediate hypersensitivity reactions, contact dermatitis, and hypersensitivity pneumonitis.
Sensitivity may be expressed as symptoms and signs in one or more organ systems.
Symptoms and signs wax and wane with exposures.
It is not necessary to identify a chemical exposure associated with the onset of the condition.
Preexistent or concurrent conditions (e.g., asthma, arthritis, somatization disorder, or depression) should not exclude patients from consideration.

1992, Association of Occupational and Environmental Clinics: Workshop on Multiple Chemical Sensitivity, Working Group on Characterizing Patients (19)

A change in health status identified by the patient.
Symptoms triggered regularly by multiple stimuli.
Symptoms experienced for at least six months.
A defined set of symptoms reported by patients.
Symptoms that occur in three or more organ systems.
Exclusion of patients with other medical conditions (psychiatric conditions are not considered exclusionary).

Table 13.1. (continued)

1993, Nethercott et al. (49)
The symptoms are reproducible with exposure. The condition is chronic. Low-level exposure results in manifestations of syndrome. Symptoms improve or resolve when incitants are removed. Responses occur to multiple, chemically unrelated substances.
1996, International Program on Chemical Safety (IPCS) (47)
An acquired disorder with multiple recurrent symptoms; associated with diverse environmental factors tolerated by the majority of people; not explained by any known medical or psychiatric disorders.

[a]Ref. 46.

- Symptoms were acquired i.e., that some time period can be defined before which the individual was documentably free of symptoms, and after which persistent symptoms were evident
- Onset coincided with an environmental or occupational health problem related to an exposure of substantial dose or setting (such as nonspecific building related illness) in which there *was* some objective evidence of toxicity or injury
- Other possible explanations, such as major depression, panic disorder, obsessive-compulsive disease or other illness, mental or physical, are excluded before a diagnosis of MCS can be made. This does not mean an MCS patient cannot be anxious or depressed, as many people with disabling illness are, but that such illness is not of magnitude and severity as to explain the clinical picture.

Although this narrower definition has been of some substantial value for research, and holds appeal to those skeptics who may view the broader definitions as too inclusive, in practice it does leave a substantial pool of individuals who have MCS-like complaints, but meet no simple diagnostic category. Practitioners in such cases have to apply judgment in management and medico-legal decision making (see below).

5 THEORIES REGARDING PATHOGENESIS OF MCS

Although almost as many theories about the pathogenesis of MCS have been promulgated as there have been investigators and commentators, the central question about the cause of MCS has focused on whether MCS is a disease of altered biology (i.e., some form of toxicologic effect of chemical exposures), or whether MCS is fundamentally a psychological or behavioral reaction to chemical exposures. Notably, based on recent advances in the underlying biology of neurobehavioral phenomena, some theory is evolving incorporating both perspectives. The following is a summary of the major theories which have

Table 13.2. Elements Common to Proposed Definitions for MCS since 1985[a]

Element	Ontario, 1985 (Ref. 39)	Cullen, 1987 (Ref. 18)	Ashford and Miller, 1991 (Ref. 48)	AAEM, 1992 (Ref. 21)	NRC, 1992 (Ref. 40)	AOEC, 1992 (Ref. 19)	Nethercott et al., 1993 (Ref. 49)	IPCS, 1996 (Ref. 47)
Multiple environmental causes	X	X	X	X		X	X	X
Time (chronicity)	X	X		X	X	X	X	
Multiorgan symptoms	X	X		X	X	X		X
Symptoms at very low levels	X	X			X	X	X	X
Symptoms affected by presence/absence of exposure	X		X		X		X	
Exclusion of other etiologies		X			X	X		X
Symptoms acquired		X				X		X
Demonstrable exposure		X						

[a]Ref. 46.

been presented, the scientific evidence which supports them and the limitations of this evidence:

5.1 Biologic Theories

5.1.1 *The Immune Hypothesis*

Starting from the days of Randolph the fundamental understanding of MCS was that it represented an atypical form of allergy, based on altered immune reactivity (20, 22, 48, 50–51). The original basis for detecting such "allergic" responses was the testing response of patients to various dermal and sublingual "challenges" with chemicals, not conceptually different from the skin testing used in modern allergy practice, but without rigorously established protocols and testing criteria. As increasing knowledge of human immunology has unfolded, newer testing modalities, including various measures of antibodies and other "markers" in blood have been proposed as evidence of immune reaction to the exogenous materials which were deemed responsible for symptoms (52–53). The AIDs epidemic also brought increasing attention to the possibility of immuno-deficiencies in addition to altered or heightened reactivity of one or another arm of the immune system. The central constructs, outlined in the statement of the Society for Human Ecology (21), however, have not changed, and are premised on both the immediate and the cumulative effects of exogenous chemicals on the function of the several limbs of the immune system. These theories have placed heavy emphasis on the large interindividual differences in innate immune function, the interrelationship between immune function and nutritional status and the possible role of subclinical infections (such as candidiasis) in modulating effects. Moreover, with the concepts of tolerance and adaptation, the theory attempts to explain different responses even in the same host to the same exposure at different times.

Although the immune hypothesis has the attractive feature of explaining the unmistakable interindividual differences in responses of workers to low level chemical exposures, and offers some explanations for the temporal variability within a single patient, there appears little else to recommend it. A detailed review of the immunologic data is beyond the scope of this chapter but, suffice it to say that currently available evidence does not support an immunologic basis for MCS. Many of the abnormalities of specific limbs of the immune system which had been reported earlier have not withstood scientific scrutiny and retesting (36, 54–55). In general, whenever groups of MCS subjects have been rigorously tested using the methods generally accepted in the medical community for immunologic function, tests have been normal (36, 54–55). Moreover, the concept of extraordinary interindividual differences in nutritional demand to support immune function has not received substantial support in immunologic research, nor has evidence emerged that infection with candida or other commensal organisms play an important role in MCS responses.

5.1.2 *Neurointoxication in MCS.*

Since many of the exposures which occur at the outset in MCS (e.g., Case 2 above) are primary neurointoxicants, and the most prominent symptomatology is often referable to the CNS, the possibility that MCS represents a response to an initial brain injury has been attractive heuristically. Evidence for this theory has been drawn from clinical series, reports of abnormal performance on neuropsychological tests and results of findings on several types of dynamic brain scans, such as the SPECT and PET scan (25, 56).

However, closer scrutiny of these data reveals a less coherent picture (57). In fact, doses of exposure to neurointoxicants have generally been far below those associated with persistent CNS effects, and the theory provides no basis for understanding why such effects should persist in the occasional subject who develops MCS. Moreover, many, if not most MCS patients develop symptoms after an initial exposure which is *not* a neurointoxicant, further complicating the theory. Carefully and blindly performed neuropsychological testing batteries on groups of MCS subjects have failed to confirm the earlier observations of profound and persistent deficits; indeed there is striking evidence that subjects perform at a high level, except perhaps when symptomatic from a recent exposure coincident with the testing (28,31,36,58). The brain scan evidence, though provocative, remains unconfirmed. In summary, although the concept of a global CNS injury akin to that seen after longstanding or overwhelming exposure to neurotoxic substances is attractive, support both theoretical and empirical is lacking.

5.1.3 Upper Respiratory Tract/Neurogenic Inflammation in MCS

This novel theory has been built on the observation of increased nasal resistance, reduced thresholds in MCS patients for nasal irritation responses and evidence of upper airway inflammation in small groups of MCS patients (59–60). Coupled with the widely reported importance of upper airway irritative symptoms in MCS patients and the importance of irritants and odors as precipitants of symptoms, the idea has emerged that MCS is caused by an acquired mucosal injury in the nose, coupled with "neurogenic" inflammation-increased responsiveness of sensory fibers to stimulate neural mediators, such as neuropeptides, and ultimately "switching" that proponents suggest is a physiologic phenomenon by which a trigger at one site could result in response, inflammatory or neural, at another.

The evidence for this theory remains fragmentary. Meggs and colleagues have demonstrated some evidence of nasal pathology involving both nerves and epithelial cells (61), though the specificity of these changes to MCS in blinded populations remains unproved. This group and colleagues involved in the role of the upper respiratory tract have delineated a series of specific testable research hypotheses (62) offering some prospect for further insight within a few years.

5.2 Psychological Theories

5.2.1 MCS as a Somatoform Illness

The failure to demonstrate physiologic alternations to explain the symptoms of MCS has prompted many to interpret these symptoms as of psychological origin, as are typically seen in panic disorder, phobias, generalized anxiety, and depression (28, 35–37, 63–67). Such physical expressions of psychiatric illness are not new, and several authors have attempted to explain MCS in terms of a new "metaphor" for expression of an old clinical problem once called hysteria or conversion disorder (68–69).

The major evidence for this theory comes from two sources. The first is the absence to find a cogent physical explanation for symptoms, the *sine qua non* of such considerations. The complimentary evidence is the high rate of anxiety and depression recognized in patients with MCS, ranging from 25–75% of the patients in most series who have been studied using either a standardized battery of tests, or standard in-depth psychiatric interviews (30, 36–37). This view has been further bolstered by epidemiologic evidence suggesting an overlap between the major risk groups for MCS and disorders of anxiety and depression, with a heavy concentration among women of high socioeconomic status in their 30s and 40s (see epidemiology of MCS, Section 6).

Unfortunately there are many gaps in knowledge which limit the certainty of this interpretation. For one, many of the case series are heavily weighted by subjects who have come to attention because of litigation, raising the question of how representative they are of MCS patients as a whole. For the most part, case definitions have relied on self-report of MCS. A second important problem is the lack of evidence that pharmacologic treatment for anxiety or depression has any demonstrable effect on the course of symptoms, in stark contrast to the impressive response of most anxious or depressed patients to available medication. Finally, there remains the undisputed evidence in each published series of a sizable proportion of patients with no prior or subsequent evidence of any discernible psychiatric pathology other than the unexplained MCS symptoms themselves. In fact, critics of this hypothesis point out that anxiety and depression are extremely common among all patients with chronic medical disorders, and perhaps the somewhat higher proportion among MCS patients could be explained by the troublesome nature of an illness which renders these individuals hostage in their own environment, and often equally hostage to skeptical employers and professionals who question the legitimacy of their impairments (70–71).

5.2.2 MCS as an Expression of Early Childhood Trauma

Selner and colleagues have published their experience with in-depth psychological evaluation and treatment of several series of MCS patients (35, 72). In their experience a very high proportion suffer from child abuse, often sexual, and they interpret the pattern of symptoms and environmental isolation as a later life response to this trauma. Importantly, they have provided encouraging evidence of therapeutic success with some patients. Although this data is anecdotal, it is unique since few other reports of cure are well described. These data are limited at present by absence of confirmation from other investigators, and a shared sense of skepticism among many who treat large numbers of MCS patients that such a mechanism could explain the unusual and highly characteristic pattern of the disorder.

5.2.3 Behavioral Models

The resemblance between MCS and phobic responses has prompted a behavioral interpretation of MCS (64–65, 73). Under this theory, subjects develop physiologic responses to an "unconditioned" stimulus (i.e., the initial adverse exposure sufficient to cause such a response) but subsequently respond to varying "conditioned" stimulae which were associated with it (e.g., smells, irritation) following classic pavlovian mechanism. The behavioral theory has received support from the extraordinary importance that odors play, which suggests the possibility of a conditioned response to smells rather than the physiologic chemical effects *per se*; some animal models are reportedly being developed. Moreover, several authors have noted evidence of an adverse response to placebos in blinded challenges, suggesting an *a priori* expectation conditioned on previous adverse reactions associated with smells. This "nocebo" effect has been tauted by some as an explanation for the occurrence of symptoms in some individuals even in circumstances where there is no definable exposure occurring (74). Importantly, there have also been some anecdotal reports of successful treatment of MCS with behavioral modification techniques, though these remain limited (75–76). Overall, although this theory is attractive heuristically, it remains untested in any sizable case series of well characterized MCS patients.

5.3 Integrated Bio–Psychological Theory

5.3.1 *The Olfactory-Limbic Model*

Bell, Miller and others have proposed a theoretical framework which attempts to bridge the evidence between biologic responses to exogenous chemical exposures and the psychological nature of symptom patterns (77). The basic concept, borrowed in part from the literature on the neurobiology of substance addiction, is that olfactory stimulation on a repeated basis could lead to "kindling" of responses by the limbic system, that portion of the forebrain responsible for affect and behavior. As has been demonstrated in the setting of addiction and withdrawal, this could lead to drastically differing dose–response relations depending on whether exposures have been ongoing or are remote, one source of symptom variability which has plagued earlier theories of biological interpretation. Moreover, the authors have speculated about the potential under such a theory for the highly "sensitized" limbic receptors to respond to differing stimulae, such as neural stimulation occurring outside the upper airway (e.g., food constituents, or exposure to physical factors such as ultraviolet light etc.). It could also account for the close clinical resemblance in some cases between MCS subjects and individuals with a range of moderately severe psychiatric disorders (which, importantly, may also be associated with limbic dysfunction). The theory could provide some basis for the observation of abnormal PET and SPECT scan results in some groups of MCS patients. Each of these tests measures regional brain metabolism and may be sensitive markers for alterations of brain function which are not associated with visible anatomic changes on MRI or CT scan.

In addition to the theoretical framework, which has the appeal suggested above, there are some animal experiments which could be interpreted as consistent with this model (78). In addition, there are a variety of components of this model which are directly testable, such as the changes induced on the metabolic scans by exposure to small chemical stimulae, or the response of affected subjects to treatment with agents which are known to blunt limbic kindling responses or seizure activity (62). In the meantime, this theoretically intriguing construct remains largely untested.

6 THE EPIDEMIOLOGY OF MCS

Many of the same problems such as lack of agreement on case definition, or lack of a simple biologic test, have also limited the ability to determine the patterns of MCS in large populations, although several attempts have been made. Notable has been the assessment of MCS within the population of Persian Gulf War veterans in the United States (41–45), and a small number of regional surveys (34, 46). Beyond these sources, most of what is known about the epidemiology of MCS has been derived from case series, which must be interpreted with caution given the diverse, and often unquantifable biases associated with patient recruitment and diagnosis. Nonetheless there are some inferences which can be derived from the available data sources.

It appears likely from several large surveys that a reasonably high fraction of the "healthy" population suffers from a tendency to find strong odors and/or chemical irritants fairly unpleasant. This may range as high as 20–40% in some subgroups (26, 79–81) such as female college students, retired elderly or office workers, depending on the survey instrument and criteria used to derive it. Although there is no uniformly applied label for such a condition, it is certainly *not* by itself sufficient for a diagnosis of MCS; indeed only 0.2% of all subjects in one survey reported physician diagnosed MCS despite a very high rate of chemical aversion (79–80). This author refers to people reporting such unpleasant reactions to chemicals in the absence of a self-reported illness as "chemically intolerant," although others have referred to them as cacosmics (people experiencing ordinary smells as unpleasant). The clinical significance, if any, of this "chemical intolerance" is unclear. It is likely that such individuals self-select for less chemically contaminated jobs and environments, do not smoke and use few, if any, scented products. It is also likely that such individuals are more likely to experience problems in the setting of an environmental problem, such as building related illness. However, whether chemical intolerance is an important predisposing risk factors for MCS remains unknown; clinical experience suggests that it is clearly evident in the premorbid background of some MCS patients and strikingly absent in others.

The prevalence of MCS in the general population is certainly lower than the prevalence of chemical intolerance, but there is a very wide range of estimates extant. A California based survey, included as part of a larger health surveillance project, estimated that 6% of the population interviewed reported physician diagnosed MCS (46). Meggs inferred a rate of 3.9% based on daily symptoms related to chemical exposures in a survey of 1027 rural North Carolinians, but was cautious about the interpretation of the figure (34). Using another instrument, investigators studying the military population in Iowa during the Persian Gulf War era demonstrated a background prevalence of MCS symptoms among personnel *not* serving in the War arena of 2.5–3%, compared with 4.7–8.6% with MCS symptoms among those with Persian Gulf War service (44). On the other hand, some clinicians, including the author, have inferred a very low prevalence of clinically apparent MCS based on the small numbers referred to treatment facilities which specialize in the care of this condition (82). It is likely from these data that clinically overt cases are somewhat rare, but that the propensity for MCS symptoms is quite common in some populations.

Whatever the population rate, most observers agree that MCS is more prevalent among women than men, most common in the 30s and 40s (although cases of all ages have been described) and apparently more common among higher socio-economic status individuals than lower (83). There appears also to be a relative excess among whites compared to nonwhites, at least in the United States, although many have raised concern about diagnostic bias as the explanation for at least some of these apparent differentials.

Consistent with this pattern of occurrence, there is now fairly good evidence that MCS bares some relationship to other reasonably prevalent and also not well understood chronic disorders affecting similar demographic groups, i.e., chronic fatigue syndrome (CFS) and fibromylagia (84). CFS, characterized by persistent and global fatigue affecting social and work life for a protracted period, has been thought to be precipitated by an infectious illness, although data are lacking. Review of series of patients seen in CFS clinics reveals that many complain of symptomatic responses to chemical exposures, and many of the features of MCS other than the precipitous onset after an environmental incident. Among patients with MCS, chronic fatigue is present in about 30% (85). Fibromyalgia is a disorder of tenderness involving muscle and associated soft tissues primarily along the axial (central) part of the body, and not corresponding to regional patterns of use. Patients typically have severe pain which is poorly relieved by antiinflammatory drugs or physical therapies. On exam, patients are found to have "trigger points" of tenderness, but laboratory tests of muscle function is normal. As with CFS, a high fraction of such patients are thought to have symptomatic responses to chemical exposures as well; likewise many MCS patients meet clinical criteria for this disorder.

The settings in which MCS begins has also been characterized in patient series, suggesting that organic solvents, pesticides and indoor air quality problems are most commonly involved. However, an enormous range of irritant and neurotoxic chemicals have been described, suggesting that specific agents may be less important than the context of exposure and clinical effects at the outset (18, 58). Once MCS symptoms have become established, the range of exposures reported to precipitate symptoms is enormous, but petroleum-derived chemicals, pesticides, tobacco smoke, and fragrances are on almost every patient's list. Moreover, the majority of patients exhibit some intolerance for all forms of sedating agents, including even small quantities of alcohol or over the counter medications.

There have been no comprehensive studies of the natural history of MCS, but some conclusions appear most consistent with published experience. On the positive side, long term follow-up of MCS patients has revealed little if any tendency towards progression of symptoms or involvement of major organs by chronic pathology. In other words, MCS is neither degenerative nor lethal, but in fact appears to have a fairly stable course, punctuated by periodic improvement and episodic exacerbations of symptoms and functioning (which patients often construe as responses to changes in their activities or therapies or both, without objective evidence). On the other hand, despite a few case reports of success with behavioral interventions or psychotropic drugs (67), and the stunning but unconfirmed reports of Staudemeyer regarding the benefits of intensive psychotherapy (35, 72), there is little basis for optimism regarding cure with any available treatment modalities. Although both conventional and nonconventional therapies (see below) are often greeted with early symptomatic gains, there is no evidence that any treatment offers more benefit than none, at least in terms of the underlying proclivity to react to common environmental factors.

7 MANAGEMENT OF MCS

In no aspect of occupational health practice is collaboration between the industrial hygienist and the clinician more sensitive and more important than in the evaluation of a worker who develops apparent problems around low-level chemical exposures. Only with the reasonable insight into possible diagnostic considerations can the hygienist provide the kind of environmental assessment which will inform the diagnostic process meaningfully; only with an understanding of the actual nature of environmental exposures and conditions, quantitatively or qualitatively as appropriate, can the physician appropriately distinguish among the diagnostic considerations and appreciate the therapeutic options. Also, only in the context of the actual diagnosis should environmental interventions be considered or undertaken. In the following sections the respective roles of hygienist and clinician in diagnosis and treatment of MCS are discussed.

7.1 Diagnosis

7.1.1 The Role of the Industrial Hygienist

Often the industrial hygienist becomes aware of symptomatic workers complaining about the work environment relatively early, and as such has a crucial role to play in ensuring that evaluation is focused and appropriate. There are several central principles in the diagnostic phase.

1. It is essential that the hygienist not fall pray to evaluating the problem of low-level exposure using the same paradigm as might be appropriate for evaluating classic toxic symptoms in the industrial setting. While it is entirely appropriate to attempt to identify the major classes of hazards which may be present in the environments which appear to be associated with symptoms, quantification for the purpose of proving that levels are "safe", i.e., far below TLV or other industrial standards, is pointless and counterproductive, as they are in assessing indoor air quality problems (see Chapt. 69).

2. Qualitative assessment of the global environment of the affected worker or workers, including such things as fragrances and personal products of co-workers, climactic factors, microenvironments which may be associated with odor or irritation locally or intermittently, sources of biologic agents such as molds and bacterial contaminants, etc., is typically more useful in the diagnostic process than quantitation.

3. Because classic allergic reactions must be initially considered in workers who are symptomatic at low level, the presence or absence of biologic or chemical hazards known to cause dermal and respiratory allergy may be particularly useful to the evaluating clinicians.

4. Whatever the evaluation may reveal, it is crucial that the industrial hygienist resist the temptation to "diagnose" the case based on the assessment of the environment, especially rendering opinions about the likelihood that any environmental finding should or should not be causing symptoms. Rather, nonjudgmental transmission of information to clinicians is ideal. When levels are very low despite complaints, summary judgments such as "these levels are far below those associated with classic toxic effects of these hazards, but may be associated with _____ in sensitive subjects" or the like may be very helpful.

5. The industrial hygienist often has more experience and knowledge about MCS and related conditions than the treating physician, especially when care is being provided outside the setting of occupational medicine practice, i.e., by personal physicians. In this

situation the hygienist can be extremely helpful by informing clinicians about the kinds of problems which have been described in low level exposure environments, to prevent the most obvious mistakes inexperienced doctors will make: incorrectly assuming toxic level exposures, inappropriately searching for unfindable "sensitizing" agents or mysterious disease; or, assuming that the patient must be crazy.

7.1.2 The Role of the Clinician.

The differential considerations in cases like those described above has already been delineated. The three most important parts of the initial workup are (1) a very careful history of the onset and nature of symptoms, and any possible association between these symptoms and the patient's various environmental exposures, i.e., at work, home, and elsewhere; (2) a careful review of the past medical history, and (*3*) an assessment of the environments in which symptoms occur, often in comparison to those which appear free of such problems. In this latter aspect the plant industrial hygienist may be extraordinarily helpful, especially when coupled with a walkthrough if the environment is not known to the physician.

Once these initial steps have been undertaken, the possibilities are usually quite limited. If the range of symptoms could be explained by any classic allergic mechanism, i.e., hay fever, rhinitis, dermatitis, asthma, urticaria, then detailed measures of immune response and lung function are appropriate, as is the search for a possible causal factor. Alternatively, if symptoms are limited to any single organ, and could be consistent with environmental exacerbation of some underlying disease such as migraine or reactive airways disease, then these possibilities need to be directly and fully evaluated. Obviously, suggestion of a major systemic disease based on exam such as a cancer, or of a severe psychologic disturbance, based on the history and interview, merit appropriate additional evaluation and/or referral.

However, in many cases none of these possibilities is strong, and the problem often boils down to whether the worker has some variant of nonspecific building related illness or MCS. The former is strongly suggested by (*1*) symptoms limited to occurrence in a single (work) environment, without problems elsewhere, and (*2*) the occurrence of similar problems in at least some co-workers. The latter, by contrast, is suggested when symptoms are occurring in *many* unrelated settings, and are *not* affecting others similarly. Where MCS seems the best explanation, the criteria discussed above should be carefully reviewed, limited testing done to evaluate particular aspects of the disorder which might be amenable to specific treatment (e.g., the patient who *also* has classic allergies, or the patient who is *also* moderately depressed, or *also* has mild asthma). There is no need to exclude the universe of possibilities in cases who meet the definition well (such as each of the cases presented above); in fact, the experience suggests that extensive testing delays institution of appropriate care and often feeds the patient's perception that they have a lethal environmental disorder.

7.2 Treatment

7.2.1 The Role of the Industrial Hygienist

Rarely can the symptomatic experience of the worker with MCS be remedied by single simple improvements in the environment. In fact, one of the hallmarks of MCS, at least in the context of building related illness, is that the MCS patient *does not* improve when steps to reduce air contamination, usually sufficient to remediate epidemic complaints, are instituted. Many are drawn to err in one of two directions: either to continue trying to improve the environment, using the MCS patient as the "canary" (which almost always fails) or presuming that nothing can help the person with MCS, and tossing in the towel. The experience with MCS patients suggests that a substantial portion can continue to function effectively in their current jobs if offered appropriate environmental controls are instituted, coupled with personnel policies which reinforce the needs of the MCS patients for control over their environment. The experienced hygienist can and often must take the lead for this to be successful.

There are aspects which must be dealt with. First, the immediate work areas (including all those in which the worker will spend predictable time) must be as well ventilated as possible, and free of established sources of offensive materials, such as machines which discharge chemicals, spraying with pesticides, cleaning with organic-solvent based detergents, renovations using new paints and materials, and new materials such as carpeting (a predictable problem), particle board furnishings etc. When such material must be used, the affected worker needs advance notice and temporary placement. In general, there are no specific quantitative guidelines for any of these measures, since patient responses differ; the subject him or herself is usually a good barometer of whether this objective has been achieved.

The second control strategy relates to exposures caused by people, in other words fragrances, tobacco products, etc. Institution of some restriction of personnel in the workspace of affected people, and the use of fragrance and smoking policies have been very effective, and surprisingly tolerable to co-workers in many settings.

Finally, since one or the other of the above measures will sometimes fail, and some exposure will occur inadvertently which precipitates unpleasant symptoms, MCS patients benefit enormously from some work flexibility which allows them to take brief strategic "walks" away from their workstation when they are overwhelmed. Obviously this has ramifications beyond those under the jurisdiction of health and safety personnel but, understanding of this aspect and its importance is often key to successful accommodation of MCS affected workers. IH personnel are in an ideal position to communicate this to managers.

Of course, sometimes these measures are either unachievable or are undertaken and nonetheless fail. At this juncture, once it has been ascertained that appropriate and feasible steps have been taken, the best approach is recognition that the work environment is not suitable even with modification, and the patient should be encouraged to look elsewhere. Further efforts to perfect the environment are rarely successful and are frustrating to all parties.

7.2.2 The Role of the Clinician

There is no established cure for MCS, nor any evidence that clinical interventions of any sort will change the natural history of the responses to chemical exposures. Although numerous radical therapies, such as intensive courses of nutrient supplements, chelation, accelerated fat turnover, radical environmental avoidance in specialized chambers, desensitization with offensive organic agents, hormonal supplements and the like have all been proposed by various fringe practitioner groups, none has been shown to be more beneficial than no treatment at all. Conversely, although anecdotal evidence of "cure" from psychotherapy, behavioral modification and pharmacologic intervention has been suggested, no evidence from even the most limited treatment trials conducted rigorously has been presented. The author has published an abysmally unsuccessful response to an otherwise very well tolerated antidepressant (57), and much clinical anecdote supports lack of efficacy of such approaches despite some theoretical reasons to suspect they should work.

There are, however, some important interventions which can assist affected workers in regaining function, if not altogether changing their symptom pattern:

1. The patient must have the nature of the illness, including the current lack of knowledge about its cause and its natural history, explained. It is particularly important that the patient understand that neither avoiding chemicals nor any other "treatment" is known to make it better, and that even left untreated it is neither progressive nor lethal, which almost all suspect it must be.

2. Once this is established, it is important to emphasize that despite lack of understanding or agreement about the basis of MCS, that the symptoms are very real, obviously disruptive and worthy of treatment. Unsympathetic, incredulous comments, suggesting "that it's all in your head" or "nothing is wrong with you" of "there's no reason you can't go back to your work" are inflammatory and hurtful to the patient, and have absolutely no therapeutic value. However much one might wish that this were true, or believe that MCS is primarily psychological in origin, such comments and suppositions are worthless in the management of workers with MCS.

3. Rather, it is essential that even as the patient is discouraged from undertaking radical, unproven, expensive and possibly harmful therapies, regular supportive care, involving both physician and some form of counseling be instituted. Where appropriate, treatable health problems such as classic allergies or depression should be treated, though without any illusion that once these are treated the MCS will disappear—it will not!

Even when these steps are undertaken the management of MCS is far from gratifying for either the clinician or the industrial hygienist. As long as the cause remains obscure so will our ability to modify the condition, or prevent it. However, ignorance is neither an excuse for ignoring the problem, nor for blaming those who are affected. MCS is, however it may all turn out, a fact of life for everyone involved in the practice of occupational and environmental health.

BIBLIOGRAPHY

1. L. Rosenstock and M. Cullen, *Textbook of Clinical Occupational and Environmental Medicine*, W. Saunders Co., Philadelphia, 1994, p. 3.
2. American Conference of Governmental Industrial Hygienists, *Documentation of the Threshold Limit Values and Biological Exposure Indices*, 6th ed., ACGIH, Cincinnati, 1998.

3. H. Gong, Jr., P. A. Lachenbruch, P. Harber, and W. S. Linn, *Toxicol. Ind. Health* **11**(5), 467–487 (1995).
4. D. B. Peden, B. Boehlecke, D. Horstman, and R. Devlin, *J. Allergy Clinical Immunol.* **100**(6, Pt 1), 802–808 (1997).
5. M. W. Frampton, J. R. Balmes, C. Cox, P. M. Krein, D. M. Speers, Y. Tsai, and M. J. Utell, *Research Report—Health Effects Institute* (78), 73–79 discussion, 81–99 (1997).
6. J. R. Balmes, R. M. Aris, L. L. Chen, C. Scannell, I. B. Tager, W. Finkbeiner, D. Christian, T. Kelly, P. Q. Hearne, R. Ferrando, and B. Welch, *Research Report—Health Effects Institute* (78), 1–37, discussion, 81–99 (1997).
7. C. Scannell, L. Chen, R. M. Aris, I. Tager, D. Christian, R. Ferrando, B. Welch, T. Kelly, and J. R. Balmes, *Amer. J. Resp. Crit. Care Med.* **154**(1), 24–29 (1996).
8. T. J. Hiltermann, J. Stolk, P. S. Hiemstra, P. H. Fokkens, P. J. Rombout, J. K. Sont, P. J. Sterk, and J. H. Dijkman, *Clinical Sci.* **89**(6), 619–624 (1995).
9. M. W. Frampton, P. E. Morrow, C. Cox, P. C. Levy, J. J. Condemi, D. Speers, F. R. Gibb, and M. J. Utell, *Environ. Res.* **69**(1), 1–14 (1995).
10. Y. Takeuchi, N. Hisanaga, Y. Ono, E. Shibata, I. Saito, and M. Iwata, *Internat. Arch. Occup. Environ. Health* **65**(1 Suppl), S227–230 (1993).
11. H. A. Peters, R. L. Levine, C. G. Matthews, S. Sauter, and L. Chapman, *Acta Pharmacol. Toxicol.* **59**(Suppl 7), 535–546 (1986).
12. M. Antti-Poika, J. Juntunen, E. Matikainen, H. Suoranta, H. Hanninen, A. M. Seppalainen, and J. Liira . *Internal. Arch. Occup. Environ. Health* **56**(1), 31–40 (1985).
13. C. A. Redlich, J. S. Sparer, and M. R. Cullen, *The Lancet* **349**, 1013–1016 (1997).
14. "Annals of Multiple Chemical Sensitivities: State-of-the-Art Science Symposium. Proceedings. Baltimore, MD, Oct. 30–Nov. 1, 1995," *Reg. Toxicol. Pharmacol.* **24**(1, Pt 2), S1–189 (1996).
15. "Proceedings of the Conference on Low-Level Exposure to Chemicals and Neurobiologic Sensitivity. Baltimore, MD, 6–7 April 1994," *Toxicol. Ind. Health* **10**(4–5), 253–669 (1994).
16. P. J. Sparks, W. Daniell, D. W. Black, H. M. Kipen, L. C. Altman, G. E. Simon, and A. I. Terr. *J Occup. Med.* **36**(7), 718–730 (1994).
17. P. J. Sparks, W. Daniell, D. W. Black, H. M. Kipen, L. C. Altman, G. E. Simon, and A. I. Terr. *J. Occup. Med.* **36**(7), 731–737 (1994).
18. M. R. Cullen, *Occup. Med.* **2**(4), 655–661 (1987).
19. Association of Occupational and Environmental Clinics, "Proceedings of the AOEC Workshop on Multiple Chemical Sensitivity, Washington DC Sept. 20–21, 1991," *Toxicol. Ind. Health* **8**(4) (1992).
20. T. G. Randolph, *J. Lab. Clin. Med.* **40**, 931–932 (1952).
21. American Academy of Environmental Medicine. *An Overview of the Philosophy of the American Academy of Environmental Medicine*, Denver, 1992.
22. American Academy of Allergy and Immunology, Executive Committee: Clinical Ecology, *J. Allergy Clin. Immunol.* **78**, 269–270 (1986).
23. American College of Physicians, Clinical Ecology, *Ann. Intern. Med.* **111**(2), 168–178 (1989).
24. M. R. Cullen, ed., *The Worker with Multiple Chemical Sensitivities*, Hanley Belfus, Philadelphia, 1987.
25. I. R. Bell, *Toxicol.* **111**(1–3), 101–117 (1996).
26. I. R. Bell, C. S. Miller, G. E. Schwartz, J. M. Peterson, and D. Amend, *Archives Environ. Health* **51**(1), 9–21 (1996).

27. I. R. Bell, E. E. Hardin, C. M. Baldwin, and G. E. Schwartz, *Environ. Res.* **70**(2), 84–97 (1995).
28. K. I. Bolla, *Reg. Toxicol. Pharmacol.* **24**(1, Pt 2), S52–54 (1996).
29. A. L. Davidoff and P. M. Keyl, *Archives Environ. Health* **51**(3), 201–213 (1996).
30. N. Fiedler, H. Kipen, J. DeLuca, K. Kelly-McNeil, and B. Natelso, *Toxicol. Ind. Health* **10**(4–5), 545–554 (1994).
31. N. Fiedler, H. M. Kipen, J. DeLuca, K. Kelly-McNeil, and B. Natelson, *Psychosomatic Med.* **58**(1), 38–49 (1996).
32. T. Hummel, S. Roscher, M. P. Jaumann, and G. Kobal, *Reg. Toxicol. Pharmacol.* **24**(1 Pt 2), S79–86 (1996).
33. H. R. Kehrl, *Environ. Health Perspectives* **105**(Suppl 2), 443–444 (1997).
34. W. J. Meggs, K. A. Dunn, R. M. Bloch, P. E. Goodman, and A. L. Davidoff, *Archives Environ. Health* **51**(4), 275–282 (1996).
35. H. Staudenmayer, *J. Allergy Clinical Immunol.* **99**(4), 434–437 (1997).
36. G. E. Simon, W. Daniell, H. Stockbridge, K. Claypoole, and L. Rosenstock, *Ann. Intern. Med.* **19**(2), 97–103 (1993).
37. D. W. Black, A. Rathe, and R. B. Goldstein, *JAMA* **26**, 3166–3170 (1990).
38. B. Hileman, *Chem. Eng. News* **69**(29), 26–42 (1991).
39. Ontario Ministry of Health. *Report of the Ad Hoc Committee on Environmental Hypersensitivity Disorders*, Toronto, 1985.
40. National Research Council, Multiple Chemical Sensitivities: a Workshop, National Academy Press, Washington, DC, 1992.
41. N. Fiedler, H. Kipen, B. Natelson, and J. Ottenweller, *Reg. Toxicol. Pharmacol.* **24**, S129–S138 (1996).
42. K. Fukuda, R. Nisenbaum, G. Stewart, W. W. Thompson, L. Robin, R. M. Washko, D. L. Noah, D. H. Barrett, B. Randall, B. L. Herwaldt, A. C. Mawle, and W. C. Reeves, *JAMA* **280**, 981–988 (1998).
43. K. C. Hyams, F. S. Wignall, and R. Roswell, *Ann. Intern. Med.* **125**, 398–405 (1996).
44. Iowa Persian Gulf Study Group, *JAMA* **277**, 238–245 (1997).
45. J. Wolfe, S. P. Proctor, J. D. Davis, M. S. Borgos, M. J. Friedman, *Amer. J. Ind. Med.* **33**, 104–113 (1998).
46. Interagency Workgroup on Multiple Chemical Sensitivities, *A Report on Multiple Chemical Sensitivities*, in press.
47. International Programme on Chemical Safety (IPCS), *Report of Multiple Chemical Sensitivities (MCS) Workshop*, February 21–23, Berlin, 1996.
48. N. A. Ashford and C. S. Miller, *Chemical Exposures: Low Levels and High Stakes*, Van Nostrand Reinhold, New York, 1991.
49. J. R. Nethercott, L. L. Davidoff, and B. Curbow, *Arch. Environ. Health* **48**(1), 19–26 (1993).
50. W. J. Rea, A. R. Johnson, G. H. Ross, J. R. Butler, E. J. Fenyves, B. Griffiths, et al., "*Considerations for the Diagnosis of Chemical Sensitivity*," in *Multiple Chemical Sensitivities*, National Academy Press, Washington, DC, 1992.
51. A. S. Levin and V. S. Byers, *Occup. Med.* **2**(4), 669–681 (1987).
52. G. Heuser, A. Wojdani, and S. Heuser, Diagnostic Markers of Multiple Chemical Sensitivity, "*Multiple Chemical Sensitivities: Addendum to Biologic Markers in Immunotoxicology*," National Academy Press, Washington, DC, 1992.

53. J. D. Thrasher, A. Broughton, and R. Madison, *Arch. Environ. Health* **45**, 217–223 (1990).
54. A. I. Terr, *Occup. Med.* **2**(4), 683–694 (1987).
55. J. F. Albright and R. A. Goldstein, *Toxicol. Ind. Health* **8**(4), 215–219 (1992).
56. H. Mayberg, *Toxicol. Ind. Health* **10**(4/5), 661–665 (1994).
57. M. R. Cullen, "Multiple Chemical Sensitivities: Is there Evidence of Extreme Vulnerability of the Brain to Environmental Chemicals," in R. L. Isaacson and K. F. Jensen, eds. *The Vulnerable Brain and Environmental Risks*, Vol. 3, *Special Hazards from Air and Water*, Plenum Press, New York, 1994, pp. 65–75.
58. N. Fiedler, C. Maccia, and H. Kipen, *J. Occup. Environ. Med.* **34**(5), 529–538 (1992).
59. R. L. Doty, D. A. Deems, R. E. Frye, R. Pelberg, and A. Shapiro, *Arch. Otolaryngol. Head Neck Surg.* **114**, 1422–1427 (1988).
60. R. Bascom, W. J. Meggs, M. Frampton, K. Hudnell, K. Kilburn, G. Kobal et al., *Environ. Health Perspect.* **105**(suppl 2), 531–537 (1997).
61. W. J. Meggs and C. H. Cleveland Jr., *Arch. Environ. Health* **48**(1), 14–18 (1993).
62. C. Miller, N. Ashford, R. Doty, M. Lamielle, D. Otto, A. Rahill, and L. Wallace, *Environ. Health Perspect.* **105**(Suppl 2), 515–519 (1997).
63. T. L. Kurt, *Clin. Toxicol.* **33**(2), 101–105 (1995).
64. D. J. Shusterman and S. R. Dager, *Occup. Med.* **6**(1), 11–27 (1991).
65. S. Siegel and R. Kreutzer, *Environ. Health Perspect.* **105**(Suppl 2), 521–526 (1997).
66. JC. Selner and H. Staudenmaer, *Toxicol. Ind. Health* **8**(4), 145–155 (1992).
67. P. Andiné, L. Rönnbäck, and B. Järvholm, *Acta Psychiatr. Scand.* **96**:**1**, 82–3 (1997).
68. C. M. Brodsky, *Psychosomatics* **24**, 731–742 (1983).
69. E. Shorter, *Scand. J. Work Environ. Health* **23**(Suppl 3), 35–42 (1997).
70. A. M. Brown-DeGagne, J. McGlone, and D. A. Santor, *J. Occup. Environ. Med.* **40**, 862–869 (1998).
71. A. L. Davidoff and L. Fogarty, *Arch. Environ. Health* **49**, 316–325 (1994).
72. H. Staudenmayer and J. C. Selner, *J. Occup. Environ. Med.* **37**(6), 704–709 (1995).
73. L. L. Davidoff, "Models of Multiple Chemical Sensitivities (MCS) Syndrome: Using Empirical Data (especially Interview Data) to Focus Investigations, Proceedings of the AOEC Workshop on Multiple Chemical Sensitivity," Washington, DC, Sept. 20–21, 1991, *Toxicol. Ind. Health* **8**(4), 229–247.
74. R. A. Hahn, *Prev. Med.* **26**, 607–611 (1997).
75. M. A. Amundsen, N. P. Hanson, B. K. Bruce, T. D. Lantz, M. S. Schwartz, and B. M. Lukach, *Reg. Toxicol. Pharmacol.* **24**(1, Pt 2), S116–118 (1996).
76. R. S. Guglielmi, D. J. Cod, and D. A. Spyker, *J. Behav. Therapy Exper. Psychiatry*, **25**(3), 197–209 (1994).
77. I. R. Bell, C. S. Miller, and G. E. Schwartz, *Biol. Psychiatry* **32**, 218–242 (1992).
78. D. H. Overstreet, C. S. Miller, D. S. Janowsky, and R. W. Russell, *Toxicology* **111**(1–3), 119–134 (1996).
79. I. R. Bell, G. E. Schwartz, J. M. Peterson, and D. Amend, *Arch. Environ. Health* **48**(1), 6–13 (1993).
80. I. R. Bell, G. E. Schwartz, J. M. Peterson, and D. Amend, *J. Amer. Coll. Nutr.* **12**(6), 693–702 (1993).

81. I. R. Bell, G. E. Schwartz, D. Amend, J. M. Petrson, and W. A. Stini, *Biol. Psychiatry* **35**, 857–863 (1994).
82. M. R. Cullen, P. E. Pace, and C. A. Redlich, *Toxicol. Ind. Health* **8**(4), 15–19 (1992).
83. S. B. Mooser, *Occup. Med.* **2**(4), 663–681 (1987).
84. M. A. Demitrack, *Psychiatric Clinics of North America* **21**(3), 671–692 (1998).
85. D. Buchwald and D. Garrity, *Arch. Intern. Med.* **154**, 2049–2053 (1994).

CHAPTER FOURTEEN

Analytical Methods

Robert G. Lieckfield, Jr., CIH

1 INTRODUCTION

The laboratory analysis of occupational health samples is an important aspect of determining occupational health exposures. The accurate determination of workplace exposures begins with the selection of an appropriate sampling device and continues through the analysis of the collected samples in the industrial hygiene laboratory. The focus of this chapter is on the methods used in the analysis of industrial hygiene samples.

Industrial hygiene chemistry covers a wide diversity of chemical substances found in modern industry with hundreds more being added yearly. The complexity of this field is compounded by the need to identify and quantify trace quantities of organic and inorganic substances in the presence of numerous potential interferences.

The industrial hygiene analytical laboratory incorporates a substantial number and wide-array of chemical and instrumental methods. The majority of industrial hygiene analyses are performed using one of the following techniques:

- Atomic Absorption Spectrophotometry (AAS)
- Electron Microscopy (Transmission and scanning microscopes)
- Gas Chromatography (GC)
- High-Pressure Liquid Chromatography (HPLC)
- Inductively Coupled Argon Plasma Spectrophotometry (ICAP)
- Ion Chromatography (IC)
- Optical Microscopy (Phase-contrast and polarized-light microscopes)
- Ultraviolet/Visible Spectrophotometry

Patty's Industrial Hygiene, Fifth Edition, Volume 1. Edited by Robert L. Harris.
ISBN 0-471-29756-9

- X-ray Diffraction
- Gravimetric

Additional methods used, although not as common, include titration, ion-selective electrode, inductively coupled argon plasma/mass spectrometry (ICP/MS), and gas chromatography/mass spectrometry (GC/MS). A basic understanding of the analytical methods and techniques is essential to the industrial hygienist, allowing for the proper interpretation of the reported laboratory results.

The early years of the industrial hygiene laboratory saw the use of classical wet chemistry methods including acid/base and indicator dye endpoint titrations, gravimetric, UV/visible spectrophotometer, and optical microscopy. Chemical contaminants such as aldehydes and hydrogen sulfide were commonly determined by iodometry titration. Gravimetric analysis of dusts is a relatively new method. The early method used in the determination of particles was the optical microscope (1). There is still evidence of optical microscopy in determining particulate in the citation of the ACGIH Threshold Limit Values (TLV) in million particles per cubic foot (mppcc).

The UV/visible spectrophotometer was a mainstay of the early industrial hygiene laboratory,used for the determination of both inorganic compounds, such as metals, and organic solvents, such as benzene. Although these methods are rarely used in the modern industrial hygiene laboratory for determination of metals and organic solvents, the chemistry involved in these determinations is clearly evident. The physical properties of metals allowed them to react with chelating agents such as sodium diethyl dithiocarbamate, dithizone, and 8-quinolonol. The extraction and concentration of metallic compounds from the sampling matrix using these chelating agents produced colored complexes that were then determined quantitatively by UV/visible spectrophotometry. Similarly, UV/visible spectrophotometry was used for analysis of aromatic organic solvents, because of the absorption characteristics of these chemicals.

The technology of laboratory instrumentation evolved through the development of both operating design and microprocessing advancements. These technology gains provided for the commercialization and the significantly lower cost of sophisticated instrumentation to the industrial hygiene community.

2 SURVEY OF INDUSTRIAL HYGIENE ANALYTICAL METHODS

A brief survey of the analytical methods used in the industrial hygiene laboratory is important for both the laboratory practitioner and the field industrial hygienist.

2.1 Spectrophotometric Methods

There are two basic analytical methods that rely on the spectrophotometric principle—ultraviolet/visible spectrophotometry and atomic absorption spectrophotometry. The basic theory behind these methods is the relationship between the absorbance of a particular compound and the compound's concentration in solution. The theory follows the mathematical relationship known as Beer's Law, shown below.

$$A = \epsilon b C$$

where A = absorbance
ϵ = absorbance constant; dependent on the compound being measured
b = cell path length
C = concentration

The relationship between absorbance and concentration is linear over a selected concentration range. Therefore, determining the concentration of an unknown solution requires construction of a calibration curve using known standard concentrations of the compound being measured.

Once the calibration curve is constructed, the absorbance of the unknown solution is plotted and the concentration calculated from the curve. Modern UV/visible and atomic absorption instrumentation perform this operation within the instrument operating software using linear regression statistics.

2.1.1 Ultraviolet/Visible Spectrophotometry

The UV/visible spectrophotometry system consists of six major components:

1. Light source (wavelength of 200 to 800 nanometers).
2. Prism (disperses white light into a spectrum).
3. Wavelength slit (selection and isolation of a particular wavelength of light).
4. Sample chamber (sample cell).
5. Detector (converts light energy into electronic signals).
6. Operating software.

The analytical wavelength is dependent on the absorbance characteristics of the compound being measured. Industrial hygiene analyses using UV/visible spectrophotometry involve the creation of colored complexes to enhance the absorbance characteristics of the compound being determined. Examples of common industrial hygiene determinations employing UV/visible methodology are shown in Table 14.1.

These are all classic industrial hygiene methods and are still frequently used as reference methods for performance of new instrumental methods. The UV/visible method for crys-

Table 14.1. Reaction Chemistries for Common Spectrophotomeric Methods

Compound	Reaction Chemistry (NIOSH Method)
Formaldehyde	Chromotropic acid (NIOSH Method 3500)
Hexavalent chromium	Diphenyl carbazide (NIOSH Method 7600)
Crystalline silica	Molybdate (NIOSH Method 7601)
Oxides of nitrogen	*N*-(1-Naphthyl) ethylenediamine dihydrochloride (NIOSH Method 6014)

talline silica, shown above, is also used for the determination of amorphous silica, the noncrystalline form of silica (SiO_2).

The UV/visible spectrophotometric methods, although continually being replaced with less labor-intensive instrumental methods of analysis, form the backbone of the industrial hygiene laboratory methods.

2.1.2 Atomic Absorption Spectrophotometry

Atomic absorption spectrophotometry (AAS) is the basic technique for the analysis of metals and metallic compounds. This method replaced the analysis of metals using chelation and colorimetric methods mentioned above. The AAS system is composed of four basic components:

1. Hollow-cathode lamp.
2. Atomization chamber.
3. Detector.
4. Operating software.

The hollow-cathode lamp is equivalent to the light source used in the UV/visible method. However, instead of requiring a prism to create a specific wavelength of light, the hollow-cathode lamp is designed to emit a single defined wavelength of light energy. The choice of hollow-cathode lamp depends on the element being determined.

The atomization chamber creates metal atoms from the sample solution that are then introduced into the light path of the hollow-cathode lamp where the metal atoms absorb at the specific wavelength characteristic of the metal. The amount of light absorbed is proportional to the metal atom concentration in the solution. There are four basic atomization techniques used in atomic absorption spectrophotometry—flame, graphite furnace, hydride generation, and cold vapor generation. The application of each atomization technique is shown below.

- Flame (common metals such as iron, lead, chromium, cadmium, nickel, and zinc).
- Graphite furnace (arsenic, selenium, beryllium, vanadium, and lead).
- Hydride generation (arsenic and selenium).
- Cold vapor generation (mercury).

The most common method is flame atomization AAS. Graphite furnace atomization provides greater sensitivity than flame AAS for a number of metallic elements commonly determined by flame AAS such as lead, cadmium, and beryllium. However, graphite furnace AAS is not applicable to all metals. Hydride generation is a relatively labor intensive method and is used predominantly for the analysis of arsenic and selenium. Cold vapor generation is used for the analysis of mercury, since elemental mercury has a high vapor pressure and can be atomized without the aid of a heat source, such as a flame or graphite furnace.

2.2 Atomic Emission Spectrophotometry–Inductively Coupled Argon Plasma (ICP) Spectrophotometry

The use of ICP, now common to the industrial hygiene laboratory, is used for the determination of metallic compounds and metals; replacing or substituting atomic absorption methods. The advantage of ICP over AAS is its applicability to all metallic elements (except mercury), simultaneous multi-element determinations, less interelement interference, smaller sample size requirement, and a faster analysis time. These features are coupled with ICP precision and accuracy comparable to AAS.

ICP operates on the principle of atomic emission spectrophotometry. Rather than *absorbing* light energy (UV/visible and AAS), the ICP utilizes the *emission* of light from metal atoms. High temperature argon plasma is used for the excitation of atoms (outer shell electrons and the corresponding release of energy in the form of light). Each metallic element has a specific emission spectrum that allows the ICP instrument to pattern and measure both qualitatively and quantitatively. The light spectrum emitted is separated into individual wavelengths by use of a diffraction grating. An element is identified by its emission wavelength and the concentration is determined by the intensity of the emission.

The ICP torch derives its power by induction from a high frequency magnetic field. An alternating current flows through an induction coil located at the end of a quartz tube. Argon, seeded with free electrons, flows through the quartz tube and passes through the induction coil. The seeded electrons interact with the magnetic field of the coil and gain sufficient energy to ionize argon atoms. The rapid movement and collision of electrons caused by the reversing electron flow between the magnetic field and argon gas produces further ionization and intense heat (6,000 to 10,000 K).

The atomic spectra generated when metal atoms are introduced into the plasma are detected by a photomultiplier tube and recorded. ICP instrumentation is available with either sequential or simultaneous detection systems. Advances in torch design and configuration (co-axial) have enhanced the sensitivity of the ICP on par with graphite furnace AAS.

The use of ICP in industrial hygiene allows for the rapid analysis of multi-element samples such as welding fumes (Ag, Cd, Cr, Cu, Fe, Ni, Mn, Pb, and Zn). ICP is more efficient and cost-effective than AAS when measuring more than one or two elements on a sample. Both NIOSH and OSHA have established methods incorporating ICP for the analysis of up to 27 metals (NIOSH 7300) on a single sample.

2.3 X-Ray Diffraction

X-ray diffraction is a powerful tool and used extensively in the qualitative and quantitative analysis of crystalline materials. However, the use of X-ray diffraction in the industrial hygiene laboratory is essentially limited to the analysis of crystalline silica. The X-ray diffraction technique is based on the principle that crystalline materials will diffract X-ray radiation at a specific angle incident to the plane of the sample. This diffraction angle is reproducible and specific in relation to the crystalline structure of the material. The diffraction pattern behaves according to the mathematical formula known as Bragg's equation.

$$m\lambda = 2d \text{ Sin } \theta$$

where m = order of diffraction
λ = wavelength of the X-ray source
d = distance between the crystal lattice
θ = diffraction angle

The x-ray diffraction method is the only industrial hygiene method for crystalline silica capable of determining the three crystalline silica polymorphs—α-quartz, cristobalite, and tridymite. Separation of the three polymorphs of crystalline silica is important when measuring crystalline silica exposures where more than one of the polymorphs may be present, since the toxicity of crystalline silica is dependent on the polymorph present.

2.4 Infrared Spectrophotometry

The infrared region of the electromagnetic spectrum extends from the red end of the visible light spectrum to the microwave region. The most useful area of the infrared spectrum in industrial hygiene chemistry is the mid-range between 200 cm^{-1} to 4000 cm^{-1}. Infrared spectroscopy involves the twisting, bending, rotating, and vibrational motions of atoms in a molecule. The absorbance of IR radiation is reproducible and unique to the material being measured. The amount of absorption is also proportional to the concentration of the crystalline material in the sample.

Infrared spectrophotometers have the same basic configuration as the UV/visible instrumentation. The light source is replaced by an infrared radiation source and the sample cell is specially made to allow infrared radiation to pass through the sample. Common cell materials are alkali halides, such as potassium bromide, sodium chloride, or silver chloride. A common sample preparation technique involves mixing the collected sample, after some preliminary preparation steps, with potassium bromide and making a pellet with a high-pressure press.

The infrared technique is used in the industrial hygiene laboratory for the analysis of crystalline silica and organic nonsoluble oils. It is also a useful technique in the determination of unknown organic materials. Usually a computer program and infrared spectral library is required for any qualitative analysis using infrared.

The infrared method for crystalline silica (NIOSH Method 7602) is commonly used, especially where the silica polymorph is known. There is some debate within the industrial hygiene laboratory profession as to the ability of the infrared method to separately quantitate the α-quartz and cristobalite polymorphs. The current NIOSH method cites the applicability of the IR method when the polymorph composition is known or inconsequential to the determination or exposure assessment.

2.5 Ion-Selective Electrodes

Ion-selective electrodes measure free ion concentration in solution. With the introduction and subsequent mainstream use of ion chromatography, the use of ion-selective electrodes has diminished. The more common electrodes in industrial hygiene chemistry are used for the determination of fluoride, chloride, cyanide, ammonia, and chlorine.

The ion-selective electrode operates by measuring the electrical activity (current) in an electrochemical solution. This is accomplished by the exchange of ions across a permeable membrane. The material used for the membrane determines the specificity of the electrode. The response of the ion selective electrode follows the Nernst equation, shown below.

$$E = \text{Constant} + [RT/F] \ln [(\text{Ion charge})\text{Internal}/(\text{Ion charge})\text{External}]$$

where E = electric potential
R = gas constant (8.314 Joules per degree-mole)
T = absolute temperature (273 + °C)
F = Faraday constant (96,487)
Ln = natural logarithm

The constant depends on the particular electrode used. The response of the ion selective electrode is logarithmic and nonlinear at lower concentrations.

Ion-selective electrodes are subject to both method and electrode interferences. Method interferences occur when some characteristic of the sample matrix prevents the electrode from sensing the ion of interest. For example, the fluoride ion forms complexes with the hydrogen ion. This fluoride complex cannot penetrate the electrode membrane and is therefore not detectable. Electrode interference occurs when the electrode responds to ions in the sample solution other than the ion being measured. Addition of buffer solution and knowledge of the sample constituents can control both of these interferences.

2.6 Chromatography Methods

There are three types of chromatographic techniques used in the industrial hygiene laboratory—gas chromatography (GC), high-pressure liquid chromatography (HPLC), and ion chromatography (IC). Gas chromatography (GC) is used for the analysis of volatile organic hydrocarbons, such as, aliphatic and aromatic hydrocarbons, alcohols, glycol ethers, ketones, and substituted hydrocarbons. High-pressure liquid chromatography (HPLC) is applicable to reactive and thermally unstable organics or high boiling organic compounds. Ion chromatography (IC) is similar to the HPLC technology and is employed for the analysis of cations and anions such as inorganic acids.

Chromatography is the science of separation of a mixture into its component parts and the subsequent measurement of the individual compounds. The chromatographic system is composed of six basic components.

1. Carrier (gas or liquid)
2. Chromatographic column
3. Column oven (GC only)
4. Injection system
5. Detection system
6. Data system

The principle behind chromatography is the distribution of sample components between a liquid or solid phase (column) and the carrier. This distribution between the two phases creates a separation of the sample components into distinct bands, detected as chromatographic peaks. The degree of separation of the sample mixture is a function of the column type, column temperature (GC only), and carrier flowrate.

After chromatographic separation of the sample mixture, the individual sample components exit the chromatographic column, and are detected by the detection system. The time between sample introduction (injection) and the detector response is called the retention time of the compound under the particular chromatographic conditions. Retention time is a characteristic of the component being measured and is commonly used as a qualitative tool for compound identification. However, it is important to note that chromatographic retention time alone is not "proof" of chemical identity.

Quantitation using chromatographic methods is performed by the analysis of known calibration standards and determining the corresponding detector response. Using the calibration data, a response curve is constructed by plotting instrument response versus concentration. The response curve is then compared to the response of unknown samples (exposure samples) and the mass of material collected is calculated. The majority of chromatographic detection systems have a linear response over a fairly wide range of concentrations. The specific linear operating range must be determined for each instrument, detection system, and compound being measured.

2.6.1 Gas Chromatography

Gas chromatography is probably the most widely used technique in the industrial hygiene laboratory. Gas chromatography is applicable to a wide range of volatile organic compounds that are thermally stable and have boiling points of <250 °C. In addition to the basic chromatographic system described above, gas chromatography employs a column oven with an operating temperature of 50 to 240 °C and an inert gas such as helium or nitrogen as a carrier.

The detection system of the GC can be selected to enhance selectivity and sensitivity, depending on the compounds being determined. Typical detection systems, along with the particular application, used with GC are shown in Table 14.2.

The most common detection system used in industrial hygiene is the flame ionization detector (FID). As noted above, the FID is applicable to a broad range of common organic solvents used in industrial applications. The FID has a wide linear operating range and sensitivities down to the low-microgram range.

The other detectors shown are used in more selective applications where either specificity or greater sensitivity is required.

2.6.2 High-Pressure Liquid Chromatography

High-pressure liquid chromatography is used for the analysis of reactive or organic compounds with boiling points generally greater than 250 °C. Examples include polyaromatic hydrocarbons (PAH), isocyanates (MDI, TDI, HDI), and organic acids.

Table 14.2. GC Detection Systems and Application

Detector	Application
Flame ionization (FID)	Aliphatic, aromatic, halogenated, and other substituted hydrocarbons
Nitrogen-phosphorus or thermionic (NPD)	Amines and nitrogen or phosphorus containing compounds
Electron capture (ECD)	Halogenated compounds such as pesticides and PCB
Flame photometric (FPD)	Sulfur and phosphorus containing compounds
Thermal conductivity (TCD)	Carbon monoxide, carbon dioxide, nitrogen, oxygen
Mass spectrometer (MS)	Qualitative analysis of volatile organic compounds

The carrier used in HPLC is a liquid passed through a chromatographic column under high pressure. In most HPLC applications, there is no need to operate the chromatographic column above ambient temperatures. It is important to maintain a relatively stable column operating temperature so that retention times do not shift unpredictably.

There are four basic detection systems used in industrial hygiene applications of HPLC—UV/visible, fluorescence, photodiode array, and refractive index. Each has advantages for the specific type of compounds being measured.

Of particular note is the use of a mass spectrometer (MS) as a detection system for the gas chromatograph or high-pressure liquid chromatograph. The mass spectrometer is a specialized detector capable of providing both qualitative and quantitative measurements of volatile organic compounds. Unlike standard GC or HPLC, which uses retention time as a means of qualitative analysis, the mass spectrometer determines the unique breakdown fingerprint pattern of an organic compound. This fingerprint pattern, or "mass spectra," is compared to a computerized mass spectral library of compounds. Through the use of a computer matching program, the mass spectra of the sample is matched to the known spectra and identification is made of the sample compound. The use of GC/MS is important when unknown solvent mixtures are being used or when investigating indoor air quality problems.

2.6.3 Ion Chromatography

Ion chromatography works on the same principle as the HPLC—a liquid carrier passed through a chromatographic column under high pressure. Because the IC is used for measurement of anions and cations, the detector used in IC measures the conductivity of the column effluent. The conductivity detector is applicable to the measurement of inorganic anions such as bromide, fluoride, chloride, nitrate, sulfate, and phosphate. These represent the principle components of the inorganic acids. Other applications include ammonia, aminoethanol compounds, nitrogen oxides, chlorine, bromine, chromic acid, sulfur dioxide, and perchlorates.

The IC incorporates two separate chromatographic columns. The first, or suppressor, column removes the background ions that would interfere with the detection of the principal ions being measured. After the background ions are scrubbed from the sample, the second, or separation, column is used in the same manner as the columns used in GC and HPLC for the separation of individual sample components.

The IC has eliminated the majority of classical "wet chemical" tests that were commonly encountered in the industrial hygiene laboratory.

2.7 Gravimetric Methods

Gravimetric methods of analysis are some of the most basic analytical methods used in the industrial hygiene laboratory. The most common application is the determination of particulate collected on a filter sample. Gravimetric methods employ either a mechanical or electro-analytical balance capable of measuring in the 1 to 10 microgram range. This measuring instrumentation is a vast improvement over the triple beam and mechanical balances used in the early industrial hygiene laboratory.

The most common use of gravimetric methods is for the determination of total or respirable particulate collected on 37-mm mixed cellulose or poly(vinyl chloride) filters. Other methods that employ gravimetric methods include measurement of benzene solubles, asphalt fume, coal tar pitch volatiles, carbon black and metalworking fluids.

2.8 Microscopy

Optical and electron microcopy are used in industrial hygiene principally for the determination of exposures to airborne fibers including asbestos, fiberglass, mineral wool, or refractory ceramics. Optical techniques include phase-contrast (PCM) and polarized-light (PLM) microscopy. Phase-contrast microscopy is used for the determination of fiber number or count on an air sample. The PCM method is nonspecific for identification and relies on fiber morphology for fiber classification. For example, the fiber morphology for asbestos is any fiber that is greater than five microns in length and has an aspect (length to wide) ratio of greater than 3 to 1. The counting criteria for commonly determined fibers are Table 14.3.

Polarized-light microscopy (PLM) is used for specific characterization of crystalline material structure and is usually used for the analysis of asbestos in bulk samples. The PLM technique uses both morphology and the optical properties of crystalline materials for mineral identification. These physical properties include refractive index, dispersion

Table 14.3. Fiber Counting Criteria

	Counting Criteria	
Fiber Type	Length	Aspect Ratio
Asbestos ("A" Rules)	>5 microns	>3 to 1
Asbestos ("B" Rules)	>5 microns	>5 to 1
Glass fiber	>10 microns	>5 to 1
Refractory ceramic	>5 microns	>5 to 1
Mineral wool	>5 microns	>3 to 1

staining, Becke line, and sign of elongation. PLM is not applicable to the analysis of individual fibers in an air sample, as in PCM fiber counting, since the magnification used is insufficient to see the fine fibers typically found in industrial hygiene exposure scenarios.

Electron microscopy methods used in industrial hygiene are transmission (TEM) and scanning (SEM) techniques. The most commonly used is the TEM for the specific identification of asbestos materials (NIOSH Method 7402). The TEM is capable of magnifications over 600,000X, although the most common magnification used in industrial hygiene applications is 15,000 to 20,000X. The TEM is typically coupled with Selected Area Electron Diffraction (SAED) instrumentation that allows the analyst to determine the specific crystalline structure of the material being examined. An Energy Dispersive Spectrometer (EDS) is also a common attachment to the TEM and provides an elemental analysis of the material.

The SEM is used in applications that do not require the magnification power of the TEM. Industrial hygiene applications for the SEM are surface texture analysis, particle sizing, and particle characterization.

3 FUTURE TRENDS IN INDUSTRIAL HYGIENE CHEMISTRY

As toxicological research continues on existing chemicals and new chemicals are created, there will be the need to develop new methods and improve existing instrumentation to meet these needs. On the horizon are new industrial hygiene applications of high technology instrumentation. These include inductively coupled argon plasma spectrophotometry coupled with mass spectrometry (ICP/MS) and high-pressure liquid chromatography and mass spectrometry (LC/MS).

The ICP/MS gives the analyst the ability to identify and quantify metallic elements at extremely low levels. The ICP/MS has detection levels equivalent to graphite furnace atomic absorption spectrophotometry while still allowing for the simultaneous analysis of numerous metals in a single analysis. Where ICP replaced the flame atomic absorption spectrophotometer, the ICP/MS is replacing the ICP and AAGF technologies. Currently, the price for the ICP/MS, in the range of $150,000 to 200,000, restricts its use on a wide commercial basis. As the price of this technology decreases, the ICP/MS will become commonplace in the industrial hygiene laboratory.

High-pressure liquid chromatography coupled with mass spectrometry, LC/MS, has the capability of measuring nanogram and picogram quantities of organic compounds. This technology is somewhat commonplace in pharmaceutical research and development. As research and development into new industrial chemicals, as well as pharmaceutical agents, continues there will be an ever-increasing need to evaluate exposures at lower and lower limits. The LC/MS will be the instrument of choice for analysis of organic chemicals. Again, as with ICP/MS, instrument cost is currently prohibitive for the typical laboratory operation.

Differential speciation of elemental and organic carbon (EC/OC) is becoming an increasingly important level of analysis particularly for diesel particulate. The speciation of elemental and organic carbon requires specially designed instrumentation following

NIOSH Method 5040. The use of the EC/OC method reduces the interferences associated with the normal gravimetric determination of diesel particulates.

In typical sampling environments, samples for elemental and organic carbon are collected on precleaned 37-mm high purity quartz filters contained in standard three-piece polystyrene sampling cassettes. The presence of other particulate matter interferes in the subsequent analysis of elemental, organic, and total carbon.

The collected samples are analyzed through evolved gas analysis using a thermal optical analyzer following NIOSH Method 5040. Both organic and elemental carbon forms are determined through a series of controlled combustion steps and chemical reactions within the thermal optical EC/OC instrumentation.

In the first stage, organic and carbonate carbon are evolved in an inert helium atmosphere as the reaction temperature is raised to 900°C. The carbon is oxidized to CO_2 using manganese dioxide and catalytically reduced to methane and quantified by flame ionization detection. In the second stage of analysis, a mixture of oxygen and helium is introduced into the combustion chamber. The oxygen mixture reacts with the elemental carbon. The resulting pyrolytically generated carbon produces an increase in the filter transmittance. Monitoring the filter transmittance during the analysis and correcting for any char produced from the first stage of analysis allows the differentiation of elemental and organic carbon species.

As technology advances and exposure limits and new chemicals are developed, the industrial hygiene chemist will incorporate more and more sophisticated instrumentation. The challenge will be to stay on the leading edge of technology development and finding practical ways to incorporate that technology into industrial hygiene analyses.

4 METHOD SELECTION

There are numerous sampling and analytical methods available to the industrial hygiene chemist. The most common reference sources are the *NIOSH Manual of Analytical Methods*, Fourth Edition (2) and the *OSHA Salt Lake City Technical Center Analytical Methods Manual* (3). In addition to the Fourth Edition, sampling and analytical methods may also be found in the second and third editions of the NIOSH methods manuals. These compilations account for over 400 individual methods. Many of the published methods are similar and some identical especially between the NIOSH and OSHA methods.

The choice of method is dependent on the following factors.

1. Available instrumentation
2. Sensitivity required
3. Interferences present
4. Sampling media
5. Sampling flowrate
6. Sampling time

Since there are multiple methods for individual contaminants, it is important to optimize both the sampling and analytical aspects of the measurement. The sampling and analytical considerations, when selecting a sampling and analytical method are shown below.

Sampling Perspective

- Sampling media
- Method detection limit
- Analytical method
- Interferences
- Applicable exposure standard

Analytical Perspective

- Compatibility of different analytes requested
- Knowledge of other compounds present in the sampling environment

Developing a sampling strategy with the laboratory will assist in the selection of the most appropriate method.

The use of the following equation is helpful in determining the necessary sample volume.

$$\text{Total Volume Required}(\text{M}^3) = \text{Analytical Detection Limit (mg)/Exposure Standard (mg/M}^3) \times 0.1$$

This equation will provide the air volume required to calculate an exposure concentration, assuming that there was no positive result, of one-tenth the targeted exposure standard. Using this simple formula helps to ensures that a non-useable analytical result is not reported.

Coupled with the reporting limits of a particular sampling and analytical method is the concept of method detection limit (MDL). This often-misunderstood concept is subject to considerable debate within the industrial hygiene laboratory community. The MDL is commonly referred to as the value that defines the point at which the analyte signal is sufficiently above the instrument background noise capable of producing a reliable result.

There are a number of ways to develop the MDL value. The most common is taken from **40** *CFR*136, Appendix B. This statistical-based procedure was developed for the analysis of environmental samples such as solids and waters. Calculating MDLs using this method requires preparation and analysis of seven laboratory spike samples using the preparation and analysis methods employed for actual samples. After the laboratory prepared samples are analyzed, the mean and standard deviation of the results are calculated. These statistics are used to calculate a 99% one-sided confidence interval using a Student's "t-test" value.

NIOSH uses a similar procedure using five or more low-level calibration standards spiked on sampling media over a range of the estimated MDL to no more than ten times the MDL. These are separately prepared spiked samples. The responses are graphed and the linear regression equation is calculated. The MDL is then determined as three times the standard regression error divided by the slope of the regression line.

NIOSH suggests that the MDL then be verified against known standards or spike samples prepared at the MDL level. This is an important practice since the MDL does not account for the recovery of the analyte from the media. Most published MDL procedures require annual development or redevelopment of MDLs.

There is further debate on whether performing MDL studies should be required by individual analysts and on individual instruments. This process can be quite cumbersome in large analytical laboratories.

There is an inherent problem with applying the MDL procedure and assuming that the value is accurate. As noted above, the current MDL process does not typically consider the analytical recovery since the statistics are based on variability due to instrument imprecision. If the recovery is less than 100% and no correction is made to the MDL, then it will be unlikely that the reported value could truly be determined at the MDL. As discussed above, it is important that the MDL be verified with the analysis of MDL spike samples.

An alternative method, is a daily check of MDL using a low-level recovery prepared and analyzed with the sample set. This method effectively validates the reported MDL with each sample set and takes into account the day-to-day variability of the instrument response. The method also accounts for the preparation and instrument techniques of various technicians performing analyses.

As an alternative, a low-level or limit of detection (LOD) standard could be used to verify measurements at the lower range of the method. This method would not take into account the preparation procedure and chemical reagents used. This method is essentially equivalent to the MDL recovery sample discussed previously, if the analytical recovery is 100%.

5 METHOD RESOURCES

There are two essential method resources for industrial hygiene sampling and analytical methods—NIOSH and OSHA. The *NIOSH Manual of Analytical Methods* is now in its Fourth Edition and encompasses over 250 methods (2). The NIOSH Fourth Edition methods are classified according to the level of validation undertaken. There are three classifications used—Full, Partial, and Unrated. Full validation covers those steps described below. Methods with a Partial evaluation have undergone a portion of the evaluation criteria or do not meet the ±25% accuracy criterion. Unrated methods have not been tested by NIOSH, but have been developed by a recognized independent source such as OSHA. The third edition of the NIOSH methods manual followed a similar validation and evaluation protocol.

The NIOSH second edition methods used a five-tier classification system depending on the degree of evaluation. A method classified as an E (Proposed) method was typically a method that was published or developed, but not subjected to any type of rigorous testing program either in the laboratory or the field. After successful use of the method on at least 15 field samples, an E class method could be upgraded to D (Operational) methodology.

A class C (Tentative) method was reserved for methods that were in general use in the industrial hygiene laboratory community. Class C methods, while in common usage in industrial hygiene laboratories, had not been evaluated under a controlled laboratory environment. The next step in the validation process was a thorough laboratory evaluation of the particular method. This laboratory validation encompassed method accuracy and precision, method recoveries, sample stability, and collection efficiencies. After successful evaluation according to this protocol, a method would be classified as a Class B (Recommended) method. The "A" (Accepted) class method was the highest order of evaluation. A Class A method had undergone both a field validation and collaborative testing.

The OSHA Manual of Analytical Methods encompasses over 200 sampling and analytical methods covering both inorganic and organic compounds (3). These methods, originating from the OSHA Salt Lake City Technical Center laboratory, include both laboratory validated and experimental "Stop-Gap" methods. The OSHA methods that undergo validation have been evaluated for precision and accuracy, recovery, stability, and collection efficiency.

6 METHOD PERFORMANCE

The performance of analytical methods is based on a combination of precision and bias. NIOSH refers to method performance as method accuracy, defined as the absolute value of the method bias plus the relative standard deviation (RSD) at the 95% confidence level. Relative standard deviation is also known as the coefficient of variation, or standard deviation divided by the mean.

The method bias is the uncorrectable relative discrepancy between the mean of the distribution of measurements from a method and the true concentration being measured. The most common means of measuring bias is the preparation of laboratory spike samples.

Precision is the relative variability of the measurements on replicate samples about the mean of the population of measurements or standard deviation. The standard deviation is derived for the 95% confidence level. The standard deviation is then divided by the mean of the sample population.

There are extensive statistical calculations published by NIOSH supporting this definition. These can be reviewed in the "Guidelines for the Air Sampling and Analytical Method Development Evaluation" (4).

The NIOSH method accuracy criterion requires that a method give a result that is within ±25% of the true concentration, at the 95% probability level, when measuring concentrations above the permissible exposure limit. At or below the permissible exposure limit the method accuracy must be ±35% and at or below the action level ±50%. The more conservative approach at the lower range of the analytical method is appropriate given the increased variability of most methods at levels nearing the lower limit of detection or measurement range.

A thorough understanding of the method accuracy, bias, and precision is important in evaluating the sampling results. Generally, most methods have been evaluated for these important statistics. As a general rule, the variability of the analytical result increases as the value approaches the method limit of detection. Conversely, as the analytical value moves up the operating range of the method, the variability will stabilize to the accuracy depicted by the defined method accuracy.

6.1 Defining Method Performance

Defining method performance, or method validation, is important in understanding the capability and limitation of the sampling and analytical method. There are various ways and degrees of conducting method validations. There is a logical progression of steps when conducting a systematic method validation that encompasses basic research, selection of analytical method and sampling media and performing laboratory analyses. Before starting the laboratory aspects of the validation, it is important to understand the method application and exposure limits that are attempting to be measured. The exposure limit is commonly used as the focal point of the validation and setting the performance ranges.

The first step is performing a literature search for the target compound. Important information gained from the literature search is physical properties, such as molecular weight, density, boiling and melting point, vapor pressure, and compound stability under various environmental conditions. In addition to physical properties, it is usually possible to obtain basic analytical methods from the chemical manufacturer. The manufacturer's methods are typically used in process quality control or synthesis and must be adapted to the lower measurement levels demanded by industrial hygiene application. As part of the initial research, information on potential uses of the compound and the industrial process can be beneficial in defining any potential interference that may be encountered and could be evaluated during the validation.

After deciding on the analytical technique and potential sampling media, the method validation sequence is defined. This encompasses the following steps.

1. Analytical method performance (linear range and precision over the range).
2. Interferences (if known).
3. Method recovery (using the sampling media selected).
4. Storage stability (using the sampling media selected).
5. Collection efficiency.
6. Field validation.
7. Shelf-life and stability of media.

The majority of industrial hygiene analytical methods have been validated to some extent for these seven steps.

Method validation is generally performed at four concentration levels corresponding to 0.1, 0.5, 1, and 2 times the target concentration, using at least six samples at each concentration level. The six trials allow for some degree of statistical significance to the final results. The more trials at each level that are performed, the more reliable the statistical evaluation of the data.

6.2 OSHA Method Validation Process

The OSHA Salt Lake City Laboratory has developed a very specific protocol for the evaluation of sampling and analytical methods (6). This approach provides a uniform and consistent means of evaluating method applicability and ruggedness. Each OSHA method write-up has the following five elements—General Discussion, Sampling Procedure, Analytical Procedure, Back-up Data, and References. Each of these elements contains significant detail into the method evaluation.

The first step in developing new sampling and analytical methods is to perform a literature review on the compound of concern; looking for existing or historical methods, compound physical properties, and toxicological data to provide a target concentration for the validation.

Once a method is proposed for evaluation a series of experiments are performed for method evaluation. These occur in a sequential process and continue until completion provided that each step in the process is completed to performance specification. If not, the method is revised and the process is restarted.

OSHA defines an acceptable method as one that produces at least a 75% recovery and has a precision of ±25% at the 95% confidence interval. These various evaluation steps are shown in Table 14.4.

6.3 New Chemical Exposure Limits (NCELs) Validation

Under the Toxic Substances Control Act (TSCA) of 1976, Section 5 established a "New Chemical Program" that regulates the risks from new chemicals before they enter the marketplace (7). This program requires manufacturers to submit a premanufacture notice (PMN) to EPA at least 90 days before the manufacture of a chemical for commercial purposes. Consistent with good industrial hygiene practice, EPA has developed performance-based sampling and analysis requirements, as an alternative to respirators, tied with performance based exposure limits or New Chemical Exposure Limits (NCELs). An important aspect of the NCELs provisions is the required validation and performance evaluation of the PMN chemical sampling and analytical method by an independent third-party laboratory.

The validation criteria under NCELs includes determination of:

- Accuracy
- Precision
- Stability
- Collection Efficiency
- Recovery Efficiency
- Breakthrough

The sampling and analytical method must clearly demonstrate that the method is accurate to within ±25% at the 95% confidence interval for concentrations of the PMN substance ranging from one-half to twice the NCEL. This can be accomplished by spiking

Table 14.4. Typical Method Evaluation Steps Used by the OSHA Laboratory

Evaluation Step	Purpose
Analytical detection limit	Ensure that the method is capable of measuring the targeted exposure concentration
Method precision	Define method variability as defined by coefficient of variation at 0.5, 1, and 2 times the target concentration using six replicates
Instrument response	Define the operating range of the method at 0.5 and 2 times the target concentration
Sampling and analytical interferences	Define the potential interferences to the method that could come from the sampling and analytical method
Collection efficiency/ Sampler capacity	Define the breakthrough concentration for sorbent tube methods and sampler collection efficiency for filter methods at 2 times the target concentration
Recovery	Define method recovery as defined by percent recovery at 0.5, 1, and 2 times the target concentration using six replicates
Storage stability	Define the effects of storage on sample stability at time zero and at three day intervals up to 15 days. Stability is determined at both room temperature and refrigeration temperatures
Method detection limit	Similar to the analytical detection limit using the actual sampling matrix and estimated air sampling volume
Reliable quantitation limit	Define the smallest amount of analyte that can be quantitated within at leas a 75% analytical recovery and a method precision of ±25%
Combined sampling and analytical method precision	Define the overall method precision through the use of the Storage studies and recovery data discussed above
Reproducibility	Define the consistency and ease of method introduction by allowing a second analyst, not familiar with the method perform the method

at least six samples at each of three levels corresponding to 0.5, 1, and 2 times the NCEL. The resulting data are then evaluated for both precision and accuracy.

Sample stability is determined through the generation of duplicate sets of samples. The manufacturer must collect two sets of six samples at the TWA concentration from either a controlled environment (gases and vapors) or direct spiking onto media (solids or liquids with a low vapor pressure). The manufacturer analyzes one set of six samples and a third-party laboratory performs analysis on the second set. If the average results of the duplicate samples are not within ±10% of the true value, the manufacturer must perform an additional sample stability study. The stability study involves collection of three sets of six samples at the TWA concentration. The manufacturer analyzes one set of six immediately after collection.

The storage stability study is performed on the remaining duplicate set of six samples. The third-party laboratory analyzes one set and the manufacturer analyzes the other set at the same time. A validated method must show the stability result in agreement of ±10% of the true value.

In addition to these specific studies, the TSCA Guidelines shown below are required method performance for each PMN chemical.

- *Linear Range*—a lower quantitation limit (LQL) defined as 5X baseline noise to an upper range of 2X NCEL must be demonstrated.
- *Instrument Calibration*—5-point calibration curve with a correlation coefficient of 0.95. The five points include standards at 0.5 NCEL, 0.5 to 1X NCEL, 1X NCEL, 1 to 2X NCEL, and 2X NCEL. Continuing calibration standards must be run every 10 samples and be within ± 25% of the initial calibration curve.
- *Sampling and Analysis Recovery*—Collection efficiency and analytical recovery must be between 75–125% as determined from the airborne generation for gases and vapors or direct spiking for particulates.
- *Sampling Device Capacity*—The sampling method must show <5% breakthrough at the challenge concentration of 2X NCEL and a relative humidity of >80%.
- *Analytical Recovery*—Analyte recovery must be determined at the NCEL using six matrix spikes. Recovery must be 75–125%.
- *Precision*—The coefficient of variation for a set of six samples collected from a controlled atmosphere at 1X NCEL, must be less than 10.5%,including an allowance of 5% sampling error.

The sampling and analytical method must be revalidated for all parameters if there are any changes in the workplace environment or method modification (from the original validation) that are reasonably likely to invalidate the initial method.

7 SAMPLING MEDIA TYPES

Sampling is an integral part of evaluating worker exposures. The overall accuracy of the analytical result can be no better than the accuracy of the sampling procedure. The sampling media must be selected to achieve the degree of specificity needed for the collection of gases, vapors, mists, particulates, and fibers. The sampler must have sufficient capacity and working range to accomplish time-weighted average and short-term exposure monitoring.

There are four types of sampling media used in industrial hygiene. Filters and sorbent tubes, both standard and chemically treated, are the most common media used in industrial hygiene. Impingers, the standard means of sampling in the 1960s and 1970s, have been almost entirely replaced by chemically treated filters and sorbent tubes. There are still many applications for the standard gas bag (Tedlar®) sampler. A relatively new means of sampling gases are evacuated stainless steel canisters (SUMMA®). These are especially useful when sampling unknown environments for analysis by gas chromatography and mass spectrometry (GC/MS).

The standard industrial hygiene filter assembly is the 37-mm filter contained in a two- or three-piece polystyrene cassette. The most common filter types for the collection of particulates and mists include mixed cellulose ester, poly(vinyl chloride) Teflon® and glass fiber. In addition to the standard filter media, numerous chemically treated media have been developed replacing the impinger and gas bag methods.

Solid sorbent tubes are the standard for the sampling of gases and vapors. The industry standard for organic solvents is the 150-mg two-section charcoal tube. Other sorbent tubes include silica gel, XAD®, Tenax®, Chromosorb®, and molecular sieve.

Impingers were the mainstays of industrial hygiene sampling and analysis for many years. The principal difficulty with impinger sampling was spillage of the impinger solution and the awkwardness of the impinger sampler on the worker. A few impinger methods are still in use today. These include formaldehyde using sodium bisulfite or water and hydrogen sulfide using a cadmium hydroxide solution.

Treated filter and sorbent tube media have been developed for numerous chemicals replacing many of the original impinger methods that were either reactive or unstable during storage. Examples include isocyanates, hydrogen fluoride, aldehydes (acetaldehyde, acrolein, formaldehyde), chlorine, hydrogen sulfide, and sulfur dioxide. A list of common treated media is shown in Table 14.5.

There are a number of sampling procedures that incorporate two and three stage sampling trains that allow the separation of the particulate and vapor phases of the chemical exposure. Sampling trains for common industrial hygiene contaminants are shown in Table 14.6.

8 PROBLEMATIC METHODS

There are a number of common industrial hygiene sampling and analytical methods that are prone to low collection efficiency, contaminated sampling media, interfering substances, and the loss of analyte during shipping and storage. These common problems must be addressed to ensure accurate evaluation of the exposure. Most of the published sampling and analytical methods note these precautions in the written methods under either the interference and/or precautions sections of the method. Another good source for this information is the analytical laboratory.

Table 14.5. Common Treated Media

Contaminant	Media Type	Chemical Treatment
Acrolein	XAD-2	2-(Hydroxymethyl) piperidine
Ethylene oxide	Charcoal tube	Hydrobromic acid
Fluorides, gaseous	Cellulose pad	Sodium carbonate
Formaldehyde	XAD-2	2-(Hydroxymethyl) piperidine
Inorganic Acids	Silica gel	Water washed
Isocyanates (MDI, TDI, HDI, IPDI)	Glass fiber filter	1-(2-Pyridyl) piperazine
Mercaptans	Glass fiber filter	Mercuric acetate
Nitrogen oxides	Molecular sieve	Triethanolamine
Sulfur dioxide	Cellulose pad	Potassium hydroxide
Triethylene tetramine	XAD-2	1-Naphthylisothiocyanate

Table 14.6. Common Industrial Hygiene Sampling Trains

Chemical	Sampling Train Configuration	Target Compound	Method Reference
Nitrogen oxides	Triethanol treated molecular sieve tube Catalyst Triethanol treated molecular sieve tube	Nitrogen dioxide (NO_2) Nitrogen oxide (NO)	NIOSH 6014
Polyaromatic hydrocarbons (PAH)	Teflon filter XAD-2	Particulate-phase PAH Vapor-phase PAH	NIOSH 5506
Cyanide	PVC Filter 0.1N KOH Impinger	Particulate cyanide Gaseous cyanide	NIOSH 7904
Polychlorinated biphenyls (PCB)	Glass fiber filter Florisil tube	Particulate PCB Vapor-phase PCB	NIOSH 5503
Sulfur dioxide	MCE filter Treated Whatman filter	Sulfate interference Sulfur dioxide	NIOSH 6004
Organophosphorous pesticides	Glass fiber filter XAD-2 Tube	Particulate Vapor-phase	NIOSH 5600
Glycols	Glass fiber filter XAD-7 Tube	Particulate Vapor-phase	NIOSH 5523

A listing of the most common problematic industrial hygiene sampling and analytical methods are provided in Table 14.7.

9 LABORATORY QUALIFICATIONS

9.1 AIHA Accreditation

The AIHA Industrial Hygiene Laboratory Accreditation Program (IHLAP) was created in 1972 through a grant from NIOSH. Over the years the AIHA has added accreditation and quality programs for field asbestos counters [Asbestos Analyst Registry (AAR)], lead [Environmental Lead Laboratory Accreditation Program (ELLAP)], and microbiology [Environmental Microbiological Laboratory Accreditation Program (EMLAP)]. In 1999, the AIHA is investigating their involvement in environmental laboratory accreditation through the National Environmental Laboratory Accreditation Program (NELAP). With the addition of these other programs the AIHA appropriately renamed their activities the AIHA Laboratory Quality Programs.

The AIHA Industrial Hygiene Laboratory Accreditation program serves a valuable function in the industrial hygiene profession by providing a standard of performance for laboratories and assisting laboratories in achieving and maintaining those performance

Table 14.7. Problematic Methods

Chemical	Method Reference	Precaution
Nitrogen oxides (NO/NO_2)	NIOSH 6014	The sampling air flow direction through the sampling train must be known and marked
Hexavalent chromium	NIOSH 7600	Samples must be collected on a PVC filter to preserve the hexavalent chromium collected
Formaldehyde	All Methods	The sampling media must be protected before and after collection to ensure that ambient formaldehyde is not inadvertently sampled.
Organic solvent mixtures (e.g., mineral spirits, naphtha, distillates)	NIOSH 1550	Bulk reference samples should be submitted with the samples and shipped in a separate package.
Hydrogen sulfide	NIOSH S4	The impinger solution must not be transferred in the field.
Polynuclear aromatic hydrocarbons	NIOSH 5506	Collection samples must be stored cold and protected form light.
Isocyanates	OSHA 42	Media must be shipped and stored cold.
Ethanolamine	NIOSH 2007	Front section of the tube is spiked after sampling with hydrochloric acid or sample has to be kept cold and analyzed within 14 days
Coal tar pitch volatiles	NIOSH 5023	Transfer filter from sampling cassette to a glass vial after sampling and protect samples from light during and after sampling.

goals. A 16-member committee offering a wide range of experience in academia, government, industry, and consulting directs the AIHA accreditation program. Through both laboratory review and performance audits, the AIHA program targets those laboratories that are providing analytical services in the evaluation of workplace exposures.

The purpose of the AIHA IHLAP is to establish and maintain standards for the industrial hygiene laboratories and assist those laboratories that do not meet IHLAP standards. The IHLAP acts as both a "Standards Enforcer" and "Guidance Counselor" to laboratories, where members are willing to assist laboratories in all matters of the accreditation process and maintenance of high standards of performance. The standards and qualifications for accreditation fall into five areas:

1. Personnel
2. Quality Assurance/Quality Control
3. Recordkeeping
4. Methods
5. Facilities

Criteria for each area is described in the operating policies of the IHLAP and patterned after ISO Guide 25 criteria. Each laboratory in the program must continually meet the program policies throughout the three-year accreditation cycle. In addition to a formal application process, the laboratory undergoes an on-site evaluation by an AIHA program site visitor. A reaccreditation application is submitted every three years. Reaccreditation follows the same criteria and site visit as the original accreditation.

A critical part of the AIHA IHLAP is the analysis of quarterly Proficiency Analytical Testing (PAT) samples that include metals, organic solvents, asbestos fibers, asbestos bulks, and silica. The AIHA is continually adding new proficiency samples as the programs expand in scope. AIHA accredited laboratories must analyze all PAT samples that are representative of their sample analysis mix.

9.2 Quality Assurance Program

A laboratory quality program is an overall process that consists of two separate but related activities; quality assurance (QA) and quality control (QC). The QA program defines the policies and overall activities that assures that the data generated by the laboratory meets defined levels of quality and defensibility. The laboratory QC program encompasses those measurements that ensure that the laboratory processes are operating correctly.

The cornerstone of a good QA program is the written QA Manual that describes the specific organizational structure and procedures that the laboratory follows in the production of analytical results. There are three basic tenets of producing quality analytical results are scientific validity, defensibility, and known precision and accuracy. The QA Manual should address the following topics (8).

- QA Manual Maintenance and Updating Procedures
- Organization and Responsibilities within the laboratory
- QA Objectives and Policies
- Personnel Qualifications and Training
- Sampling Materials and Procedures
- Chain of Custody/Sample Receiving Procedures
- Reagents and Standards
- Equipment Calibration and Maintenance Procedures
- Analytical Methods
- Data Reduction, Validation and Reporting
- Internal Quality Control Procedures
- Performance and System Audits
- Corrective Action
- Quality Assurance Reports
- Documentation and Recordkeeping
- Sample Retention and Disposal

The QA Manual should be reviewed annually and updated as the laboratory operation evolves.

Quality assessments are an essential part of the QA program and encompass both performance and system audits. A performance audit is a quantitative evaluation of the laboratory measurement system. A system audit is a qualitative evaluation of all the components of the measurement system.

The laboratory QC program encompasses those procedures and measurements that ensure that the laboratory processes are operating correctly. The most common form of quality control is the introduction of laboratory control samples (blank media spiked with a known amount of analyte) and external reference samples. The QC samples are analyzed with each sample set and used to determine the accuracy and precision of the analyses performed at a particular time interval.

All QC measurements must be evaluated against acceptance criteria that are developed by the individual laboratory. Performing QC measurements without corresponding acceptance criteria are unproductive. Statistical based acceptance criteria are recommended whenever possible. A minimum of 20 QC data points is considered adequate for evaluating results using statistical methods. There may be exceptions where a laboratory procedure is not performed regularly and lacks sufficient data to generate statistical based criteria. In these cases, the laboratory must either use the data available (at least seven points) or set acceptance limits using professional judgement.

The most common means of statistically evaluating QC data is the development of control charts. Control charts are graphical plots of QC tests results with respect to time or sequence and are used for both evaluating data and determining performance trends. The control limits are statistical boundaries that define the range of expected fluctuation in analytical data. The control chart is constructed with five defined levels as follows.

Mean (X)
Upper warning limit (X $\pm$ 2SD)
Lower warning limit (X $\pm$ 2SD)
Upper control limit (X $\pm$ 3SD)
Lower control limit (X $\pm$ 3SD)

The mean (X) is generally calculated using at least 15 to 20 data points generated from prepared laboratory spike samples. The upper and lower warning limits are defined at the 95% confidence interval (2 standard deviations from the mean). The upper and lower control limits are set at the 99% confidence interval (3 standard deviations form the mean). Once constructed the control chart is then used to assess the current laboratory performance against historical practice (control chart data). Results outside of either the upper or lower warning limits are within the control limits but require investigation for possible errors or system problems. Results outside of the control limits are considered "out of control" and require corrective action and reanalysis prior to reporting the analytical results.

The best use of the control chart is determining trends and taking corrective action prior to producing out of control results. Charting QC data by analytical methods, instrument type, and analysts creates an efficient means of problem solving and more importantly problem prevention.

9.3.1 *Data Quality Documentation*

The degree of defensibility of laboratory data is an important aspect of laboratory analysis. Without defensible results the exposure measurements derived have little value in decisionmaking. The QA program discussed above is the foundation for data quality. The defensibility relies on appropriate documentation that the QA/QC program was performed. Documentation should include records pertaining to analyst training, instrument calibration, support equipment calibration, batch recovery or blind spike samples, laboratory blanks, and data review process. Table 14.8 shows the basic QA/QC processes and the rationale for each step.

The overall defensibility of data relies on the presence of acceptance criteria for all quality measurements. There is little or no value to a quality measurement that is not comparable to acceptance criteria. For example, the simple process of calibrating an analytical balance, a process performed in virtually every laboratory every single day, has no inherent value unless the person performing the calibration is able to determine if the resulting measurement is within normal expected operating range. Defining the normal operating range is best accomplished through the statistical evaluation of historical data, including mean and standard deviation. As noted above, the use of control charts for this function are useful in plotting trends and alerting the analyst of the need for preventative actions.

9.3.2 *Assessing Laboratory Performance*

Assessing a laboratory's performance requires a basic understanding of laboratory operations and quality assurance and quality control principles. There are four basic areas for

Table 14.8. QA/QC Processes

QA/QC Principle	Purpose	Defensibility Record
Analyst training	Ensure that the analyst performing the analysis understands and has demonstrated a basic comprehension of the procedure	Education and training records, demonstrated capability for the particular analytical procedure
Instrument calibration	Ensure instrument operation	Calibration data with performance criteria
Equipment calibration	Ensure equipment operation	Calibration data with acceptance limits
Batch recovery samples	Monitor method and analyst performance	Recovery data and statistical acceptance limits
Laboratory blanks	Monitor matrix and laboratory-based contamination	Blank data and acceptance criteria
Data review process	Verification of procedures followed and quality check of all data	Accountability of data through sign-off and approval of quality records

review. The assessment process should include both a review of paper records and on-site assessment of the laboratory operation. The first is the qualifications and experience of the laboratory management. This can only be gleaned from personal interviews with supervisory staff. Second is the existence of a written Quality Assurance program that is specific to the laboratory. There are many laboratories that copy or create more of a generic QA Manual that does not specifically address their particular operation. A simple test consists of taking a section of the QA Manual and requesting that the laboratory staff demonstrates that they practice what is written. Third is a review of the documentation that supports the QA program as noted in the key points shown in Table 14.8 above. The fourth area of assessment is record retention procedures and policies. There is no set rule for the required length of time records must be retained or what records need to be retained. At a minimum, all laboratory records supporting the analytical procedure must be retained. Examples are provided in Table 14.9.

The on-site assessment, mentioned above, will usually require about 6 to 8 hours on-site, depending on the size of the laboratory operation. Although there is no "best way" of conducting an on-site assessment, a typical assessment would include the following elements.

The most important aspect of an assessment is detailed interviews with senior management and supervisory staff. The senior management sets the capabilities of the operations, so these discussions will provide the management approach and attitudes about technical, quality, and operational characteristics of the laboratory operation. Of particular importance is acquiring an understanding of the operation of the quality program. This can be gleaned from either the laboratory director or the quality assurance coordinator. A technique that proves valuable in obtaining factual information is having the management staff describe the operation of the quality program without the use of notes or simple quotations from the quality manual. The management description of the program can then be compared to what is written in the Quality Assurance manual, described above, and the actual practices followed by the laboratory staff.

After obtaining an operational overview from management staff, the next step is performing a general laboratory walkthrough gaining information on housekeeping and ge-

Table 14.9. Laboratory Records

Analytical Procedure	Supporting Record
Calibration	Instrument data printouts and chromatograms
Sample analysis	Identification of analyst
	Method reference
	Preparation data
	Instrument data printouts and chromatograms
Quality control	Recovery sample results
	Acceptance criteria
	Control charts
	Corrective actions
Peer and management review of data, procedures, and training	Review sheet with signature and dates

neric safety procedures. After this walkthrough is completed, the next step is randomly selecting finished report from the files and retracing the entire process that the samples, represented by the selected report, followed through the laboratory. This is a detailed, laborious process and will include the sample chain of custody and tracking (typically referred to as the login process), discussions with laboratory analysts that performed the work, the quality control and data review procedures, and final report review and authorization.

It is important during this stage of the on-site assessment that the laboratory records be complete and capable of recreating, through laboratory documentation, the entire analytical process. The laboratory staff must be able to demonstrate what was done to the samples and the quality control procedures that were incorporated into their analytical regimen. Specific questions related to how the instrument calibration, preparation of standards, batch quality control performed, data review, and final report authorization must be addressed. All records must have supporting documentation that can be used as verification that these steps were indeed performed.

Once a laboratory is selected for use, it is important to verify performance on a continuing basis. This can be accomplished through review of laboratory performance in interlaboratory proficiency programs, such as the AIHA Proficiency Analytical Testing Program (PAT). Another method, that is probably a more genuine indicator of laboratory performance, is submission of "double blind" quality control samples. These QC samples are prepared and disguised as regular samples intermingled with an actual set of field samples.

10 SUMMARY

The defensibility and quality of industrial hygiene exposure data relies on the overall quality programs of both the sample collector and the industrial hygiene laboratory. While this chapter has focused on the laboratory aspects of industrial hygiene, the exposure assessment program must encompass all areas, from sampling strategy, calibration and functionality of the sampling equipment, chain-of-custody and laboratory analysis, to the final review of the exposure data against exposure standards. With careful attention to quality details, the resulting data and corresponding decisions can be made with confidence. Useful general references are listed in bibliography (9–15).

ACKNOWLEDGMENTS

The author wishes to express his appreciation to Laura McMahon, Edith Liptow, and Allen Schinsky, CIH of Clayton Laboratory Services for their assistance in reviewing this manuscript.

BIBLIOGRAPHY

1. S. A. Roach, E. J. Baier, H. E. Ayer, and R. L. Harris, "Testing Compliance with Threshold Limit Values for Respirable Dusts," *Am. Ind. Hyg. Assoc. J.* 543–49 (Nov.–Dec. 1967).

2. "*NIOSH Manual of Analytical Methods*," 4th ed., U.S. Department of Health and Human Services, NIOSH Publication No. 94-113, Aug. 1994.
3. "*OSHA Manual of Analytical Methods, OSHA Methods Evaluation Branch*," OSHA Salt Lake City Laboratory, Salt Lake City, Utah, Oct. 1989.
4. "*Guidelines for Air Sampling and Analytical Method Development and Evaluation*," U.S. Department of Health and Human Services, NIOSH Technical Report, PB96-134564, May 1995, DHHS/PUB/NIOSH-96-117.
5. "*Occupational Exposure Sampling Strategy Manual*," U.S. Department of Health and Human Services, NIOSH Publication No. 77-173, Jan. 1977.
6. "*OSHA Manual of Analytical Methods, OSHA Methods Evaluation Branch*," OSHA Salt Lake City Laboratory, Salt Lake City, Utah, Oct. 1989.
7. "*Guidance for New Chemical Exposure Limits (NCELs) Development and Validation of Analytical Methods for Workplace Monitoring Under Section 5 of TSCA*," USEPA Office of Pollution Prevention and Toxics, May 1995-Draft.
8. "*Laboratory Quality Assurance Program Policies*," AIHA Laboratory Quality Assurance Program, Jan. 1999.
9. *The Industrial Environment—Its Evaluation and Control*, U.S. Department of Health, Education, and Welfare, Public Health Service, Center for Disease Control, National Institute for Occupational Safety and Health, 1973.
10. S. R. Di Nardi, ed., *The Occupational Environment—Its Evaluation and Control*, AIHA Press, Fairfax, VA, 1997.
11. H. H. Willard, L. L. Merritt, Jr., and J. A. Dean, "*Instrumental Methods of Analysis*," 5th ed., Van Nostrand Company, New York, 1974.
12. H. H. Willard, L. L. Merritt, Jr., J. A. Dean, and F. A. Settle Jr., " *Instrumental Methods of Analysis*," 7th ed., Van Nostrand Company, New York, 1988.
13. "*AIHA Quality Assurance Manual for Industrial Hygiene Chemistry*," AIHA Analytical Chemistry Committee, 1988 and 1994.
14. AIHA, "Understanding EPA's New Chemical Exposure Limits Program," *The Synergist*, (April 1995).
15. American Chemical Society, "A History of Instrumentation," *Today's Chemist at Work*, **8**(3) (March 1999).

CHAPTER FIFTEEN

Calibration

Morton Lippmann, Ph.D., CIH

1 INTRODUCTION

Proper interpretation of any environmental measurement depends on an appreciation of its accuracy, precision, and whether it is representative of the condition or exposure of interest. This chapter is concerned primarily with the accuracy and precision of industrial hygiene measurements. Although considerations of the location, duration, and frequency of measurements may be equally or even more important in the evaluation of potential hazards, they require knowledge of the process variables, the kinds of hazard and/or toxic effect that may result from exposures, and their temporal variations. Such considerations, which require the exercise of professional judgment, are beyond the scope of this chapter.

The accuracy of a given measurement depends on a variety of different factors including the sensitivity of the analytical method, its specificity for the agent or energy being measured, the interferences introduced by cocontaminants or other radiant energies, and the changes in response resulting from variations in ambient conditions or instrument power levels. In some cases the influence of these variables can be defined by laboratory calibration, providing a basis for correcting a field sample or instrument reading response. In cases such as the effect of variable line voltage on an instrument's response, they can be avoided by modifications in the circuitry or by the addition of a constant voltage transformer. When the effects of the interferences cannot be controlled or well defined, it may still be desirable to make field measurements, especially in range-finding and exploratory surveys. The interpretation of any such measurements is greatly aided by an appreciation of the extent of the uncertainties. Laboratory calibrations can provide the basis for such an appreciation.

It is important to document the nature and frequency of calibrations and calibration checks to meet legal as well as scientific requirements. Measurements made to document

Patty's Industrial Hygiene, Fifth Edition, Volume 1. Edited by Robert L. Harris.
ISBN 0-471-29756-9

the presence or absence of excessive exposures are only as reliable as the calibrations upon which they are based. Formalized calibration audit procedures established by federal agencies provide a basis for quality assurance where they apply. They can also provide a systematic framework for developing appropriate calibration procedures for situations not governed by specific reporting requirements.

State and local air monitoring networks that are collecting data for compliance purposes are required to have an external performance audit on an annual basis (1). The audit also summarizes the performance of the instruments. In the case of ozone, for example, this would include the records of the weekly multipoint calibrations at 0.1, 0.2, and 0.4 ppm.

The *NIOSH Manual of Analytical Methods* (2) recommends that sampling pumps should be calibrated with each use, and that this calibration be performed with the sampler in line. It also recommends that records of calibration be recorded with each unit.

2 TYPES OF CALIBRATIONS

Occupational health problems can arise from exposure to airborne contaminants, heat stress, excessive noise, vibration, ionizing radiations, and nonionizing electromagnetic radiations. Each of these types of exposure involves a different set of measurement variables and calibration considerations, and they are considered separately. Other types of measurements requiring calibration are associated with the evaluation of ventilation systems used to control exposures to airborne contaminants.

2.1 Air Sampling Instruments

Air samples are collected to determine the concentrations of one or more airborne contaminants. To define a concentration, the quantity of the contaminant of interest per unit volume of air must be ascertained. In some cases the contaminant is not extracted from the air (i.e., it may simply alter the response of a defined physical system). An example is the mercury vapor detector, where mercury atoms absorb the characteristic ultraviolet radiation from a mercury lamp, reducing the intensity incident on a photocell. In this case the response is proportional to the mercury concentration, not to the mass flow rate through the sensing zone; hence concentration is measured directly.

In most cases, however, the contaminant either is recovered from the sampled air for subsequent analysis or is altered by its passage through a sensor within the sampling train, and the sampling flow rate must be known to be able ultimately to determine airborne concentrations. When the contaminant is collected for subsequent analysis, the collection efficiency must also be known, and ideally it should be constant. The measurements of sample mass, of collection efficiency, and of sample volume are usually done independently. Each measurement has its own associated errors, and each contributes to the overall uncertainty in the reported concentration.

The sampled volume measurement error often is greater than that of the contaminant mass measurement. The usual reason is that the volume measurement is typically made in the field with devices designed more for portability and light weight than for precision and accuracy. Flow-rate measurement errors can further affect the determination if the collection efficiency depends on the flow rate.

Each element of the sampling system should be calibrated accurately before initial field use. Protocols should also be established for periodic recalibration, because the performance of many transducers and meters changes with the accumulation of dirt, as well as with corrosion, leaks, and misalignment due to vibration or shocks in handling, and so on. The frequency of such recalibration checks should be high initially, until experience indicates that it can be safely reduced.

2.1.1 Flow and/or Volume

If the contaminant of interest is removed quantitatively by a sample collector at all flow rates, the sampled volume may be the only airflow parameter that need be recorded. On the other hand, when the detector response depends on both the flow rate and sample mass, as in many length-of-stain detector tubes, both quantities must be determined and controlled. Finally, in many direct-reading instruments, the response depends on flow rate but not on integrated volume.

In most sampling situations the flow rates are, or are assumed to be, constant. When this is so, and the sampling interval is known, it is possible to convert flow rates to integrated volumes, and vice versa. Therefore flow-rate meters, which are usually smaller, more portable, and less expensive than integrated volume meters, are generally used on sampling equipment even when the sample volume is the parameter of primary interest. Little additional error is normally introduced in converting a constant flow rate into an integrated volume, because the measurement and recording of elapsed time generally can be performed with good accuracy and precision.

Flowmeters can be divided into three groups on the basis of the type of measurement made: integrated volume meters, flow-rate meters, and velocity meters. The principles of operation and features of specific instrument types in each group are discussed in succeeding pages. The response of volume meters, such as the spirometer and wet test meter, and flow-rate meters, such as the rotameter and the orifice meter, are determined by the entire sampler flow. In this respect they differ from velocity meters such as the thermoanemometer and the Pitot tube, which measure the velocity at a particular point of the flow cross section. Because the flow profile is rarely uniform across the channel, the measured velocity invariably differs from the average velocity. Furthermore, because the shape of the flow profile usually changes with changes in flow rate, the ratio of point-to-average velocity also changes. Thus, when a point velocity is used as an index of flow rate, there is an additional potential source of error, which should be evaluated in laboratory calibrations that simulate the conditions of use. Despite their disadvantages, velocity sensors are sometimes the best indicators available, for example, in some electrostatic precipitators, where the flow resistance of other types of meters cannot be tolerated. Velocity sensors are also used in measurements of ventilation airflow and to measure one of the components of heat stress.

2.1.2 Calibration of Collection Efficiency

A sample collector need not be 100% efficient to be useful, provided its efficiency is known and consistent and is taken into account in the calculation of concentration. In practice, acceptance of a low but known collection efficiency is reasonable procedure for most types of gas and vapor sampling, but it is seldom, if ever, appropriate for aerosol sampling. All the molecules of a given chemical contaminant in the vapor phase are essentially the same size, and if the temperature, flow rate, and other critical parameters are kept constant, the molecules will all have the same probability of capture. Aerosols, on the other hand, are rarely monodisperse. Because most particle capture mechanisms are size dependent, the collection characteristics of a given sampler are likely to vary with particle size. Furthermore, the efficiency will tend to change with time because of loading; for example; a filter's efficiency increases as dust collects on it, and electrostatic precipitator efficiency may drop if a resistive layer accumulates on the collecting electrode. Thus aerosol samplers should not be used unless their collection is essentially complete for all particle sizes of interest.

2.1.3 Recovery from Sampling Substrate

The collection efficiency of a sampler can be defined by the fraction removed from the air passing through it. However, the material collected cannot always be completely recovered from the sampling substrate for analysis. In addition, the material sometimes is degraded or otherwise lost between the time of collection in the field and recovery in the laboratory. Deterioration of the sample can be particularly severe for chemically reactive materials. Sample losses may also be due to high vapor pressures in the sampled material, exposure to elevated temperatures, or reactions between the sample and substrate or between different components in the sample.

Laboratory calibrations using blank and spiked samples should be performed whenever possible to determine the conditions under which such losses are likely to affect the determinations desired. When it is expected that the losses would be excessive, the sampling equipment or procedures should be modified as much as feasible to minimize the losses and the need for calibration corrections.

2.1.4 Sensor Response

When calibrating direct-reading instruments, the objective is to determine the relation between the scale readings and the actual concentration of contaminant present. In such tests the basic response for the contaminant of interest is obtained by operating the instrument sensor in known concentrations of the pure material over an appropriate range of concentrations. In many cases it is also necessary to determine the effect of such environmental cofactors as temperature, pressure, and humidity on the instrument response. Also, many sensors are nonspecific, and atmospheric cocontaminants may either elevate or depress the signal produced by the contaminant of interest. If reliable data on the effect of such interferences are not available, they should be obtained in calibration tests. Procedures for establishing known concentrations for such calibration tests are discussed in detail in Section 4.6 and Section 4.7.

2.2 Ventilation System Measurements

2.2.1 Air Velocity Measurements

Most ventilation performance measurements are made with anemometers (i.e., instruments that measure air velocities) and, with the exception of the Pitot tube, all require periodic calibration. Instruments based on mechanical or electrical sensors are sensitive to mechanical shocks and/or may be affected by dust accumulations and corrosion. Operational characteristics of anemometers are indicated in Table 15.1.

Anemometers are usually calibrated in a well-defined flow field that is relatively large in comparison to the size of meter being calibrated. Such flow fields can be produced in wind tunnels, which are discussed in Section 4.5.

2.2.2 *Pressure Measurements*

Although the standard Pitot tube and a water-filled U-tube manometer may not require calibration, Pitot tubes and other flowmeters may be used with pressure gauges that do. Many direct-reading gauges can give false readings because of the effects of mechanical shocks and/or leakage in connecting tubes or internal diaphragms. For pressures of approximately 1 in. H_2O or greater, a liquid-filled laboratory manometer should be an adequate reference standard. For lower pressures, it may be necessary to use a reference whose calibration is traceable to a National Institute of Standards and Technology (NIST) standard. Characteristics of pressure gauges are given in Table 15.2.

2.3 Heat Stress Measurements

The four environmental variables used in heat stress evaluations are air temperature, humidity, radiant temperature, and air velocity. The measurement of air velocity is discussed in detail elsewhere in this chapter. Liquid-in-glass thermometers used to measure dry-bulb, wet-bulb, or globe thermometer temperatures should have calibrations traceable to certified NIST standards, and should not need recalibration. Other temperature sensors will require periodic recalibration.

2.4 Electromagnetic Radiation Measurements

The electromagnetic spectrum is a continuum of frequencies whose effects on human health are discussed in Chapters 21 and 22. Most of these effects are frequency dependent, and some are attributable to narrow bands of frequency. Thus many of the instruments used to measure frequency and/or intensity are designed to operate over specific frequency regions. Other instruments, known as bolometers, measure the total incident radiant flux over a wide range of frequencies.

The literature on calibration techniques for the measurement of electromagnetic radiation is too extensive to summarize adequately in the space available. Instead, the reader is directed to other chapters of this book and to the reference works cited in those chapters.

2.5 Noise-Measuring Instruments

The accuracy of sound-measuring equipment may be checked by using an acoustic calibrator (Fig. 15.1) consisting of a small, stable sound source that fits over a microphone and generates a predetermined sound level within a fraction of a decibel. The acoustic calibration provides a check of the performance of the entire instrument, including micro-

Table 15.1. Characteristics of Flow Instruments

Instrument	Range (fpm)	Hole Size (for ducts) (in.)	Range Temp.[a]	Dust, Fume Difficulty	Calibration Requirements	Ruggedness	General Usefulness and Comments
Pilot tubes with inclined manometer	600 up	3/8	Wide	Some	None	Good	
Standard	600 up	3/8	Wide	Some	None	Good	Good except at low velocities
Small size	600 up	3/16	Wide	Yes	Once	Good	Good except at low velocities
Double	500 up	3/4	Wide	Small	Once	Good	Special
Swinging vane anemometers	25–10,000	1/2–1	Medium	Some	Frequent	Fair	Good
Rotating vane anemometers							
Mechanical	30–10,000	Not for duct use	Narrow	Yes	Frequent	Poor	Special; limited use
Electronic	25–200 25–500 25–2,000 25–5,000	Not for duct use	Narrow	Yes	Frequent	Poor	Special; can record; direct reading

[a]Temperature range: Narrow, 20–150°F; medium, 20–300°F; wide, 0–800°F.

Source: *Industrial Ventilation*, 22nd ed., Courtesy American Conference of Governmental Industrial Hygienists.

Table 15.2. Characteristics of Pressure Measuring Instruments: Static Pressure, Velocity Pressure, and Differential Pressure

Instrument	Range (in. H_2O)	Manufacturer's Stated Precision (in. H_2O)	Comments
Liquid manometers			
Vertical U-tube	No limit	0.1	Portable; needs no calibration
Inclined 10:1 slope	Usually up to 10	0.005	Portable; needs no calibration; must be leveled
Hook gauge	0–24	0.001	Not a field instrument; tedious, difficult to read; for calibration only
Micromanometer (Meriam model 34FB2TM)	0–10	0.001	Heavy; must be located on vibration-free surface; not difficult to read; uses magnifier
	0–20		
Micromanometer (Vernon Hill type C)	0.001–1.2	0.0004	Small; portable, uses magnifier; need experience to read to manufacturer's precision; calibration needed
Micromanometer (Electric Microtonic, F.W. Dwyer, Mfg.)	0–2	0.0003	Portable; needs vibration-free mount; no magnifier, slow to use; no eyestrain; no calibration needed
Diaphragm and mechanical			
Diaphragm-magnehelic gauge	0–0.5	0.01	Calibration recommended; no leveling; no mounting needed, direct reading
	0–1	0.02	
	0–4	0.10	
Swinging vane anemometer	0–0.5	5% scale	Calibration recommended; no leveling; no mounting needed; use manufacturer's exact recommendation for size of SP hole
	0–20		
Pressure transducers	0.05–6	0.3%	Must be calibrated; remote reading responds to rapid change in pressure

Source: *Industrial Ventilation*, 20th Ed., Courtesy American Conference of Governmental Industrial Hygienists.

Figure 15.1. Acoustic calibrator.

phone and electronics. Some sound level meters have internal means for calibration of electronic components only. Sound level calibrators should be used only with the microphones for which they are intended. Manufacturers' instructions should be followed regarding the use of calibrators and indications of malfunction of instruments.

An ANSI standard for acoustic calibrators (ANSI S1.40-1984) stipulates that calibrators "normally include a sound source which generates a known sound pressure level (SPL) in a coupler into which a microphone is inserted. A diaphragm or piston inside the coupler is driven sinusoidally and generates a specified SPL and frequency within the coupler. The calibrator presents to the inserted microphone of a sound level meter or other sound measuring system a reference or known acoustic signal so one can verify the system sensitivity or set the system to indicate the correct SPL at some frequency."

Many acoustic calibrators provide two or more nominal SPLs and operate at two or more frequencies. Multiple levels and multiple frequencies are useful for checking the linearity and, in a limited way, the frequency response of a measuring system, respectively. The latter is useful for gross checks of microphones and weighting filters for failure. An acoustic calibrator that produces multiple frequencies may also be used to determine a single-number composite calibration for a broad-band sound through calibration at several frequencies. Additional signals such as tone bursts may be provided for use in checking some important electroacoustic characteristics of sound level meters (SLMs) and other acoustical instruments. Such signals may also be useful in checking performance characteristics of instruments to measure sound exposure level or time-period average sound level.

The ANSI standard specifies tolerances for SPLs produced by calibrators. These range from ±0.3 dB for calibration of microphones expected to be used with types 0 and 1 SLM instruments to ±0.4 dB for calibration of type 2 instruments including dosimeters.

One should not place blind trust in any calibrator or instrument. A simple procedure, when several calibrators and/or SLMs are available, is to perform a cross check among them. This should produce results within the published tolerance limits.

Coupler-type calibrators should be used only with microphones for which they are intended. Instructions supplied by the manufacturer on instrument use and corrections for barometric pressure and temperature should be carefully followed. Use of single-frequency calibrations may result in overlooking damage to microphones that is manifested at frequencies other than the calibrator frequency.

3 CALIBRATION STANDARDS

Calibration procedures generally involve a comparison of instrument response to a standardized atmosphere or to the response of a reference instrument. Hence the calibration can be no better than the standards used. Reliability and proper use of standards are critical to accurate calibrations. Reference materials and instruments available from or calibrated by NIST should be used whenever possible. Information on calibration aids available from NIST is summarized in Table 15.3.

Test atmospheres generated for purposes of calibrating collection efficiency or instrument response should be checked for concentration using reference instruments or sampling and analytical procedures whose reliability and accuracy are well documented. The best procedures to use are those that have been refereed or panel tested, that is, methods that have demonstrably yielded comparable results on blind samples analyzed by different laboratories. Organizations that publish standard calibration methods are listed in Table 15.4.

Calibrations of flow rate or sampled volume are generally made by comparing the indicated rate or volume passing through the sampler with that passing through or into a calibration standard in series with the sampler. The standards used are classified as primary, intermediate, and secondary.

Primary standards provide direct and unequivocal calibration because they are based on direct and easily measurable linear dimensions, such as the length and diameter of a cylinder of displaced fluid. Secondary standards are instruments or meters that have been calibrated against primary standards. Intermediate standards are secondary standards that can provide an accuracy comparable to primary standards, for example, 1%. Examples include wet test meters and dry gas meters. Secondary standards must be handled with care, and properly stored and maintained when not in use, and they should be periodically recalibrated against a suitable primary standard.

4 INSTRUMENTS AND TECHNIQUES

4.1 Cumulative Air Volume

Many air sampling instruments utilize an integrating volume meter for measurement of sampled volume. Most of them measure displaced volumes that can be determined from

Table 15.3. Categories of National Institute of Standards and Technology (NIST) Standard Reference Materials (SRMs)* used for Industrial and Environmental Hygiene Calibrations

Category	Descriptions
105	Clinical Laboratory Materials (gas, liquid, and solid forms) Various SRMs are available for calibrating apparatus and validating analytical methods used in clinical and pathology laboratories.
105	Materials on Filter Media These SRMs consist of potentially hazardous materials deposited on filters to be used to determine the levels of these materials in industrial atmospheres.
105	Trace Constituent Elements in Blank Filters SRMs 2678 (cellulose actetate membrane—47 mm, 0.45 μm pore size) and 2681 (ashless—42.5 mm) are for use in evaluating the performance of air sampling filter methods with two different filter types or sizes commonly used in air sampling of industrial atmospheres. For both SRMs, either certified values (in μg), or limits of detection (X_D), for each of 30 constituent elements as well as six leachable anions and cations are provided.
105	Respirable Silica (powder form) SRMs 1878a (alpha quartz) and 1879a (cristobalite) are crystalline silica materials with particles in the respirable range. They are intended for use in determining by x-ray diffraction, the levels of respirable silica in an industrial atmosphere according to NIOSH Analytical Method P&CAM 259 or equivalent methods. NOTE: These SRMs are not certified for particle size.
105	Lead in Paint, Dust, and Soil (powder and sheet forms) These SRMs and RM have been developed in conjunction with the U.S. EPA to monitor paint, soil, and dust sources of lead. SRMs 2570 through 2576 consist of one Mylar™ sheet per unit. Each sheet, 7.6 cm × 10.2 cm, is coated with a single uniform paint layer for use with portable x-ray fluorescence analyzers. SRM 2579a consists of a set of six Mylar™ sheets, one each of SRMs 2570 through 2575, SRMs 2580, 2581, 2582, and 2589 consist of paint that has been ground and homogenized into a powder, 99+ % of which passes a 100 μm sieve. SRM 2583 consists of dust, 99+ % of which passes a 100 μm sieve, that was collected in vacuum cleaner bags during routine cleaning of dwelling interiors. SRM 2583 is certified for arsenic, chromium, cadmium, lead, and mercury. (Also see Category 106) SRMs 2584, 2586, and 2587 are dust or soil matrices containing lead from paint. RM 8680 consists of a 10.2 cm wide × 15.2 cm long × 1.3 cm thick section of painted fiberboard and is intended for use in the evaluation of destructive and nondestructive methods of measuring lead in paint on fiberboard.

105 Asbestos

These SRMs are for use in identifying and quantifying asbestos types. SRM 1866a consists of a set of three common bulk mine-grade asbestos materials; chrysotile, grunerite (Amosite), riebeckite (Crocidolite), and one glass fiber sample. SRM 1867 consists of a set of three uncommon mine-grade asbestos materials; antophyllite, tremolite, and actinolite. The optical properties of SRMs 1866a and 1867, as observed by polarized light microscopy (PLM), have been characterized so that they may serve as primary calibration standards for the identification of asbestos types in building materials.

SRM 1868 consists of a set of two common bulk mine-grade asbestos materials; chrysotile and grunerite (Amosite), contained in matrices simulating building materials (calcium carbonate and glass fiber), in quantities at just below the U.S. EPA regulatory limit of 1%. This SRM is certified by weight for the quantity of each asbestos material present.

SRM 1876b is intended for use in evaluating the techniques used to identify and count chrysotile asbestos fibers by transmission electron microscopy (TEM). A unit consists of sections of mixed-cellulose—ester filters containing chrysotile asbestos fibers deposited by an aerosol generator.

RM 8411 consists of a section of collapsed mixed-cellulose—ester filters with a high concentration (138 fibers/0.01 mm^2) of chrysotile asbestos and a medium concentration (43 fibers/0.01 mm^2) of grunerite (Amosite) asbestos. It is intended for use in evaluating the techniques used to identify and count asbestos fibers by transmission electron microscopy (TEM).

106 Metal Constituents in Natural Matrices (liquid and solid forms)

These SRMs and RM are for analysis of materials of health or environmental interest. (Also see Categories 105 and 111)

106 Thin Films for X-Ray Fluorescence

SRM 1833 is for standardizing x-ray spectrometers. It may be useful in elemental analysis of particulate matter collected on filter media, and where x-ray spectrometer calibration functions are determined using thin film standards. Each SRM unit is individually certified and consists of a silica base glass film (0.5 μm thick) deposited on a 47 mm diameter polycarbonate filter mounted on an aluminum ring.

106 Carbon Modified Silica (powder form)

SRM 1216 is intended for th calibration of instruments used to measure total elemental carbon. The SRM consists of three, 1-g bottles of chemically modified microparticulate silica.

106 Trace Elements (solid from)

SRM 1648 is for analysis of trace elements in urban particulate matter.

Table 15.3. (continued)

Category	Descriptions
107	Primary Gas Mixtures. These SRMs are for calibrating equipment and apparatus used to measure various components of gas mixtures and atmospheric pollutants. The typical gas mixture is supplied in a DOT 3AL specification aluminum (6061 alloy) cylinder with a nominal pressure exceeding 12.4 mPa that provides the user with approximately 0.73 m^3 of usable mixture. Due to increasing customer demand, these primary gas mixtures are in short supply and may not be readily available for sale. In such cases, a NIST traceable reference gas described below may be substituted. A NIST Traceable Reference Material (NTRM) is a reference material produced by a commercial supplier with a well-defined traceability to NIST. This traceability is established via criteria and protocols defined by NIST that are tailored to meet the needs of the meterological community to be served. The NTRM concept was established to allow NIST to respond to the increasing needs for high quality reference materials by leveraging its relatively fixed human and financial resources with secondary reference material producers. Reference material producers adhering to NIST defined protocol requirements are allowed to use the "NTRM" trademark to identify their product. The gas NTRM program was established in 1992 in partnership with the U.S. EPA and specialty gas companies as a means for providing end-users with the wide variety of certified gas standards needed to implement the "Emissions Trading" provision of the 1990 Clean Air Act. Gas NTRMs are produced and distributed by specialty gas companies with NIST oversight of the production and maintenance, and direct involvement in the analysis. NTRMs can be developed for any pollutant, concentration, and balance gas combination for which a NIST primary standard or SRM exists. The gas standards prepared according to this program are related, within known limits of *uncertainty,* to specific gaseous primary standards maintained by NIST. NTRMs are available from commercial suppliers. A supplier list can be obtained upon request from the SRM Program Sales Office.
107	Permeation Devices These SRMs are primarily intended for use in calibrating air pollution monitoring apparatus and for calibrating air pollution analytical methods and procedures. Each tube is individually calibrated and certified according to NIST procedures and protocols.

109	GC/MS and LC System Performance (liquid form) These SRMs and RM are for evaluating the sensitivity of gas chromatography/mass spectrometry (GC/MS) instrumentation and for characterizing liquid chromatography (LC) column selectivity.
109	Organic Constituents (liquid and solid forms) These SRMs and RMs are for calibrating or measuring organic contaminants found in a variety of environmental matrices. SRMs are available for PAHs, Pesticides, PCBs, methylmercury, and mercury. The SRMs are grouped according to application—calibration or natural matrix measurement. The calibration SRMs are useful for validating the chromatographic separation step while the natural matrix SRMs, which are similar to actual environmental samples, can be used to validate all the steps of an analytical procedure. SRMs and RMs certified for organic components in matrices such as oil, iso-octane, and methanol are also available.
301	Sizing: Particle Size (powder and solid forms) These SRMs are for evaluating and calibrating specific types of particle size measuring instruments, including light scattering, electrical zone flow-through counters, optical and scanning electron microscopes, sedimentation systems, and wire cloth sieving devices. SRM 659 consists of equiaxed silicon nitride particles with a minimal amount of large agglomerates. SRMs 1003b, 1004a, 1017b, 1018b and 1019b each consist of soda-lime glass beads covering a particular size distribution range. SRMs 1690, 1691, 1691, 1692 and 1963 are commercially manufactured latex particles in a water suspension.
301	Cement Turbidimetry and Fineness SRM 114 p is for calibrating the Blaine fineness meter according to the latest issue of Federal Test Method Standard 158, Method 2101 or ASTM C 204 to calibrate the Wagner turbidimeter according to ASTM C 115 and to determine sieve residue according to ASTM C 430. Each set consists of 20 sealed laminated film pouches, each containing approximately 10 g of cement.

[a]For current information on availability of specific SRMs, contact the NIST SRM Program by telephone (301-975-2021), fax (301-926-4342), e-mail (nancy.trahey@nist.gov) or visit the NIST web-site (http://ts.nist.gov/srm).

Table 15.4. Organizations Publishing Recommended or Standard Methods and/or Test Protocols Applicable to Air Sampling Instrument Calibration

Abbreviation	Full Name and Address
ANSI	American National Standards Institute, Inc. 11 W. 42nd Street, 13 Floor New York, NY 10036
ASTM	American Society for Testing and Materials D-22 Committee on Sampling and Analysis of Atmospheres, and E-34 Committee on Occupational Health and Safety 100 Barr Harbor Drive West Conshohocken, PA 19428
AWMA	Air and Waste Management Association 1 Gateway Center, Third Floor Pittsburgh, PA 15222
EPA/NERL	U.S. Environmental Protection Agency National Exposure Research Laboratory Quality Assurance Division Research Triangle Park, NC 27711
NIOSH	National Institute for Occupational Safety and Health Editor, *NIOSH Manual of Analytical Methods* (NMAM®) Division of Physical Sciences and Engineering 4676 Columbia Parkway Cincinnati, OH 45226

linear measurements and geometric formulas. Such measurements usually can be made with a high degree of precision.

4.1.1 Water Displacement

Figure 15.2 is a schematic drawing of a Mariotte bottle. When the valve at the bottom of the bottle is opened, water drains out of the bottle by gravity, and air is drawn by way of a sample collector into the bottle to replace it. The volume of air drawn in is equal to the change in water level multiplied by the cross section at the water surface.

4.1.2 Spirometer or Gasometer

The spirometer (Fig. 15.3) is a cylindrical bell with its open end under a liquid seal. The weight of the bell is counterbalanced so that the resistance to movement as air moves in or out of the bell is negligible. The device differs from the Mariotte bottle in that it measures displaced air instead of displaced liquid. The volume change is calculated in a similar manner (i.e., change in height times cross section). Spirometers are available in a wide variety of sizes and are frequently used as primary volume standard (1).

4.1.3 "Frictionless" Piston Meters

Cylindrical air displacement meters with nearly frictionless pistons are frequently used for primary flow calibrations. The simplest version is the soap bubble meter illustrated in

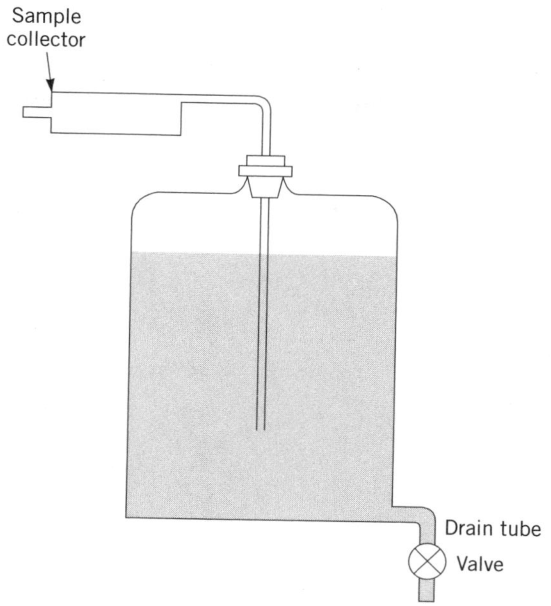

Figure 15.2. Mariotte bottle. From Ref. 3.

Figure 15.4. It utilizes a volumetric laboratory burette whose interior surfaces are wetted with a detergent solution. If a soap-film bubble is placed at the left side, and suction is applied at the right, the bubble is drawn from left to right. The volume displacement per unit time (i.e., flow rate) can be determined by measuring the time required for the bubble to pass between two scale markings that enclose a known volume.

Soap-film flowmeters and mercury-sealed piston flowmeters are available commercially from several sources (4). In the mercury-sealed piston, most of the cylindrical cross section is blocked off by a plate that is perpendicular to the axis of the cylinder. The plate is separated from the cylinder wall by an O-ring of liquid mercury, which retains its toroidal shape because of its strong surface tension. This floating seal has a negligible friction loss as the plate moves up and down.

4.1.4 Wet Test Meter

A wet test meter (Fig. 15.5) consists of a partitioned drum half-submerged in a liquid (usually water), with openings at the center and periphery of each radial chamber. Air or gas enters at the center and flows into an individual compartment, causing the drum to rise, thereby producing rotation. This rotation is indicated by a dial on the face of the instrument. The volume measured depends on the fluid level in the meter, because the liquid is displaced by air. There is a sight gauge for determining fluid height, and the meter is leveled by screws and a sight bubble provided for this purpose.

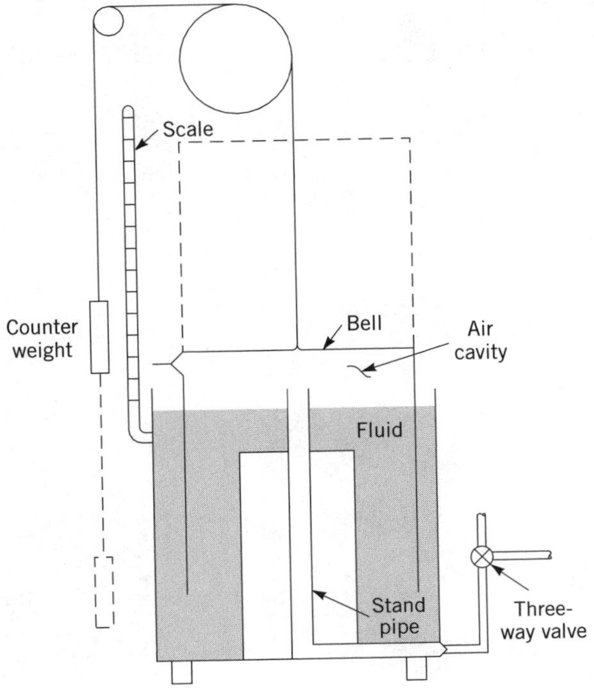

Figure 15.3. Spirometer.

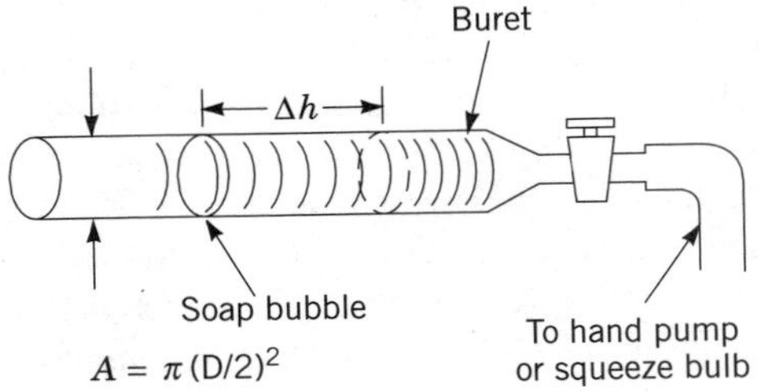

Figure 15.4. Bubble meter. From Ref. 3.

Several potential errors are associated with the use of a wet test meter. The drum and moving parts are subject to corrosion and damage from misuse, there is friction in the bearings and the mechanical counter, and inertia must be overcome at low flows (<1 rpm), whereas at high flows (>3 rpm) the liquid might surge and break the water seal at the inlet or outlet. In spite of these factors, the accuracy of the meter usually is within 1 percent when used as directed by the manufacturer.

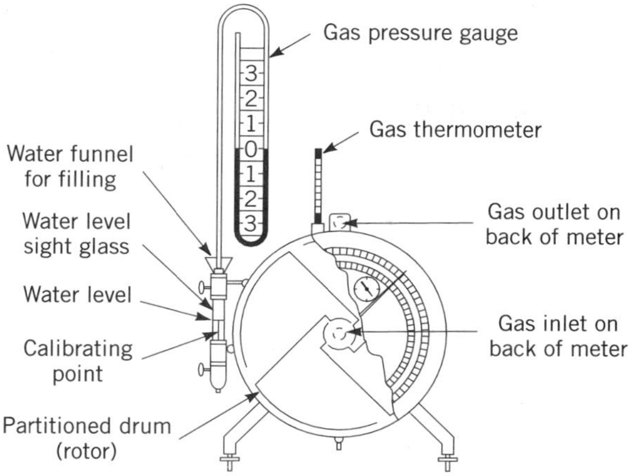

Figure 15.5. Wet test meter. From Ref. 3.

4.1.5 Dry Gas Meter

The dry gas meter shown in Figure 15.6 is very similar to the domestic gas meter. It consists of two bags interconnected by mechanical valves and a cycle-counting device. The air or gas fills one bag while the other bag empties itself; when the cycle is completed, the valves are switched, and the second bag fills while the first one empties. Any such

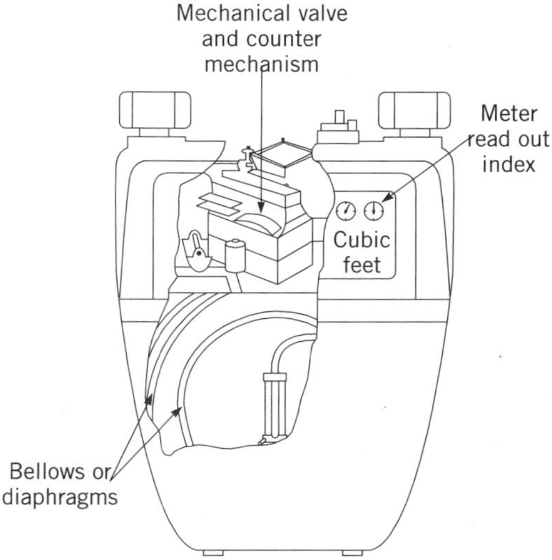

Figure 15.6. Dry gas meter.

device has the disadvantages of mechanical drag, pressure drop, and leakage; however, the advantage of being able to use the meter under rather high pressures and volumes often outweighs the disadvantages created by the errors, which can be determined for a specific set of conditions. The alternate filling of two chambers is also used as the basis for volume measurement in twin-cylinder piston meters. Such meters can also be classified as positive displacement meters.

4.1.6 Positive Displacement Meters

Positive displacement meters consist of a tight-fitting moving element with individual volume compartments that fill at the inlet and discharge at the outlet parts. Another multicompartment continuous rotary meter uses interlocking gears. When the rotors of such meters are motor driven, these units become positive displacement air movers.

4.2 Volumetric Flow Rate

The volume meters discussed in the preceding paragraphs were all based on the principle of conservation of mass; specifically, the transfer of a fluid volume from one location to another. The flow rate meters in this section all operate on the principle of the conservation of energy; more specifically, they utilize Bernoulli's theorem for the exchange of potential energy for kinetic energy and/or frictional heat. Each consists of a flow restriction within a closed conduit. The restriction causes an increase in the fluid velocity and therefore an increase in kinetic energy, which requires a corresponding decrease in potential energy (i.e., static pressure). The flow rate can be calculated from a knowledge of the pressure drop, the flow cross section at the constriction, the density of the fluid, and the coefficient of discharge, which is the ratio of actual flow to theoretical flow and makes allowance for stream contraction and frictional effects.

Flowmeters that operate on this principle can be divided into two groups. The larger group, which include orifice meters, venturi meters, and flow nozzles, have a fixed restriction and are known as variable head meters because the differential pressure head varies with flow. The other group, which include rotameters, are known as variable area meters, because a constant pressure differential is maintained by varying the cross section for fluid flow.

4.2.1 Variable Area Meters (Rotameters)

A rotameter consists of a "float" that is free to move up and down within a vertical tapered tube that is larger at the top than the bottom (Fig. 15.7). The fluid flows upward, causing the float to rise until the pressure drop across the annular area between the float and the tube wall is just sufficient to support the float. The tapered tube is usually made of glass or clear plastic, and the flow-rate scale is etched directly on it. The height of the float indicates the flow rate. Floats of various configurations are used. They are conventionally read at the highest point of maximum diameter, unless otherwise indicated.

Most rotameters have a range of 10:1 between their maximum and minimum flows. The range of a given tube can be extended by using heavier or lighter floats. Tubes are made in sizes from about ⅛ to 6 in. (3 mm to 15 cm) in diameter, covering ranges from

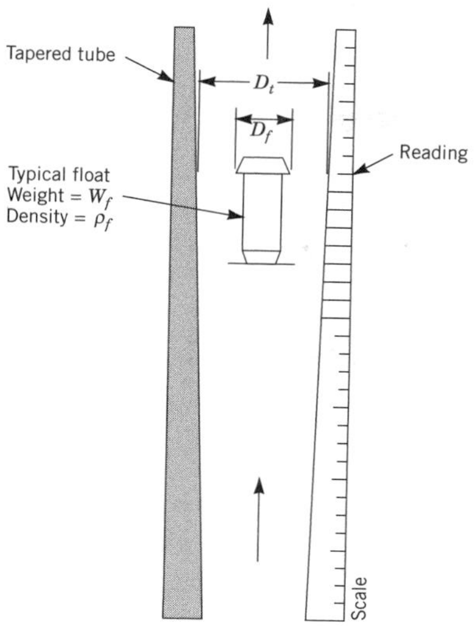

Figure 15.7. Schematic of rotameter.

a few cubic centimeters per minute-to more than 1000 cfm (28 m^3/min). Some of the shaped floats achieve stability by having slots that make them rotate, but these are less commonly employed than previously. The term "rotameter" was first used to describe such meters with spinning floats but now serves generally for tapered metering tubes of all types.

Rotameters are the most commonly used flowmeters on commercial air samplers, especially on portable samplers. For such sampler flowmeters, the most common material of construction is acrylic plastic, although glass tubes may also be used. Because of space limitations, the scale lengths are generally no more than 4 in. (10 cm), and most commonly nearer 2 in. (5 cm). Unless they are individually calibrated, the accuracy is unlikely to be better than $\pm 25\%$. When individually calibrated, $\pm 5\%$ accuracy may be achieved. It should be noted, however, that with the large taper of the bore, the relatively large size of the float, and the relatively few scale markers on these rotameters, the precision of the readings may be a major limiting factor.

Calibrations of rotameters are performed at an appropriate reference pressure, usually atmospheric. Because good practice dictates that the flowmeter be located downstream of the sample collector or sensor, however, the flow is actually measured at a reduced pressure, which may also be a variable pressure if the flow resistance changes with loading. If this resistance is constant, it should be known; if variable, it should be monitored, to permit adjustment of the flow rate as needed, and the making of appropriate pressure corrections for the flowmeter readings.

For rotameters with linear flow-rate scales, the actual sampling flow will approximately equal the indicated flow rate times the square roots of the ratios of absolute temperatures and pressures of the calibration and field conditions (5). The ratios increase when the field pressure is less than the pressure in the calibration laboratory or the field temperature is greater than that in the laboratory. Thus, if the flowmeter was accurate at ambient pressure and the flow resistance of the sampling medium was relatively low, for example, 30 mm Hg for a flow rate of 11 L/min., the flow rate indicated on the rotameter would be 11 $\times$ (730/760)½ = 10.8 L/min., a difference of only 1.8%. On the other hand, for a 25-mm diameter AA Millipore filter with a 3.9 cm^2 filtering area and a sampling rate of 11 L/min., the flow resistance would be ~190 mm Hg, and the indicated flow rate would be 11 $\times$ (570/760)½ = 9.5 L/min., a difference of 14%.

A further correction is needed when the sampling is done at atmospheric pressures and/or temperatures that differ substantially from those used for the calibration. For example, at an elevation of 5000 ft (1500 m) above sea level, the atmospheric pressure is only 83% of that at sea level. Thus the actual flow rate would be 9.6% greater than that indicated on a rotameter scale, based upon the altitude correction alone. If the temperature in the field was 35°C while the meter was calibrated at 20°C, the actual flow rate in the field would be (308/293)½ $\times$ 100 = 2.5% greater than that of standard air.

In a situation where there were corrections needed for the pressure drop of the sampler, high altitude, and high temperature, the overall correction could be, for the examples cited, 1.14 $\times$ 1.096 $\times$ 1.025 = 1.28 or 28%.

Craig (6) has shown that the change in calibration with air density cannot be made by simple computation, especially for the small-diameter rotameter tubes and floats commonly used on air sampling equipment. Figure 15.8 gives experimental calibration data at various suction pressures for a specific glass rotameter. Clearly it is not practical to generate such a family of empirical calibration curves for each rotameter. Craig recommends: (*1*) that a pressure gauge be used at the inlet to the rotameter; (*2*) that the flow rates used be those that give as low a pressure drop as possible; and (*3*) that the meter size be selected to give readings near the upper end of the scale.

4.2.2 *Variable Head Meters*

When orifice and venturi meters are made to standardized dimensions, their calibration can be predicted with ±10% accuracy using standard equations and published empirical coefficients. The general equation (7) for this type of meter is:

$$W = q_1 \rho_1 = KYA_2\sqrt{2g_c(P_1 - P_2)\rho_1} \tag{1}$$

where $K = C/(1 - \beta^4)^{1/2}$
C = coefficient of discharge (dimensionless)
A_2 = cross-sectional area of throat (ft^2 or m^2)
g_c = 32.17 ft/sec^2 for English units or 1 for metric units
P_1 = upstream static pressure (lb/ft^2 or Pa)
P_2 = downstream static pressure (lb/ft^2 or Pa)
q_1 = volumetric flow at upstream pressure and temperature (ft^3/s or m^3/s)
W = weight rate of flow (lb/sec) for English units
= mass rate of flow (kg/s) for metric units
Y = expansion factor (see Fig. 15.11)

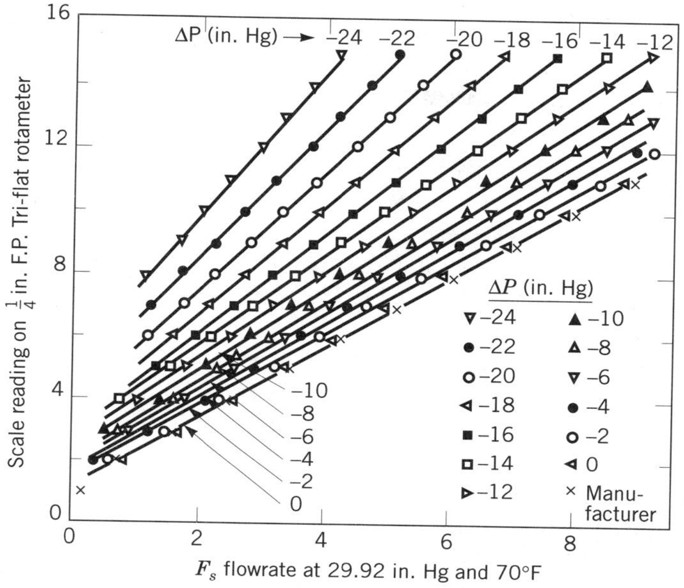

Figure 15.8. Rotameter reading versus airflow rate under standard conditions for various pressure gauge readings at rotameter.

β = ratio of throat diameter to pipe diameter (dimensionless)
ρ_1 = density at upstream pressure and temperature (lb/ft^3 or kg/m^3)

4.2.2.1 Orifice Meters. The simplest form of variable head meter is the square-edged or sharp-edged orifice illustrated in Figure 15.9. It is also the most widely used because of its ease of installation and low cost. If it is made with properly mounted pressure taps, its

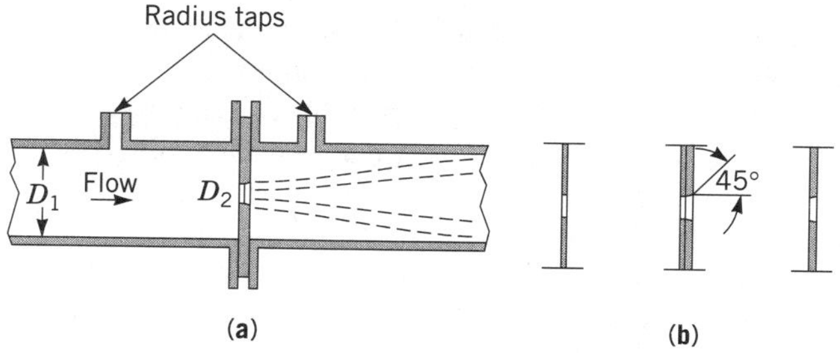

Figure 15.9. Square-edged or sharp-edged orifices. The plate at the orifice opening must not be thicker than 1/30 of pipe diameter, 1/8 of the orifice diameter, or 1/4 of the distance from the pipe wall to the edge of the opening. (**a**) Pipeline orifice. (**b**) Types of plate.

calibration can be determined from Equation 1 and Figure 15.10 and Figure 15.11. However, even a nonstandard orifice meter can serve as a secondary standard, provided it is carefully calibrated against a reliable reference instrument.

Although the square-edged orifice can provide accurate flow measurements at low cost, it is inefficient with respect to energy loss. The permanent pressure loss for an orifice meter with radius taps can be approximated by $(1-\beta^2)$ and often exceeds 80%.

4.2.2.2 Venturi Meters. Venturi meters have optimal converging and diverging angles of about 25° and 7°, respectively, which means that they have high pressure recoveries; that is, the potential energy that is converted to kinetic energy at the throat is reconverted to potential energy at the discharge, with an overall loss of only about 10%.

For air at 70°F (21°C) and 1 atm and for $\frac{1}{4} < \beta < \frac{1}{2}$, a standard venturi would have a calibration described by:

$$Q = V_n \beta^2 D^2 (\Delta h)^{1/2} \tag{2}$$

where V_n = 21.2 for English units, or 58.4 for metric units
Q = flow (cfm or L/min)
β = ratio of throat to duct diameter (dimensionless)
D = duct diameter (in or cm)
Δh = differential pressure (in H_2O, or cm H_2O)

4.2.2.3 Other Variable Head Meters. The characteristics of various other types of variable head flowmeters (e.g., flow nozzles, Dall tubes, centrifugal flow elements) are described in standard engineering references (7, 9). In most respects they have similar properties to the orifice meter, the venturi meter, or both.

One type of variable head meter that differs significantly from all the foregoing is the laminar flowmeter. These devices are seldom discussed in engineering handbooks because they are used only for very low flow rates. Because the flow is laminar, the pressure drop is directly proportional to the flow rate. In orifice meters, venturi meters, and related devices, the flow is turbulent and flow rate varies with the square root of the pressure differential.

Laminar flow restrictors used in commercial flowmeters consist of egg-crate or tube bundle arrays of parallel channels. Alternatively, a laminar flowmeter can be constructed in the laboratory using a tube packed with beads or fibers as the resistance element. Figure 15.12 illustrates this kind of homemade flowmeter. It consists of a "T" connection, pipette or glass tubing, cylinder, and packing material. The outlet arm of the "T" is packed with a porous plug, and the leg is attached to a tube or pipette projecting into the cylinder filled with water or oil. A calibration curve of the depth of the tube outlet below the water level versus the rate of flow should produce a linear curve. Saltzman (10) has used tubes filled with mineral fiber to regulate and measure flowrates as low as 0.01 cm^3/min.

4.2.2.4 Pressure Transducers. All the variable head meters require a pressure sensor, sometimes referred to as the secondary element. Any type of pressure sensor can be used,

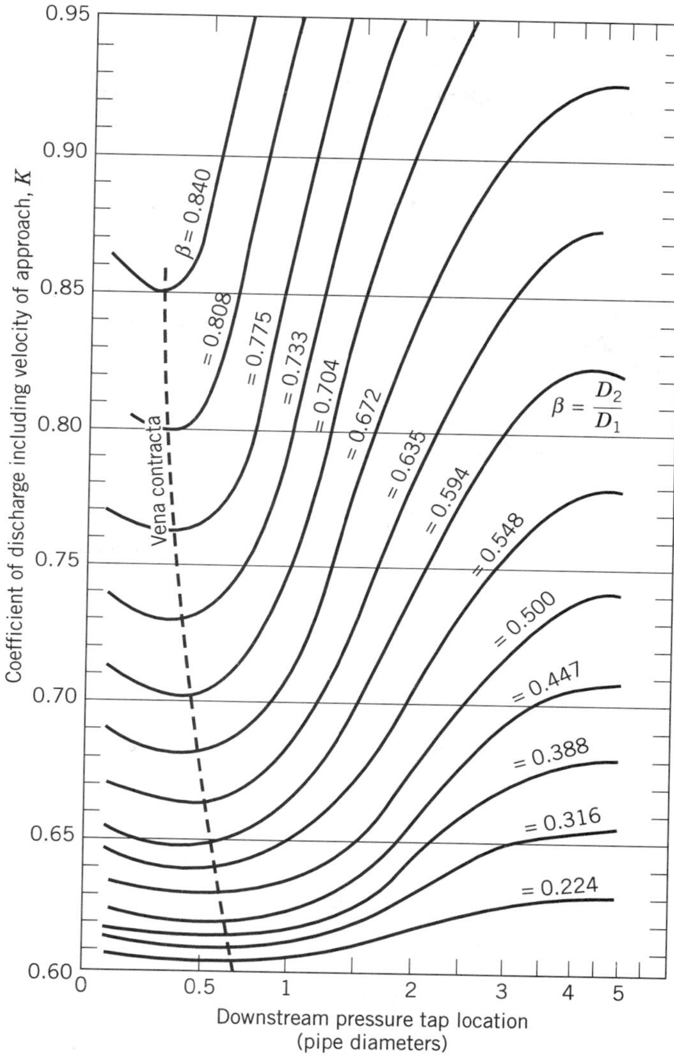

Figure 15.10. Downstream pressure tap location in pipe diameters. Coefficient of discharge for square-edged circular orifices for $N_{Re2} > 30{,}000$ with the upstream tap location between one and two pipe diameters from the orifice position. From Ref. 8.

although high cost and fragility usually rule out many electrical and electromechanical transducers.

Liquid-filled manometer tubes are sometimes used, and if they are properly aligned, and the density of the liquid is accurately known, the column differential provides an

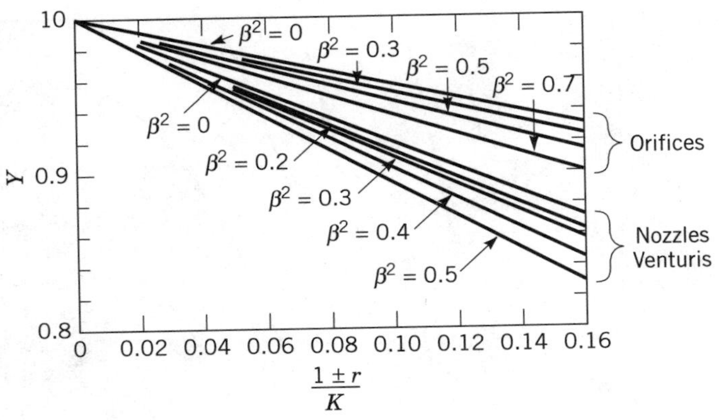

Figure 15.11. Values of expansion factor Y for orifices, nozzles, and venturis.

unequivocal measurement. In most cases, however, it is not feasible to use liquid-filled manometers in the field, and the pressure differentials are measured with mechanical gauges with scale ranges in centimeters or inches of water. For these low-pressure differentials the most commonly used gauge is the Magnehelic®, illustrated in Figure 15.13. These gauges are accurate to ± 2 percent of full scale and are reliable, provided they and their connecting hoses do not leak and their calibration is periodically rechecked.

4.2.2.5 Critical Flow Orifice. For a given set of upstream conditions, the discharge of a gas from a restricted opening increases with a decrease in the ratio of absolute pressures P_2/P_1, where P_2 is the downstream pressure and P_1 the upstream pressure, until the velocity through the opening reaches the velocity of sound. The value of P_2/P_1 at which the maximum velocity is just attained is known as the critical pressure ratio. The pressure in the throat will not fall below the pressure at the critical point, even if a much lower downstream pressure exists. When the pressure ratio is below the critical value, therefore, the rate of flow through the restricted opening depends only on the upstream pressure.

It can be shown (7) that for air flowing through rounded orifices, nozzles, and venturis, when $P_2 < 0.53\ P_1$ and $S_1/S_2 > 25$, the mass flow rate w is constant.

$$w = 0.533\ \frac{C_v S_2 P_1}{T_1}\ \text{lb/sec} \tag{3}$$

where C_v = coefficient of discharge (normally ~1)
S_1 = duct or pipe cross section (in.^{2})
S_2 = orifice area (in.^{2})
P_1 = upstream absolute pressure (psi)
T_1 = upstream temperature (°R).

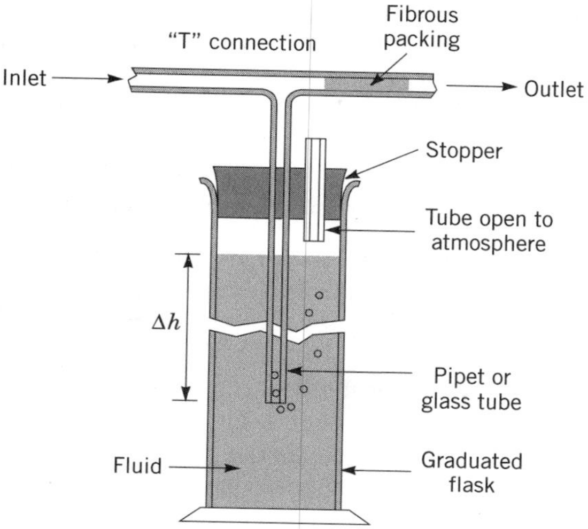

Figure 15.12. Packed plug flowmeter. From Ref. 3.

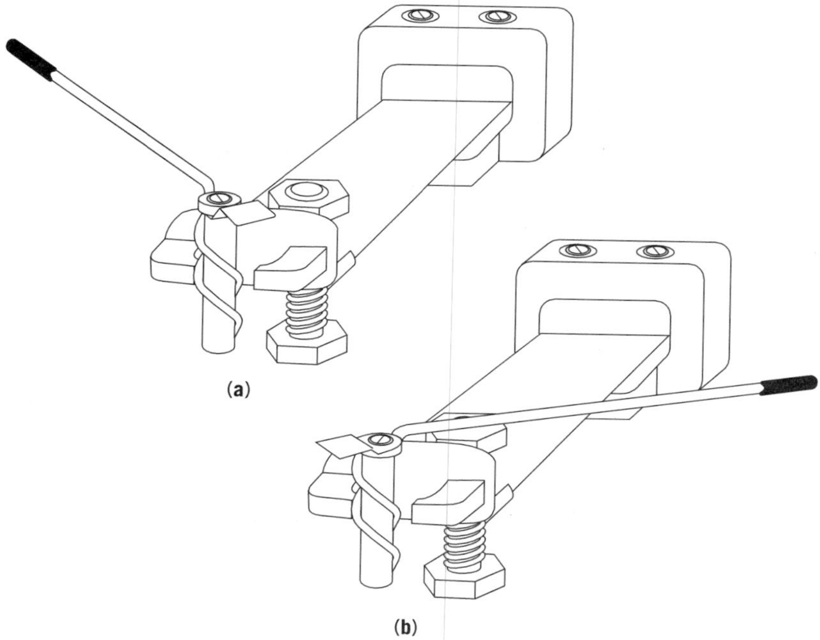

Figure 15.13. Workings of the magnetic linkage. Photograph courtesy of Dwyer Instruments, Inc., Michigan City, Ind., Bulletin No. A-20.

Critical flow orifices are widely used in industrial hygiene instruments such as the midget impinger pump and squeeze bulb indicators. They can also be used to calibrate flowmeters by using a series of critical orifices downstream of the flowmeter under test. The flowmeter readings can be plotted against the critical flows to yield a calibration curve.

The major limitation in their use is that the orifices are extremely small when they are used for flows of 1 cfm (28.3 L/min.) or less. They become clogged or eroded in time and therefore require frequent examination and/or calibration against other reference meters.

4.2.2.6 Bypass Flow Indicators. In most high-volume samplers, the flow rate depends strongly on the flow resistance, and flowmeters with a sufficiently low flow resistance are usually too bulky or expensive. A commonly used metering element for such samplers is the bypass rotameter, which actually meters only a small fraction of the total flow; a fraction, however, that is proportional to the total flow. As shown schematically in Figure 15.14, a bypass flowmeter contains both a variable head element and a variable area element. The pressure drop across the fixed orifice or flow restrictor creates a proportionate flow through the parallel path containing the small rotameter. The scale on the rotameter generally reads directly in cubic feet per minute or liters per minute of total flow. In the versions used on portable high-volume samplers, there is usually an adjustable bleed valve at the top of the rotameter that should be set initially, and periodically readjusted in laboratory calibrations so that the scale markings can indicate overall flow. If the rotameter tube accumulates dirt, or the bleed valve adjustment drifts, the scale readings may depart greatly from the true flows.

4.3 Mass Flow and Tracer Techniques

4.3.1 Thermal Meters

A thermal meter measures mass air or gas flow rate with negligible pressure loss. It consists of a heating element in a duct section between two points at which the temperature of the

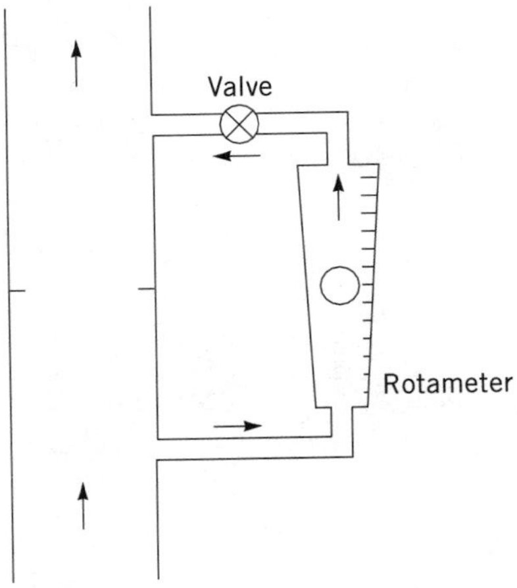

Figure 15.14. Bypass flow indicators.

air or gas stream is measured. The temperature difference between the two points depends on the mass rate of flow and the heat input.

4.3.2 *Mixture Metering*

The principle of mixture metering is similar to that of thermal metering. Instead of adding heat and measuring temperature difference, a contaminant is added and its increase in concentration is measured; or clean air is added and the reduction in concentration is measured. This method is useful for metering corrosive gas streams. The measuring device may react to some physical property such as thermal conductivity or vapor pressure.

4.3.3 *Ion-Flow Meters*

In the ion-flow meter illustrated in Figure 15.15 ions are generated from the central disk and flow radially toward the collector surface. Airflow through the cylinder causes an axial displacement of the ion stream in direct proportion to the mass flow. The instrument can measure mass flows from 0.1 to 150 scfm (3–4.250 L/min) and velocities from 1 to 12,000 fpm (5 cm/sec to 60 m/sec).

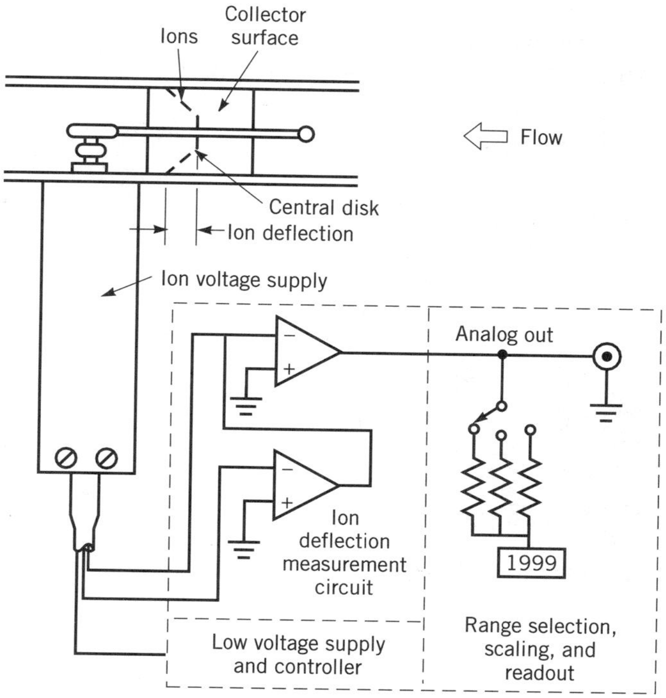

Figure 15.15. Ion-flow mass flowmeter. Schematic courtesy of TSI, St. Paul, Minn., Leaflet No. TSI-54100671.

4.4 Air Velocity Meters (Anemometers)

Air velocity is a parameter of direct interest in heat stress evaluations and in some ventilation evaluations. Though it is not the parameter of interest in sampling flow measurements, it may be the only feasible parameter to measure in some circumstances, and it usually can be related to flow rate, provided the sensor is located in an appropriate position and is suitably calibrated against overall flow.

4.4.1 *Velocity Pressure Meters*

4.4.1.1 Pitot Tube. The Pitot tube is often used as a reference instrument for measuring the velocity of air. A standard Pitot device, carefully made, will need no calibration. It consists of an impact tube whose opening faces axially into the flow, and a concentric static pressure tube with eight holes spaced equally around it in a plane that is eight diameters from the impact opening. The difference between the static and impact pressures is the velocity pressure. Bernoulli's theorem applied to a Pitot tube in an airstream simplifies to the dimensionless formula:

$$V = (2g_c P_v)^{1/2} \tag{4}$$

where V = linear velocity
g_c = gravitational constant
P_v = pressure head of flowing fluid or velocity pressure

Expressing V in linear feet per minute (fpm), P_v in inches of water, that is, (h_v), and with:

$$g_c = 32.17 \frac{(\text{lb} - \text{mass})(\text{ft})}{(\text{lb} - \text{force})(\text{sec}^2)}$$

equates to:

$$V = 1097\left(\frac{h_v}{\rho}\right)^{1/2} \tag{5}$$

where ρ = the density of air or gas (lb/ft^3).

If the Pitot tube is to be used with air at standard conditions (70°F and 1 atm), Equation 5 reduces to:

$$V = 4005(h_v)^{1/2} \tag{6}$$

where V = velocity (fpm)
h_v = velocity pressure (in H_2O)

There are several serious limitations to Pitot tube measurements in most sampling flow calibrations. It may be difficult to obtain or fabricate a small enough probe, and the velocity pressure may be too low to measure at the velocities encountered. Thus at 1000 fpm (5.1 m/sec), $h_v = 0.063$ in H_2O (1.6 mm H_2O), a low value even for an inclined manometer.

4.4.1.2 Other Velocity Pressure Meters. There are several means of utilizing the kinetic energy of a flowing fluid to measure velocity besides the Pitot tube. One way is to align a jeweled-bearing turbine wheel axially in the stream and count the number of rotations per unit time. Such devices are generally known as rotating vane anemometers. Some are very small and are used as velocity probes. Others are sized to fit the whole duct and become indicators of total flow rate; sometimes these are called turbine flowmeters.

The Velometer, or swinging vane anemometer, is widely used for measuring ventilation airflows, but it has few applications in sample flow measurement or calibration. It consists of a spring-loaded vane whose displacement indicates velocity pressure.

4.4.2 Heated Element Anemometers

Any instrument used to measure velocity can be referred to as an anemometer. In a heated element anemometer, the flowing air cools the sensor in proportion to the velocity of the air. Instruments are available with various kinds of heated element, including heated thermometers, thermocouples, films, and wires. They are all essentially nondirectional (i.e., with single element probes); they measure the air speed but not its direction. They all can accurately measure steady state air speed, and those with low mass sensors and appropriate circuits can also accurately measure velocity fluctuations with frequencies above 100,000 Hz. Because the signals produced by the basic sensors depend on ambient temperature, as well as air velocity, the probes are usually equipped with a reference element that provides an output that can be used to compensate or correct errors due to temperature variations. Some heated element anemometers can measure velocities as low as 10 fpm (50 cm/sec) and as high as 8000 fpm (41 m/sec).

4.5 Procedure for Calibrating Velocity; Flow and Volume Meters

It is not possible to describe here all the techniques available, or to go into great detail on those commonly used. This discussion is limited to selected procedures that should serve to illustrate recommended approaches to some commonly encountered calibration procedures.

4.5.1 Producing Known Velocity Fields

Known flow fields can be produced in wind tunnels of the type illustrated in Figure 15.16. The basic components needed have been described by Hama (11) as follows:

1. *A Satisfactory Test Section.* Because this is the location of the probe or sensing element of the device being calibrated, the gas flows must be uniform, both perpendicular and axial to the plane of flow. Streamlined entries and straight runs of duct are essential to eliminate pronounced vena contracta and turbulence.

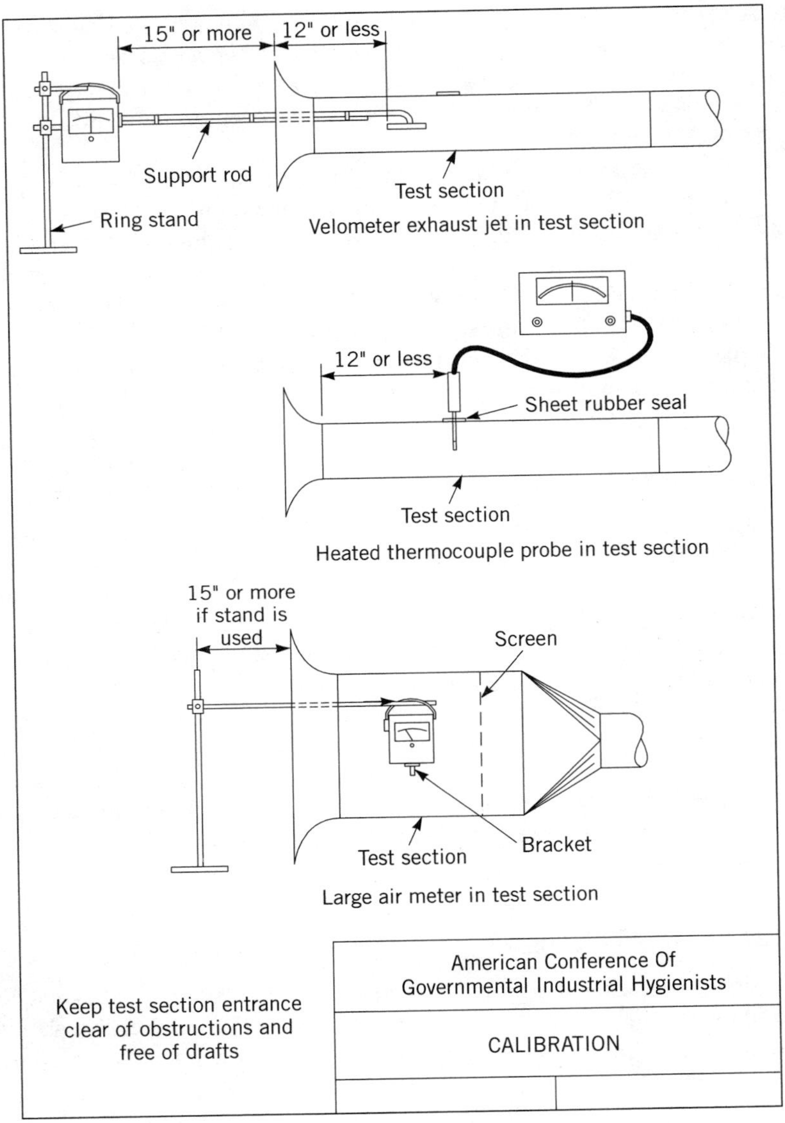

Figure 15.16. Wind tunnel and its uses for calibration of anemometers. From *Industrial Ventilation,* 19th ed. Courtesy of American Conference of Governmental Industrial Hygienists.

2. *A Satisfactory Means of Precisely Metering Airflow.* A meter with adequate scale graduations to give readings of $\pm 1\%$ is required. Venturi and orifice meters represent optimal choices, because they require only a single reading.

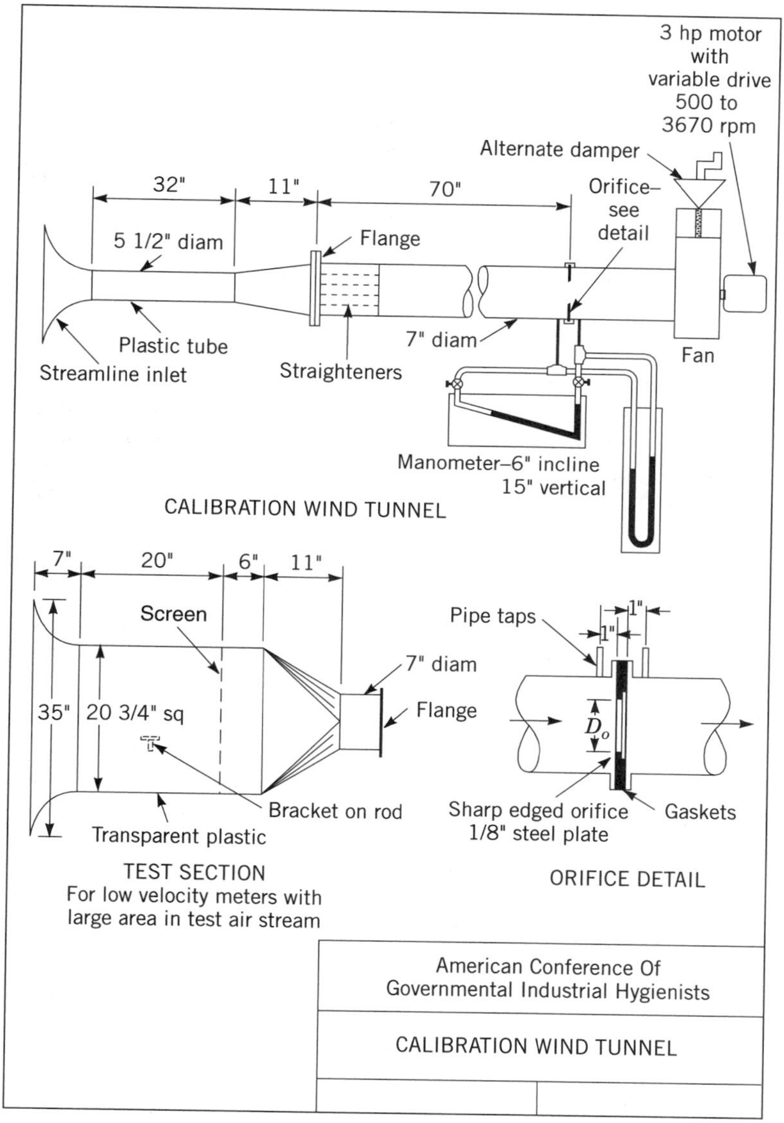

Figure 15.16. (continued).

3. *A Means of Regulating Airflow.* A wide range of flows is required. A suggested range is 50 to 10,000 fpm (2.5 cm/sec to 51 m/sec); therefore the fan must have sufficient capacity to overcome the static pressure of the entire system at the maximum velocity required. A variable drive provides for a means of easily and precisely attaining a desired velocity.

Meters must be calibrated in a manner accurately reflecting their use in the field. Vane-actuated devices should be set on a bracket inside a large test section with a streamlined entrance. Low-velocity, probe-type devices may be tested through appropriate openings in the same type of tunnel. High-velocity ranges of probe-type devices and impact devices should be tested through appropriate openings in a circular duct at least 8.5 diameters downstream from any interference. If straighteners (Fig. 15.16) are used, this requirement can be reduced to seven diameters.

NOTE: Devices must be calibrated at multiple velocities throughout their operating range.

4.5.2 *Comparison of Primary and Secondary Standards*

Figure 15.17 presents an experimental setup for checking the calibration of a secondary standard (in this case, a wet test meter) against a primary standard (a spirometer). The first step should be to check all the system elements for integrity, proper functioning, and interconnections. Both the spirometer and the wet test meter require specific internal water levels and leveling. The operating manuals for each should be examined, because they usually outline simple procedures for leakage testing and operational procedures.

After all connections have been made, it is a good policy to recheck the level of all instruments and determine that all connections are clear and have minimum resistance. If compressed air is used in a calibration procedure, it should be cleaned and dried.

Actual calibration of the wet test meter in Figure 15.17 is accomplished by opening the bypass valve and adjusting the vacuum source to obtain the desired flow rate. The optimal range of operation is between 1 and 3 revolutions/min (rpm). Before actual calibration is initiated, the wet test meter should be operated for several hours in this setup to stabilize the meter fluid with respect to temperature and absorbed gas, and to work in the bearings and mechanical linkage. After all elements of the system have been adjusted, zeroed, and stabilized, several trial runs should be made. During these runs, the cause of any indicated difference in pressure should be determined and corrected. The actual procedure would be

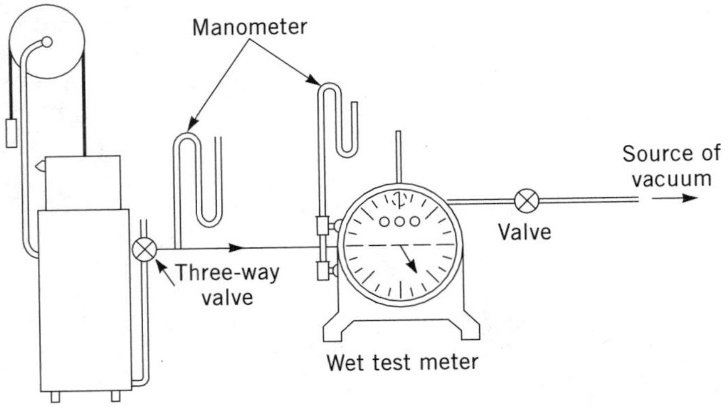

Figure 15.17. Calibration setup for calibrating a wet test meter.

to divert the air instantly to the spirometer for a predetermined volume indicated by the wet test meter (minimum of one revolution), or to near capacity of the spirometer, then return to the bypass arrangement. Readings, both quantity and pressure of the wet test meter, must be taken and recorded while the device is in motion, unless a more elaborate system is set up. In the case of a rate meter, the interval of time that the air is entering the spirometer must be accurately measured. The bell should then be allowed to come to equilibrium before displacement readings are made. Enough different flow rates are taken to establish the shape or slope of the calibration curve, and the procedure being repeated three or more times for each point. For an even more accurate calibration, the setup should be reversed so that air is withdrawn from the spirometer. In this way any imbalance due to pressure differences would be canceled.

A permanent record should be made, consisting of a sketch of the setup and a list of data, conditions, equipment, results, and personnel associated with the calibration. All readings (volume, temperatures, pressures, displacements, etc.) should be legibly recorded, including trial runs or known faulty data, with appropriate comments. The identifications of equipment, connections, and conditions should be complete, enabling another person, solely by use of the records, to replicate the same setup, equipment, and connections.

After all the data have been recorded, the calculations (e.g., correction for variations in temperature, pressure, and water vapor) are made, using the ideal gas laws:

$$V_s = V_1 \times \frac{P_1}{760} \times \frac{273}{T_1} \tag{7}$$

where V_s = volume at standard conditions (760 mm and 0°C)
V_1 = volume measured at conditions P_1 and T_1
T_1 = temperature of V_1 (K)
P_1 = pressure of V_1 (mm Hg)

In most cases the water vapor portion of the ambient pressure is disregarded. Also, the standard temperature of the gas is often referred to normal room temperature (i.e., 21°C rather than 0°C). The instruments, data reading and recording, calculations, and resulting factors or curves should be manipulated with extreme care. If a calibration disagrees with previous calibrations or the supplier's calibration, the entire procedure should be repeated and examined carefully to assure its validity. Upon completion of any calibration, the instrument should be tagged or marked in a semipermanent manner to indicate the calibration factor, where appropriate, the date, and the identity of the calibrater.

4.5.3 Reciprocal Calibration by Balanced Flow System

It is impractical to remove the flow-indicating device for calibration in many commercial instruments. This may be because of physical limitations, characteristics of the pump, unknown resistance in the system, or other limiting factors. In such situations it may be necessary to set up a reciprocal calibration procedure: that is, a controlled flow of air or gas is compared first with the instrument flow, then with a calibration source. A further complication is often introduced by the static pressure characteristics of the air mover in the instrument. In such instances supplemental pressure or vacuum must be applied to the system to offset the resistance of the calibrating device. An example of such a system appears in Figure 15.18.

The instrument is connected to a calibrated rotameter and a source of compressed air. Between the rotameter and the instrument an open-ended manometer is installed. The connections, as in any other calibration system, should be as short and resistance-free as possible.

In the calibration procedure, the flow through the instrument and rotameter is adjusted by means of a valve or restriction at the pump until the manometer indicates "0" pressure difference to the atmosphere. When this condition is achieved, both the instrument and the rotameter are operating at atmospheric pressure. The indicated and calibrated rates of flow are then recorded, and the procedure is repeated for other rates of flow.

4.5.4 Dilution Calibration

Gas dilution techniques are normally employed for instrument response calibrations; however, several procedures (11–13) have been developed whereby sampling rates of flow can be determined. The principle is essentially the same except that different unknowns are involved. In airflow calibration a known concentration of the gas (e.g., carbon dioxide) is contained in a vessel. Uncontaminated air is introduced and mixed thoroughly in the chamber to replace that removed by the instrument to be calibrated. The resulting depletion of the agent in the vessel follows the theoretical dilution formula:

$$C_t = C_0 e^{bt} \tag{8}$$

where C_t = concentration of agent in vessel at time t
C_0 = initial concentration at $t = 0$
e = base of natural logarithms
b = air changes in the vessel per unit time
t = elapsed time

The concentration of the gas in the vessel is determined periodically by an independent method. A linear plot should result from plotting concentration of agent against elapsed time on log paper. The slope of the line indicates the air changes per minute b, which can

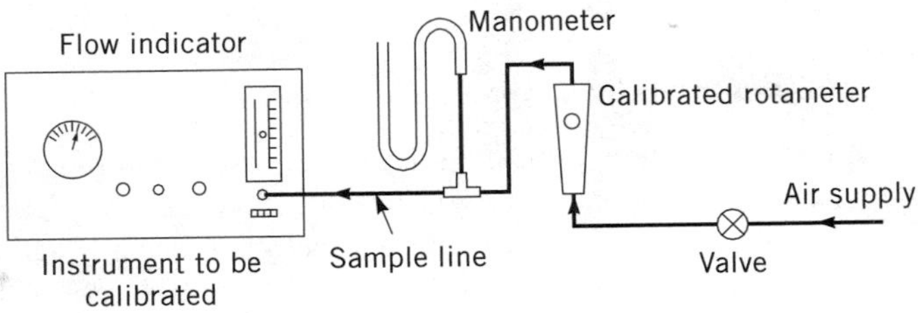

Figure 15.18. Setup for balanced flow calibration.

be converted to the rate Q of air withdrawn by the instrument from the relationship $Q = bV$, where V is the volume of the vessel.

This technique is advantageous in that virtually no resistance or obstruction is offered to the airflow through the instrument; however, it is limited by the accuracy of determining the concentration of the agents in the air mixture.

4.6 Production of Known Vapor Concentrations

4.6.1 *Introduction and Background*

Gaseous pollutants should be generated specifically for the calibration of air sampling instruments. The accurate analysis of pollutant concentration, whether it be through direct-reading instrumentation, wet chemical techniques, or indirect methods, is only as good as the calibration system. In order to test the collection efficiency of a sampler for a given contaminant, it is necessary either: (*1*) to conduct the test in the field using a proven reference instrument or techniques as a reference standard; or (*2*) to reproduce the expected field atmosphere in a calibration chamber or flow system. Techniques and equipment for producing such test atmospheres are discussed here and in detail in various other sources.

Test atmospheres that are generated for the purpose of calibrating instrument response and collection efficiency should be checked for accuracy by using mass balance relations or, where applicable, by using reference instruments or sampling and analytical procedures whose reliability and accuracy are well documented. Professional groups and governmental agencies that publish such recommendations, guidelines, and standards are listed in Table 15.4.

Methods of producing known concentrations are usually divided into two general classes: (*1*) static or batch systems; and (*2*) dynamic or continuous flow systems. With static systems, a known amount of gas is mixed with a known amount of air to produce a known concentration, and samples of this mixture are used for calibration. Static systems are limited by two factors, loss of vapor by surface adsorption and the finite volume of the mixture. In a dynamic calibration system, air and gas or vapor are continuously metered in proportions that will produce the final desired concentration. They provide an unlimited supply of the test atmosphere, and wall losses become negligible after equilibration has taken place.

In the field of industrial hygiene, gas or vapor concentrations are usually discussed in terms of parts per million (ppm). In this case, "parts per million" refers to a volume-to-volume relationship (i.e., so many liters of contaminant per liter of air). Thus by definition, both 1 μL of SO_2 per liter of air and 1 mL of SO_2 per cubic meter of air are equal to 1 ppm of SO_2. In the field of air pollution these concentrations may also be discussed as parts per 100 million or parts per billion, also based on a volume-to-volume ratio.

Occasionally with direct-reading instruments, and more frequently with chemical analysis of the atmosphere, confusion arises in converting milligrams (mg) or micrograms (μm) per cubic meter (m^3) to ppm or parts per billion (ppb). Dimensional analysis is very useful in avoiding these errors. Thus, if one has a concentration in mg/m^3 of air, it must be converted to millimoles (mm) per m^3, and to (mL) per m^3 or ppm:

$$\left(\frac{\text{mg}_x}{\text{m}^3}\right)\left(\frac{\text{mM}_x}{\text{mg}_x}\right)\left(\frac{22.4\ \text{mL}_x}{\text{mM}_x}\right)(F_t)(F_p) = \frac{\text{mL}_x}{\text{m}^3} = \text{ppm} \tag{9}$$

where F_t and F_p are the pressure and temperature conversion factors from the well-known gas laws, and the subscript x refers to a trace contaminant. Conversely:

$$\left(\frac{\text{mL}_x}{\text{m}^3}\right)\left(\frac{\text{mM}_x}{22.4\ \text{mL}_x}\right)\left(\frac{\text{mg}_x}{\text{mM}_x}\right)(F_t)(F_p) = \frac{\text{mg}_x}{\text{m}^3} \tag{10}$$

Chemical analysis of atmospheric samples is further complicated by procedures that call for some fixed volume of absorbing or reacting solution and sometimes for dilution. In this case it is convenient to determine the concentration of contaminant in solution and, by multiplying by the volume of solution, calculate the total amount of contaminant collected. This is then related to the volume of air sampled and converted to ppm. For example, after bubbling 5 L of air at 25°C and 755 mm Hg through 25 mL of an appropriate absorbing solution (with 100% collection efficiency) it was determined that the SO_2 (molecular weight = 64) concentration in solution was 5 μg/mL.

The total amount of SO_2 measured was

$$\frac{5\ \mu\text{g}}{\text{mL}} \times 25\ \text{mL} = 125\ \mu\text{g} \tag{11}$$

The volume of 1 μM of SO_2 at 25°C and 755 mm Hg is found as follows:

$$1\ \mu\text{M} \times \frac{22.4\ \mu\text{L}}{\mu\text{m}} \times \frac{298}{273} \times \frac{760}{755} = 24.6\ \mu\text{L} \tag{12}$$

The concentration in ppm is

$$\frac{125\ \mu\text{g SO}_2}{5\ \text{L}} \times \frac{\text{mol SO}_2}{64\ \mu\text{g SO}_2} \times \frac{24.6\ \mu\text{L SO}_2}{\mu\text{M SO}_2} = \frac{9.6\ \mu\text{L SO}_2}{\text{L}} = 9.6\ \text{ppm} \tag{13}$$

When producing test atmospheres, many factors may interfere with the contaminant gas or the instrument or analytical procedure. These include: (*1*) specificity of reagents or instrument being used to measure the particular contaminant; and (*2*) loss of the contaminant by reaction with, or adsorption onto, other trace contaminants in the carrier gas or elements of the system. Thus before establishing test atmospheres, the dilution gas should be purified. In addition, the nature and chemistry of the material to be analyzed, as well as the detection principle, must be thoroughly understood.

4.6.2 *Static Systems: Batch Mixtures*

In static systems, a known amount of material is introduced into a container, either rigid or flexible, then diluted with an appropriate amount of clean air. If flexible systems are desired, bags of Mylar, aluminized Mylar, Teflon, Tedlar, or polyethylene can be used. They provide the advantage that the entire volume of the bag is usable. Polyethylene is simple to use, but many pollutants either diffuse through it or are adsorbed onto the walls (12). Mylar and aluminized Mylar are less permeable (13), and because they are inelastic, offer the additional advantage of filling to a constant volume. Usually these bags come with a valved inlet port that can accept some type of gas chromatograph septum. The O-rings in the valve and the septum must have very little or no affinity for the components of the calibration atmosphere.

The bag should first be evacuated as thoroughly as possible prior to metering in the dilution gas and injection of calibration gas or liquid. Calibrated syringes provide a simple method for injection of these materials. Just prior to use, the syringe should be flushed several times with the component of interest.

Side-port needles should be used to penetrate septa that have an inert surface, like Teflon, coating a material such as rubber, that might otherwise absorb the calibration gas or vapor. The side port will ensure that no piece of the latter material is cut out and injected into the bag along with the gas or liquid to be vaporized. The actual injection should be performed by gently depressing the syringe.

A rigid container such as a bottle can be modified to function as a collapsible bag by insertion of a balloon. Inflation and deflation of the balloon allows loading and unloading of the calibration atmosphere with no dilution. The fittings necessary for controlling the sampling can be attached directly to the exposure-container, separate from the air source for the balloon.

Rigid containers such as 5-gal bottles (Fig. 15.19) have been used for static systems. The bottles are usually equipped with an inlet tube, an inlet valve, and an outlet tube. A third inlet or pass-through port for introduction of the contaminant may also be provided. In practice, after the mixture has come to equilibrium, samples are drawn from the outlet side while replacement air is allowed to enter through the inlet tube. Thus, the mixture is being diluted while it is being sampled.

Under ideal conditions, the concentration remaining is a known function of the number of air changes in the bottle. If one assumes instantaneous and perfect mixing of the in-

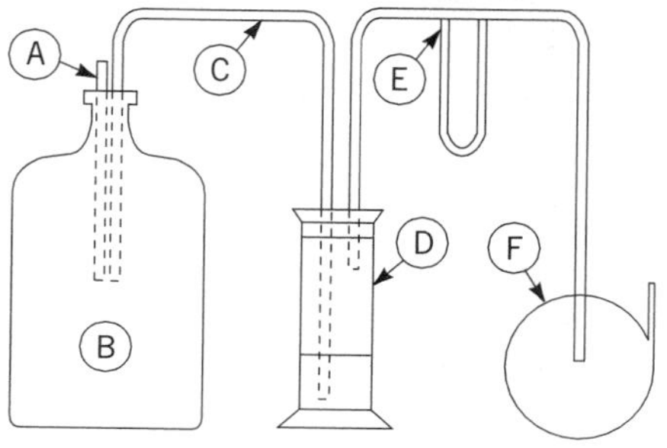

Figure 15.19. Five-gallon bottle for static calibration. A, intake tube; B, 5-gal. bottle; C, withdrawal tube; D, collecting device of direct reading instrument; E, flowmeter; F, suction pump. From Ref. 14.

coming air with the entire sample volume, the concentration change, as a small volume is withdrawn, is equal to the concentration times the percentage of the volume withdrawn:

$$dC = C \frac{dV}{V_0} \tag{14}$$

which integrates to:

$$C = C_o e^{-(V/V_0)} \text{ or } 2.3 \log \frac{C_o}{C} = \frac{V}{V_0} \tag{15}$$

where C = the concentration at any time
V = total volume of sample withdrawn
C_o = original concentration
V_0 = the volume of the chamber

Thus, if one-tenth the volume is removed, one can write:

$$2.3 \log \frac{C_o}{C} = 0.1 \tag{16}$$

$$\log \frac{C_o}{C} = \frac{0.1}{2.3} = 0.0435 \tag{17}$$

$$\frac{C_o}{C} = 1.1053 \quad \text{or} \quad \frac{C}{C_o} = 0.9047 \tag{18}$$

The average concentration of the sample withdrawn is 0.9524. If instantaneous mixing does not occur, and the inlet and outlet port are separated, the average concentration may be even higher.

If one were interested in a maximum of 5% variation from the average concentration, only about 10% of the sample could be used. Setterlind (15) has shown that this limitation can be overcome by using two or more bottles of equal volume (V_0) in series, with the initial concentration in each bottle being the same. When the mixture is withdrawn from the last bottle, it is not displaced by air but by the mixture from the preceding bottle. If, as previously, a maximum of 5% variation in concentration can be tolerated, two bottles in series provides a usable sample of 0.6 V_0. With five bottles, the usable sample will increase to about 3 V_0. A table in Setterlind's paper gives both residual concentration and average concentration of the withdrawn sample as a function of the number of volumes withdrawn for each of five bottles in series.

As noted earlier, a rigid system can also be modified to give greater usable volumes by attaching a balloon to the intake side inside the bottle. Air from the bottle can then be displaced without any dilution by merely expanding the balloon.

4.6.2.1 Introduction of Material into Static Systems. Calibrated syringes provide a simple method for introduction of materials, either gaseous or liquid, into static systems. A wide variety of both gas and liquid syringes are available down to the microliter range (1). A second method is to produce glass ampuls containing a known amount of pure contaminant and break them within the fixed volume of the static system. Setterlind (15) has discussed the preparation of ampuls in detail. Other devices such as gas burettes, displacement manometers, and small pressurized bombs have been used successfully (16). Gaseous concentrations can also be produced by adding stoichiometrically determined amounts of reacting chemicals.

Finally, a standard cylinder can be evacuated, filled with a measured volume of gas or liquid, and repressurized with compressed air or other carrier gas to produce the concentrations required. This mixture can be used with further dilution if necessary. The techniques for filling cylinders have been discussed by Cotabish et al. (17). A number of various gases and vapors are available in different concentrations from several manufacturers.

Measuring the weight of compound that is placed inside the standard cylinder is the most accurate method of producing a calibration atmosphere. Standard cylinders from a commercial source usually provide gravimetric analyses of the weight of the cylinder before and after the pure compound was introduced. The cylinder is repressurized with a known number of moles of diluent gas such as air or nitrogen. The actual cylinders should always be checked by an independent analysis procedure, because the trace gas may not be adequately mixed or may be partially lost due to wall adsorption. The mixture of the gas produced in cylinders cannot be assumed to be uniform. In fact, the cylinders should be rolled, or otherwise treated, in order to ensure that the injected gases are thoroughly mixed.

In the partial pressure method, the relative concentrations of the different component gases of the calibration atmosphere are determined by the partial pressure used to introduce each gas into the cylinder. The cylinder or cylinders are filled in sequence beginning with the component that will have the lowest partial pressure (concentration) in the gas mixture.

Standard cylinders can also be filled with known volumes of contaminant gas and diluent gas by having known flows of each of these gases supplying the input line to the compressor. The concentration in the standard cylinder will be the same as that determined by the ratio of these flows.

4.6.3 Dynamic Systems: Continuous Flow

In dynamic systems the rate of airflow and the rate of addition of contaminant to the airstream are carefully controlled to produce a known dilution ratio. Dynamic systems offer a continuous supply of material, allow for rapid and predictable concentration changes, and minimize the effect of wall losses as the contaminant comes to equilibrium with the interior surfaces of the system. Both gases and liquids can be used with dynamic systems. With liquids, however, provision must be available for conversion to the vapor state.

4.6.3.1 Gas Dilution Systems. A simple schematic of a gas dilution system appears in Figure 15.20. Air and the contaminant gas are metered through restrictions and mixed. The output can be used as is or further diluted in a similar system. In theory this process

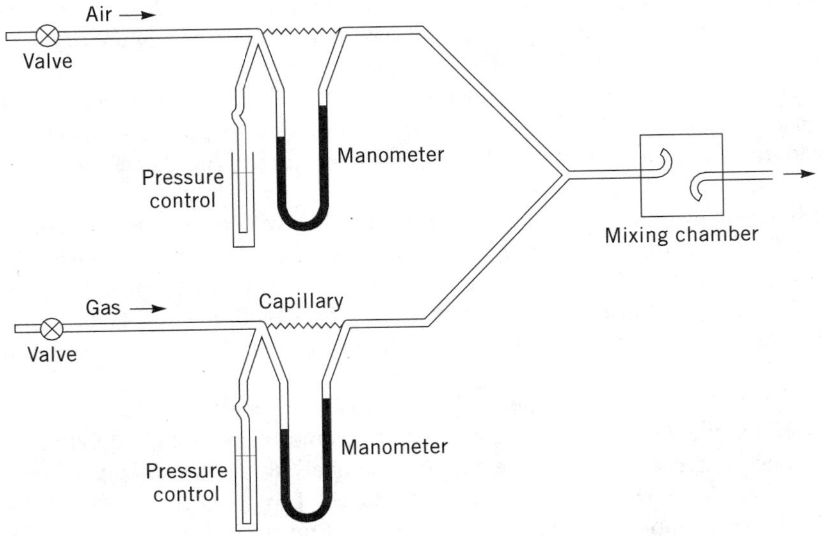

Figure 15.20. Continuous mixer for dynamic gas concentrations. From Ref. 17. Courtesy of Williams & Wilkins Company, Baltimore, and Mine Safety Appliances Company, Pittsburgh.

can be repeated until the necessary dilution ratio is obtained. In practice, series dilution systems are subject to a variety of instabilities that make them difficult to control.

Saltzman (10, 18) has described two flow dilution devices; Figure 15.12 shows a porous plug flowmeter, a device that assures that a restricting fiber mat-plugged capillary receives a constant pressure of the contaminant gas. The contaminant gas flow, a function of the pressure, is controlled by the height of the column of water or oil. A second device (Fig. 15.21) minimizes back pressure and includes a mixing chamber, because the airstream is

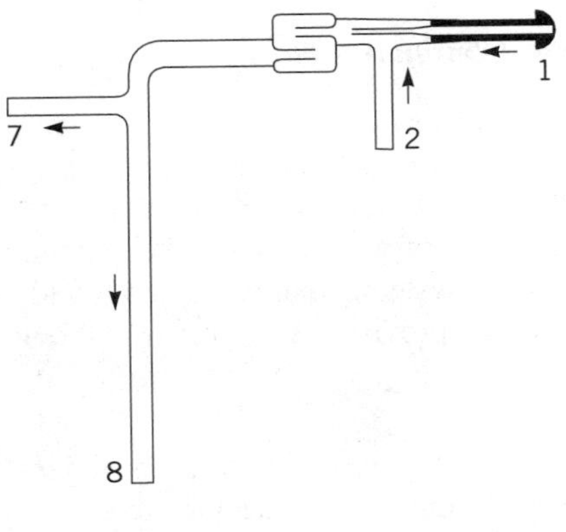

Figure 15.21. All-glass system. From Ref. 10.

split. The majority of the gas can be piped to waste through the larger tube. Immersing the end of this tube in water provides a slightly positive pressure at the smaller sidearm delivery tube.

Cotabish et al. (17) have described a system originally patented by Mase for compensation of back pressure (Fig. 15.22) in which both the air and contaminant gas flow are regulated by the height of a water column, which in turn is controlled by the back pressure of the calibration system. Thus an increase in back pressure causes an increase in the delivery pressure of both air and contaminant gas.

Calibrated Instruments, Inc., has two instruments available for calibration purposes: the ppm Maker (Fig. 15.23) and the Stack Gas Calibrator (Fig. 15.24). The ppm Maker consists of a four-output, positive displacement pump and two mechanized four-way stopcocks with single-bore plugs. The bore is normally aligned with the carrier gas flow. When activated, the stopcock is rotated 180°, momentarily aligning the bore with the contaminant gas airflow and delivering a precise volume to the carrier gas. A mixing chamber downstream mixes the carrier gas and the contaminant. The mixture is then pumped through a second identical system. By varying the flow rates of the carrier gas, dilution ratios in the order of $1{:}10^9$ can be achieved. The stepwise increments of the pumps and the stopcocks provide more than 10,000 different concentration ratios.

The Stack Gas Calibrator traps a fixed amount of gas by a series of valves and tubing and releases this volume into the carrier gas. The number of volumes released can be varied from 1 to 10 per minute. Depending on the size of the volume, three dilution ranges of 10 steps are available, 200 to 2000 ppm, 660 to 6600 ppm, and 1320 to 13,200 ppm.

Another gas dilution system, the Dyna-Blender of Matheson Gas Products, is represented schematically in Figure 15.25. Still another device for constant delivery of a pol-

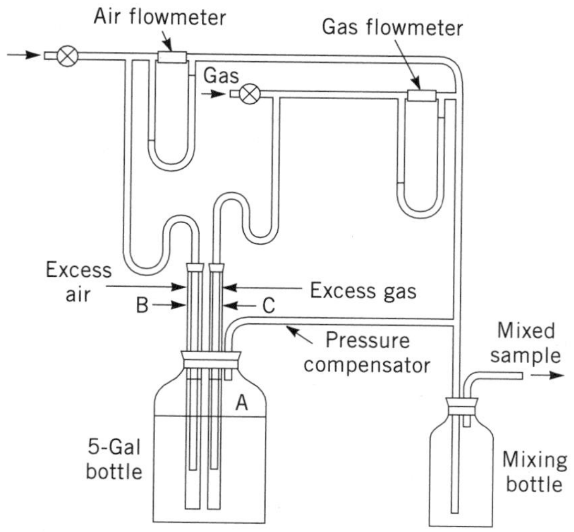

Figure 15.22. Modified Mase gas mixer for compensation of back pressure. From Ref. 17. Courtesy of Williams & Wilkins Company, Baltimore, and Mine Safety Appliances Company, Pittsburgh.

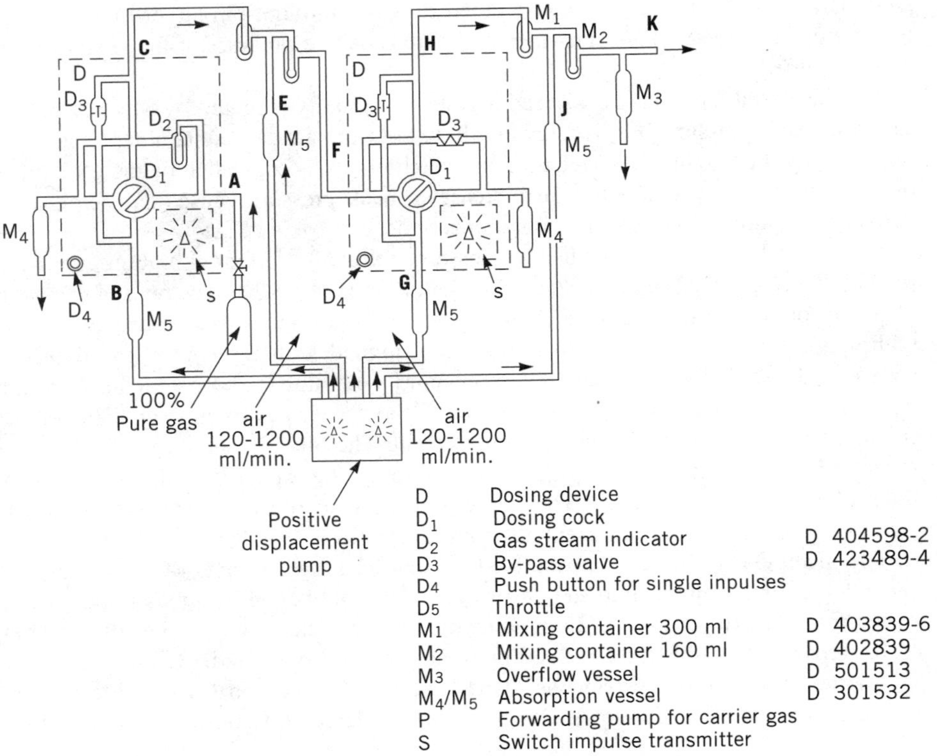

Figure 15.23. The ppm Maker. Schematic courtesy of Calibrated Instruments, Incorporated.

lutant gas has been described by Goetz and Kallai (19) (Fig. 15.26). It consists of a large, gastight syringe with a centrifugal rotor attached to the piston so that the piston rotates around its axis. The rotation, caused by a jet of air directed tangentially toward the rotor, is nearly friction-free and induces a constant pressure in the gas. The outlet of the syringe is connected on one side of a glass T-tube. Dilution air is piped into the base of the T, and the mixture exits the T-tube from the outer sidearm.

4.6.3.2 Liquid Dilution Systems. When the contaminant is a liquid at normal temperatures, a vaporization step must be included. One procedure is to use a motor-driven syringe (16, 17) and meter the liquid onto a wick or a heated plate in a calibrated airstream. Nelson and Griggs (20) have described a calibration apparatus that makes use of this principle (Fig. 15.27 and Fig. 15.28). The system consists of an air cleaner, a solvent injection system, and a combination mixing and cooling chamber. A large range of solvent concentrations can be produced (2 to 2000 ppm). The device permits rapid changes in the concentrations and is accurate to about 1%. It can also be used to produce gas dilutions with an even wider range of available concentrations (0.05 to 2000 ppm).

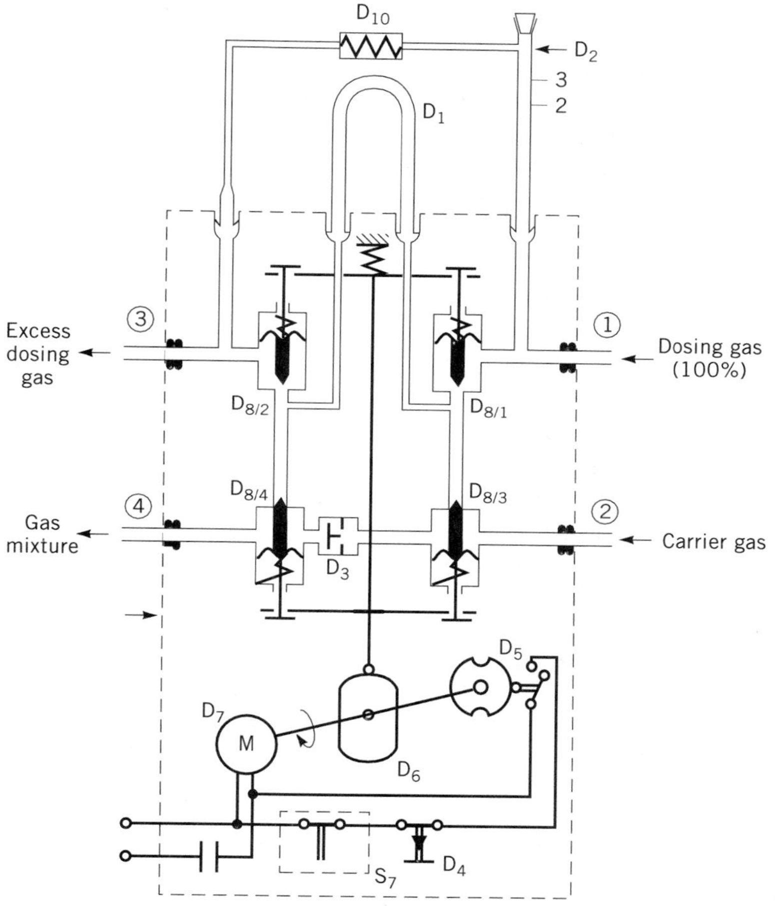

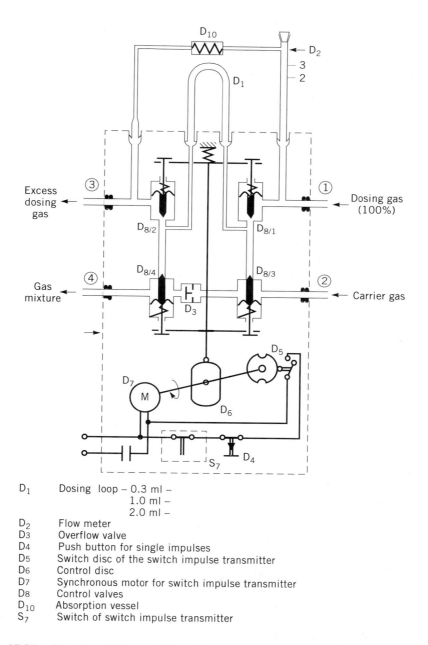

D_1 Dosing loop – 0.3 ml –
1.0 ml –
2.0 ml –
D_2 Flow meter
D_3 Overflow valve
D_4 Push button for single impulses
D_5 Switch disc of the switch impulse transmitter
D_6 Control disc
D_7 Synchronous motor for switch impulse transmitter
D_8 Control valves
D_{10} Absorption vessel
S_7 Switch of switch impulse transmitter

Figure 15.24. Stack Gas Calibrator. Schematic courtesy of Calibrated Instruments, Incorporated.

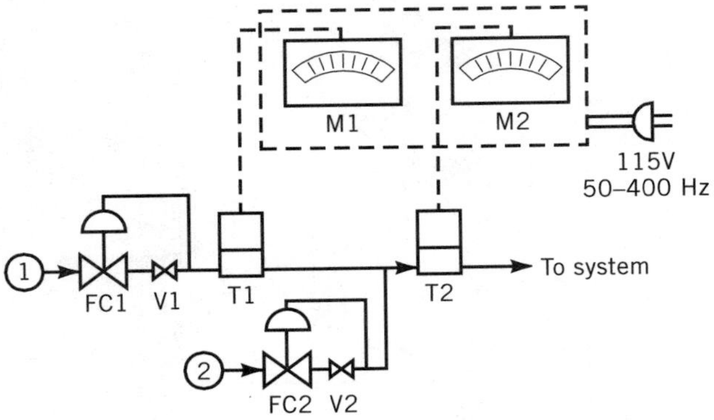

1 "Concentrate" stream

2 Carrier stream

T1 - Flow transducer for concentrate stream

T2 - Flow transducer for total flow (carrier & concentrate)

V1 - Concentrate flow control valve

V2 - Carrier flow control valve

FC1 - Flow controller–concentrate

FC2 - Flow controller–carrier

M1 - Readout meter–concentrate flow

M2 - Readout meter total flow

Figure 15.25. Matheson Dyna-Blender. Schematic courtesy of Matheson Gas Products, Division of Will Ross, Inc.

A second vapor generation method is to saturate an airstream with vapor and dilute to the desired concentration with makeup air. The amount of vapor in the saturated airstream depends on both the temperature and the vapor pressure of the contaminant and can be precisely calculated. A simple vapor saturator is illustrated in Figure 15.29. The inert carrier gas passes through two gas-washing bottles in series, and these contain the liquid to be volatilized. The first bottle is kept at a higher temperature than the second one, which is immersed in a constant temperature bath. By using the two bottles in this fashion, saturation of the exit gas is assured. A filter is sometimes included to remove any droplets entrained in the airstream as well as any condensation particles. A mercury vapor generator using this principle has been described by Nelson (21).

Diffusion cells have been used to produce known concentrations of gaseous vapors (22, 24). All such devices have the basic design shown in Figure 15.30. Liquid is placed in a reservoir that is connected to a mixing zone by a tube of known length and cross section. The concentration in the outlet of the diffusion cell is a function of the diffusion rate of the vapor, q_d, and the total flow, Q_T, in the system.

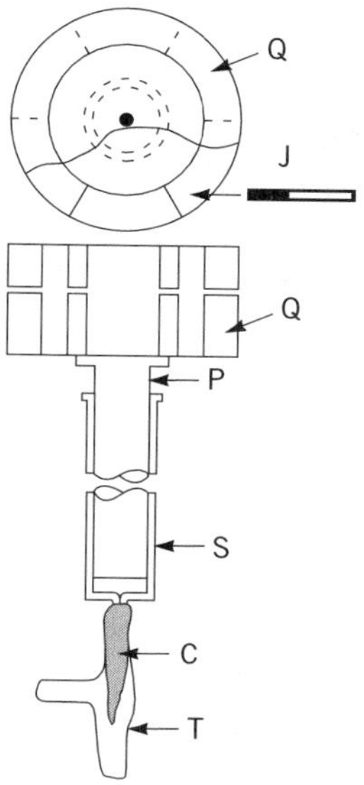

Figure 15.26. Schematic of spinning syringe calibrator assembly: Q, fan vanes; J, air jet; P, glass piston; S, large glass syringe; C, capillary tube; T, T-tube. From Ref. 19. Courtesy of the Air Pollution Control Association.

$$C(\text{ppm}) = \frac{q_d}{Q_T} 10^6 \tag{19}$$

When designing a diffusion cell for a specific application or output rate, the diffusion rate of the vapor produced above the liquid in the reservoir can be calculated from the environmental conditions in the system, the diffusion coefficient of the vapor, and the length and cross sectional area of the diffusion tube. If the diffusion coefficient is unknown, it can be estimated from the molecular weights of the vapor and the diluent gas (22). In general, these calculations for estimating the diffusion rate are subject to assumptions that cause a significant difference between the calculated and measured values.

In practice, the diffusion rate of the vapor through the specific geometry of the diffusion tube can be estimated for a given temperature by measuring the change in weight of the reservoir or by measuring the output concentration with calibrated chemical detectors. In operating such devices, the accuracy and stability of the output concentrations are directly proportional to the ability to control temperature in the reservoir, diffusion tube, and mixing zone.

4.6.3.3 Diffusion of Vapors: Permeation Devices. Permeation methods for producing controlled atmospheres used in calibrating air sampling instruments have been reviewed by Nelson (20), O'Keefe and Ortman (24), and Lodge (25). In general, the permeation of molecules of a source material through plastics can be used to reproducibly generate a

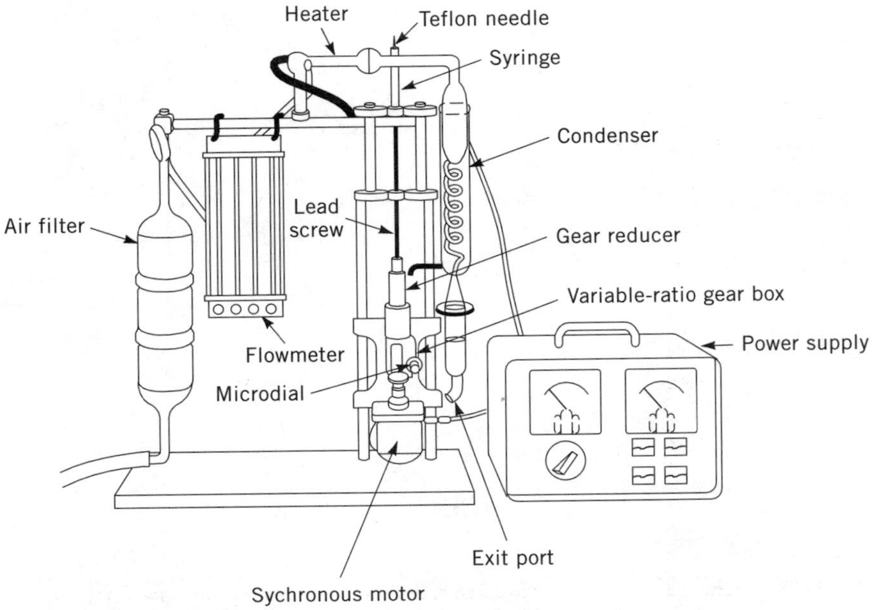

Figure 15.27. Syringe drive calibration assembly. Reprinted from U.C.R.L.-70394, courtesy of Lawrence Radiation Laboratory and the U.S. Atomic Energy Commission.

controlled atmosphere provided the critical temperature of the source material is above 20 to 25°C. Plastics such as fluorinated ethylene propylene (FEP Teflon), tetrafluoroethylene (TFE Teflon), polyethylene, poly(vinyl acetate) and poly(ethylene terephthalate) (Mylar) are a few of the materials that have been used in permeation devices (22). In the operation of these devices, the source material will usually dissolve in the plastic and permeate through it. The rate of permeation is primarily a function of the thickness of the plastic material, the total internal area exposed to the source material, and the temperature. These devices are sensitive to the temperature to the extent that a 0.1°C change in temperature can result in a 1% change in the permeation rate through the plastic container.

All existing permeation devices have one of three basic formats (Fig. 15.31): source liquid in a permeation tube, permeation tube in the source liquid, and source liquid reservoir with permeable and impermeable wall. In all cases, diluent air at flow rate Q_0 (L/min.) passes over or through the device and the source liquid held at constant temperature. The concentration in the output flow is proportional to the ratio of the permeation rate of the source material, q_p, to the output flow rate, Q_T.

Like diffusion devices, the output of a permeation device can be estimated from the thickness of the wall material, the type of the wall material, the source liquid, the magnitude of the permeable areas exposed to the source material, the pressure drop, and the temperature (22). These approximations can be accurate to within 10% and are useful mainly in designing a device with specific output capabilities.

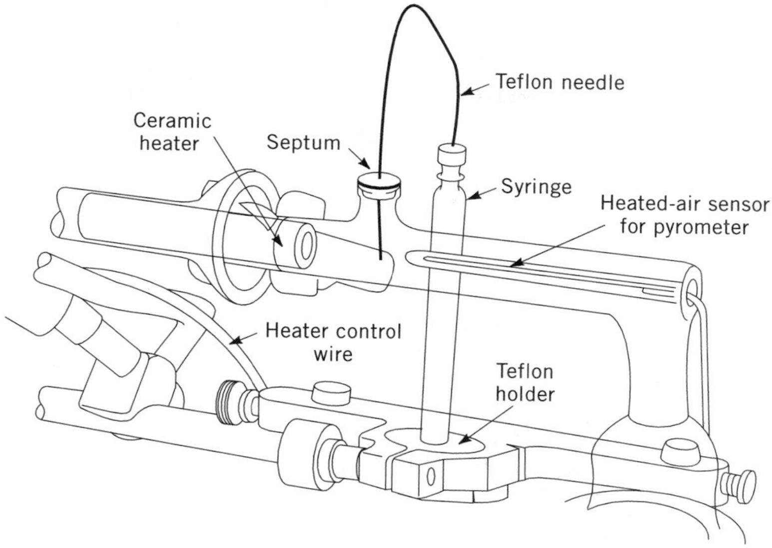

Figure 15.28. Detailed view of heating system and injection port. Reprinted from U.C.R.L.-70394, courtesy of Lawrence Radiation Laboratory and the U.S. Atomic Energy Commission.

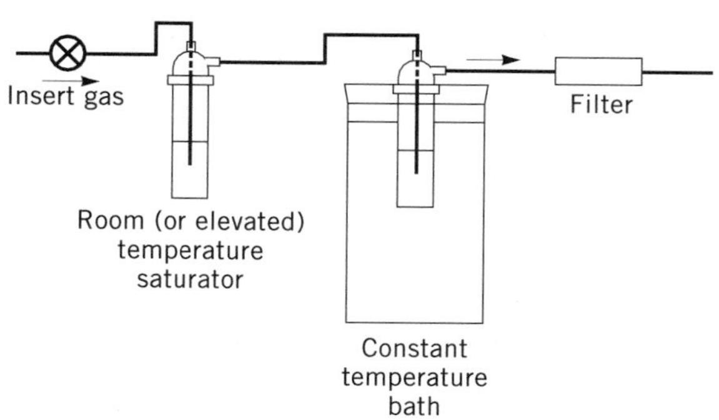

Figure 15.29. Vapor saturator. From Ref. 17. Courtesy of Williams & Wilkins Company, Baltimore, and Mine Safety Appliances Company, Pittsburgh.

For permeation devices to be used in calibration of air sampling instruments, they must be either weighed before and after use, calibrated with chemical detectors just prior to use, or purchased as precalibrated units. The advantage of permeation devices are that they are extremely simple and, under adequate temperature control, can be highly precise. A wide range of concentrations in the high (hundreds of ppm) to very low (hundredths of ppm)

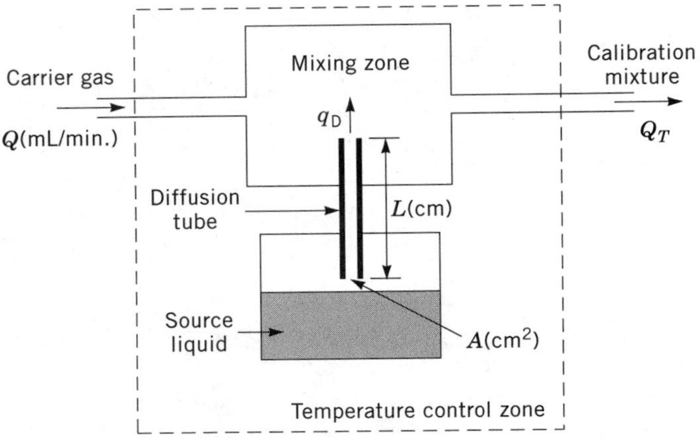

Figure 15.30. Basic design of diffusion cells.

range can be produced. However, the total output rate and, thus, the achievable concentrations from these permeation devices, is low.

4.7 Dynamic Calibration of Aerosol Samplers

Calibration systems for air sampling instruments may be as simple as a single delivery line from the generator to the inlet of the air sampling instrument, or they may be as complex as an exposure system that includes a chamber of sufficient size to allow several air sampling instruments to be completely immersed in the calibration atmosphere. Sampling conditions encountered in the industrial environment can often be duplicated by placing instruments inside a calibration chamber or wind tunnel. When such facilities are not available, the instruments are tethered about a gas distribution system in a manner much like that used for nose-only exposure of animals to test atmosphere. Techniques, insights, and pitfalls common in whole body and nose-only exposure of animals to test atmospheres are directly applicable to completing accurate air sampling instrument calibrations (26).

Regardless of whether a calibration duct or calibration chamber is used, the operator should have a basic understanding of the air flow through the calibration system, the assumptions contained in the equations used to predict concentration at different points in the calibration system as a function of time, and the influence of deviations from optimal operating conditions on the stability of the calibration atmosphere. These considerations have been discussed in detail by Moss (27).

4.7.1 Components of Aerosol Samplers

An aerosol sampler usually consists of a sampling inlet, a detection or collection section, a flow metering device, an air mover, and a flow controller. The sampling inlet, which is

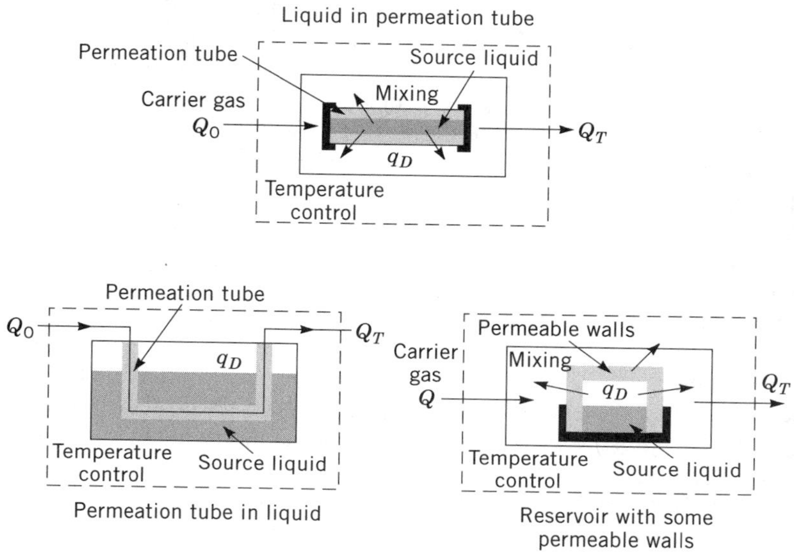

Figure 15.31. Three basic formats of permeation devices.

the entrance to the instrument, is connected to the detection section with a short transport line. The air mover (usually a pump) draws air into the sampler, and flow controllers control flow rates. Personal samplers and some area samplers, including impactors and filters, have separate pumps and flowmeters. A few passive samplers, such as personal photometers, do not have the air mover and flowmeters. They rely on air current in the atmosphere to bring the aerosol into the detector. Most direct-reading aerosol instruments, such as optical counters, condensation nuclei counters (CNCs), and photometers, include the sampling inlet, transport line, detector, pump, and flowmeter in a single unit. Each unit is indeed a complete system. For other systems such as filter samplers, individual components must be assembled.

It is important to know that there can be significant particle losses in the sampling inlet (aspiration efficiency) and transport lines (transport losses), especially for very large (>5 µm) and very small particles (<0.01 µm). Also, every instrument has finite detection limits and detection efficiency. The flowmeter controls a constant volumetric flow rate so that the instrument can be operated properly, and accurate sampling volume and, therefore, aerosol concentration can be determined.

4.7.2 Measured Parameters

Depending on the function of the instrument, the measured parameter can be separated into particle concentration (number or mass) and particle size distribution. For size-selective samplers, mass concentrations for inspirable, thoracic, and respirable or fine fractions will be determined. Often, both parameters have to be considered in the instrument calibration. For example, in impactor calibration, the collection efficiency as a function of particle size has to be determined, taking into account the sampling and transport losses. It is recommended that the investigator measure the detection efficiency of the whole system, including the aspiration efficiency of the inlet, losses in the transport lines, and efficiencies of the detector or sensor.

Each instrument has a finite detection range, and is therefore only useful in that range. The applicable size range is based on the sampling principle, the detector efficiency, and inlet design. For example, inertial-type instruments such as impactors usually collect particles between 0.5 and 15 μm; for diffusion batteries, the size ranges between 0.005 and 0.5 μm. For optical instruments, the lower detection limit is about 0.1 μm. Therefore, the instrument influences the selection of the test aerosol. In most calibrations, particles having the size in the applicable range of the instrument are required to establish the calibration curves. Also, most instruments have minimum and maximum detection limits in aerosol concentration. For example, instruments that are based on the detection of scattered light for single particles have very low maximum concentration limits (in the order of 100 particles/cm^3). At higher concentrations, increased coincidence errors occur, due to the simultaneous presence of two or more particles in the sensing volume of the detector. For aerosol collectors, overloading the substrates causes sampling errors; therefore, appropriate sampling time and mass concentration have to be considered.

4.7.3 Sampling Environments

Depending on the wind speed in an environment, the sampling procedure can be classified as calm air sampling or sampling in the flow stream. Calm air sampling generally refers to a wind speed less than 50 cm/sec, and applies to indoor environments, including residential units, offices, and factories. Flow stream sampling refers to environments with higher wind speed, such as ambient atmosphere, or inside ventilation ducts and stacks. The air flow pattern in an environment affects particle movement and, therefore, is an important parameter for the inlet aspiration efficiency. Criteria for calm air and flow stream sampling have been discussed by Davies (28), Hinds (29), and Brockmann (30). To stimulate various flow conditions in the environment, test facilities with different capabilities should be considered. Instrument chambers with uniform, low air speeds are suitable for testing samplers under calm air conditions, and aerosol wind tunnels are required for testing samplers under flow stream conditions.

4.7.4 Test Programs

Aerosol instruments can be tested on several levels for performance. The decision on the appropriate test program is largely driven by regulatory and/or scientific needs. The three levels of test programs include:

1. Flow calibration and system integrity.
2. Single point check.
3. Full-scale calibration.

The next level of testing involves instrument response for a single point. For an aerosol sizing instrument, a test aerosol (usually polystyrene latex (PSL) particles) of a defined size is used, and the response is compared to an existing calibration curve. This procedure assumes that a full-response calibration curve is available, and the user performs the test to make sure it is functioning normally. A full-scale calibration requires testing the instrument response over its full operational range. Therefore, the calibration curve in terms of response as a function of particle size or concentration can be established. The efforts, equipment, and test facility needed to perform these programs increase substantially from the simple flow calibration to the full-scale test.

Full-scale calibration is needed to establish a calibration curve for each new instrument. While it is usually assumed that the manufacturer or instrument developer should provide such data, there are some commercial instruments supplied without calibration data. In many cases, independent investigators provide careful evaluation and calibration of such instruments, and their results have often been published in the open literature.

A user needs to obtain aerosol measurement data of high quality in order to meet scientific guidelines or regulatory standards established by government agencies. The user must ensure that the instrument performs according to its specifications. The flow and system integrity should be checked whenever possible, and a single-point check should be considered for scientific validation. To a large extent, decisions on full-scale calibration depend on the scientific justifications and regulatory requirements for each study. Aerosols are usually measured for scientific research, regulatory compliance for health protection purposes, and toxicity testing.

For scientific research, calibration data provided by the manufacturer or published in a scientific journal can be used, when the instrument is used under normal conditions. If the instrument is used under different ambient pressures or flow rates, then it may need to be recalibrated under the actual operating conditions. At a minimum, the flowmeter must be calibrated and a single-point check be performed to see whether the instrument responds differently than the original calibration.

Regulations often specify standard sampling methods or equivalent methods that follow the same performance specifications. For example, the U.S. EPA performance specifications and test procedures applicable to a size-selective instrument, PM_{10}, are contained in 40 *CFR* Part 53-Ambient Air Monitoring Reference and Equivalent Methods (31). The PM_{10} samplers should be tested in a wind tunnel with liquid particles (10 sizes ranging from 3 to 25 μm in aerodynamic diameter) at wind speeds of 2, 8, and 24 km/h. The 50% cutoff determined for each speed must be 10 ± 0.5 μm. The precision for determination of concentration and flow stability is also specified (32).

4.7.5 Calibration Standards

Calibration curves are generated by comparing responses from the test instrument to those of a calibration standard. Several standard methods are now available for calibrating aerosol instruments. The primary standard method for particle size and number concentration determination is the microscopic examination, whereas the gravimetric method is the primary standard for mass concentration determination. However, several secondary standard methods have been developed and frequently used because they are often easier to use than the primary method. Table 15.5 lists aerosol instruments and test standards that have been used for their calibration. Further discussion of such standards has been provided by Cheng and Chen (33).

4.7.6 *Calibration Systems and Test Facilities*

Figure 15.32 is a schematic diagram of a typical calibration apparatus for aerosol instruments. It includes an aerosol generator, conditioning devices, a flow mixer, a test chamber, pressure and air flow monitoring equipment, the instrument to be calibrated, and the calibration standard. The aerosol from the generator can be monodisperse or polydisperse, solid or liquid, wet or dry, charged or uncharged, or spherical or nonspherical. Generally, aerosols require conditioning before use. For an aerosol containing volatile vapors or water droplets, a diffusion dryer with desiccant and/or charcoal is commonly used to remove the solvent. In some cases, a heat treatment using a high-temperature furnace is required for the production of a test aerosol (34, 35). The heat treatment involves either sintering or fusing the particles to reach the desired particle morphology and chemical form, or initiating particle evaporation and subsequent condensation to produce monodisperse particles. Because aerosol particles are usually charged by static electrification during formation, a neutralizer containing a bipolar ion source (e.g., ^{85}Kr, ^{241}Am, or ^{63}Ni) is often used in the aerosol treatment. This reduces the number of charges on the particles and results in an aerosol with charge equilibrium (36). In addition, a size-classifying device is often used in the aerosol treatment to segregate particles of a similar size or of a desired size fraction (37–39). Furthermore, a concentrator or a dilutor is often used to adjust the aerosol concentration (40, 41).

An extensive description of test facilities for aerosol calibrations has been provided by Cheng and Chen (33).

4.7.7 *Test Aerosol Generation*

Test aerosols contain either monodisperse or polydisperse, spherical or nonspherical, solid, or liquid particles (29, 42–44). The characteristics of an ideal aerosol generator are a constant and reproducible production of monodisperse and stable particles whose size and concentration can be easily controlled. To calibrate an instrument for a specific environment, the test aerosol should have similar physical and chemical properties to those of the aerosol of interest. In addition, the environment in which the instrument is to be operated must be considered when selecting the test aerosol. For example, if the instrument is to be operated in a high-temperature environment, the desired test aerosol could be a refractory metal oxide, because of its thermal stability and chemical inertness. Generally, as long as the desired aerosol is determined, the appropriate method of generation can be identified. Table 15.6 lists test aerosols that are frequently used for instrument calibration. Monodisperse aerosols containing spherical particles are the most widely used. Particles with nonspherical shapes are sometimes used in calibration to study the possible effect of shape on the instrument response. Polydisperse dust particles have also been used in calibrating dust monitors. This is important, because most real aerosols contain nonspherical particles of different sizes and densities.

Table 15.5. Calibration Standards of Aerosol Instruments

Instrument	Measured Parameter	Particle Size Range (μm)	Calibration Standard
		Size Measurement	
Cascade impactor	Flow rate, gas medium, physical dimension in and around the nozzle	0.05–30	Monodisperse, spherical particles with a known size and density
Aerodynamic particle sizing instrument	Flow rate, pressure, gas medium	0.5–20	Monodisperse, spherical particles with a known size, shape, and density
Optical particle counter	Wavelength of the light source, range of scattering angles, sensitivity of detector	0.3–15	Monodisperse, spherical particles with a known size and refractive index
Electrical mobility analyzer	Flow rate, charging mechanism, electric field strength	0.001–0.1	Monodisperse, spherical particles with a known size
Diffusion battery	Flow rate, temperature, deposition surface	0.001–0.1	Monodisperse, spherical particles with a known size
		Number Concentration Measurement	
Condensation nuclei counter	Flow rate, saturation ratio, temperature gradient	0.001–0.5	Electrical classifier with electrometer
		Mass Concentration Measurement	
Photometer	Wavelength of the light source, range of scattering angles, sensitivity of detector	0.3–1.5	Gravimetric measurement of filter samples
β-Attenuation monitor	Uniformity of particle deposit	1–15	Gravimetric measurement of filter samples
Quartz crystal mass balance	Sensitivity of the sensor	0.02–10	Gravimetric measurement of filter samples

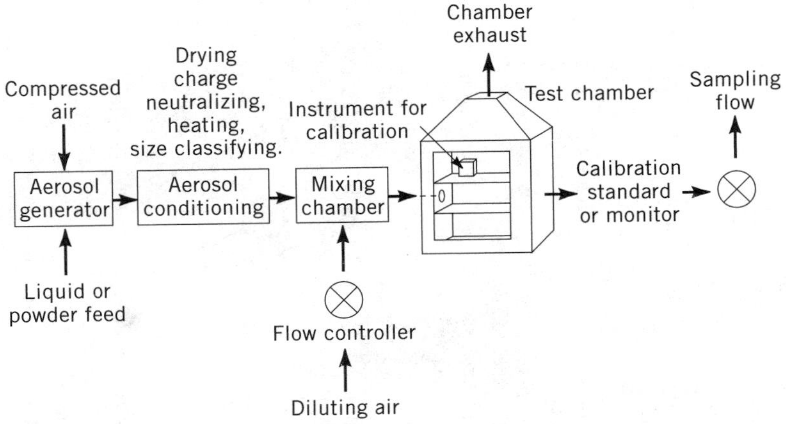

Figure 15.32. Diagram of a typical calibration system for aerosol samplers.

The size distribution and concentration of a test aerosol depend on the characteristics of the generator and the feed material. The information given in this section is intended to assist in the selection of appropriate generation techniques. The actual size distribution in each application should always be measured directly with the appropriate instruments.

Laboratory aerosol generators that have available or have been described in detail in previous reviews by Kerker (45), Fraser et al. (46), Silverman (16), and Axelrod and Lodge (12). A detailed review of techniques and equipment for producing monodisperse aerosols was prepared by Fuchs and Sutugin (47). From these and other sources, a condensed summary of techniques for generating monodisperse test aerosols has been constructed (Table 15.7). Sources of commercially available devices for producing test aerosols are tabulated in Air Sampling Instruments (4).

Aerosol generators can be divided into two types: those that produce condensation aerosols, and those that produce dispersion aerosols. In the former type, the material to be aerosolized is dispersed in the vapor phase and allowed to condense on airborne nuclei.

4.7.8 Generation of Monodisperse Condensation Aerosols

In an isothermal supersaturated environment, vapor molecules diffuse to and condense on airborne nuclei. Wilson and LaMer (48) demonstrated that the surface area of the resulting droplets increases linearly with time. Thus as the droplets become large in comparison to the nuclei, the size range becomes quite narrow, even when the nuclei on which the droplets grew may have varied widely in size.

The LaMer-Sinclair (49) aerosol generator, illustrated schematically in Figure 15.33 was based on these considerations. Improvements in the basic LaMer-Sinclair design have been described by Muir (50). Further refinements were made by Huang et al. (51). Rapaport and Weinstock (52) have described a condensation aerosol generator that is simpler and less expensive to produce, and requires less critical control of temperature and flow rate

Table 15.6. Test Aerosols and Generation Methods Used for Instrument Calibration

Test Aerosol	Particle Morphology	Size Range[a,b] VMD (μm)	σ_g	Density (g/cm^3)	Refractive Index[c]	Generation Method	Aerosol Output (particles/cm^3)
PSL (PVT)	Spherical solid	0.01–30	≤1.02	1.05 (1.027)	1.58	Nebulization	$<10^4$
Fluorescent uranine	Irregular, solid	<8	1.4–3	1.53	—	Nebulization	$<10^9$
Dioctyl phthalate	Spherical, liquid	0.5–40	≤1.1	0.99	1.49	Vibrating atomization	$<10^5$
Oleic acid	Spherical, liquid	0.5–40	≤1.1	0.89	1.46	Vibrating atomization	$<10^5$
Ammonium fluorescein	Spherical, solid	0.5–50	≤1.1	1.35	—	Vibrating atomization	$<10^5$
Fused ferric oxide	Spherical, solid	0.2–10	≤1.1	2.3	—	Spinning disc (top) atomization	$<10^7$
Fused aluminosilicate	Spherical, solid	0.2–10	≤1.1	3.5	—	Spinning disc (top) atomization	$<10^7$
Fused cerium oxide	Spherical, solid	0.2–10	≤1.1	4.33	—	Spinning disc (top) atomization	$<10^7$
Sodium chloride	Irregular, solid	0.002–0.3	≤1.2	2.17	1.54	Evaporation/condensation	$<10^6$
Silver	Irregular, solid	0.002–0.3	≤1.2	10.5	0.54	Evaporation/condensation	$<10^6$
Coal dust	Irregular, solid	~3.3	~3.2	1.45	1.54–0.5i	Dry powder dispersion	<30 mg/m^3
Arizona road dust	Irregular, solid	~3.8	~3.0	2.61	—	Dry powder dispersion	<30 mg/m^3

[a]Aerosol treatment of drying, charge neutralization, and size classification is generally used.
[b]VMD = volume median diameter; σ_g = geometric standard deviation.
[c]— indicates refractive index (RI) unknown; i indicates imaginary RI for absorption coefficient.

Table 15.7. Techniques for Generating Monodisperse Test Aerosols

Generator	Principles	Types of Aerosols	Aerosol Flow Rate (L/min)	Particle Size Range (μm)
Electrostatic classifier	Electrical mobility	Liquid and solid	2–4	0.005–1
Constant output generators	Controlled condensation	Liquid and solid	1–4	0.04–8
Spinning-top aerosol generator	Centrifugal spray	Liquid and solid	120	1–100
Vibrating-orifice aerosol generator	Jet spray	Liquid and solid	100	1–40

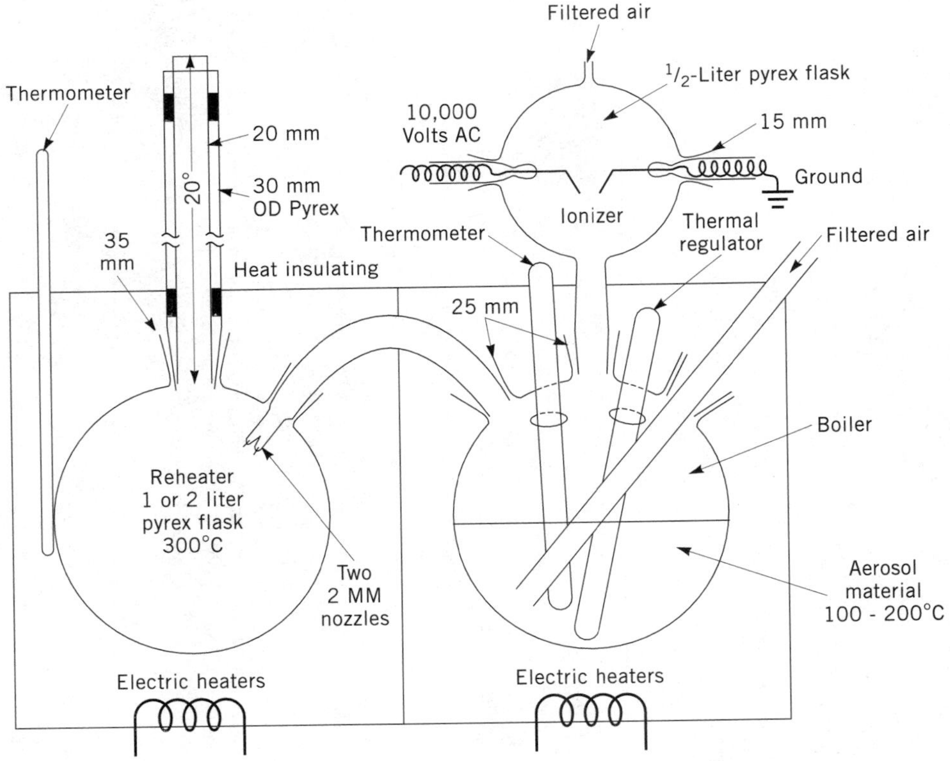

Figure 15.33. LaMer–Sinclair type of condensation aerosol generator. From U.S. Atomic Energy Commission *Handbook on Aerosols.*

for the production of monodisperse aerosol. A more sophisticated version of this generator has been described by Liu and Lee (53). This type of generator is capable of producing high-quality aerosols of high-temperature boiling, low vapor pressure liquids (e.g., dioctyl phthalate, triphenylphosphate, and sulfuric acid in the size range of about 0.03 to 1.3 μm). Prodi (54) has described a modified LaMer-Sinclair generator for monodisperse particles in the 0.2 to 8 μm size range at a concentration of $\sim 100/cm^3$. Apparatus for producing monodisperse condensation aerosols of lead, zinc, cadmium, and antimony using a high-frequency induction furnace has been described by Homma (55) and Movilliat (56). Matijevic et al. (57) and Kitani and Ouchi (58) have produced monodisperse condensation aerosols of sodium chloride. The particles produced by condensation generators are liquid and spherical unless the material vaporized has a melting point above ambient temperature. In this case the particles solidify and, if crystalline, may form nonspherical shapes. Summaries of techniques for producing radioactively labeled monodisperse condensation aerosols, including organic compounds and inorganic materials have been presented by Spurny and Lodge (59) and Kerker (45).

4.7.9 Generation of Dry Dispersion Aerosols

Dry dispersion generators comminute a bulk solid or packed powder by mechanical means, usually with the aid of an air jet. They often include an impaction plate at the outlet for removal of oversize particles and for breaking up aggregates. The aerosol particles produced are typically composed of solid, irregularly shaped particles having a broad range of sizes. Also, the rate of generation is usually not perfectly uniform, because it depends on the uniformity of hardness, or friability, of the bulk material being subdivided, as well as on the uniformity of the feed-drive mechanism and air jet pressure.

The characteristics of a variety of types of dry dust generators have been described by Ebens (60), including the widely used Wright Dust Feed (61) illustrated in Figure 15.34. Among the more difficult kinds of dry dust aerosol to generate are plastics that develop high electrostatic charges. Laskin et al. (62) have described two types of generator for such materials. One uses a high-speed fan to create a stable fluidized bed from which aerosol can be drawn; the second uses a high-speed grinder to comminute a block of solid material.

Other generator designs developed for "problem" dusts include those by Dimmock (63) for viable dusts, by Brown et al. (64) for deliquescent dust, and by Timbrell et al. (65) and Holt and Young (66) for fibrous dust. A thorough review of dry-dispersion generators has been provided by Chang and Chen (33). Table 15.8 from their review summarizes the operating parameters of some commercially available dry powder aerosol generators.

Useful aerosols of dry particles of metal and metal oxides are also produced with electrically heated (67, 68) or exploded wires (69, 70). These techniques have some disadvantages because of the very broad size distributions of the resulting particles and because of the tendency of particles to coalesce. There are applications, however, for this type of aerosol, and it is possible with a wire-heating method to produce spherical particles of many different metals or their oxides. Aerosols of very small particles have also been produced by arc vaporization (71).

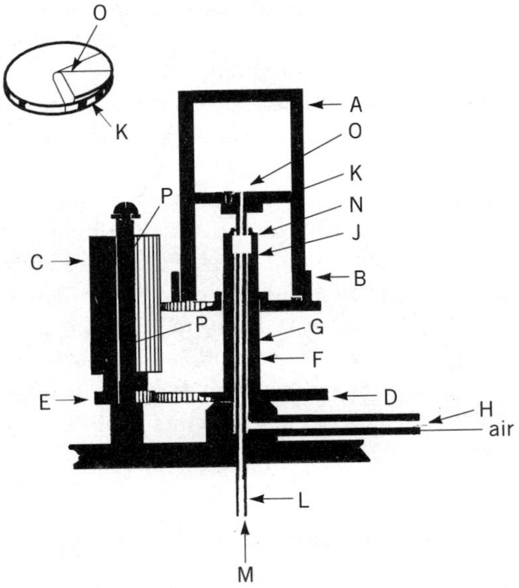

Figure 15.34. The Wright Dust Feed: A, dust cylinder; B, cap, with peripheral gear; C, pinion; D, wheel; E, pinion; F, threaded tube; G, tube, connected to H, compressed air line; J, small tube, carrying scraper head K, which communicates with jet L, which is above impaction plate M for breaking up aggregates; O is a spring disk with cutting edge. From Ref. 61.

4.7.10 Wet Dispersion Aerosol Generators

Wet dispersion generators break up bulk liquid into droplets. If the liquid is nonvolatile, the resulting aerosol is a mist or fog. If a volatile liquid is aerosolized, the resulting particles are composed of the nonvolatile residues in the feed liquid and are much smaller than the droplets dispersed from the generator. Solid particles can be produced by nebulizing salt or dye solutions or particle suspensions. Aqueous solutions, of course, produce water-soluble particles that may be hygroscopic. This may be an important factor, because the aerodynamic size for such aerosols varies with ambient humidity.

A variety of techniques can be used to subdivide bulk liquid into airborne droplets. In most cases the liquid is accelerated by the application of mechanical pneumatic, or centrifugal forces and drawn into filaments or films that break up into droplets because of surface tension. Centrifugal pressure nozzles and fan spray nozzles use hydraulic pressure to form a sheet of liquid that breaks up into droplets, but these generally have high liquid feed rates and produce very large droplets. They are seldom employed for producing aerosols for instrument calibration.

A commonly used type of aerosol generator is the two-fluid nozzle, which uses pneumatic energy to break up the liquid. Several laboratory-scale compressed air-driven nebulizers have been described in detail by Mercer et al. (77). Table 15.9 summarizes the

Table 15.8. Operating Parameters of Dry Powder Aerosol Generators

	Wright Dust Feed	Fluidized Bed	Small-Scale Powder Disperser	Jet-O Mizer
Type of operation	Scraping the packed plug and dispersing it with air	Feeding the powder to the bed on a conveyor and air fluidizing it	Using rotating plate to deliver the powder and dispersing it with Venturi suction	Using Venturi suction to feed the powder into a fluid energy mill in which centrifugal force and air velocity are used to break up the agglomerate and disperse the powder
Air flow rate, L/min	8.5–4.0	5–20	12–21	14–113
Feed flow rate mm^3/min	0.24–210	1.2–36	0.9–2.5	2000–30,000
Output mass concentration, g/m^3 ($\rho = 1$ g/cm^3)	0.012–11.5	0.13–4.0	0.0003–0.04	10–1500

Table 15.9 Operating Parameters of Air-Blast and Ultrasonic Nebulizers

Nebulizer	Operating Conditions: Orifice Diameter (mm)	Air Pressure (psig)	Frequency (mHz)	Flow Rate[a] (L/min)	Aerosol Output (μL/L)	Droplet Size Distribution: VMD (μm)	σ_g
			Airblast Type				
Collison	0.35	15		2.0	8.8	2.5–3	—
		25		2.7	7.7	1.9–2	—
DeVilbiss[b]	0.84	15		12.4	15.5	4.2	1.8
D-40		30		20.9	12.1	2.8	1.9
DeVilbiss	0.76	15		9.4	23.2	4.0	—
D-45		30		14.5	22.9	3.4	—
Lovelace	0.26	20		1.5	40	5.8	1.8
		50		2.3	27	2.6	2.3
Retec	0.46	20		5.0	46	5.7	1.8
X-70/N		50		9.7	47	3.2	2.2
			Ultrasonic Type				
De Vilbiss	(2)[c]		1.35	41.0	54	5.7	1.5
880	(4)[c]		1.35	41.0	150	6.9	1.6
Sono-Tek			0.025–0.12	10^{-6}–0.44	—	18–80	—

[a]Output per orifice.
[b]Vent closed.
[c]Power settings.

operational characteristics of six such nebulizers, including the Lovelace design, which is illustrated in Figure 15.35. The DeVilbiss No. 40 is made of glass, which not only makes it fragile but also limits its precision of manufacture and reproducibility. Ready reproducibility led Whitby to select the British Collison (73) nebulizer for his atomizer-impactor aerosol generator (74). Other commercially available nebulizers, including those of Wright (75) (Fig. 15.36) and Dautrebande (76), were machined to close tolerance from plastic materials.

Nebulizers produce droplets of many sizes, and resultant aerosol particles after evaporation are therefore polydisperse, although relatively narrow size dispersions can be obtained with Whitby's atomizer-impactor (74) and Dautrebande's D-30 (76). The droplet distributions described for nebulizers are the initial distributions at the instant of formation; droplet evaporation begins immediately, even at saturation humidity, because the vapor pressure on a curved surface is elevated (77). The rate of evaporation depends on many factors, including solute concentration, the hygroscopicity of the solute (78, 79), the presence of immiscible liquids or evaporation inhibitors (80); and the size of the droplets.

Evaporative losses cause an increase in the concentration of the solution or of the suspended particles, resulting in an increase in the size of the dry particles formed when

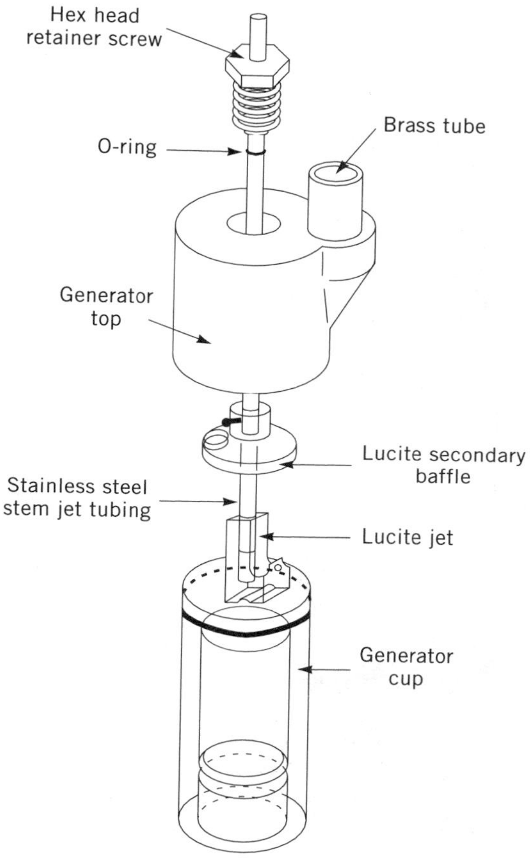

Figure 15.35. The Lovelace nebulizer, which operates with a liquid volume of ~4 mL and incorporates a jet baffle similar to that of Wright (95). Schematic courtesy of Dr. Otto G. Raabe.

the liquid evaporates. Evaporation occurs both from the surface of the liquid and from the droplets, which evaporate slightly and hit the wall of the nebulizer to be returned to the reservoir; it is most important in nebulizers with small reservoirs but large volumetric airflows.

Rotary atomizers, such as the spinning disk, utilize centrifugal force to break up the liquid, which undergoes an acceleration as it spreads from the center to the edge of the disk. The liquid leaves the edge of the disk as individual droplets or as ligaments that disintegrate into droplets. Walton and Prewett (81) demonstrated that these atomizers can produce monodisperse aerosols when operated with low liquid feed rates and high peripheral speeds. A spinning disk generator designed specifically for the production of monodisperse test aerosols with radioactive tags has been described by Lippmann and Albert (81) and is illustrated in Figure 15.37.

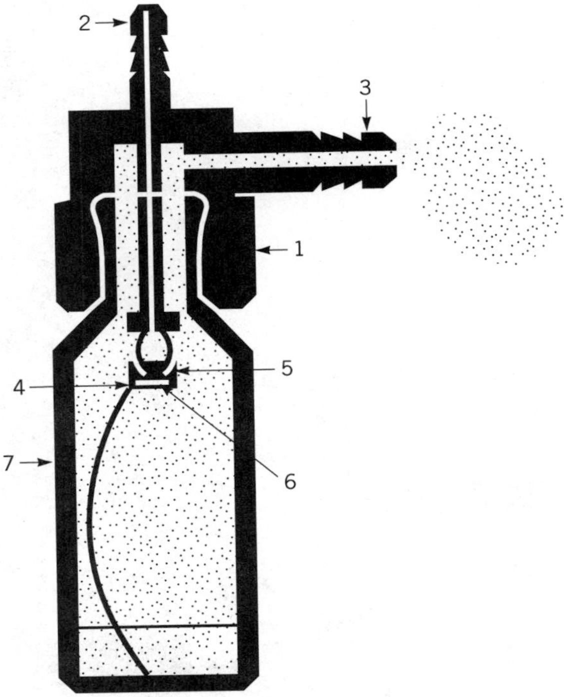

Figure 15.36. The Wright nebulizer. It consists of a solid cap 1, into which can be screwed any suitable bottle; 2, inlet connection; 3, outlet connection. The inlet connection communicates with a fine jet 4 onto which is screwed a knurled nozzle 5. The nozzle carries a circular baffle plate 6 mounted on an eccentric pillar through which passes a flexible feed tube 7. As the air jet passes through the nozzle 5, a vacuum is created that draws liquid up the feed pipe 7. The resulting spray impacts against the baffle plate and the coarser droplets (more than about 8 μm diameter) are trapped, coalesce, and fall back into the liquid. From Ref. 75.

Monodisperse test aerosols can also be produced by a variety of techniques that break up a laminar liquid jet into uniform droplets. Most of them vibrate a capillary at high speed with a variety of transducers and types of motion. Dimmick's (83) generator, for example, uses transverse vibrations, whereas Strom's (84) uses axial vibrations. Wolf (85) uses a vibrating reed, wetted to a constant length by passage through a liquid reservoir, to create the droplet stream. The generator described by Raabe (86) has an air jet above the orifice and uses an ultrasonic transducer to convert a high-frequency power signal into mechanical axial vibrations of the orifice. The vibrating orifice generator of Berglund and Liu (87), which is illustrated in Figure 15.38 uses a cylindrical piezoelectric ceramic to vibrate a thin orifice plate, with holes from 3 to 22 μm in diameter, producing droplet diameters from about 10 to 50 μm. The particles produced by these generators can be made to vary less than 1% in volume, less than the variation in size of aerosols generated by a spinning disk device.

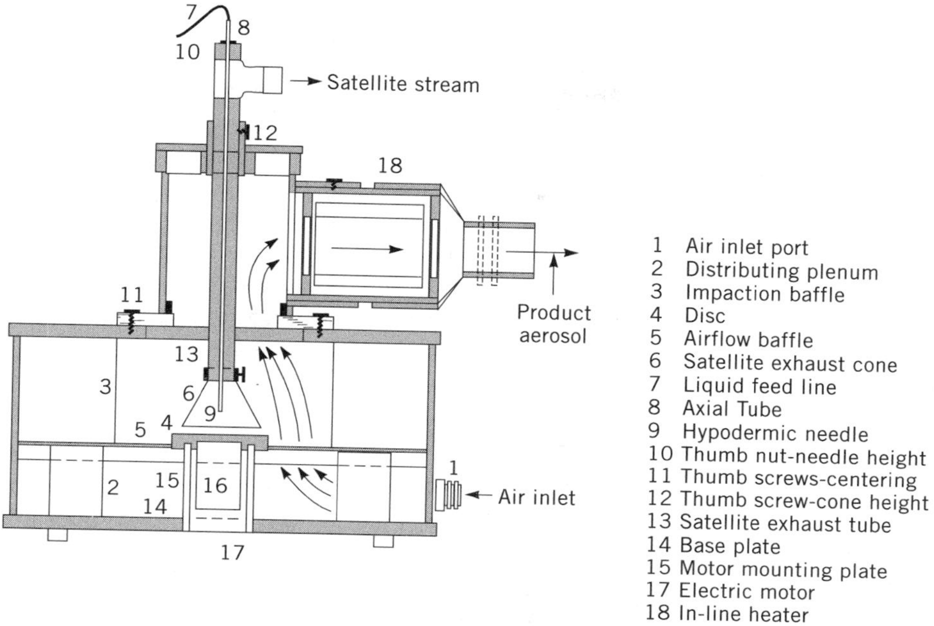

Figure 15.37. Electric motor driven spinning disk aerosol generator of Lippmann and Albert (82). 1, Air inlet port; 2, distributing plenum; 3, impaction baffle; 4, disc; 5, airflow baffle; 6, satellite exhaust cone; 7, liquid feed line; 8, axial tube; 9, hypodermic needle; 10, thumb nut-needle height; 11, thumb screws-centering; 12, thumb screw-cone height; 13, satellite exhaust tube; 14, base plate; 15, motor support tube; 16, motor mounting plate; 17, electric motor; 18, in-line heater.

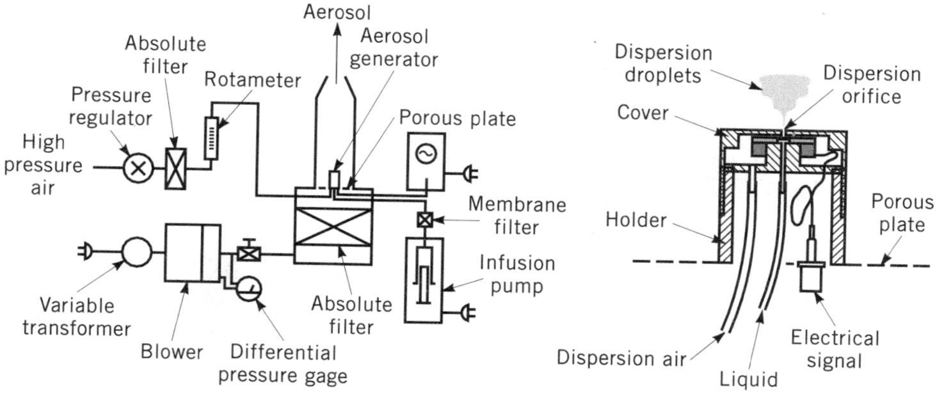

Figure 15.38. Vibrating orifice monodisperse aerosol generator. *Left:* schematic of system. *Right:* generator head. From Ref. 88.

Electrostatic atomization can also produce monodisperse aerosols. Electric charges on a liquid surface act to decrease the surface tension. Liquid flowing through a capillary at high voltage is drawn into a narrow thread that breaks up into very small droplets (89–91).

Liu and Pui (92) developed a generator (Fig. 15.39) for quite monodisperse submicrometer aerosol particles in which the polydisperse output of a compressed air nebulizer is classified electrostatically. A solution or colloid is aerosolized in a Collison atomizer, mixed with dry air to form a solid aerosol, and brought to a state of charge equilibrium with the aid of a ^{85}Kr source. This aerosol, which is polydisperse (geometric standard deviation ≅ 2.7, median particle diameter ranging from 0.009 to 0.65 μm), is introduced into a differential mobility analyzer, which functions as a particle size classifier, based on the electrical mobility of the different size categories. This apparatus consists of an inner cylindrical electrode along which flows a sheath of clean air surrounded by an outer concentric sheath of the aerosol. Depending on the voltage of the electrode and the flow rate and the geometry, the more mobile particles drift through the clean air sheath to the electrode, where they are discharged and adhere, thereby being removed from the aerosol

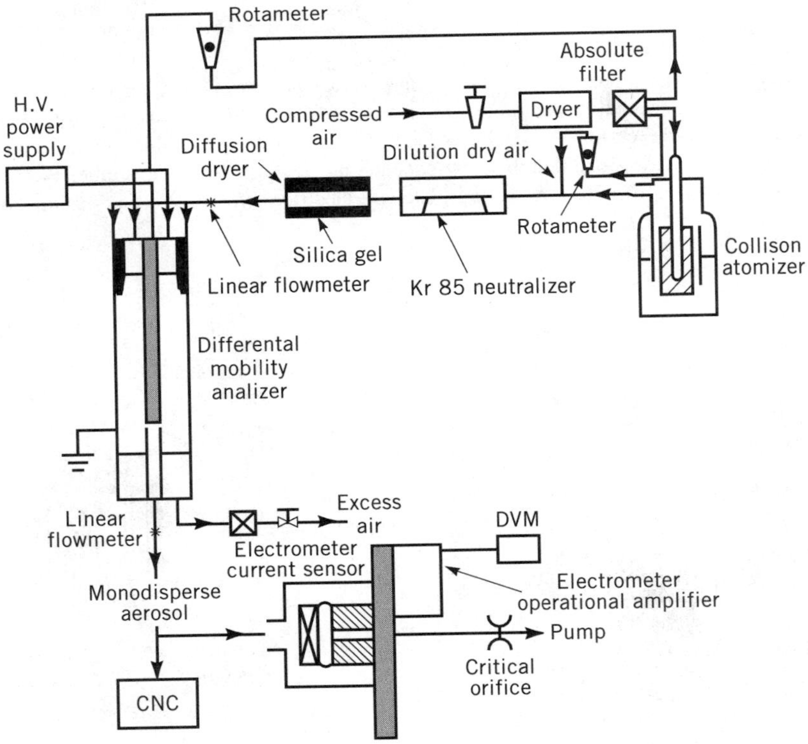

Figure 15.39. Apparatus for generating submicron aerosol standards.

stream. Under a given set of operating conditions, a particular class of particles drift to a particular position, where they can be vented. These particles make up the monodisperse aerosol. For the particular design described by Liu and Pui, there was a coefficient of variance of .04 to .08 in particle size for singly charged particles. The concentration of aerosol is measured by collection on a filter, where the electrostatic charge is discharged through an electrometer. The presence of doubly charged particles increases greatly for particles larger than 0.3 μm. Thus above this size, the mobility no longer defines the particle size.

Commercially available ultrasonic aerosol generators can vibrate a liquid surface at a frequency high enough to result in the disintegration of the surface liquid into a polydisperse droplet aerosol. For mass median droplet diameters below 5 μm, the transducer must vibrate at a frequency greater than 1 MHz. The output characteristics of ultrasonic nebulizers (Fig. 15.40) have been described by Raabe (86).

4.7.11 Generation of Solid Insoluble Aerosols with Wet Dispersion Generators

Solid insoluble aerosols can be produced by nebulizing particle suspensions. One technique for producing monodisperse test aerosols is to prepare a uniform suspension of the particles

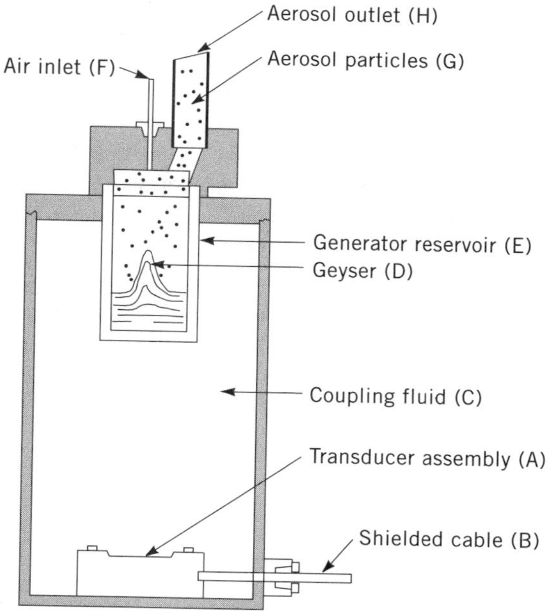

Figure 15.40. Sectional view of an operating Ultrasonic Aerosol Generator: transducer assembly A receives power through shielded cable B, generates an acoustic field in the coupling fluid C, creating an ultrasonic geyser D in the generator reservoir E, and air entering at F carries away aerosol G through outlet H. Figure by G. J. Newton, reproduced from Raabe (86).

(latex, bacteria, etc.) in which the concentration is sufficiently dilute in the liquid phase that the probability of more than one particle being present in each droplet is acceptably small (93–95). This will result in a high vapor to particle ratio, thus limiting the mass concentration of aerosol produced. Another approach is to use a colloid as the feed liquid. In this case the diameter of the colloid particles can be orders of magnitude smaller than the particles in the resulting aerosol. Thus the volume of the droplet, and the size of the dried aggregate particles, is determined by the solids content of the sol.

An aerosol with chemical properties different from those of the feed material can be produced by utilizing suitable gas phase reactions such as polymerization or oxidation. Kanapilly et al. (96) describe the generation of spherical particles of insoluble oxides from aqueous solutions with heat treatment of the aerosols. This procedure involves: (a) nebulizing a solution of metal ions in chelated form; (b) drying the droplets; (c) passing the aerosol through a high-temperature heating column to produce the spherical oxide particles; and (d) cooling the aerosol with the addition of diluting air. Another example of aerosol alteration is the production of spherical aluminosilicate particles with entrapped radionuclides by heat fusion of clay aerosols (97). This method involves: (a) ion exchange of the desired radionuclide cation into clay in aqueous suspension and washing away of the unexchanged fraction; (b) nebulization of the suspension yielding a clay aerosol; and (c) heat fusion of clay aerosol, removing water and forming an aerosol of smooth solid spheres.

4.7.12 Characterizing Aerosols

4.7.12.1 Size Dispersion. The size dispersion of a test aerosol produced by a laboratory generator is determined by the characteristics of the generator and feed materials. The data on size included in the preceding discussion on generator characteristics, and in Table 15.7 indicate the approximate range obtainable in normal operation. The actual size distribution in a given case should always be measured with appropriate techniques and instrumentation. Sampling for particle size analysis has been discussed by Knutson (3). Sampling and analytical techniques have also been reviewed by Raabe (86) and Giever (98).

The distribution of droplets produced by nebulizers and some dry dust generators can usually be described by assuming that the logarithms of size are normally distributed. This log-normal distribution of sizes allows for simple mathematical transformation (99) and usually describes volume distributions satisfactorily (100). The characteristic parameters of a log-normal distribution are the median (or geometric mean) and the geometric standard deviation σ_g. The median of a distribution of diameters is called the count median diameter CMD; the median diameter based on the surface area is called the surface median diameter SMD; the median of the mass or volume distribution of the droplets or particles is called either the mass median diameter MMD or the volume median diameter VMD. These are related as follows:

$$\ln(\text{SMD}) = \ln(\text{CMD}) + 2\ln^2\sigma_g$$
$$\ln(\text{MMD}) = \ln(\text{CMD}) + 3\ln^2\sigma_g$$

in which ln designates the natural logarithm. A representative log-normal distribution appears in Figure 15.41 for a CMD equal to 1 μm and a σ_g equal to 2.

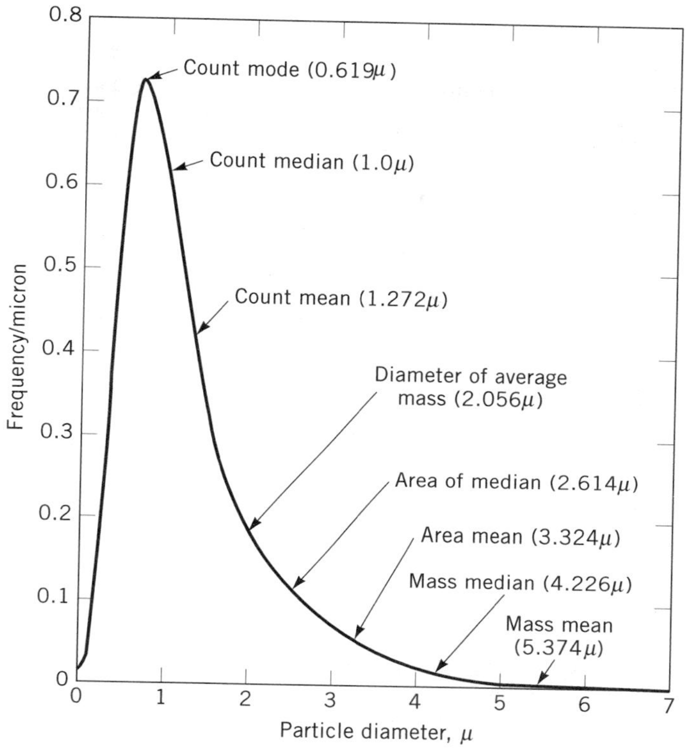

Figure 15.41. An example of the log-normal distribution function in normalized form of CMD = 1.0 μm and σ_R = 2.0 μm, showing the mode, median, and mean diameters, the mass distribution median and mean diameters, and the diameter of average mass. Graph courtesy of Dr. Otto G. Raabe (86).

When particles are classified on the basis of their airborne behavior, a parameter called aerodynamic diameter is often used. It refers to the size of a unit density sphere having the same terminal settling velocity as the particle in question. For radioactive particles, ICRP task groups (101, 102) have used the parameter aerodynamic mass activity diameter (AMAD), which is the aerodynamic median size for airborne particulate activity.

4.7.12.2 Physical and Chemical Properties. An aerosol of a pure material having the desired physical and chemical characteristics can be prepared by dispersing that material into the air by any appropriate technique previously described. It is also possible to produce aerosols that differ in physical and/or chemical properties from the feed material. For example, particle size can be varied by dissolving or suspending the material in a suitable volatile solvent that evaporates in the air to leave residue particles smaller than the nebulized droplets.

Solid aerosols resulting from droplet evaporation are generally spherical, but not always. Too rapid solvent evaporation, low pH and the presence of impurities may cause the dried particles to be wrinkled or to assume various shapes (103).

Aerosols produced from aqueous solutions (and some other methods) are charged by the random imbalance of ions in the droplets as they form (104). After evaporation, aerosol particles can be relatively highly charged; this may cause a small evaporating droplet to break up if the Rayleigh limit (105) is reached because of the repelling forces of the electrostatic charges overcoming the liquid surface tension (106). In some cases the net charge on a particle may be tens or even hundreds of electronic charge units, which will affect both the aerosol stability and behavior. Therefore a reduction in the net charges on aerosols produced by nebulization is desirable and in some experiments may be imperative. This can be accomplished either by mixing the aerosol with bipolar ions (107) or by passing it through a highly ionized volume near a radioactive source (108).

4.7.13 Detection of Aerosol Particles and Tagging Techniques

For many applications, such as efficiency testing of aerosol samplers or filters, it is often necessary to be able to measure concentrations that differ by several orders of magnitude. This type of testing can be done with untagged particles, such as polystyrene latex, using sensitive light-scattering photometers for concentration measurements. When particles other than the test aerosol are present, however, as in many field test situations, this method should not be used. Also, light-scattering techniques can be used over only a limited range of particle size, and the equipment is relatively expensive. Another approach to efficiency testing entails a microscopic count and/or size analyses of upstream and downstream samplers. However, this procedure is so tedious and time-consuming that it is seldom the method of choice.

Particle detection is often facilitated by incorporating dye or radiosotope tags in the particles in their production. Test aerosols composed of or containing fluorimetric dyes that can be analyzed in solutions containing as little as 10^{-10} g/cm^3 have been used for such applications (74). The particles are soluble in water or alcohol, and can be quantitatively leached from many types of filters and collection surfaces for analysis. Colorimetric dyes such as methylene blue, which is used in the British Standard Test for Respirator Canisters (73), can be used in similar fashion when extremes of sensitivity are not required.

Radioisotope tags have been used in many forms and can usually be detected at extremely low concentrations. Spurny and Lodge (59) have discussed a variety of techniques for preparing radioactively labeled aerosols, including: (*1*) preparation by means of neutron activation of aerosols in a nuclear pile or other neutron source; (*2*) labeling by means of decay products of radon and thoron; (*3*) preparation by means of radioactively labeled elements and compounds (condensation aerosols, disperse aerosols, and plasma aerosols); and (*4*) preparation by means of radioactively labeled condensation nuclei.

Method 2 refers to a process in which the particle surface is tagged while the particle is airborne. Procedures for surface tagging of polystyrene latex particles with isotopes in liquid suspensions by emulsion-polymerization reactions have been described by Black and Walsh (109), Bogen (110), and Singer et al. (111). Flachsbart and Stöber (112) have described a technique for growing uniform silica particles in suspension and incorporating various radioactive tags.

Other insoluble test aerosols containing nonleaching radioisotope tags have been produced by several techniques. The technique of heat fusion of ion-exchange clays (97) was discussed in the preceding section on insoluble aerosols. Techniques for producing insoluble spherical aggregate particles by nebulizing colloidal suspensions and plastics in solution have been described (82, 103). These aerosols made from colloids can be tagged with radioisotopes by mixing the nonradioactive colloid with a much lower mass concentration of an insoluble radioactive colloid before nebulization. The plastic particles can be tagged with radioisotopes in chelated form, dissolved in the plastic solution (103, 113–115).

4.8 Calibration of Sampler's Collection Efficiency

4.8.1 Use of Well-Characterized Test Atmospheres

To test the collection efficiency of a sampler for a given contaminant, it is necessary either (*1*) to conduct the test in the field using a proved reference instrument or technique as a reference standard or (*2*) to reproduce the atmosphere in a laboratory chamber or flow system. Techniques and equipment for producing such atmospheres are discussed in Section 4.6 and Section 4.7.

4.8.2 Analysis of Sampler's Collection and Downstream Total Collector

The best approach to use in the analysis of a sampler's collection is to operate the sampler under test in series with downstream total collector, as illustrated in Figure 15.42. The sampler's efficiency is then determined by the ratio of the sampler's retention to the retention in the sampler and downstream collector combined. This approach is not always feasible, however. When the penetration is estimated from downstream samples there may be additional errors if the samples are not representative.

4.8.3 Analysis of Sampler's Collection and Downstream Samples

It is not always possible or feasible to collect quantitatively all the test material that penetrates the sampler being evaluated. For example, a total collector might add too much flow resistance to the system or be too bulky for efficient analysis. In this case, the degree of penetration can be estimated from an analysis of a sample of the downstream atmosphere, as illustrated in Figure 15.43. When this approach is used, it may be necessary to

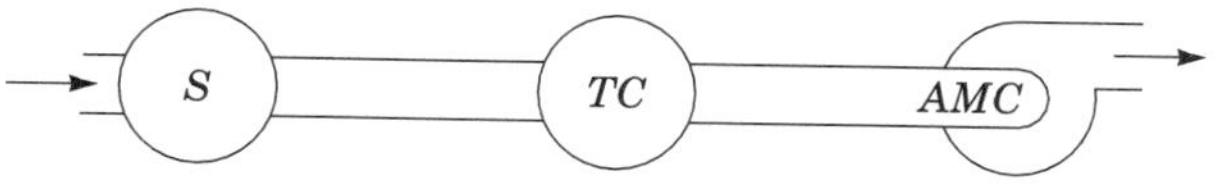

Figure 15.42. Sampler efficiency evaluation with downstream total collector; analysis of collections in sample under test *S* and total collector *TC*; *AMC*, air mover, flowmeter, and flow control.

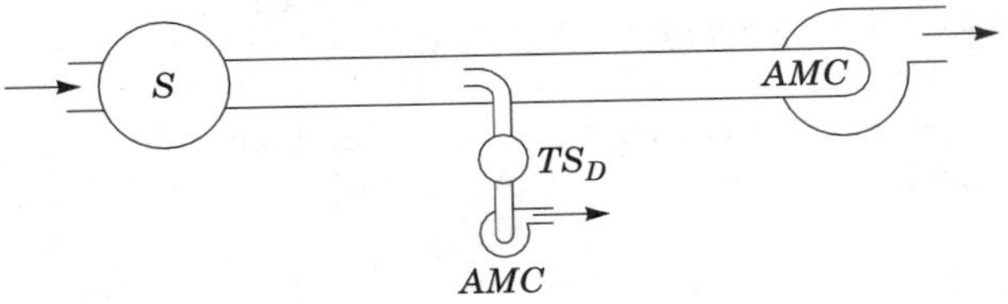

Figure 15.43. Sample efficiency evaluation with downstream concentration sampler: analysis of collections in sample under test S and downstream sampler total collector TS_D.

collect a series of samples across the flow profile, rather than a single sample, to obtain a true average concentration of the penetrating atmosphere.

4.8.4 Analysis of Upstream and Downstream Samples

In some cases it is not possible to recover or otherwise measure the material trapped within elements of the sampling train such as sampling probes. The magnitude of such losses can be determined by comparing the concentrations upstream and downstream of the elements in question, as schematized in Figure 15.44.

4.9 Determination of Sample Stability and/or Recovery

The stability and the recovery of trace contaminants from sampling substrates are difficult to predict or control. Thus these factors are best explored by realistic calibration tests.

If the sample is divided into a number of aliquots that are analyzed individually at periodic intervals, it is possible to determine the long-term rate of sample degradation, or any tendency for reduced recovery efficiencies with time. These analyses would not, however, provide any information on losses that may have occurred during or immediately after collection because they had different rate constants. Such losses should be investigated using spiked samples.

4.9.1 Analysis of Spiked Samples.

If known amounts of the contaminants of interest are intentionally added to the sample substrate, subsequent analysis of sample aliquots will permit calculation of sample recov-

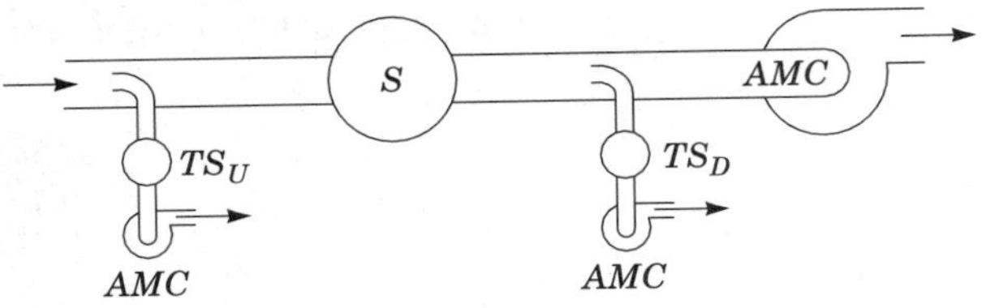

Figure 15.44. Sampler efficiency evaluation with upstream and downstream concentration samplers: Analysis of collections in upstream and downstream samplers, total collection, TS_U and TS_D.

ery efficiency and rate of deterioration. These results will be valid only insofar as the added material is equivalent in all respects to the material in the ambient air. There are two basic approaches to spiked sample analyses: (*1*) the addition of known quantities to blank samples; and (*2*) the addition of radioactive isotopes to either blank or actual field collected samples.

When the material being analyzed is available in tagged form, the tag can be added to the sample in negligible or at least known low concentrations. If there are losses in sample processing or analysis, the fractional recovery of the tagged molecules will provide a basis for estimating the comparable loss that took place in the untagged molecules of the same species.

4.10 Calibration of Sensor Response

Direct-reading instruments are generally delivered with a direct reading panel meter, a set of calibration curves, or both. The unwary and inexperienced user tends to believe the manufacturer's calibration, and this often leads to grief and error. Any instrument with calibration adjustment screws should of course be suspect, because such adjustments can easily be changed intentionally or accidentally, for example, in shipment.

All instruments should be checked against appropriate calibration standards and atmospheres immediately upon receipt and periodically thereafter. Procedures for establishing test atmospheres are discussed earlier in this chapter. Verification of the concentrations of such test atmospheres should be performed whenever possible using analytical techniques that are referee tested or otherwise known to be reliable.

With these techniques, calibration curves for direct-reading instruments can be tested or generated. When environmental factors such as temperature, ambient pressure, and radiant energy may be expected to influence the results, these effects should be explored with appropriate tests whenever possible. Similarly, the effects of cocontaminants and water vapor on instrument response should also be explored.

5 ESTIMATION OF ERRORS

5.1 Sources of Sampling and Analytical Errors

The difference between the air concentration reported for an air contaminant on the basis of a meter reading or laboratory analysis, and the true concentration at that time and place represents the error of the measurement. The overall error is often due to a number of smaller component errors rather than to a single cause. To minimize the overall error, it is usually necessary to analyze each of its potential components, concentrating one's efforts on reducing the component error that is largest. It would not be productive to reduce the uncertainty in the analytical procedure from 10 to 1.0% when the error associated with the sample volume measurement is ± 15 percent.

Sampling problems are so varied in practice that it is possible only to generalize on the likely sources of error to be encountered in typical sampling situations. In analyzing a particular sampling problem, consideration should be given to each of the following:

1. Flow rate and sample volume.
2. Collection efficiency.
3. Sample stability under conditions anticipated for sampling, storage, and transport.
4. Efficiency of recovery from sampling substrate.
5. Analytical background and interferences introduced by sampling substrate.
6. Effect of atmospheric cocontaminants on samples during collection, storage, and analyses.

5.2 Cumulative Statistical Error

The most probable value of the cumulative error E_e can be calculated from the following equation:

$$E_e = [E_1^2 + E_2^2 + E_3^2 + \ldots + E_n^2]^{1/2}$$

For example, if accuracies of the flow rate measurement, sampling time, recovery, and analysis are ±15, 2, 10, and 10% respectively, and there are no other significant sources of error, the cumulative error would be:

$$Ee = [15^2 + 2^2 + 10^2 + 10^2]^{1/2} = [429]^{1/2} = \pm 20.7\%$$

It should be remembered that this provides an estimate of the deviation of the measured concentration from the true concentration at the time and place the sample was collected. As an estimate of the average concentration to which a worker was exposed in performing a given operation, it would have additional uncertainty, depending on the variability of concentration with time and space at the work station.

6 SUMMARY AND CONCLUSIONS

Determinations of the concentrations of trace level contaminants in air and of heat stress, noise, and radiant energies are subject to numerous variables, many of them difficult to control. Thus it is prudent to perform frequent calibration checks on all industrial hygiene instruments. Such calibrations should be based on realistic simulations of the conditions encountered in the field.

The production of test atmospheres in the range of occupational threshold limits is often difficult. This chapter provides a review of available techniques for the production of test atmospheres of gases, vapors, and aerosols, with sketches of many of the more useful techniques.

Extreme care should be exercised in performing all calibration procedures. The following guidelines should be followed:

1. Use standard or reference atmospheres, instruments, and devices with care and attention to detail.

2. Check all standard materials and instruments and procedures periodically to determine their stability and/or operating condition.
3. Perform calibrations whenever a device has been changed, repaired, received from a manufacturer, subjected to use, mishandled, or damaged, and at any time when a question arises with respect to its accuracy.
4. Understand the operation of an instrument before attempting to calibrate it, and use a procedure or setup that will not change the characteristics of the instrument or standard within the operating range required.
5. When in doubt about procedures or data, make certain of their validity before proceeding to the next operation.
6. Keep all sampling and calibration train connections as short and free of constrictions and resistance as possible.
7. Exercise extreme care in reading scales, timing, adjusting, and leveling if needed, and in all other operations involved.
8. Allow sufficient time for equilibrium to be established, inertia to be overcome, and conditions to stabilize.
9. Obtain enough points or different rates of flow on a calibration curve to give confidence in the plot obtained. Each point should be made up of more than one reading whenever practical.
10. Maintain a complete permanent record of all procedures, data, and results. This should include trial runs, known faulty data with appropriate comments, instrument identification, connection sizes, barometric pressure, and temperature.
11. When a calibration differs from previous records, determine the cause of change before accepting the new data or repeating the procedure.
12. Identify calibration curves and factors properly with respect to conditions of calibration, device calibrated and what it was calibrated against, units involved, range and precision of calibration, date, and name of the person who performed the actual procedure. Often, it is convenient to indicate where the original data are filed and to attach a tag to the instrument indicating the items just listed.

BIBLIOGRAPHY

1. *Code of Federal Regulations*, Title 40, Part 58, Ambient Air Quality Surveillance, Appendix D-Network Design for State and Local Air Monitoring Stations (SLAMS) and National Air Monitoring Stations (NAMS), U.S. Government Printing Office, Washington DC, 1981, pp. 149–159.
2. P. M. Eller, ed., *NIOSH Manual of Analytical Methods*, 3rd ed., US DHHS, CDC, NIOSH, Cincinnati, OH, 1984.
3. C. H. Powell and A. D. Hosey, eds., *The Industrial Environment—Its Evaluation and Control*, 2nd ed., Public Health Service Rd. No 614, U.S. Government Printing Office, Washington, DC, 1965.
4. American Conference of Governmental Industrial Hygienists, *Air Sampling Instruments*, 9th ed., ACGIH, Cincinnati, OH, 1999.

5. N. A. Leidel, K. A. Busch, and J. R. Lynch, *Occupational Exposure Sampling Strategy Manual*, US DHEW, CDC, NIOSH, Cincinnati, OH, 1977.
6. D. Craig, *Health Phys* **21**, 328–332 (1971).
7. J. H. Perry et al., eds., *Chemical Engineering Handbook*, 4th ed., McGraw-Hill, New York, 1963.
8. Spitzglass, *Trans.* ASME **44**, 919 (1922).
9. American Society of Mechanical Engineers, "Flow Measurement by Means of Standardized Nozzles and Orifice Plates," ASME Power Test Code (PTC 19.5.4-1959), ASME, New York, 1959.
10. B. E. Saltzman, *Anal. Chem.* **33**, 1100–1112 (1961).
11. G. Hama, *Air Eng.* **9**, 18 (1967).
12. H. D. Axelrod and J. P. Lodge, in A. C. Stern, ed., *Air Pollution*, Vol. 3, 3rd ed., Academic Press, New York, 1976, pp. 145–182.
13. W. D. Conner and J. S. Nader, *Am. Ind. Hyg. Assoc. J.* **25**, 291–297 (1964).
14. F. M. Stead and G. J. Taylor, *J. Ind. Hyg. Toxicol.* **29**, 408–412 (1947).
15. A. N. Setterlind, *Am. Ind. Hyg. Assoc. Quart.* **14**, 113–120 (1953).
16. L. Silverman, in P. L. Magill, F. R. Holden, and C. Ackley, eds., *Air Pollution Handbook*, McGraw-Hill, New York, 1956, pp. 12:1–12:48.
17. H. N. Cotabish, P. W. McConnaughey, and H. C. Messer, *Am. Ind. Hyg. Assoc. J.* **22**, 392–402 (1961).
18. B. E. Saltzman and A. F. Wartburg, Jr. *Anal. Chem.* **37**, 1261 (1965).
19. A. Goetz and T. Kallai, *J. Air Pollut. Control Assoc.* **12**, 437–443 (1962).
20. G. O. Nelson and K. S. Griggs, *Rev. Sci. Instrum.* **39**, 927–928 (1968).
21. G. O. Nelson, *Rev. Sci. Instrum.* **41**, 776–777 (1960).
22. G. O. Nelson, *Gas Mixtures Preparation and Control*, Lewis Publishers, Chelsea, MI, 1992.
23. A. P. Altshuller and L. R. Chohe, *Anal. Chem.* **32**, 802 (1960).
24. A. E. O'Keefe and G. O. Ortman, *Anal. Chem.* **38**, 760–763 (1966).
25. J. P. Lodge, *Methods of Air Sampling and Analysis*, 3rd ed., Lewis Publishers, Inc., Chelsea, MI, 1989.
26. R. O. McClellan and R. F. Henderson, *Concepts in Inhalation Toxicology*, Hemisphere Publishing Corp., New York, 1989.
27. O. R. Moss, "Calibration of Gas and Vapor Samplers," in B. S. Cohen and S. V. Hering, eds., *Air Sampling Instruments*, 8th ed., American Conference of Governmental Industrial Hygienists, Cincinnati, 1995, Chapt. 8 pp. 151–163.
28. C. N. Davies, *Proc. Roy Soc.* **279a**, 413 (1964).
29. W. Hinds, *Aerosol Technology*, 2nd ed., John Wiley & Sons, New York, 1999.
30. J. E. Brockmann, in K. Willeke and P. A. Baron eds., *Aerosol Measurement: Principles, Techniques, and Applications*, Van Nostrand Reinhold, New York, 1993.
31. U.S. Environmental Protection Agency, *Code of Federal Regulations* **40**, Part 53 (1987).
32. M. B. Ranade, M. C. Wood, F. L. Chen, et al., *Aerosol Sci. Technol.* **13**, 54 (1990).
33. Y. S. Chang and B. T. Chen, "Aerosol Sampler Calibration" in B. S. Cohen and C. McCammon, eds., *Air Sampling Instruments*, 9th ed., American Conference of Governmental Industrial Hygienists, Cincinnati, OH (in press).
34. G. M. Kanapilly, O. G. Raabe, and G. J. Newton, *J. Aerosol Sci.* **1**, 313 (1970).

35. B. T. Chen, Y. S. Cheng, and H. C. Yeh, *Aerosol Sci. Technol.* **12**, 278 (1990).
36. W. John, in K. Willeke, ed., *Generation of Aerosols and Facilities for Exposure Experiments*, Ann Arbor Science, Ann Arbor, MI, 1980.
37. B. Y. H. Liu and D. Y. H. Pui, *J. Colloid Interface Sci.* **47**, 155 (1974).
38. B. T. Chen, H. C. Yeh, and M. A. Rivero, *J. Aerosol Sci.* **19**, 137 (1988).
39. F. J. Romay-Novas and D. Y. H. Pui, *Aerosol Sci. Technol.* **9**, 123 (1988).
40. E. B. Barr, M. D. Hoover, G. M. Kanapilly, et al., *Aerosol Sci. Technol.* **2**, 437 (1983).
41. H. C. Yeh, Y. S. Cheng, and R. L. Carpenter, *Am. Ind. Hyg. Assoc. J.* **44**, 358 (1983).
42. T. T. Mercer, *Aerosol Technology in Hazard Evaluation*, Academic Press, New York, 1973.
43. O. G. Raabe, in B. Y. H. Liu, ed., *Fine Particles: Aerosol Generation, Measurement, Sampling, and Analysis*, Academic Press, New York, 1976.
44. B. T. Chen, H. C. Yeh, and C. H. Hobbs, *Aerosol Sci. Technol.* **19**, 109 (1993).
45. M. Kerker, *Adv. Colloid Interface Sci.* **5**, 105–172 (1975).
46. D. A. Fraser, R. E. Bales, M. Lippmann, and H. E. Stokinger, *Exposure Chambers for Research in Animal Inhalation*, Public Health Monograph No. 57, Public Health Service Publication No. 662, U.S. Government Printing Office, Washington, DC, 1959.
47. N. A. Fuchs and A. G. Sutugin, in C. N. Davies, ed., *Aerosol Science*, Academic Press, London, 1966, pp. 1–30.
48. B. Wilson and V. K. LaMer, *J. Ind. Hyg. Toxicol.* **30**, 265–280 (1948).
49. V. K. LaMer and D. Sinclair, *An Improved Homogeneous Aerosol Generator*, OSRD Report No. 1668, Department of Commerce, Washington, DC, 1943.
50. D. C. F. Muir, *Ann. Occup. Hyg.* **8**, 233–238 (1965).
51. C. M. Huang, M. Kerker, E. Matijevic, and D. D. Cooke, *J. Colloid Interface Sci.* **33**, 244 (1970).
52. E. Rapaport and S. G. Weinstock, *Experimentia* **11**(9), 363 (1955).
53. B. Y. H. Liu and K. W. Lee, *Am. Industr. Hyg. Assoc. J.* **36**, 861–865 (1975).
54. V. Prodi, in T. T. Mercer, P. E. Morrow, and W. Stöber, eds., *Assessment of Airborne Particles*, Charles C. Thomas, Springfield, IL, 1971, pp. 169–181.
55. K. Homma, *Ind. Health* **4**, 129–137 (1966).
56. P. Movilliat, *Ann. Occup. Hyg.* **4**, 275 (1962).
57. E. Matijevic, W. F. Espenscheid, and M. Kerker, *J. Colloid Interface Sci.* **18**, 91–93 (1963).
58. S. Kitani and S. Ouchi, *J. Colloid Interface Sci.* **23**, 200–202 (1967).
59. K. R. Spurny and J. P. Lodge, Jr., *Atmos. Environ.* **2**, 429–440 (1968).
60. R. Ebens, *Staub* **29**, 89–92 (1969).
61. B. M. Wright, *J. Sci. Instrum.* **27**, 12–15 (1950).
62. S. Laskin, S. Posner, and R. Drew, paper presented at the annual meeting of the American Industrial Hygiene Association, St. Louis, May 1968.
63. R. L. Dimmock, *AMA Arch. Ind. Health* **20**, 8–14 (July 1959).
64. J. R. Brown, J. Horwood, and E. Mastromatteo, *Ann. Occup. Hyg.* **5**, 145–147 (1962).
65. V. Timbrell, A. W. Hyett, and J. W. Skidmore, *Ann. Occup. Hyg.* **11**, 273–281 (1968).
66. P. F. Holt and D. K. Young, *Ann. Occup. Hyg.* **2**, 249–256 (1960).
67. J. C. Couchman, Metallic Microsphere Generation, EG&G, Inc., Santa Barbara, CA, 1966.
68. M. Polydorova, *Staub (Engl. transl.)* **29**, 38 (1969).

69. F. G. Karioris and B. R. Fish, *J. Colloid Sci.* **17**, 155–161 (1962).
70. M. Tomaides and K. T. Whitby, *Proceedings of the Seventh International Conference on Condensation and Ice Nuclei*, Academia, Prague, 1969.
71. J. D. Holmgren, J. O. Gibson, and C. Sheer, *J. Electrochem. Soc.* **3**, 362–369 (1964).
72. T. T. Mercer, M. I. Tillery, and H. Y. Chow, *Am. Ind. Hyg. Assoc. J.* **29**, 66–78 (1968).
73. British Standards Institute, "Methylene Blue Particulate Test for Respiratory Canister," B.S. No. 2577, British Standards Institute, London, 1955.
74. K. T. Whitby, D. A. Lundgren, and C. M. Peterson, *Int. J. Air Water Pollut.* **9**, 263–277 (1965).
75. B. M. Wright, *Lancet* 24–25 (1958).
76. L. Dautrebande, *Microaerosols*, Academic Press, New York, 1962.
77. V. K. LaMer and R. Gruen, *Trans. Faraday Soc.* **48**, 410–415 (1952).
78. C. Orr, F. K. Hurd, and W. J. Corbett, *J. Colloid Sci.* **13**, 472–482 (1952).
79. M. J. Pilat and R. J. Charlson, *J. Rech. Atmos.* **2**, 165–170 (1966).
80. C. C. Snead and J. T. Zung, *J. Colloid Interface Sci.* **27**, 25–31 (1968).
81. W. H. Walton and W. C. Prewett, *Proc. Phys. Soc. (London)*, **62**, 341–350 (1949).
82. M. Lippmann and R. E. Albert, *Am. Ind. Hyg. Assoc. J.* **28**, 501–506 (1967).
83. N. A. Dimmick, *Nature*, **166**, 686–687 (1950).
84. L. Strom, *Rev. Sci. Instrum.* **40**, 778–782 (1969).
85. W. R. Wolf, *Rev. Sci. Instrum.* **32**, 1124–1129 (1961).
86. O. G. Raabe, in M. G. Hanna, P. Nettesheim and J. R. Gilbert, eds., Inhalation Carcinogenesis, CONF-691001, Clearinghouse for Federal Scientific and Technical Information, NBS, U.S. Department of Commerce, Springfield, VA, April 1970.
87. R. N. Berglund and B. Y. H. Liu, *Environ. Sci. Technol.* **7**, 147 (1973).
88. B. Y. H. Liu, *Air Pollut Control Assoc. J.*, **24**, 1170–1172 (Dec. 1974).
89. M. A. Nawab and S. G. Mason, *J. Colloid Sci.* **12**, 179–187 (1958).
90. E. P. Yurkstas and C. J. Meisenzehl, "Solid Homogeneous Aerosol Production by Electrical Atomization," University of Rochester Atomic Energy Report No. UR-652, Rochester, NY, Oct. 30, 1964.
91. V. A. Drozin, *J. Colloid Sci.* **10**, 158 (1955).
92. B. Y. H. Liu and D. Y. H. Pui, *J. Colloid Interface Sci.* **10**, 158, (1955).
93. P. C. Reist and W. A. Burgess, *J. Colloid Interface Sci.* **24**, 271–273 (1967).
94. O. G. Raabe, *Am. Ind. Hyg. Assoc. J.* **29**, 439–443 (1968).
95. S. C. Stern, J. S. Baumstark, A. I. Schekman, and R. K. Olson, *J. Appl. Phys.* **30**, 952–953 (1959).
96. G. M. Kanapilly, O. G. Raabe, and G. J. Newton, *Am. Ind. Hyg. Assoc. J.* **30**, 125 (1969) (abstract).
97. G. M. Kanipilly, O. G. Raabe, and G. J. Newton, *Aerosol Sci.* **1**, 313 (1970).
98. P. M. Giever in A. C. Stern, ed., *Air Pollution*, Vol. 3, 3rd ed., Academic Press, New York, 1976, pp. 3–50.
99. T. Hatch and S. P. Choate, *J. Franklin Inst.* **207**, 369–387 (1929).
100. T. T. Mercer, R. F. Goddard, and R. L. Flores, *Ann. Allergy*, **23**, 314–326 (1967).
101. P. E. Morrow, *Health Phys.* **12**, 173–208 (1966).

102. ICRP Task Group, "Human Respiratory Tract Model for Radiological Protection," *Ann. ICRP*, **24**, 1–482 (1994).
103. R. E. Albert, H. G. Petrow, A. S. Salam, and J. R. Spiegelman, *Health Phys.* **10**, 933–940 (1964).
104. T. T. Mercer, *Health Phys.* **10**, 873–887 (1964).
105. L. Rayleigh, *Phil. Mag.* **14**, 184–186 (1882).
106. K. T. Whitby and B. Y. H. Liu, in C. N. Davies, ed., *Aerosol Science*, Academic Press, New York, 1966, pp. 59–86.
107. K. T. Whitby, *Rev. Sci. Instrum.* **32**, 351–355 (1961).
108. S. L. Soong, M. S. thesis, University of Rochester, Rochester, NY, 1968.
109. A. Black and M. Walsh, *Ann. Occup. Hyg.* **13**, 87–100 (1970).
110. D. C. Bogen, *Ann. Industr. Hyg. Assoc. J.* **31**, 349–352 (May–June 1970).
111. M. Singer, C. J. Van Oss, and W. Wanderhoff, *J. Reticuloendothel. Soc.* **6**, 281–286 (1969).
112. H. Flachsbart and W. Stöber, *J. Colloid Interface Sci.* **30**, 568–573 (1969).
113. R. E. Albert, J. Spiegelman, M. Lippmann, and R. Bennett, *Arch. Environ. Health* **17**, 50–58 (July 1968).
114. J. R. Spiegelman, G. D. Hanson, A. Lazarus, R. J. Bennett, M. Lippmann, and R. E. Albert, *Arch. Environ. Health* **17**, 321–326 (1968).
115. D. V. Booker, A. C. Chamberlain, J. Rundo, D. C. F. Muir, and M. L. Thomson, *Nature* **215**, 30–33 (1967).
116. T. T. Mercer, R. F. Goddard, and R. L. Flores, *Ann. Allergy* **26**, 18–27 (1968).
117. K. E. Lauterbach, A. D. Hayes, and M. A. Coelho, *AMA Arch. Ind. Health* **13**, 156–160 (1956).
118. K. R. May, *J. Aerosol Sci.* **4**, 235–243 (1973).
119. O. G. Raabe in B. Y. H. Liu, ed., *Fine Particles*, Academic Press, New York, 1976, pp. 60–110.

CHAPTER SIXTEEN

Quality Control

William E. Babcock

1 INTRODUCTION

Why do we need quality control? This question has likely been asked countless times of quality managers by laboratory managers, laboratory analysts, and clients eager to save time or money. Certainly there are costs associated with a quality control effort and the results may be difficult to quantify.

There are a number of answers to this question. One needs quality control to evaluate data quality, determine compliance with standard procedures, provide legally defensible data, provide data on which effective decisions can be made, and to evaluate analyst performance. Quality control is needed to assure the integrity of test data and to provide a consistently good product.

How do we know if analytical data are any good? We cannot see, hear, smell, taste, or feel it. The laboratory client may be pleased with the test result of a defective analysis and unhappy with the result of an excellent analysis. The only way to accomplish an adequate evaluation of the data is if the analytical procedure is of a known and acceptable accuracy and precision.

Laboratory sample analyses are measurements of actual quantities. As with any measurement process, there will be variability in the system. The reported laboratory result is almost certainly not exactly the actual quantity. There is really a range of possible results of various probabilities upon which various professionals will make a variety of regulatory, legal, and occupational health decisions. The effectiveness of these decisions is dependent on the quality of the laboratory measurement. Quality control enables us to make sound inferences about analytical results and to act rationally upon those inferences.

The laboratory quality system is designed to reduce the variability and to measure the variability of the test data.

Patty's Industrial Hygiene, Fifth Edition, Volume 1. Edited by Robert L. Harris.
ISBN 0-471-29756-9

2 INTERNATIONAL CONSENSUS STANDARD

Globalization has been a driving force in quality systems in manufacturing. As international consensus standards have developed their use has spread to the United States. The International Organization for Standardization (ISO) Guide 25 is the international standard for the general requirements for the competence of calibration and testing laboratories (1).

The ISO/IEC Guide 25 1990 revision was the 3rd revision of the standard. It is undergoing another revision and will subsequently be known as the ISO DIS 17025. This document was available in draft form in late 1998 and a final issue date had not been determined at that time.

Guide 25 addresses a number of areas for which laboratory management must develop policies, documented procedures, or arrangements to assure an appropriate quality system. These areas are

1. Organization and management.
2. Quality system, audit, and review.
3. Personnel.
4. Accommodation and environment.
5. Equipment and reference materials.
6. Measurement traceability and calibration.
7. Calibration and test methods.
8. Handling of calibration and test items.
9. Records.
10. Certificates and reports.
11. Subcontracting of calibration or testing.
12. Outside support services and supplies.
13. Complaints.

The ISO Guide 25 was developed so that national authorities could adopt or adapt it to regulate or accredit testing laboratories within their jurisdiction. In the United States, accreditation of industrial hygiene testing laboratories is performed by the American Industrial Hygiene Association. Laboratories seeking to establish a good quality system can obtain the ISO Guide 25 and begin to address the individual requirements.

3 NATIONAL LABORATORY ACCREDITATION STANDARD

The American Industrial Hygiene Association (AIHA) laboratory accreditation program began in 1974. Two years earlier, AIHA and The National Institute for Occupational Safety and Health (NIOSH) entered a joint agreement to develop and implement an accreditation program for IH laboratories which would include proficiency test samples. The initial program began by evaluating government and government contract laboratories which were performing lead, silica, and asbestos testing for the Occupational Safety and Health

Administration (OSHA) Target Health Hazard Program. By July 1974 there were 26 accredited laboratories (2).

The AIHA accreditation program focused on five evaluation factors. These were proficiency testing, personnel, quality control, facilities, and records.

"The primary purpose of the AIHA Laboratory Quality Assurance Programs is to establish and maintain the highest possible standards of performance for laboratories analyzing samples to support the evaluation of workplace and environmental exposures to hazardous agents." (3)

In recent years, AIHA accreditation policies have been revised to be more consistent with ISO Guide 25 and the policies now state that "the Industrial Hygiene Laboratory Accreditation Program (IHLAP) complies with the recommendations of the International Standards Organization (ISO/IEC) 25 and 58 Guides."

Applicant laboratories now go through initial accreditation from the AIHA and reaccreditation every three years. The accreditation process involves submitting a completed application and payment of relevant fees, application review, proficiency testing requirements, a laboratory site visit, and board approval. The reaccreditation process is very similar.

Requirements for accreditation include policies regarding:

1. Organization and function.
2. Facilities.
3. Equipment.
4. Personnel.
5. Analytical methods.
6. Quality assurance.
7. Safety and health.
8. Proficiency testing.
9. Site visit.
10. Accreditation maintenance.

A number of these categories might reasonably be considered to have quality assurance implications. Specific quality assurance requirements are

1. QA Manual.
2. Organization and responsibility.
3. QA objective and policies.
4. Personnel qualification and training.
5. Sampling materials and procedures.
6. Chain of custody and samples receiving.
7. Reagents and standards.
8. Equipment calibration and maintenance.
9. Data reduction validation and reporting.

10. Internal QC procedures.
11. Performance and system audits.
12. Corrective actions.
13. QA reports.
14. Documentation and recordkeeping.
15. Sample retention and disposal.

As with ISO Guide 25, this list of quality requirements for AIHA accreditation gives any laboratory a good starting point for developing its own quality system.

4 QUALITY SYSTEM

4.1 General

The laboratory quality system is a set of policies and procedures to assure quality products. As previously mentioned, either the ISO Guide 25 or AIHA accreditation standards can serve as an outline for the types of policies and procedures which the laboratory should address.

The quality system is the method by which laboratory management specifies their philosophy and commitment to quality products. Involvement of management throughout the development, implementation, review, and revision of the quality system is of integral importance. This assures that management understands and endorses the policies and procedures in place for quality.

The five steps important to instituting an effective quality system are

1. Design.
2. Document.
3. Implement.
4. Audit.
5. Review.

The design phase may involve adapting current laboratory practice or establishing new policies which comply with national or international standards. Document the policies and practices in a laboratory quality assurance manual. Implement the policies and practices that have been designed and documented. Determine how they will be carried out in practice. The audit phase examines quality practices in the laboratory to determine whether or not practice and policy are harmonized. Management review addresses those areas where the audit determines that policy and practice conflict. This is an iterative process to focus on continuous improvement and assure that quality policy and practice are harmonized.

If there existed a motto for this process it would be “say what you do and do what you say”. Introducing an elegant system full of wonderful policies and procedures is of no value at all if it not followed in laboratory practice. Indeed, following careful laboratory procedures may be of limited value if not properly documented. Good laboratory practices followed by one analyst may not be followed by others. Or training of new personnel may be inconsistent without proper documentation. Or analysts with a particular skill and no trained back-up may leave. There are any number of potential flaws in undocumented systems.

4.2 Design

Design of the quality system should be based on current laboratory practice where practical. Inclusion of analysts in development of the quality system will only strengthen the program. Any new program requirements must be supported by management and staff or they will probably fail.

4.3 Documentation

The quality assurance manual (QAM) is the document which defines the quality system and the policies and procedures which comprise the system. Although the system is a management system and the QAM is a management document, it should have considerable input from technical personnel. They are often the closest to and most familiar with the various technical issues in developing the quality system.

4.4 Implementation

Implementation issues may be the most difficult phase of the quality system process. This is particularly true with established laboratories with long-term employees in which significant changes are required. This is why it is important to build upon current laboratory practice and to involve technical employees at all stages of the process.

Where new procedures are required, implementing in incremental fashion is usually the best alternative. Individual issues can be implemented and audited without making large changes in work practices all at once.

4.5 Audit

Internal audits should be conducted at specified intervals. These intervals may be determined by laboratory management depending upon laboratory circumstances, but should be documented at least annually. More frequent audits are recommended for laboratories which are just introducing and implementing a quality system or which perpetually have numerous and repeated nonconformances during internal audits.

Audits should be conducted by personnel trained in such activities and which are independent of the operation to be audited. In IH laboratories of sufficient size to have designated QA personnel, they should be trained in auditing principles and conduct audits of laboratory operations. Such training is available from a variety of sources.

Auditors should audit to specific standards which are clearly stated. These standards may be the laboratory QAM or other consensus standards such as AIHA IHLAP or ISO Guide 25. Instances where laboratory practice conflicts with a documented standard should be expressed in terms of that standard. If the conflict cannot be expressed in terms of the standard, then there is no nonconformance. Ideally, this would preclude arguments between supervisors of the audited operation and the auditor. If the operation is being done in a manner that is nonconforming to a required standard, then corrective action must be taken to align the policy and practice. Issues are not always so clearly defined, however.

One method to assure consistency and that all relevant areas are covered is the use of a checklist. Various checklists have been developed for laboratory use (4) or the laboratory could generate its own based on its QAM or other requirements such as AIHA Policies or ISO Guide 25. While this does have advantages, care must be used in developing the check list to avoid yes-or-no answers without further evaluation or analysis. Especially for internal audits, this can lead to a quality system that is superficially adequate, but with no depth. For example, the checklist may include the presence of a QA Manual. If one exists, no matter how defective, this would elicit a positive response. Further exploration of the issue is critical. More experienced auditors may be able to depart from the checklist or even eliminate it completely in the course of their audit. Their investigation may take on the semblance of a Socratic dialogue—asking a series of relatively easy questions which lead to an ultimate answer, conclusion, or realization which the auditor was seeking. Or each answer may lead to a new inquiry or a new direction to determine the facts. Without regard to how the audit proceeds, it is important to focus on the fact-finding, not fault-finding, nature of the audit process.

Audit schedules and operations to be audited should be determined in advance. Annual audit planning allows for coordination of audits to assure that each operation is covered in a one year audit cycle (if multiple audits per year are conducted). Audits may be specific to certain parts of a laboratory facility or specific to certain standards. Examples of these are: comprehensive audit of a specific operation such as chromatography with all standards evaluated; comprehensive audit of specific standards across all laboratory operations such as auditing calibrations for chromatography, spectroscopy, X-ray diffraction. Another audit style is to select individual samples and follow them cradle-to-grave to assure compliance with all required standards. Yet another internal audit is to follow-up action items from external audits to assure that any policy changes have been implemented into laboratory practice.

Following the actual audit, the auditors should document their findings and meet with managers of the audited operations. Audit findings are presented and the auditors and managers can discuss any nonconformances and time frames for required corrective action to be completed and documented.

Following completion and documentation of required corrective action a final audit report should be submitted to the laboratory director. This report can serve as a springboard to discussion for the management review process. Areas in which policy and practice have diverged should be discussed.

4.6 Management Review

Management review of the system should be conducted at least annually. This review is designed to assure that quality policies reflect the intentions of management and that those policies are being implemented as intended.

Audit findings may reveal discrepancies between policy and practice, ambiguous policies, or the existence of no policy at all regarding program elements which must be addressed. It is the responsibility of management to harmonize policy and practice. Where the two are in conflict management must take steps to change either policy or practice or both. The management review process enables this to occur.

This process allows managers to "get on the same page" regarding quality policies. To focus on what they intend by the various quality policies.

The review should be documented and revisions to the QAM should be signed and documented.

This can be viewed as an endlessly repeated cycle—a continuous improvement process to identify areas of weakness and address them through a management system.

5 ADMINISTRATIVE AND MANAGEMENT ISSUES

Quality issues need to be integrated into the total function of the laboratory. This includes cost accounting, planning, budgeting, staffing, and other administrative and management issues.

5.1 Organization and Responsibility

Organizing and defining quality responsibilities throughout the organization is a function of senior management. Demonstrated commitment to quality is most visible through budget and staffing considerations. In issues such as quality and safety, management must not only talk-the-talk, but walk-the-walk. Commitment in thought, word, and deed is essential to give employees the message that quality is important.

The most demonstrable action senior management can take is the appointment of a quality assurance manager or coordinator. The functions of such a position are development of laboratory quality policies (with management and staff), administration and evaluation of quality programs, and reporting on quality issues to laboratory management. This would include developing statistical protocols for accuracy and precision, conducting corrective action report and internal audits, maintaining accreditation status and contacts, and coordinating any external proficiency testing or auditing. This would usually be considered a staff, rather than line, position, and thus not have responsibility for the actual implementation of the quality system throughout the laboratory.

Laboratory supervisors are responsible for implementing the quality system in the laboratory. They help define training requirements and needs and assure that adequate training of personnel occurs before they are assigned new procedures. Adequate documentation of this process is essential.

Laboratory analysts are responsible to follow quality policies and provide input to laboratory management regarding improving quality policies and procedures.

5.2 Quality Objectives and Policies

The objectives and policies of the QA Program should be clearly stated in the QA Manual. The management should be actively involved in the development and clearly support the objectives and policies.

Objectives should concentrate on why the quality system is implemented. What benefits does the laboratory hope to gain by its investment in the quality system? Some possible objectives are stated in the introduction. Some possible objectives are

To produce test results that are scientifically and legally defensible.

To produce test results of known and acceptable accuracy and precision.

To establish and maintain AIHA accreditation status.

To evaluate analyst, instrument, and methodology performance.

To monitor and improve the quality of laboratory analyses.

Policies state how the laboratory intends to accomplish the objectives. A good starting point for developing policies is with any international or national consensus standards or accreditation criteria. Some possible policies are

To use validated sampling and analytical methods to produce test data.

To conduct internal audits to accepted consensus standards such as ISO Guide 25.

To calibrate each instrument with each use against a National Institute for Science and Technology (NIST) traceable standard.

To participate in external proficiency testing programs such as the PAT Program.

To establish statistical data for accuracy and precision of each analytical technique.

To use chain-of-custody procedures to ensure sample security and integrity.

To use corrective action reports to document identified deficiencies.

In addition to objectives and policies, the QAM may contain procedures or refer to documented procedures. Procedures are the nuts-and-bolts of how policies are implemented. They are a step-by-step guide on how things are done. For example, if sampling and analytical methods are validated in the laboratory, a standard operating procedure will need to be developed or adopted which specifies the steps of method validation and how they are to be accomplished.

5.3 Personnel Requirements

Personnel requirements must be determined by laboratory management for each job category. Guidance regarding qualifications can be found in Guide 25 and the AIHA IHLAP standards. Combinations of education and professional experience are the most common requirements. Technical personnel will usually be required to have a relevant college degree coupled with some professional experience. Professional certification is a valuable asset as well for many technical positions.

5.4 Training

Training requirements will be determined on a continuous basis to enable personnel to maintain technical currency, proficiency, and to expand laboratory capabilities into new areas. Training documentation should be well maintained in a central location. Training and qualification records should focus on demonstrating defined job requirements and how individual laboratory personnel have met those requirements. Training programs may include specific training courses in technical areas and mentoring or on-the-job training at the laboratory. In addition, competency to perform specific analytical tasks should be demonstrated by acceptable performance on internal quality assurance reference samples and external proficiency testing samples.

Specific areas for which training criteria and documentation should be established include: asbestos fiber counting, atomic absorption spectroscopy, gas chromatography (identify each detector system, e.g., flame ionization detector), glassware washing, high performance liquid chromatography, ion chromatography, portable field instrumentation, sample receiving, sample media, X-ray diffraction. Whatever individual tasks are identified by laboratory staff which require training should be so documented, both as to the training criteria and to the individuals as they receive and complete each training module.

6 SAMPLING AND ANALYTICAL METHODS

Laboratory policy should include the use of validated sampling and analytical methods wherever feasible. Methods such as those validated by the National Institute for Occupational Safety and Health (NIOSH) or the Occupational Safety and Health Administration (OSHA) are readily available to IH laboratories in a variety of formats.

Validated methods ensure that designated guidelines or evaluation criteria have been followed to assure acceptable performance for sampling and analysis of a particular contaminant (5).

These criteria should include relevant background information, such as toxic effects, workplace exposures, and physical properties. The sampling procedure should be well-documented including such items as sampler capacity, desorption or extraction efficiency, storage effects at ambient and reduced temperatures, and possible collection interferences. The analytical procedures should also be clearly documented. This would include the apparatus, analytical SOP, safety precautions, detection limit, accuracy and precision information, and possible interferences.

The review and approval process for the method should be clearly described in the QA Manual. This is important not only if the laboratory develops methods internally, but also for the acceptance of externally developed methods.

Methods should be in written form and available to all analysis. Any deviations from the written method should be documented in the laboratory notebook or whatever record of the sample analysis process is used. Deviations from written methods are a matter of laboratory policy and dependent on the skill and expertise of the analysts or supervisor responsible for approving such deviations.

7 SAMPLE HANDLING

7.1 Sample Media

Prior to sampling, the person performing the sampling must acquire the sample media. This may be done independently of the laboratory, with advice from the laboratory, or directly from the laboratory. The closer the communication and coordination between the person sampling and the laboratory, the more likely to have a sample taken consistent with the validated sampling and analytical method. This will assure the best analysis possible and that re-sampling is not needed because of a defective sample.

Commercially available media is readily obtainable for most common sampling situations. These are generally suitable and the most practical alternative, but should be evaluated individually to assure against prohibitive expense, poor quality or instability issues. Manufacturers expiration dates should be checked to assure that media are used or disposed by the expiration date.

Care should be taken to assure that blank samples are included to be shipped to the laboratory. Appropriate shipping considerations should be followed including compliance with Department of Transportation regulations and the International Air Transport Association (IATA) guidelines.

7.2 Sample Acceptance and Receiving

The laboratory should have procedures in place to assure that incoming samples are reviewed to assure the laboratory has appropriate facilities and resources before accepting or commencing work.

When samples are received, appropriate procedures should be in place to assure the integrity of samples is verified and maintained. One method of maintaining sample integrity from the field to the laboratory is the use of sample seals. Once the samples have been collected in the field, the seal is applied in such a manner that the seal cannot be removed without breaking it. This prevents tampering and ensures sample integrity if the seal is intact when received in the laboratory. The condition of the seals should be noted and documented by sample receiving personnel.

Samples should be immediately transported to a designated receiving and logging area. This area should be secured in appropriate ways such as restricted access, visitors accompanied by laboratory personnel, locked drawers or storage areas, and locked doors during non-business hours.

Sample receiving personnel should check each sample for signs of tampering or damage, including the condition of seals, as previously mentioned. Any such visual evidence should be noted and the samples logged into the laboratory sample tracking system. Each sample receives a unique identifier number. Chain-of-custody documentation should then be maintained throughout the sample process until disposal. Sample documentation from whomever submits the samples should be examined for completeness and accuracy. Minimum information should include the name of the person sampling, the chemical which was sampled for, the sampling medium, the air volumes sampled, the sampling and shipping date, the establishment from which the samples were taken, possible interferences, and atmospheric conditions if relevant.

7.2.1 Chain of Custody

Chain of custody refers to evidence of an unbroken record of possession from the time a sample is collected until the completion of the analysis (4). In some cases, it may extend into a court case, but in most industrial hygiene applications the sample will be used during the analysis or stored for a short time before disposal. This is further addressed in the section on sample retention.

Each person responsible for the sample during this process should sign the appropriate documentation regarding the dates in which the sample is in their custody, beginning with sample collection. If samples must be sent by mail or other delivery service, appropriate certification or documentation should be acquired and retained. Sample receiving personnel and laboratory analysts are also part of this chain.

Possession is determined if the sample is actually in the custody or visual contact of the custodian, or is otherwise secured such as in a locked storage room, drawer, or refrigerator.

The purpose of chain of custody documentation is primarily legal or regulatory. The primary reasons for maintaining this record is to demonstrate that the sample analyzed is in fact the same sample that was collected and that it has not been tampered with, altered, or damaged in any way.

7.3 Sample Rejection

If samples are accepted under certain conditions, it follows that they are also rejected under certain conditions. Even laboratories which typically do not reject samples for analysis will have certain conditions under which samples are rejected.

These may include samples that are compromised in some way. In these cases the laboratory client should be advised of the compromised nature of the results and a decision reached whether or not to continue. Examples of these instances include incorrect sampling medium, overloaded sampling medium, bulk and air samples in the same shipping container, or visual evidence of tampering. In these cases it may be useful to continue with the analysis for screening or other purposes even though the use of results may not be acceptable for legal or regulatory purposes.

In addition to cases where samples may have been compromised the laboratory may have other restrictions or justifications for rejecting samples. Samples of human fluid or tissue may not be appropriate for laboratories which have not been licensed under the Clinical Laboratories Improvement Act (CLIA). Shipping containers holding liquid samples or sampling reagents which have been damaged and leaked in shipment may be rejected either because no sample is left or handling the resultant mess is too hazardous.

7.4 Sample Security

Sample security issues will be decided based on the needs of individual laboratories and the importance or costs of keeping individual samples secure. Security measures which laboratories may wish to take include visitor logbooks and visitor tags in laboratory areas, locked storage drawers or areas for samples received, stored, or in process, restricted access to laboratory areas through the use of keypads or other restricted access devices to allow only authorized personnel to access test areas. Security during business hours and after hours should be addressed.

7.5 Analytical Sample Handling

Analytical sample handling will largely be addressed in either the sampling and analytical method or by laboratory SOP. Appropriate measures should be taken to assure sample integrity throughout the analytical process, including the use of quality assurance reference samples to verify or validate that sample integrity was maintained.

7.6 Sample Retention and Disposal

Sample retention and disposal policy addresses the question of what to do with the sample material when the analytical work is complete. This will depend on the nature of the laboratory and the samples, the client requirements, and cost considerations of retention or disposal.

Samples that may need to be reanalyzed or that may be subject to litigation or regulatory action may be retained for some period of time consistent with anticipated needs. These may be subjective on a case-to-case basis, but a written policy should be developed that explains the process for determining this decision.

If there is no need to retain completed samples, they should be disposed of in a proper manner. This includes complying with applicable laws or regulation regarding disposition of potentially hazardous materials. There are a number of commercial vendors capable of disposing of hazardous materials and providing advice or assistance on appropriate control measures.

8 REAGENTS AND STANDARDS

Analytical reagents and standards should be of the purity specified in the sampling and analytical method. Analytical standards must be traceable the National Institute for Science and Technology (NIST) or other national or international reference standard where possible. In the event that no traceable reference standard is available, other methods of assuring quality should be undertaken. These might include the use of internal quality assurance reference samples prepared from another source material or the use of proficiency or external collaborative testing.

All standards must be properly labeled as specified by the laboratory or required by AIHA IHLAP. Labels or other documentation should include manufacturer, purity, date prepared, chemical name, and disposal or re-evaluation date. Chemicals should either be retested or disposed by such a date. Any standards which are retested and found to be still valid should be relabeled with appropriate dates.

Reagents and standards must be stored in appropriate locations with safety precautions as required. This may include flammable or corrosive storage cabinets, refrigerators, and freezers.

9 EQUIPMENT CALIBRATION AND MAINTENANCE

Calibration is the act of determining instrument response to a sample of known or traceable content. Instrument response to the sample to be analyzed is then compared to the response for the known amount to determine the amount present in the sample. For IH laboratory analyses, calibration standards should be traceable to NIST or other national or international standard when practical.

Several points relevant to IH laboratory analytical calibration should be observed. Calibration frequency should be determined by laboratory policy. With most instrumental methods this will mean that calibration is accomplished each day or with each use. Although single point calibration is sometimes used, such as with flame ionization gas chromatographic methods, a multipoint calibration should be performed with each instrument use. The use of zero and one additional calibration point are sometimes used also, but are less suitable generally than true multipoint calibration.

Multipoint calibration can be accomplished by diluting a stock standard of known content to several different concentration levels, analyzing the standard at each level, and recording instrument response. This can be graphed (or electronically recorded) with instrument response on the y axis and concentration on the x axis. The response of the sample is then interpolated onto the graph and a corresponding concentration (or amount) is determined. All sample responses should be bracketed by standards. Generally, extrapolation beyond the highest or lowest concentration standard is not advisable. Laboratory policy should determine if and when extrapolation is acceptable. For example, extrapolations of small magnitude (e.g., 10% or less) may be considered acceptable or some extrapolations may be permitted for detectors which have been demonstrated to have linear responses over large concentration ranges.

In addition to analyzing calibration standards at the beginning and end of each analysis, calibration solutions should also be analyzed frequently during sample analysis to verify calibration. The criteria for frequency depends upon a number of factors, including how many samples are to be analyzed, the stability of instrument response, and the cost of reanalyzing samples if the instrument loses calibration during the course of the analysis.

Calibration criteria should also be developed for ancillary or support equipment which is used in the production of test data. These types of equipment include analytical balances and pipettes. Acceptance or performance limits should be clearly listed in the calibration criteria documentation and instruments should be appropriately labeled with calibration status or have logbooks which record calibration status. Calibration with each use or at specified intervals may be acceptable.

In addition to calibration, appropriate instrument maintenance and repair is essential to producing good quality data. The first steps to appropriate equipment maintenance are appropriate facilities and appropriate training. Sensitive instruments should not be subjected to unusual conditions or dramatic changes which can affect not only their longevity but their performance as well. Conditions which the laboratory should be aware of include dust, vibration, corrosive fumes, radiation, humidity, and excessive temperature fluctuations. Temperature changes during the course of the day or night can significantly affect retention times of high performance liquid chromatographs, for example. Once appropriate accommodations and utilities are assured, analyst training in the proper use of instruments is needed. Unauthorized or untrained personnel should not be allowed to operate equipment with which they are not familiar. Training either by instrument manufacturer representatives or other in-house analysts or managers familiar with the instrument should suffice. Operating manuals for each instrument should be readily available nearby.

Maintenance logs should be maintained for each instrument. The logs can be at each instrument or a centralized location within the laboratory. Entries should be made whenever any type of maintenance work is performed, including the date, person performing the maintenance, and descriptions of what was actually done. Preventive maintenance schedules determined by laboratory management should be documented, followed, and recorded in maintenance logs when actually performed.

Repairs should be included in the maintenance log or in a separate log.

10 INTERNAL REFERENCE SAMPLES AND STATISTICAL QUALITY CONTROL

10.1 General

The use of internal reference samples and statistical techniques for evaluation are required by AIHA IHLAP. Most laboratories, whether or not they are accredited will have a quality objective to produce data that is scientifically defensible or to produce data of known and acceptable accuracy and precision. This can only be accomplished by an effective internal quality reference sample program. Statistical programs for laboratory quality control have been developed and described (6).

Some of the common terms of this program should be clarified. Accuracy is a measurement of how close the analytical result comes to the true or expected value. The accuracy measure most commonly used is the arithmetic mean. The arithmetic mean is simply the sum of all results divided by the number of observations. The difference between the mean and true value is sometimes called bias. Precision is a measure of dispersion of results around the mean recovery. The measure of dispersion most common in the IH laboratory is standard deviation.

10.2 Reference Samples

A simple program of spiked samples and blanks can be effective in evaluating analytical performance. Spiking an aliquot of the compound of interest onto the appropriate sampling medium provides a reasonable simulation of field samples. This can be accomplished using an adjustable volume microliter pipette or syringe, commercially available or internally prepared stock standards, and the medium used for field sampling.

Alternatives to liquid spiking samples are the use of reference standards or the use of air-generating the compound of interest onto the sampling medium. Spiking is widely used because the use of standards only does not allow for evaluation of how the field sample handling procedure (e.g., desorption, extraction, digestion) affects the samples and air-generating quality assurance reference samples is too time-consuming, costly, or otherwise unsuitable for many laboratories.

Reference samples should be prepared using sources independent from those used in the analysis and should be traceable to NIST or other national or international standard. They should also be prepared by someone other than the analyst. Thus these reference samples are prepared by a different person and from a different source, effectively eliminating those possible sources of error if the results are consistent with the theoretical load level.

Reference samples should also be prepared to bracket the range of interest such as the OSHA Permissible Exposure Limit or ACGIH Threshold Limit Value. Samples should not always be prepared at the same level and spiking levels should be blind to the analyst if possible. One way to accomplish this is to analyze multiple reference samples at different spike levels. This also provides additional information if corrective action is required. Blunders which affect only one sample are quite easily distinguished from other systemic errors which affect the entire analysis and thus all the reference and field samples.

Analyses should be conducted using validated methods where possible and the reference samples should be handled in the same manner as the field samples.

10.3 Data Evaluation and Control Limits

Once the analysis is completed and the results are calculated, the reference sample results should be reported and evaluated. Results are compared to theoretical results and the standardized recovery then compared to established control limits. Numerous statistical processes are available for this evaluation.

After a suitable number of data points have been determined, usually about 20, statistical treatments can be applied. Most commonly each data point is normalized or standardized to either percent recovered or decimal fraction recovered by dividing the amount found by the theoretical amount spiked (times 100 if percent recovery is used). Then using a calculator or statistical software a mean recovery and standard deviation for the data set are calculated.

Control limits are the statistical extremes within which we expect the data to fluctuate. Control limits are set as desired, usually three standard deviations from the mean recovery. Warning limits are occasionally used also. These are commonly set at two standard deviations from the mean.

Establishing control limits at three standard deviations from the mean implies a 99.7% confidence that any data outside those limits is due to a determinate error. That is, if only random effects are occurring that we expect 99.7% of the data to be within control limits. So when outliers occur, there is just a 3-in-1000 chance that it is due to random error.

Excursions outside warning limits are intended to alert the laboratory to a possible or impending problem for which some level of action may be taken. Excursions outside control limits alert the laboratory to an immediate problem that is almost certainly due to some determinate cause, not to random error. In any event, the data represent a statistically very unlikely result. Data results outside control limits should always result in the established corrective action process occurring.

The procedure for establishing and updating control limits should be established in the QAM. A sufficient number of data points should be obtained before the initial determination of control limits. Approximately 15 to 20 data points should suffice. As more data are generated they may be added to the data base so that more data points are considered in updating control limits. The use of 50 data points, when available, is not uncommon. At this point, as new data are added the initial data are deleted so that the most recent 50 data points only are considered. The control limits can be updated on a real-time basis with each new data point, or established and fixed for a period of time so that control limits are relatively static. Both approaches are used and each has certain merits.

Although data points outside control limits set at three standard deviations from the mean are statistically very unlikely events and therefore probably due to some determinate error, the reverse is not true. That is, data points inside the control limits are not necessarily free from determinate error. However, that error is within the limits that we have defined as acceptable, or that we are willing to live with for the particular analysis.

In addition to warning limits, there are published various general rules regarding how to handle multiple or consecutive data points at 1 or 2 standard deviations from the mean, and the various probabilities can be calculated fairly easily once laboratory management has determined limits of acceptability (7).

10.4 Control Charts

As quality reference sample data are generated, they are entered on a control chart. Control charts are a graphical representation of quality control data and provide a valuable method to assess internal reference samples (7–10).

According to Kelley, et al., the control charting technique was invented by Dr. Walter Shewart of Bell Labs to assess manufacturing systems for process stability and described in his book, *Economic Control of Quality Manufactured Product*, published in 1931 (7). The value of the control chart is in distinguishing random error from determinate error. Random error is the normal variability associated with the procedure. Determinate error has an assignable cause which investigators may or may not be able to identify. Examples of determinate error include deteriorated analytical standard or incorrect standard, poor or incorrect calibration, and calculation or data entry error.

The control charting technique is based on three assumptions to be valid. These are variation, randomness, and normality (10). All processes contain variation and analytical determinations in the IH laboratory are no different. No matter how many controls are implemented, repeated test results will give answers of some degree of variation. This variation should be random. That is, there is no determinate error being applied. If analyst calculations, invalid calibration standards, or any other error factor is occurring, the data will not be random. Finally, the distributions of random variations in most IH analyses will be normal or Gaussian (bell-shaped). This is not the case with asbestos analyses and other techniques must be applied (11).

For use in the IH laboratory control charts should be developed for a specific compound of interest, sampling medium, and analytical technique. Data should not be mixed for any of these parameters. For example, there should be a unique control chart for zinc collected on mixed-cellulose ester filter and analyzed by atomic absorption spectroscopy.

Charts are constructed with recovery (preferably normalized to percent or fraction recovered) on the y-axis and sample number on the x-axis). Horizontal lines representing the mean recovery (or alternatively, 100% recovery) and the upper and lower control limits are drawn. Data is then charted and may be evaluated.

Data points outside the control limit lines are obviously outliers and will need to be investigated for cause through whatever corrective action process the laboratory has designed.

Control charts are most useful, however, for the ability to look for trends or other indications of system breaking down through error sources before it becomes an out-of-control situation.

Since random error should produce a random pattern on the control chart, this can most easily be accomplished by evaluating runs (9). Data runs are consecutive results that have the same characteristic such as increasing or decreasing recovery, above the mean or below the mean recovery. Each of these characteristics has only two possible outcomes each of which is equally likely in a random distribution of data.

Since there are only two possible outcomes, each of which is equally likely, there is a 50–50 chance with each new data point that it will be specifically one of the two characteristics, much like flipping a coin with heads or tails as the two possible equally likely outcomes. Identifying runs such as consecutive data points above the mean are analogous to consecutive coin flips landing on heads.

With each succeeding like result it gets more unlikely that the results obtained are due to random effects. Thus, it becomes increasingly likely that such results are due to some determinate effect which we have not yet identified. By the time data points have run 10 consecutive times with the same characteristic, the odds are less than 1 in 1000 that it is due to random causes.

Results consecutively above the mean or below the mean may be due to differences in analytical standards between the analytical group and the quality assurance group. Steps should be taken to evaluate or rationalize any differences that appear.

Results consecutively increasing or decreasing may be due to deteriorating standards by either laboratory function. They may also indicate deteriorating performance of the analytical instrument.

These evaluations may point directly to a problem or give an indication of a developing problem which needs to be investigated and any needed corrective action taken.

10.5 Corrective Action

Corrective action should be initiated whenever established control limits are exceeded. This action should consist of an investigation by analyst and quality assurance personnel to determine the cause or source of outlier, if possible, propose remedial action to eliminate the root cause of the error, identify opportunities to improve the analytical or quality system through training or changes in SOPs, and document the entire incident.

No sample results should be reported for the analysis in question until appropriate resolution. This may include determining the cause of the outlier and correcting the error, demonstrating the outlier to be a random event that does not affect the sample data, or reliably demonstrating the sample analytical results to be less than the detection or quantitation limit.

Strategies for the corrective action sequence should be determined and documented in the QAM. The nature of the outlier data, however, may suggest particular strategies or causes which may be more likely than others to have occurred.

Analytical personnel should consider calculation checks, calibration checks, reproducibility data, standard preparation, and instrumental performance.

Quality assurance personnel should check calculations, appropriate control limits, standard preparation and stability, and sample preparation procedures.

The entire process should be documented in some fashion. The most common and effective manner is the Corrective Action Report (CAR). Contents of the CAR may be adapted to suit the particular laboratory, but should generally contain a description of the problem, corrective steps taken by those involved, and a conclusion or resolution to the event.

Also considered should be the impact on the client samples and disposition of data. Problems which may have impacted client sample results but cannot be corrected should be reported to the client with a statement of how the samples would likely be impacted.

Disposition of the quality assurance reference sample data should be based on predetermined criteria. The determination to replace or delete possibly erroneous data should not be taken lightly. This decision will impact future control limits.

Data may be rejected for cause or on a statistical basis. Rejection for cause is usually a straightforward exercise. If a problem occurred in the analysis which can be identified, it should be corrected if possible and the correct data inserted for the quality assurance reference samples and the client samples. If an identifiable problem occurred, but cannot be corrected, the QA data should be rejected and the sample results reported to the client with the appropriate qualifier information.

Numerous statistical tests exist to evaluate outlier data (10). These tests include the Dixon test, the Grubb test, and the Youden test. These or other valid statistical tests can be used to accept or reject suspect QC data for which no assignable cause of error can be found.

Another test which is sometimes mentioned is the rule of huge error. This simply assigns some large multiple of the standard deviation, such as 4 or 5, and rejects such data simply because it is statistically extremely unlikely to occur due to random error.

Data rejection criteria should be applied judiciously so data that is suspect, but nonetheless valid, is not rejected. Indiscriminate rejection of data will also lead to ever tightening control limits, followed by increasing outliers for which no assignable cause exists.

10.6 Replicates

Replicate analyses may be used to establish method precision and demonstrate the analytical system can reproduce results for identical spikes from the same sample. This should be done with both quality assurance reference samples and field samples. Even though the theoretical or actual amount on the field samples is unknown, the ability to replicate whatever result is obtained is the critical point of replicate sample analysis.

Control charts should also be maintained for replicate samples. These charts often plot the range obtained in replicate analysis of the same sample. With horizontal control limits at zero (the lowest possible range of two values) and at the upper control limit, the data is plotted as it is obtained. As with the accuracy control charts discussed earlier, excursions outside the established control limits should initiate an established corrective action procedure to investigate, correct, and document any problems discovered.

10.7 Blanks

Laboratories should incorporate the use of blanks into the quality system by using them in the on-site sampling process and into the quality assurance reference sample process.

Field blanks are sampling media that are subjected to the same handling as the field samples, except no air is pulled through the sampler. Since field blanks have not sampled the occupational atmosphere, they should contain no air contaminants from employee breathing zones. The purpose is to show that samplers handled in the same manner have no hazardous materials present and thus any materials present in the field samples are due to sampling the occupational atmosphere.

Blank values are usually none detected or very low, and customarily would be treated by subtracting the blank value from any value reported for the field sample. This becomes more problematic if relatively large amounts are found on the field blank. One common method to handle this situation is to not blank subtract and to report the actual amount found on the blank sample. This may be due to inadvertent contamination that did not adversely affect the field samples, or it may be due to some other factor that did indeed impact the sample results for the field samples.

Laboratory blanks as part of the quality assurance program are also media blanks with none of the compound of interest added. The instrument response should be zero or very nearly so. This demonstrates that nothing in the laboratory process contributed to the positive test results. Again, problems in data handling for the blank arise in the event a large amount of the compound of interest is found.

Although finding large blank values is a relatively rare occurrence, it does happen. Laboratory management and technical personnel should develop policies and procedures for evaluating blank sample results, document the process, and assure that employees are adequately trained in how to handle the results of blank samples.

10.8 Blind Samples

Another evaluation tool which can be used in the quality control program is the blind sample, sometimes referred to as double-blind. Unlike the quality assurance reference sample (or single-blind sample), which is known to the analyst as a quality control sample but with an unknown amount spiked on it, the blind sample is also unknown to the analyst as a quality control sample (8).

This provides the laboratory with a check of routine performance on client samples, because the analyst is not inclined to provide extra effort that might accompany a known quality control or proficiency test sample.

There are, however, disadvantages to using these type samples. One major disadvantage is that these samples do not easily replicate field samples. Samples spiked in the laboratory are detected by analysts because of their clean appearance or clean chromatography with relatively high quantities detected.

Although blind samples do provide valuable information on the routine level of performance, they should be used sparingly and evaluated carefully, focusing on the entire quality system and not as a punitive measure for analyst performance.

10.9 Nonroutine Procedures

For nonroutine procedures many of the evaluation criteria will not be relevant or will be impractical. There will often not be a validated or standard method. Some sort of mini-validation procedure may be necessary to provide an indication of sample stability and recovery.

In addition, there will not be sufficient data for statistical quality control procedures to be performed. There may not be sources for independent standards or for NIST-traceable standards.

Replicate analysis or sample-splitting will provide some insight into the reproducibility of the procedure. In addition, spiking reference samples provides valuable information on the quality of the analysis. Control limits can be arbitrarily established based on results obtained from similar compounds or at some alternative acceptance level at which the laboratory is comfortable issuing sample results.

10.10 Asbestos Quality Control

Because of the unique nature of the asbestos analysis—it is not a normal distribution, for example—it requires statistical treatments different than those we have already discussed. Specific programs to establish quality control programs for asbestos are available (7).

In addition, special attention is needed because of regulation. The OSHA Asbestos Standard **29** *CFR* 1910.1001 Appendix A requires certain quality control procedures. Training requirements that all asbestos analysts must have the NIOSH course or equivalent training are mandated. Blind recounts of 10% of samples are also required (12).

10.10.1 Intralaboratory Program

Laboratories with more than one microscopist must establish a statistical program of quality assurance with blind recounts and comparisons between microscopists. In addition, for companies with multiple laboratories the program must include variability between the various sites.

10.10.2 Interlaboratory Program

A round-robin testing program must be established which includes at least two other independent laboratories and testing must be performed at least semi-annually. Each laboratory must submit representative slides and the evaluation of performance must be statistically based. Laboratories are also required to participate in the AIHA PAT Program or Asbestos Registry.

11 PERFORMANCE AND SYSTEM AUDITS

Various aspects of performance and system audits have been discussed throughout this chapter. For documentation of the laboratory quality system, they should be aggregated into one section of the QAM.

Internal audits, participation in proficiency testing programs, and audits by accreditation agencies or other outside groups should all be addressed in the section. Laboratory policy regarding each relevant area should be addressed. This would include frequency, scope, personnel, and reporting requirements for internal audits. Participation in proficiency testing or in other types of external, collaborative, or round robin testing should be addressed. Any regular external audit or accreditation site visits could be addressed if applicable.

12 CORRECTIVE ACTION

Appropriate corrective action procedures need to be incorporated into the QAM. Corrective action should be initiated whenever there is a departure from documented policies or whenever there is a testing discrepancy. There are two separate processes which may be referred to as corrective actions.

In the internal audit process, corrective actions refer to measures taken to address a nonconformance. In this case, there is a departure from documented policies or other standards. The audit report will list a nonconformance and a required corrective action. These types of corrective action are addressed further in Section 4.

In the internal quality assurance reference sample process, corrective actions refer to measures taken to investigate, correct, and document a statistical outlier. In this case, there is a testing discrepancy. These types of corrective action are addressed further in Section 10.

13 QA REPORTS

QA reports should be designed as needed or required. Many of the reports have already been discussed in previous sections and will be mentioned only briefly.

13.1 Management Review

The annual management review process should be documented in a report. Specific issues discussed and the resolution of those issues should be addressed. Any changes to the QAM should be made by the appropriate authority (usually the QA manager), signed by the appropriate authorities (usually the QA manager and laboratory director), the superannuated QAM filed, any controlled copies called in, and new QAM copies distributed.

13.2 Quarterly Report on Quality Issues

This is a requirement for AIHA accredited laboratories, but an excellent idea for those that are not. At least quarterly the QA manager gathers relevant information regarding the quality system with specific emphasis on problems encountered or notices regarding possible impending issues. These might include detailing or summarizing specific corrective action investigations, listing or describing specific analyses for which the accuracy and precision have exceeded predetermined limits or are approaching those limits, a brief summary of any internal audits or external performance audits conducted, any QAM modifications, and any new issues regarding the quality system.

13.3 Report on Control Limits, Accuracy, and Precision

This should be done as frequently as updates occur. If real-time updates are performed daily, then a weekly or monthly update would be sufficient. These should be available to managers and staff so they can get an idea of the relative performance of each analytical procedure.

13.4 Control Charts

Control charts should be updated with each analysis, or as frequently as possible. Trends should be evaluated on a continuous basis to avert problems before they lead to an out-of-control situation.

13.5 Quality Assurance Reference Sample Report Data

This is the report that is generated with each quality assurance reference sample analysis. It should list any relevant information. Such information might include the analyst, preparer, date prepared, compound of interest, date analyzed, analytical technique or instrument, amount found, amount theoretical, percent or fraction recovered, in/out of control, and control limits. Any other relevant parameters could also be included.

13.6 Other Reports

Other general laboratory reports could be considered part of the quality system including sample submission forms, log-in forms or books, raw data files, and final reports of results.

14 DATA REDUCTION, VALIDATION, REPORTING, AND DOCUMENTATION

Data reduction, validation, reporting, and documentation should be done in accordance with quality requirements as determined by the laboratory. Plots and printouts from the analytical instrumentation should be reviewed and processed to determine an accurate result for the compound of interest. Validation by peer-review should be accomplished prior to reporting of sample results. Reports to the appropriate client and comprehensive documentation of the analytical process complete the task.

Data validation is the process of assuring that analytical results have been determined in an appropriate manner following the validated sampling and analytical method and approved policies and procedures of the laboratory. This can be accomplished by having another analyst familiar with the analytical procedure check or peer-review the entire process. This check would likely include verification of calculations, data transfer and data entry, appropriate analytical techniques, instrumentation, and standards.

Data documentation is a multistep procedure throughout the laboratory process. Sample receiving initiates the laboratory portion of data documentation by including whatever additional information is determined to be necessary to the documentation which accompanies the samples. Some basic requirements include the date the sample was received, by whom it was received, the condition of samples or sample seals, and a unique laboratory identification number.

Sample receiving personnel input or log-in the relevant data from the sampling documentation. This may include the date sampled, the chemical sampled, identifier numbers, the establishment from which the sample was taken, the shipping date, the person performing the sampling, the number of samples, the sampling time and air volumes, the sampling medium, and identifiers for blank samples.

The next documentation level is the custody transfer to the analyst. This information would include the identification of the analyst and the date assigned or transferred.

Finally, the analyst enters the relevant information to report and complete the sample documentation. This would include: the analytical technique and instrument, the date analysis was completed or reported, identification of checker or peer-reviewer who verified the analytical results, the analytical results, any general comments or specific comments regarding the analysis, and identification of supervisor or other management official who verified or approved the final results or report.

Comments which may be included in the report are almost limitless, but anything which qualifies the analysis in some way is appropriate. This may include comments about detection or quantitation limit, appropriateness of sample medium, interferences encountered, verification by alternate instrument such as mass spectrometry, appropriateness of air volumes taken, or any sample integrity issues such as leaks, loose or missing sample seals.

Raw data and other information which can be used to reconstruct the analysis if necessary should be filed and retained for some time period deemed appropriate by laboratory management. Three to five years is a common time period. Information included in this file would include all relevant information on the instrument and analytical parameters, calibration standards, calibration curves, data calculations, quality assurance reference sample results, corrective actions, and any deviations from the validated sampling and analytical method.

15 EXTERNAL PROFICIENCY TESTING

External proficiency testing provides the laboratory with the opportunity for external validation of test data.

Laboratories which are accredited by the AIHA are required to perform in the Proficiency Analytical Testing (PAT) Program (13). In addition, many laboratories which are not seeking AIHA accreditation participate in this program for the value of the independent check on their test results. Clients of IH laboratories should be aware of the distinction between an AIHA accredited laboratory which participates in the PAT Program as part of their accreditation and a laboratory which purchases PAT participation as an external quality check. These laboratories have not been through the accreditation process detailed previously.

PAT Program samples are available to IH laboratories for organic vapors, metals, fibers, silica, diffusive samplers, and bulk asbestos. The diffusive samplers program (with analytes benzene, toluene, and xylene) and bulk asbestos have either different schedules or different performance criteria than the traditional PAT Program.

Organic vapors included in the PAT Program rotation are benzene, *n*-butyl acetate, chloroform, 1,2-dichloroethane, *p*-dioxane, ethyl acetate, isopropanol, methanol, methyl ethyl ketone, methyl isobutyl ketone, tetrachloroethylene, toluene, 1,1,1-trichloroethane, trichloroethylene, and *o*-xylene.

Metals included in the PAT rotation are cadmium, chromium, lead, and zinc.

Asbestos/fibers included in the PAT rotation are amosite, chrysotile, and man-made fibers.

The PAT Program operates on a quarterly basis. Laboratories analyze PAT samples in the same manner as client samples would be analyzed. Laboratories receive four samples and a blank for each contaminant for which they participate. Individual results, not averages from multiple runs or analysts, are reported. Statistical treatment of the data is performed based on the results of preselected reference laboratories. Target values are assigned and performance limits are established at three standard deviations from the target value (or mean result). Any individual result outside the performance limit is considered an outlier.

Proficiency ratings are established for each analyte category for each participating laboratory. Proficiency is defined as all acceptable results for the previous two rounds or 75% acceptable results for the previous four rounds. In addition, an overall proficiency rating is assigned to each laboratory. Laboratories which do not meet the proficiency criteria for any performance element are rated nonproficient.

In addition, to the AIHA PAT Program, laboratories may wish to consider participation in other round-robin type programs. These might include non-PAT analytes of interest to the laboratory, such as ions (e.g., chloride, sulfate) or other metals or organic vapors. Asbestos analysis laboratories are required by OSHA standard to participate in round-robin type programs.

16 SUBCONTRACTING

When the analytical laboratory does not have the capability to perform a requested analysis, due to lack of proper facilities, equipment, or trained analysts, it may be necessary to contract the analysis to another laboratory with appropriate capabilities (4). Laboratories should conduct a review of their capabilities before accepting new work and advise clients that samples will be contracted to a qualified laboratory. Any report issued should clearly state that the work was performed by another laboratory.

Selection of subcontract laboratories should follow criteria determined and documented in the QA Manual. Possible selection criteria may include accreditation status with a relevant accreditation body, quality control procedures and quality system documentation, proficiency testing results, and of course, cost.

It is the responsibility of the laboratory to assure that adequate quality measures are taken. To do so may include the use of quality assurance reference samples along with the client samples or a site visit to determine satisfaction with the contract laboratory and their quality system.

17 COMPLAINTS

The laboratory QA Manual should contain the documented process for clients to press complaints and what procedures the laboratory is to use in response. The process should include a definition of what will be considered a complaint.

Once a complaint is lodged, the process should begin with a central point of contact, such as the QA Manager or other management designee. All client contact regarding complaints should be directed through this designee, who gathers all relevant information from the client and laboratory personnel and conducts an investigation into the validity of the complaint.

Any continuing or concluded investigations should be summarized in the quarterly QA report to management. A more thorough report on each individual complaint would also be prepared for appropriate management personnel as the investigation is concluded (4).

18 CONCLUSION

Adequate quality systems are essential to validate data obtained in the IH laboratory. Without a knowledge of the measurement uncertainty, there is no reasonable way to evaluate test data provided by the laboratory. Users of analytical services can perform a basic check of accreditation status, internal and external proficiency test results, internal audit reports, and accuracy and precision of the analytical procedure. Such a check is critical to appropriate decision making regarding analytical laboratory results. Laboratories which need to establish an adequate quality system can start with ISO Guide 25 and AIHA IHLAP criteria to develop and implement adequate systems.

BIBLIOGRAPHY

1. International Organization for Standardization, *Guide 25: General Requirements for the Competence of Calibration and Testing Laboratories*, Geneva, Switzerland, 1990.
2. American Industrial Hygiene Association, *Sound Data Smart Decisions* brochure, Fairfax, VA, 1998.
3. American Industrial Hygiene Association, *Laboratory Quality Assurance Program Policies*, Fairfax, VA, 1999.
4. T. A. Ratliff, Jr., *The Laboratory Quality Assurance System: A Manual of Quality Procedures with Related Forms*, Van Nostrand Reinhold, New York, 1990.
5. Occupational Safety and Health Administration, *OSHA Manual of Analytical Methods*, ACGIH, Cincinnati, OH, 1990.
6. J. A. Burkart, L. M. Eggenberger, J. H. Nelson, and P. R. Nicholson, "A Practical Statistical Quality Control Scheme for the Industrial Hygiene Chemistry Laboratory," *Am. Ind. Hyg. Assoc. J.* **45**(6), 386–392 (June 1984).

7. W. D. Kelley, T. A. Ratliff, Jr., and C. Nenadic, *Basic Statistics for Laboratories: A Primer for Laboratory Workers*, Van Nostrand Rheinhold, New York, 1992.
8. J. P. Dux, *Handbook of Quality Assurance for the Analytical Chemistry Laboratory*, Van Nostrand Rheinhold, New York, 1990.
9. E. L. Grant, R. S. Leavenworth, *"Statistical Quality Control,"* McGraw Hill, New York, 1972.
10. J. K. Taylor, *Quality Assurance of Chemical Measurements*, Lewis Publishers, Chelsea, MI, 1987.
11. G. Kateman and L. Buydens, *Quality Control in Analytical Chemistry*, John Wiley & Sons, New York, 1993.
12. **29** *CFR* 1910.1001 Appendix A, 1997.
13. C. A. Esche and J. H. Groff, "Proficiency Analytical Testing (PAT) Program November 29, 1997," *Am. Ind. Hyg. Assoc. J.*, **59**(4), 292–293 (April 1998).

CHAPTER SEVENTEEN

Odors: Measurement and Control

Amos Turk, Ph.D., Angela Merlo, MD, Ph.D. and Samuel Cha, MS

1 INTRODUCTION

The scope and applications of odor measurement and control are sometimes confused by various uses of the word *odor*. In the past, odor has been used to mean either (*1*) the perception of smell, referring to the sensation, or (*2*) that which is smelled, referring to the stimulus. To eliminate confusion, *odor* is best used only for the former meaning, and *odorant* should be defined as any odorous substance. Odor intensity, then, is the magnitude of the olfactory sensation produced on exposure to an odorant. *Odor control* is a term that can be used to describe any process that makes olfactory experiences more acceptable to people. The perceptual route to this objective is usually, but not always, the reduction of odor intensity. An alternative route is the change of odor quality in some way that is considered to be an improvement.

When the reduction of odor intensity is accomplished by removal of odorant from the atmosphere, the process is equivalent to gas and vapor abatement, or to air cleaning, but with some special considerations. These include (*1*) problems related to the need to attain very low concentrations, often approaching threshold levels, (*2*) uncertainties with regard to the reliability of sensory or chemical analyses (see Section 4), and (*3*) difficulties associated with diffuse or sporadic sources.

When an odorous atmosphere is improved by the addition of another (usually pleasant) odorant under conditions in which chemical reactions are not involved, the process is called odor modification, referring to modification of the odor perception, not of the odorant. These and other methods of odor control are considered in some detail in Section 6.

Patty's Industrial Hygiene, Fifth Edition, Volume 1. Edited by Robert L. Harris.
ISBN 0-471-29756-9

2 OLFACTORY PERCEPTION

2.1 Anatomy of the Olfactory System

The olfactory system (Figure 17.1) is a unique sensory system in which the cells responsible for the receipt and transduction of odorous stimuli also transmit encoded information to the brain. Olfactory receptor cells act to process and not merely relay information. In humans, these cells can number in the billions. They are interspersed among supporting and basal cells in a yellow-pigmented epithelium called the *olfactory mucosa*, located in the upper part of the nasal cavity. Research on other species suggests that basal cells divide

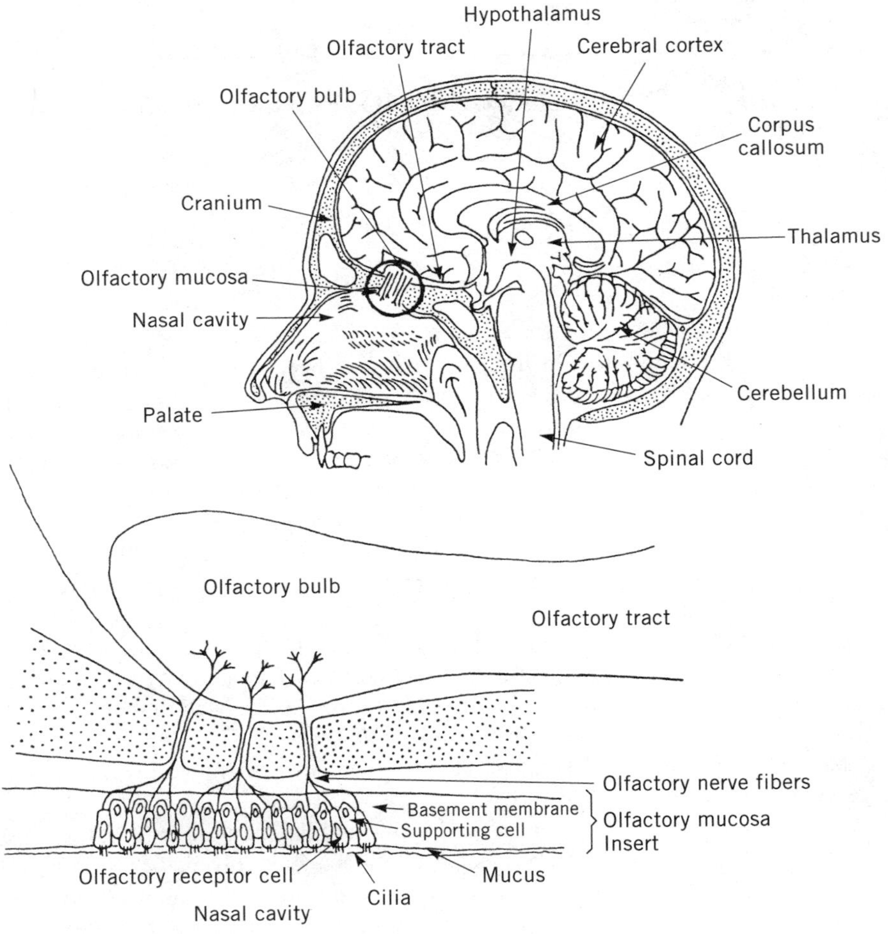

Figure 17.1. Schematic of human olfactory system in sagittal view showing location of olfactory receptor cells.

to form precursors of new olfactory receptor cells, allowing for turnover of the receptor cell population. Olfactory receptor cells are poised above the path of the main air currents that enter the nose with normal inspiration. Odorant molecules reach the olfactory mucosa by diffusion or are drawn up by sniffing.

Olfactory receptor cells are bipolar neurons with two main processes. The dendrite of each cell penetrates the surface of the mucosa, and several hairlike projections called *cilia* extend through the layer of mucus that bathes the mucosa. The cilia are typically 0.3 μm in diameter and 50 to 150 μm long; they house the molecular transduction apparatus responsible for receiving odorant stimuli. The basal process of each olfactory receptor cell forms an unmyelinated axon about 0.2 μm in diameter. This diameter is small, so propagation of nerve impulses is slow. The axons bundle together as they pass out of the olfactory mucosa, forming the *olfactory nerve*, or first cranial nerve. In humans, the olfactory nerve is only a few millimeters long. It courses centrally, enters the cranium through a series of perforations called the cribiform plate, and terminates in the surface layers of a region of brain called the *olfactory bulb*. The olfactory bulb is the first synaptic relay center in the olfactory pathway.

The olfactory bulb comprises a complex pattern of connections about which little is known. It is here that olfactory information is processed. Relay neurons have axons that leave the olfactory bulb in the *olfactory tract* to transmit olfactory information to other parts of the brain. Olfactory information participates in complex behaviors such as feeding and reproduction, and in emotional responses such as fear, pleasure, and excitement. The olfactory bulb is only two or three relays from nearly every center in the brain. Information from other brain areas is likewise relayed to the olfactory bulb. There is thus tremendous potential for the modulation of activity in the olfactory system.

The *trigeminal nerve*, or fifth cranial nerve, also innervates the nasal passages and relays information on odorants. The nasal trigeminal system contributes to the perception of odor intensity and perhaps also odor quality, although its role has not been well delineated.

2.2 Sensory Characteristics of Odors

An odor can be characterized by its absolute threshold, its intensity, its quality, and its affective tone or pleasantness-unpleasantness dimension. Determinations of thresholds of odor perception and measurements of odor intensity are considered in Sections 2.4 and Section 2.5, respectively. Discussion of measurements of odor quality and odor surveys are reserved for Section 4.4. It is important to recognize throughout that any program of odor measurement should be related to realistic analytical objectives. For example, if the odor from a landfill operation is to be controlled by masking agents, it is irrelevant to appraise odor intensities or threshold levels; a more appropriate measurement would describe the changed odor quality or character and would assay its acceptability to the people in the community. On the other hand, a nonselective method of odor reduction, such as ventilation or activated carbon adsorption, may well be monitored by measurements of intensity or by threshold dilution techniques.

2.3 Psychophysics and the Measurement of Odor Perception

G. T. Fechner is generally considered to be the founder of psychophysics, which he defined as the exact theory of the functionally dependent relations of body and soul, or more generally, of the physical and the psychological worlds (1). Fechner developed the concept of estimating or measuring the magnitude of sensations by assigning numerical values, reasoning that the relative magnitudes of sensations would be mathematically related to the magnitudes of the corresponding physical stimuli. In the years since Fechner's work, psychophysicists have endeavored to establish the laws relating physical stimuli to the resulting conscious sensations of the human observer. These efforts have concerned the responses of subjects to a number of basic questions about physical stimuli. Referring specifically to odors, these questions and the types of sensory evaluation needed to answer them can be tabulated (Table 17.1).

Each except the last of the problems named in Table 17.1 has a counterpart in quantitative neurophysiology. Indeed, applications of psychophysical techniques in some fields of research have aided in the investigation of underlying neural mechanisms. However, electrophysiological signals, even if it were convenient to monitor them routinely from human subjects, are not direct expressions of subjective experience. The magnitudes of sensations are therefore measured along psychophysical scales, and for olfactory sensations the reference points on such scales consist of the perceptions experienced on exposure to standard odorants.

Odor identification depends on accessing olfactory information in memory as well as in perception. Fluctuations in performance have been attributed to problems with semantic memory, that is, forming the association between the odor and its label (2). The situation becomes more complicated when odorants exist in mixtures. Mixing odorants alters perceived quality as well as intensity. The perceived intensity of a mixture has been found to closely approximate that of the stronger of the separate components, whereas the quality of a mixture composed of equally strong unmixed components lies between those of the components (3). There may be active suppression as well as hypoadditivity.

2.4 Decision Processes and Thresholds of Odor Perception

The word *threshold* means a boundary value, a point on a continuum that separates values that produce a physiological or psychological effect from those that do not. The upper

Table 17.1. Basic Questions for Sensory Evaluation of Odor Stimuli

Questions about an Odor Stimulus	Relevant Category of Sensory Evaluation
Is an odor present?	Determination of sensitivity to the detection of odor
Is a particular odor (e.g., that of phenol) present?	Determination of sensitivity to the recognition of odor
Is this odor different from that?	Determination of sensitivity to the discrimination between odors
How strong (intense) is this odor?	The scaling of odor intensity
What does this odor smell like, or what is the quality of this odor?	The scaling of the degree of similarity or dissimilarity between a given odor and each of a set of different odors
How pleasant (or unpleasant) is this odor?	The scaling of the hedonic, or like-dislike, response to an odor

threshold is the odorant concentration above which further increases do not produce increases in perceived intensity. The detection or absolute threshold is the minimum odorant concentration that can be distinguished from an environment free of that odor. Correspondingly, the recognition threshold is the minimum concentration at which an odorant can be individually identified. The recognition threshold of a particular odorant is never lower than its absolute threshold.

Thresholds are not firmly fixed values. Sensitivity fluctuates irregularly, and a certain odorant concentration will elicit a response at one time but not at another. Classical psychophysicists recognized the instability inherent in sensitivity, and they defined the absolute threshold statistically as that odorant concentration that is detected as often as not over a series of presentations. The probability of detection of such a stimulus is 50%. This definition of threshold has prevailed in the literature of air pollution and industrial hygiene. The theory of signal detection, originally engineered for telephone and radio communication systems by Shannon and Weaver (4) and later translated into a more general theory by Swets and others (5), contributed significantly to the development of modern psychophysics by specifying the dependence of threshold determinations on certain experimental variables. Detection theory has revealed the criteria employed by subjects in making perceptual judgments.

To illustrate how odor thresholds differ from other sensory evaluations of odor, refer to the questions about odor tabulated in Table 17.1 and note that the first three questions, which deal with thresholds, can be answered only by "yes" or "no." Because either answer can be right or wrong, four possibilities emerge when a subject is asked one of these questions. These possibilities are set out in a response matrix in Table 17.2.

In such an experimental situation, the subject must decide whether, in an observation period of set duration, there is only noise (i.e., background interference either introduced by the experimenter or inherent in the sensory process), or whether a signal (i.e., the designated odor stimulus) is present as well. Two kinds of error can be made by the subject in such a situation: a miss (i.e., the failure to detect an odor stimulus when one is present) and a false alarm (i.e., the report of perception of an odor stimulus when one is absent). Detection theory emphasizes the relation between the occurrence of these two types of error. According to detection theory (6), a positive response by a subject is favored (*1*) by positive expectations that are enhanced by a high rate of stimulus presentation or suggested by the experimenter's instructions, and (*2*) by reluctance to miss the presence of an odorant in accordance with rewards obtained for hits and/or punishments incurred for misses. Conversely, a negative response is favored (*1*) by an expectation, perhaps resulting from experience, that no odor stimulus is present, and (*2*) by reluctance to score a false alarm,

Table 17.2. Response Matrix

Odorant	Response: Yes	Response: No
Present	Hit	Miss
Absent	False alarm	Correct rejection

perhaps resulting from rewards obtained for correct rejections and/or punishments incurred for false alarms. These personal biases can be so strong that it is possible, in laboratory situations, to manipulate a subject's responses by manipulation of the variables influencing the decision process. Thus the experimental paradigm can drive the value determined for the odor threshold to a particular target. For example, a subject can avoid false alarms by not saying "yes"; this behavior generates a high value for the threshold. Conversely, a subject will not miss by not saying "no"; the consequence of such a decision is the generation of a low value for the threshold. Young and Adams (7), Steinmetz et al. (8), and Johansson et al. (9) describe such effects of the manipulation of experimental variables on odorant threshold determinations.

It is against this background of psychological biases that determinations of the limits of sensory perception must be evaluated. The absolute thresholds of different odorants determined in the same investigation can vary over 6 orders of magnitude, for example, 0.00021 ppm for trimethylamine and 214 ppm for methylene chloride (10). Such wide ranges may reflect real differences among the odors of different substances, yet since some reported threshold concentrations are very low, analytical and calibration errors may be large, and the accuracy of the threshold values is therefore suspect. Furthermore, the signal detection model of threshold determinations suggests that the variance can reflect decision as well as sensory processes. This suggestion could account for the wide range of scatter found in values of the absolute threshold generated for the same odorant in different investigations, for example, 3.2×10^{-4} to 10 ppm for pyridine (11). Another reasonable explanation for such scatter is the possibility that different samples of nominally identical odorants may really be very dissimilar, owing to different contents of odorous impurities. For example, phosphine, for which values of the detection threshold reported in the literature range from 0.2 to 3 ppm, has been described as odorless when pure (12), the reported odors being due to impurities in the form of organic phosphine derivatives. Similarly, it has been demonstrated (13) that ultrapurification engenders radical changes in the odor of nominally pure samples. Interindividual differences in sensitivity are yet another possible cause of large differences among values of the absolute threshold generated for the same odorants in different investigations. Interindividual variation can be high, but typically this factor is ignored in data analyses for the determination of odorant thresholds. There is also the problem of diversity in experimental procedures, for example, the mode of odorant presentation and sampling, which could introduce adaptation effects (Section 2.7), odor masking (Section 6.7), and errors in values reported for odorant concentration (14).

2.5 Factors Relating Odor Intensity to Odorant Concentration

A major objective of psychophysics is to measure the dynamic properties of sensory systems. Odorant concentration is the one property of the olfactory stimulus that can be varied somewhat systematically, although its effective control at the level of the olfactory receptor cell or even the olfactory mucosa has yet to be demonstrated. In the past, odorant concentration has been estimated or measured at some distance from the olfactory receptor cells. Psychophysicists have strived to characterize the functional dependence of odor intensity on such values of stimulus magnitude.

Psychophysical scaling attempts to determine the mathematical form of the relation between odor intensity and odorant concentration. *Direct scaling* refers to methods in which direct assessments of psychological quantities are made on an equal-interval scale (i.e., a graduated series with a constant unit) or on a ratio scale (i.e., a graduated series with a constant unit and a true zero). Direct scaling procedures include estimation methods in which stimuli are manipulated by the experimenter and judged by subjects, and production methods in which stimuli are manipulated by the subject to achieve a defined relation. In category estimation a given segment of a sensation continuum (usually predefined by stimuli supplied as anchors by the experimenter) is partitioned by subjects into a predetermined number of perceptually equal intervals, and subsequently presented stimuli are distributed by subjects among these categories. In ratio estimation the apparent ratio corresponding to a pair of stimulus magnitudes is numerically estimated. Magnitude estimation is a method in which various magnitudes of a stimulus are individually presented, and subjects assign a number to each signal in proportion to perceived intensity. For a complete discussion of psychophysical scaling methods, refer to work by Engen (15).

Stevens (16) proposed the psychophysical relation in which perceived intensity is a power function of stimulus magnitude, specifically the relation $R = cS^n$, where R is the perceived intensity, S is the stimulus magnitude, and c and n are constants referring to the intercept and the slope, respectively, when R and S are plotted on log-log coordinates. This yields a linear relation between the logarithm of the magnitude of the stimulus and the logarithm of the magnitude of the sensation. This relation has been verified for a wide range of sensory continua and has been proposed to be the fundamental psychophysical law. Hyman (14) has tabulated values of the exponent of the psychophysical function for odor intensity for 33 odorants generated in different investigations. The values are typically less than 1, with some exceptions for some subjects. They range from 0.03 for 1-decanol to 0.82 for 1-hexanol. Some of the results of the scaling of perceived intensity for the same and even for different odorants are quite similar, whereas considerable disagreement exists among the results of many other investigations of the intensity of different and even of the same odorants. Hyman (14) has discussed the factors capable of influencing the outcome of olfactory psychophysical scaling in the hope of elucidating possible sources of discrepancies. In brief, the factors concerning stimulus magnitude are (*1*) the use in different investigations of different physical scales that are not linearly related; (*2*) the use in some investigations of excessively strong stimuli or intertrial intervals so short that adaptation effects are introduced (see Section 2.7); (*3*) the use in some investigations of stimuli in concentrations so low that analytical and calibration errors are large; (*4*) the adsorption of vapors on the walls of gas-sampling vessels, which introduces errors in reported odorant concentrations; (*5*) the evaporation of volatiles from mixtures, which alters their composition during a given study; and (*6*) deviations from ideal behavior as given by Raoult's law, which is used to predict the vapor pressure of a dissolved substance on the basis of proportionality to solute concentration.

Published values of the exponent of the psychophysical function for odor intensity are generally based on odor intensity assessments collected from a group of subjects, then pooled to form one regression plot. The assumption implicit in such a pooling procedure is that subjects are interchangeable psychophysical transducers, or that they represent a normal distribution of such transducers and that the distribution is the same for all odorants.

But variation across individuals can be high (17, 18), and individual subjects can be consistent in the values of the exponent they generate for a particular odorant over repeated test sessions. It therefore seems improper to regard subjects as interchangeable for the purpose of pooling data in the indiscriminate manner traditionally employed for the calculation of odor intensity function parameters. Indeed, interindividual variation can be as large as differences between values of the exponent of the psychophysical function for different odorants derived from odor intensity assessments pooled from groups of subjects. Group measures may be obscuring important attributes of the olfactory system that these investigations are endeavoring to elucidate. Not only can the value of the exponent of the psychophysical function for odor intensity for any particular odorant vary greatly between subjects, but large intraindividual differences can exist between values of the exponent for different odorants. Thus it seems appropriate to direct investigations to individual behavior over a variety of stimulus and response modes within the olfactory sensory continuum while concurrently striving to minimize variation in experimental procedures.

2.6 Pure and Impure Odor Perceptions

The experience of smell can be taken to mean either any perception that results from nasal inspiration or only the sensations perceived by way of the olfactory receptor cells. An odorant that is sensed only by the olfactory receptor cells (e.g., air containing 0.01 ppm vanillin) is said to be psychologically pure. An odorant that stimulates both olfactory receptor cells and other sensitive cells (the so-called common chemical sense) is said to be psychologically impure; an example is air containing 50 ppm propionic acid. The common chemical sense includes sensations of heat, cold, pain, irritation, and dimensions of pungency described by our language inadequately. The trigeminal nerve, or fifth cranial nerve, is the primary mediator of common chemical sensations. The receptors responsible for the transduction of these perceptions remain incompletely characterized. The concept of psychological purity implies nothing about the chemical composition of the stimulus.

Differences between psychologically pure and impure odors are neglected in many odor measurements, especially in industrial hygiene and community air pollution applications, where irritants and stenches are grouped together as objectionable atmospheric contaminants that ought to be removed. It is not always operationally feasible to make distinctions among odors based on degrees of psychological purity. In establishing sensory measurement scales, however, it is important to recognize that *irritating odor* is not necessarily an extension in magnitude of *strong odor* but refers to a different type of sensation.

The transitions in perceived quality that can accompany changes in the perceived intensity of an odorant may be purely olfactory in origin, or they may be due to a change from purely olfactory stimulation at low odorant concentrations to olfactory plus trigeminal stimulations at high odorant concentrations. Evidence has accumulated in support of the notion that the trigeminal nerve contributes to the overall intensity of odorants. For human subjects with unilateral destruction of the trigeminal nerve, the magnitudes of perceived intensities for certain odorants are consistently lower through the deficient nostril, with the most dramatic differences at high odorant concentrations (19). Laska and coworkers (20) reported the detection of odorous vapors by what they considered to be anosmic human observers; such detection presumably occurs by way of the trigeminal system. The olfactory and trigeminal systems can have different psychophysical functions for a particular odorant, and psychophysical scaling data reported for odor intensity could actually be some combination of the two. There is usually no effort to eliminate the trigeminal component from assessments of odor intensity. It is therefore plausible that excitation of trigeminal receptors is responsible, at least in part, for differences among values of the exponent of the psychophysical function for odor intensity and for departures from that relation reported for some odorants in some investigations.

2.7 Olfactory Adaptation

Olfactory adaptation is the decrement in sensitivity to an odorant following exposure to what is termed an adapting odorous stimulus. The rate and degree of loss in sensitivity and subsequent recovery depend on the adapting stimulus and on its concentration (21). Self-adaptation to an odorant affects the perceived intensity of the same odorant, whereas cross-adaptation affects the perceived intensity of other odorants. For example, workers with daily exposure exhibit both a loss of sensitivity to the odorant and a decrease in their perception of other odorants (22). Olfactory adaptation is common, yet its physiological mechanism has not yet been elucidated.

Olfactory adaptation has been studied by measuring the time required for the odor of a continuously presented stimulus to disappear, and by determining absolute threshold values before and after exposure to an odorant for a given period. An increase in odorant concentration usually increases the time required for disappearance of an odor, and for a particular experimental design, absolute threshold values increase with an increase in the magnitude of the adapting odorous stimulus.

Self-adaptation affects the magnitude of the olfactory sensation as a function of stimulus magnitude. Values of the exponent of the psychophysical function for odor intensity are significantly greater after self-adaptation. High concentrations of self-adapting stimuli generate steeper psychophysical functions for odor intensity than do self-adapting stimuli of low concentrations (23). The use of weak stimuli closely spaced in magnitude, or long periods of exposure to an adapting odorous stimulus can permit resolution of the influence of duration of an adapting stimulation on the psychophysical function for odor intensity. However, without consideration of these factors, increasing the duration of self-adaptation produces only minor effects on the odor intensity function.

Olfactory adaptation may be responsible for some of the differences among values of the exponent of the psychophysical function for odor intensity reported for the same and even for different odorants. Adaptation effects can be introduced in an experiment by excessively strong stimuli, short intertrial intervals, the presence of odors apart from those generated by stimuli in intertrial intervals, and the presentation of a standard reference stimulus immediately before test concentrations (24).

In the context of industrial hygiene and in air pollution situations, two important questions arise. Does prolonged exposure to odors produce a permanent loss of ability to smell? And to what extent does temporary adaptation interfere with the sensory measurements of odor intensity carried out in the workspace or in the outside community? With regard to the first question, there is simply no evidence on which a reliable answer can be based. In view of the many difficulties involved in quantifying sensory odor measurements under controlled test conditions, it would seem hopelessly unreliable to attempt any studies that depended on retrospective estimates of exposures to odors. To answer the second question, we must consider patterns of recovery from adaptation and the opportunities for such recovery during typical odor survey programs. Koster (25) studied recovery times from adaptation to the odors of benzene and various alkylbenzenes, as well as to isopropyl alcohol, dioxane, cyclopentanone, and β-ionone. Typically, 60%–70% recovery of sensory response to the odors was realized in about 2 minutes. The implication of these findings, which is in accord with practical experience, is that if a subject is downwind from a point source such as a stack and is exposed to the odor intermittently depending on changes in wind direction, he can make perfectly good sensory evaluations when the plume comes his way again after a lapse of 5 min or more. However, a person in an odorous enclosure, who is subjected to unrelieved exposure to odor, will indeed become adapted to a point that can invalidate any sensory judgment.

Cognitive factors such as context and expectations can modulate the overall sensory perception of odorants and even override mechanisms of sensory adaptation. Information about potential toxicity of an odorant can alter the perception of odor intensity, as illustrated by reports of heightened sensitivity to odorants after prolonged exposure (26).

2.8 Theories of Olfactory Mechanism

Theories of olfaction attempt to delineate the mechanism by which odorant stimuli elicit responses in olfactory receptor cells, and correlate the sensory characteristics of odors with physicochemical properties of the odorants. The objective is to understand the physiological basis of olfactory discrimination.

The perception of smell begins with the impingement of odorant molecules upon the olfactory receptor. Humans can perceive thousands of different odorants, with differences in perceived quality often related to subtle differences in molecular structure. The recent identification and cloning of genes encoding odorant receptors (27, 28) has greatly enhanced our understanding in this area. Odorant receptors are located in the cilia of the olfactory cells. They are G-protein-coupled molecules that transverse the cell membrane and bind odorant molecules in special areas that exhibit tremendous diversity. The extremely large number of odorant receptor genes suggests that receptors are specific in their recognizing and binding with only one or a small set of structurally similar odorants. The diversity among receptors allows the population to interact with a large number of structurally diverse odorant molecules. Odorant receptors can be grouped into subfamilies with similar divergent regions. The subfamilies of receptors are thought to be able to detect subtle differences between structurally similar odorants. Odor discrimination, then, seems to depend heavily on the initial interaction of receptor and odorant.

The expression of odorant receptor genes is organized onto distinct topographical zones within the olfactory mucosa (29, 30). The zones are bilaterally symmetrical in the two nasal cavities. There are many different odorant receptors within each zone, although a given odorant receptor is confined to a single zone. Neurons expressing a particular odorant receptor gene may be widely distributed throughout the zone. The topographical organization of odorant receptors is preserved in the axonal projections to the olfactory bulb, suggesting an important role in the encoding of olfactory sensory information.

Odorant-receptor binding triggers generation of action potentials in the olfactory receptor cells through an adenyl cyclase second-messenger system (28, 31). The biochemical cascade that leads to the opening of cation channels in the cilia membrane and, in turn, depolarization of the cell is now well understood. The system is a classic cAMP-based system with certain remarkable characteristics such as the ability to rapidly generate large amounts of cAMP upon odorant exposure and then quickly return to the quiescent state. There are several amplification steps in the pathway between odorant binding and signal generation, so that only a few odorant molecules are needed to generate neuronal action potentials. This amplification system accounts for the low thresholds of olfactory detection.

Negative feedback processes have been identified within the olfactory receptor neuron that act to truncate an ongoing response in the continued presence of an odorant stimulus and thereby mediate adaptation (32, 33). Electrophysiological studies demonstrate a transient neural response to a sustained odorant stimulus. Calcium and cAMP have been identified as the negative feedback messengers. Intracellular concentrations of calcium rising as a result of the opening of cation channels following odorant-receptor binding inhibit the opening of the channel; and cAMP produced by odorant-receptor binding activates an enzyme that in turn inactivates the transduction process. Thus, the rise of intracellular calcium and cAMP following odorant stimulation cause the olfactory receptor cell to adapt to further stimulation.

Innovative biochemical, molecular biological, and electrophysiological studies have provided tremendous insight into the operating characteristics of the olfactory sensory system. Odor detection, odor discrimination, and adaptation all have correlates in basic cellular mechanisms.

3 CHARACTERISTICS OF ODOROUS SUBSTANCES

3.1 Odor Sources

A vented storage tank being filled with liquid ethyl acrylate from a delivery truck discharges to the atmosphere a volume of air saturated with the acrylate vapor at the prevailing temperature of the liquid. A system of this type is a simple example of a confined odor source: the location, molecular aggregation, composition, concentration, and volumetric discharge of the source can be quantitatively specified. Such definite characterizations facilitate the establishment of relationships between the odor source and the odor measurements made in the work space or the community. In general, an odor source may be said to be confined when its rate of discharge to the atmosphere can be measured and when the atmospheric discharge is amenable to representative sampling and to physical or chemical processing for purposes of odor abatement. For meteorological diffusion calculations, the location at which the odor is discharged into the atmosphere is assumed to be a point in space.

The characterization of a confined odor source should include (*1*) the volumetric rate of gas discharge (gas volume/time); (*2*) the temperature at the point of discharge; (*3*) the moisture content; (*4*) the location, elevation, and the area and shape of the stack or vent from which the discharge is emitted; and (*5*) a description of the state of aggregation (gas, mist, etc., including particle size distribution) of the discharge. Such information is useful when meteorological factors relating to odor reduction by dispersion and dilution are being evaluated. The characterization of an odor source in terms of chemical composition will depend on the relationships to be established between sensory and chemical measurements. If there are no technical or legal uncertainties regarding the source of a given community odor, a detailed chemical analysis is not needed; instead, a comparative method of appraising the effect of control procedures will suffice. On the other hand, chemical characterization may be helpful in tracing a community odor to one of several alleged sources, in relating variations in odor from a given source to changes in process conditions, or in appraising the effectiveness of chemical procedures that are designed for odor abatement.

The direct sensory characterization of a confined odor source is frequently impossible because extremes of temperature or of concentration of noxious components make the source intolerable for human exposure. Methods suitable for the dilution and cooling for sensory evaluation of odor sources are described in Section 4.2.

A drainage ditch discharges odorous vapors to the atmosphere along its length. The composition of the contents of the ditch changes from place to place, and the rate of discharge of odorants to the atmosphere is affected by wind and terrain. No single air sample is likely to be representative of the odor source at the time the sample is taken. Such a configuration is an unconfined odor source. Other examples are garbage dumps, settling lagoons, and chemical storage areas. An unconfined source may be represented by an imaginary emission point for the purpose of meteorological diffusion calculations. Such assignment may be made on the basis of supposing that if all the odor from the unconfined area were being discharged from the emission point, the dispersion pattern would just include the unconfined source. Figure 17.2 schematically illustrates this procedure.

3.2 Odorous Gases and Vapors

In general, gases and vapors are odorous. The relatively few odorless or practically odorless exceptions include oxygen, nitrogen, hydrogen, steam, hydrogen peroxide, carbon monoxide, carbon dioxide, methane, and the noble gases. No precise relationship has been established between the olfactory quality of a substance and its physicochemical properties, although attempts have been made for many years. This topic is treated in more detail in

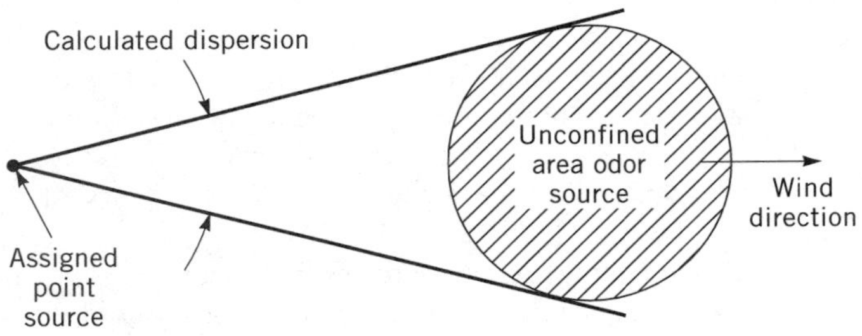

Figure 17.2. Assignment of emission point to an unconfined area odor source.

Section 4.1. Most odorants encountered in the work space or in the outside air are complex mixtures, however, and it cannot be assumed that the odor of the mixture is that of its major component. Thus a phenolic odor from the curing of a phenolic resin is not the same as the odor of pure phenol, and the pungent odor of burning fat is not the same as that of acrolein, though it is often so characterized. Even more dramatic are instances in which the ultrapurification of a supposedly pure substance produces a considerable change in odor (13).

3.3 Odors Associated with Airborne Particulate Matter

It is often assumed that all odorants are gases or vapors. However, there is evidence that some particles can stimulate the sense of smell because they are volatile or because they are desorbing a volatile odorant (34). There is also speculation that some particulate matter is capable of stimulating the sense of smell. Regardless of the mechanism by which particles may stimulate the olfactory sense, they definitely appear to be involved.

The idea that odors are associated with particles is supported by observations that filtration of particles from an odorous airstream can reduce the odor level. It has been shown (35) that the removal of particulate matter from diesel exhaust by thermal precipitation effects a marked reduction in odor intensity. The precipitation method was selected because it provides minimal contact between the collected particles and the gaseous components of the diesel exhaust stream. Thus effects that could be produced by a filter bed, such as removal of odorous gases by adsorption in the filter cake, are eliminated. Therefore the observed odor reduction must have resulted directly from the removal of particulate matter.

3.3.1 Volatile Particles

Liquid or even solid aerosols may be sufficiently volatile that their vaporization on entering the nasal cavity produces enough gaseous material to be detected by smell. Such aerosols may be relatively pure substances such as particles of naphthalene or camphor, or they may be mixtures that release volatile components. The retention of the odorous properties of volatile aerosols depends on the prevailing temperature and on the length of time they are dispersed in air. In a cold atmosphere, the relatively greater temperature rise accompanying inhalation accelerates the production of gaseous odorant.

3.3.2 Desorption of Odorous Matter from Particles

The interaction between gas molecules and the surface of airborne particles has been discussed by Goetz (36). The theoretical considerations were directed to the question of transfer of toxicants by particles, but they are also applicable to odors. Even if a given aerosol is intrinsically odorless, it can act as an odor intensifier if the sorptive capacity of the aerosol particles for the odorant were smaller than the affinity of the odorant for the nasal receptor, and if at the same time, the sorptive capacity of the aerosol particles were large enough to produce an accumulation of odorant on the particle surface. Such aerosol particles would concentrate odorous molecules on their surfaces, but the odorous matter would be transferred to olfactory receptors when the aerosol entered the nasal cavity. The odorous matter would then be present at the receptor sites in concentrations higher than in the absence of the aerosol. The resulting effect would be synergistic. If an odorant is more strongly adsorbed by the aerosol particles than by the olfactory receptors, transfer of the odorant to the receptors would be impeded and the particles would actually attenuate the odor.

3.3.3 *Odorous Particles*

No study has rigorously defined the upper limit of particle size for airborne odorous matter. Particles up to about 10^{-3} μm (10 Å) in diameter are considered to be molecules that can exist in equilibrium with a solid or liquid phase from which they escape by vaporization. The vapor pressure decreases as the molecular weight increases, and particles above about 10 Å do not generally exist in significant concentrations in equilibrium with a bulk phase; hence we do not consider them to be vapors. Nonetheless it is possible that odorant properties do not disappear when particle sizes exceed those of vapor molecules. Our knowledge about particles in the size range of 10 to 50 Å (up to about the size of small viruses) is relatively meager, and we do not know whether they can be intrinsically odorous. Larger particles may also be intrinsically odorous, although their more significant role may be to contribute to odor by absorbing and desorbing odorous gases and vapors.

3.3.4 *Odor Threshold and Related Concepts*

Among the sensory characteristics of odors (i.e., threshold, intensity, quality, and affective tone or pleasantness), threshold is directly related to the concentration of the odorant and is often used in practical engineering studies.

Listings of threshold values for pure chemicals are common in many environmental resource references. However, because threshold values are not physical constants, but rather statistically derived perception values, reported values from different laboratories or even from the same laboratory can vary by as much as several orders of magnitude. This extreme scatter results from differences in experimental methodologies, in the design of odor sample presentation apparatus, in the sensory evaluation decision process as discussed in Section 2.4, and in odor judge panel size, training, and selection, as discussed in Section 4.3. A scrutiny of methods of determining odor thresholds served to critique the certainty of published experimental values. The review and a compilation of critiqued threshold values were published by American Industrial Hygiene Association (AIHA) for 183 chemicals with occupational health standards (37). Table 17.3 is a summary of critiqued odor threshold values of 38 compounds selected from the AIHA publication.

For environmental odors, which are generally complex mixtures of many compounds, it is difficult to express threshold values as chemical concentration levels. Instead, the odor level of the odorous mixture can be expressed as a dilution-to-threshold ratio. Thus, a sample of odorous air that must be greatly diluted to reach its odor threshold is said to have a high *odor level*, in the sense of being highly pervasive. This ratio is therefore defined as V_t/V_o, or assuming no change in odorant mass on dilution, C_o/C_t, where subscripts o and t refer to the original and threshold concentration levels or sample volumes. The two terms most commonly used to express the dilution-to-threshold ratio are *D*/*T* and

Z, both dimensionless. Then $D/T = Z = C_o/C_t = V_t/V_o$; the Z is for Zwaardermaker who was the pioneer in using dilution ratios for odor measurement (38).

An older, somewhat awkward related term is the *odor unit*, defined as one volume of odorous air at the odor threshold, often expressed in cubic feet. Then the ratio odor units/cu ft = volume of sample diluted to threshold (cu ft)/original volume of sample (cu ft) becomes the same as D/T or Z. The odor unit is a unit of volume and is not a synonym of D/T or Z.

A derived term is *odor emission rate*, expressed in unit volume per unit time (cu ft/min); it is the product of the D/T value of the odorous air and the air emission rate (cfm or its equivalent). The odor emission rate is used to describe the pervasiveness of an odor source that may travel to downwind locations.

4 ODOR MEASUREMENT

4.1 Relation Between Sensory and Physicochemical Measurements

No one can accurately predict the odor of a compound from its molecular structure, nor is it possible to specify the molecular structure of a substance that will yield a predicted odor. In the context of odorous air pollutants, this means that attempts to predict odor nuisance from a knowledge of sources and processes can serve as a guide to signal possible problems but not as a substitute for direct sensory evaluation. Older statistical attempts at multidimensional scaling of such attributes (39) attempted to explain the degree of similarity between different odors in terms of physicochemical variables. The information generated by such methods, however, seems to be of the type that is already common knowledge, for example, that compounds differing greatly in molecular weight are likely to have very different odors, or that sulfides do not smell like aldehydes. Because most organic chemists believe they can identify the functional group (alcohol, amine, ester, etc.) in a compound by smell, it is interesting to determine the degree to which such attempts are successful. Brower and Schafer (40) conducted such a study and found that for most representative compounds the functional group was correctly identified in 45% of the cases. The performance was poor for alcohols, ethers, and halides, and excellent for amines, sulfur compounds, esters, phenols, and carboxylic acids. When the subjects missed the functional group, they used the labels alcohol, ester, and ketone twice as often as average. The label *sulfur compound* was misapplied in only 1% of all cases. Bulky hydrocarbon groups near the functional group can weaken or obliterate the odor quality, but aliphatic amines and sulfur compounds are very resistant to steric hindrance. On the other hand, they are greatly weakened by electron-withdrawing groups. Aliphatic compounds bearing a multiplicity of methyl groups have the odor of camphor or menthol.

Of course, most odor problems are produced by complex mixtures of odorants, which severely complicates the task of predicting sensory effects. The literature from 1958 through March 1969 was surveyed for instances of instrumental-sensory correlation studies under the sponsorship of ASTM Committee E-18 on Sensory Evaluation of Materials and Products (41). Of the several thousand articles reviewed in 65 major technical journals, including journals devoted to air pollution, only 45 were judged to have any direct value

Table 17.3. Selected Critiqued Odor Threshold Values[a]

Compound	Geometric Mean Odor Threshold Value (ppm)[b]	Type of Threshold	Range of Acceptable Values (ppm)[c]	Range of All Referenced Values (ppm)
Acetaldehyde	0.067	Detection	0.067	0.0028–1000
Acetic acid	0.074	Detection	0.037–0.15	0.010–31
Acetone	62	Detection	3.6–653	0.40–800
Ammonia	17	Detection	17	0.43–53
Amyl acetate	0.052	Detection	0.052	0.0075–7.3
Aniline	2.4	Detection	0.58–10	0.012–10
Benzene	61	Detection	34–119	0.78–160
	97	Recognition	97	
Bromine	NA[d]		NA	<0.0099–0.46
Bromoform	NA		NA	0.19–15
Butane	NA		NA	1262–5048
n-Butyl acetate	0.31	Detection	0.063–7.4	0.0063–368
	0.68	Recognition	0.038–12	
n-Butyl alcohol	1.2	Detection	0.12–11	0.05–990
	5.8	Recognition	1–20	
Camphor	0.079	Detection	0.079	0.0026–0.96
Carbon disulfide	NA		NA	0.016–0.42
Carbon tetrachloride	252	Detection	140–584	1.6–706
	250	Recognition	250	
Chlorine	0.080	Detection	0.080	0.021–3.4
Chloroform	192	Detection	133–276	0.6–1413
m-Cresol	0.00060	Detection	0.00005–0.0079	0.000011–0.0068
Ethyl alcohol	180	Detection	49–716	0.34–40,333
	100	Recognition	100	
Ethyl mercaptan	0.00035	Detection	0.000098–0.003	0.000098–18
	0.00040	Recognition	0.00040	
Hydrogen sulfide	0.0094	Detection	0.001–0.13	0.00007–1.4
	0.0045	Recognition	0.0045	

Isobutyl alcohol	NA		NA	0.028–0.072
Isopropyl alcohol	43	Detection	37–610	1.0–610
	19	Recognition	19	
Methyl alcohol	160	Detection	4.2–5960	3.3–198,656
	690	Recognition	53–8940	
Methyl amine	4.7	Detection	4.7	0.0009–4.7
Methyl ethyl ketone	16	Detection	2–85	0.25–85
	17	Recognition	5.4–55	
Methyl isobutyl ketone	0.88	Detection	0.1–7.8	0.1–16
	2.1	Recognition	0.27–16	
Methyl mercaptan	0.00054	Detection	0.0000002–0.041	0.0000002–0.56
	0.001	Recognition	0.001	
Naphthalene	0.038	Detection	0.038	0.0095–0.64
Phenol	0.06	Detection	0.06	0.0045–1
Propionic acid	0.066	Detection	0.026–0.17	0.00099–1.5
	0.033	Recognition	0.033	
Styrene	0.14	Detection	0.017–1.9	0.0047–61
	0.15	Recognition	0.15	
Sulfur dioxide	2.7	Detection	2.7	0.33–5.0
	4.4	Recognition	4.4	
Toluene	1.6	Detection	0.16–37	0.021–69
	11	Recognition	1.9–69	
Trimethylamine	NA		NA	0.00011–0.87
o-Xylene	5.4	Detection	5.4	0.18–5.4
m-Xylene	0.62	Detection	0.62	0.081–0.55
p-Xylene	2.1	Detection	2.1	0.12–2.1

[a]Geometric mean based on critiqued and accepted values.
[b]Geometric mean based on critiqued and accepted values.
[c]Single value indicates only a single accepted value.
[d]NA, not available.

Source: Selected from "Odor Thresholds for Chemicals with Established Occupational Health Standards," and used with permission of the American Industrial Hygiene Association.

toward progress in establishing standards for instrumental-sensory correlations. Of these, some 40 were concerned with foods, and the remainder with pure chemicals and other topics. None dealt with odorous air pollutants per se. Although the power and sensitivity of analytical methods continue to improve, more recent attempts to relate chemical composition to perceived odor of complex mixtures such as sewage effluents are only partly successful (42).

4.2 Sampling for Odor Measurements

In the case of nuisance odors in the work space or in the community, sensory judgments are usually made from evidence gained on direct exposure of human subjects to the odorous atmosphere in question. Where odor intensities are low enough to be tolerable and where they vary from time to time in accordance with atmospheric turbulence, changes in indoor air currents or outdoor wind direction, and changes in the odor source, it is much better for the observers to expose themselves directly to the odorous atmosphere rather than to a previously collected sample. Furthermore, when odor levels are variable, there are often enough intervals of low odor intensity to provide adequate recovery from odor fatigue. Under such circumstances, it would require considerable effort to collect samples that were representative of all the experiences a roving observer could accumulate in a single session of an hour or so.

In some instances, however, it is advantageous to collect a sample of odorant before presenting it for sensory measurements. Such instances occur when the odorant must be diluted, concentrated, warmed, cooled, or otherwise modified before people can be exposed to it; when it is necessary to have a uniform sample large enough to be presented to a number of judges; or when the samples must be transported to another location at which sensory evaluations are made.

4.2.1 Grab Sampling

A grab sample places a volume of odorant at barometric pressure into a container from which it can subsequently be presented to judges for evaluation. The general principles of sampling for gases and vapors are described in Chapter 8 of this volume. The subject has also been reviewed by Weurman (43). However, the objective of preserving the integrity of a sample for odor measurement may present special problems because mass concentrations of odorants are often very low. As a result, adsorption or absorption by the container walls may alter the odor properties of the sample. It is therefore important to use containers that are known to be inert to the odorant and are large enough to minimize wall effects. A number of container materials have been used, including glass, stainless steel, and various inert plastics.

When the grab sample is to be evaluated by human subjects, it must be expelled from the container into the space near the judge's nose. This expulsion may be effected by displacement of the sample with another fluid (usually water) or by collapsing the container. It must be recognized that when a person smells a jet of air from a small orifice, the aspiration of ambient air around the jet creates some additional dilution before the odorant reaches the subject. Springer (44) has designed a conical funnel that is positioned in front of the subject's nose to prevent this effect.

4.2.2 *Sampling with Dilution*

Dilution procedures for sampling odor sources such as oven exhausts can reduce concentrations and temperatures to levels suitable for human exposure without permitting condensation of the odorant material.

A suitable device for sampling of hot odor sources with dilution is a stainless steel tank fitted with valves, a vacuum gauge, and a pressure gauge. The tank is first evacuated, then filled through a metal probe to a desired pressure P_1, which must be low enough to prevent condensation when the gas warms to room temperature. This dilution is $Z_1 = 1$ atm/P_1(atm). The tank is then pressurized with odor-free air to a new pressure, P_2, which is above atmospheric pressure. The diluted, pressurized sample may then be sniffed by one or more judges for odor evaluation. The dilution that occurs when pressurized gas is released to the atmosphere is $Z_2 = P_2(\text{atm})/1$ atm. The overall dilution ratio (38) is

$$Z = Z_1 \times Z_2 = \frac{1\ \text{atm}}{P_1(\text{atm})} \times \frac{P_2(\text{atm})}{1\ \text{atm}} = \frac{P_2}{P_1}$$

For example, if the evacuated tank is first filled to 0.1 atm, then pressurized to 2 atm (gauge), $P_1 = 0.1$ atm, and $P_2 = (2 + 1)$ atm (absolute), and $Z = 3/0.1 = 30$.

4.2.3 *Sampling with Concentration*

A dilute odorant may have to be concentrated before it can be adequately characterized by chemical analysis. Such circumstances are likely to arise when a community malodor is to be traced to one or more possible sources. The ratio of concentrations between the odor source and the odorant outdoors where it constitutes a nuisance may be in the range of 10^3 to 10^5. Under such circumstances the comparison of the odorant with the alleged source is greatly facilitated if the concentration is increased to approximate that of the source.

The most widely used methods of concentration are freezeout trapping and adsorptive sampling. Chapter 8 elaborates on both these methods.

When the concentration ratio must be high, the accumulation of water in a cold trap is a serious drawback, and adsorptive sampling is usually preferred. Activated carbon has been used, but it is so retentive to organic vapors that recovery of the sample often requires special methods. Various porous polymers have been found to be both effective and convenient, especially when the concentrated sample is to be injected into a gas chromatograph.

4.3 Sensory Evaluation

4.3.1 *General Conditions*

Sensory testing requires attention and high motivation on the part of human judges. Therefore, outside interferences such as noise, extraneous odors, and any other environmental distractions or discomforts must be kept to a minimum. Odor control is generally achieved by use of air conditioning combined with activated carbon adsorbers. The testing room should be under slight positive pressure to prevent infiltration of odors. All materials and equipment inside the room should be either odor free or have a low odor level. If highly odorous products are to be examined or high humidities are anticipated, partitions may be sprayed with an odorless, strippable, soft-colored coating that can be replaced if it becomes contaminated.

The panel moderator must exert every precaution to avoid bias with respect to any of the tests. Ideally, the moderator, like the judges, should be unaware of the identity of the samples, so that the test is double blind.

4.3.2 Selection and Training of Judges

The general criteria for selection of judges, as outlined in an ASTM manual (45), involve the individuals' natural sensitivity to odors, their motivation, and their ability to work in a test situation. Wittes and Turk (46) have pointed out that three variables characterize the efficacy of any screening procedure: (*1*) the cost, as determined by the number of sensory tests; (*2*) the proportion of potentially suitable candidates rejected by the screen; and (*3*) the proportion of potentially unsuitable candidates accepted by the screen. These variables are functionally dependent; that is, specification of any two determines the value of the third. Therefore a screening procedure can be designed to favor any two of these variables, but not all three.

It is generally best, especially with novices, to start with familiar substances such as food flavors. The transition to malodors can come later. Screening tests are based on the ability of a judge to identify the odd sample in a group of three odors of which two are the same (triangle test), or to order samples in a sequence of increasing intensity, or to identify the components of a mixture. Details are given by Wittes and Turk (46).

After a group of judges has been selected, they must be trained. The screening tests should be repeated, and this time any errors should be discussed and analyzed. The judges are then introduced to the types of sensory tests they will carry out. If possible, the original measurements should be made on known standards.

4.4 Odor Inventories and Surveys

First, we define odor inventories as assessments of odors at potential sources or at the origins of the odorous emission, and odor surveys as assessments of odors at the impacted area. As shown in Section 3.3.4, odor emission rate is the product of the dilution-to-threshold (Z or D/T) and the air flow rate and is a convenient means of quantifying the potential of an odor source that may impact the downwind locations. In general, the higher the odor emission rate, the more severe the odor problem potential. For a multisources situation, odor emission rates from different sources are additive if the odor qualities of the odors from different sources are identical or similar. For multisources with different odor qualities, odor emission rates are not always additive because they may differ in their objectionability and in the exponents of their psychophysical functions, which determines their pervasiveness. For example, in a coating operation, odor emission rates of the spray booth stacks should not be combined with the odor emission rates of drying oven stacks.

An approach to odor surveying that distinguishes among sources of different qualities was used as early as the 1950s in Louisville, Kentucky (47). In that study a kit of 14 reference odorants was chosen to "represent the principal odors expected to occur" in the area being surveyed. Some of the reference standards were actual samples of presumed odorant sources, such as a creosote mixture used as a wood preservative in an operation that was known to generate an odor nuisance. In the actual survey, the observers reported daily the nature (by identification with the reference samples), the intensity (category scale), and the location of the odors they smelled. In later studies odor reference standards have been used in conjunction with improved methods of intensity scaling to yield quality-intensity inventories of odor sources of various kinds (48).

Assessments of odor intensity and quality, however, do not serve as reliable predictors of human affective responses to unpleasant odors. There have been two approaches to direct surveying of such reactions. In one approach, developed by Springer and co-workers (42) for diesel exhaust odors, the odor source to be surveyed is set up in a transportable facility, under controlled conditions of dilution and presentation to judges. Responses are elicited from a large number of randomly selected, untrained people. For the diesel odor, the exhaust from an engine was diluted and piped into compartments in a trailer for sensory evaluations. The source-trailer combination was then used in several sites in various cities in the United States, to constitute a national survey. The questionnaire submitted to each judge was designed around the cartoon scale of Figure 17.3. Note that the facial expressions, the bodily actions, and the descriptive adjectives are all mutually reinforcing.

A second approach to measuring human affective responses to odors is a survey of annoyance reactions. Jonsson (49) has reviewed human reactions studied in different odor surveys and concludes that responses to questions about annoyance are more reliable than other indexes such as willingness to sign a petition or to take direct action to modify the environment (e.g., by using household deodorants or installing air conditioning). A detailed survey procedure is described in "A Study of the Social and Economic Impact of Odors,

Figure 17.3. Cartoon scale for odor testing.

Phase II," also known as the Copley Report (50). An important feature of this procedure is the use of an odor-free area as a control to account for the fact that in some instances annoyance to odor is expressed when no odor source exists. Even in an odorless area, complainants may call attention to the odor problem, but the opinions are not generally typical of the majority of the community. Consumer and social research studies have found repeatedly that the likes and dislikes of persons who volunteer their opinions are different from the likes and dislikes of their neighbors who must be solicited for their opinions. The purpose of the survey, then, is to compare the attitudes of people residing in a community thought to be an odor problem area, with attitudes of similar people residing in an odor-free area. Attitudes of both groups are determined by conducting interviews by telephone with residents of both areas. The survey proceeds in the following sequence:

1. The first task is to define the possible odor problem area.
2. Next, a matching odor-free area is located.
3. Utilizing a street address (reverse order) telephone directory, a list of telephone numbers in each area is made, and a sample of these telephone numbers is selected at random.
4. Utilizing the questionnaire provided, telephone interviews are conducted with the adult occupant of the house for each telephone number included in the sample.
5. The total number of responses to key questions asked in both areas is tabulated and compared for problem identification.
6. If an odor problem is found, an odor problem index number is calculated.

The questionnaire itself explores the respondent's length of residence in the community, attitude to various categories of complaints about environmental problems, personal experiences of odor, degree of annoyance by odors, and opinion with regard to the origin of the odors; place of employment is also ascertained for all members of the household. The final calculated odor index expresses the degree to which it can be confidently stated that there is a statistically significant difference between the test area and the control area; such a difference is said to constitute an odor problem identification.

During the last 20 years, the use of atmospheric dispersion models in environmental impact studies has been improved. Practitioners often input dilution-to-threshold values, source air emission rates, meteorological conditions (wind speed, wind direction, stability class), as well as physical configurations of the odor source (stack height, stack diameter, flue temperature, etc.) and the surround structures, into dispersion models to study the odor impacts at the downwind locations. More discussion about the odor dispersion modeling can be found in Section 6.2.2.

5 SOCIAL AND ECONOMIC EFFECTS OF ODORS

5.1 Social Effects

For convenience in definition, social effects can be said to differ from economic effects in that the former cannot be measured directly in monetary terms (51). The social effects of odors include (*1*) interference with the everyday activities of the exposed individuals, (*2*) feelings of annoyance caused by offensive smell, (*3*) physical symptoms of physiological changes, (*4*) actual complaints to an authority, and (*5*) various forms of direct individual action to modify the environment other than through complaints.

All these effects are very difficult to quantify; for example, the average homeowner cannot say with assurance how many times an unpleasant odor has prevented him from using his yard to entertain friends. As described in the preceding section, the principal tool used to assess social effects has been the attitude survey, and the results yield only a statistical measure of the confidence that an odor problem is correctly identified, not a rating of the intensity of human reactions.

5.2 Economic Effects

Assuming that unpleasant odors annoy people, can we say that they also cause economic loss? If so, how can this loss be determined? Again the most direct approach is to survey the affected area by asking a representative sample of persons in the community how much they would be willing to pay to get rid of the odors. Though theoretically valid, this approach runs into the practical difficulty of separating what people say they would pay from what they actually will pay if asked to comply with their own responses.

A more fruitful approach lies in determining what people actually have paid to obtain an odor-free environment. This requires recognition that the economic impact of odor pollution is most likely to manifest itself in reduced property values, reduced productivity of industrial companies, and reduced sales in commercial areas.

Economic theory states that if odors are bothersome, people should be willing to pay more to live in an odor-free area. Thus two similar properties in similar neighborhoods should sell for different prices if one area is affected by odors and the other is not. Thus we are assuming that all economic losses due to the presence of odors are capitalized negatively into property values and that buyers need only know that they prefer some properties to others and be willing to pay more for them.

Odors may affect industrial property values in somewhat the same way they affect residential areas. However, another form of loss to commercial and industrial establishments is possible, namely, odors may reduce the productivity of employees because of induced illness or distraction from work assignments. Productivity losses are likely to be particularly noticeable if such odors are intermittent as well as strong. Such a situation would be offensive, while also tending to inhibit persons from becoming adapted to their work environment.

Commercial areas may suffer economic losses from odors in the form of a general loss of customers and reduced sales per customer. The one serious attempt to measure such economic effects was carried out in Los Angeles in 1969 as part of the Copley Report (50). Unfortunately, the attempt was not successful, either because the methods were not sensitive enough to isolate economic effects caused by odors, or because transitory odors are not capitalized into property values.

6 ODOR CONTROL METHODS

6.1 Process Change and Product Modification

The first step in controlling odorous emissions from chemical or other industrial operations should be a reconsideration of the process itself. Such review should include the process conditions, choice of raw materials, and perhaps even a modification of the product. Following is a suggested sequence.

Start with the actions that can be taken without delay and that can only be helpful. Many of these will be low in cost and may actually save money. Leaks in process equipment, spills that are not promptly dealt with, and uncontrolled transfers of materials are recurrent sources of fugitive emissions that can be readily minimized or eliminated. Upgrade the quality of valves, pumps, drainlines, and other potential sources of leaks. Set up a rigorous and monitored maintenance schedule for all such potential sources. Keep buckets of granular activated carbon in laboratory hoods and other trouble spots for instant deodorization of even minor odorous releases (52).

A second step which, like the first, involves only internal actions, would be a review of process conditions to minimize the creation of odorous byproducts. This is obviously not a trivial effort for a manufacturing facility, since it involves simultaneous monitoring of the effects of the process changes on the product, on the operating costs, and on the formation and release of unwanted odorants. The potential benefit may include an overall improvement in the efficiency of the operation.

A third step would be a reconsideration of the raw materials and other feedstocks used in the operations. Some of these changes, such as a switch from organic- to aqueous-based solvents, may be attractive, but of course must be evaluated as carefully as any other process modification. Also, the constancy of feedstocks purchased from suppliers should not be taken for granted, even if they always conform to specified properties. Set up a pilot program that mimics the full-scale manufacturing process with various substituted feedstocks, and that evaluates the resulting formation of malodorous byproducts.

Finally, what about changing the product? No such option can be considered lightly, but, in any case, it may be balanced against the cost of investment in odor abatement equipment.

6.2 Dispersal and Dilution Techniques

6.2.1 Ventilation Systems in Enclosed Spaces

A time-honored approach to controlling odors in enclosed spaces has been to exhaust the malodorous air to the outdoors. There are several fundamental limitations to such a remedy. First, the exhausted air may transfer its nuisance to the outdoors. Such instances are particularly troublesome in congested areas where people may be exposed to the exhaust, or where one building's exhaust becomes another's intake. Under some conditions such atmospheric short-circuiting may even occur between the vents of the same structure. Second, the exhausted air must be replaced, and if the makeup air requires heating or cooling, large consumptions of energy may be required. Moreover, the odorant concentration is not reduced to zero; rather, it approaches an equilibrium level in which generation and removal rates are equal, and $C_\infty = G/Q$, where C_∞ is the concentration of odorant at equilibrium (mg/m^3), G is the rate of generation of odor (mg/min), and Q is the ventilation rate (m^3/min). This approach occurs at an exponentially decreasing rate (53, 54). In addition, depending on the mixing characteristics of the space, there may be areas in which people are exposed to greater than average odorant concentrations. Finally, the perceived odor intensity does not decrease linearly with the decrease of odorant concentration (see Section 2), but more slowly, because the intensity exponents are typically less than unity.

6.2.2 *Validity of Outdoor Dispersion Models*

Some of the problems of ventilation described in the preceding section have their counterparts in outdoor dispersal methods. Thus, exhausted ventilation air must be replaced, often with heating or cooling; and the intensities of outdoor odors are not reduced to match the physical dilution. In addition, the use of conventional diffusion models to predict the extent of odor is not always successful. Early efforts (55) revealed extreme discrepancies in which odors were experienced at far greater distances than were predicted. Later approaches by Hogstrom (56) established a model to predict the frequency of occurrence, as a function of distance from the source, of instantaneous concentrations equal to or above an absolute odor threshold level. Hogstrom reports fairly good agreements between observed and predicted frequencies up to several kilometers from the source, but at distances between 5 and 20 km the observed frequencies are larger than those calculated by a factor of 2 or 3. The reasons for these discrepancies are not well understood, but they may involve failure of the dispersion model to account for peak concentrations of short duration, irregularities in threshold responses, or participation of particulate matter. These considerations are discussed in a review by Turk and Shapira (57).

6.3 Adsorption Systems

6.3.1 *General Principles*

Any gas or vapor will adhere to some degree to any solid surface. This phenomenon is called adsorption. When adsorbed matter condenses in the submicroscopic pores of an adsorbent, the phenomenon is called capillary condensation. Adsorption is useful in odor control because it is a means of concentrating gaseous odorants, thus facilitating their disposal, their recovery, or their conversion to innocuous or valuable products. When an odorous airstream is passed through a fresh adsorbent bed, almost all the odorant molecules that reach the surface are adsorbed, and desorption is very slow. Furthermore, if the bed consists of closely packed granules, the distance the molecules must travel to reach some point on the surface is small, and the transfer rate is therefore high. In practice, the half-life of airborne molecules streaming through a packed adsorbent bed is of the order of hundredths of a second, which means that very high removal efficiencies can be realized with thin beds of fresh activated carbon, as in gas masks.

An adsorbent bed becomes progressively loaded with adsorbate, and thus exhausted, in the direction of the air or other gas flowing through it. In Figure 17.4 the lower solid curve represents the reduction of odorant concentration when a gas stream passes through a fresh adsorbent bed. The critical bed depth, l_c, of the fresh adsorbent is the minimum required

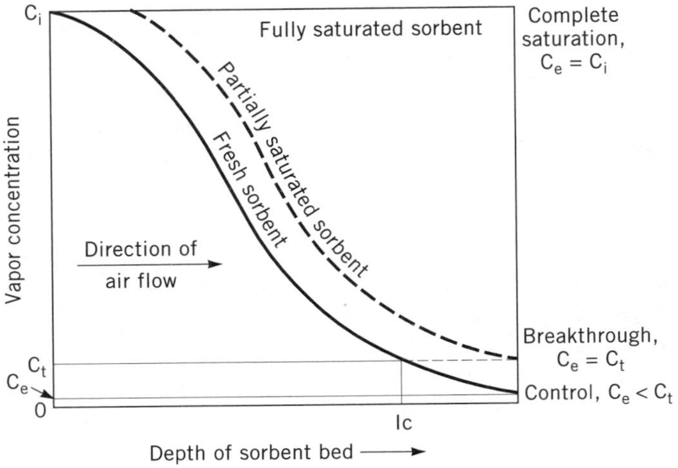

Figure 17.4. The adsorption wave.

to reduce the odorant concentration to a target, or threshold, level, C_t. Breakthrough occurs when the effluent reaches that level, as exhibited by the upper dashed curve.

When the adsorbent is exhausted, it may be discarded or separated from its adsorbed odorants and recovered for reuse. The odorants, too, can be recovered if they are valuable.

6.3.2 Adsorbents

The most widely used adsorbent for general-purpose odor removal is activated carbon. The reason for this preference is that most odorants that must be removed from indoor air or from effluents to the outdoors are organic substances that occur as byproducts or wastes in commerce and industry, or as products of putrefaction or decomposition. These substances are, for the most part, significantly higher in molecular weight than the normal components of air (including CO_2 at MW 44) and insoluble or only slightly soluble in water. Activated carbon, consisting as it does largely of only carbon atoms, presents an essentially nonpolar surface for adsorption that rejects water molecules in favor of the much less polar organic ones. It is for this reason that activated carbon can remove odorants from moist air streams, in contrast to oxygenated adsorbents such as activated alumina or silica gel, which preferentially retain water. There are, however, important exceptions, such as hydrogen sulfide, methyl mercaptan, ammonia, trimethylamine, formaldehyde, phosphine, and arsine; for none of these is physical adsorption an optimal choice. One alternative that has been used, especially for removal of formaldehyde, is activated alumina impregnated with potassium permanganate, which serves as an oxidant, converting formaldehyde largely to CO_2. The advantage of oxidation over adsorption is that it is irreversible in air, whereas an adsorbed gas can be desorbed when the temperature rises or when it is displaced by competing gases or vapors. However, activated carbon can be impregnated with various chemicals that serve as catalysts for air oxidation of some odorants, or as chemical reagents for removal of specific gases.

The pore size distributions of activated carbons are important determinants of their adsorptive properties. Pores less than about 20 Å (2 nm) in diameter are generally designated as micropores; those between 20 and 500 Å (2 to 50 nm) are called mesopores, and larger ones are macropores (58). The distinction is important because most molecules of concern in air pollution range in diameter from about 4 to about 8.5 or 9 Å. If the pores are not much larger than twice the molecular diameter, opposite-wall effects play an important role in the adsorption process by facilitating capillary condensation. Maximum adsorption capacity is determined by the liquid packing that can occur in such small pores.

6.3.3 Equipment and Systems

When odorant concentrations are low, thin-bed adsorbers often provide a useful service life while offering the advantage of low resistance to airflow. Flat or pleated bed shapes are retained by perforated metal or plastic sheets (Figure 17.5) and typically handle about 1000 to 2000 cfm.

Thick-bed adsorbers are used when large capacity is needed and sometimes when on-site regeneration is used. Bed depths are usually in the range of 1 to 3 ft (0.3 to 0.9 m). Linear airflows through such beds are typically about 50 ft/min (15 m/min). Figure 17.6

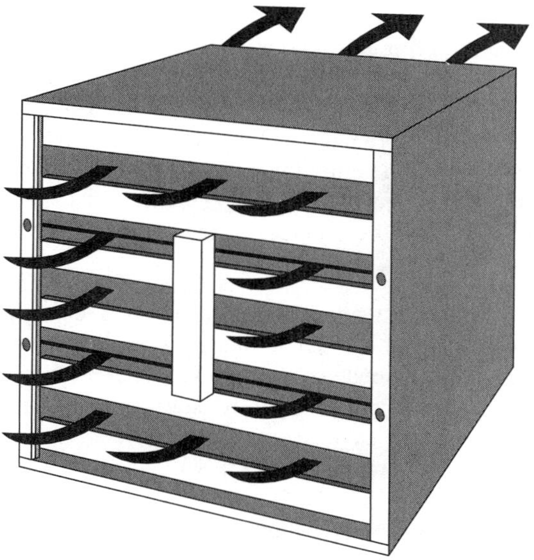

Figure 17.5. Thin-bed activated carbon adsorption cell. The small test element affixed to the inlet side of the cell contains carbon to be analyzed for degree of exhaustion after some time of service, thereby allowing estimation of the remaining capacity of the cell.

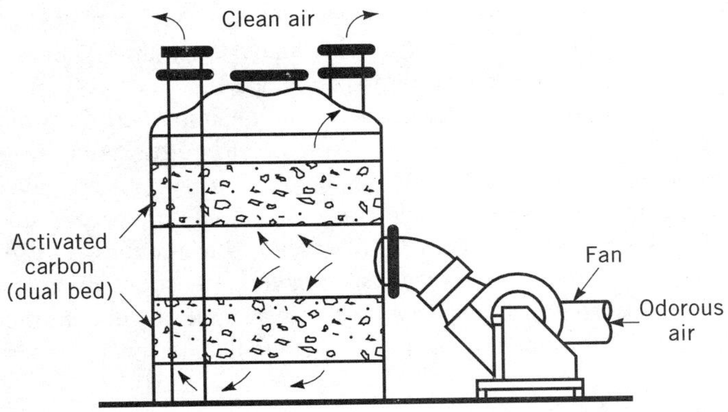

Figure 17.6. Dual-bed activated carbon adsorber.

shows a dual-bed adsorber, typically 12 ft (3.7 m) in diameter, with 3 ft (0.9 m) thick carbon beds, such as is often used for odor control in sewage treatment plants. Solvent recovery systems, which require on-site regeneration, necessarily involve multiple adsorbers (usually two but sometimes four), so that adsorption and regeneration cycles can be exchanged between beds. Other mechanical systems, such as fluidized, rotating, and falling bed adsorbers, have been described but are not common.

The service life of the carbon is limited by its capacity and by the contaminating load. Provisions must therefore be made for determining when the carbon is saturated and for renewing it. When the carbon is used for odor control, it is effectively exhausted when the odor intensity of the effluent air becomes unacceptable. There are several approaches to estimating the degree of exhaustion of the carbon, but such information does not translate directly to an assessment of its remaining life, because there is no guarantee that the concentration of the odorant challenging the carbon will be constant. For thick beds provided with multiple sampling ports (Figure 17.6) simultaneous analyses at several points can provide a helpful picture of the progress of exhaustion. Following are the available analytical methods:

1. Change in dry density. As the carbon progresses toward exhaustion, its density increases with the adsorption of organic matter and cycles with the gain and loss of moisture. A reasonable estimate of the carbon's decreasing capacity for adsorption of organic odorants can be obtained by measuring its increase in dry density. This value can be approximated by measuring the density (59) of the carbon after it is dried at a relatively low temperature (90 to 95°C) so as not to remove too much of the organic adsorbate.
2. Loss of mass on heating (60). A sample of the carbon is heated in a muffle furnace to 950°C for 7 minutes. Its volatile content is calculated from this loss of mass less its moisture content as determined from a separate sample.
3. Heat of immersion (61). Since adsorption is exothermic, the temperature rise of a sample of carbon on exposure to an organic adsorbate is an approximate indication

of its residual capacity. The measurement is made by inserting the carbon into a simple device consisting of a standard bottle containing a thermometer and an organic liquid such as hexane.

In many cases a schedule for renewal of adsorbent is determined by actual deterioration of performance (odor breakthrough), or it may be based on a time schedule calculated from previous performance history.

When the adsorbent is saturated it must be removed and replaced, or regenerated in place. Thin-bed adsorbers, which are used for light odorant loads and thus are expected to have long service lives, are normally replaced when they are exhausted. For thick-bed adsorbers and heavy contaminant loads of valuable solvents, it is generally economical to regenerate the adsorbent by on-site stripping with superheated steam with recovery of the desorbed matter.

Activated carbon may be impregnated to enhance its capacity for some malodorants that are too low-boiling to be economically removed by physical adsorption. For example, carbon impregnated with a strong base such as NaOH or KOH, or injected with ammonia gas (62) is used to remove even mildly acidic odorous gases such as hydrogen sulfide and methyl mercaptan. Phosphoric acid impregnation is used for enhanced removal of basic light gases such as mono- and trimethyl amine. However, impregnants for such light malodorants necessarily occupy some of the surface area and pore volume that offer capacity for physical adsorption of heavier malodorous gases that are also generally present, and therefore impregnations for odor control are often a mixed blessing.

In general, activated carbon adsorption is the method of choice for deodorizing at ambient temperature an odorous airstream whose vapor concentrations are low (ppm range or below). At higher temperatures and concentrations other methods, as described in the following sections, become progressively more attractive, and the choice of activated carbon usually must be justified by some additional benefit such as recovery of a valuable solvent. When a less efficient but cheaper method can serve to remove the bulk of contaminant organic matter from an airstream, an activated carbon adsorbent may be used as a final polishing stage to advance the cleanup to a condition of complete deodorization.

For low emission rates of malodorous air streams, earth filters, known as *biofilters* (63), may be suitable. This technology utilizes a filter bed based on materials such as peat, soil, leaf compost, or wood chips, which provides a physical environment for microorganisms and renders the entire medium biologically active. Odorous organic compounds, as well as hydrogen sulfide and ammonia, can serve as nutrients and be oxidized to odorless or nearly odorless products. The bed must be continuously moist, and the temperature moderate. The pH, which is critical for the organisms and odorants involved, is generally neutral to slightly alkaline, but some organisms thrive in acidic media. The physical condition of the bed should be monitored to maintain reasonably high void fractions and reasonably low pressure drop. Under ideal conditions, a biofilter maintains itself with little required attention.

6.4 Oxidation by Air at Elevated Temperature

The complete oxidation of most odorants in air results in deodorization. Some final products are odorless (H_2O, CO_2), and others (SO_2, SO_3, NO, NO_2) have much higher odor thresholds than their precursors. When malodorous waste gases containing halogens are oxidized, however, the products may include the free halogens, the halogen acids, or other toxic halogen compounds such as phosgene. All such substances must be removed by scrubbing before the gas stream is discharged to the atmosphere.

In addition, partial oxidation of hydrocarbons and oxygenates often yields intermediate products that are more highly odorous than their precursors. Unsaturated aldehydes or ketones, and unsaturated carboxylic acids, all having highly pungent odors, are frequently encountered.

The system to be used depends on the reactivity of the contaminants with oxygen, their heat content, and the concentration of oxygen in the gas stream. Many malodorous vapors, especially those formed in decomposition reactions, are relatively easy to oxidize. These include such odorants as rendering plant emissions, cooking vapors, and coffee roasting effluents. On the other hand, hydrocarbons, especially aromatic ones, are much more resistant to oxidation.

The heat content of the oxidizable vapors determines the temperature rise of the gas stream during oxidation. This rise, ΔT, can be conveniently expressed in terms of the lower explosive limit (LEL) of the vapor (64). Such a relationship is convenient because combustible gas meters are scaled directly in percentage of the LEL. The expression is

$$\Delta T(^{\circ}\mathrm{C}) = 16 \times (\text{percentage of the LEL})$$

The choice of mode of operation depends on the various factors outlined previously. In flame incineration, air and fuel are used to sustain a flame. This is, in effect, an enclosed flare, and it operates at temperatures of 2500°F (1371°C) or higher. The fuel cost is so high that the method is practical only when the combustible gas concentration exceeds 50 percent of the LEL, thus contributing at least about 1500°F (816°C) temperature rise. Such high vapor loadings often suggest that solvent recovery by activated carbon may be the better choice.

In thermal incineration the operating temperature is about 1200 to 1400°F (649 to 760°C); the half-life of reactive odorants is about 0.1 sec, and that of the more stable hydrocarbons is slightly longer. Consequently, a detention time of about 0.5 sec is sufficient for most odor control objectives. Figure 17.7 schematically represents a thermal incinerator.

Catalytic incineration is designed to give performance like that of a thermal incinerator but at a lower temperature and faster detention time. The advantages result from the action of a solid catalyst that consists of a noble metal alloy or in some cases a metallic oxide mixture. Operating temperatures are typically in the 600 to 900°F (316 to 482°C) range. The loss of catalyst activity, which determines catalyst life, and hence equipment maintenance costs, is related to three major factors: (*1*) the presence of catalyst poisons (such as metallic or organometallic vapors or sulfur compounds) in the odorous air, (*2*) the obstruction of catalyst surface by deposit of inorganic materials such as silicates from silicone resins, and (*3*) the mechanical loss of catalyst through abrasion by solid particles in the air stream.

For air free from particles and metal-containing vapors, a long catalyst life may be realized, and some installations have been reported (65) to give more than 23,000 hr of

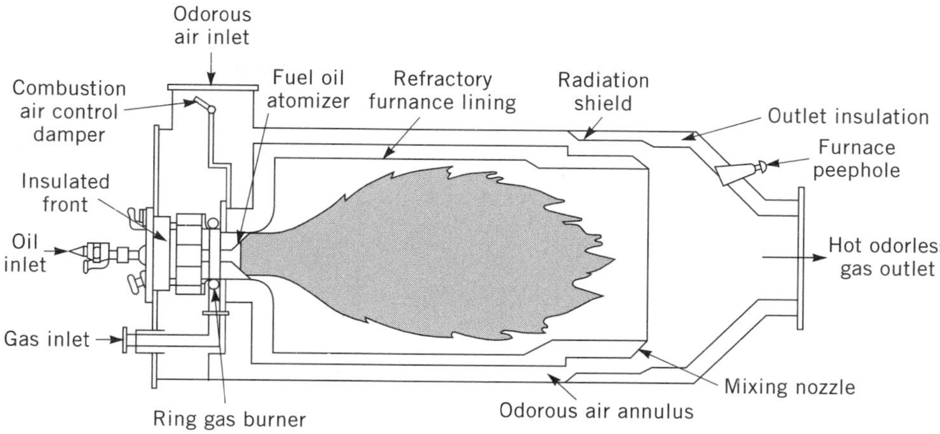

Figure 17.7. Direct-fired air heater. Schematic courtesy of Peabody Engineering Co.

service without catalyst regeneration. In other cases, however, loss of activity may be quite rapid. A pilot run before installing full-scale equipment is generally advisable.

6.5 Liquid Scrubbing

Liquid scrubbing is widely used for odor control. The mechanisms for its action include (*1*) solution of the odorous vapors into the scrubber liquid, (*2*) condensation of odorous vapors by the cooling action of the liquid, (*3*) chemical reaction of the odorants with the scrubber liquid to yield an innocuous product, and in some cases, (*4*) adsorption of odorant onto particles suspended in the scrubber liquid.

The physical actions of solution and condensation generally approach equilibrium conditions that still involve a significant partial pressure of the odorant vapors. Therefore such actions are only partially effective in deodorizing gas streams. For example, the water scrubbing of a gas stream containing ammonia and other nitrogenous odorants will remove much of the ammonia but only a small portion of some of the organic nitrogen compounds, which may be extremely odorous.

Chemical conversions in scrubbers are generally oxidations or acid-base neutralizations. Because the latter category involves very rapid proton exchange, the important determinant of effective action is the equilibrium condition. Thus soluble acidic odorants such as hydrogen sulfide and phenol are effectively scrubbed by basic solutions, and basic odorants such as ammonia and soluble amines can be neutralized by acids.

Reagents for chemical oxidation include potassium permanganate, sodium hypochlorite, chlorine dioxide, and hydrogen peroxide. In general, such oxidations are much slower than flame reactions and require considerably more residence time for effective odor control. Furthermore, the absolute rate of vapor removal is virtually independent of the chemical nature of the odorants in the case of physical adsorption, moderately dependent in the case of incineration, and extremely dependent in the case of ambient temperature oxidation. These differences are significant because a ratio of, say, 5:1 in the rates of removal of two vapors has only marginal significance when dealing with hundredths of seconds in a granular bed; it has more significance in the ranges of tenths of seconds in a flame; and it can well be of overriding importance in the much slower, ambient-temperature reactive systems.

The various oxidants cited earlier differ in their reaction pathways for different odorants, and it is meaningless to specify which one is best. Potassium permanganate (66) is generally used under mildly alkaline conditions (pH $\sim$ 10), where its reduction product is the insoluble manganese dioxide (MnO_2), which poses a waste disposal problem. However the MnO_2 slurry acts as an oxidation catalyst, and thus makes it possible for air to participate in the oxidation of some very reactive odorants, such as mercaptans and some amines. Sodium hypochlorite has been found to react more rapidly with rendering plant emissions and has been recommended for such applications (67).

Figure 17.8 represents a typical scrubber installation for control of rendering plant odors. The malodorous gases enter the first stage of the system, where they pass through

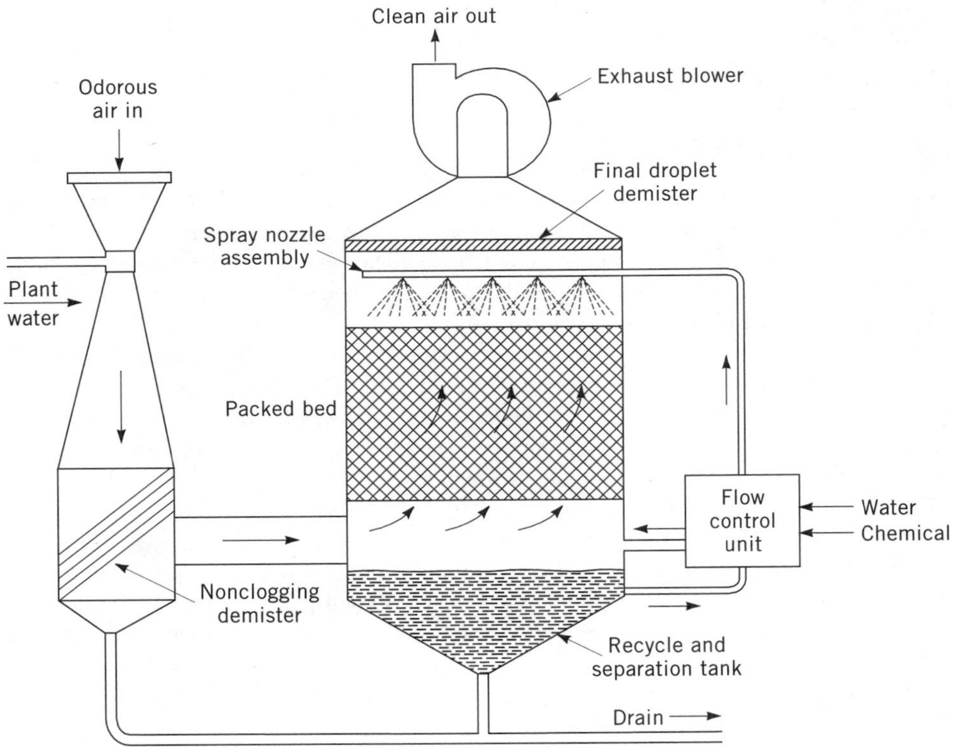

Figure 17.8. Two-stage chemical scrubbing system. Schematic courtesy of Environmental Research Corp., St. Paul, Minnesota.

water in a venturi scrubber. This step removes particulate matter and cools and saturates the gas stream. The gases are then passed through a packed bed, where they contact a countercurrent stream of scrubbing liquid containing a hypochlorite or other oxidant. Malodorous gases are absorbed and oxidized. The scrubbed gas stream leaves the packed bed, flows through a mist elimination section, and is exhausted to the atmosphere. The depleted scrubbing liquid is collected and recycled to the scrubber. A portion of the depleted scrubbing solution is continuously removed from the recycle stream and replaced with makeup water and chemicals. The bleed stream is combined with the wastewater from the venturi scrubber and sent to a sewage treatment facility.

6.6 Ozonation

Ozone is a reactive ambient-temperature oxidizing agent that has been used for gas-phase conversion of malodorants to less offensive products. The toxicity of ozone renders it unfit for use in occupied spaces, however. Before they were prohibited by regulations, ozone-producing devices were offered for indoor use, but they generate such low concentrations that their effect in controlling malodorants is nil and the injury they cause is probably too little to be evident.

For controlling odorants before they are discharged to the outdoors, ozone is introduced into the odorous airstream in concentrations of 10 to 30 ppm, and a reaction time of 5 sec or more is provided during passage of the stream through a stack or special detention chamber. Ozonation is chemically selective and is not equivalent to thermal incineration or catalytic oxidation, which can convert malodorants to their ultimate oxidation products. Ozone reacts with mercaptans to produce sulfones and sulfoxides, with amines to produce amine oxides, and with unsaturated hydrocarbons to produce aldehydes, ketones, and carboxylic acids. Mercaptans and amines are so highly odorous that even any partial oxidation is a great improvement. For unsaturated hydrocarbons or oxygenates, however, odor control performance is less reliable. For example, ozonation of a mixture of styrene and vinyl toluene from resin operations yielded a mixture of aromatic aldehydes with a distinct cherrylike odor, and attempts to deodorize acrylic esters by ozonation during the detention time available in typical industrial stacks have been unsuccessful (68).

6.7 Odor Masking and Counteraction

When a mixture of odorants is smelled, the odor qualities of the components may be perceived separately or may blend into one quality such that the individual components cannot be recognized. The odor mixture may be perceived as stronger than, equal to, or less than the sum of the odor intensities of the components. Likewise, the odor intensity of any single component of such a mixture may be stronger than, equal to, or less than the odor intensity of that component smelled alone. When referring to the mixture as a whole, these effects are designated hyperaddition, complete addition, and hypoaddition, respectively. When referring to any individual component, the effects are called synergism, independence, and antagonism, respectively.

Interaction effects on odor intensity have been studied for some two-component mixtures (69). For example, the perceived intensity of vapor-phase mixtures of various concentrations of pyridine and a second component such as linalyl acetate, linalool, or lavandin oil is less than the sum of the perceived intensities of the two components smelled alone. The addition of the second component to a relatively weak stimulus of pyridine causes an increase in overall odor intensity, but the addition of the same amount of the second component to a relatively intense stimulus of pyridine causes a reduction in overall odor intensity. Mixtures of 1-propanol and *n*-amyl butyrate have been reported to interact similarly (70). These data suggest the existence of complex interactions in the perceived intensity of odorous mixtures.

When mixtures of odor components are perceived as a single blend, a vector summation model of odor interaction has been suggested as a means of predicting the odor intensity of mixtures of malodorants such as dimethyl disulfide, dimethyl monosulfide, hydrogen sulfide, methyl mercaptan, and pyridine (71–73). For components equal in perceived intensity when smelled alone, a direct proportionality has been reported between odor intensity of the mixtures and the arithmetic sum of the odor intensities of the components.

The interpretation of the application of these phenomena to practical odor control objectives presents difficulties, and the common industrial terminology does not make matters easier. Counteraction has been used to connote reduction of intensity, although it is not always clear whether this refers to the blend or to the malodorant alone. Cancellation means reduction to zero intensity, a phenomenon that has never been convincingly documented. Masking refers to a change in odor quality that makes the malodorant unrecognizable; the connotation of concealment has made the term unpopular. In spite of this variety of terminology, the odor control practices to which the words refer are operationally indistinguishable. The materials used are selected from industrially available high intensity odorants, often from by-product sources. They may be applied in undiluted form or as an aqueous emulsion. They may be incorporated into the process or product that constitutes the malodorous source, sprayed into a stack or over a stack exit, or vaporized over a large outdoor area.

The general method has the important practical advantages of low initial equipment costs, negligible space requirements, and greater freedom from the necessity of confining the atmosphere into a closed space for treatment. It is not applicable when irritation or toxicity accompany odor.

Clearly it is very difficult to estimate the effectiveness of this category of odor control methods. Not the least of the problems is that of choosing a criterion for evaluation. Furthermore, industrial or commercial installations are not designed to be controlled experiments. Instead, these methods are generally combined with other beneficial actions, such as improvements in sanitation and general housekeeping, to maximize the opportunities for odor reduction. As a result, information concerning the performance of such systems consists entirely of descriptions of actual operations and other anecdotal reports.

6.8 Epilogue

Odor control has long been considered to be more or less equivalent to the reduction of gaseous emissions that happen to be odorous. Furthermore, if the criterion of odor control is the reduction of odorant concentration to the threshold level, the efficiency required of a control device can be calculated if a reliable threshold level is known. The ratio of high source concentrations to very low odor threshold levels may lead to rather unprecedented requirements for the performance of gas-cleaning devices, so the usual remedy is to count on atmospheric dispersal to help solve the problem. Then, in the schematic diagram of Figure 17.9, we assume an odorant concentration C_s at a source that is treated by an abatement device (dashed lines) of efficiency E_a to discharge the abated concentration C_a to the atmosphere. Atmospheric dispersal of efficiency E_d further reduces the odorant concentration to the target or threshold value C_t before it reaches ground level. Then,

$$E_a = \frac{C_s - C_a}{C_s}$$

$$E_d = \frac{C_a - C_t}{C_a}$$

and the overall efficiency, E, is

$$E = E_a + E_d - E_aE_d$$

Now, if we assume that the required overall efficiency can be measured by human judges exposed to source samples of progressively greater dilution, and that the atmospheric dilutions can be predicted by dispersion models, we can readily calculate the required design efficiency of the abatement equipment to be applied to the source.

We have pointed out the many deficiencies in this sequence, such as the inadequacies of dispersion models, the inconstancy of the odor threshold, the variability of the exponent of odor intensity functions, and the influence of odor quality on human reactions to odor. The technical and anecdotal literature offers many instances in which this strategy fails to predict human responses to unpleasant odors, in terms of either the distance at which odors can be detected or the affective reactions or overt social initiatives they elicit.

The recognition that odor control effectiveness is most truly manifested by its reduction of human annoyance has also, unfortunately, been of little help. Although such recognition

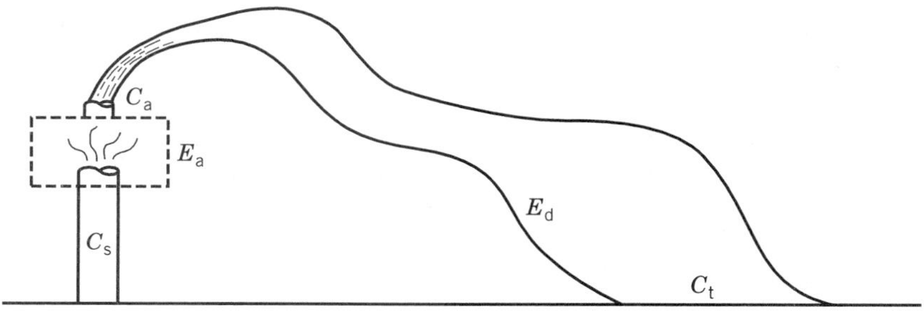

Figure 17.9. Overall efficiency of control of gaseous emission is $E = E_a + E_d - E_aE_d$, where E_a is the efficiency of a vapor control device and E_d is the efficiency of vapor dilution by atmospheric dispersal.

does allow for the consideration of odor masking and counteraction methods that other strategies do not accommodate, it fails to offer a basis for establishing design criteria for odor abatement systems. Furthermore, the measurement of economic effects of odors, such as the depression of property values or the reduction in work efficiency, has not been successfully accomplished in any precise manner.

On the other hand, methods of controlling gaseous emissions are being improved continually by means such as the design of more efficient scrubbers, the development of new catalysts, and advances in the effectiveness of systems for heat recovery and the regeneration of adsorbent beds. In addition, the continued refinement of instrumented monitoring systems serves all these abatement methods. The overall result has been that odor control strategies are determined in large measure by the available technology, which is often very good. Attempts to predict how the control of gaseous emissions reduces odors are described by two terms currently in vogue. The first is *dose-response relationships*, which purport to relate odorous emissions at the source to odor problems in the community or the workspace. However, we have seen that such relationships have not been quantitatively established. Instead, they are generally approximated in the give-and-take fashion that is typical of political and social processes in an open society—by persuasion, by negotiation, and sometimes by legal adjudication. The second popular term is *cost-benefit analysis*, which purports to predict how much odor control each dollar will buy. It should be obvious that human benefit from odor control is no less difficult to quantify than human response. However, it is much easier to compare costs of alternative abatement methods that yield the same reductions of odorant concentrations, and such comparisons are valid and important. The main findings of the latter approaches have been that as the costs of energy continue to increase, the search for more sophisticated approaches to odor control, starting with a full review of the role of the process itself, becomes ever more urgent.

BIBLIOGRAPHY

1. G. T. Fechner, *Elemente der Psychophysik*, Breitkopf and Harterl, Leipzig, 1860. English translation of Vol. 1 by H. E. Adler (D. H. Howes and E. G. Boring, Eds.), Holt, Rinehart and Winston, New York, 1966.
2. W. Cain, R. de Wijk, C. Lulejian, F. Schiet, and L. See, *Chem. Senses,* **23**, 309–326 (1998).
3. W. Cain, F. Schiet, M. Ilsson, and R. de Wijk, *Chem. Senses*, **20**, 625–637 (1995).
4. C. E. Shannon and W. Weaver, *The Mathematical Theory of Communication*, University of Illinois Press, Urbana, 1949.
5. J. A. Swets, W. P. Tanner, Jr., and T. G. Birdsall, *Psychol. Rev.*, **68**, 301 (1961).
6. T. Engen, "Psychophysics, 1. Discrimination and Detection," in J. W. Kling and L. A. Riggs, Eds., *Woodworth and Schlosberg's Experimental Psychology,* 3rd ed., Holt, Rinehart and Winston, New York, 1971, pp. 11–46.
7. F. A. Young and D. F. Adams, in *Proceedings of the 74th Annual Convention of the American Psychological Association*, New York, 1966, p. 75.
8. G. Steinmetz, G. T. Pryor, and H. Stone, *Percept. Psychophys.*, **6**, 142 (1969).
9. B. Johansson, B. Drake, B. Berggren, and K. Vallentin, *Lebensm.-Wiss. Technol.*, **6**, 115 (1973).

10. G. Leonardos, D. Kendall, and N. Barnard, *J. Air Pollut. Control Assoc.*, **19**, 91 (1969).
11. E. M. Adams, "Physiological Effects," in *Air Pollution Abatement Manual*, Manufacturing Chemists Association, Washington, DC, 1951, Chapter 5.
12. E. Fluck, *J. Air Pollut. Control Assoc.*, **26**, 795 (1976).
13. A. Turk and J. Turk, "The Purity of Odorant Substances," in D. G. Moulton, A. Turk, and J. W. Johnston, Eds., *Methods in Olfactory Research*, Academic Press, New York, 1975, pp. 63–73.
14. A. M. Hyman, *Sensory Processes*, **1**, 273 (1977).
15. T. Engen, "Psychophysics, II. Direct Scaling Methods," in J. W. Kling and L. A. Riggs, Eds., *Woodworth and Schlosberg's Experimental Psychology*, 3rd ed., Holt, Rinehart and Winston, New York, 1971, pp. 47–86.
16. S. S. Stevens, in G. Stevens, Ed., *Psychophysics: Introduction to Its Perceptual, Neural and Social Prospects*, Wiley, New York, 1975.
17. B. Bergland, U. Bergland, G. Ekman, and T. Engen, *Percept. Psychophys.*, **9**, 379 (1971).
18. M. J. Mitchell and R. A. M. Gregson, *J. Exp. Psychol.*, **89**, 314 (1971).
19. W. S. Cain, *Ann. N. Y. Acad. Sci.*, **237**, 28 (1974).
20. M. Laska, H. Distel, and R. Hudson, *Chem. Senses*, **22**, 447–456 (1997).
21. G. Steinmetz, G. T. Pryor, and H. Stone, *Percept. Psychophys.*, **8**, 327 (1970).
22. P. Dalton and C. Wysocki, *Percept. Psychophys.*, **58**(5), 781–792 (1996).
23. W. S. Cain, *Percept. Psychophys.*, **7**, 271 (1970).
24. W. S. Cain and T. Engen, "Olfactory Adaptation and the Scaling of Odor Intensity," in C. Pfaffmann, Ed., *Olfaction and Taste,* Vol. 3, Rockefeller University Press, New York, 1969, pp. 127–141.
25. E. P. Koster, *Adaptation and Cross-Adaptation in Olfaction*, Bronder-Offset, Rotterdam, 1971.
26. P. Dalton, *Chem. Senses*, **21**, 447–458 (1996).
27. L. Buck and R. Axel, *Cell*, **65**, 175–187 (1991).
28. S. Firestein, R. F. Margolskee, and S. Kinnamon, "Molecular Biology of Odor and Taste," in G. Siegel, Ed., *Basic Neurochemistry: Molecular, Cellular, and Medical Aspects*, 6th ed., Lippincott-Raven, 1999, pp. 985–1006.
29. K. Ressler, S. Sullivan, and L. Buck, *Cell*, **73**, 597–609 (1993).
30. Vasser, J. Nagai, and R. Axel, *Cell*, **74**, 309–318 (1993).
31. H. Breer, *Semin. Cell Biol.*, **5**, 25–32 (1994).
32. S. Firestein, G. Shepherd, and F. Werblin, *J. Physiol.*, **430**, 135–158 (1990).
33. T. Chen and K. Yau, *Nature*, **368**, 545–548 (1994).
34. W. R. Roderick, *J. Chem. Educ.*, **43**, 510 (1966).
35. A. T. Rossano and R. R. Ott, "The Relationship Between Odor and Particulate Matter in Diesel Exhaust," paper presented at the Pacific Northwest Section Meeting of the Air Pollution Control Association, Portland, Ore., November 5–6, 1964.
36. A. Goetz, *Int. J. Air Water Pollut.*, **4**, 168 (1961).
37. AIHA, *Odor Thresholds for Chemicals with Established Occupational Health Standards*, AIHA Press, Fairfax, VA, 1989.
38. A. Turk, *Atmosph. Environ.*, **7**, 967 (1973).
39. S. S. Schiffman, *Science*, **185**, 112 (1974).

40. K. R. Brower and R. Schafer, *J. Chem. Educ.*, **52**, 538 (1975).
41. American Society for Testing and Materials, *Reviews of Correlations of Objective-Subjective Methods in the Study of Odors and Taste*, Special Technical Publication No. 451, ASTM, Philadelphia, 1969.
42. A. Turk, S. Yousef, S. E. Ellis, J. H. Worthington, E. Miseo, T. Aciukewicz, T. Stolki, and R. Steeves, "The Organic Content of Odorous Sewage Gases," in C. H. McGinley and J. R. Swanson, Eds., *Odors: Indoor and Environmental Air*, Air and Waste Management Assn., Pittsburgh, PA, 1995, pp. 133–142.
43. C. Weurman, "Sampling in Airborne Odorant Analysis," in A. Turk, J. W. Johnston, and D. G. Moulton, Eds., *Human Responses to Environmental Odors*, Academic Press, New York, 1974, pp. 263–328.
44. K. Springer, "Combustion Odors," in A. Turk, J. W. Johnston, and D. G. Moulton, Eds., *Human Responses to Environmental Odors*, Academic Press, New York, 1974, pp. 227–262.
45. American Society for Testing and Materials, *Manual on Sensory Testing Methods*, Special Publication No. 434, ASTM, Philadelphia, 1968.
46. J. Wittes and A. Turk, "The Selection of Judges for Odor Discrimination Panels," in *Correlation of Subjective-Objective Methods in the Study of Odors and Taste*, American Society for Testing and Materials Special Publication No. 440, ASTM Philadelphia, 1968, pp. 49–70.
47. *The Air Over Louisville*, Technical Report of the Public Health Service, U.S. Department of Health, Education and Welfare, Washington, DC, 1956–1957.
48. A. Turk, J. T. Wittes, L. R. Reckner, and R. E. Squires, *Sensory Evaluation of Diesel Exhaust Odors*, Publication No. AP-60, National Air Pollution Control Administration, Washington, DC, 1970.
49. E. Jonsson, "Annoyance Reactions to Environmental Odors," in A. Turk, J. W. Johnston, and D. G. Moulton, Eds., *Human Responses to Environmental Odors*, Academic Press, New York, 1974, pp. 330–333.
50. Copley International Corp., *A Study of the Social and Economic Impact of Odors, Phase I, 1970; Phase II, 1971, Phase III, 1973*, Report of Contract No. 68-02-0095, U.S. Environmental Protection Agency, Washington, DC.
51. R. D. Flesh and A. Turk, "Social and Economic Effects of Odors," in P. N. Cheremisinoff and R. A. Young, Eds., *Industrial Odor Technology Assessment*, Ann Arbor Science Publishers, Ann Arbor, MI, 1975, pp. 57–74.
52. A. Turk, H. Karamitsos, K. Mahmood, J. Mozaffari, R. Loewi, and V. Tola, J. *Chem. Educ.*, **69**, 929–932 (1992).
53. A. Turk, *J. ASHRAE*, October, 1963.
54. A. Turk, "Concentrations of Odorous Vapors in Test Chambers," in *Basic Principles of Sensory Evaluation*, American Society for Testing and Materials Special Publication No. 433, ASTM, Philadelphia, 1968, pp. 79–83.
55. H. C. Wohlers, *Int. J. Air Water Pollut.*, **7**, 71 (1963).
56. U. Hogstrom, "Transport and Dispersal of Odors," in A. Turk, J. W. Johnston, and D. G. Moulton, Eds., *Human Responses to Environmental Odors*, Academic Press, New York, 1974, pp. 164–198.
57. R. K. Shapira and A. Turk, "Assessing Industrial Odors: Dispersion Models or Community Complaints," in *Environmental Quality and Ecosystem Stability*, Vol. IVA, Israel Soc. Ecology and Environ. Health Sci. Pub., Jerusalem, Israel, 1989, pp. 167–175.

58. American Society for Testing and Materials, *Standard Terminology Related to Activated Carbon*, D 2652, ASTM, Philadelphia, 1994.
59. American Society for Testing and Materials, *Standard Method for Apparent Density of Activated Carbon*, D 2852, ASTM, Philadelphia, 1996.
60. American Society for Testing and Materials, *Standard Test Method for Volatile Content of Activated Carbon Samples*, D 5832, ASTM, Philadelphia, 1998.
61. A. Bagreev, T. Bandosz, R. K. Shapira, and A. Turk, *Monitoring Carbon Saturation by Heat of Immersion*, Intl. Activated Carbon Conf., PACS, Inc., Coraopolis, PA, 1999.
62. A. Turk, E. Sakalis, J. Lessuck, H. Karamitsos, and O. Rago, *Environ. Sci. Technol.* **23**, 1242 (1989).
63. H. L. Bohn, *J. Air Pollut. Control Assoc.*, **25**, 953 (1975).
64. R. J. Ruff, "Catalytic Method of Measuring Hydrocarbon Concentrations in Industrial Exhaust Fumes," in *American Society for Testing and Materials Special Publication No. 164*, ASTM, Philadelphia, 1954, p. 13.
65. R. J. Ruff, *Am. Ind. Hyg. Assoc. Quart.*, **14**, 183 (1953).
66. H. S. Posselt and A. H. Reidies, *Ind. Eng. Chem. Prod. Res. Develop.*, **4**, 48 (1965).
67. T. R. Osag and G. B. Crane, *Control of Odors from Inedibles-Rendering Plants*, Publication No. 450/1-74-006, U.S. Environmental Protection Agency, Washington, DC, 1974.
68. A. Turk, personal communication.
69. W. S. Cain and M. Drexler, *Ann. N. Y. Acad. Sci.*, **237**, 427 (1974).
70. W. S. Cain, *Chem. Sens. Flav.*, **1**, 339 (1975).
71. B. Berglund, U. Berglund, and T. Lindvall, *Acta Psychol.*, **35**, 255 (1971).
72. B. Berglund, U. Berglund, T. Lindvall, and L. T. Svensson, *J. Exp. Psychol.*, **100**, 29 (1973).
73. B. Berglund, *Ann. N.Y. Acad. Sci.*, **237**, 35 (1974).

CHAPTER EIGHTEEN

Interpreting Levels of Exposures to Chemical Agents

Stephen M. Rappaport, Ph.D., CIH

1 INTRODUCTION

The past half-century has witnessed great advances in the characterization of exposures to toxic chemicals in the workplace. With the development of robust personal monitors (1), it is now a relatively simple matter to measure the daily air levels among various workers in an observational group (a group of workers sharing observable factors such as job, location, department, etc.). [Although Ref. 1 was published in 1960, personal sampling did not become widely applied until well into the 1970s. In fact, Symanski et al. observed that essentially all published exposure data prior to 1972 (from 696 data sets) involved either area or short-term breathing zone sampling (2, 3).] Likewise, statistical tools that assume lognormally distributed exposures (4, 5) open many avenues for characterizing air levels. And finally, recognition that exposures vary both within and between workers in observational groups (6–9) permits assessments of exposure, risk and control options to be integrated.

Unfortunately, the interpretation of levels of exposure has not achieved the same degree of sophistication. In fact, decisions still tend to rely upon one-to-one comparisons of the highest measured air concentration with occupational exposure limits (OELs), much as they did 50 years ago (see, for example, Refs. 10–14). If the highest air concentration is less than the OEL, then exposure is acceptable (in compliance) and vice versa. While such interpretations would be reasonable if occupational exposures were relatively constant, they seem remarkably naïve when considered in light of daily exposure levels which vary between 10 and 4000 fold. [Based upon estimated within- and between-worker variance components compiled by Kromhout et al. (15).]

Patty's Industrial Hygiene, Fifth Edition, Volume 1. Edited by Robert L. Harris.
ISBN 0-471-29756-9

1.1 Exposure as a Random Process

In order to interpret exposures to environmental contaminants it is helpful to consider exposure as a random process with numerous sources of variability. One can visualize many random variables that affect the magnitude of exposure at a particular time. These include location, source of contamination, type of control, temperature, wind speed and direction, tasks, practices, etc. The multiplicative interaction of these variables tends to produce the extraordinary ranges of exposures observed in occupational studies.

The scatter plot shown in Figure 18.1 represents a sample of occupational exposure data published about 20 years ago (16). These 177 personal measurements (daily average levels) of inorganic lead were obtained from six workers in a factory producing tetra-alkyl lead compounds. The range of air levels was 1.3–62.6 μg/m^3, about 50-fold, which is typical of continuous indoor exposures in major industries (15). When formed into a histogram, as depicted in Figure 18.2(**a**), these air levels show the characteristic skewness of exposure data, with generally small values punctuated by a few very large observations. Since such skewness is a property of the lognormal distribution, it is now common practice to treat occupational exposure data as lognormally distributed for statistical purposes (5, 9, 17). If air levels are lognormally distributed, then the logged values should be normally distributed and, therefore, appropriate for standard statistical models and hypothesis tests. Referring again to the lead data, after taking natural logarithms, the histogram assumes an approximate bell-shaped curve as shown in Figure 18.2(**b**).

1.2 Compliance Testing

As noted above, the traditional approach for interpreting levels of exposure involves direct comparisons of minimal numbers of measurements with an OEL. Although the numbers

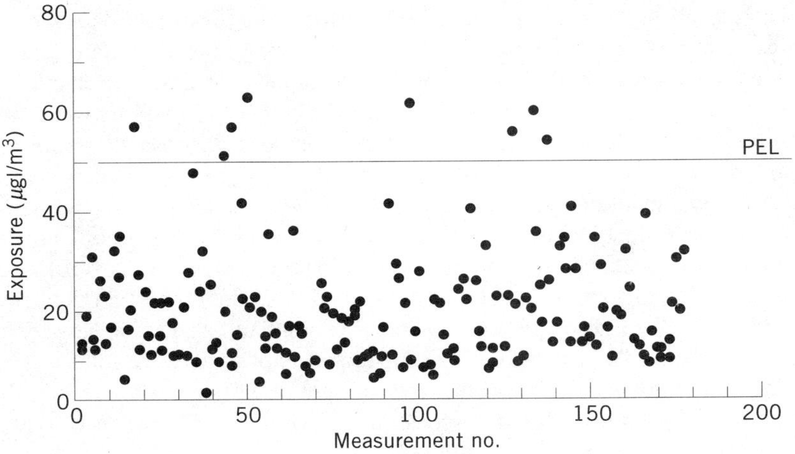

Figure 18.1. Scatter plot of exposures to inorganic lead (N = 177) for an observational group of workers in a tetra-alkyl lead manufacturing plant (Group 1). Data from Ref. 16.

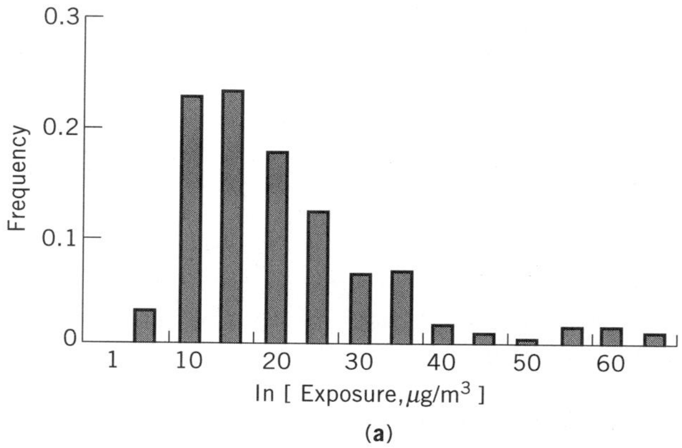

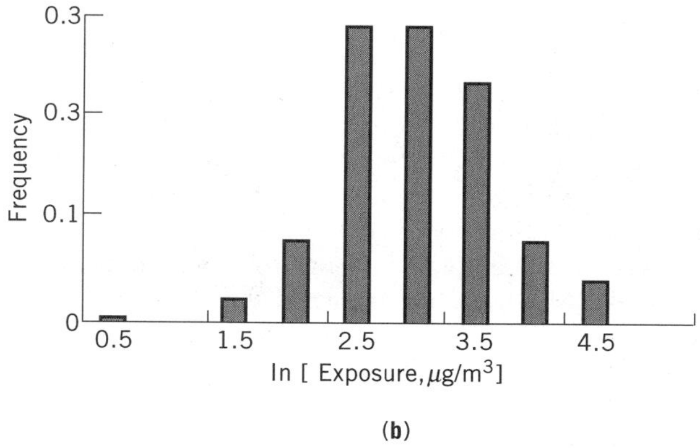

Figure 18.2. Histograms of exposures to inorganic lead depicted in Figure 18.1. (**a**) Exposure distribution showing characteristic skewness. (**b**) Distribution of the natural logarithms of the exposures.

of measurements obtained in industrial surveys are not generally available, a large study of data from the nickel industry indicates that only one or two are collected about ⅔ of the time (14). This practice will henceforth be termed compliance testing for obvious reasons. In some countries, including the United States, compliance testing can result in legal sanctions against the employer when evidence indicates that exposures exceed the OEL. However, even in situations where the legal basis for monitoring is unclear, the practice is common because of its strong historical precedence and seductive simplicity. Thus, compliance testing should be viewed as a general mechanism for assessing exposures that transcends the legal milieu.

The major problem with compliance testing is that it cannot deal effectively with exposure variability. It is known that exposures vary to such an extent that a typical person exposed at one tenth of the OEL on one day can be exposed above the OEL on the next, and that a pair of workers in the same job typically have average exposures differing by 4 fold. Referring again to Figure 18.1, a few of the daily measurements were above the operative OEL of 50 $\mu g/m^3$ [the permissible exposure limit (PEL) of the U.S. Occupational Safety and Health administration (OSHA)]. Indeed, measurements above OELs inevitably occur among workers in primary industries and the likelihood of documenting such an event depends critically upon the sample size (13). For example, if only one or two measurements had been obtained from the lead workers, it is likely that all observations would have been below the OEL (in compliance). However, because many measurements were performed, a few exceeded the OEL (out of compliance). Some of the vagaries of compliance testing have been explored (9, 13, 14) and will be addressed later in the chapter.

1.3 Variation of Exposures Within and Between Workers

The scatter plot illustrated in Figure 18.1 is difficult to interpret because it was compiled from repeated measurements from six different workers. As such it provides no insight into the effect that each worker might have had upon the exposure distribution. In Figure 18.3(**a**), the logged measurements are sorted by person and then plotted along with the worker-specific average values. Daily levels varied greatly *within workers* from day to day while the average levels varied only marginally *between workers* (among the six individual averages). These data can be compared with three other data sets, also depicted in Figure 18.3, representing exposures to benzene in the petroleum-refining industry [Figs. 18.3(**b**) and 18.3(**c**); data from Ref. 8] and to styrene in the reinforced-plastics industry [Fig. 18.3(**d**); data from Refs. 18, 19]. In aggregate, these datasets point to situations in which the between-worker variability was small [Figs. 18.3(**a**) and 18.3(**b**)], moderate [Fig. 18.3(**d**)] and large [Fig. 18.3(**c**)], consistent with results from much larger databases (14, 15).

The concept of between-worker variability is central to the assessment and control of exposure (9). If there were no between-worker variation, then variability would occur only within workers (i.e., from day to day), all workers would be exposed to the same level on average, and the group could be regarded as uniformly exposed (9, 20). [This designation of uniform exposure was developed to avoid semantic difficulties (9).] In such cases, workers are at equal risk to the effects of chronic exposure and engineering or administrative controls (which affect everyone more-or-less equally) would be required to reduce unacceptable levels. However, if substantial variation exists between workers, the distribution of risk across the population becomes heterogeneous and control options change markedly. Suppose, for example, that factors affecting the personal environments of a few workers produced exposures well in excess of those experienced by others in the same job. In such situations, the few highly exposed persons would bear the brunt of elevated health risk and might not benefit from engineering or administrative controls applied at the group level. Indeed, it could well be more effective to identify these persons so that the individual factors giving rise to the higher exposures might be understood and modified.

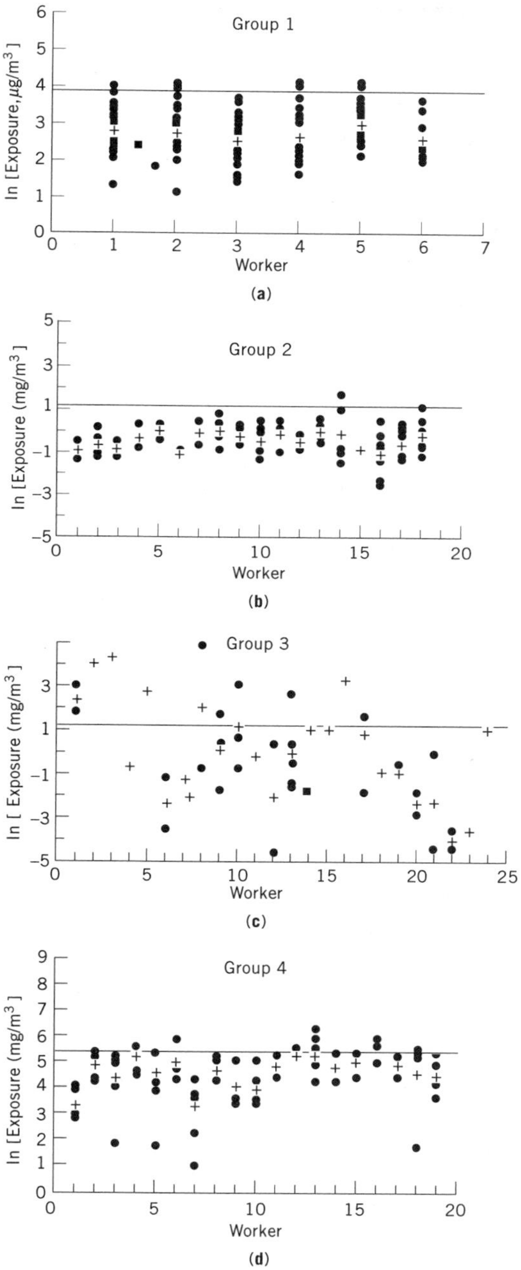

Figure 18.3. Scatter plots of logged exposures of four observational groups of workers defined in Table 18.1. Each point represents the daily exposure and each (+) represents the average exposure for a worker. Horizontal lines represent the logged PELs (at the time of monitoring).

1.4 Statistical Sampling Strategies

The variability inherent in exposure and the consequent pitfalls of compliance testing have motivated occupational hygienists to consider statistical strategies for assessing exposures. However, most strategies still assume that workers in each observational group are uniformly exposed. Indeed, this notion is so deeply ingrained in current practice that observational groups are often referred to as homogeneous-exposure groups (HEGs) (21).

Most sampling strategies have focused upon either the probability that a typical worker would be exposed above the OEL during a single work shift (designated as the exceedance) (10, 22–27) or the probability that the group mean exposure would be greater than the OEL (21, 27–33). Both of these approaches have improved the quality of exposure assessments, not only because they recognize the inherent variability in exposure, but also because they foster the collection of more data and of random sampling. However, the statistical methods currently used with these strategies can legitimately be applied only with single measurements from each of several persons, and as such cannot differentiate sources of variability within and between workers. [Replicate measurements from the same worker tend to be positively correlated and are, therefore, not independent as required for standard statistical tests.] It follows, therefore, that these strategies only lend themselves to interventions at the group level (i.e., engineering and administrative controls).

A sampling strategy recognizing between-worker variability has recently been proposed and provides methods suitable for samples containing repeated measurements from the same persons (34–36). Since this approach accepts that workers' average exposures can vary within a group, assessments relate directly to the associated risks of disease. Furthermore, by coupling interventions to the between-worker variability, this strategy integrates the assessment and control functions of occupational hygiene to a much greater extent than do those assuming uniform exposure. For example, a small amount of variability between workers would suggest the need for group-based (engineering or administrative) controls whereas a large amount would encourage individual-based interventions (based upon tasks, locations, equipment, practices, etc.).

1.5 Purpose of this Chapter

Because exposures must be evaluated in the face of great variability within and between workers in observational groups, statistical methods are central to proper assessment procedures. Likewise, since monitoring is motivated by, and must ultimately relate to, exposure limits, the impact of exposure variability on the interpretation of OELs is important. The purpose of this chapter is to integrate statistical methods and the philosophical bases for exposure limits within a conceptual framework for interpreting exposure data. The methods will be illustrated with the four sets of exposure data depicted in graphical form in Figure 18.3. The sources and characteristics of these measurements are summarized in Table 18.1 and all data are listed in Appendix A for those who wish to work through the various examples in the text.

Table 18.1. Descriptions of Four Observational Groups of Workers[a]

Group	Chemical	Industry	Job	N	k	Units	Fold Range
1	Inorganic lead	Alkyl-lead manufacturing	Unknown	177	6	$\mu g/m^3$	48.2
2	Benzene	Petroleum refining	Refining operator	90	18	mg/m^3	63.9
3	Benzene	Petroleum refining	Transfer/movement operator	48	24	mg/m^3	12,400
4	Styrene	Boat manufacturing	Sprayer/laminator	103	19	mg/m^3	141

[a] N represents the number of measurements; k represents the number of workers; fold range is the ratio of the highest to the lowest measured level.

2 SAMPLING OCCUPATIONAL EXPOSURES

2.1 Establishing Observational Groups

The first step in assessing exposures is to place workers into observational groups, using common factors related to the job. Such schemes probably originated with retrospective investigations of occupational disease based upon qualitative descriptors of the job and workplace. (Studies of pneumoconioses in the first half of the 20th century provide particularly relevant examples, e.g., see Refs. 37 and 38.) Over time, observational grouping became part of the fabric of occupational hygiene for prospective purposes, being codified by Corn and Esmen in the late 1970s (23, 39).

Since observational groups are assigned on the basis of inspection, the grouping process is open-ended in the sense that observation can be extended to an ever-expanding array of processes, environments and tasks. As such, this process can be sufficiently time consuming as to undercut the monitoring of exposures *per se*. That is, available resources can be devoted to qualitative characterization of factors thought to give rise to exposure rather than to quantitative exposure assessment. It is, therefore, recommended that observational grouping be restricted to job title and location in most cases, consistent with extensive applications of analysis-of-variance (ANOVA) methods (14, 15). If it is necessary to combine several job titles to include sufficient numbers of workers, grouping should generally be based upon location (e.g., room, building or department) so that all workers will share the same process-related and environmental factors.

2.2 Measurements of Exposure

After assigning workers into observational groups, sufficient data should be gathered to test exposures relative to limits. Each measurement should be obtained by personal sampling over the full work shift. (For practical purposes, the monitoring period should be at least 4 h to minimize statistical problems associated with combining measurements of different averaging times.) The goal is to obtain one or more random measurements from each member of a representative sample of workers in the observational group. (Repeated measurements from the same workers are necessary to obtain information about the within- and between-worker variance components in the group.) The preferred sampling design involves balanced data, where the same number of measurements is obtained from each worker (n measurements per worker) in the sample. However, the ideal of balanced data is difficult to achieve in practice and statistical methods can be applied to either balanced or unbalanced samples. In any event, sampling should be carried out over a sufficient period (one year is recommended) to cover the full range of operational and environmental conditions (9, 34, 40, 41). Sample sizes in the range of 10–20 measurements per group (i.e., two measurements from 5–10 workers) should generally be sufficient for initial assessments of exposure.

2.3 Random vs. Worst-Case Sampling

Prior to the development of personal sampling, it was necessary to restrict the sampling effort to a handful of short-term area or breathing-zone measurements during each survey.

(For insight into early applications see Refs. 37 and 42.) Because monitoring was so difficult in those days, occupational hygienists attempted to identify highly exposed individuals and to ascertain whether their exposures were in the acceptable range, thereby placing an upper bound on the exposure for the entire group. This bias towards high levels, referred to as worst-case sampling, became so deeply rooted in professional practice that it persisted after the development of personal monitors (10, 43), and is still encouraged in some quarters (32, 33). The practice probably continues because worst-case sampling is expedient within the confines of compliance testing.

Although worst-case sampling has some merit in the context of governmental inspections (14, 44), it should be resisted more generally for several reasons. First, there is little evidence that hygienists can consistently identify highly exposed workers solely on the basis of observation. [For example, occupational hygienists could predict jobs with high, medium and low exposures in some cases but not in others (6, 45). Also, governmental inspectors (who apparently targeted worst cases) reported higher levels on average than obtained otherwise (46, 47); however, the differences depended upon the contaminant being measured and the two distributions were highly overlapping.] Second, selection of the worst case depends upon the particular task(s) to be performed; and tasks are also subject to great variability because of differences in duration, process, environment, and worker (48–53). Third, the intentional biasing of results invalidates the use of statistical tools for assessment purposes (54). Finally, biased data provide no information concerning the population as a whole, thereby obscuring exposure-response relationships (47, 55) and invalidating risk assessments. For these reasons, it is highly recommended that random measurements be obtained to the extent feasible in a particular situation.

2.3.1 Random Selection of Workers and Days

All workers within a group should be eligible for monitoring. Each person should be assigned a number and the particular sample of k persons chosen randomly. If a worker refuses to cooperate or leaves the group, then another individual should be randomly selected.

After selecting workers, days should also be randomly sampled. This is important to avoid problems arising from situations in which measurements obtained from the same person on different days are autocorrelated (autocorrelation of exposures will be discussed later). If exposures are autocorrelated, then non-random selection of days (as, for example, collecting all measurements within one week or monitoring only on Monday of each week) could provide measurements from the same worker that are not statistically independent.

Irrespective of the way in which days are chosen for monitoring, it is recommended that all k workers in the sample be repeatedly monitored until n measurements have been obtained from each. If a particular worker is absent from the workplace on one day, then the same individual should be sought out on another occasion to complete the collection of data. It would also be useful to obtain information about tasks as the measurements are made, if this can be done relatively easily. (See Refs. 56, 57 for good examples in the rubber industry.) This information will be useful at later stages to determine the impact of tasks upon exposure, to regroup the workers, or to evaluate options for controls.

On any given day it is recommended that only one measurement be made per group because measurements collected simultaneously among members of the group can be correlated, thereby creating statistical problems. Such temporal correlation could arise from scheduling of activities or production. For example, by conducting repairs of the same piece of equipment on a given day several workers in a maintenance job could be exposed to similar levels. Several different groups can be monitored on the same day as long as only one measurement is obtained per group.

If occupational hygienists are routinely available on site, then the above recommendations about random sampling should not present major problems. However, in cases where hygiene staff is not routinely on site, serious consideration should be given to enlisting the assistance of other types of on-site personnel (e.g., safety engineers, nurses, supervisory staff, or specially-trained workers) to carry out the monitoring in consultation with the occupational hygienist. After an initial period of training in the use and calibration of sampling equipment, it should be a relatively simple matter for these individuals to contribute productively to the sampling effort. (See Ref. 58 for a good example.)

In some cases monitoring cannot be conducted on site routinely. Then, it becomes difficult to randomly select among workers and days and the only practical solution may well be for the occupational hygienist to make all measurements during a discrete campaign of a few days time. This approach can still lead to valid inferences if exposures are not highly autocorrelated, if exposures among persons in the group are not highly correlated on given days, and if the full range of activities giving rise to exposure was covered during the sampling campaign. However, occupational hygienists should recognize that campaign sampling casts doubts on exposure assessments.

2.4 Measurements Below the Limit of Detection

Occasionally, measurements are reported as less than the analytical limit of detection (LOD) for the method of measurement. Such undetected measurements are referred to as (left) censored values. It is not uncommon to encounter situations where more than 10% of the values are left censored. For example, among 44 benzene-exposed groups in the petroleum-refining industry, Spear et al. (8) found between 0 and 64% censoring with a median value of 12%. Replacing censored values with fixed values leads to biased estimates of the mean and variance of the distribution; e.g., substitution of the LOD for censored values overestimates of the mean exposure. Hornung and Reed (59) showed that replacing censored values with $\mathrm{LOD}/\sqrt{2}$ minimizes this bias for lognormally distributed exposures. More complex methods employ the assumed underlying distribution to impute values for the censored observations. These range from rather simple graphical procedures (60) to complex iterative methods (61, 62).

3 STATISTICAL CONSIDERATIONS

The object of a prospective sampling program is to allow inferences to be drawn concerning the degree of exposure of the individual or group over periods of months or years or, in the narrower context mentioned earlier, to determine whether exposures are acceptable relative to particular OELs. This section focuses on the statistical issues related to characterizing distributions of exposure.

3.1 Stationarity

The concept of stationarity will not be dealt with in detail. It implies that the statistical parameters of the underlying process that gives rise to exposure, namely, the mean, variance, and autocorrelation functions, do not change over the time period of interest. The work of Symanski and coworkers offers some insight into the stationarity of occupational exposures (2, 3, 41, 63). Regarding long-term exposures (years to decades), the authors identified pervasive trends, generally toward lower levels (78% of 696 datasets), with rates of reduction ranging between 4% and 14% per year (median 8% reduction per year) (2, 3). Several factors were found to influence the rate of reduction, including the industrial sector and the type of contaminant (gas or vapor). On the other hand, relatively little change in exposure was detected in surveys less than one year apart (significant changes were detected in 2 of 25 data sets collected within a year but in 10 of 28 datasets obtained over more than one year) (41) and most time series of consecutive daily measurements obtained from the same workers within one month appeared to be stationary (63).

Based upon these results, hygienists should view exposure assessment as an ongoing activity in which inferences about the levels of airborne chemicals are made periodically. Since most trends appear to be towards lower exposure, a finding of low exposure during one period (e.g., year) would suggest that even lower levels will follow; thus, it would be reasonable to reduce the sampling effort (numbers of measurements, etc.) in subsequent periods.

3.2 The One-Way Random Effects Model

In order to use relatively few measurements to make inferences about the underlying distributions of exposures, it is necessary to have a model with which to summarize the data and to develop appropriate tests. A simple model that allows exposure to vary within and between persons in a group is the one-way random effects model (hereafter simply random effects model). This model has been used extensively to evaluate occupational exposures to chemical and physical agents (6, 8, 9, 14, 15, 20, 27, 34–36, 41, 44, 56, 57, 64–76).

Let X_{ij} represent the exposure level (e.g., mg/m^3) for the i-th person on the j-th day (we assume that each exposure covers a full workday) and let Y_{ij} represent the natural logarithm of X_{ij}, that is $Y_{ij} = \ln(X_{ij})$. We designate the mean and variance of X_{ij} as μ_x and σ_x^2, respectively, and those of Y_{ij} as μ_y and σ_y^2, respectively. Thus, the random-effects model is specified by the following expression:

$$Y_{ij} = \ln(X_{ij}) = \mu_y + \beta_i + \epsilon_{ij} \quad \text{for } (i = 1, 2, \ldots, k \text{ persons}) \text{ and } (j = 1, 2, \ldots, n_i \text{ days})$$

where μ_y represents the fixed mean (logged) exposure of the group, β_i represents the random effect of the i-th person (given by the deviation of the i-th person's mean (logged) exposure, $\mu_{yi} = \mu_y + \beta_i$, from μ_y) and ϵ_{ij} represents the random deviation on the j-th day for person i from μ_{yi} ($\epsilon_{ij} = Y_{ij} - \mu_{yi}$). It is assumed under the model that both β_i and ϵ_{ij} are normally distributed, with a mean of zero and variances of σ_B^2 and σ_W^2, respectively, and that the β_is and ϵ_{ij}s are mutually independent. (This assumption of independence of the β_is and ϵ_{ij}s has important implications that were alluded to earlier). Thus, the parameters σ_W^2 and σ_B^2 represent the within- and between-person components of the total variance, given by $\sigma_y^2 = \sigma_W^2 + \sigma_B^2$, and Y_{ij} is normally distributed with mean μ_y and variance σ_y^2.

The following assumptions are inherent in the random effects model:

1. The expected value of Y_{ij} is μ_y for all i and j. In fact, the $\{Y_{ij}\} = \{Y_{11}, Y_{12}, \ldots, Y_{kn_k}\}$ constitute a random sample from a normal distribution of logged exposures received by the i-th person, such that the expected value of the $\{Y_{ij}|\mu_{yi}\}$ equals μ_{yi} (indicating that the expected value of each Y_{ij} conditional upon the i-th person is equal to μ_{yi}).

2. The variance of Y_{ij}, i.e., σ_y^2 is equal to $\sigma_W^2 + \sigma_B^2$ for all i and j. Also, the variance of the conditional distribution of exposures for the i-th person $\{Y_{ij}|\mu_{yi}\}$ is equal to σ_W^2 (indicating that daily exposures from person i vary about μ_{yi} with variance σ_W^2).

3. The within-person variance component σ_W^2 does not vary with i (i.e., there is an assumption of homogeneous variance so that the intraperson variability is the same for all individuals).

4. For $j \neq j'$, the covariance of $[Y_{ij}, Y_{ij'}]$ (a pair of measurements from the i-th person on different days) is equal to σ_B^2, so that the correlation of $[Y_{ij}, Y_{ij'}] = \rho = \sigma_B^2/\sigma_y^2$ (this is the so-called intraclass correlation where "class" = person.) This means that only positive correlation is assumed between pairs of exposures from the same person and that the correlation is the same for any pair of exposures (i.e., the exposures are not autocorrelated).

These assumptions have been evaluated and tested with databases of occupational exposures (14, 15, 34–36, 77). Using graphical procedures to evaluate the normality assumption of the random-person effects (based upon Refs. 78, 79), Tornero-Velez et al. (14) indicated that the random effects model adequately fit data from 220 of 252 (87%) observational groups. This suggests that this model is generally appropriate for applications involving occupational exposure data.

Figure 18.4 illustrates various hypothetical probability distribution functions (PDFs) under the random effects model. In this case, the area under the PDF curve refers to the theoretical relationship between the logged exposure and the probability of observing such a level in the exposure distribution. The curve labeled 'All' in Figure 19.4(**a**) refers to the logged exposures for a hypothetical group where Y_{ij} is normally distributed with mean $\mu_y = 2.3$ and variance $\sigma_y^2 = 0.693$. It is assumed that the within- and between-person components of variance are equal for this group; thus, $\sigma_W^2 = \sigma_B^2 = 0.693/2 = 0.346$. The numbered curves in Figure 18.4(**a**) indicate the conditional distributions (i.e., the $\{Y_{ij}|\mu_{yi}\}$) of a sample of 5 workers (designated 1 to 5), each normally distributed with mean μ_{yi} for $i = 1, 2, \ldots, 5$ and variance σ_W^2. Under the random effects model it is implicit that the $\{\mu_{yi}\} = \{\mu_{y1}, \mu_{y2}, \ldots, \mu_{yk}\}$ are derived from a normal distribution with mean μ_y and variance σ_B^2. Figure 18.4(**b**) shows this distribution of individual means (values of μ_{yi}) across the population.

3.2.1 The Lognormal Distribution of Exposures

Although the random effects model is applied to the logged exposures represented by the set of $\{Y_{ij}\}$, it provides a valuable tool for making inferences about the underlying log-

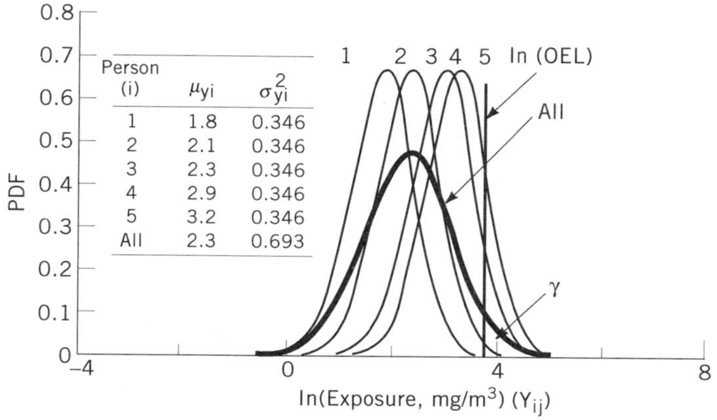

(a)

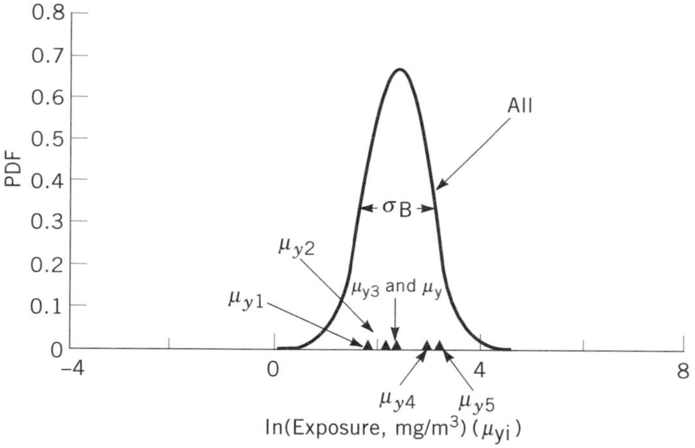

(b)

Figure 18.4. Hypothetical normal distributions of logged exposures for an observational group of workers. (**a**) Distributions of exposures received from day to day. Curves 1–5 represent a sample of five workers from the group while the curve labeled "All" represents the hypothetical population of daily exposures. (**b**) The curve labeled "All" represents the distribution of individual mean exposures (μ_{yi}) from the population shown in (**a**). Points labeled μ_{y1}–μ_{y5} represent the individual means of the distributions labeled 1–5 in (**a**). For explanation of terms, see Nomenclature.

normal distribution of exposure levels, i.e., the $\{X_{ij}\} = \{X_{11}, X_{12}, \ldots, X_{kn_k}\}$. This is because it is implicit in the random effects model that $X_{ij} = e^{(\mu_y + \beta_i + \epsilon_{ij})}$ is lognormally distributed with mean $\mu_x = e^{(\mu_y + 0.5\sigma_y^2)}$ and variance $\sigma_x^2 = \mu_x^2(e^{\sigma_y^2} - 1)$, in keeping with current knowledge of occupational exposures (9). Therefore, even though the model pertains to the logged values $\{Y_{ij}\}$, it also establishes a mechanism for relating the corresponding exposures $\{X_{ij}\}$ with dose and with the subsequent risk of disease (9, 80).

Assuming that the $\{X_{ij}|\mu_{xi}\}$ constitute a random sample of exposures from the i-th person, this conditional distribution of X_{ij} has an expected value of μ_{xi} and variance of $\sigma_{xi}^2 = \mu_{xi}^2(e^{\sigma_W^2} - 1)$; furthermore the $\{X_{ij}|\mu_{xi}\}$ are assumed to be conditionally independent and to all have the same underlying distribution. Here, σ_W^2 is the parameter reflecting within-person variability just as it was with the logged values. Figure 19.5(**a**) illustrates the conditional distributions of exposures for the five hypothetical persons previously shown in Figure 19.4(**a**). Note that the variances of the conditional distributions (i.e., the values of σ_{xi}^2) increase with μ_{xi}. The curve labeled "All" again refers the hypothetical distribution for the population, just as for the logged values in Figure 18.4(**a**).

It is also implicit in the random effects model that the mean exposures $\{\mu_{xi}\} = \{\mu_{x1}, \mu_{x2}, \ldots, \mu_{xk}\}$ of the k persons in a group come from a lognormal distribution with overall (group) mean $\mu_x = e^{[\mu_y + 0.5(\sigma_W^2 + \sigma_B^2)]}$ and variance $\sigma_x^2 = \mu_x^2(e^{\sigma_B^2} - 1)$. Thus, under the random effects model, an observational group is necessarily a monomorphic group [a group described by a lognormal distribution for μ_{xi} (9, 34)], which is appropriate for making statistical inferences about the unobservable individual mean exposures, i.e., the $\{\mu_{xi}\}$. Figure 18.5(**b**) illustrates the lognormal distribution of $\{\mu_{xi}\}$. Here, σ_B^2 is the parameter reflecting between-person variability just as it was for the logged values.

3.2.2 *Estimating Parameters*

The random-effects model can be applied to an observational group after measuring n_i exposures for each member of a random sample of k persons. Let x_{ij} represent the j-th measurement from the i-th person and y_{ij}, the corresponding logged measurement. The following standard ANOVA methods are applied to estimate the parameters under the model (81). Note that the hat ("^") symbol above a letter refers to an estimated value:

$$\bar{Y}_i = \frac{1}{n_i} \sum_{j=1}^{n_i} y_{ij}$$

is the sample mean of the logged measurements for the i-th person;

$$\bar{Y} = \frac{1}{N} \sum_{i=1}^{k} \sum_{j=1}^{n_i} y_{ij}$$

is the overall sample mean based upon all $N = \Sigma_{i=1}^{k} n_i$ logged measurements;

$$\hat{\sigma}_W^2 = \text{MSW}$$

estimates σ_W^2, the within-person variance component; and

$$\hat{\sigma}_B^2 = \frac{(k - 1)[MSB - MSW]}{\left(N - \sum_{i=1}^{k} n_i^2/N\right)}$$

estimates σ_B^2, the between-person variance component, where

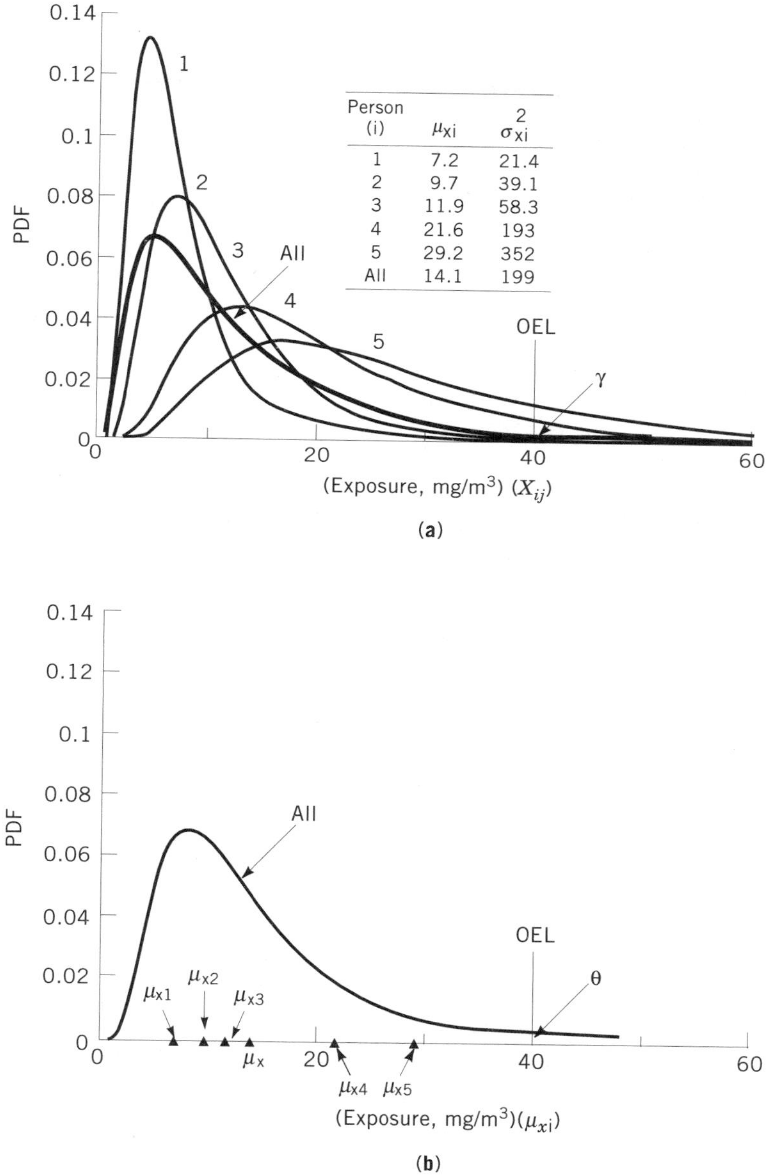

Person (i)	μ_{xi}	σ^2_{xi}
1	7.2	21.4
2	9.7	39.1
3	11.9	58.3
4	21.6	193
5	29.2	352
All	14.1	199

Figure 18.5. Hypothetical lognormal distributions of exposures for an observational group of workers. (**a**) Distributions of exposures received from day to day. Curves 1–5 represent a sample of five workers from the group while the curve labeled "All" represents the hypothetical population of daily exposures. (**b**) The curve labeled "All" represents the distribution of individual mean exposures (μ_{xi}) from the population shown in (**a**) Points labeled μ_{x1}–μ_{x5} represent the individual means of the distributions labeled 1–5 in (**a**). For explanation of terms, see Nomenclature.

$$MSW = \frac{\sum_{i=1}^{k} \sum_{j=1}^{n_i} (y_{ij} - \bar{Y}_i)^2}{(N - k)} = \frac{SSW}{(N - k)} \text{ and}$$

$$MSB = \frac{\sum_{i=1}^{k} n_i(\bar{Y}_i - \bar{Y})^2}{(k - 1)}$$

are the mean squares within persons and between persons, respectively, obtained from the ANOVA table;

$$\hat{\mu}_y = \frac{\sum_{i=1}^{k} (\bar{Y}_i/(\hat{\sigma}_B^2 + \hat{\sigma}_W^2/n_i))}{\sum_{i=1}^{k} (1/(\hat{\sigma}_B^2 + \hat{\sigma}_W^2/n_i))}$$

estimates the group mean μ_y for all logged measurements, and

$$\hat{\sigma}_y^2 = \hat{\sigma}_B^2 + \hat{\sigma}_W^2$$

is the ANOVA estimate of the group variance σ_y^2 for all logged measurements.

Because MSW can be greater than MSB in some instances, the estimate of the between-person variance (i.e., $\hat{\sigma}_B^2$) is occasionally negative, particularly when k and N are small (81). In the event that $\hat{\sigma}_B^2 < 0$, $\hat{\sigma}_B^2$ should be set to zero for estimating σ_y^2.

For the special case where only one exposure is measured for each person ($j = 1$), then $N = k$ and the above estimates revert to the familiar forms:

$$\hat{\mu}_y = \bar{Y} = \frac{1}{k} \sum_{i=1}^{k} y_i \text{ and}$$

$$\hat{\sigma}_y^2 = \frac{1}{(k - 1)} \sum_{i=1}^{k} (y_i - \bar{Y})^2$$

3.2.3 *The ANOVA Table*

Following application of the random effects model with any standard statistical package, an ANOVA table should be available that includes the following components:

Source	Sum of Squares	Degrees of Freedom (d.f.)	Mean Squares	Parameter Estimated
Between-person	SSB	$k - 1$	$\frac{SSB}{(k-1)}$	$\sigma_W^2 + n_o\sigma_B^2$
Within-person (error)	SSW	$N - k$	$\frac{SSW}{(N-k)}$	σ_W^2

where $n_o = [N - (\Sigma_{i=1}^{k} n_i^2/N)]/(k - 1)$. Assuming the model holds, Searle et al. (81) provide expressions for the variances and covariances of σ_B^2 and σ_W^2 that allow confidence intervals and tests of hypotheses of parameters to be constructed.

Table 18.2 lists the parameter estimates for Groups 1–4 following application of the random effects model to the data. From the table we see that both estimated variance components covered a wide range; in fact, extremely large estimates were observed for Group 3. The last column shows the estimated intraclass correlation represented by the ratio of the estimated between-person variance component to the variance ($\hat{\varrho} = \hat{\sigma}_B^2/\hat{\sigma}_y^2$), which ranged from 0.058 for Group 1 to 0.441 for Group 3. These values are consistent with results from a large database of occupational exposures from 220 observational groups where the median value of $\hat{\varrho}$ was 0.22, 25% of groups had $\hat{\varrho} \leq 0.04$ and 25% had $\hat{\varrho} \geq 0.41$ (14). Since the intra-class correlation is generally significant, it follows that repeated measurements obtained from the same persons in occupational groups tend to be positively correlated. As mentioned previously, statistical methods that do not recognize such correlation can lead to biased or otherwise invalid results (e.g., variances may be overly precise).

3.2.4 Relative Measures of Variability

The ranges of exposures indicated by σ_B^2 and σ_W^2 are often difficult to gauge because they relate to the logged values. To make things more intuitive, Rappaport (9) defined a scale-independent measure of exposure variability, $R_{0.95}$, representing the fold range containing 95% of the exposures. That is,

$$R_{0.95} = \frac{97.5\text{-th Percentile exposure}}{2.5\text{-th Percentile exposure}}$$

This measure of the fold range can also be applied to either the distribution of worker-specific means $\{\mu_{xi}\}$ or the conditional distribution of daily exposures $\{X_{ij}|\mu_{xi}\}$. For the distribution of μ_{xi} the fold-range containing 95% of the values is designated ${}_BR_{0.95} = \mu_{x,0.975}/\mu_{x,0.025}$, where $\mu_{x,0.975} = e^{[(\mu_y + 0.5\sigma_W^2) + 1.96\sigma_B]}$, and $\mu_{x,0.025} = e^{[(\mu_y + 0.5\sigma_W^2) - 1.96\sigma_B]}$, so that ${}_BR_{0.95} = e^{3.92\sigma_B}$. For example, if ${}_BR_{0.95} = 4$ then 95% of the lognormal distribution of individual mean exposures would lie within a 4-fold range; this occurs when $\sigma_B = 0.353$.

Table 18.2. Parameter Estimates for Groups 1–4 (ANOVA estimates from application of the random effects model to the data listed in Appendix A)[a]

Group	$\hat{\mu}_y$	$\hat{\sigma}_B^2$ (${}_B\hat{R}_{0.95}$)	$\hat{\sigma}_W^2$ (${}_W\hat{R}_{0.95}$)	$\hat{\sigma}_y^2$ ($\hat{R}_{0.95}$)	$\hat{\mu}_x$	$\hat{\varrho} = \frac{\hat{\sigma}_B^2}{\hat{\sigma}_y^2}$
1	2.73	0.023 (1.8)	0.377 (11.1)	0.400 (11.9)	18.7 μg/m³	0.058
2	−0.424	0.032 (2.0)	0.466 (14.5)	0.499 (15.9)	0.840 mg/m³	0.071
3	9.23×10^{-2}	2.38 (421)	3.01 (901)	5.39 (8960)	16.2 mg/m³	0.441
4	4.62	0.292 (8.3)	0.513 (16.6)	0.805 (33.7)	152 mg/m³	0.362

[a]See Nomenclature for explanation of terms.

Likewise, for variation within persons we can visualize the corresponding percentiles of the conditional distribution of $\{X_{ij}|\mu_{xi}\}$; that is, $X_{0.975}|\mu_{xi} = e^{(\mu_{yi}+1.96\sigma_W)}$ and $X_{0.025}|\mu_{xi} = e^{(\mu_{yi}-1.96\sigma_W)}$ so that ${}_WR_{0.95} = e^{3.92\sigma_W}$. For example, if ${}_WR_{0.95} = 15$ then 95% of the lognormal distribution of daily exposures experienced by any particular individual would have a 15-fold range; this occurs when $\sigma_B = 0.691$.

Values of $R_{0.95}$ can be estimated by substituting the ANOVA estimates of σ_B^2 and σ_W^2 into the above relationships. These estimates are designated ${}_B\hat{R}_{0.95} = e^{3.92\hat{\sigma}_B}$, ${}_W\hat{R}_{0.95} = e^{3.92\hat{\sigma}_W}$, and $\hat{R}_{0.95} = e^{3.92\hat{\sigma}_y}$, respectively. Table 18.2 lists values of ${}_B\hat{R}_{0.95}$, ${}_W\hat{R}_{0.95}$, and $\hat{R}_{0.95}$ along with the corresponding variance components for Groups 1–4.

3.3 Converting among Parameters

It was shown above that the lognormal distributions of exposures (i.e., the $\{X_{ij}\}$ and the $\{\mu_{xi}\}$) were related to the corresponding normal distributions of logged exposures (i.e., the $\{Y_{ij}\}$ and the $\{\mu_{yi}\}$). In fact, the parameters representing the means and variances of these distributions can easily be converted from one to another and can also be used to derive the corresponding geometric mean and geometric standard deviation (designated μ_g and σ_g) if desired. Table 18.3 gives the useful relationships for doing so (based upon Refs. 10, 82). The most important of these is $\mu_x = e^{(\mu_y+0.5\sigma_y^2)}$, which indicates that the mean of the lognormal distribution μ_x is always greater than the geometric mean $\mu_g = e^{\mu_y}$ and that the difference between the two increases with σ_y^2. Another useful relationship in Table 19.3 is $\sigma_y^2 = \ln[1 + (\sigma_x^2/\mu_x^2)]$, which allows the variance of the normal distribution to be related to the squared coefficient of variation (CV) of the lognormal distribution. Table 18.3 can also be used with the ANOVA estimates of the corresponding parameters for applications with samples of data.

Table 18.3. Formulas for Converting among Parameters of a Lognormal Distribution[a]

Given	To Obtain	Use
μ_y	$\mu_g =$	$\exp(\mu_y)$
μ_x, σ_x	$\mu_g =$	$\mu_x^2/\sqrt{\mu_x^2 + \sigma_y^2}$
σ_y	$\sigma_g =$	$\exp(\sigma_y)$
μ_x, σ_x	$\sigma_g =$	$\exp\sqrt{\ln[1 + (\sigma_x^2/\mu_x^2)]}$
μ_y, σ_y	$\mu_x =$	$\exp[\mu_y + \frac{1}{2}\sigma_y^2]$
μ_g, σ_y	$\mu_x =$	$\mu_g \exp[\frac{1}{2}\sigma_y^2]$
μ_y, σ_y	$\sigma_x =$	$\sqrt{[\exp(2\mu_y + \sigma_y^2)][\exp(\sigma_y^2) - 1]}$
μ_g, σ_j	$\sigma_x =$	$\sqrt{\mu_g^2[\exp(\sigma_y^2)][\exp(\sigma_y^2) - 1]}$
μ_g	$\mu_y =$	$\ln(\mu_g)$
μ_x, σ_y	$\mu_y =$	$\ln(\mu_x) - \frac{1}{2}\sigma_y^2$
σ_g	$\sigma_y =$	$\ln(\sigma_g)$
μ_x, σ_x	$\sigma_y =$	$\sqrt{\ln[1 + (\sigma_x^2/\mu_x^2)]}$

[a]See Glossary for explanation of terms.

3.4 Probabilities of Exceeding Particular Levels

The parameters of a lognormal distribution can be used to calculate various probabilities, including the likelihood that exposure would exceed an OEL. Two different probabilities will be considered in this context, namely the exceedance, γ, representing the probability that a randomly selected worker's exposure would exceed the OEL on any day and overexposure, θ, representing the probability that a worker's average exposure would exceed the OEL. The reasons for differentiating between these two probabilities have been elucidated by Tornero-Velez et al. (14) and will be discussed later.

3.4.1 *Calculating Exceedance*

The exceedance is related to the OEL and parameters of the exposure distribution as follows:

$$\gamma = P\{X_{ij} > OEL\} = P\{Y_{ij} > \ln(OEL)\} = 1 - \Phi\left\{\frac{\ln(OEL) - \mu_y}{\sigma_y}\right\}, \qquad (1)$$

where $\Phi\{z_{(1-\gamma)}\}$ denotes the probability that a standard normal variate Z would fall below the value $z_{1-\gamma} = \{[\ln(OEL) - \mu_y]/\sigma_y\}$. Following the hypothetical example shown in Figures 18.4 and 18.5, where $\mu_y = 2.3$, $\sigma_y^2 = 0.693$, and $OEL = 40$ mg/m^3, one finds:

$$\gamma = P\{Y_{ij} > \ln(OEL)\} = 1 - \Phi\left\{\frac{\ln(40) - 2.3}{\sqrt{0.693}} = 1.67\right\} = 0.048,$$

from which we infer that there is about a 5% chance that a random exposure would exceed the OEL. Note that this probability refers to the group as a whole. From Figures 18.4(**a**) and 18.5(**a**), however, it is clear that individual workers can have exceedances greater or less than 5%, depending upon the magnitude of the between-person variance component. Tables of the standard normal distribution are available in most statistics textbooks and values can be obtained as needed from common microcomputer software.

3.4.2 *Calculating Overexposure*

The probability that a randomly selected worker in a monomorphic group would be exposed on average above the OEL is given by:

$$\theta = P\{\mu_{xi} > OEL\} = 1 - \Phi\left\{\frac{\ln(OEL) - \mu_y - \dfrac{\sigma_W^2}{2}}{\sigma_B}\right\} \qquad (2)$$

where $\Phi\{z_{1-\theta}\}$ denotes the probability that a standard normal variate Z would fall below the value $z_{(1-\theta)} = \{[\ln(OEL) - \mu_y - (\sigma_W^2/2)]/\sigma_B\}$. Extending the earlier example where $OEL = 40$, $\mu_y = 2.3$, $\sigma_y^2 = 0.693$, and $\sigma_B^2 = \sigma_W^2 = 0.346$, then

$$\theta = P\{\mu_{xi} > OEL\} = 1 - \Phi\left\{\frac{\ln(40) - 2.3 - \frac{0.346}{2}}{\sqrt{0.346}} = 2.07\right\} = 0.020.$$

Thus, there is a 2% probability that a worker from this group would have an average exposure (μ_{xi}) above the OEL.

3.4.3 *Estimating γ and θ*

The above relationships can be used with the estimated parameters of the exposure distributions to estimate γ and θ from samples of data. That is:

$$\hat{\gamma} = 1 - \Phi\left\{\frac{\ln(OEL) - \hat{\mu}_y}{\hat{\sigma}_y}\right\} \text{ and } \hat{\theta} = 1 - \Phi\left\{\frac{\ln(OEL) - \hat{\mu}_y - \frac{\hat{\sigma}_W^2}{2}}{\hat{\sigma}_B}\right\}$$

where $\hat{\mu}_y$, $\hat{\sigma}_B^2$, and $\hat{\sigma}_W^2$, refer to the ANOVA estimates of μ_y, σ_B^2, and σ_w^2, respectively. Table 18.4 shows the estimates for γ and θ for Groups 1–4 (with OELs shown at the time of sampling; two OELs were operative for lead and benzene). Note that the relative magnitudes of exceedance and overexposure do not always coincide; that is, $\hat{\theta} \ll \hat{\gamma}$ for Groups 1 and 2; $\hat{\theta} \cong \hat{\gamma}$ for Group 4; and $\hat{\theta} > \hat{\gamma}$ for Group 3.

3.4.4 *Relationship to the Mean Exposure*

Because the mean μ_x and the variance σ_x^2 of a lognormal distribution are not functionally independent, the levels of exceedance and overexposure are maximal at a given value of OEL/μ_x (85). By substituting $\mu_x = e^{(\mu_y + 0.5(\sigma_W^2 + \sigma_B^2))}$ into Equations 1 and 2, and solving for

Table 18.4. Estimates of Exceedance (γ) and Overexposure (θ) for Groups 1–4 (Based upon Equations 1 and 2 using ANOVA Estimates of the Parameters of the Exposure Distributions)

Group	OEL[a]	$\hat{\gamma}$	$\hat{\theta}$
1	50 μg/m³ (PEL)	0.031	<0.001
	30 μg/m³ (AL)	0.144	0.001
2	3.2 mg/m³ (PEL)	0.012	<0.001
	1.6 mg/m³ (AL)	0.103	<0.001
3	3.2 mg/m³ (PEL)	0.322	0.610
	1.6 mg/m³ (AL)	0.435	0.767
4	213 mg/m³ (PEL)	0.204	0.184

[a]OELs refer to OSHA's Permissible Exposure Limits (PELs) and Action Levels (ALs, if available) at the time that data were collected.

OEL/μ_x one finds that

$$\frac{\text{OEL}}{\mu_x} = e^{0.5(z^2_{(1-\gamma)} - (z_{(1-\gamma)} - \sigma_y)^2)} \text{ and} \tag{3}$$

$$\frac{\text{OEL}}{\mu_x} = e^{0.5(z^2_{(1-\theta)} - (z_{(1-\theta)} - \sigma_B)^2)} \tag{4}$$

Because OEL/μ_x is maximal when $z_{(1-\gamma)} = \sigma_y$ or $z_{(1-\theta)} = \sigma_B$, there is a corresponding maximum probability representing exceedance and overexposure, designated as γ_{max} and θ_{max}, respectively, given by:

$$\left(\frac{OEL}{\mu_x}\right)_{max} = e^{(0.5z^2_{(1-\gamma_{max})})} \text{ and} \tag{5}$$

$$\left(\frac{\text{OEL}}{\mu_x}\right)_{max} = e^{(0.5z^2_{(1-\theta_{max})})} \tag{6}$$

From Equations 5 and 6 we see that, at a given value of OEL/μ_x, there exists a maximum probability that the *OEL* will be exceeded, over either days or individuals, regardless of the magnitudes of the within- and between-person variance components. Table 18.5 lists values of γ_{max} and θ_{max} that correspond to given values of $(OEL/\mu_x)_{max}$. The table shows, for instance, that when $(OEL/\mu_x)_{max} = 2$, γ_{max} and $\theta_{max} = 0.12$. In other words, if the mean is less than half of the OEL, then not more than 12% of the exposures from a lognormal distribution (of either days or individual mean values) can exceed the *OEL*. Various implications of this property of the lognormal distribution have been discussed (22, 83) and will be explored further in this chapter.

3.5 Autocorrelated Exposure Series

It has been shown that the distribution of exposures received by a worker can be summarized by the conditional exposure distribution $\{Y_{ij}|\mu_{yi}\}$, which is assumed under the random effects model to be distributed normally with mean μ_{yi} and variance σ^2_W. However,

Table 18.5. Maximum Levels of Exceedance and Overexposure at a Given Ratio of the OEL to the Mean Exposure $(\mu_x)^a$

$\left(\frac{\text{OEL}}{\mu_x}\right)_{max}$	$z_{(1-\gamma \text{ or } \theta)}$	γ_{max} or θ_{max}
2	1.177	0.120
4	1.665	0.048
8	2.039	0.021
16	2.355	0.009
32	2.633	0.004

[a] γ_{max} is the maximum exceedance; θ_{max} is the maximum overexposure; $z_{(1-\gamma \text{ or } \theta)}$ is the value from the standard normal distribution associated with γ_{max} or θ_{max}.

these parameters do not provide any information concerning the correlation of exposures measured at different times. This is given by a third characteristic of the distribution, the autocorrelation function, which defines the relationship between exposures separated by a fixed interval of time, referred to as the lag. Under the random effects model it was assumed that the correlation of any pair of $[Y_{ij}, Y_{ij'}]$ (for $j \neq j'$) was equal to $\rho = \sigma_B^2/\sigma_y^2$, regardless of when they were collected. If the correlation were, in fact, related to the lag, then this assumption would be corrupted and could lead to biased estimates of the parameters of the exposure distribution (e.g., see Refs. 41, 84). For example, the practice of campaign sampling, where all measurements are obtained during consecutive days, could pose problems.

Given the importance of unbiased estimates of the parameters of the exposure distribution, the question of autocorrelation of occupational exposures should be addressed. Unfortunately, a reasonable estimate of the autocorrelation function over the first 10–12 lags requires the measurement of at least 50 sequential exposures (85). Since occupational hygienists rarely collect such large numbers of measurements, relatively little information is currently available to allow autocorrelation functions to be analyzed. This paucity of time-series data has motivated investigators to rely upon models to gain insight into the potential importance of autocorrelation on the estimation of exposure. The most popular of these has been the first-order autoregressive process [AR(1) process], which depicts the current exposure as a weighted fraction of the previous exposure plus a random input (63, 84, 86–92).

In the context of day-to-day exposures, there are only anecdotal suggestions that significant autocorrelation exists. For example, Esmen (93) noted a strong correlation between production rate and exposure, particularly for batch operations; Ulfvarson (55) commented upon seasonal variations and multiyear trends in exposure; and Buringh and Lanting (94) observed that the variances of data sets with $3 \leq n \leq 6$, obtained from a variety of industries, were smaller when all measurements were collected within a single week than otherwise. However, in the only studies to investigate the question directly (with time series of daily exposures), relatively little evidence of autocorrelation was observed, and when found, the first-lag coefficients tended to be rather small (63, 84, 92, 95). [In the most extensive of these investigations, Symanski and Rappaport (63) questioned the validity of the AR(1) process for day-to-day exposure series since sample autocorrelation functions were inconsistent with the AR(1) model in about half the time series demonstrating significant correlation at lag one.] If these results are typical of the day-to-day correlation observed in most workplaces, then it may well be possible to estimate the distribution of shift-long measurements of exposure on the basis of discrete campaigns of a few days duration. However, as noted earlier, potential problems can be avoided by random sampling.

Very little work has been published concerning the levels of autocorrelation of intraday exposures. Coenen (40, 86) noted that dust concentrations and vinyl-chloride concentrations measured continuously at fixed locations in manufacturing facilities were highly autocorrelated. Kumagai et al. (95) confirmed this finding for short-term personal samples obtained from 16 worker-chemical combinations and also presented evidence that the AR(1) process was appropriate for such time series. Roach (87), Spear et al. (89) and Rappaport and Spear (91) modeled an AR(1) process based upon air exchange rates to predict short-term autocorrelation that might exist in occupational environments (where mass transport of the contaminant is governed by turbulent diffusion). These models suggest that short-term exposures would be highly autocorrelated under realistic scenarios. Collectively, these studies indicate that periods of hours can be required between measurements to obtain relatively uncorrelated data. Thus, the occupational hygienist wishing to predict the frequencies of brief excursions to short-term-exposure limits (STELs) should be wary of doing so on the basis of sequential brief measurements. Recent advances in the development of personal monitors that measure short-term exposures and that store the data over an entire shift should allow actual autocorrelation functions to be investigated.

4 VARIABILITY WITHIN AND BETWEEN WORKERS

A random effects model was presented to characterize exposures in observational groups. In illustrating the estimation of parameters under this model, it was shown that the within- and between-person variance components for Groups 1–4 covered a wide range. Results of broad application of this model to observational groups will be summarized in this section, so that the magnitudes of the variance components can be gauged more generally. Then the implications of the between-person variance component will be considered in light of traditional interpretations of exposure among groups of workers.

4.1 Ranges of Variance Components

In an extensive investigation of variability of occupational exposures, (based upon 13,945 daily exposures (personal measurements) from 1,574 workers) Kromhout et al. (15) reported values of $\hat{\sigma}_W^2$ and $\hat{\sigma}_B^2$ for 165 groups of workers classified by job and location (factory). The cumulative distributions of these variance components (reported as values of $\hat{R}_{0.95}$, representing the fold-ranges containing 95% of the values) are shown in Figure 18.6. The figure shows that exposure tends to vary more within than between persons (${}_W\hat{R}_{0.95}$: median $=$ 15.1, ${}_B\hat{R}_{0.95}$: median $=$ 4.06), a finding confirmed with smaller data sets (68, 71, 76). However, estimates of ${}_W\hat{R}_{0.95}$ and ${}_B\hat{R}_{0.95}$ both covered extremely wide ranges (from less than two to greater than 1000), suggesting many sources of variability operating at levels of the group and the individual worker.

4.2 Sources of Variability in Observational Groups

As part of their study, Kromhout et al. (15) evaluated the influence of several environmental and process-related covariates upon $\hat{\sigma}_W^2$ and $\hat{\sigma}_B^2$. This subset analysis was conducted with data compiled from 87 (of 165) groups, having complete information. Univariate analyses showed that several of these factors were significantly associated with increasing variability. The following comparisons indicated higher values of both $\hat{\sigma}_W^2$ and $\hat{\sigma}_B^2$ in the directions shown: intermittent processes $>$ continuous processes ($p < 0.001$); outdoor exposures $>$ indoor exposures ($p < 0.01$); general ventilation $>$ local-exhaust ventilation ($p < 0.01$); mobile workers $>$ stationary workers ($p < 0.05$); and local sources $>$ general

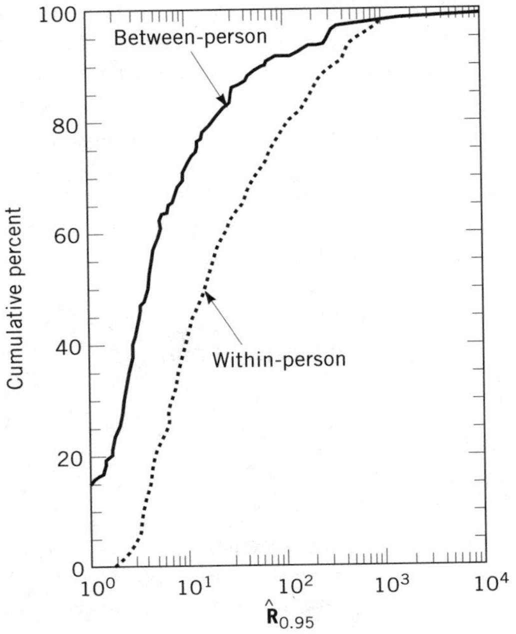

Figure 18.6. Cumulative distributions of within- and between-person sources of variation in exposure for 165 observational groups of workers (from Ref. 15). Expressed as the estimated fold-range $\hat{R}_{0.95}$ containing 95% of the observations of the respective distributions.

sources ($p < 0.05$). The authors commented that these associations were all in the expected directions.

Multivariate models were then used to investigate the combined effects of the environmental and process-related factors upon $\hat{\sigma}_W^2$ and $\hat{\sigma}_B^2$ (15). The results, summarized in Table 18.6, show that the final model for $\hat{\sigma}_W^2$ contained only two effects (continuous vs. intermittent process and indoor vs. outdoor exposure, $p = 0.0001$) while that for $\hat{\sigma}_B^2$ contained only a single effect (continuous vs. intermittent process, $p = 0.0005$). Each of the variance components listed in the table represents the model prediction for the particular combination of covariates shown. Note that only three combinations of covariates were available for predictions of σ_B^2 because no continuous outdoor processes were represented in the 87 occupational groups used to develop the model.

These results suggest that variability within persons is largely dictated by factors related to the process and the environment, a finding confirmed by Peretz et al. (71). Since such factors would affect all persons in the group more or less equally, the fact that they explain 41% of the variability of $\hat{\sigma}_W^2$ (Table 18.5) is, perhaps, not surprising. On the other hand, these same factors explained only 13% of the variation in $\hat{\sigma}_B^2$ and the fit of the model was rather poor. Thus, at present we can only speculate about the factors that contribute to different exposures between persons in the same job and factory. Certainly the particular

Table 18.6. Multivariate Evaluation of Covarates upon Within- and Between-person Variance Components for 87 Occupational Groups[a,b]

Dependent Variable	Covariate(s)	R^2 (%)	Model Prediction	
Within-person variance component	Process and environment	41	σ_W^2	$_W R_{0.95}$
	Continuous and indoors		0.320	9.2
	Intermittent and indoors		1.30	87.6
	Intermittent and outdoors		1.60	142
Between-person variance component	Process	13	σ_B^2	$_B R_{0.95}$
	Continuous		0.053	2.5
	Intermittent		0.320	9.2

[a]From Ref. 15.
[b]σ_W^2 and σ_B^2 represent the within- and between-worker components of variance; $R_{0.95}$ represents the fold range containing 95% of the values from the corresponding lognormal distribution.

mix of tasks, locations, equipment and work practices could differ greatly within a group. Elucidation of such causative factors is an important goal for future investigations.

The predicted variance components in Table 18.6 also indicate that intermittent processes are typically much more variable than continuous processes. This has implications for sample sizes required to test exposures relative to limits, as will be shown.

4.3 Implications to the Homogeneous Exposure Group

The wide range of σ_B^2 reported in the above studies casts doubt upon the traditional view that exposure is uniform within observational groups (HEGs). Since many current sampling strategies assume uniform exposure (e.g., see Ref. 21), the implications of between-worker variability deserve special attention.

4.3.1 A Model for the HEG

Let X_j represent the exposure of a randomly selected person within a HEG on the j-th day. Then, the following model can be used to define the presumed sources of exposure variability:

$$Y_j = \ln(X_j) = \mu_y + \epsilon_j \quad \text{for } (j = 1, 2, \ldots, N \text{ daily exposures})$$

where, Y_j represents the logged value of the j-th exposure, μ_y represents the mean (logged) level for the group, and ϵ_j is the error term, which is assumed to be normally distributed with mean zero and variance σ_y^2. As with the random effects model, this 'HEG model' assumes that the $\{Y_j\}$ are normally distributed with mean μ_y and variance σ_y^2 and that the $\{X_j\}$ are lognormally distributed with mean $\mu_{yx} = e^{(\mu_y + 0.5\sigma_y^2)}$ and variance $\sigma_x^2 = \mu_x^2(e^{\sigma_y^2} - 1)$. However, by ignoring the effect of the worker, the HEG model also assumes that the $\{Y_j\text{s}\}$ are statistically independent of each other even if they are obtained from the same person on different days; i.e., it is assumed that for $j \neq j'$, the covariance of $[Y_j, Y_{j'}]$

equals 0. Based upon our current knowledge, this is unreasonable because, when σ_B^2 is large, the covariance of $[Y_j, Y_{j'}]$ is significantly greater than zero. Thus, without *a priori* knowledge that σ_B^2 is trivial, the HEG model can legitimately be applied only with single measurements from a sample of workers or with repeated measurements from a single worker (in the latter case the HEG would consist of one person).

4.3.2 Uniformity of Exposure

As noted previously, uniform exposure occurs when the variability between workers is very small (i.e., σ_B^2 approaches 0 or ${}_BR_{0.95}$ approaches one). As a rule of thumb one can regard a group where ${}_BR_{0.95} \leq 2$ (occurring when $\sigma_B^2 \leq 0.036$) as uniformly exposed because 95% of the workers in this group would have mean exposures within a two-fold range (9). Referring now to the between-person variability of Groups 1–4 (Table 18.2), Groups 1 and 2 provide evidence of uniform exposure since ${}_B\hat{R}_{0.95} = 1.8$ and 2.0, respectively. However, Groups 3 and 4, with ${}_B\hat{R}_{0.95} = 421$ and 8.3, respectively, provide evidence of highly variable exposures among workers. Thus, even though the four groups are legitimate HEGs, in the sense that all members share the same job and location, they are vastly different in terms of the uniformity of exposure.

The uniformity of exposure can be considered more generally with the compilation of between-person variance components reported by Kromhout et al. (15) for 165 observational groups. Values of ${}_B\hat{R}_{0.95}$ ranged from 1 to 2000 with a median value of 4.06 (see Fig. 18.6). This indicates that workers in a typical HEG experienced about a four-fold range of individual mean exposures. Only about 20% of these groups had values of ${}_B\hat{R}_{0.95} < 2$, suggesting uniform exposure, while an equal percentage had values of ${}_B\hat{R}_{0.95} > 18$, suggesting highly variable exposure among workers.

With evidence suggesting that about ⅘ of observational groups are not uniformly exposed, hygienists should be wary of relying upon a model (the HEG) that cannot accurately explain the variability of occupational exposures. As noted in the introduction, such an interpretation presents intractable problems regarding the interpretation of occupational exposures. In particular, the HEG model hampers any systematic investigation of differences in tasks, practices and personal environments, which might be used to assess and reduce exposures at the level of the individual worker. For these reasons, it is strongly suggested that occupational hygienists move away from the HEG model and toward the more general random effects model, which can even accommodate the special case where exposure is uniform ($\sigma_B^2 \cong 0$).

5 EXPOSURE LIMITS

Since most prospective assessments of exposure have been motivated by the need to comply with OELs, one cannot consider the development of sampling strategies without discussing exposure limits. The following analysis considers the implications of exposure variability upon the interpretation of exposure limits used in the United States.

5.1 Threshold Limit Values

Prior to passage of the Occupational Safety and Health Act (OSH Act) of 1970 (96) limits of airborne exposure in the United States generally reflected consensus standards developed by various groups, most notably the Threshold Limit Values (TLVs) of the American Conference of Governmental Industrial Hygienists (ACGIH). The TLVs have had a profound influence on the practice of occupational hygiene everywhere in the world (97). The United States government (OSHA), several states within the United States, and many other countries have adopted some or all of the TLVs as their official limits.

5.1.1 Long-Term vs. Short-Term TLVs

The ACGIH defines TLVs according to criteria for monitoring over both the long term and the short term. The long-term limit, referred to as the TLV–TWA (time-weighted average), is defined as (98):

> . . . the time-weighted average concentration for a conventional 8-hour workday and 40-hour workweek to which it is believed that nearly all workers may be repeatedly exposed, day after day, without adverse effect.

By defining a TLV-TWA as the level to which a worker can be repeatedly exposed over the long term, the ACGIH seems to suggest that this is the average level to which a worker could be exposed. Logically, a 'conventional' workday or week would reflect average conditions and not those associated with unusually high or low exposures. Likewise the wording "repeatedly exposed, day after day" suggests a cumulative exposure (i.e., average exposure times duration of exposure). However, the ACGIH has not provided guidance as to the exact meaning of the TLV-TWA when confronted with variability within workers from day to day. Indeed, the above interpretation is controversial and inconsistent with the ACGIH's recommendations regarding short-term excursions of the TLV-TWA (33). That is, the ACGIH states that " . . . worker exposure levels may exceed three times the TLV–TWA for no more than a total of 30 minutes during a work-day, and under no circumstances should they exceed five times the TLV–TWA, provided that the TLV–TWA is not exceeded" (98).

The short-term exposure limit or TLV-STEL, is defined as an air concentration averaged over 15 min, which represents (98)

> . . . the concentration to which workers can be exposed continuously for a short period of time without suffering from (*1*) irritation, (*2*) chronic or irreversible tissue damage, or (*3*) narcosis . . ., provided that the daily TLV-TWA is not exceeded. It is not a separate independent exposure limit, rather it supplements the TWA limit where there are recognized acute effects from a substance whose toxic effects are primarily of a chronic nature.

Two of the three criteria, that is, irritation and narcosis, are consistent with the accepted perception that STELs relate to effects of transient high levels that might arise even though exposures are chronic in nature and the average concentration is at or below the TLV–TWA (99–101). Likewise, the fact that irreversible tissue damage, such as neural death associated with anoxia, might be associated with periods of intense exposure, provides a clear rationale for STELs. However, the notion that long-term damage might depend on the rate of exposure at a given TLV–TWA is more provocative. This suggests that transmission of the contaminant to the target tissues is rapid and that the relationship between the tissue burden and damage is nonlinear (curving upward) (9, 102). Little, if any, evidence of such behavior has been reported for situations in which the average exposure is less than the TLV–TWA.

A third type of limit designated the TLV–C represents a ceiling value that should not be exceeded even instantaneously. A review of the 1999 list of TLVs (98) suggests that ceiling limits are applied to substances that only produce effects following acute exposures.

5.1.2 Health Basis of TLVs

The ACGIH's definition of its TLV–TWAs as levels that protect " . . . nearly all workers repeatedly exposed, day after day, without effects . . ." implies that these limits are based primarily on health considerations. For many agents the paucity of exposure-response data available to or reported by, the ACGIH has caused this interpretation to be questioned (99, 101, 103, 104). The documentation supporting a particular TLV provides a relatively brief review and listing of reference material; although it contains no formal analysis of exposure–response relationships, the documentation may comment upon such analyses from other sources (105).

Some TLVs are based on animal experiments, others on reports of human experience both in the workplace and in controlled experiments with volunteers. Referring to human experience, Roach and Rappaport reported that the risks inherent in certain TLVs (from the 1976 and 1986 lists) were surprisingly large, averaging between 14 and 17 per 100 individuals exposed at or below these levels (104). They further observed that the TLVs were highly correlated with the levels of exposure reported in the studies. They speculated, therefore, that the TLVs historically reflected the perception that more stringent limits were unrealistic given the current state of control. This interpretation is supported by statements of prominent industrial toxicologists in the 1950s and 1960s and is consistent with staged reductions in TLVs (97), which have occurred in step with general trends towards lower exposures. [Interestingly, Symanski et al. presented evidence that the rate of reduction in exposure from year to year was *less* for substances having one or more changes in the TLV (2, 3).] Because the TLVs are updated rather infrequently the notion that many of them represent conditions in the distant past should be troubling to occupational hygienists. [The median age of TLVs on the 1991–1992 list was 16.5 years (97).]

5.1.3 Recognition of Exposure Variability

Although the ACGIH's TLV committee has acknowledged short-term fluctuations in air concentrations and has used this variation to justify STELs, ceiling limits, and excursion factors, the variability of daily exposures has not been addressed. This neglect of variability of daily exposures is difficult to justify in light of current evidence, but it undoubtedly conditioned past thinking concerning exposure assessment. That is, the TLVs were intended for and used by occupational hygienists as reference points in a world where data were few and limited to area and occasional breathing-zone measurements. Because measurements could not easily be related to actual exposures, area concentrations were equated with long-term exposures of groups of workers in the vicinity. This practice, and the fact that some prominent TLVs included little if any margins of safety, led hygienists to conclude that any measurement above the TLV pointed to a hazardous situation.

The above analysis suggests that compliance testing in the United States resulted primarily from two factors: (*1*) the inability to measure the individual exposures of large numbers of workers; and (*2*) the interpretation of TLVs as concentrations not to be exceeded over any full day. The technological leap from stationary and hand-held samplers to personal sampling equipment in the 1960s and 1970s made it possible for occupational hygienists to measure daily exposures directly and, therefore, to recognize the importance of variability over days and between individuals. Unfortunately, the introduction of OSHA standards in 1971–1972 appears to have codified compliance testing in the United States just at the time when it was possible for more sophisticated approaches to evolve.

5.2 OSHA Standards

The OSH Act of 1970 established OSHA as the official standard-setting body in the United States (96). The PELs established by OSHA for chemical agents fall into two categories: those originally adopted as existing standards in 1971 under a provision of the OSH Act, and those subsequently developed as new standards. Those adopted as existing standards included PELs for about 400 chemicals, most of which had originally been issued as TLVs by the ACGIH in 1968. OSHA has developed new exposure limits for 13 substances. As shown in Table 18.7, the new standards often included a second exposure criterion called an Action Level (AL, an exposure concentration of roughly one-half of the PEL, which triggers certain requirements), and sometimes a STEL.

5.2.1 Risk and Feasibility

Court decisions over the last three decades have determined that OSHA may issue a new standard after demonstrating that workers are at "significant risk" of adverse effect at the existing PEL. Significant risk has been interpreted by OSHA, in situations involving carcinogens, as an individual lifetime risk of at least 1 per 1000 (106). Having demonstrated a significant risk at an existing limit, OSHA establishes a new PEL representing the lowest level thought to be economically and technologically feasible in at least one industrial sector (97). If the residual risk at this new PEL still exceeds 1 per 1000, then the door is left open for OSHA to institute further reductions in the future. This explicit estimation of risk and feasibility differentiates OSHA's new PELs from TLVs and apparently from most OELs promulgated elsewhere as well (99, 101).

The two-tiered process of setting new PELs is illustrated in Figure 18.7 using the benzene standard (97, 106) as an example; this process has been used for 8 of the 13 new PELs, as indicated in Table 18.6. In its benzene standard, OSHA devoted a large section of the preamble to assessing the levels of risk at the old and new PELs. Using the results from three epidemiological studies to arrive at an exposure–response relationship, OSHA applied a linear cancer model to assess the risk inherent in the old and new PELs. Assuming

Table 18.7. Exposure Limits Set by OSHA in its New Standards

Substance	PEL[a]	AL[b]	STEL[c]	Risk Assess.[d]
Acrylonitrile	2 ppm	1 ppm	10 ppm	No
Arsenic	10 μg/m³	5 μg/m³		Yes
Asbestos	0.1 fiber/cm³		1.0 fiber/cm³	Yes
Benzene	1 ppm	0.5 ppm	5 ppm	Yes
Cadmium	5 μg/m³	2.5 μg/m³		Yes
Coke oven emissions	150 μg/m³			No
Cotton dust	200 μg/m³			No
1,2-Dibromo-3-chloropropane	1 ppb			Yes
Ethylene oxide	1 ppm	0.5 ppm	5 ppm	Yes
Formaldehyde	0.75 ppm	0.5 ppm	2 ppm	Yes
Lead	50 μg/m³	30 μg/m³		No
Methylenedianiline	10 ppb	5 ppb	100 ppb	Yes
Vinyl chloride	1 ppm	0.5 ppm	5 ppm	No

[a]Permissible Exposure Limit expressed as an 8-hr TWA air concentration.
[b]Action level expressed as an 8-hr TWA air concentration.
[c]Short Term Exposure Limit expressed as a 0.25-hr air concentration, except for asbestos where it is expressed as a 0.5-hr air concentration.
[d]Indicates whether or not OSHA performed a formal risk assessment.

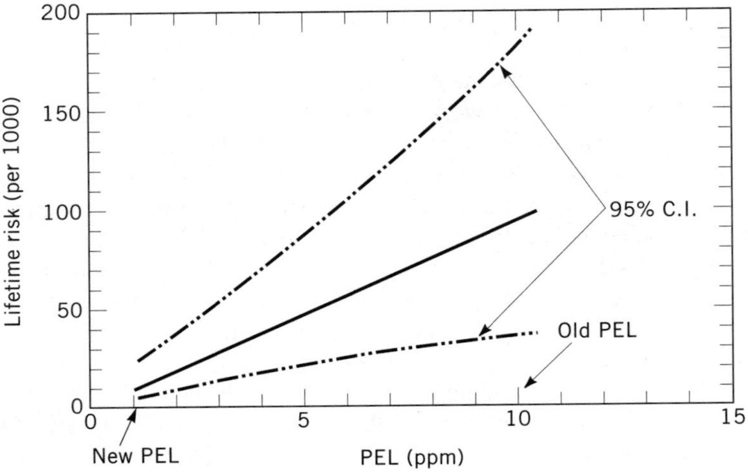

Figure 18.7. Exposure–response relationship and 95% confidence interval used by OSHA to establish the risk of leukemia at the old and new PELs for benzene. From Ref. 97.

a cumulative exposure to benzene at the old PEL (10 ppm), under a regimen of 8 hr/day, 5 day/week for 45 years, OSHA estimated the risk to be 95 excess deaths from leukemia per 1000 workers, with a 95% confidence interval (CI) of between 37 and 186 deaths. Clearly, such a large risk would be well above the 1 per 1000 level that is needed to justify a new PEL.

Having demonstrated the need for a new standard, OSHA then endeavored to determine the lowest level of benzene exposure that was feasible based upon on both technologic and economic factors. The preamble to OSHA's benzene standard contained a lengthy discussion of the levels of exposure in various industrial sectors, i.e., petrochemical production, petroleum refining, coke- and coal-chemical production, tire manufacturing, bulk terminals and plants, and transportation (via tank trucks). OSHA subsequently determined that, in fact, only petrochemicals and coke- and coal-production could not feasibly achieve reductions in exposure below 1 ppm (97, 106). Thus, it seems that OSHA developed its new benzene PEL of 1 ppm relative to the two worst segments of industry affected by the standard. This conclusion is supported by analyses indicating that, at the time the benzene standard was set, petrochemicals and coke- and coal-production had overall mean exposures of 0.7 and 1.4 ppm, respectively, levels close to the new PEL of 1 ppm (97). Surprisingly, OSHA's data indicated that these two sectors contained only 2.2% of all workers exposed to benzene and only 9.1% of all workers exposed above 1 ppm (97). Thus, the two smallest industrial sectors dictated the official limit for the vast majority of workers exposed to benzene. Note from Figure 18.7 that the risk inherent in this new PEL is 9.5 per 1000 (95% CI: 4–22 per 1000), well above the 1 per 1000 threshold needed to justify a new standard.

5.2.2 Interpretation of PELs

Since OSHA equates risk with cumulative exposure over 45 years and considers feasibility in the context of average conditions, each new PEL represents the average exposure received by the individual worker over 45 years in those segments of industry considered least able to control exposures. Thus, the philosophy used to establish a new PEL is generally consistent with the idea that the risk to the *i*-th worker is proportional to μ_{xi} over a fixed period of time (e.g., 45 years). The validity of this concept is reinforced by toxicokinetic analyses showing that the long-term tissue dose under linear kinetics is proportional to cumulative exposure (9, 80, 107, 108). The assumption of linear kinetics seems reasonable for exposures that are low relative to saturable processes involved with metabolism and repair. This should be the usual situation involving occupational exposures today in the U.S., but one should be wary of scenarios, which might invalidate the assumption of linear kinetics at expected levels of exposure (109–111). OSHA also noted that some segments of industry affected by the benzene standard could consistently maintain exposures below the new PEL and the workers would experience proportionally lower risk (106); here again, individual risk under the linear model would be proportional to μ_{xi}.

Notwithstanding the fact that the standard-setting process used by OSHA equates risk with cumulative exposure, the definition used by the agency for enforcement is ambiguous.

For example, the benzene standard defined the PEL in terms of the onus to the employer, as follows (106):

> The employer shall assure that no employee is exposed to an airborne concentration of benzene in excess of 1 ppm as an 8-hour time-weighted average.

Although this statement can be taken to mean that no employee should be exposed to more than one ppm of benzene *on the average*, such an interpretation was initially rejected (106). However, OSHA subsequently indicated that it will not issue a citation on the basis of a single measurement above the PEL, but rather requires that exposures be less than the PEL " . . . *in most operations most of the time* . . ." (italics added), consistent with the risk assumptions in its new standards. [This quotation was taken from a particularly lucid discussion of OSHA's standard for inorganic lead that arose from legal arguments in the U.S. Court of Appeals (112). Interestingly, in that case, OSHA argued that the most appropriate interpretation of its PEL was that of the geometric mean exposure level.]

5.3 Working Limits

If statistical methods are to be adopted for assessing long-term exposures in the workplace, it is essential that industrial hygienists become aware of the underlying premises and shortcomings of the OELs that they employ. As suggested above, some limits, including many TLVs, contain little or no margins of safety. Likewise, new OSHA standards, which are based on the feasibility of achieving a level in the worst segment(s) of industry, are not sufficiently protective in sectors where exposures can be better controlled. In either case, it is prudent to adopt working limits that allow for uncertainties in the underlying risk to health and for reduction of exposure to the lowest levels feasible in all segments of industry. OSHA's promulgation of action levels (ALs) imposes a working limit of one half of the PEL in many segments of industry affected by its new standards.

Each safety factor should be chosen on the basis of a reasoned judgment concerning the biological effects, the quality of data supporting the OEL, and the circumstances under which exposures arise (113–115). Large companies in the United States, for example, are capable of maintaining exposures at most facilities well below a new PEL and could choose a working limit of, say, a half to a fourth of the PEL on the basis of feasibility alone. Many have adopted corporate exposure guidelines that do this. More marginal facilities, which might be unable to achieve the desired levels of control, should strive for incremental reductions of exposure over time. In such situations it might initially be necessary to test relative to the PEL. Then annual testing could be performed, with concurrent adoption of controls, until the target level of, say, PEL/2, is achieved.

6 TESTING RELATIVE TO LIMITS

Some strategies will now be considered for determining whether exposures are acceptable relative to OELs. Four alternatives will be evaluated, namely, compliance testing (observing a measurement above the OEL), testing exceedance (the probability that a daily exposure would exceed the OEL), testing the group mean exposure, and testing overexposure (the probability that a worker's mean exposure would exceed the OEL). Although the analysis will focus primarily upon measurements over the whole workday, some evaluations relative to STELs will also be mentioned.

6.1 Criteria for Evaluating Strategies

Since each strategy approaches the problem from a different perspective, objective criteria are needed to evaluate the relative strengths and weaknesses. An ideal test would have the following characteristics:

1. It would provide for rigorous evaluation in the sense that it would test a specific hypothesis, relating some characteristic(s) of the exposure distribution with the appropriate OEL. To discourage the practice of declaring situations acceptable by default, it would assume that exposures are unacceptable till proven otherwise (9, 25, 28, 30, 32, 34, 43).
2. It would lead to decisions that relate to the degree of hazard. In other words, the likelihood of declaring exposures unacceptable should be directly related to the risk of adverse effect.
3. It would provide information suitable for selecting among options for controlling unacceptable situations.

6.2 Outcome vs. Risk

In order to relate the outcome of a test with the risk of disease, it is helpful to differentiate between risks associated with acute and chronic exposures.

6.2.1 Risks of Acute Exposure

Regarding fast-acting substances (which can lead to either death or serious effects following a single exposure), it seems clear that risk is primarily related to the likelihood that a very large air concentration would be encountered. Thus, it is theoretically possible to assess acute exposures by focusing upon the extreme right tail of the distribution (116). However, it is naïve to suppose that one can accurately predict the likelihood of exposures at levels that are, perhaps, 1000 times greater than the mean exposure without very large numbers of measurements as well as assumptions regarding the underlying distribution of exposures (9, 116). Thus, it seems unwise to consider sampling strategies in terms of personal monitoring of individuals exposed to fast acting substances. Rather it would make more sense to continuously monitor either the source of exposure or the air in the vicinity of the worker and to implement practices that would protect the individual from accidental releases of these contaminants. It follows from this that ceiling limits can be implemented rigorously only through continuous monitoring of air contaminants over time scales of seconds to minutes.

6.2.2 *Risk of Chronic Exposure*

When considering chronic exposures, it seems unlikely that risk would be unduly influenced by transient exposures of one day or less but rather should be related to the long-term exposures received from year to year (9, 34, 117). Indeed, OSHA adheres to this reasoning in the development of new standards where it is implicit that the risk of disease is strictly proportional to the cumulative exposure over 45 years. In this context, an individual's risk can be stated simply as

$$Risk_{i,t} = K(CE_{i,t}) = K(\mu_{xi} \times t)$$

where $Risk_{i,t}$ represents the risk to the i-th individual after t years of exposure, K is a constant of proportionality and $CE_{i,t}$ is the individual's cumulative exposure, which is equivalent to the product of the mean exposure and time.

Since the upper limit of acceptable risk would be proportional to $OEL \times t$, it follows that acceptable exposure (over t years) exists when $\mu_{xi} < OEL$, as implicitly assumed by OSHA in promulgation of its standards. Although this assumption strictly holds only when the distribution describing $\mu_{xi}(t)$ does not vary over the entire t years, acceptable risk is still demonstrable as long as $\mu_{xi} < OEL$ at each successive stage of evaluation throughout the entire period. Since it is impractical to measure the exposure of all persons in the group, inferences about the likelihood that $\mu_{xi} > OEL$ must be made on the basis of samples of persons representing the unobservable $\{\mu_{xi}\}$ in the population. Thus, there is always a non-zero probability that a worker would be overexposed, i.e., would have $\mu_{xi} > OEL$, and the key to successful risk management is to define acceptable exposure in terms of this likelihood.

Previously the probability of overexposure was defined as $\theta = P\{\mu_{xi} > OEL\}$ and was related to the parameters of the exposure distribution under the random-effects model. Let A represent the maximum acceptable probability of overexposure; acceptable exposure is, therefore, defined by the condition where $\theta < A$ (34). Although it makes sense that A should be small, there is no consensus as to what the value should be. For illustration purposes, we will use a value of $A = 0.10$ following limited precedence (34–36); other values can obviously be used.

6.3 Compliance Testing

The usual test for compliance involves direct comparison of the largest of a few measured exposures with an OEL, which has been interpreted in this context as an air concentration not to be exceeded on any day. The essential elements of the test, as well as particular applications to OSHA standards, have been discussed (9, 10, 12–14, 24, 118) and will be commented upon here.

6.3.1 *The Probability of Compliance*

Two characteristics determine the outcome of compliance testing, namely, the probability that a given exposure would exceed the limit (i.e., the exceedance γ), and the number of measurements, N. Assuming independent data, the probabilities of compliance P(Comp.)

and of noncompliance *P*(Noncomp.) are easily obtained from the following expressions (13):

$$P(\text{Comp.}) = (1 - \gamma)^N, \tag{7}$$

$$P(\text{Noncomp.}) = 1 - \text{P(Comp.)} \tag{8}$$

These expressions are illustrated in Figure 18.8 for sample sizes between 1 and 20. Although the OEL would generally be a shift-long limit, these relationships can also be used to evaluate compliance of short-term exposures relative to STELs. If an action level (AL) is also operative, additional rules and monitoring requirements essentially require that the probability of exceeding the AL (usually half of the PEL) be used in the above expressions. This arises from increases in sample size that are required by OSHA whenever an initial measurement is found between the AL and the PEL (13).

It is important to recognize that compliance testing assumes acceptable exposure by default and requires evidence that exposures are out of compliance. Since compliance can be maximized by minimizing the number of measurements [i.e., $P(\text{Comp.}) = 1$ for $N = 0$] this creates a disincentive for monitoring (13). Since $0 < \gamma < 1$, Equations 7 and 8 also make explicit what is known intuitively, that any population of exposures can be declared out of compliance given a large enough sample. Indeed, Figure 18.8 shows that sample size is often the most important single determinant of compliance. Since *N* is totally unrelated to the health hazard, this illustrates a critical flaw in the strategy.

To illustrate the importance of sample size on compliance testing, consider the data for Groups 1–4. The estimated values of γ are summarized in Table 18.4 for all four groups given the operative OELs. The probabilities of noncompliance can easily be estimated by

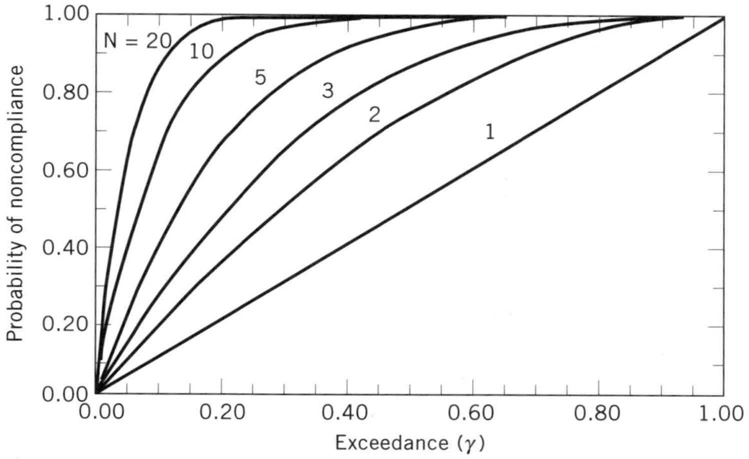

Figure 18.8. Relationship between the exceedance (γ) and the probability of noncompliance for sample sizes (*N*) between one and 20 measurements. From Equation 8.

substituting values of $\hat{\gamma}$ into Equations 7 and 8. For example, for Group 1 ($\hat{\gamma}$ (AL) = 0.144), *P*(Noncomp.) = 0.14, 0.37, 0.54, and 0.96 for N = 1, 3, 5 and 20, respectively. So an assessment with five measurements would more likely than not turn up a value above the AL; then, additional monitoring would eventually lead to a measurement above the PEL. A decision as to whether exposures of members of Group 1 are in compliance would, therefore, be reduced to the following incredible statement: if an initial assessment included fewer than five measurements, then compliance should be the outcome and if it included five or more, noncompliance is more likely. In the case of Group 2 [$\hat{\gamma}$ (AL) = 0.102], 7 measurements would be required for noncompliance to be more likely; for Group 3 [$\hat{\gamma}$ (AL) = 0.435], two measurements would be required; and, finally, for Group 4 [$\hat{\gamma}$ (PEL) = 0.204], three measurements.

Because of the critical importance of sample size (N) upon compliance testing, this strategy is necessarily constrained to very small numbers of measurements. Although sample sizes used for compliance testing are not generally known, Tornero-Velez et al. (14) reported the numbers of measurements employed for annual surveys of the nickel industry world wide. This database contained 15,576 personal measurements of inorganic nickel in 4,864 surveys of 1,282 observational groups. Their results indicate that N was typically quite small. Indeed 49% of the groups had $N = 1$, 67% had $N \leq 2$, 77% had $N \leq 3$, and 85% had $N \leq 4$. This suggests that the typical determination of compliance is based upon one or two (and rarely more than 4) measurements per group.

6.3.2 Compliance vs. Risk

Since compliance is determined by the probability that a measurement will exceed the OEL at a given sample size (according to Equation 7) and risk is related to the probability of overexposure θ, the relationship between compliance and risk is complicated (13, 14, 44). However, because compliance testing is typically applied to highly skewed exposure distributions with very small sample sizes ($N \leq 2$), acceptable exposure is very likely to be the outcome for any given test. What, then, is the likelihood that a group with a large probability of overexposure would go undetected? In order to sort this out, we require information about the likelihood that θ would be large for any given group. Relying upon the work of Tornero-Velez et al. (14), summarized in Figure 18.9, it appears that about ⅕ of 179 observational groups had values of $\hat{\theta}$ greater than 0.10, suggesting unacceptable conditions. Because $\hat{\gamma}$ tends to be equal to or less than $\hat{\theta}$ for these groups (14), Equation 7 can be used to infer that compliance testing would find these groups acceptable quite often. For example, if $\gamma = 0.2$ (suggesting that $\theta \geq 0.2$) and $N = 2$, then compliance would be declared 64% of the time. Thus, compliance testing can lead to serious underestimation of risk for realistic scenarios of industrial exposure.

6.3.3 Options for Control

With very small numbers of measurements (typically one measurement from 1–4 persons), little insight can be gained from compliance testing as to the selection of controls. Indeed,

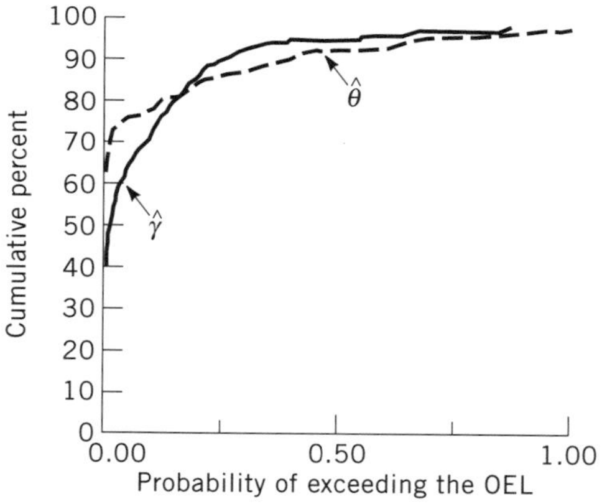

Figure 18.9. Cumulative distributions of estimates of exceedance ($\hat{\gamma}$) and overexposure ($\hat{\theta}$) for 179 observational groups of workers. From Ref. 14.

any recommendations about control would be restricted to the few persons and conditions monitored.

6.3.4 Summary

Several shortcomings make compliance testing a poor choice for evaluating long-term exposures. First, hypotheses can neither be postulated nor tested since the sample size is not controlled and compliance is accepted by default. Second, since the probability of compliance decreases as the number of measurements increases, decisions tend to be based upon very small sample sizes. Third, when N is small, compliance can easily be declared despite large probabilities of overexposure. Fourth, compliance testing would rarely provide sufficient data with which to weigh options for control. Finally, compliance testing artificially increases the importance of assay errors since decisions hinge on comparison of a single (largest) value with the OEL (10). Thus, even though assay error term is usually a trivial component of total variation, it might appear important when the largest air concentration is only slightly greater than the OEL. As an example, if the CV of exposure equals 1.0 (equivalent to $\sigma_y^2 = 0.693$) and the corresponding CV of the assay (CV_a) equals 0.10 ($\sigma_a^2 = 0.00995$), then the imprecision of the assay only contributes [0.00995/(0.693 + 0.00995) = 0.014] 1.4% to the total variance.

6.4 Testing the Exceedance

A more sophisticated testing structure emerges by defining an acceptable exposure distribution as one in which the exceedance γ is less than E, where E represents some small probability, usually 0.05 (11, 21, 24–27). This approach retains γ as a decision-making element, and can, therefore, be regarded as an outgrowth of compliance testing. However, rigorous tests of hypotheses can be conducted via one-sided tolerance limits (25, 27). For example, the following testing structure can be used:

$$H_0\colon \gamma \geq E \text{ (unacceptable)}$$

versus

$$H_1\colon \gamma < E \text{ (acceptable)}$$

where H_0 and H_1 represent the null and alternative hypotheses. If one is able to reject the null hypothesis in favor of the alternative, that γ is less than E, then exposure can be declared acceptable at a given level of significance. Note that this pair of hypotheses assumes unacceptable exposure by default and requires evidence to declare exposures acceptable.

Following Lyles and Kupper, the null and alternative hypotheses can be rewritten as

$$H_0\colon \mu_y + z_{(1-E)}\sigma_y \geq \ln(OEL) \text{ (unacceptable)}$$

versus

$$H_1\colon \mu_y + z_{(1-E)}\sigma_y < \ln(OEL) \text{ (acceptable)}$$

where $z_{(1-E)}$ is the $100(1 - E)^{\text{th}}$ percentile of the standard normal distribution (27). This pair of hypotheses allows a determination to be made as to whether $X_{(1-E)} < OEL$, where $X_{(1-E)} = \exp(\mu_y + z_{(1-E)}\sigma_y)$ is the $100(1 - E)^{\text{th}}$ percentile of the distribution of X_{ij}.

Because the statistical test requires independent observations, only one measurement per worker can be used to evaluate exceedance. Assuming a sample of single measurements (x_{ij} for $j = 1$) from k workers, the test statistic is the one-sided tolerance limit (25, 27), defined as

$$T_u = \hat{\mu}_y + \left(\frac{-1}{\sqrt{k}}\right) t'_{k-1,\alpha}(\lambda)\hat{\sigma}_y \tag{9}$$

where $\hat{\mu}_y$ and $\hat{\sigma}_y$ are the mean and square root of the variance of the logged measurements (y_{ij} for $j = 1$), $t'_{k-1,\alpha}$ is the $100(\alpha)^{\text{th}}$ percentile of the noncentral t distribution with $(k - 1)$ degrees of freedom and noncentrality parameter $\lambda = -\sqrt{k}z_{(1-E)}$, and α is the significance level of the test. If $T_u < ln(OEL)$, then H_0 is rejected at an α level of significance and exposure is declared acceptable.

6.4.1 Sample Size Requirements

The test of γ encourages monitoring since the power of the test increases with sample size (in this context the power of the test is the probability of declaring exposures acceptable when $\gamma < E$); this is a direct consequence of the hypothesis-testing structure. Lyles and Kupper (27) provide a relationship to compute sample sizes at a given power, as well as values of k required for a range of parameters. Table 18.8 reproduces sample size requirements from (27) for $E = 0.05$, $\alpha = 0.05$ and power $= 0.80$. It shows that many mea-

Table 18.8. Number of Workers (k) Required for the Test of the Exceedance when E = 0.05, α = 0.05 and power = 0.80[a,b]

$OEL/X_{(1-E)}$	$\sigma_y^2 = 0.50$	$\sigma_y^2 = 1.0$	$\sigma_y^2 = 1.5$	$\sigma_y^2 = 2.0$	$\sigma_y^2 = 2.5$	$\sigma_y^2 = 3.0$
1.5	58	107	154	202	249	295
2.0	24	42	59	76	93	109
2.5	16	27	37	47	57	67
3.0	13	20	28	35	42	49

[a]From Ref. 27, assuming one measurement per worker.
[b]$X_{(1-E)}$ is the $100(1 - E)^{\text{th}}$ percentile of the distribution of exposures; σ_y^2 is the variance of the logged exposures (Y_{ij} for j = 1 measurement per person).

surements (workers) are required to achieve an 80% level of power when $OEL/X_{(1-E)}$ is less than about two or three depending upon the variance.

For example, using the estimated 95th percentiles and variances Group 1 ($PEL/\hat{X}_{(1-E)} = 1.2$; $\hat{\sigma}_y^2 = 0.400$) would require $k > 58$ and Group 2 ($PEL/\hat{X}_{(1-E)} = 1.5$; $\hat{\sigma}_y^2 = 0.499$) would require $k = 58$. Also, Groups 3 and 4 could not be declared acceptable more than $100(\alpha)\%$ of the time because $\hat{X}_{(1-E)} > OEL$ in both cases ($PEL/\hat{X}_{(1-E)}$ = 0.06 and 0.48, respectively).

More generally, the variance components predicted by Kromhout et al. (15) (Table 18.6) suggest that a typical intermittent process (where $1.3 < \hat{\sigma}_y^2 < 2.0$, depending upon whether exposure occurs indoors or outdoors) would require substantially more measurements to declare exposures acceptable than would a comparable continuous processes (where $\hat{\sigma}_y^2 < 0.5$).

6.4.2 *Outcome vs. Risk*

The test of exceedance considers only the likelihood that exposure would exceed the OEL on a single workday. It does not evaluate either the magnitude or the distribution of such excursions among the workers in the group. Thus, without evidence that the group is uniformly exposed, some workers can experience exposures many times the OEL rather frequently and others rarely [see Fig. 18.5(**a**)]. Risk, on the other hand, is related to the individual mean exposure μ_{xi} that ultimately determines a worker's cumulative exposure and absorbed dose. Because the probability of overexposure θ captures this property, the question logically arises as to the relationship between γ and θ.

In fact, γ can be either larger or smaller than θ depending upon the particular within- and between-worker components of variance for the group in question. The empirical relationship between $\hat{\theta}$ and $\hat{\gamma}$ is shown in Figure 18.10 for 179 observational groups of workers (14); $\hat{\gamma}$ was greater than $\hat{\theta}$ in 93 cases, $\hat{\gamma}$ was less than $\hat{\theta}$ in 30 cases, and $\hat{\gamma}$ and $\hat{\theta}$ were both effectively zero (<0.001) in 56 cases.

Figure 18.10 also reveals a peculiar feature of the relationship between θ and γ. Values of $\hat{\theta}$ tend to be less than $\hat{\gamma}$ when $\hat{\gamma}$ is small and greater than $\hat{\gamma}$ when $\hat{\gamma}$ is large. The two vertical lines in the figure represent the transition zone where γ ceases to be a conservative surrogate for θ and becomes anticonservative instead. The point of transition depends upon

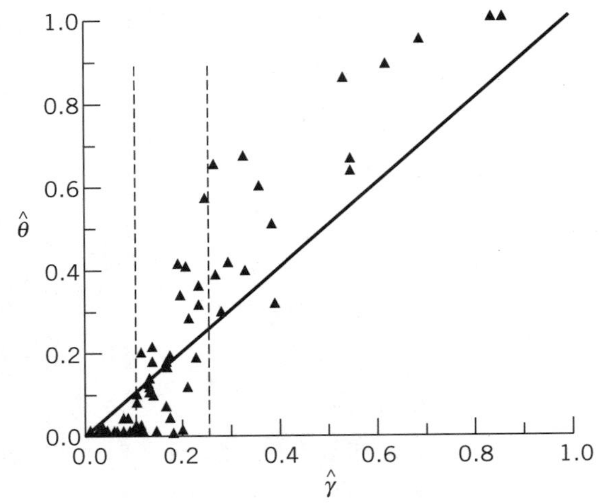

Figure 18.10. Estimates of overexposure ($\hat{\theta}$) versus exceedance ($\hat{\gamma}$) for 179 observational groups of workers. Data from Ref. 14. Dashed lines represent the transition zone where exceedance changes from a conservative to an anticonservative surrogate for overexposure. The solid line represents the case where $\hat{\theta} = \hat{\gamma}$.

the variance $\hat{\sigma}_y^2$ and the correlation ρ of the particular group (14). Given the ranges of $\hat{\sigma}_y^2$ and $\hat{\varrho}$ represented by these groups, the transition zone covers the approximate range of $0.10 < \hat{\gamma} < 0.25$. From theoretical considerations, a test of $\gamma \geq E = 0.05$ will similarly constrain θ only for $\hat{\sigma}_y^2 \leq 2.59$ (14). Since $\hat{\sigma}_y^2$ exceeded 2.59 in 11% of the 179 groups, a test of $\gamma \geq E = 0.05$ would *not* serve as an upper limit for θ in at least 11% of the tests. Because there is no theoretical upper limit to the variance that a group can possess, there is likewise no theoretical lower limit of γ that one would adopt a priori as a 'conservative' bound for θ.

Furthermore, within the transition zone, the intra-class correlation has a profound influence on the rate at which γ changes from an upper to a lower bound for θ (14). Indeed, groups with very little correlation demonstrate abrupt changes in values of θ for small changes in γ, suggesting that the consequences of incorrectly specifying γ can be large. Thus, the selection of a universally conservative value of *E* becomes problematic. By adopting smaller and smaller values of *E*, it becomes increasingly difficult to declare exposures acceptable, even for very small values of θ. In fact, we see evidence of this at the level of $E = 0.05$, where for Groups 1 and 2 it is unlikely that exposure would be declared acceptable even though $\hat{\theta} < 0.001$ in both cases.

6.4.3 Application to STELs

Since probabilities with which exposure limits can be exceeded over short periods are either explicitly or implicitly built into STELS (83), there is a rationale for testing the exceedance of short-term exposures. In this context, the test could be applied either to a random sample of persons, each contributing one short-term measurement, or to the conditional distribution of exposures received by a single worker, in which case all measurements would be obtained from that individual. Caution must be exercised concerning the timing of short-term measurements, however, because exposures are likely to have significant autocorrelation within shifts, particularly in continuous, indoor operations (95). Since such correlation of serial values would disrupt the assumption of independence, it can be necessary to collect short-term data over several days to achieve the desired sample sizes.

6.4.4 *Options for Control*

Because the test of exceedance can only be applied with single measurements from each worker in a sample, it provides little insight into the differences in exposure between workers in the group. As such, unacceptable situations would be dealt with via engineering or administrative controls that affect all persons more or less equally. Any information concerning tasks, which had been gathered during the surveys, could be used to focus controls on particular activities at the group level.

6.4.5 *Summary*

The test of exceedance provides a mixed bag to the occupational hygienist. On the one hand, it allows acceptable exposure to be rigorously defined and requires evidence to make such a declaration. But on the other hand, exceedance has only a tenuous connection to the long-term risk associated with exposure. This muddles the rationale for arriving at an acceptable level of exceedance and can lead to overly conservative testing schemes in many cases. The necessity to constrain the sampling to single measurements from each person causes practical problems, particularly in situations where many persons are required to declare exposures acceptable. Likewise, the inability to gather repeated measurements limits control options to engineering and administrative solutions.

6.5 Testing the Group Mean Exposure

The notion that exposure should be evaluated in terms of the mean air concentration of the observational group is certainly not new. It was argued in Great Britain in the early 1950s that, because of the cumulative effect of coal dust on respiratory function, exposure to coal dust should be assessed in terms of the mean exposure received over time (4, 38, 119, 120). During the subsequent 10–15 years, attention shifted to statistical methods to estimate means of lognormal exposure distributions (40, 121–124) and to the consideration of exposures to radioactive particles, for which cumulative exposure is also important (e.g., Refs. 125, 126). More recent work has focused on the appropriateness of the group mean exposure for determining acceptable conditions in general (9, 12, 83, 127) and upon statistical methods for testing the means of lognormal distributions relative to limits (27–32). Strategies for evaluating mean exposures relative to OELs have been applied to underground mines in the United States (128, 129) and for monitoring long-term exposures to hazardous substances in Germany (130, 131).

6.5.1 Hypothesis Testing

The following method can be used for testing hypotheses concerning means from a log-normal distribution,

$$H_0\text{: } \mu_x \geq OEL \text{ (unacceptable)}$$

versus

$$H_1\text{: } \mu_x < OEL \text{ (acceptable)}$$

(Again, note that evidence is required to declare exposure acceptable). Various test statistics have been proposed to perform this test at an α level of significance, notably those due to Rappaport and Selvin (30), to Lyles and Kupper (27), and to Land (132). Although uniformly the most powerful, the test of Land requires special tables and/or graphs to perform the calculations and often involves various forms of interpolation as well. The other two methods are much simpler to apply. Extensive simulation has shown that the method of Lyles and Kupper is more powerful than that of Rappaport and Selvin and performs essentially as well as that of Land with 10 or more observations (27, 32). For these reasons, the method of Lyles and Kupper (27) is recommended for routine use and will be illustrated here.

As with the test of exceedance, this test of the group mean exposure requires that all observations be independent. Thus, we will assume a sample of single measurements from k persons (x_{ij} for $j = 1$). The test statistic, representing the upper $100(1 - \alpha)^{\text{th}}$ percentile confidence bound of μ_x is given by the following expression,

$$T_m = \hat{\mu}_y + \hat{e}\hat{\sigma}_y$$

where $\hat{\mu}_y$ and $\hat{\sigma}_y$ are the mean and square root of the variance of the logged measurements (y_{ij} for $j = 1$), $\hat{e} = [-\hat{\delta}\sqrt{(k - 1)/k}/\chi_{k-1,\alpha}] + t_{k-1,1-\alpha}/\sqrt{k}$ is a constant, incorporating the estimated noncentrality parameter $\hat{\delta} = (-\sqrt{k}\hat{\sigma}_y)/2$, $\chi_{k-1,\alpha}$ is the square root of the $100(\alpha)^{\text{th}}$ percentile of the chi square distribution with $k - 1$ degrees of freedom, and $t_{k-1,1-\alpha}$ is the $100(1 - \alpha)^{\text{th}}$ percentile of the t distribution with $k - 1$ degrees of freedom, and α is the significance level of the test.

6.5.2 Sample Size Requirements

Lyles and Kupper (27) presented a relationship to estimate sample sizes required for their test as well as sample sizes required to declare the group mean exposure acceptable with $\alpha = 0.05$ and power $= 0.80$ for situations where the OEL/μ_x is between 1.5 and 4.0 (reproduced as Table 18.9). The sample sizes indicate that it should be possible to apply the test to situations in which μ_x is less than a fourth to a half of the OEL, except in some cases where $\hat{\sigma}_y^2 > 1.0$, as observed for intermittent processes (Table 18.6).

For example, using the estimated parameters of Groups 1–4 it appears that about 12 measurements (workers) would be required to reject H_0 (unacceptable exposure) with $\alpha = 0.05$ and power $= 0.80$ for Group 1 (PEL/$\hat{\mu}_x = 2.7$, $\hat{\sigma}_y^2 = 0.400$) and about 7 mea-

Table 18.9. Number of Workers (k) Required for the Test of the Group Mean Exposure (μ_x) when $\alpha = 0.05$ and power $= 0.80$[a,b]

OEL/μ_x	$\sigma_y^2 = 0.35$	$\sigma_y^2 = 0.50$	$\sigma_y^2 = 1.0$	$\sigma_y^2 = 2.0$	$\sigma_y^2 = 3.0$	$\sigma_y^2 = 4.0$
1.5	27	38	83	191	322	473
2.0	12	16	32	71	118	171
4.0	5	7	11	22	35	49

[a]From Ref. 27, assuming one measurement per worker.
[b]σ_y^2 is the variance of the logged exposures (Y_{ij} for $j = 1$ measurement per person).

surements (workers) for Group 2 (PEL/$\hat{\mu}_x = 3.8$ $\hat{\sigma}_y^2 = 0.499$). In the case of Group 3, where PEL/$\hat{\mu}_x = 0.20$, it would not be possible to declare the group mean acceptable with probability greater than α. Finally, Table 18.9 indicates that about 83 measurements (workers) would be required from Group 4 in order to declare exposure acceptable (PEL/$\hat{\mu}_x = 1.4$, $\hat{\sigma}_y^2 = 0.805$).

6.5.3 *Outcome vs. Risk*

The test of μ_x relates to the likelihood that the mean exposure of the group would exceed the OEL, without considering the distribution of individual μ_{xi} among the workers. Thus, without evidence that the group is uniformly exposed, some workers can have much greater probabilities of overexposure than others [see Figure 18.5(b)]. Since individual risk depends upon θ, the simple assurance that the group mean μ_x is less than the OEL is not sufficient to ensure an acceptable level of risk for all members of the group.

An alternative application of this strategy takes advantage of the lognormality of the individual mean exposures $\{\mu_{xi}\}$ to demonstrate indirectly an acceptable level of overexposure, i.e., that $\theta < A$. In Equation 6 it was shown that $\theta_{\max}$ is related to a particular ratio of the OEL to μ_x; for example, $\theta_{\max} = 0.10$ when OEL/$\mu_x = 2.3$. So rejection of a test of H_0: $\mu_x \geq$ OEL/2.3 versus H_1:$\mu_x <$ OEL/2.3 would lead to the valid inference that $\theta < A = 0.10$ at an α level of significance (83). Since such a test would be accomplished without repeated measurements, it might be useful in initial screening studies where it is anticipated that μ_x would be very small. Let L represent an arbitrary fraction of the OEL. To estimate sample sizes for such an application, Table 18.9 can be used by substituting the value of $L =$ OEL/2.3 for OEL. For example, 7 measurements (workers) would be needed to demonstrate acceptable exposure with 80% power when $L/\mu_x = 4$ (equivalent to OEL/$\mu_x = 9.2$) and $\sigma_y^2 = 0.5$ at an $\alpha = 0.05$ level of significance.

6.5.4 *Application to STELs*

Since the mean from a lognormal distribution is independent of the averaging time of measurements (89), it is theoretically possible to evaluate the frequency with which short-term exposures exceed a STEL based solely upon knowledge of μ_x (83). For example, if it can be demonstrated that $\mu_x <$ STEL/4, then no more than 5% of exposures are expected to exceed the STEL over 15 min (or other) periods (see Table 18.5). Since μ_x can be estimated with measurements covering a full shift, it should be possible to evaluate transient excursions of the STEL without collecting short-term data. In fact, because autocorrelation should generally decrease with increasing averaging time (89), long-term measurements can arguably allow more valid inferences to be drawn about the probabilities of peak exposures. For example, daily measurements tend to be much more independent than short term measurements obtained during a single shift.

The idea that long-term monitoring data can be used to evaluate short-term exposures is potentially useful. Rappaport et al. (83) explored this notion with exposure data obtained from 41 workers exposed to toluene diisocyanate in seven factories that manufactured polyurethane foams. In each case that μ_x was demonstrably less than 0.005 ppm (the TLV–TWA) fewer than 5% of 15-min. exposures exceeded the TLV–STEL of 0.02 ppm = 4(TLV–TWA), as predicted by theory.

It is, therefore, implicit in any pair of long- and short-term limits that, insofar as exposures are lognormally distributed and μ_x is less than the long-term exposure limit (designated LTEL, e.g., TLV–TWA), the maximum probability that the STEL can be exceeded is

$$\gamma_{\text{STEL}} = P\{X_{ij} > STEL\} = 1 - \Phi\left\{\left(2 \times \ln\left[\frac{STEL}{LTEL}\right]\right)^{1/2}\right\}$$

where $\Phi\{z_{(1-\gamma STEL)}\}$ denotes the probability that a standard normal variate Z would fall below the value $z_{(1-\gamma STEL)} = \{(2 \times \ln[STEL/LTEL])^{1/2}\}$ and X_{ij} refers here to the j-th short-term measurement. For example, if $\mu_x <$ TLV–TWA = 0.5 ppm for benzene (98), then the maximum probability that the TLV-STEL = 2.5 ppm would be exceeded is

$$\gamma_{\text{STEL}} = 1 - \Phi\left\{\left(2 \times \ln\left[\frac{2.5}{0.5}\right]\right)^{1/2} = 1.79\right\} = 0.037$$

6.5.5 *Options for Control*

Because the test of the group mean exposure can only be applied with single measurements from samples of workers, it provides little with which to differentiate levels of exposure between workers. Thus, engineering or administrative controls remain the only valid options in situations deemed to be unacceptable. As noted in the last section, any information obtained during the surveys about particular tasks can be used to partially focus control efforts.

6.5.6 *Summary*

As with exceedance, the test of the group mean exposure allows acceptable exposure to be rigorously tested and does not assume acceptable exposure by default. Unfortunately, demonstration that $\mu_x < OEL$ does not ensure that the probability of overexposure is also acceptably low (i.e., that $\theta < A$). However, a variant of the test can be used to make inferences about θ by demonstrating that $\mu_x < L$, where L is again an arbitrary constant (e.g., if $L = OEL/2.3$ then $\theta_{\max} < 0.10$). Unfortunately existing test statistics require independent observations; thus, only single measurements from a sample of persons can be used and many workers can be needed in some cases. Because repeated measurements from the same workers are not obtained, control options remain limited to engineering and administrative solutions.

6.6 Testing Overexposure

The definition of θ is the probability that a randomly selected worker from an observational group would have a value of $\mu_{xi} > OEL$ (34). Acceptable exposure can, therefore, be defined by the condition where $\theta < A$. This strategy differs fundamentally from those discussed previously because its application requires information about the within- and between-person components of variance. Thus, repeated measurements are required from a sample of workers in the observational group and the random effects model should be applied to estimate the parameters of the distribution.

As originally described, the test of overexposure was part of a structured protocol that assisted the occupational hygienist in obtaining samples of data, determining whether exposures were acceptable, and using the data to focus options for controlling unacceptable exposures (34). The following discussion will focus primarily upon the testing component of the protocol.

6.6.1 Hypothesis Testing

Following previous convention (where unacceptable exposure is assumed by default) the test of overexposure applies the following null and alternative hypotheses:

$$H_0\colon \theta \geq A \text{ (unacceptable)}$$

versus

$$H_1\colon \theta < A \text{ (acceptable)}$$

A condition equivalent to $\theta < A$ is $(\mu_y + 0.5\sigma_w^2 + \sigma_B z_{1-A}) < \ln(OEL)$ where z_{1-A} represents the $100(1 - A)^{\text{th}}$ percentile of the standard normal distribution. Thus, by defining the quantity,

$$R = \mu_y + 0.5\sigma_w^2 + \sigma_B z_{1-A} - \ln(OEL)$$

an equivalent test of $\theta < A$ can be constructed as (34):

$$H_0\colon R \geq 0 \text{ (unacceptable)}$$

versus

$$H_1\colon R < 0 \text{ (acceptable)}$$

Since σ_B^2 and σ_W^2 are unknown, Lyles and coworkers (34–36) developed a Wald-type test of H_0 versus H_1. Under the random effects ANOVA model, with n_i measurements from the i-th person in a sample of k workers, the Wald-type statistic for testing H_0 versus H_1 can be written in terms of the estimated values of R and its variance as follows:

$$\hat{W} = \frac{\hat{R}}{\sqrt{\hat{V}[\hat{R}]}}, \text{ for}$$

$$\hat{R} = \hat{\mu}_y + 0.5\hat{\sigma}_w^2 + \hat{\sigma}_B z_{1-A} - \ln(OEL), \text{ and}$$

$$\hat{V}[\hat{R}] = \hat{a} + \frac{\hat{b}}{4} + \frac{z_{1-A}}{4\hat{\sigma}_B}\left(2\hat{c} + \frac{\hat{d}z_{1-A}}{\hat{\sigma}_B}\right)$$

where $\hat{\sigma}_W^2$ and $\hat{\sigma}_B^2$ represent the ANOVA estimates of the within- and between-worker components of variance, and the algebraic constants are defined as follows:

$$\hat{a} = \left(\sum_{i=1}^{k} \frac{n_i}{(n_i\hat{\sigma}_B^2 + \hat{\sigma}_W^2)}\right)^{-1}$$

$$\hat{b} = 2\hat{\sigma}_W^4/(N - k)$$

$$\hat{c} = \frac{-2\hat{\sigma}_W^4(k - 1)}{(N - k)(N - S_2/N)}, \text{ and}$$

$$\hat{d} = \frac{2N}{(N^2 - S_2)}\left(\frac{N(N - 1)(k - 1)}{(N - k)(N^2 - S_2)}\hat{\sigma}_W^4 + 2\hat{\sigma}_W^2\hat{\sigma}_B^2 + \frac{\hat{\sigma}_B^4(N^2S_2 + S_2^2 - 2NS_3)}{N(N^2 - S_2)}\right),$$

where $S_2 = \Sigma_{i=1}^k n_i^2$ and $S_3 = \Sigma_{i=1}^k n_i^3$

Under H_0 the statistic $\hat{W}$ has an approximate standard normal distribution. Thus, H_0 can be rejected in favor of H_1 (and exposure declared acceptable) when $\hat{W} \leqq z_\alpha$, where z_α is the value from the standard-normal distribution associated with an α-level of significance.

The Wald-type test has been studied in detail for balanced sets of exposure data (36). The authors showed that the methodology compared favorably with more complicated test statistics, with the exception of situations in which the population value of the ratio σ_B^2/σ_W^2 was relatively large. To account for such behavior (likely due to the reduced accuracy of the normal approximation for small sample sizes), they suggested the following simple adjustment to the rejection rule when applying the Wald-type test: if $\hat{\sigma}_B^2/\hat{\sigma}_W^2 \geqq 0.5$, then reject H_0 at the $\alpha/2$ level of significance.

Finally, in situations where a negative estimate of σ_B^2 is obtained, the Wald-type statistic, behaves inappropriately (i.e., if $\hat{\sigma}_B^2$ is set equal to zero, then $\hat{W} = 0$ regardless of the values of the other parameter estimates). In such cases the authors recommended an alternative test assuming that the N observations comprise a random sample of data. (When the estimated between-person variance component is negative, the intra-class correlation would be close to zero; thus, it would be reasonable to assume independent data even if repeated measurements are obtained from the same persons.) The alternative test is similar to that described for the test of the mean exposure in the last section, for the special case where H_0: $\mu_x \geqq L$ versus H_1: $\mu_x < L$ (where $L = OEL/2.3$ when $A = 0.10$). To apply that test, simply replace the number of workers k by the total number of measurements N for all calculations.

6.6.2 Sample Size Requirements

Rappaport et al. (34) provided a relationship to estimate sample sizes required at a given power to reject the null hypothesis. The authors commented that the limited power of the Wald-type test made it difficult to declare exposure acceptable for situations where μ_x is less than about a third of the OEL ($OEL/\mu_x < 3$). This is illustrated in Table 18.10 which used the published relationship to estimate sample sizes ($N = kn$ for $n = 2$ measurements from each of k persons) for various exposure distributions with 80% power and a significance level of 0.05. The values of the variance and the intra-class correlation used for the calculations were approximately equal to the 25th, 50th and 75th percentiles of the cumulative distributions of $\hat{\sigma}_y^2$ and $\hat{\varrho} = \hat{\sigma}_B^2/\hat{\sigma}_y^2$ estimated by Tornero-Velez et al. (14) from 220 observational groups. The table shows that when $OEL/\mu_x \leq 2.0$ very large sample sizes would be required to reject H_0 with 80% power. However, as the group mean becomes smaller, sample sizes are generally achievable, particularly in cases where $\sigma_y^2 \leqq 1$, as is generally observed for continuous processes (Table 18.6).

Table 18.10 is based upon $n = 2$ measurements per person. Other values for n resulted in somewhat different sample sizes at a given power (results not shown), depending upon the particular parameters of the exposure distribution. The optimal values of n were consistently between 2 and 4 for the combinations of parameters given in Table 18.10.

Returning now to the data for the four groups, the sample sizes listed in Table 18.10 suggest that Group 1 ($\hat{\theta} < 0.001$; $PEL/\hat{\mu}_x = 2.7$, $\hat{\sigma}_y^2 = 0.400$; $\hat{\varrho} = 0.058$) and Group 2 ($\hat{\theta} < 0.001$; $PEL/\hat{\mu}_x = 3.8$, $\hat{\sigma}_y^2 = 0.499$; $\hat{\varrho} = 0.064$) would both be declared acceptable using the Wald-type test with as few as four measurements (for $A = 0.1$ and $\alpha = 0.05$). On the other hand, Group 3 ($\hat{\theta} = 0.610$; $PEL/\hat{\mu}_x = 0.20$, $\hat{\sigma}_y^2 = 5.39$; $\hat{\varrho} = 0.441$) could not be declared acceptable with probability greater than α and Group 4 ($\hat{\theta} = 0.184$; $PEL/\hat{\mu}_x = 1.4$, $\hat{\sigma}_y^2 = 0.805$; $\hat{\varrho} = 0.362$) is unlikely to be declared acceptable at achievable sample sizes.

Assuming that it is desirable to make at least 10 measurements with at least two measurements per person, then all workers would typically be monitored in groups containing five or fewer members. The inability to monitor more workers in these groups (i.e., to increase k above 5) has obvious power implications, which can be only partially offset by

Table 18.10. Number of Measurements (N) Required for the Test of Overexposure when $\alpha = 0.05$ and power $= 0.80$[a]

	$\sigma_y^2 = 0.50$			$\sigma_y^2 = 1.0$			$\sigma_y^2 = 2.0$		
OEL/μ_x	$\rho = 0.05$	$\rho = 0.25$	$\rho = 0.50$	$\rho = 0.05$	$\rho = 0.25$	$\rho = 0.50$	$\rho = 0.05$	$\rho = 0.25$	$\rho = 0.50$
2.0	178	138	334	444	764	>1000	>1000	>1000	>1000
3.0	4	28	34	4	74	108	272	246	424
4.0	4	16	16	4	34	42	142	92	118
8.0	4	4	8	4	12	14	4	26	28

[a] μ_x is the group mean exposure; σ_y^2 is the variance of the logged exposures (Y_{ij} for $j = 2$ measurements per person); ρ is the intra-class correlation (σ_B^2/σ_y^2).

increasing the number of measurements per person. A possible solution would be to combine several small groups that still share a common factor (e.g., department or building), but do not have the same job. However, this is likely to increase the between-person variation and, therefore, would make it more difficult to declare exposure acceptable.

6.6.3 *Outcome vs. Risk*

The test of overexposure was motivated by the reasonable assumption that the risks associated with long-term exposures to chemical agents should be related to the cumulative exposures of the individual persons, that is to $\mu_{xi} \times t$ (9, 80, 102, 117). As noted above, this assumption is inherent in OSHA's risk assessments. It is also supported by theoretical and empirical studies indicating that toxicokinetic processes tend to be linear when exposures are low relative to saturable biochemical pathways (9, 107, 108, 117). Because this should be the usual situation involving occupational exposures today (at least in countries with well-developed programs for occupational health) the test of overexposure should be generally appropriate for applications to chronic exposures. Nonetheless, one should be wary of situations invalidating the assumption of linear kinetics at expected levels of exposure (79, 109–111).

Although cumulative exposure should be a valid measure of long-term risk, problems can be encountered in the extreme. Obviously, it would be inappropriate for a worker to be exposed to a lifetime's exposure or even a year's exposure on one day. For this reason, Rappaport et al. (34) encouraged hygienists to demonstrate that $\theta < A$ at each stage of evaluation (e.g., yearly). They also suggested that any evidence of individual overexposure (when, for example, the predicted value of an individual worker's mean exposure was greater than the OEL) should be dealt with directly.

6.6.4 *Options for Control*

A great advantage of the strategy based upon overexposure is the ability to use the exposure data when necessary to select among control options. If the group is uniformly exposed, then engineering or administrative controls should be instituted. However, if the group is not uniformly exposed, the goal should be to determine why exposures differ among the workers. Then interventions can be considered relative to the particular sources of exposure in a more effective manner.

Large interindividual differences in exposure suggest either that the workers are performing different tasks or that the individual personal environments differ substantially. (The term 'personal environment' includes all factors related to the individual worker, e.g., differences in equipment, location, and work practices.) If tasks vary among persons, then the group can be divided into two or more subgroups (based upon sets of tasks), each of which would presumably be more uniformly exposed. Then, additional measurements would be made (if required) and the test repeated for each of the subgroups. Since it is likely that some subgroups would have lower exposures than others, the result of regrouping could well be to demonstrate that one or more subgroups has acceptable exposures. If regrouping is not a viable option, then it is necessary to intervene by investigating and altering personal environmental factors that contribute to large exposures. Then, the testing process would be repeated with a new sample of data.

To illustrate the use of exposure data for control purposes, consider Groups 3 and 4, both of which would probably be found to have unacceptable exposures. The scatterplots, shown in Figure 18.3, suggest different situations between the two groups. Exposures among members of Group 3, consisting of Transfer/Movement Operators in a petroleum refinery, were extremely variable both within and between workers. Indeed, the variability was so great as to suggest that the group contains at least two subgroups (with distinct tasks), each of which includes a complicated mix of individual personal environments among the workers. Thus, a reasonable plan to reduce exposures for Group 3 would be to (*1*) assess the mix of tasks involved (with the goal of defining subgroups) and (*2*) evaluate the particular personal factors which gave rise to the highest and lowest daily exposures. Obviously, any information about tasks and activities that had been noted on the days that measurements were performed would be extremely useful at this stage.

Regarding Group 4, consisting of Sprayers and Laminators in a boat factory, Figure 18.3 suggests a somewhat different situation. Here, much of the between-worker variability can be traced to two persons (Workers 1 and 7), both of which had relatively low exposures. Indeed, when the random effects model was applied to the data after removing measurements from these two persons, the variance and correlation were altered greatly ($\hat{\sigma}_y^2$ dropped from 0.805 to 0.547 and $\hat{\varrho}$ dropped from 0.362 to 0.107), suggesting that these two workers were quite influential. Here, it would make sense to ascertain why these two individuals were exposed to much lower levels than the others (again tasks or personal factors could be important). After excluding Workers 1 and 7, the other 18 workers in Group 4 had ${}_B\hat{R}_{0.95} = 3.51$, suggesting a general problem ultimately requiring some form of engineering solution. Because the author is aware that the exposures took place in a large naturally-ventilated building with diffuse sources of styrene, this is a reasonable conclusion.

6.6.5 *Summary*

The test of overexposure (Section 6.6) comes closest to the ideal exposure assessment procedure mentioned in Section 6.1 on testing relative to limits. Overexposure can be tested rigorously in a manner that motivates collection of data to declare exposure acceptable. The connection between long-term risk and the outcome of the test is straightforward and consistent with assumptions inherent in risk assessment and sound toxicological principles. Methods allow (indeed require) multiple measurements from the same persons so that the within- and between-person components of variance can be estimated and used to guide options for controlling exposure.

Although the Wald-type test used to evaluate overexposure is fairly robust, it is not particularly powerful. This can cause problems when group mean exposures are greater than about a third of the OEL due to the large sample sizes required to declare exposures acceptable.

7 CONCLUSIONS

This chapter has described a complicated milieu in which occupational hygienists interpret exposures to chemical agents. The complexity stems largely from the great variability in exposure that occurs both within and between workers performing the same job. With recognition of these sources of variability, increasingly sophisticated statistical methods have evolved to assist in making decisions. The major thrust of this chapter has been to present these methods within the context of defining acceptable exposure and testing exposures relative to limits. Although some of the statistics are more complex than those used by occupational hygienists in the past, the difficulties in applying them to the problem are not as great as they might appear initially. The methods lend themselves to microcomputer software including common spreadsheet programs. In fact, all calculations describing Groups 1–4 were performed with Microsoft Excel©.

7.1 Increases in Sample Size

A sample size of 10–20 observations (two measurements each from 5–10 workers) has been recommended for initial assessments of exposure in an observational group. Although samples of this magnitude are certainly larger than those used for traditional compliance testing, they are still rather modest when considered in light of the enormous variability in exposure levels, which currently exists (14, 15). Since exposures can be evaluated over a year, these sample sizes should not present a great barrier to organizations with full-time hygiene staff. However, smaller or highly decentralized organizations can be challenged to obtain such samples and should seek mechanisms for increasing numbers by whatever means possible. For example, the use of passive monitors for gaseous contaminants require little professional training and can easily increase the number of measurements by 10-fold over pump-and-collector systems. Also, any compromise between sample size and precision of monitoring should come down heavily on the the side of larger numbers of measurements because environmental variability is invariably much greater than the imprecision of the assays (133). In fact, there is a critical need for inexpensive assays of modest precision (e.g., assay coefficient of variation $CV_a \leq 50\%$) that require minimal calibration and are suitable for use by individuals without professional qualifications (9).

The increased sampling demands, required to deal effectively with exposure variability, will necessarily alter the role of the professional hygienist in assessing exposure (9). Because it will often be impractical for hygienists to make all of the measurements themselves, the physical process of obtaining exposure data must be transferred to individuals with less formal training who can be on site at the times dictated by the sampling schedule (9). The hygienist then becomes the individual who designs the sampling program, facilitates the measurement process, designs follow-up studies, and interprets the data appropriately to allow decisions to be made regarding the need for and types of controls.

7.2 Stationary Behavior and Independence

Most statistical testing assumes that the distribution does not change with time. Based upon current information, it appears reasonable to regard the exposure distribution as stationary for measurements collected within a year but not for longer periods (2, 3, 41). Thus, it is prudent to view exposure assessment as an ongoing activity with periodic tests of the environment. Likewise, occupational hygienists should be reluctant to base judgments entirely upon data collected during brief campaigns (regarding either intra- or inter-day sampling) owing to the possibility of significant autocorrelation of exposures.

Regarding independence, it has been emphasized that the tests of exceedance and the group mean exposure implicitly assume independent data. Since current knowledge indicates pervasive between-worker variability in observational groups, it is unreasonable to assume *a priori* that repeated measurements from the same workers would be independent (i.e., that the intra-class correlation ρ would be zero). Thus, as currently conducted, these tests (of exceedance and the group mean exposure) can only legitimately be applied to a group with one measurement per worker, a major constraint in many situations.

7.3 Assessment of Peak Exposures

The major focus of this chapter had been upon long-term exposures rather than upon short-term or so-called peak exposures. This emphasis was motivated by the recognition that the risk of exposure to fast-acting toxicants is related to the very small probability of encountering extremely large (peak) concentrations. Since it is a great challenge to model rare events (associated with accidents or gross stupidity, for example) on the basis of small samples of exposure data, it seems important to monitor the source(s) of these substances rather than the individual workers (9).

Chemicals that are not fast acting produce effects after years of exposure. In such cases individual risk should be primarily related to the cumulative exposure over relatively long time scales. Indeed, in situations where the individual mean exposure is demonstrably below the OEL, it is unlikely that peak exposures would introduce nonlinear effects into the exposure-response relationship (9, 80). However, if evidence suggests that peak exposures are etiologically important in the long term, then sampling strategies should be developed to discern differences in the exposure regimen among workers. Since such strategies will be even more resource-intensive, their development poses a formidable challenge. The conceptual framework for the problem has been described (79, 109–111) and should provide a starting point for assessing exposures under such situations.

7.4 Acceptable Probabilities of Exceedance or Overexposure

The issue of defining an acceptable probability of exceedance or overexposure is important. Although the values of $E = 0.05$ or $A = 0.10$, used in this chapter, have limited precedence they should provide suitable protection if the OEL is well founded. [In fact, the use of $E = 0.05$ is likely to be extremely conservative in most cases because in typical groups $\gamma > \theta$, suggesting that the corresponding value of A would be much less than 0.05.] Indeed, the range of accessible sample sizes would rarely allow a declaration of acceptable exposure under such conditions when either γ or θ is greater than about 0.01.

Regardless of the value of E or A, the notion of allowable exceedance or overexposure has notable implications. These quantities limit, respectively, the probability that the exposure on a randomly selected workday or (on average) from a randomly selected worker, would exceed the OEL. As such, the probabilities are rather impersonal because they relate to an unidentifiable person in the population represented by the sample. However, in applying the tests, there would be instances where there is strong evidence that particular workers were exposed above the OEL. Provision should be made for investigating such documented cases, regardless of the outcome for the population as a whole.

Regulatory groups that set and enforce exposure limits have been reluctant to embrace the concept of allowable exceedance or overexposure, even though it is essential for statistical testing. Such reluctance has been understandable because the notion appears to legitimize situations where workers are more highly exposed than desirable. In practice, however, this well-intentioned inflexibility encourages minimal sampling to avoid breaches of the OEL (13), with the results that exposures remain poorly characterized, some individuals remain overexposed, and inappropriate control measures are sometimes chosen. Hopefully, regulators will recognize that their interest in protecting workers is not served by the mere appearance of compliance, but rather by encouraging long-term assessments of exposure.

7.5 Methods for Testing Exposure

Four approaches were evaluated for testing exposures relative to limits: compliance testing, testing exceedance, testing the group mean exposure and testing overexposure. Of these, compliance testing was found wanting by all criteria, save expediency. Since minimizing data maximizes the probability of compliance, sample sizes remain small and decisions become highly capricious; indeed, given current sample sizes ($n \leq 4$) compliance testing is very likely to allow hazardous conditions to go unnoticed. The other tests are more useful than compliance testing because they can be rigorously conducted and encourage the collection of more and better data. However, the tests of exceedance and the group mean exposure have only limited utility because they implicitly assume that the group is uniformly exposed. Thus, these tests must be limited to single measurements from each worker; outcomes of tests are only marginally related to individual risk; and data provide little insight into proper selection of controls. The test of overexposure (section 6.6) overcomes these disadvantages by recognizing the inherent variability both within and between persons. Thus, the outcome of a test of overexposure is consistent with the individual risk of long-term exposure, and the data can be used to optimize controls.

APPENDIX A

Exposure data for the four groups of workers described in Table 18.1 of the text. Air concentrations are given in $\mu g/m^3$ for Group 1 and mg/m^3 for Groups 2–4.

Group	Worker	Conc.	Group	Worker	Conc.
1	1	10.40	1	1	20.20
1	1	11.40	1	1	15.90
1	1	17.10	1	1	9.00
1	1	28.90	1	1	9.50
1	1	12.40	1	1	19.10
1	1	10.00	1	1	9.00
1	1	24.60	1	1	25.70
1	1	21.50	1	1	46.90
1	1	11.30	1	1	7.90
1	1	15.00	1	1	22.20
1	1	30.40	1	2	30.80
1	1	25.40	1	2	1.30
1	1	34.00	1	2	23.80
1	1	3.90	1	2	10.10
1	1	14.40	1	2	11.40
1	1	18.60	1	2	7.50
1	1	56.40	1	2	50.10
1	1	25.60	1	2	18.00
1	1	10.00	1	2	56.50
1	1	21.90	1	2	9.70
1	1	12.90	1	2	13.70
1	1	8.60	1	2	40.90
1	1	19.60	1	2	20.80
1	1	13.00	1	2	62.60
1	1	10.20	1	2	19.20
1	1	19.50	1	2	21.00
1	2	3.20	1	4	5.80
1	2	17.70	1	4	6.80
1	2	13.10	1	4	6.90
1	2	34.00	1	4	5.00
1	2	16.70	1	4	20.00
1	2	13.50	1	4	19.60
1	2	10.10	1	4	13.20
1	3	9.70	1	4	9.40
1	3	5.10	1	4	9.90
1	3	15.20	1	4	7.90
1	3	34.80	1	4	22.50
1	3	8.70	1	4	22.60
1	3	14.70	1	4	24.60
1	3	13.30	1	4	20.10
1	3	6.60	1	4	39.30
1	3	5.20	1	4	24.40

Group	Worker	Conc.	Group	Worker	Conc.
1	3	7.70	1	4	14.00
1	3	7.70	1	4	10.60
1	3	23.70	1	4	31.90
1	3	18.80	1	4	6.30
1	3	21.40	1	4	10.10
1	3	7.10	1	4	6.90
1	3	17.80	1	4	21.30
1	3	10.20	1	4	10.30
1	3	16.50	1	4	10.80
1	3	11.20	1	4	21.50
1	3	16.00	1	4	55.00
1	3	16.80	1	4	19.90
1	3	17.70	1	4	7.60
1	3	7.70	1	5	8.80
1	3	19.70	1	5	20.80
1	3	8.80	1	5	18.90
1	3	9.20	1	5	59.70
1	3	9.30	1	5	35.00
1	3	4.30	1	5	23.20
1	3	4.90	1	5	15.70
1	3	8.50	1	5	53.50
1	3	14.50	1	5	24.40
1	3	40.40	1	5	11.50
1	3	9.20	1	5	15.70
1	3	27.80	1	5	31.40
1	4	25.10	1	5	33.50
1	4	6.10	1	5	26.90
1	4	19.90	1	5	39.90
1	4	61.10	1	5	11.30
1	4	8.10	1	5	26.90
1	4	13.80	1	5	11.80
1	4	26.50	1	5	14.70
1	5	12.40	2	5	1.09
1	5	13.10	2	5	1.15
1	5	33.10	2	5	0.70
1	5	11.00	2	5	1.28
1	5	27.50	2	6	0.38
1	5	18.60	2	6	0.29
1	5	15.00	2	7	0.54
1	5	8.90	2	7	1.47
1	5	18.10	2	7	0.51
1	5	16.70	2	7	1.53
1	5	16.70	2	8	2.27

Group	Worker	Conc.	Group	Worker	Conc.
1	5	31.00	2	8	0.73
1	5	22.90	2	8	1.37
1	5	22.90	2	8	1.15
1	6	12.00	2	8	0.42
1	6	10.90	2	8	1.02
1	6	8.60	2	9	0.51
1	6	38.20	2	9	0.51
1	6	7.30	2	9	1.18
1	6	13.80	2	9	0.54
1	6	10.80	2	9	1.34
1	6	10.30	2	10	1.15
1	6	8.10	2	10	1.63
1	6	12.00	2	10	0.51
1	6	8.10	2	10	0.86
1	6	19.40	2	10	0.38
1	6	29.00	2	10	0.27
1	6	18.30	2	10	0.42
1	6	30.60	2	10	0.45
2	1	0.48	2	11	1.05
2	1	0.32	2	11	1.76
2	1	0.38	2	11	0.80
2	1	0.27	2	11	0.38
2	1	0.54	2	12	0.80
2	2	0.54	2	12	0.42
2	2	1.09	2	13	1.82
2	2	1.18	2	13	1.28
2	2	0.67	2	13	0.57
2	2	0.32	2	13	0.70
2	2	0.28	2	14	2.68
2	2	0.57	2	14	0.25
2	3	0.42	2	14	0.35
2	3	0.28	2	14	5.75
2	3	0.61	2	14	0.45
2	4	1.28	2	15	0.45
2	4	0.57	2	16	0.61
2	4	40.45	2	16	1.69
2	5	0.99	2	16	0.83
2	16	0.51	3	12	0.13
2	16	0.25	3	13	0.26
2	16	0.10	3	13	0.61
2	16	0.42	3	13	0.22
2	16	0.09	3	13	13.74
2	16	0.35	3	13	1.31

Group	Worker	Conc.	Group	Worker	Conc.
2	17	1.18	3	14	2.75
2	17	0.35	3	15	2.72
2	17	1.47	3	16	23.32
2	17	0.73	3	17	0.16
2	17	0.93	3	17	5.11
2	17	0.29	3	18	0.42
2	17	0.51	3	19	0.32
2	17	0.48	3	19	0.57
2	17	0.30	3	19	0.38
2	18	1.63	3	19	0.32
2	18	3.19	3	20	0.16
2	18	0.57	3	20	0.06
2	18	1.02	3	21	0.01
2	18	0.83	3	21	0.96
2	18	0.31	3	22	0.03
2	18	0.54	3	22	0.01
2	18	0.45	3	23	0.03
3	1	18.53	3	24	2.84
3	1	5.85	4	1	19.39
3	2	48.87	4	1	18.73
3	3	70.60	4	1	21.56
3	4	0.51	4	1	56.20
3	4	0.51	4	1	48.00
3	5	14.79	4	2	66.87
3	6	0.03	4	2	75.08
3	6	0.29	4	2	81.14
3	7	0.22	4	2	184.21
3	7	0.32	4	2	220.50
3	8	6.48	4	2	222.22
3	8	0.51	4	2	155.30
3	8	123.61	4	3	60.52
3	9	0.89	4	3	131.06
3	9	1.34	4	3	185.05
3	9	0.19	4	3	84.66
3	9	5.43	4	3	178.88
3	10	19.23	4	3	6.64
3	10	1.92	4	3	157.30
3	10	0.51	4	4	206.25
3	11	0.80	4	4	303.68
3	12	1.47	4	4	166.63
3	12	0.01	4	4	108.61
4	4	96.78	4	11	209.56
4	4	114.87	4	12	157.58

Group	Worker	Conc.	Group	Worker	Conc.
4	4	264.47	4	12	233.61
4	5	67.86	4	12	181.40
4	5	50.66	4	12	244.71
4	5	210.81	4	12	188.01
4	6	107.50	4	12	194.63
4	6	118.46	4	13	241.17
4	6	133.33	4	13	384.00
4	6	82.65	4	13	139.03
4	6	138.27	4	13	204.40
4	6	153.63	4	13	284.59
4	6	361.13	4	13	71.01
4	7	2.78	4	14	74.20
4	7	10.15	4	14	221.67
4	7	31.64	4	15	89.93
4	7	47.05	4	15	179.08
4	7	72.61	4	15	208.47
4	7	82.00	4	16	189.83
4	8	74.42	4	16	213.78
4	8	179.75	4	16	153.32
4	8	86.23	4	16	176.79
4	8	82.29	4	16	391.33
4	8	163.53	4	16	287.04
4	9	32.08	4	17	108.77
4	9	58.54	4	17	188.63
4	9	39.52	4	17	89.47
4	9	49.02	4	17	123.10
4	9	66.61	4	17	175.80
4	9	69.23	4	17	131.26
4	9	150.06	4	17	169.31
4	10	30.60	4	18	178.20
4	10	56.00	4	18	6.32
4	10	35.09	4	18	256.08
4	10	42.45	4	18	275.59
4	10	70.52	4	19	69.57
4	10	64.07	4	19	43.57
4	10	154.14	4	19	140.49
4	11	122.41	4	19	66.71
4	11	87.96	4	19	199.22

NOMENCLATURE

α The significance level of a statistical hypothesis test.

β_i The random deviation of the i-th worker's mean logged exposure from μ_y.

γ	The exceedance; the probability that a daily exposure would exceed the OEL.
$\gamma_{\max}$	The maximum value of γ from a lognormal distribution at a given ratio of the OEL to the mean exposure.
$\hat{\gamma}$	The sample estimate of γ.
$\hat{\delta}$	Estimated noncentrality parameter used in the calculation of T_m for the test of the group mean exposure.
λ	Noncentrality parameter used in the calculation of T_u for the test of exceedance.
ρ	The intra-class correlation coefficient under the one-way random effects ANOVA model ($\varrho = \sigma_B^2/(\sigma_B^2 + \sigma_W^2)$.
μ_x	The mean of X_{ij}.
$\hat{\mu}_x$	The sample estimate of μ_x.
μ_g	The geometric mean of X_{ij}.
μ_{xi}	The mean exposure for the i-th worker.
μ_y	The mean of Y_{ij}.
$\hat{\mu}_y$	The sample estimate of μ_y.
μ_{yi}	The mean logged exposure for the i-th worker.
θ	The probability that a randomly selected worker's mean exposure is greater than the OEL.
$\hat{\theta}$	The sample estimate of θ.
$\theta_{\max}$	The maximum value of θ from a lognormal distribution at a given ratio of the OEL to the group mean exposure.
σ_B^2	The between-worker component of variance of the logged exposure variable Y_{ij}.
$\hat{\sigma}_B^2$	The sample estimate of σ_B^2.
σ_g	The geometric standard deviation of X_{ij}.
σ_W^2	The within-worker component of variance of the logged exposure variable Y_{ij}.
$\hat{\sigma}_W^2$	The sample estimate of σ_W^2
σ_x^2	The variance of X_{ij}.
σ_y^2	The variance of Y_{ij}.
$\hat{\sigma}_y^2$	The sample estimate of σ_y^2.
A	An arbitrarily assigned acceptable probability that a randomly selected worker's mean exposure is greater than the OEL.
ACGIH	American Conference of Governmental Industrial Hygienists. A private organization of industrial hygienists employed by federal, state and local governments or by educational institutions, in the United States. A committee of the ACGIH sets TLVs.
AL	(Action Level) An OEL for an air contaminant established by OSHA at about one half of the PEL.

$\hat{a}$, $\hat{b}$, $\hat{c}$, $\hat{d}$, S_2, S_3	Algebraic constants used in calculations of $\hat{V}[\hat{R}]$ for the test of overexposure.
E	An arbitrarily assigned acceptable probability that a randomly selected worker's exposure is greater than the OEL on a randomly selected day.
$\hat{e}$	Algebraic constant used in calculation of T_m for the test of the group mean exposure.
$CE_{i,t}$	The cumulative exposure of the i-th worker after t years.
CV	Coefficient of variation; σ_x/μ_x.
CV_a	Coefficient of variation of the assay for measuring exposure levels.
HEG	(Homogeneous Exposure Group) An observational group assumed to be uniformly exposed.
k	The number of workers in a group or sample.
K	Proportionality constant between individual risk and cumulative exposure.
L	An arbitrary fraction of the OEL.
LTEL	Long term exposure limit (an average exposure level used in connection with limiting the exceedance of STELs) with the group mean test.
MSB	The mean square between workers obtained from the ANOVA table.
MSW	The mean square within workers obtained from the ANOVA table.
n	The number of measurements obtained from each worker in a balanced set of data.
n_i	The number of measurements obtained from the i-th worker.
N	The total number of measurements obtained from a group of workers.
OEL	(Occupational Exposure Limit) A standard or guide for an air contaminant.
OSHA	Occupational Safety and Health Administration, an agency of the United States government which sets and enforces PELs.
P(Comp.)	Probability of compliance; the likelihood that all measurements from a sample of given size would be less than the OEL
P(Noncomp.)	Probability of noncompliance; the likelihood that at least one measurement from a sample of given size would exceed the OEL.
PEL	(Permissible Exposure Limit) An OEL for an air contaminant established by OSHA.
R	A quantity which is used to develop the test of overexposure.
$\hat{R}$	The sample estimate of R.
$R_{0.95}$	Ratio of the 97.5th and 2.5th percentiles of the lognormally distributed daily exposures $\{X_{ij}\}$ of a group of workers; this is equivalent to a factor containing 95% of the $\{X_{ij}\}$.
$\hat{R}_{0.95}$	Sample estimate of $R_{0.95}$.

${}_{B}R_{0.95}$	Ratio of the 97.5th and 2.5th percentiles of the lognormally distributed mean exposures $\{\mu_{xi}\}$ of a group of workers; this is equivalent to a factor containing 95% of the $\{\mu_{xi}\}$ derived from a lognormal distribution (monomorphic group).
${}_{B}\hat{R}_{0.95}$	Sample estimate of ${}_{B}R_{0.95}$.
${}_{W}R_{0.95}$	Ratio of the 97.5th and 2.5th percentiles of the lognormally distributed daily exposures of the i-th worker $\{X_{ij}\|\mu_{xi}\}$; this is equivalent to a factor containing 95% of the $\{X_{ij}\|\mu_{xi}\}$.
${}_{W}\hat{R}_{0.95}$	Sample estimate of ${}_{W}R_{0.95}$.
$Risk_{i,t}$	Risk of disease arising from cumulative exposure to the *i*-th person over *t* year.
t	Time; the duration of exposure.
T_m	Test statistic for determining whether the group mean exposure from a lognormal distribution is less than the OEL.
T_u	Test statistic for determining whether less than 100(E)% of exposures from a lognormal distribution exceed the OEL (test of exceedance).
TLV	(Threshold Limit Value) OEL for an air contaminant established by the ACGIH.
$\hat{V}[\hat{R}]$	The estimated variance of $\hat{R}$.
$\hat{W}$	The Wald-type test statistic used for the test of overexposure.
X_{ij}	The daily exposure received by the *i*-th worker on the *j*-th day in an observational group.
x_{ij}	The daily personal measurement of exposure received by the *i*-th worker on the *j*-th day in an observational group.
$X_{(1-E)}$	The $100(1 - E)^{\text{th}}$ percentile of the distribution of X_{ij}.
$\hat{X}_{(1-E)}$	The sample estimate of $X_{(1-E)}$.
$[X_{ij}\|\mu_{xi}]$	The value of X_{ij} conditional on the *i*-th worker.
$[Y_{ij}\|\mu_{yi}]$	The value of Y_{ij} conditional on the *i*-th worker.
Y_{ij}	The natural logarithm of X_{ij}.
y_{ij}	The natural logarithm of x_{ij}.
$\bar{Y}$	The overall sample mean of the logged measurements (based upon N measurements).
$\bar{Y}_i$	The mean of the logged measurements for the *i*-th worker (based upon n_i measurements).
Z	A standard-normal variate.
$z_{1-\gamma}$	A value from the standard-normal distribution such that $P\{Z > z_{1-\gamma}\} = \gamma$.
$z_{1-\theta}$	A value from the standard-normal distribution such that $P\{Z > z_{1-\theta}\} = \theta$.
z_{1-A}	A value from the standard-normal distribution such that $P\{Z > z_{1-A}\} = A$.
z_{1-E}	A value from the standard-normal distribution such that $P\{Z > z_{1-E}\} = E$.

ACKNOWLEDGMENTS

The author appreciates the helpful comments of Mark Weaver, Rogelio Tornero-Velez, and Peter Egeghy based upon a draft of the manuscript. He also acknowledges the stimulating work with S. Selvin, L. Kupper, E. Symanski, H. Kromhout, and R. Tornero-Velez that motivated much of this chapter.

BIBLIOGRAPHY

1. R. J. Sherwood and D. M. S. Greenhalgh, "A Personal Air Sampler," *Ann. Occup. Hyg.* **2**, 127–132 (1960).
2. E. Symanski, L. L. Kupper, I. Hertz-Picciotto, and S. M. Rappaport, "Comprehensive Evaluation of Long-Term Trends in Occupational Exposure: Part 2. Predictive Models for Declining Exposures," *Occup. Environ. Med.* **55**, 310–316 (1998).
3. E. Symanski, L. L. Kupper, and S. M. Rappaport, "Comprehensive Evaluation of Long-Term Trends in Occupational Exposure: Part 1. Description of the Database;" *Occup. Environ. Med.* **55**, 300–309 (1998).
4. P. Oldham and S. A. Roach, "A Sampling Procedure for Measuring Industrial Dust Exposure." *Br. J. Ind. Med.* **9**, 112–119 (1952).
5. P. Oldham, "The Nature of the Variability of Dust Concentrations at the Coal Face," *Br. J. Ind. Med.* **10**, 227–234 (1953).
6. H. Kromhout, Y. Oostendorp, D. Heederik, and J. S. Boleij, "Agreement between Qualitative Exposure Estimates and Quantitative Exposure Measurements," *Am. J. Ind. Med.* **12**, 551–562 (1987).
7. S. M. Rappaport, R. C. Spear, and S. Selvin, "The Influence of Exposure Variability on Dose–Response Relationships," *Ann. Occup. Hyg.* **32 S1**, 529–537 (1988).
8. R. C. Spear, S. Selvin, J. Schulman, and M. Francis, "Benzene Exposure in the Petroleum Refining Industry," *Appl. Ind. Hyg.* **2**, 155–163 (1987).
9. S. M. Rappaport, "Assessment of Long-Term Exposures to Toxic Substances in Air," *Ann. Occup. Hyg.* **35**, 61–121 (1991).
10. N. A. Leidel, K. Busch, and J. Lynch, *Occupational Exposure Sampling Strategy Manual (NIOSH 77-173)* U.S. Government Printing Office, Washington, DC, 1977.
11. U. Ulfvarson, "Statistical Evaluation of the Results of Measurement of Occupational Exposure to Air Contaminants," *Scand. J. Work Environ. Health* **3**, 109–115 (1977).
12. J. Rock, "A Comparison between OSHA-Compliance Criteria and Action-Level Decision Criteria," *Am. Ind. Hyg. Assoc. J.* **43**, 297–313 (1982).
13. S. M. Rappaport, "The Rules of the Game: An Analysis of OSHA's Enforcement Strategy," *Am. J. Ind. Med.* **6**, 291–303 (1984).
14. R. Tornero-Velez, E. Symanski, H. Kromhout, R. C. Yu, and S. M. Rappaport, "Compliance versus Risk in Assessing Occupational Exposures" [published erratum appears in *Risk Anal.* **17**(5), 657 (Oct. 1997)]. *Risk Anal* **17**, 279–292 (1997).
15. H. Kromhout, E. Symanski, and S. M. Rappaport, "A Comprehensive Evaluation of Within- and Between-Worker Components of Occupational Exposure to Chemical Agents," *Ann. Occup. Hyg.* **37**, 253–270 (1993).

16. R. Cope, B. Panacamo, W. E. Rinehart, and G. L. Ter Haar, "Personal Monitoring for Tetra-alkyllead in an Alkyllead Manufacturing Plant," *Am. Ind. Hyg. Assoc. J.* **40**, 372–379 (1979).
17. N. Esmen and Y. Hammad, "Lognormality of Environmental Sampling Data," *J. Environ. Sci. Health, A12* 29–41 (1977).
18. J. W. Yager, W. M. Paradisin, and S. M. Rappaport, "Sister-Chromatid Exchanges in Lymphocytes are Increased in Relation to Longitudinally Measured Occupational Exposure to Low Concentrations of Styrene," *Mutat. Res.* **319**, 155–165 (1993).
19. S. M. Rappaport, K. Yeowell-O'Connell, W. Bodell, J. W. Yager, and E. Symanski, "An Investigation of Multiple Biomarkers among Workers Exposed to Styrene and Styrene-7,8-oxide," *Cancer Res.* **56**, 5410–5416 (1996).
20. S. M. Rappaport, H. Kromhout, and E. Symanski, "Variation of Exposure between Workers in Homogeneous Exposure Groups," *Am. Ind. Hyg. Assoc. J.* **54**, 654–662 (1993).
21. N. C. Hawkins, S. K. Norwood, and J. C. Rock, *A Strategy for Occupational Exposure Assessment*, American Industrial Hygiene Association, Akron, OH, 1991.
22. N. Esmen, "A Distribution-Free Double Sampling Method for Exposure Assessment," *Appl. Occup. Environ. Hyg.* **7**, 613–621 (1992).
23. M. Corn and N. A. Esmen, "Workplace Exposure Zones for Classification of Employee Exposures to Physical and Chemical Agents," *Am. Ind. Hyg. Assoc. J.* **40**, 47–57 (1979).
24. R. M. Tuggle, "Assessment of Occupational Exposure using One-sided Tolerance Limits," *Am. Ind. Hyg. Assoc. J.* **43**, 338–346 (1982).
25. S. Selvin, S. M. Rappaport, R. C. Spear, J. Schulman, and M. Francis, "A Note on the Assessment of Exposure using One-Sided Tolerance Limits," *Am. Ind. Hyg. Assoc. J.* **48**, 89–93 (1987).
26. AIHA *LOGAN Workplace Exposure Evaluation System User's Manual*. American Industrial Hygiene Association, Akron, OH, 1990.
27. R. H. Lyles and L. L. Kupper, "On Strategies for Comparing Occupational Exposure Data to Limits," *Am. Ind. Hyg. Assoc. J.* **57**, 6–15 (1996).
28. W. Coenen and G. Riediger, "Die Schatzung des zeitlichen Konzentrationsmittelwertes gefahrlicher Arbeitsstoffe in der Luft bei stichprobenartigen Messungen (The estimation of the time related mean air concentrations of harmful substances in random samples)," *Staub Reinhalt. Luft* **38**, 402–409 (1978), in German.
29. H. G. Galbas, "Ein beschranktes sequentielles Testverfahren zur Beurteilung von Schadstoffkonztrationen am Arbeitsplatz (A limited sequential method to control concentrations of noxious substances in the workplace)," *Staub Reinhalt. Luft* **39**, 463–467 (1979), in German.
30. S. M. Rappaport and S. Selvin, "A Method for Evaluating the Mean Exposure from a Lognormal Distribution," *Am. Ind. Hyg. Assoc. J.* **48**, 374–379 (1987).
31. J. S. Evans and N. C. Hawkins, "The Distribution of Student's *t*-Statistic for Small Samples from Lognormal Exposure Distributions," *Am. Ind. Hyg. Assoc. J.* **49**, 512–515 (1988).
32. P. Hewett, "Mean Testing: II. Comparison of Several Alternative Approaches," *Appl. Occup. Environ. Hyg* **12**, 347–355 (1997).
33. P. Hewett, "Mean Testing: I. Advantages and Disadvantages," *Appl. Occup. Environ. Hyg.* **12**, 339–346 (1997).
34. S. M. Rappaport, R. H. Lyles, and L. L. Kupper, (1995) "An Exposure-Assessments Strategy Accounting for Within- and Between-Worker Sources of Variability," *Ann. Occup. Hyg.* **39**, 469–495 (1995).

35. R. H. Lyles, L. L. Kupper, and S. M. Rappaport, "A Lognormal Distribution-Based Exposure Assessment Method for Unbalanced Data," *Ann. Occup. Hyg.* **41**, 63–76 (1997).
36. R. H. Lyles, L. L. Kupper, and S. M. Rappaport, "Assessing Regulatory Compliance of Occupational Exposures via the Balanced One-Way Random Effects ANOVA Model," *J. Ag. Biol. Environ. Stat.* **2**, 64–86 (1997).
37. P. Drinker and T. Hatch, *Industrial Dust*, McGraw Hill, New York, 1936.
38. S. A. Roach, "A method of Relating the Incidence of Pneumoconiosis to Airborne Dust Exposure," *Brit. J. Ind. Med.* **10**, 220–226 (1953).
39. N. Esmen, *On Estimation of Occupational Health Risks*, Princeton Scientific Publishers, Princeton, NJ, 1984.
40. W. Coenen, "Measurement Assessment of the Concentration of Health-Impairing, especially Silicogenic Dusts at Work Places of Surface Industries," *Staub. Reinhalt. Luft* **31**, 16–23 (1971).
41. E. Symanski, L. L. Kupper, H. Kromhout, and S. M. Rappaport, "An Investigation of Systematic Changes in Occupational Exposure," *Am. Ind. Hyg. Assoc. J.* **57**, 724–735 (1996).
42. F. Patty, *Sampling and Analysis of Atmospheric Contaminants*, Interscience, New York, 1958.
43. S. A. Roach, E. J. Baier, and H. E. Ayer, and L., H. R. "Testing Compliance with Threshold Limit Values for Respirable Dusts," *Am. Ind. Hyg. Assoc. J.* **28**, 543–553 (1967).
44. R. C. Spear and S. Selvin, "OSHA's Permissible Exposure Limits: Regulatory Compliance versus Health Risk," *Risk Anal.* **9**, 579–586 (1989).
45. W. Post, H. Kromhout, D. Neeperik, D. Noy, and R. S. Duijzentku, "Semiquantite Estimates of Exposure to Methylene Chloride and Shepene: The Influence of Quantitative Exposure Data," *Appl. Occup. Environ. Hyg.* **6**, 197–204 (1991).
46. E. Olsen, B. Laursen, and P. S. Vinzents, "Bias and Random Errors in Historical Data of Exposure to Organic Solvents," *Am. Ind. Hyg. Assoc. J.* **52**, 204–211 (1991).
47. E. Olsen, "Effect of Sampling on Measurement Errors," *Analyst* **121**, 1155–1161 (1996).
48. E. Olsen and B. Jensen, "On the Concept of the "Normal" Day: Quality Control of Occupational Hygiene Measurements," *Appl. Occup. Environ. Hyg.* **9**, 245–255 (1994).
49. E. Olsen, "Analysis of Exposure using a Logbook Method," *Appl. Occup. Environ. Hyg.* **9**, 712–722 (1994).
50. M. Nicas and R. C. Spear, "A Task-Based Statistical Model of a Worker's Exposure Distribution: Part I—Description of the Model," *Am. Ind. Hyg. Assoc. J.* **54**, 211–220 (1993).
51. M. Nicas and R. C. Spear, "A Task-Based Statistical Model of a Worker's Exposure Distribution: Part II—Application to Sampling Strategy," *Am. Ind. Hyg. Assoc. J.* **54**, 221–227 (1993).
52. M. J. Nieuwenhuijsen, C. P. Sandiford, D. Lowson, R. D. Tee, K. M. Venables, and A. J. Newman-Taylor, "Peak Exposure Concentrations of Dust and Flour Aeroallergen in Flour Mills and Bakeries" [see comments]. *Ann. Occup. Hyg.* **39**, 193–201 (1995).
53. J. W. Cherrie, "Are Task-Based Exposure Levels a Valuable Index of Exposure for Epidemiology?" [letter; comment], *Ann. Occup. Hyg.* **40**, 715–722 (1996).
54. E. Olsen, "Effect of Sampling on Measurement Errors," *Analyst* **121**, 1155–1161 (1996).
55. U. Ulfvarson, "Limitations to the use of Employee Exposure Data on Air Contaminants in Epidemiologic Studies," *Int. Arch. Occup. Environ. Health* **52**, 285–300 (1983).
56. H. Kromhout, P. Swuste, and J. S. Boleij, "Empirical Modelling of Chemical Exposure in the Rubber-Manufacturing Industry," *Ann. Occup. Hyg.* **38**, 3–22 (1994).

57. H. Kromhout and D. Heederik, "Occupational Epidemiology in the Rubber Industry: Implications of Exposure Variability," *Am. J. Ind. Med.* **27**, 171–85 (1995).

58. S. M. Rappaport, M. Weaver, D. Taylor, L. Kupper, and P. Susi, "Application of Mixed Models to Assess Aerosol Exposures Measured by Construction Workers during Hot Processes," *Ann. Occup. Hyg.* **43**, 457–469 (1999).

59. R. W. Hornung and L. D. Reed, "Estimation of Average Concentration in the Presence of Nondetectable Values," *Appl Occup. Environ. Hyg.* **5**, 132–141 (1990).

60. L. L. Travis and M. L. Land, "Estimating the Mean Data Sets with Nondetectable Values," *Environ. Sci. Technol.* **24**, 961–962 (1990).

61. A. P. Dempster, N. M. Laird, and D. B. Rubin, "Maximum Likelihood from Incomplete Data via the EM Algorithm," *J. Royal Stat. Soc., Series B* **39**, 1–38 (1977).

62. K. Lange, *Numerical Analysis for Statisticians*, Springer-Verlag, New York, 1999.

63. E. Symanski and S. M. Rappaport, "An Investigation of the Dependence of Exposure Variability on the Interval between Measurements," *Ann. Occup. Hyg.* **38**, 361–372 (1994).

64. D. Heederik, J. S. M. Boleij, H. Kromhout, and T. Smid, "Use and Analysis of Exposure Monitoring Data in Occupational Epidemiology: An Example of an Epidemiological in the Dutch Animal Food Industry," *Appl. Occup. Environ. Hyg.* **6**, 458–464 (1991).

65. S. Lagorio, I. Iavarone, N. Iacovella, I. R. Proietto, S. Frselli, L. T. Baldassarri, and A. Carere, "Variation of Benzene Exposure among Filling Station Attendants," *Occup. Hyg.* **4**, 15–30 (1998).

66. H. Kromhout, D. P. Loomis, G. J. Mihlan, L. A. Peipins, R. C. Kleckner, R. Iriye, and D. A. Savitz, "Assessment and Grouping of Occupational Magnetic Field Exposure in Five Electric Utility Companies," *Scand. J. Work Environ. Health* **21**, 43–50 (1995).

67. H. Kromhout, D. P. Loomis, R. C. Kleckner, and D. A. Savitz, "Sensitivity of the Relation between Cumulative Magnetic Field Exposure and Brain Cancer Mortality to Choice of Monitoring Data Grouping Scheme," *Epidemiology* **8**, 442–445 (1997).

68. H. Kromhout and D. P. Loomis, "The Need for Exposure Grouping Strategies in Studies of Magnetic Fields and Childhood Leukemia" [letter; comment], *Epidemiology* **8**, 218–219 (1997).

69. S. Kumagai, Y. Kusaka, and S. Goto, "Cobalt Exposure Level and Variability in the Hard Metal Industry of Japan," *Am. Ind. Hyg. Assoc. J.* **57**, 365–369 (1996).

70. M. J. Nieuwenhuijsen, D. Lowson, K. M. Venables, and A. J. Newman-Taylor, "Flour Dust Exposure Variability in Flour Mills and Bakeries," *Ann. Occup. Hyg.* **39**, 299–305 (1995).

71. C. Peretz, P. Goldberg, E. Kahan, S. Grady, and A. Goren, "The Variability of Exposure Over Time: A Prospective Longitudinal Study," *Ann. Occup. Hyg.* **41**, 485–500 (1997).

72. L. Preller, H. Kromhout, D. Heederik, and M. J. Tielen, "Modeling Long-term Average Exposure in Occupational Exposure–Response Analysis," *Scand. J. Work Environ. Health* **21**, 504–512 (1995).

73. L. Preller, D. Heederik, H. Kromhout, J. S. Boleij, and M. J. Tielen, "Determinants of Dust and Endotoxin Exposure of Pig Farmers: Development of a Control Strategy using Empirical Modelling," *Ann. Occup. Hyg.* **39**, 545–557 (1995).

74. S. M. Rappaport, E. Symanski, J. W. Yager, and L. L. Kupper, "The Relationship between Environmental Monitoring and Biological Markers in Exposure Assessment," *Environ. Health Perspect.* **103**, 49–53 (1995).

75. N. S. Seixas and L. Sheppard, "Maximizing Accuracy and Precision Using Individual and Grouped Exposure Assessments," *Scand. J. Work Environ. Health* **22**, 94–101 (1995).

76. S. R. Woskie, P. Shen, E. A. Eisen, M. H. Finkel, T. J. Smith, R. Smith, and D. H. Wegman, "The Real-Time Dust Exposures of Sodium Borate Workers: Examination of Exposure Variability," *Am. Ind. Hyg. Assoc. J.* **55**, 207–217 (1994).

77. S. M. Rappaport, L. L. Kupper, and R. H. Lyles, "Re.: A Log-Normal Distribution-Based Exposure Assessment Method for Unbalanced Data" [Letter], *Ann. Occup. Hyg.* **42**, 417–422 (1998).

78. N. Lange and L. M. Ryan, "Assessing Normality in Random Effects Models," *Ann. Stat.* **17**, 624–642 (1989).

79. A. P. Dempster and L. M. Ryan, "Weighted Normal Plots," *J. Am. Stat. Assoc.* **80**, 845–850 (1985).

80. S. M. Rappaport, "Biological Considerations in Assessing Exposures to Genotoxic and Carcinogenic Agents," *Int. Arch. Occup. Environ. Health* **65**, S29–35 (1993).

81. S. R. Searle, G. Casella, and C. E. McCulloch, *Variance Components*, John Wiley & Sons, 1992, New York.

82. J. Aitchison and J. A. C. Brown, *The Lognormal Distribution*, Cambridge University Press, London, 1987.

83. S. M. Rappaport, S. Selvin, and S. Roach, "A Strategy for Assessing Exposure with Reference to Multiple Limits," *Appl. Ind. Hyg.* **3**, 310–315 (1988).

84. M. Francis, S. Selvin, R. Spear, and S. Rappaport, "The Effect of Autocorrelation on the Estimation of Workers' Daily Exposures," *Am. Ind. Hyg. Assoc. J.* **50**, 37–43 (1989).

85. G. E. P. Box and G. M. Jenkins, *Time Series Analysis: Forecasting and Control*. Holden-Day, San Francisco, 1976.

86. W. Coenen, "Beschreibung des zeitlichen Verhaltens von Schadstoffkonzentrationen durch einen stetigen Markow-Process," *Staub Reinhalt. Luft* **36**, 240–248 (1976), in German.

87. S. A. Roach, "A Most Rational Basis for Air Sampling Programmes," *Ann. Occup. Hyg.* **20**, 65–84 (1977).

88. A. Koizumi, T. Sekiguchi, M. Konno, and M. Ikeda, "Evaluation of the Time Weighted Average of Air Contaminants with Special References to Concentration Fluctuation and Biological Half Time," *Am. Ind. Hyg. Assoc. J.* **41**, 693–699 (1980).

89. R. C. Spear, S. Selvin, and M. Francis, "The Influence of Averaging Time on the Distribution of Exposures," *Am. Ind. Hyg. Assoc. J.* **47**, 365–368 (1986).

90. B. Preat, "Application of Geostatistical Methods for Estimation of the Dispersion Variance of Occupational Exposures," *Am. Ind. Hyg. Assoc. J.* **48**, 877–884 (1987).

91. S. M. Rappaport and R. C. Spear, "Physiological Damping of Exposure Variability during Brief Periods," *Ann. Occup. Hyg.* **32**, 21–33 (1988).

92. D. K. George, M. R. Flynn, and R. L. Harris, "Autocorrelation of Interday Exposures at an Automobile Assembly Plant," *Am. Ind. Hyg. Assoc. J.* **56**, 1187–1194 (1993).

93. N. Esmen, "Retrospective Industrial Hygiene Surveys," *Am. Ind. Hyg. Assoc. J.* **40**, 58–65 (1979).

94. E. Buringh and R. Lanting, "Exposure Variability in the Workplace: Its Implications for the Assessment of Compliance [see comments]." *Am. Ind. Hyg. Assoc. J.* **52**, 6–13 (1991).

95. S. Kumagai, I. Matsunaga, and Y. Kusaka, "Autocorrelation of Short-Term and Daily Average Exposure Levels in Workplaces," *Am. Ind. Hyg. Assoc. J.* **54**, 341–350 (1993).

96. Anonymous "Occupational Safety and Health Act of 1970, PL 91-596, S.2193." U.S. Government Printing Office, Washington, DC, 1970.
97. S. M. Rappaport, "Threshold Limit Values, Permissible Exposure Limits, and Feasibility: The Bases for Exposure Limits in the United States," *Am. J. Ind. Med.* **23**, 683–694 (1993).
98. ACGIH, *1999 Threshold Limit Values and Biological Exposure Indices for Chemical Substances and Physical Agents*, ACGIH, Cincinnati, OH, 1999.
99. D. Henschler, "Exposure Limits: History, Philosophy and Future Developments," *Ann. Occup. Hyg.* **28**, 79–92 (1984).
100. U. Ulfvarson, "Assessment of Concentration Peaks in Setting Exposure Limits for Air Contaminants at Workplaces, with Special Emphasis upon Narcotic and Irritative Gases and Vapors," *Scand. J. Work Environ. Health* **13**, 389–398 (1987).
101. R. L. Zielhuis, P. C. Noordam, H. Roelfzema, and A. A. E. Wibowo, "Short-term Occupational Exposure Limits: A Simplified Approach," *Int. Arch. Occup. Environ. Health* **61**, 207–211 (1988).
102. S. M. Rappaport, "Smoothing of Exposure Variability at the Receptor: Implications for Health Standards," *Ann. Occup. Hyg.* **29**, 201–214, (1985).
103. D. H. Halton, "A Comparison of the Concepts used to Develop and Apply Occupational Exposure Limits for Ionizing Radiation and Hazardous Chemical Substances," *Reg. Tox. Pharmacol.* **8**, 343–355 (1988).
104. S. A. Roach and S. M. Rappaport, "But They Are Not Thresholds: A Critical Analysis of the Documentation of Threshold Limit Values," *Am. J. Ind. Med.* **17**, 727–753 (1990).
105. ACGIH, *Documentation of the Threshold Limit Values and Biological Exposure Indices*, American Conference of Governmental Industrial Hygienists, Cincinnati, OH, 1990.
106. OSHA, "Occupational Exposure to Benzene." *Fed. Reg.* **52**, 34460–34579 (1987).
107. W. H. Olson and R. B. Cumming, "Chemical Mutagens: Dosimetry, Haber's Rule and Linear Systems," *J. Theor. Biol.* **91**, 383–395 (1981).
108. D. Hattis, "Pharmacokinetic Principles for Dose–Rate Extrapolation of Carcinogenic Risk from Genetically Active Agents," *Risk Anal.* **18**, 303–316 (1998).
109. H. Checkoway and C. H. Rice, "Time-Weighted Averages, Peaks, and Other Indices of Exposure in Occupational Epidemiology," *Am. J. Ind. Med.* **21**, 25–33 (1992).
110. T. Smith, "Exposure Assessment for Occupational Epidemiology," *Am. J. Ind. Med.* **126**, 249–268 (1987).
111. T. Smith, "Occupational Exposure and Dose Over Time: Limitations of Cumulative Exposure," *Am. J. Ind. Med.* **21**, 35–51 (1992).
112. United States Court of Appeals, *American Iron and Steel Institute and Bethlehem Steel Corporation v. Occupational Safety and Health Administration*, Federal Reporter 2nd Series (DC Cir. 1991) **939**, 975–1010 (1991).
113. R. L. Zielhuis and F. W. van der Kreek, "The Use of a Safety Factor in Setting Health Based Permissible Levels for Occupational Exposure I. A Proposal," *Int. Arch. Occup. Environ. Health* **42**, 191–201 (1979).
114. R. L. Zielhuis and F. W. van der Kreek, "Calculation of a Safety Factor in Setting Health Based Permissible Levels for Occupational Exposure II. Comparison of Extrapolated and Published Permissible Levels," *Int. Arch. Occup. Environ. Health* **42**, 203–215 (1979).
115. M. L. Dourson and J. F. Stara, "Regulatory History and Experimental Support of Uncertainty (Safety) Factors," *Reg. Tox. Pharmacol.* **3**, 224–238 (1983).

116. S. M. Rappaport, S. Selvin, R. C. Spear, and C. Keil, "Air Sampling in the Assessment of Continuous Exposures to Acutely-Toxic Chemicals. Part I—Strategy," *Am. Ind. Hyg. Assoc. J.* **42**, 831–838 (1981).
117. S. Rappaport, "Selection of The Measures of Exposure for Epidemiology Studies," *Appl. Occup. Environ. Hyg.* **6**, 448–457 (1991).
118. R. M. Tuggle, "The NIOSH Decision Scheme," *Am. Ind. Hyg. Assoc. J.* **42**, 493–498 (1981).
119. W. M. Long, "Airborne Dust in Coal Mines: The Sampling Problem," *Br. J. Ind. Med.* **10**, 241–244 (1953).
120. B. M. Wright, "The Importance of the Time Factor in the Measurement of Dust Exposure," *Brit. J. Ind. Med.* **10**, 235–240 (1953).
121. R. C. Tomlinson, "A Simple Sequential Procedure to Test whether Average Conditions Achieve a Certain Standard," *Appl. Statistics* **6**, 198–207 (1957).
122. J. Juda and K. Budzinski, "Fehler bei der Bestimmung der mittleren Staubkonzentration als Funktion der Anzahl der Einzelmessungen," *Staub Reinhalt. Luft* **24**, 283–287 (1964), in German.
123. J. Juda and K. Budzinski, "Determining the Tolerance Range of the Mean Value of Dust Concentration," *Staub Reinhalt. Luft* **27**, 12–16 (1967).
124. W. Coenen, "The Confidence Limits for the Mean Values of Dust Concentration," *Staub Reinhalt. Luft* **26**, 39–45 (1966).
125. R. J. Sherwood, "On the Interpretation of Air Sampling for Radioactive Particles," *Am. Ind. Hyg. Assoc. J.* **27**, 98–109 (1966).
126. W. A. Langmead, "Air Sampling as Part of an Integrated Programme of Monitoring of the Worker and His Environment," *Inhaled Part. Vapors* **2**, 983–995 (1970).
127. P. Hewett, "Interpretation and Use of Occupational Exposure Limits for Chronic Disease Agents," *Occup Med* **11**, 561–590 (1996).
128. M. Corn, P. Breysse, T. Hall, G. T. R. Chen, and D. L. Swift, "A Critique of MSHA Procedures for Determination of Permissible Respirable Coal Mine Dust Containing Free Silica," *Am. Ind. Hyg. Assoc. J.* **46**, 4–8 (1985).
129. M. Corn, "Strategies of Air Sampling," *Scand. J. Work Environ. Health* **11**, 173–180 (1985).
130. G. Heidermanns, G. Kuhnen, and G. Riediger, "Messung and Beurteilung gesundheitsgefährlicher Staube am Arbeitsplatz," *Staub Reinhalt. Luft* **40**, 367–373, (1980) in German.
131. G. Riediger, "Die Anwendung von Maximalen Arbeitsplatzkonzentrationen (MAK) nach der TRgA 402," *Staub Reinhalt. Luft* **46**, 182–186 (1986), in German.
132. C. Land, in E. S. K. Crow, ed., *Lognormal Distributions*, Marcel Dekker, New York, 1988, pp. 87–112.
133. M. Nicas, B. P. Simmons, and R. C. Spear, "Environmental Versus Analytical Variability in Exposure Measurements," *Am. Ind. Hyg. Assoc. J.* **52**, 553–557 (1991).

Index